Advances in
Carbohydrate Chemistry and Biochemistry

Volume 38

Advances in Carbohydrate Chemistry and Biochemistry

Editors

R. STUART TIPSON

DEREK HORTON

Board of Advisors

LAURENS ANDERSON
STEPHEN J. ANGYAL
GUY G. S. DUTTON
ALLAN B. FOSTER
DEXTER FRENCH

BENGT LINDBERG
HANS PAULSEN
NATHAN SHARON
MAURICE STACEY
ROY L. WHISTLER

Volume 38

1981

ACADEMIC PRESS

A Subsidiary of Harcourt Brace Jovanovich, Publishers

New York   London   Toronto   Sydney   San Francisco

COPYRIGHT © 1981, BY ACADEMIC PRESS, INC.
ALL RIGHTS RESERVED.
NO PART OF THIS PUBLICATION MAY BE REPRODUCED OR
TRANSMITTED IN ANY FORM OR BY ANY MEANS, ELECTRONIC
OR MECHANICAL, INCLUDING PHOTOCOPY, RECORDING, OR ANY
INFORMATION STORAGE AND RETRIEVAL SYSTEM, WITHOUT
PERMISSION IN WRITING FROM THE PUBLISHER.

ACADEMIC PRESS, INC.
111 Fifth Avenue, New York, New York 10003

United Kingdom Edition published by
ACADEMIC PRESS, INC. (LONDON) LTD.
24/28 Oval Road, London NW1 7DX

LIBRARY OF CONGRESS CATALOG CARD NUMBER: 45-11351

ISBN 0-12-007238-6

PRINTED IN THE UNITED STATES OF AMERICA

81 82 83 84   9 8 7 6 5 4 3 2 1

CONTENTS

LIST OF CONTRIBUTORS ix
PREFACE xi

Emil Hardegger (1913–1978)

HASSAN S. EL KHADEM

Text 1
Appendix I 4
Appendix II 10

Carbon-13 Nuclear Magnetic Resonance Spectroscopy of Polysaccharides

PHILIP A. J. GORIN

I. Introduction 13
II. Configurationally Dependent, ^{13}C Parameters 15
III. Identification of ^{13}C Signals of Monosaccharides: O-Substitution Shifts . . . 19
IV. Identification of ^{13}C-N.m.r. Signals of Oligosaccharides and
    Polysaccharides 22
V. Molecular Motion 25
VI. ^{13}C-N.m.r. Spectra of Polysaccharides 37

Photochemical Reactions of Carbohydrates

ROGER W. BINKLEY

I. Introduction 105
II. Alkenes 106
III. Carbonyl Compounds 122
IV. Acetals 142
V. Unprotected Alditols, Aldohexoses, and Related Compounds 147
VI. Sulfur-Containing Compounds 150
VII. Nitrogen-Containing Compounds 160
VIII. Iodo Compounds 186
IX. Organometallic Compounds 190
X. Phosphates 192

Fluorinated Carbohydrates

ANNA A. E. PENGLIS

I. Introduction 195
II. Synthesis of Fluorinated Carbohydrates 199
III. Physical Methods 253
IV. Some Biological Applications of Fluorinated Carbohydrates 281
V. Addendum 284

CONTENTS

The Gulono-1,4-lactones: A Review of Their Synthesis, Reactions, and Related Derivatives

THOMAS C. CRAWFORD

I. Introduction	287
II. Synthesis	288
III. Crystal and Solution Structure, and Spectroscopic Properties of the Gulono-1,4-lactones and Derivatives	296
IV. Reactions of the Lactone Moiety in the Gulono-1,4-lactones	298
V. Reactions of the Hydroxyl Groups in the Gulono-1,4-lactones	302
VI. Deoxy and Anhydro Derivatives of the Gulono-1,4-lactones	305
VII. Amino Derivatives of the Gulono-1,4-lactones	308
VIII. Miscellaneous Derivatives of the Gulono-1,4-lactones	310
IX. Oxidation of the Gulono-1,4-lactones and Derivatives	314
X. Reduction of the Gulono-1,4-lactones and Derivatives	315
XI. Nucleophilic Additions to the Carbonyl Group of Derivatives of the Gulono-1,4-lactones	317
XII. Analytical Procedures	318
XIII. Biological Role of the Gulono-1,4-lactones and Gulonic Acids	320
XIV. Addendum	321

The Chemistry and Biological Significance of 3-Deoxy-D-*manno*-2-Octulosonic Acid (KDO)

FRANK M. UNGER

I. Introduction	324
II. Occurrence, Location, and Linkages of KDO Residues in Bacterial Polysaccharides	326
III. Elucidation of the Structure of KDO	357
IV. Chemical Synthesis and Monosaccharide Chemistry of KDO	365
V. Enzymes of KDO Metabolism	378
VI. The Design of Inhibitors of KDO Metabolism	387

Methylation Techniques in the Structural Analysis of Glycoproteins and Glycolipids

HEIKKI RAUVALA, JUKKA FINNE, TOM KRUSIUS, JORMA KÄRKKÄINEN, AND JOHAN JÄRNEFELT

I. Introduction	389
II. Permethylation of the Sample	390
III. Analysis of Permethylated Carbohydrates without Depolymerization of the Sample	392
IV. Degradation of the Permethylated Sample, and Analysis of the Partially Methylated Sugars	396
V. Methylation Analysis Combined with Specific Degradation	407

Bibliography of Crystal Structures of Carbohydrates, Nucleosides, and Nucleotides: 1977 and 1978

GEORGE A. JEFFREY AND MUTTAIYA SUNDARALINGAM

I. Introduction 417
II. Data for Carbohydrates 418
III. Data for Nucleosides and Nucleotides 485
IV. Preliminary Communications 526

AUTHOR INDEX. 531
SUBJECT INDEX. 554
ERRATA 577

LIST OF CONTRIBUTORS

Numbers in parentheses indicate the pages on which the authors' contributions begin.

ROGER W. BINKLEY, *Department of Chemistry, Cleveland State University, Cleveland, Ohio 44115* (105)

THOMAS C. CRAWFORD, *Central Research, Chemical Process Research, Pfizer Inc., Eastern Point Road, Groton, Connecticut 06340* (287)

HASSAN S. EL KHADEM, *Department of Chemistry and Chemical Engineering, Michigan Technological University, Houghton, Michigan 49931* (1)

JUKKA FINNE, *Department of Medical Chemistry, University of Helsinki, Siltavuorenpenger 10 A, SF-00170 Helsinki 17, Finland* (389)

PHILIP A. J. GORIN, *Prairie Regional Laboratory, National Research Council of Canada, Saskatoon, Saskatchewan S7N OW9, Canada* (13)

JOHAN JÄRNEFELT, *Department of Medical Chemistry, University of Helsinki, Siltavuorenpenger 10 A, SF-00170 Helsinki 17, Finland* (389)

GEORGE A. JEFFREY, *Department of Crystallography, University of Pittsburgh, Pittsburgh, Pennsylvania 15260* (417)

JORMA KÄRKKÄINEN, *Department of Medical Chemistry, University of Helsinki, Siltavuorenpenger 10 A, SF-00170 Helsinki 17, Finland* (389)

TOM KRUSIUS, *Department of Medical Chemistry, University of Helsinki, Siltavuorenpenger 10 A, SF-00170 Helsinki 17, Finland* (389)

ANNA A. E. PENGLIS,* *Department of Chemistry, Queen Elizabeth College, Campden Hill Road, London W8 7AH, United Kingdom* (195)

HEIKKI RAUVALA, *Department of Medical Chemistry, University of Helsinki, Siltavuorenpenger 10 A, SF-00170 Helsinki 17, Finland* (389)

MUTTAIYA SUNDARALINGAM, *Department of Biochemistry, University of Wisconsin, Madison, Wisconsin 53706* (417)

FRANK M. UNGER, *Sandoz Forschungsinstitut Wien, Brunner Strasse 59, A-1235 Vienna, Austria* (323)

* Present address: The Dyson Perrins Laboratory, University of Oxford, South Parks Road, Oxford OX1 3QY, United Kingdom.

PREFACE

In Volume 38, the thirtieth in this series that has been edited by the Senior Editor, a discussion of the ^{13}C-nuclear magnetic resonance spectroscopy of polysaccharides is provided by P. A. J. Gorin (Saskatoon). This technique has opened up a major, new avenue for determination of the structure of these macromolecular carbohydrates, and has proved much more fruitful than ^{1}H-n.m.r. spectroscopy, a methodology discussed largely in the context of simple sugars, by B. Coxon in Volume 27 and L. D. Hall in Volume 29. R. W. Binkley (Cleveland) complements and updates the article by G. O. Phillips in Volume 18 on the photochemical reactions of carbohydrates, with particular emphasis on organic-chemical aspects. A comprehensive treatment of fluorinated carbohydrates is contributed by A. A. E. Penglis (Oxford); this is a subject touched upon briefly in previous articles (by L. J. Haynes and F. H. Newth in Vol. 10; T. G. Bonner, F. Micheel, and A. Klemer in Vol. 16; J. E. G. Barnett in Vol. 22; and H. Paulsen in Vol. 26). A review of the synthesis and reactions of the gulono-1,4-lactones and related derivatives is provided by T. C. Crawford (Groton). F. M. Unger (Vienna) contributes an article on the biological significance of 3-deoxy-D-*manno*-2-octulosonic acid (KDO) and the chemical approaches for synthesis of this important compound. J. Järnefelt and coworkers (Helsinki) have written a detailed description of the use of the time-honored methylation technique, as adapted for modern, microscale work in the structural analysis of glycoproteins by J. Montreuil in Volume 37. A continuing article in the series of bibliographic surveys of crystal-structure work on carbohydrates, nucleosides, and nucleotides, covering the literature thereon for 1977–1978, is provided by G. A. Jeffrey (Pittsburgh) and M. Sundaralingam (Madison); these surveys not only present crystallographic results in a pictorial format readily comprehended by chemists and biochemists, but they correct errors found in the original-literature interpretations. H. S. El Khadem (Houghton) has written an interesting account of the life and work of Emil Hardegger, a Swiss sugar chemist whose work has been, perhaps, less well recognized than it might have been; a useful feature of this article is the citation of unpublished research that is available in doctoral dissertations.

Kensington, Maryland R. STUART TIPSON
Columbus, Ohio DEREK HORTON
June, 1980

1913 – 1978

EMIL HARDEGGER*

1913–1978

Emil Hardegger was born on November 25, 1913, in St. Gallen, Switzerland, to Emil and Anna Hardegger, both of whom were teachers in that city's school system. Soon after Emil's birth, his father was elected to the National Assembly, and he occupied a seat there for many years: he was also elected to the City Council of St. Gallen. He died at the age of 93, only a year before his son's premature death. Young Emil attended primary school and then the Kantonschule in St. Gallen. After passing the Maturität examination in 1934, he went to London, and spent a year there, learning the English language.

Early in his youth, Emil Hardegger showed a great interest in chemistry; he used to conduct chemical experiments in the attic of his home, much to his parents' distress. During one such experiment, fortunately conducted in an open field, an explosion occurred that burnt his hair and narrowly missed his eyes. It was therefore no surprise to his parents when Emil decided to study chemistry at the Eidgenössiche Technische Hochschule (ETH).

In 1936, two years before receiving therefrom the degree of Diplom Ingenieur Chemiker, he met Olga Lüdi, the daughter of Albert and Marie Lüdi of Berne. They were married on February 16, 1940, after he had been named an assistant at the ETH. The Hardeggers lived in a furnished room until he received the doctorate in 1941, and then they rented a house on Toblerstrasse, overlooking the University. Their first son, Ivar, was born in 1947, and, three years later, they had a second son, Béat. As the house was "becoming too small," they rented another house, in Lufingen, 15 miles from Zürich. In 1948, after his Habilitation, he bought his first car, a small Volkswagen, which he would drive nonstop to Spain on vacations. Then, when Ivar started to attend school in Zürich, the Hardeggers bought a house in Gockhausen in order to be near his school. In 1968, they bought a country house in Toggenburg, near woods and meadows, and they spent all their week-ends and holidays there until Professor Hardeg-

---

* The author is indebted to Mrs. Emil Hardegger for her help and advice during the preparation of this article.

ger's sickness. Emil Hardegger loved to work on the land; he chopped wood and tilled the soil. Another hobby of his was collecting all types of engines. His pride and joy were two old Coronado jet-propulsion engines, which he used to operate in an abandoned factory-building on the property.

Emil Hardegger was very proud of his two sons. When they grew up, they used to visit him in Toggenburg on week-ends and holidays. At first, they came with their girl-friends; later, with their wives, and then, children. Ivar became a pharmaceutical wholesaler and married a bank employee, Marlys; they have two sons, Raoul and Gilles. Béat, who received a doctorate in pharmacy at the ETH, married Janine, a pharmacy student. They opened a pharmacy on Oberengstringen near Zürich, and have two daughters, Prisca and Andrea.

When Emil Hardegger enrolled as a student in the ETH, the most active research group in the Department of Chemistry was headed by Professor Leopold Ruzicka, who, at that time, was actively involved in the synthesis of sex hormones, work that earned him the Nobel Prize in 1939. It was not surprising that young Emil Hardegger was drawn to this group, and chose to work for Professor Ruzicka, both for his Diplomarbeit and his Doktorarbeit. His Diplomarbeit dealt with the synthesis of 1,10-dimethylpicene; it was published by Ruzicka and Hoffman, and Hardegger's name appeared in the Experimental section.

After Hardegger received his degree of Diplom Ingenieur Chemiker in the Autumn of 1938, he joined Ruzicka's sex-hormone research-team. His doctoral thesis comprised two parts; the first dealt with the elimination products of anhydro-oxyprogesterone in human urine, and the second, with the preparation and reactions of Δ^4-17,20-oxypregnen-3-one. Despite lengthy interruptions in his research, caused by active duty during the war years, he finished his thesis in the winter of 1941, three years after he started it.

Hardegger, who had decided to make a career of teaching and research at the ETH, worked extremely hard to win an appointment at this prestigious institution and often stayed in the laboratory until 2 o'clock in the morning. The following three years were spent on research related to his doctoral work, carried out in collaboration with Professors Ruzicka and P. A. Plattner.

As Oberassistent, Hardegger supervised graduate students jointly with Professor Ruzicka. Klemens Scholz and Leo Blunschy were among his first graduate students, but Blunschy died before he could write his thesis.

Later, after Professor T. Reichstein moved to Basel, Dr. Emil Hardegger was asked to take charge of the carbohydrate research in the

ETH, and he moved to Reichstein's laboratory in the basement of the chemistry building. Among the first group of his graduate students working in the carbohydrate field were Montavon, DePasqual, El Heweihi, and the present writer. This multinational group worked on a variety of topics, ranging from the partial p-toluenesulfonylation of D-glucose to 1,2-anhydrides, dithioacetals, and osazones. In 1948, Emil Hardegger presented his published work for the habilitation, and was appointed Privatdozent. At that time, Jucker, Robinet, Spitz, and Leemann joined his research group. Jucker worked on the p-toluenesulfonylation of D-glucose, continuing Montavon's project; Robinet, on the glycosides of steroids and triterpenes; and Spitz, on D-glucuronic acid; Leemann continued the work started by DePasqual on 1,2-anhydro sugars. Then, Schreier, Kreis, and Stöhr joined the group, and all three worked on anhydro-osazones, continuing the present writer's project.

The years 1945 to 1952 were mainly devoted to carbohydrate chemistry. During this period, Hardegger published some 25 papers, of which only two were on noncarbohydrate topics. In 1952, when Professor Plattner left the ETH, Emil Hardegger assumed part of his teaching responsibilities, and supervised research in the areas of alkaloids and antibiotics. The graduate students joining the Hardegger research group during the period from 1952 to 1966 were involved with such problems as the stereochemistry of tropine alkaloids, the synthesis of lysergic acid, and structural studies on plant and mushroom degradation-products, as, for example, lycomarasmine, muscarine, fusaric acid, and prostaglandins. A large team, composed of ten graduate students, was at one time working on the synthesis of the last-mentioned compounds. This period saw the promotion of Emil Hardegger to Ausserordentlich Professor in 1957, and to Ordentlich Professor in 1966.

During the following ten years, Emil Hardegger returned to his true love, the carbohydrates. He started to emphasize such topics as urea derivatives of 2-amino-2-deoxy-D-glycosides, the synthesis of monosaccharides, thio sugars, the electrochemical oxidation and reduction of sugars, the oxidation of D-glucose with oxygen, and the epimerization of aldonic acids.

Emil Hardegger was a man of average height and build, 172 cm tall and weighing 68 kilograms. He had dark-brown eyes and hair, the latter turning gray in later years. He was a very humane person of great warmth and kindness, but was never affiliated formally with a particular church. In December, 1975, at the peak of his scientific productivity, Emil Hardegger's health suddenly deteriorated, and his condition was diagnosed as a cirrhosis of the liver. Diazomethane poisoning was suspected; he had worked for several years with this toxic reagent, which must have affected his system. After a short recovery, he soon relapsed, and had to undergo surgery. Finally, his condition worsened, and he died on the 18th of May, 1978.

During his short, but fruitful, academic career, from 1939 to 1975, Emil Hardegger published 123 papers and supervised 46 Doctoral dissertations. The results of some of his later students' work were never published, due to the decline of his health, and therefore the titles, authors, and year of doctoral theses he supervised are given in Appendix II.

Emil Hardegger always had a very good relationship with his many research students. They, in turn, admired not only his scientific knowledge, but also his overall warmth and human qualities. Emil Hardegger will long be remembered in the scientific community as one of the great carbohydrate and natural-products chemists.

HASSAN S. EL KHADEM

APPENDIX I

The following is a chronological list of the scientific publications of the late Professor Emil Hardegger.

"Umwandlung von 17-Aethinyl-androstendion-(3,17) in Pregnadienol-(3)," by L. Ruzicka, M. W. Goldberg, and E. Hardegger, *Helv. Chim. Acta*, 22 (1939) 1294–1300.

"Ueber die bei Verabreichung von Anhydro-oxy-progesteron im Harn ausgeschiedenen Steroide," by M. W. Goldberg and E. Hardegger, *Schweiz. Med. Wochenschr.*, 71 (1941) 1041–1042.

"Oxydation von $\Delta^{4,17}$-Pregnadienon-(3) mit Phthalmonopersäure," by L. Ruzicka, M. W. Goldberg, and E. Hardegger, *Helv. Chim. Acta*, 25 (1942) 1297–1305.

"Die Umlagerung von Δ^4-17,20-Oxido-pregnen-3-on durch Eisessig," by L. Ruzicka, M. W. Goldberg, and E. Hardegger, *Helv. Chim. Acta*, 25 (1942) 1680–1689.

"17α-Oxy-20-keto-Verbindungen der Pregnen- und der Allopregnan-Reihe," by M. W. Goldberg, R. Aeschbacher, and E. Hardegger, *Helv. Chim. Acta*, 26 (1943) 680–686.

"Zur Kenntnis der unverseifbaren Lipoide aus arteriosklerotischen Aorten," by E. Hardegger, L. Ruzicka, and E. Tagmann, *Helv. Chim. Acta*, 26 (1943) 2205–2221.

"Herstellung des β-(*trans-p*-Oxy-cyclohexyl)-$\Delta^{\alpha,\beta}$-butenolids," by E. Hardegger, P. A. Plattner, and F. Blank, *Helv. Chim. Acta*, 27 (1944) 793–800.

"Ueber die $\Delta^{5,16}$-3β-Oxy-ätiocholadiensäure und einige ihrer Umwandlungsprodukte," by L. Ruzicka, E. Hardegger, and C. Kauter, *Helv. Chim. Acta*, 27 (1944) 1164–1173.

"Ueber den sterischen Verlauf der Hydrierung von Doppelbindungen in 17,20-Stellung der Steroide," by P. A. Plattner, H. Bucher, and E. Hardegger, *Helv. Chim. Acta*, 27 (1944) 1177–1184.

"Ueber zwei weitere Homologe der digitaloiden Aglucone," by P. A. Plattner, E. Hardegger, and H. Bucher, *Helv. Chim. Acta*, 28 (1945) 167–173.

"Versuche zur Herstellung von 4,13-Dioxychrysen-Derivaten," by E. Hardegger, D. Redlich, and A. Gal, *Helv. Chim. Acta*, 28 (1945) 628–637.

"Neue Derivate und Umwandlungsprodukte von Δ^5-17-Aethinylandrosten-diol-(3β,17α) und Δ^5-Pregnen-20-on-diol-(3β,17α)," by E. Hardegger and C. Scholz, *Helv. Chim. Acta*, 28 (1945) 1355–1360.

"Ueber Δ^2-Androsten-dion-(6,17) und Δ^2-Cholestenon-(6)," by L. Blunschy, E. Hardegger, and H. L. Simon, *Helv. Chim. Acta*, 29 (1946) 199–204.

"Herstellung des β-(cis-p-Oxy-cyclohexyl)-$\Delta^{\alpha,\beta}$-butenolids," by E. Hardegger, H. Heusser, and F. Blank, *Helv. Chim. Acta*, 29 (1946) 477–483.

"Herstellung weiterer Modell-Lactone für digitaloide Aglucone," by E. Hardegger, *Helv. Chim. Acta*, 29 (1946) 1195–1198.

"Ueber die Herstellung von α- und β-Chinovose-tetraacetat aus Glucose," by E. Hardegger, *Helv. Chim. Acta*, 29 (1946) 1199–1203.

"Versuche zur Herstellung von 3,4-Diacetyl-d-chinovosan-α-1,2;1,5," by E. Hardegger and R. M. Montavon, *Helv. Chim. Acta*, 30 (1947) 632–638.

"Osotriazole aus l-Xylose-, l-Rhammose- und d-Chinovose-phenylosazon," by E. Hardegger and H. El Khadem, *Helv. Chim. Acta*, 30 (1947) 900–904.

"p-Tolylosotriazole einiger Monosaccharide," by E. Hardegger and H. El Khadem, *Helv. Chim. Acta*, 30 (1947) 1478–1483.

"Glucoside und β-1,3,4,6-Tetraacetyl-glucose aus Triacetyl-glucosan-α-1,2;1,5," by E. Hardegger and J. de Pascual, *Helv. Chim. Acta*, 31 (1948) 281–286.

"Herstellung und Hydrogenolyse von Benzhydrylestern," by E. Hardegger, Z. El Heweihi, and F. G. Robinet, *Helv. Chim. Acta*, 31 (1948) 439–445.

"Di-tosylierung der Glucose," by E. Hardegger, R. M. Montavon, and O. Jucker, *Helv. Chim. Acta*, 31 (1948) 1863–1867.

"Umwandlungsprodukte von α-1,3,4-Triacetyl-2,6-ditosyl-glucose," by E. Hardegger, O. Jucker, and R. M. Montavon, *Helv. Chim. Acta*, 31 (1948) 2247–2251.

"Derivate der 3,6-Anhydro-glucose und des Glucose-6-jodhydrins," by E. Hardegger and O. Jucker, *Helv. Chim. Acta*, 32 (1949) 1158–1162.

"Herstellung von Derivaten des α-Methyl-D-glucuro-pyranosids," by E. Hardegger and D. Spitz, *Helv. Chim. Acta*, 32 (1949) 2165–2170.

"Herstellung und Derivate des β-Methyl-D-glucuro-pyranosids," by E. Hardegger and D. Spitz, *Helv. Chim. Acta*, 33 (1950) 337–342.

"β-D-Quinovosides du cholesterol, de l'acide Δ^5-hydroxy-3-cholenique et leurs produits d'hydrogenation," by E. Hardegger and F. G. Robinet, *Helv. Chim. Acta*, 33 (1950) 456–462.

"Kristallisierte Mercaptale von D-Ribose, von D-Lyxose und Derivate des Galactose- und Glucose-dibenzyl-mercaptals," by E. Hardegger, E. Schreier, and Z. El Heweihi, *Helv. Chim. Acta*, 33 (1950) 1159–1164.

"β-D-Quinovoside de l'acide oleanolique," by E. Hardegger and F. G. Robinet, *Helv. Chim. Acta*, 33 (1950) 1871–1876.

"D-Arabohexose-β-naphthyl-osotriazol und p-Bromphenyl-osotriazole einiger Monosaccharide," by E. Hardegger, H. El Khadem, and E. Schreier, *Helv. Chim. Acta*, 34 (1951) 253–257.

"Synthesen mit 3,4,6-Triacetyl-glucosan (Brigl-Anhydrid)," by E. Hardegger and H. Leeman, *Chimia*, 5 (1951) 108–109.

"Zur Kenntnis der D-, L- und DL-Erythron- und Threon-säurelactone," by E. Hardegger, K. Kreis, and H. El Khadem, *Helv. Chim. Acta*, 34 (1951) 2343–2348.

"Die Konstitution des Diels'schen 'Anhydro-phenylosazons'," by E. Hardegger and E. Schreier, *Helv. Chim. Acta*, 35 (1952) 232–247.

"Oxydation einiger Mono- und Disaccharide mit Alkali und Sauerstoff," by E. Hardegger, K. Kreis, and H. El Khadem, *Helv. Chim. Acta*, 35 (1952) 618-623.

"Ueber das 3,6- und das 5,6-Anhydro-D-fructose-phenylosotriazol," by E. Hardegger and E. Schreier, *Helv. Chim. Acta*, 35 (1952) 623-631.

"Glucoside von Oleoanolsäure-estern und glykosidische Bindung im Zuckerrüben-Saponin," by E. Hardegger, H. J. Leemann, and F. G. Robinet, *Helv. Chim. Acta*, 35 (1952) 824-829.

"Oxydativer Abbau von Uronsäuren und Derivate der Trioxy-glutarsäuren," by E. Hardegger, K. Kreis, and D. Spitz, *Helv. Chim. Acta*, 35 (1952) 958-963.

"Ueber die Umwandlung des D-Fructose-phenylosazons in das Diels-Anhydro-osazon," by H. El Khadem, E. Schreier, G. Stöhr, and E. Hardegger, *Helv. Chim. Acta*, 35 (1952) 993-999.

"Beweis der Konfiguration des Pseudo-tropins, bezw. des Tropins," by E. Hardegger and H. Ott, *Helv. Chim. Acta*, 36 (1953) 1186-1189.

"Anhydroderivate der Tagatose- bezw. Sorbose-phenyl-osazone und -phenylosotriazole," by E. Schreier, G. Stöhr, and E. Hardegger, *Helv. Chim. Acta*, 37 (1954) 35-41.

"Abbau von vier 3,6-Anhydro-hexose-phenylosotriazolen zu kristallisierten antipoden Dialdehyd-hydraten," by E. Schreier, G. Stöhr, and E. Hardegger, *Helv. Chim. Acta*, 37 (1954) 574-583.

"Umsetzung von Nor-ψ-tropin und Nor-tropin mit Aldehyden," by E. Hardegger and H. Ott, *Helv. Chim. Acta*, 37 (1954) 685-690.

"Herstellung des Indol-4-aldehyds," by E. Hardegger and H. Corrodi, *Helv. Chim. Acta*, 37 (1954) 1826-1827.

"Derivate des Diäthanolamins und des N-Oxyäthyl-morpholins," by E. Hardegger and H. Ott, *Helv. Chim. Acta*, 38 (1955) 213-215.

"Konfiguration des Cocains und Derivate des Ecgoninsäure," by E. Hardegger and H. Ott, *Helv. Chim. Acta*, 38 (1955) 312-320.

"Herstellung und Umsetzungen der 4-Halogen-indole," by H. Corrodi and E. Hardegger, *Chimia*, 9 (1955) 116.

"Umsetzungen des o-Nitrophenylessigesters und des 2-Chlor-6-nitrophenyl-brenztraubensäureesters," by A. Romeo, H. Corrodi, and E. Hardegger, *Helv. Chim. Acta*, 38 (1955) 463-467.

"Umwandlungsprodukte der 4-Chlorindol-2-carbonsäure und des 4-Chlorindols," by E. Hardegger and H. Corrodi, *Helv. Chim. Acta*, 38 (1955) 468-473.

"Die Konfiguration des Colchicins und verwandter Verbindungen," by H. Corrodi and E. Hardegger, *Helv. Chim. Acta*, 38 (1955) 2030-2033.

"Die Konfiguration des Apomorphins und verwandter Verbindungen," by H. Corrodi and E. Hardegger, *Helv. Chim. Acta*, 38 (1955) 2038-2043.

"Fusarinsäure-[carboxyl-^{14}C] und quaternäre Derivate der Fusarinsäure," by E. Hardegger and E. Nikles, *Helv. Chim. Acta*, 39 (1956) 223-229.

"Untersuchungen in der Pyridin-Reihe; ein neuer Weg zur Fusarinsäure," by E. Hardegger and E. Nikles, *Helv. Chim. Acta*, 39 (1956) 505-513.

"Herstellung von 4-Jod-oxindol, 4-Jod-indol, 4-Jod-skatolyl-aceton und verwandten Verbindungen," by E. Hardegger and H. Corrodi, *Helv. Chim. Acta*, 39 (1956) 514-517.

"Die Konfiguration des natürlichen (+)-Laudanosins, sowie verwandter Tetrahydro-isochinolin-, Aporphin- und Tetrahydro-berberin-alkaloide," by H. Corrodi and E. Hardegger, *Helv. Chim. Acta*, 39 (1956) 889-897.

"Notiz über die β-Carboxy-γ-methylamino-buttersäure und den Methylamino-malonester," by E. Hardegger and H. Corrodi, *Helv. Chim. Acta*, 39 (1956) 980-983.

"Notiz über einige dem Tryptophan nahestehende Verbindungen und über das 4-

Cyan-indolyl-(3)-acetonitril," by E. Hardegger and H. Corrodi, *Helv. Chim. Acta*, 39 (1956) 984-986.

"Herstellung des racemischen Colchicins und des un-natürlichen (+)-Colchicins," by H. Corrodi and E. Hardegger, *Helv. Chim. Acta*, 40 (1957) 193-199.

"Synthese von Stereoisomeren des Muscarins," by H. Corrodi, E. Hardegger, F. Kögl, and P. Zeller, *Experientia*, 13 (1957) 138-141.

"Synthese von S(+)-6,7-Dihydroxy-3-tropanon," by E. Hardegger and H. Furter, *Helv. Chim. Acta*, 40 (1957) 872-874.

"Diaporthin, ein Welketoxin aus Kulturen von *Endothia parasitica* (Murr.) And.," by A. Boller, E. Gäumann, E. Hardegger, F. Kugler, S. Naef-Roth, and M. Rosner (18. Mitteilung, Welkstoffe und Antibiotika), *Helv. Chim. Acta*, 40 (1957) 875-880.

"Eine neue Synthese der Fusarinsäure," by E. Hardegger and E. Nikles (19. Mitteilung, Welkstoffe und Antibiotika), *Helv. Chim. Acta*, 40 (1957) 1016-1021.

"Ueber induzierte Abwehrstoffe bei Orchideen I.," by A. Boller, H. Corrodi, E. Gäumann, E. Hardegger, H. Kern, and N. Winterhalter-Wild (20. Mitteilung, Welkstoffe und Antibiotika), *Helv. Chim. Acta*, 40 (1957) 1062-1066.

"Präparative Herstellung der 2-Desoxy-D-ribose," by E. Hardegger, M. Schellenbaum, R. Huwyler, and A. Züst, *Helv. Chim. Acta*, 40 (1957) 1815-1818.

"Die absolute Konfiguration des (+)-Catechins," by E. Hardegger, H. Gempeler, and A. Züst, *Helv. Chim. Acta*, 40 (1957) 1819-1822.

"Ueber Muscarin, 7. Mitteilung. Synthese und absolute Konfiguration des Muscarins," by E. Hardegger and F. Lohse, *Helv. Chim. Acta*, 40 (1957) 2384-2389.

"Die Reaktion von 3-Butyl-pyridin mit Methylradikalen," by E. Hardegger and E. Nikles, *Helv. Chim. Acta*, 40 (1957) 2421-2427.

"Herstellung der Fusarinsäure aus 2,5-Lutidin," by E. Hardegger and E. Nikles (21. Mitteilung, Welkstoffe und Antibiotika), *Helv. Chim. Acta*, 40 (1957) 2428-2433.

"Ueber Muscarin, 8. Mitteilung. Herstellung von racemischen Allomuscarin," by H. Corrodi, E. Hardegger, and F. Kögl, *Helv. Chim. Acta*, 40 (1957) 2454-2461.

"Ueber Muscarin, 9. Mitteilung. Ueber die Synthese von racemischen Muscarin, seine Spaltung in die Antipoden und die Herstellung von (−)-Muscarin aus D-Glucosamin," by H. C. Cox, E. Hardegger, F. Kögl, P. Liechti, F. Lohse, and C. A. Salemink, *Helv. Chim. Acta*, 41 (1958) 229-234.

"Ueber Muscarin, 10. Mitteilung. Herstellung des un-natürlichen $2R,3S,5R$-(−)-Muscarins aus 2-Desoxy-D-ribose," by E. Hardegger, H. Furter, and J. Kiss, *Helv. Chim. Acta*, 41 (1958) 2401-2410.

"Hydroxy-morphinane, 12. Mitteilung. Die Konfiguration der Morphinane," by H. Corrodi, J. Hellerbach, A. Züst, E. Hardegger, and O. Schnider, *Helv. Chim. Acta*, 42 (1959) 212-217.

"Relative Configurations of the Two 1-Keto-6-hydroxy*spiro*[4.4]nonanes and the Three 1,6-Dihydroxy*spiro*[4.4]nonanes," by E. Hardegger, E. Maeder, H. M. Semarne, and D. J. Cram, *J. Am. Chem Soc.*, 81 (1959) 2729-2737.

"Beiträge zur Stereochemie der Chitarsäure," by J. Kiss, E. Hardegger, and H. Furter, *Chimia*, 13 (1959) 336-338.

"2-Deoxy-D-ribose," by E. Hardegger, *Methods Carbohydr. Chem.*, 2 (1960) 177-179.

"Die absolute Konfiguration des (−)-Epicatechins," by A. Züst, F. Lohse, and E. Hardegger, *Helv. Chim. Acta*, 43 (1960) 1274-1279.

"Verwendung radioaktiver Fusarinsäure und Synthese von Fusarinsäure-[2,5'-^{14}C]," by H. Biland, F. Lohse, and E. Hardegger (22. Mitteilung, Welkstoffe und Antibiotika), *Helv. Chim. Acta*, 43 (1960) 1436-1440.

"Ueber die Isolierung von Culmomarasmin, einem peptidartigen Welkstoffe aus dem Kulturfiltrat von *Fusarium culmorum* (W. G. Sm.) Sacc.," by J. Kiss, S. Naef-Roth, E. Hardegger, A. Boller, F. Lohse, E. Gäumann, and P. A. Plattner (23. Mitteilung, Welkstoffe und Antibiotika), *Helv. Chim. Acta*, 43 (1960) 2096–2101.

"Ueber Muscarin, 11. Mitteilung. Synthese von bisquaternären, dem Muscarin ähnlich Verbindungen," by J. Kiss, H. Furter, F. Lohse, and E. Hardegger, *Helv. Chim. Acta*, 44 (1961) 141–147.

"Konfigurationsbeweis für L-Tryptophan durch erschöpfende Ozonisierung," by E. Hardegger and H. Braunschweiger, *Helv. Chim. Acta*, 44 (1961) 1125–1127.

"Ueber Muscarin, 12. Mitteilung. Synthese des DL-2-Methyl-muscarons, des DL-2-Methyl-muscarins und des DL-2-Methyl-epimuscarins," by H. Corrodi, K. Steiner, N. Halder, and E. Hardegger, *Helv. Chim. Acta*, 44 (1961) 1157–1162.

"Ueber Muscarin, 13. Mitteilung. Synthese von Salzen des DL-Muscarons und DL-Allomuscarons über die Norbasen," by E. Hardegger, H. Corrodi, and N. Chariatte, *Helv. Chim. Acta*, 44 (1961) 1193–1198.

"Die absolute Konfiguration des Aglucons im Neosperidin," by E. Hardegger and H. Braunschweiger, *Helv. Chim. Acta*, 44 (1961) 1413–1417.

"Die Konstitution des Lycomarasmins," by E. Hardegger, P. Liechti, L. M. Jackman, A. Boller, and P. A. Plattner (24. Mitteilung, Welkstoffe und Antibiotika), *Helv. Chim. Acta*, 46 (1963) 60–74.

"Präparative Herstellung von Krist. 6-Desoxy-6-amino-D-glucose-hydrochlorid und Konstitution einiger Zwischenprodukte," by E. Hardegger, G. Zanetti, and K. Steiner, *Helv. Chim. Acta*, 46 (1963) 282–287.

"Synthese der natürlichen Dehydrofusarinsäure," by K. Steiner, U. Graf, and E. Hardegger (25. Mitteilung, Welkstoffe und Antibiotika), *Helv. Chim. Acta*, 46 (1963) 690–694.

"Synthese der Anhydrolycomarasminsäure," by E. Hardegger, J. Seres, R. Andreatta, F. Szabo, W. Zankowska-Jasinska, A. Romeo, C. Rostetter, and H. Kindler (Vorläufige 26. Mitteilung, Welkstoffe und Antibiotika), *Helv. Chim. Acta*, 46 (1963) 1065–1067.

"Ueber induzierte Abwehrstoffe bei Orchideen II," by E. Hardegger, M. Schellenbaum, and H. Corrodi (27. Mitteilung, Welkstoffe und Antibiotika), *Helv. Chim. Acta*, 46 (1963) 1171–1180.

"Synthese von 2,4-Dimethoxy-6-hydroxy-phenanthren und Konstitution des Orchinols," by E. Hardegger, H. R. Biland, and H. Corrodi (28. Mitteilung, Welkstoffe und Antobiotika), *Helv. Chim. Acta*, 46 (1963) 1354–1360.

"Synthese von 2,4-Dimethoxy-6-hydroxy-9,10-dihydrophenanthren," by E. Hardegger, N. Rigassi, J. Seres, C. Egli, P. Müller, and K. O. Fitzi (29. Mitteilung, Welkstoffe und Antibiotika), *Helv. Chim. Acta*, 46 (1963) 2543–2551.

"Synthese des 5,6,7-Trimethoxy-indols und des 6,7-Dimethoxy-bufotenin methyläthers," by E. Hardegger and H. Corrodi, *Pharm. Acta Helv.*, 39 (1964) 101–107.

"Ausgangsprodukte zur Totalsynthese des Javanicins," by E. Hardegger, K. Steiner, E. Widmer, H. Corrodi, T. Schmidt, H. P. Knopfel, W. Rieder, H. J. Meyer, F. Kugler, and H. Gempeler (30. Mitteilung, Welkstoffe und Antibiotika), *Helv. Chim. Acta*, 47 (1964) 1996–2017.

"Bicyclische Zwischenprodukte zur Synthese des Javanicins," by E. Hardegger, K. Steiner, E. Widmer, and T. Schmidt (31. Mitteilung, Welkstoffe und Antibiotika), *Helv. Chim Acta*, 47 (1964) 2017–2026.

"Synthese von Javanicin-5,8-dimethyläther und Konstruktion von Javanicin und Fusarubin," by E. Hardegger, K. Steiner, E. Widmer, and A. Pfiffner (32. Mitteilung, Welkstoffe und Antibiotika), *Helv. Chim. acta*, 47 (1964) 2027–2030.

"Selektion Spaltung der Aether Chelierter Phenole und Synthese von Javanicin," by

E. Hardegger, E. Widmer, K. Steiner, and A. Pfiffner (33. Mitteilung, Welkstoffe und Antibiotika), *Helv. Chim. Acta,* 47 (1964) 2031–2037.

"Synthese von Isojavanicin und vereinfachte Synthese von Javanicin," by E. Widmer, H. J. Meyer, A. Walser, and E. Hardegger (34. Mitteilung, Welkstoffe und Antibiotika), *Helv. Chim. Acta,* 48 (1965) 538–553.

"Ueber Muscarin, 14. Mitteilung. Konfigurative Zusammenhänge," by E. Hardegger, N. Chariatte, and N. Halder, *Helv. Chim. Acta,* 49 (1965) 580–586.

"Konstitution des Diaporthins und Synthese der Diaporthinsäure," by E. Hardegger, W. Rieder, A. Walser, and F. Kugler (35. Mitteilung, Welkstoffe und Antibiotika), *Helv. Chim. Acta,* 49 (1966) 1283–1290.

"Ueber Muscarin, 15. Mitteilung. Reduktion Normuscaron mit Borwasserstoff und bicyclische Derivate der Muscarinreihe," by E. Hardegger and N. Halder, *Helv. Chim. Acta,* 50 (1967) 1275–1283.

"Synthese von Verbindungen der Lycomarasmin-Reihe," by E. Hardegger, R. Andreatta, F. Szabo, W. Zankowska-Jasinska, C. Rostetter, and H. Kindler (36, Mitteilung, Welkstoffe und Antibiotika), *Helv. Chim. Acta,* 50 (1967) 1540–1545.

"Synthese der DL-Form eines natürlichen Prostaglandins (Vorläufige Mitteilung)," by E. Hardegger, H. P. Schenk, and E. Broger, *Helv. Chim. Acta,* 50 (1967) 2501.

"Zwischenprodukte zur Synthese von Verbindungen der Lycomarasmin-Reihe aus Glycin und L-Serin," by E. Hardegger, F. Szabo, P. Liechti, C. Rostetter, and W. Zankowska-Jasinska (37. Mitteilung, Welkstoffe und Antibiotika), *Helv. Chim. Acta,* 51 (1968) 78–85.

"Zur Synthese der Anhydro-lycomarasminsäure," by E. Hardegger, C. Rostetter, J. Seres, and R. Andreatta (38. Mitteilung, Welkstoffe und Antibiotika), *Helv. Chim. Acta,* 52 (1969) 873–880.

"Eine präparative Synthese von Streptozoticin," by E. Hardegger, A. Meier, and A. Stoos, *Helv. Chim. Acta,* 52 (1969) 2555–2560.

"Synthese der 2-Thio-D-glucose und einiger 3-Thio-D-altrose-Derivate," by E. Hardegger and W. Schüep, *Helv. Chim. Acta,* 53 (1970) 951–959.

"Herstellung der 1-Thio-D-chinovose," by W. Schüep and E. Hardegger, *Helv. Chim. Acta,* 53 (1970) 1336–1339.

"Fusarinolsäure," by K. Steiner, U. Graf, and E. Hardegger (39. Mitteilung, Welkstoffe und Antibiotika), *Helv. Chim. Acta,* 54 (1971) 845–851.

"Synthese von 2-Thioaldosen über α,β-ungesättigte Nitrokörper," by P. Wirz and E. Hardegger, *Helv. Chim. Acta,* 54 (1971) 2017–2026.

"Synthese von D- und L-S-Auro-1-thio-xylit," by P. Wirz, J. Staněk, Jr., and E. Hardegger, *Helv. Chim. Acta,* 54 (1971) 2027–2030.

"Salicaceae, 7-O-Methylaromadendrin from *Populus alba*," by A. Stoessl, A. Toth, E. Hardegger, and H. Kern, *Phytochemistry,* 10 (1971) 1972.

"Versuche zur Synthese der 4-Thio-D-glucose," by L. Vegh and E. Hardegger, *Helv. Chim. Acta,* 56 (1973) 1792–1799.

"Notiz über 1,6-Anhydro-2-S-benzyl-2-thio-β-D-idopyranose," by L. Vegh and E. Hardegger, *Helv. Chim. Acta,* 56 (1973) 1961–1962.

"Zur Kenntnis der 4-Thioglucose," by L. Vegh and E. Hardegger, *Helv. Chim. Acta,* 56 (1973) 2020–2025.

"4-Desoxy-1-thio-β-D-glucose und deren Auro-(1:1)-Komplex," by L. Vegh and E. Hardegger, *Helv. Chim. Acta,* 56 (1973) 2079–2082.

"Synthesen von Dehydroorchinol-methyläther und Dehydroorchinol," by P. Müller, J. Seres, K. Steiner, S. E. Helali, and E. Hardegger (40. Mitteilung, Welkstoffe und Antibiotika), *Helv. Chim. Acta,* 57 (1974) 790–795.

"2,4-Dimethoxy-8-hydroxy-, 2,4-Dimethoxy-7-hydroxy-, und 1,3,4-Trimethoxy-6-hydroxy-9-methyl-phenanthren," by C. Egli, J. Seres, K. Steiner, S. E. Helali, and E. Hardegger (41. Mitteilung, Welkstoffe und Antibiotika), *Helv. Chim. Acta*, 57 (1974) 796–802.

"Zur Synthese des Orchinols," by K. Steiner, C. Egli, N. Rigassi, S. E. Helali, and E. Hardegger (42. Mitteilung, Welkstoffe und Antibiotika), *Helv. Chim. Acta*, 57 (1974) 1137–1141.

"Modifizierte Streptozotocine," by A. Meier, F. Stoos, D. Martin, G. Büyük, and E. Hardegger, *Helv. Chim. Acta*, 57 (1974) 2622–2626.

"Die Benzoin-Phenol-Synthese," by C. Egli, S. E. Helali, and E. Hardegger, *Helv. Chim. Acta*, 58 (1975) 104–110.

"Varianten im Zuckerteil des Streptozotocins," by N. Gassmann, F. Stoos, A. Meier, G. Büyük, S. E. Helali, and E. Hardegger, *Helv. Chim. Acta*, 58 (1975) 182–185.

"Fusarinonsäure und die enantiomeren Fusarinolsäuren," by G. Büyük and E. Hardegger (43. Mitteilung, Welkstoffe und Antibiotika), *Helv. Chim. Acta*, 58 (1975) 682–687.

APPENDIX II

The following is a chronological list of doctoral theses* supervised by Professor Emil Hardegger.

"Synthese des Cyclohexano-α-pyrons. Umwandlungen von Δ^5-3,17-Dioxy-17-aethinyl-androsten," by K. Scholz (1946).

"Tosylation Partielle du Glucose. Quinovose et Diacetylquinovosane," by R. Montavon (1949).

"Analytische Untersuchungen an Mono- und Disacchariden," by H. S. El Khadem (1950).

"A. Mercaptale und Xanthogenate der Kohlenhydrate und Derivate der 6-Thioglucose. B. Herstellung und Hydrogenolyse von Benzhydrylestern," by Z. El Heweihi (1950).

"Di-Tosylierung der α-D-Glucose und Versuche zur Herstellung von 5-Thiomethyl-Ribose," by O. Jucker (1950).

"Ueber die D-Glucuronsäure und den Zuckerteil der Glyzyrrhizinsäure," by D. Spitz (1951).

"Saponosides Stéroïdes et Triterpéniques de Synthèse," by F.-G. Robinet (1951).

"Synthese von Glucosiden und Glucose-1-estern mit Triacetylglucosan-α(1,2)-β(1,5)," by H. Leeman (1951).

"Die Konstitution des Diels'schen Anhydro-D-glucose-phenylosazons," by E. Schreier (1953).

"Oxydativer Abbau reduzierender Zucker und Diels'scher Anhydro-hexosazone," by K. Kreis (1953).

"Zur Kenntnis der Diels'schen Anhydro-hexosazone," by G. Stöhr (1953).

"Der räumliche Bau der Tropa-alkaloide," by H. Ott (1955).

"Versuche zur Totalsynthese der Lysergsäure," by H. R. Corrodi (1955).

"Versuche zur Totalsynthese der Lysergsäure," by M. Gerecke (1957).

"Ueber Lycomarasmin," by P. Liechti (1958).

---

* The doctoral theses of the Eidgenössiche Technische Hochschule are printed, and are available at the ETH library.

"Synthese von (−)-Muscarin aus 2-Desoxy-D-ribose. Synthese von Muscarin-ähnlichen Verbindungen und ein neuer Konfigurationsbeweis für Chitarsäure," by H. Furter (1958).
"Synthese und absolute Konfiguration des Muscarins und seines Antipoden," by F. Lohse (1958).
"Die relative Konfiguration der drei 1,6-Dihydroxy-spiro[4.4]nonane und der beiden 1-Keto-6-hydroxy-spiro[4.4]nonane," by E. Maeder (1958).
"Neue Synthesen der Fusarinsäure," by E. Nikles (1958).
"Synthesen der 2,5-Bisdesoxy-D-ribose und verwandter Verbindungen," by R. Huwyler (1959).
"Isolierung und Konstitutionsaufklärung des Orchinols," by M. Schellenbaum (1959).
"Die absolute Konfiguration des Catechins, Epicatechins und des Dromorans," by A. Züst (1960).
"Hydrolyse des Kanamycins und Versuche zur präparativen Synthese der Spaltprodukte," by G. Zanetti (1962).
"Isolierung und Konstitutionsermittlung von Marticin und Isomarticin, zwei neuen Welktoxinen aus *Fusarium martii*," by A. Pfiffner (1963).
"Diaporthin und Synthese von Diaporthinsäure," von W. Rieder (1963).
"Synthese von Iso-orchinol und verwandten Verbindungen," by N. Rigassi (1963).
"A. Synthesen in der Furanreihe. B. Synthese von Dehydroorchinol," by P. Müller (1964).
"Zwischenprodukte zur Synthese der Prostaglandine," by W. Graf (1965).
"Beiträge zur Synthese der Prostaglandine," by H. A. Kindler (1966).
"Versuche zur Synthese der Prostaglandine," by F. Näf (1967).
"Totalsynthese von Prostaglandinen," by H. Schenk (1967).
"Versuche zur Totalsynthese der Prostaglandine," by J. Vonarburg (1967).
"Beiträge zur Synthese der Prostaglandine," by F. Leuenberger (1968).
"Totalsynthese von prostaglandinähnlichen Verbindungen," by H. Hoffmann (1968).
"Versuche zur Synthese der Prostaglandine," by S. E. Helali (1969).
"Versuche zur Totalsynthese der Prostaglandine," by U. Koelliker (1969).
"Ueber Harnstoffderivate des D-Glucosamins," by F. Stoos (1969).
"Versuche zu neuartigen Glucosidsynthesen," by K. Meyer (1970).
"Zur präparativen Herstellung von Monosacchariden," by P. Guldbrandsen (1970).
"Synthesen von Albocorticin," by W. Mueller (1970).
"Synthese von Thiozuckern," by P. Wirz (1971).
"Elektrochemische Reduktionen und Oxidationen in der Zucker-reihe," by B. Bruttel (1972).
"Beiträge zur Synthese mittlerer Ringe," by J. Ott (1973).
"Cystein-Derivate der D-Glucose," by E. Winkelmann (1973).
"Versuche zu Totalsynthesen in der Prostaglandinreihe," by P. de Roche (1973).
"Oxydation von D-Glucose und von Diaceton-L-sorbose mit Sauerstoff," by F. Bacher (1976).
"Epimerisierungen von Aldonsäuren und Aldonsäure-Isopropyliden-derivaten," by R. D. Uribe-Echevarria (1976).

CARBON-13 NUCLEAR MAGNETIC RESONANCE SPECTROSCOPY OF POLYSACCHARIDES

By Philip A. J. Gorin

Prairie Regional Laboratory, National Research Council of Canada, Saskatoon, Saskatchewan S7N 0W9, Canada

I. Introduction	13
II. Configurationally Dependent, ^{13}C Parameters	15
1. Chemical Shifts	15
2. Single-Bond, ^{13}C–H Coupling	17
3. Three-Bond, ^{13}C–H Coupling	17
III. Identification of ^{13}C Signals of Monosaccharides: O-Substitution Shifts	19
IV. Identification of ^{13}C-N.m.r. Signals of Oligosaccharides and Polysaccharides	22
V. Molecular Motion	25
1. Quantitation of Signal Intensities: Spectral Quality	25
2. T_1, T_2, and n.O.e. Values of Solutions and Gels	26
VI. ^{13}C-N.m.r Spectra of Polysaccharides	37
1. α-D-Glucans	37
2. β-D-Glucans	48
3. Linear D-Mannans	52
4. Branched-Chain D-Mannans	56
5. Heteropolymers Having D-Mannose-Containing Main-Chains	60
6. Furanose-Containing Structures	68
7. Sulfated Polysaccharides	72
8. O-Phosphonohexopyranans	81
9. A Teichoic Acid	88
10. Polysaccharides Containing Sialic Acid	89
11. Polysaccharides Containing 3-Deoxy-D-*manno*-octulosylonic Acid Residues	91
12. Alginic Acid	94
13. D-Xylose-Containing Polysaccharides	95
14. Miscellaneous Heteropolymers	96
15. Synthetic Derivatives of Polysaccharides	97

I. Introduction

The advent of studies of polysaccharides by carbon-13 nuclear magnetic resonance (^{13}C-n.m.r.) spectroscopy coincided, not surprisingly, with that of instrumentation incorporating Fourier transform. Although, in some instances, the ^{13}C technique is complementary to proton magnetic resonance (p.m.r.) spectroscopy, it does in fact fill a large gap left by the older technique. Allerhand[1] pointed out that the amount of information provided by proton spectra diminishes as the size and complexity of the molecule increases, and that the "resonance density" is comparatively high. This factor is aggravated by the tendency, of protons in polymers, towards low values of the spin–spin relaxation-time (T_2) and marked broadening of signals.

The well-defined, ^{13}C spectra obtained under conditions of proton decoupling have provided a wealth of information on chemical and physicochemical properties of polysaccharides. The expansion in this field is evident from a review by Komoroski and coworkers[2] covering the literature on the ^{13}C-n.m.r. spectroscopy of polymers, up to 1974, that contained only 11 references to work conducted specifically on polysaccharides. Shaskov and Chizhov[3] published a detailed article on the ^{13}C-n.m.r. spectroscopy of all carbohydrates. Jennings and I. C. P. Smith[4] reviewed methods for the assignment of signals, sequencing, and conformation, using selected polysaccharides as examples, and provided data on experimental techniques used in obtaining proton-decoupled, ^{13}C-n.m.r. spectra. Perlin has discussed carbon-13 n.m.r. spectroscopy of carbohydrates, including oligo- and poly-saccharides,[5] and glycosaminoglycans in particular.[6]

The present article covers literature published to the end of 1978 on aspects of ^{13}C-n.m.r. spectroscopy relating to polysaccharides. The ^{13}C-n.m.r. spectra of polysaccharides can serve a number of useful purposes, and of particular interest is the information given on structural sequences in simple polymers. For example, a hexopyranan that is homogeneous in structure or has a two-unit, or three-unit, repeating sequence can give 6, 12, or 18 resonances, respectively. The extent of long-range effects of substitution, namely, those shift effects more distant than α and β carbon nuclei, is not yet clear, and a spectrum containing 24 signals, corresponding to a four-unit repeating-structure, has not been found. Values of $J_{C,H}$ can indicate glycosidic configuration and, sometimes, the macromolecular conformation, as

(1) A. Allerhand, *Pure Appl. Chem.*, 41 (1975) 247–273.
(2) R. A. Komoroski, I. R. Peat, and G. C. Levy, *Top. Carbon-13 NMR Spectrosc.*, 2 (1976) 179–267.
(3) A. S. Shashkov and O. S. Chizhov, *Bioorg. Khim.*, 2 (1976) 437–497; English version, pp. 312–368.
(4) H. J. Jennings and I. C. P. Smith, *Methods Enzymol.*, 50 Part C (1978) 39–50.
(5) A. S. Perlin, *MTP Int. Rev. Sci. Org. Chem.*, Ser. Two, 7 (1976) 1–34.
(6) A. S. Perlin, *Fed. Proc.*, 36 (1977) 106–109.

expressed by glycosidic torsion-angles. Also, information at the macromolecular level can be gained through values of T_2, particularly in gel structures, and other insights into molecular motion and chemical structure may be obtained through measurements of the spin–lattice relaxation-time, T_1.

^{13}C-n.m.r. spectroscopy is useful in monitoring the purity of polysaccharide preparations. Similarly, in a multicomponent, biological system, such as exists in fungal cell-walls, the varying proportion of each polysaccharide can be gauged, once characteristic signals have been identified.

As with p.m.r. spectroscopy, the ^{13}C technique has the potential of detecting unusual or unexpected structures within polysaccharides. Location of such substituents as acetate, malonate, phosphate, or sulfate groups is often simple. However, as in determining the position of O-glycosylation, it is first necessary to correlate each signal to a given carbon atom. As compared with monosaccharides, fewer routes are available for direct, unambiguous assignments of polysaccharide resonances. The commonest strategy, which is at least applicable with spectra having well separated signals, is indirect. It is assumed that the chemical shifts of the polysaccharide signals are the same as those of the component monosaccharide having the appropriate configuration, and that a strong, downfield displacement of the O-glycosylated resonance (α-effect) takes place, with smaller upfield displacements of the signal(s) of adjacent carbon atoms (β-effect). More-distant effects, of >1 p.p.m., are not common.

One confusing aspect of the literature on ^{13}C-n.m.r. spectroscopy is the large number of different systems for signal reference that have been used, leading to a multitude of different, chemical-shift values. These include external carbon disulfide, internal sodium 2,2,3,3-tetradeuterio-4,4-dimethyl-4-silapentanoate (TSP), and sodium 4,4-dimethyl-4-silapentane-1-sulfonate (DSS), and the preferred, external tetramethylsilane. In the present article, the original literature values are given, and, although correction factors are available for different reference-standards, and for temperature and pH, they have been avoided for fear of error.

II. Configurationally Dependent, ^{13}C Parameters

1. Chemical Shifts

The C-1 resonances of the pyranoid forms of glucose, xylose, galactose, arabinose, methyl glucoside, and methyl xyloside were shown by Hall and Johnson[7] to be sensitive to the anomeric configuration. This is also true of the aldofuranoid series.[8–10] In subsequent Sections, it will be seen that the C-1 signals of some oligosaccharides or polysaccharides fall into two different ranges of shifts, depending on whether they contain α- or β-linkages. Small, internal variations may occur, due to different O-substitution α-shifts, which can sometimes be diagnostically useful. Frequently, resonances of other carbon atoms are configurationally dependent; for example, a glucuronoxylan containing α- and β-D-xylopyranosyl residues gives two C-1 and two C-5 signals.[11] One of the C-5 signals undergoes an upfield, β-shift of 1.6 p.p.m. relative to the C-5 resonance of methyl α-D-xylopyranoside, so that 4-O-substituted α-D-xylopyranosyl residues were indicated (see Section VI, 13). Although C-1 resonances of α- and β-D-mannopyranosyl residues are often close to each other,[12] the respective C-2, C-3, and C-5 signals are readily distinguishable.

The C-1 chemical-shifts of furanosides are generally at lower field than those of their anomeric counterparts in the pyranose series.[13] Sometimes, an immediate identification may be made when very low-field signals of δ_c 107 or more are present, for example, for β-galactofuranoside[14–17] and α-arabinofuranoside.[18] Generally, characteristic signals of furanoside-ring carbon atoms are present at low field (δ_c 80 –85) also.

(7) L. D. Hall and L. F. Johnson, *J. Chem. Soc. Chem. Commun.*, (1969) 509–510.
(8) A. S. Perlin, N. Cyr, H. J. Koch, and B. Korsch, *Ann. N. Y. Acad. Sci.*, 222 (1973) 935–942.
(9) T. Usui, S. Tsushima, N. Yamaoka, K. Matsuda, K. Tuzimura, H. Sugiyama, S. Seto, K. Fujieda, and G. Miyajima, *Agric. Biol. Chem.*, 38 (1974) 1409–1410.
(10) R. G. S. Ritchie, N. Cyr, B. Korsch, H. J. Koch, and A. S. Perlin, *Can. J. Chem.*, 53 (1975) 1424–1433.
(11) L. Mendonça-Previato, P. A. J. Gorin, and J. O. Previato, *Biochemistry*, 18 (1979) 149–154.
(12) P. A. J. Gorin, *Can. J. Chem.*, 51 (1973) 2375–2383.
(13) A. S. Perlin, B. Casu, and H. J. Koch, *Can. J. Chem.*, 48 (1970) 2596–2606.
(14) M. Ogura, T. Kohama, M. Fujimoto, A. Kuniaka, H. Yoshino, and H. Sugiyama, *Agric. Biol. Chem.*, 38 (1974) 2563–2564.
(15) P. A. J. Gorin and M. Mazurek, *Carbohydr. Res.*, 48 (1976) 171–186.
(16) L. Mendonça, P. A. J. Gorin, K. O. Lloyd, and L. R. Travassos, *Biochemistry*, 15 (1976) 2423–2431.
(17) J.-P. Joseleau, M. Lapeyre, M. Vignon, and G. G. S. Dutton, *Carbohydr. Res.*, 67 (1978) 197–212.
(18) J.-P. Joseleau, G. Chambat, M. Vignon, and F. Barnoud, *Carbohydr. Res.*, 58 (1977) 165–175.

2. Single-Bond, ^{13}C–H Coupling

By using D-glucose enriched with carbon-13, Perlin and Casu[19] showed, by ^{13}C-n.m.r. and p.m.r. spectroscopy, that the $J_{C-1,H-1}$ values of the α and β anomer are 169 and 160 Hz, respectively, and a difference of ~10 Hz was observed for several anomeric pairs of pyranoses and derivatives,[20–23] and for oligosaccharides and polysaccharides.[9,15,24–26] This general, distinguishing feature is attributable to the fact that the equatorial C–H bond of the α anomer is gauche to the two lone-pair orbitals of the ring-oxygen atom, whereas the axial C–H bond of the β anomer has one trans and one gauche interaction.[27] Coupling constants of ~175 and 172.5 Hz, respectively, have been reported for the α and β anomer of methyl D-galactofuranoside,[28] and, although these values have been used to identify the β-D-galactofuranosyl residues in a polysaccharide,[17] the difference is small in comparison to the line width. Broadening often occurs through longer-range, ^{13}C–H coupling, and, in this case, a chemical-shift approach to the identification of configuration appears preferable.

Configurational dependence of $^1J_{C-1',C-2'}$, values has been observed with acetates of [1'-^{13}C]-labelled cellobiose, laminarabiose, maltose, and nigerose, the α- and β-linked disaccharides respectively giving[29] the values of ~46.5 and 49.2 Hz.

The quality of a proton-coupled, ^{13}C-n.m.r. spectrum may be improved by use of gated decoupling, whereby most of the nuclear Overhauser enhancement (n.O.e) is retained.[30]

3. Three-Bond, ^{13}C–H Coupling

Geminal, and vicinal, ^{13}C–H coupling-constants are 8 Hz or less, and difficulties have been reported in observing splittings in the ^{13}C-

---

(19) A. S. Perlin and B. Casu, *Tetrahedron Lett.*, (1969) 2921–2924.
(20) J. A. Schwarcz and A. S. Perlin, *Can. J. Chem.*, 50 (1972) 3667–3676.
(21) K. Bock, I. Lundt, and C. Pedersen, *Tetrahedron Lett.*, (1973) 1037–1040.
(22) K. Bock and C. Pedersen, *J. Chem. Soc. Perkin Trans. 2*, (1974) 293–297.
(23) K. Bock and C. Pedersen, *Acta Chem. Scand.*, 29 (1975) 258–264.
(24) F. R. Taravel and P. J. A. Vottero, *Tetrahedron Lett.*, (1975) 2341–2344.
(25) G. K. Hamer and A. S. Perlin, *Carbohydr Res.*, 49 (1976) 37–48.
(26) D. Y. Gagnaire and M. R. Vignon, *Makromol. Chem.*, 178 (1977) 2321–2333.
(27) A. S. Perlin, *Isot. Org. Chem.*, 3 (1977) 171–235.
(28) G. Chambat, J.-P. Joseleau, M. Lapeyre, and A. Lefebvre, *Carbohydr. Res.*, 63 (1978) 323–326.
(29) D. Y. Gagnaire, R. Nordin, F. R. Taravel, and M. R. Vignon, *Nouv. J. Chim.*, 1 (1977) 423–430.
(30) O. A. Gansow and W. Schittenhelm, *J. Am. Chem. Soc.*, 93 (1971) 4294–4295.

n.m.r. spectra of monosaccharides[22] and polysaccharides.[26] Vicinal coupling, observed in a ^{13}C–X–X–H system, is of particular interest, as the coupling depends on the dihedral angle formed by the three bonds joining these nuclei in the manner of the proton Karplus curve.[19,20,31] For glycosidic bonds, it is of interest to determine the ^3J value of C-1 coupled to the proton of the O-substituted carbon atom, and that of the O-substituted carbon atom and the glycosidic H-1 atom. The respective, dihedral angles may then be used to express torsion angles ϕ and ψ, respectively, where ϕ = the dihedral angle of O_{ring}–C-1–O_{glyc}–C_{subst}, and ψ = the dihedral angle of C-1–O_{glyc}–C_{subst}–C in the glycosidic bond of an oligosaccharide. Coupling measurements have generally been made by using p.m.r. spectroscopy of ^{13}C-enriched compounds, so that the ^{13}C satellite proton signals may be observed.[19,20,31] $^3J_{C-1',H}$ values were obtained for [1'-^{13}C]-labelled peracetates of cellobiose, laminarabiose, maltose, and nigerose, thus giving information on their ψ torsion angles.[32] However, Perlin and coworkers[33,34] were able to determine ^3J values of disaccharides and oligosaccharides from natural-abundance-^{13}C spectra. This was easier for the α-D-glucopyranose series, as C-1 of α-D-glucose gives a doublet having ^1J 170 Hz, with apparent line-broadening of ~2 Hz due[34] to coupling with H-5. On the other hand, the C-1 signal for α-D-glucose exhibits coupling with H-2 of −5.5 Hz, in addition to line broadening. Cyclohexaamylose[34] showed C-1',H-4 coupling by virtue of a split C-1' signal (estimated ^3J value of 3.5–4.0 Hz), an interaction that was confirmed by selective, proton decoupling. Occurrence of C-4,H-1' coupling was evidenced by line broadening of ~4 Hz. The peracetate gave unusually broad, single signals, and respective couplings of ~4.5 and ~4.0 Hz were estimated. The suggested torsion-angles ϕ and ψ are consistent with conformations in aqueous solution that are more staggered than those present in the solid state. In the acetate series, ϕ appears to be smaller, and ψ larger, than for maltose.

O-Phosphonomannans containing phosphoric diester bridges lend themselves to a similar approach, as ^{13}C–C–O–^{31}P couplings may be interpreted in the same way. This subject is discussed in Section VI,8.

(31) R. U. Lemieux, T. L. Nagabhushan, and B. Paul, *Can. J. Chem.*, 50 (1972) 773–776.
(32) G. Excoffier, D. Y. Gagnaire, and F. R. Taravel, *Carbohydr. Res.*, 56 (1977) 229–238.
(33) A. S. Perlin, N. Cyr, R. G. S. Ritchie, and A. Parfondry, *Carbohydr. Res.*, 37 (1974) c1–c4.
(34) A. Parfondry, N. Cyr, and A. S. Perlin, *Carbohydr. Res.*, 59 (1977) 299–309.

Configurationally dependent effects of Gd^{3+} on the ^{13}C-n.m.r. spectra of uronic acids and derivatives are considered in Section VI,7.

III. IDENTIFICATION OF ^{13}C SIGNALS OF MONOSACCHARIDES: O-SUBSTITUTION SHIFTS

The methods available for correct assignment of monosaccharide signals have been assessed by Shashkov and Chizhov.[3] Briefly, in the first attempt of this kind, Dorman and Roberts[35] assumed that, on going from an equatorial to an axial hydroxyl group, the shifts of α, β, and γ-carbon atoms are independent and constant. However, the method proved untenable in the light of later work by Perlin and coworkers,[13] who made the more-reliable assumption that, in such an epimerization, the changes in chemical shift may be summarized in the form of independent contributions from non-bonded interactions. This resembled the method of Angyal,[36] who estimated differences in free energy of various carbohydrate isomers. However, difficulties arose in distinguishing signals having very close chemical shifts, and unequivocal assignments were based on the ^{13}C-n.m.r. spectra of compounds specifically labelled with isotopes.

Replacement of a ring proton by a deuteron causes (a) disappearance of the ^{13}C-signal of the appended carbon atom, as in [3-^{2}H], [5-^{2}H], and [5,6,6-^{2}H$_3$] derivatives of glucose,[37] the results leading to reassignment[13] of the ^{13}C signals of glucose,[35] or (b) conversion into a triplet, ~0.4 p.p.m. upfield, as observed in the spectra of [5-^{2}H] derivatives of pentoses or [6-^{2}H] derivatives of 6-deoxyhexoses.[38] Simultaneous, upfield shifts of 0.04–0.12 p.p.m. were reported by Gorin and Mazurek[38,39] for β-carbon signals. The size of the shift varied as the number of attached deuterons, and was consistently less with more-deshielded resonances.[15] With two, suitably deuterium-labelled sugars, the complete ^{13}C-n.m.r. spectrum of a pentose, hexose, or derivative could be rationalized. The studies were useful in correcting earlier signal assignments for, amongst others, fucose and galactose.[13] Confirmation was educed by Matwiyoff and coworkers,[40]

---

(35) D. E. Dorman and J. D. Roberts, *J. Am. Chem. Soc.*, 92 (1970) 1355–1361.
(36) S. J. Angyal, *Angew. Chem. Int. Ed. Engl.*, 8 (1969) 157–166.
(37) H. J. Koch and A. S. Perlin, *Carbohydr. Res.*, 15 (1970) 403–410.
(38) P. A. J. Gorin and M. Mazurek, *Can. J. Chem.*, 53 (1975) 1212–1223.
(39) P. A. J. Gorin, *Can. J. Chem.*, 52 (1974) 458–461.
(40) T. E. Walker, R. E. London, T. W. Whaley, R. Barker, and N. A. Matwiyoff, *J. Am. Chem. Soc.*, 98 (1976) 5807–5813.

who determined long-range, $^{13}C-H$ couplings in [1-^{13}C]-enriched monosaccharides.

The aforementioned deuterated derivatives were prepared by way of reduction of a ketone, aldehyde, or ester with sodium borodeuteride, or by deuteroboration of an alkene. An interesting reaction, perhaps eventually applicable to direct deuteration of polysaccharides, was reported by Koch and Stuart[41a] and by them and their coworkers,[41b] who found that treatment of methyl α-D-glucopyranoside with Raney nickel catalyst in deuterium oxide results in exchange of protons attached to C-2, C-3, C-4, and C-6. In other compounds, some protons of CHOH groups are not replaced, but the spectra may nevertheless be interpreted with the aid of α- and β-deuterium effects.

Shifts occurring on O-alkylation of hydroxyl groups bear a resemblance to those produced on O-glycosylation, in that strong, downfield α-shifts occur with the substituted carbon resonance,[7,42] with weaker, upfield β-shifts for the adjacent carbon atom.[42] For O-methylation of 6-membered-ring systems, these displacements are +7 to +11 and −1 to −2 p.p.m., respectively, and small downfield β-shifts sometimes occur.[43,44] Methylation of an equatorially (e)[42] or axially (a) attached group[44−46] results in a stronger shift of the β-carbon resonance (−3 to −5 p.p.m.) if the β-carbon–oxygen bond is axial. Dorman and Roberts[47] considered that, in the OMe-e–OH-a system, the upfield shift of 4.5 p.p.m. is due to the favored, gauche disposition of the methoxyl group, where it has a nonbonded interaction (γ-effect) with the carbon atom bearing the axial substituent (1). In the OMe-a–OH-a system, as in 1,4-di-O-methyl-*chiro*-inositol, as compared with 4-O-methyl-*chiro*-inositol, a gauche interaction takes place as in 2, and the shift of the β signal at C-6 is[46] −3.5 p.p.m.

The values of a series of α- and β-shifts have been determined for mono-O-methylglucoses and mono-O-methylmannoses, in order to aid in assignments of ^{13}C signals of oligosaccharides and polysaccha-

---

(41) (a) H. J. Koch and R. S. Stuart, *Carbohydr. Res.*, 59 (1977) c1–c6; (b) F. Balza, N. Cyr, G. K. Hamer, A. S. Perlin, H. J. Koch, and R. S. Stuart, *ibid.*, 59 (1977) c7–c11.
(42) D. E. Dorman, S. J. Angyal, and J. D. Roberts, *J. Am. Chem. Soc.*, 92 (1970) 1351–1354.
(43) P. Colson, K. N. Slessor, H. J. Jennings, and I. C. P. Smith, *Can. J. Chem.*, 53 (1975) 1030–1037.
(44) P. A. J. Gorin, *Carbohydr. Res.*, 39 (1975) 3–10.
(45) P. A. J. Gorin and J. F. T. Spencer, *Can. J. Microbiol.*, 18 (1972) 1709–1715.
(46) J. W. Blunt, M. H. G. Munro, and A. J. Paterson, *Aust. J. Chem.*, 29 (1976) 1115–1118.
(47) D. E. Dorman and J. D. Roberts, *J. Am. Chem. Soc.*, 93 (1971) 4463–4472.

from OMe to O-glycosyl if they both exist as rotamer **3**, which lacks a strong gauche-interaction. α-Sophorose and α-maltose have such a preponderant rotamer, as their C-1' shift is close to that of methyl α-D-glucopyranoside.[48] Equal shifts are, however, not so common, and, for estimation of O-glycosyl shifts, O-isopropyl derivatives are preferable as models.

In the furanose series, O-methylation α- and β-shifts of +9 to +11 p.p.m. and −1 to −3 p.p.m., respectively, were observed, with smaller shifts occurring on O-substitution with isopropyl, which was suggested, as just mentioned, as a more-accurate, model substituent for gauging O-glycosylation shifts.[15] The O-isopropylation displacements were found to be independent of steric distortion, or variation of population of conformational components, or both.

IV. IDENTIFICATION OF ^{13}C-N.m.r. SIGNALS OF OLIGOSACCHARIDES AND POLYSACCHARIDES

The foregoing approach, although satisfactory for the assignment of polysaccharide signals that are widely spaced, is unsuitable for signals separated by 1 p.p.m. or less, as small, longer-range, substituent effects become important, and there is still a lack of satisfactory, model derivatives for comparison. An alternative method is direct substitution of the polysaccharide, or structurally related oligosaccharide, in a known position. For example, replacement of the OH-6 group of glucopyranose by chlorine results in α- and β-shifts of −17 and −1.2 p.p.m., respectively, so that C-5 resonances may be identified.[43]

Signals of ^{13}C nuclei having two attached protons, such as C-6 of hexose units, appear as triplets when obtained with off-resonance decoupling. Identification is simple if the signal is in an uncrowded region of the spectrum. Assignment of ^{13}C signals of polysaccharides can be achieved unambiguously by carbon-13, proton heteronuclear decoupling, provided that the proton spectrum is simple and readily interpretable.[26,50,51]

Direct labelling with deuterium is another approach, and three promising methods have been reported. One, described by Koch and Stuart and coworkers,[41] is based on the specific exchange of hydrogen attached to hydroxylated carbon atoms by deuterium atoms that occurs on treatment with Raney nickel–D_2O. Although no exchange was observed with soluble starch or glycogen, the H-2, H-3, and H-6 atoms of

(50) D. Bassieux, D. Gagnaire, and M. Vignon, *Carbohydr. Res.*, 56 (1977) 19–33.
(51) D. Gagnaire and M. Vincendon, *Bull. Soc. Chim. Fr.*, (1977) 479–482.

rides consisting of glucose[48] and mannose,[44,45] respectively. In only one instance were the O-methylation and O-glycosylation shifts found to be markedly different. 2-O-Methylation of β-D-mannose gives a C-1, β-shift of +0.4 p.p.m.,[44] whereas the difference in shift of C-1 atoms of nonreducing end-groups and internal residues of a β-D-(1→2)-linked D-mannopyranan is[11] +2.0 p.p.m., probably attributable to different glycosidic torsion-angles, ϕ and ψ, between those of the (nonreducing) end-groups and those of the 2-O-substituted residues, which would be modified in order to relieve steric compression.

Replacement of hydroxyl by benzyloxy gives α- and β-shifts similar to those of methoxyl, and resonances of the ^{13}C-n.m.r. spectrum of methyl 2,3-di-O-benzyl-α-D-galactofuranoside[28] have been found to be sufficiently close to those of a polysaccharide containing 2,3-di-O-substituted α-D-galactofuranosyl residues as to make possible a structural confirmation.[17]

Koch and coworkers[49] found that the C-1 shifts of isopropyl and tert-butyl α-D-glucopyranoside respectively lie 4 and 7.7 p.p.m. to higher field than that of the methyl glycoside; this was assumed to be wholly due to the γ-effect between methyl groups of the aglycons and the anomeric centers. Isopropyl α-D-glucopyranoside can exist as three rotamers **3**, **4**, and **5**, and it may be seen that the last two contribute to

upfield shifts; this serves as a model for O-glycosylation α-shifts, in the sense that no upfield displacements would take place on going

(48) T. Usui, N. Yamaoka, K. Matsuda, K. Tuzimura, H. Sugiyama, and S. Seto, *J. Chem. Soc. Perkin Trans. 1*, (1973) 2425–2432.
(49) K. F. Koch, J. A. Rhoades, E. W. Hagaman, and E. Wenkert, *J. Am. Chem. Soc.*, 96 (1974) 3300–3305.

cyclohexaamylose were replaced,[41] and earlier assignments[48] for the C-2 and C-5 signals were reversed (see Section VI,1). Exchange was also observed for inulin.[41(a)] Gagnaire and Vincendon[52(a)] found that, in dimethyl sulfoxide solution, hydroxylated ^{13}C nuclei of compounds having their OH groups partly exchanged with O^{2}H give rise to doublets by virtue of an upfield β-shift of ~0.08 p.p.m. on going from ^{13}C-O-H to ^{13}C-O-^{2}H. The spectrum of methyl β-cellobioside was interpreted thus, but in those of amylose, cellulose, cycloheptaamylose, laminaran, and pseudonigeran,[52(b)] the signals of carbinol groups were merely broadened, values for amylose being from Δν 15 Hz to 20–25 Hz.This method, like the Raney-nickel approach, may be more applicable to oligosaccharide models than to the parent polysaccharides. The latter have line widths greater than the deuterium shift, and diminution of line widths by increase of temperature is not applicable, as the doublets collapse at 80° because of rapid OH–O^{2}H exchange. This problem has now been overcome by use of ^{2}H$_2$O and H$_2$O solutions in a dual, coaxial, n.m.r. cell.[52(c)] Gorin[44] reported relatively direct incorporation of deuterium into a yeast mannan containing alternate (1→3)- and (1→4)-linked β-D-mannopyranosyl residues. When [6-^{2}H$_2$], [5-^{2}H], and [3-^{2}H] derivatives of D-glucose were used as precursors in the medium, C-6, C-5, and C-3 signals were virtually eliminated in respective experiments, although they were still detectable, because of a little randomization of the respective tracer. As the half-height line-width was ~6 Hz, the β-deuterium shift (−0.12 p.p.m.) was observable in the C-5 signal only when the dilabelled, [6-^{2}H$_2$] precursor was used. In a similar experiment leading to the formation of a fungal rhamnomannan, only the label of D-[6-^{2}H$_2$]glucose was sufficiently unrandomized that signal diminution could be detected.[53]

D-[1-^{13}C]Glucose and [2-^{13}C]-enriched D-glucose may have promise as precursors in assigning ^{13}C signals of polysaccharides grown in culture media.[53] The ^{13}C-n.m.r. spectrum of the rhamnomannan contained enlarged C-1 and C-6, or C-2 and C-5, signals, respectively, depending on whether a [1-^{13}C]- or a [2-^{13}C]-labelled precursor was used. For the former, where the level of label was high, there was a possibil-

(52) (a) D. Gagnaire and M. Vincendon, *J. Chem. Soc. Chem. Commun.*, (1975) 509–510; (b) D. Gagnaire, D. Mancier, and M. Vincendon, *Org. Magn. Reson.*, 11 (1978) 344–349; (c) P. E. Pfeffer, K. M. Valentine, and F. W. Parrish, *J. Am. Chem. Soc.*, 101 (1979) 1265–1274.
(53) P. A. J. Gorin, R. H. Haskins, L. R. Travassos, and L. Mendonça-Previato, *Carbohydr. Res.*, 55 (1977) 21–33.

ity of observing $^{13}C-^{13}C$ coupling[54] of ~45 Hz, but, in the crowded spectrum, the doublet was obscured by adjacent resonances.

Once assignments of polysaccharide signals are known, they may be used as a basis in determination of the position of substitution by such groups as acetate, malonate, phosphate, and sulfate, whose α- and β-shifts may be estimated by referral to suitable monosaccharide models. For phosphate, the phosphated carbon atoms and adjacent resonances may be identified, as they give coupled signals and are subject to lanthanide-induced shifts. These data are described in Section VI, 8.

The shifts ocurring on variation of pH or pD are due to changes in conformation at the polymeric level; this effect was noted with amylose[47] and other polysaccharides,[44,55] and is sometimes useful in improving spectral resolution. Hydrogen-bond modification plays an important role, but rules have not as yet been formulated so that all of the signal displacements may be interpreted in terms of signal assignment.

The ^{13}C-n.m.r. spectra of polysaccharides having mixed linkages can sometimes be interpreted by reference to the spectra of homopolymers representing each type of linkage. For example, the spectrum of lichenan, which contains one β-D-(1→3) per two β-D-(1→4) units, is a virtual composite of those of laminaran and cellulose in a 1:2 ratio.[51] More-poorly resolved spectra of glucans containing α-D-(1→4) and α-D-(1→6) units,[55] α-D-(1→3), α-D-(1→4), and α-D-(1→6) units,[55] and β-D-(1→3) and β-D-(1→6) units[56] may be interpreted similarly.

There appears to be some potential in using T_1 values in distinguishing and assigning ^{13}C resonances in ^{13}C-n.m.r. spectra of branched-chain polysaccharides, as the side-chain units should exhibit more segmental motion than those of the main chain. The branched mannan from *Saccharomyces rouxii*, which consists of a main chain of (1→6)-linked α-D-mannopyranosyl residues substituted at O-2 by O-α-D-mannopyranosyl-(1→2)-O-α-D-mannopyranosyl side-chains, contains C-1 atoms having T_1 values at 70° of 0.20 s (nonreducing end-unit), 0.13 s (adjacent, side-chain unit), and 0.09 s (main-chain unit).[57] This contrasts with the linear mannan from *Hansenula capsulata*, which contains a repeating sequence of α-D-(1→2), α-D-(1→2),

(54) G. J. Karabatsos, J. D. Graham, and F. M. Vane, *J. Am. Chem. Soc.*, 84 (1962) 37–40.
(55) P. Colson, H. J. Jennings, and I. C. P. Smith, *J. Am. Chem. Soc.*, 96 (1974) 8081–8087.
(56) H. Saito, T. Ohki, N. Takasuka, and T. Sasaki, *Carbohydr. Res.*, 58 (1977) 293–305.
(57) P. A. J. Gorin and M. Mazurek, *Carbohydr. Res.*, 72 (1979) c1–c5.

and α-D-(1→6) units that each has a T_1 value for C-1 atoms of 0.14 s; observations were made at 25.2 MHz. In branched-chain mannans, T_2 values of C-1 atoms of side-chain units, as reflected by line widths, were noticeably longer than those of the main chain, at 30°. At higher temperatures, the difference in line width was difficult to detect. No correlation of n.O.e. values and segmental motion was observed. Similar variations of T_1 values were observed for branched-chain dextrans.[58]

V. MOLECULAR MOTION

1. Quantitation of Signal Intensities: Spectral Quality

The integral intensities of signals of polysaccharides, obtained with a modern spectrometer under the usual operating conditions, are proportional or quasi-proportional, to the number of ^{13}C nuclei present. This has been observed, in particular, in the case of linear hexopyranan structures containing one type,[11,47,51,55,56,58–61] or two equal types, of linkage,[62] or branched-chain polymers having two linkage types,[53,58] where the resonances are readily resolved. In such cases, the T_1 values would be low, 0.2 s or less, and the n.O.e. values would be approximately equal. However, few actual determinations of these values have been made, and extrapolation of such assumptions to more-complex polysaccharide structures is not recommended, as outlined in Section V,2.

In general, however, quantitative spectra may be conveniently obtained if n.O.e. effects are suppressed, as in an anti-gated, coupling technique in which the accumulation of data is made when the decoupler is on, but off during a delay of a least 5 times the T_1 value of the resonance observed (90° pulse).[4] In such an experiment, integration, rather than peak-height measurement, is preferable, as line widths may vary in a given spectrum, and each resonance must be defined by a sufficient number of data points. Improved resolution may be obtained by increasing the temperature of the solution, which

(58) F. R. Seymour, R. D. Knapp, and S. H. Bishop, *Carbohydr. Res.*, 72 (1979) 229–234.
(59) H. Saito, T. Ohki, and T. Sasaki, *Biochemistry*, 16 (1977) 908–914.
(60) A. Allerhand, R. F. Childers, R. A. Goodman, E. Oldfield, and X. Ysern, *Am. Lab.*, 4 (1972) 19–20.
(61) J. O. Previato, L. Mendonça-Previato, and P. A. J. Gorin, *Carbohydr. Res.*, 70 (1979) 172–174.
(62) J. F. T. Spencer and P. A. J. Gorin, *Biotechnol. Bioeng.*, 15 (1973) 1–12.

leads to narrower line-widths, and by using higher magnetic fields,[63,64] which provide a decreased spectral-density. However, it is suspected that high fields lead to increased T_1 values, and it is not as yet known whether they are disadvantageous in quantitative analysis, as distinct from spectral quality.

Aspects of n.m.r. instrumentation, and methods for the preparation of solutions for ^{13}C-n.m.r. spectroscopy, are not dealt with herein, as they have already been covered.[4] However, one particularly interesting technique is the use of 1.7-mm-diameter capillary-tubes with which, for example, 1 milligram of lactose can give rise to a spectrum (25.2 MHz) having a signal to noise (s/n) ratio of 7:1 after an overnight experiment.[65] At present, however, with an ordinary capillary tube, difficulties are encountered at high temperatures, as the solvent evaporates, and condenses at the top of the tube. Methods for preventing this phenomenon and for readily introducing solutions into suitable tubes would be welcome.

The use of 20-mm-diameter tubes for increased spectral sensitivity is possible when large samples are available.[60]

2. T_1, T_2, and n.O.e. Values of Solutions and Gels

The quantitation of ^{13}C spectra, which involves T_1 and n.O.e. values, is often interrelated with the spin–spin relaxation-time (T_2), which is short for polysaccharides and can lead to broad lines and, thus, lack of resolution. Thus, a description of each parameter is necessary from the standpoints of quantitation, and knowledge, of molecular motions in solutions and gels.

In a large molecule, such as that of a polysaccharide, relaxation takes place almost entirely through dipole–dipole interactions. The spin–lattice relaxation-time (T_1) describes the time required for the spin system to return to thermal equilibrium after application of a radiofrequency field at resonance, which perturbs equilibrium populations so that their spins are promoted to the higher energy-level. The relevant, relaxation mechanism is through the lattice (that is, environment), which undergoes an energy exchange with the spins. The speed at which this process takes place is dependent on a number of factors.

(63) J. M. Berry, G. G. S. Dutton, L. D. Hall, and K. L. Mackie, *Carbohydr. Res.*, 53 (1977) c8–c10.
(64) P. Colson, H. C. Jarrell, B. L. Lamberts, and I. C. P. Smith, *Carbohydr. Res.*, 71 (1979) 265–272.
(65) P. A. J. Gorin and M. Mazurek, unpublished results.

The theory of ^{13}C relaxation has been dealt with in detail by Komoroski and coworkers[2] and Lyerla and Levy.[66] For the purposes of this article, the phenomenon of T_1 may be described by the equation[2]

$$1/T_1 = 0.1\hbar^2\gamma_H^2\gamma_C^2\ r_{CH}^{-6}n[f(\omega_H - \omega_C) + 3f(\omega_C) + 6f(\omega_H + \omega_C)], \quad (1)$$

where $f(\omega) = \tau_C/(1 + \omega^2\tau_C^2)$, $\hbar$ = Planck's constant/2π, γ_H and γ_C = gyromagnetic constants for protons and ^{13}C nuclei, respectively, r_{CH} = length of the carbon to proton bond, n = the number of protons attached to the carbon atom, τ_C = the molecular rotational, correlation time, and ω_H and ω_C = the Larmor frequency of protons and ^{13}C nuclei, respectively.

Without an oscillating, radiofrequency field, spins precess about the axis of an external field with the phases at random, so that there is no transverse magnetization. On application of a radiofrequency field at the resonance of the spins, a transverse magnetization arises, and the component, magnetic moments are polarized along the axis of the rotating field. The magnetization decays to zero, as there is a spread of Larmor frequencies across the sample, and the spins dephase at a rate expressed by the constant T_2, the transverse or spin–spin relaxation time. This may vary according to the equation[2]

$$1/T_2 = 0.05\ \hbar^2\gamma_H^2\gamma_C^2 r_{CH}^{-6}n[4\ \tau_C + f(\omega_H - \omega_C) + 3f(\omega_C) \\ + 6f(\omega_H + \omega_C)]. \quad (2)$$

In small molecules having some correlation-times (τ_C), ($\omega_H + \omega_C)\tau_C \ll 1$, and $T_1 = T_2$, but, with higher correlation times of > 10 ns, equations 1 and 2 indicate that T_1 and T_2 diverge, as may be seen in Fig. 1.

In a system where ^{13}C couples with protons that are then irradiated, the nuclear Overhauser enhancement is important. This phenomenon is a consequence of the non-Boltzmann distribution of energy levels of ^{13}C nuclei on saturation of the proton resonances; the theory has been dealt with in detail elsewhere.[3] In molecules of low molecular weight, its value is 3, regardless of the number of attached protons, but it decreases[2] to 1.1 for polymers having $\tau_C \geq 10$ ns (see Fig. 2).[67] Two terms are currently used to express enhancements, namely, (a) n.O.e., which is the ratio of intensity of decoupled and coupled signals, and (b) the nuclear Overhauser enhancement factor (n.O.e.f. = η), which is (n.O.e. $-$ 1). N.O.e. values can be measured by comparing the signal to noise (s/n) ratio of a given signal obtained under conventional decoupling-conditions with that obtained by an anti-gated experiment, all other spectral parameters being identical.

(66) J. R. Lyerla and G. C. Levy, *Top. Carbon-13 NMR Spectrosc.*, 1 (1974) 81–147.
(67) D. Doddrell, V. Glushko, and A. Allerhand, *J. Chem. Phys.*, 56 (1972) 3683–3689.

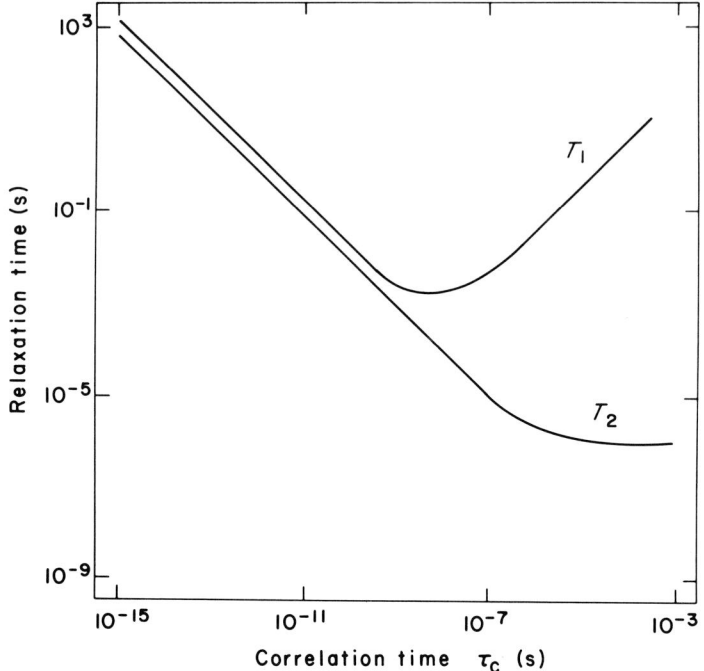

Fig. 1.—Calculated Dependence of Relaxation Times T_1 and T_2 on Correlation Time τ_c.

As may be seen from equations 1 and 2, T_1 and T_2 may be dependent on the Larmor frequency, and Fig. 3 shows that, at low correlation-times, the applied magnetic field has a large effect on T_1 values, in contrast to T_2 values, which are less affected.[2] This marked increase of

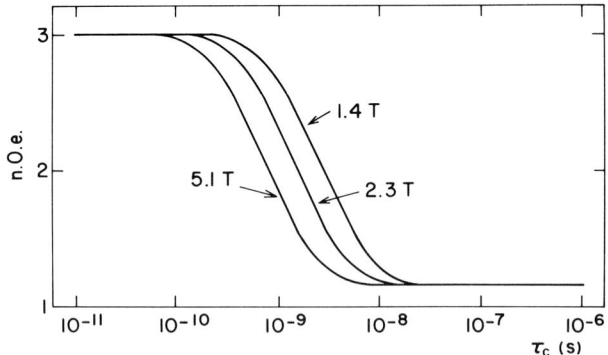

Fig. 2.—Dependence of Nuclear Overhauser Enchancement (n.O.e.) on Correlation Time τ_c at Various Magnetic Fields, T (tesla).

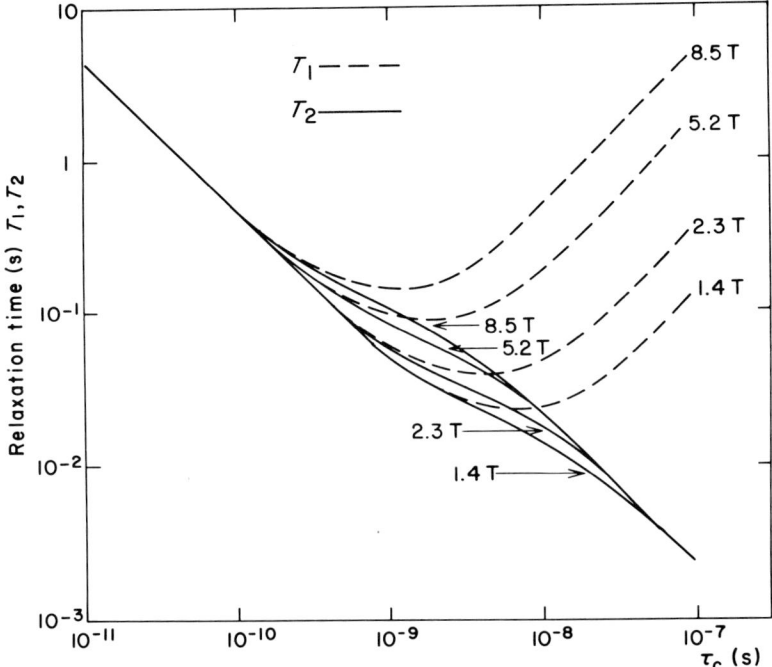

FIG. 3.—Calculated Dependence of Relaxation Times T_1 and T_2 on the Correlation Time τ_c for Various Magnetic Fields, T, for a ^{13}C Nucleus in a CH group.

T_1 values may occur with such native biopolymers as ribonuclease A, deoxyribonucleic acid, and collagen, whose molecular motions are restricted, but, as yet, high T_1 values have not been observed for polysaccharides in solution, or for gels, in which these motional-restriction effects may be equivalent, or less marked. However, an extensive relaxation-study by Levy and coworkers[68] on poly(n-alkyl methacrylates) may serve as a model for future experiments on polysaccharides, as this type of molecule has a main chain and side chains, albeit more mobile than those in polysaccharides.

In the n-butyl derivative **6**, T_1 values of the side-chain carbon atoms

(68) G. C. Levy, D. E. Axelson, R. Schwartz, and J. Hochmann, *J. Am. Chem. Soc.*, **100** (1978) 410–424.

TABLE I

^{13}C Spin–Lattice Relaxation-Times and n.O.e. Values of Resonance of Poly(n-butyl methacrylate) as 50% (w/w) Solution in Toluene-^{2}H$_8$

Temp. (°C)	C-1	C-2	C-3	C-4	CH$_3$	CH$_2$	C
			T_1 (s) at 67.9 MHz				
6	0.35	0.23	0.39	1.0	0.06	0.21	2.3
55	0.33	0.69	1.5	2.9	0.11	0.10	1.7
111	0.35	1.4	3.4	6.2	0.26	0.11	1.6
			T_1 (s) at 22.6 MHz				
10	0.13	0.20	0.45	1.1	0.03	—	0.69
49	0.09	0.33	0.89	1.7	0.05	0.03	0.54
105	0.15	0.67	1.4	2.8	0.13	0.04	0.70
			n.O.e. at 67.9 MHz				
14	2.4	2.7	—	2.8	1.71	1.67	1.47
64	1.61	2.3	—	2.4	2.3	1.44	1.59
101	1.42	1.98	—	2.3	2.4	1.36	1.59
			n.O.e at 22.6 MHz				
11	1.47	2.5	—	2.6	1.99	—	1.89
63	1.79	2.9	—	2.3	2.3	1.71	1.75
104	2.1	2.3	—	2.4	2.4	1.59	2.1

(C-1 to C-4) increase, as expected, as a function of the distance from the anchoring main-chain, and increase with temperature, thus reflecting their ease of motion (see Table I). Unexpectedly, the n.O.e. values are less than their maximum, and the effect of field intensity is not marked. However, almost all T_1 values, both those of main and side chains, increase on going from 22.6 to 67.9 MHz. Unfortunately, T_2 values could not be measured in this study, because of line-broadening due to tacticity.

A number of pulse-sequence methods are available for measurement of T_1 values, and those most commonly used are the methods of saturation recovery (s.r.F.t.),[69,70] progressive saturation (p.s.F.t.),[71] inversion recovery (i.r.F.t.),[72] and the Freeman–Hill modification of in-

(69) J. L. Markley, W. J. Horsley, and M. P. Klein, *J. Chem. Phys.*, 55 (1971) 3604–3605.
(70) G. G. McDonald and J. S. Leigh, Jr., *J. Magn. Reson.*, 9 (1973) 358–362.
(71) R. Freeman and H. D. W. Hill, *J. Chem. Phys.*, 54 (1971) 3367–3377.
(72) R. L. Vold, J. S. Waugh, M. P. Klein, and D. E. Phelps, *J. Chem. Phys.*, 48 (1968) 3831–3832.

version recovery (FH-i.r.F.t.).[73] Their relative merits for nuclei having long and short T_1 values have been weighed by Levy and Peat.[74] For sugars that have short T_1 values (<2 s) the i.r.F.t. methods are preferable, and the most accurate, as such instrumental instabilities as magnet homogeneity and drift are less critical. The Freeman–Hill modification is advantageous, as an improved s/n ratio is obtained. However control of temperature is still extremely important in each method, as, for example, on going from 10 to 66°, the T_1 values of pure glycerol increase[75] from ~0.05 to ~0.4 s and, with 2:1 (v/v) 1,4-dioxane–D_2O, T_1 increases[76] at the rate of 2.5% per degree from 30 to 50°. Dissolved oxygen has an effect on the relaxation rates, and, on degassing the solution, T_1 values of carbonyl carbon atoms of glucose derivatives increase[77] by 20%.

Although few determinations of T_1 have been made on polysaccharides, it is of interest to compare these values with those of oligosaccharides, and to use the data in anticipating the T_1 values of branched-chain polysaccharide structures. In the first, T_1 study on a carbohydrate, Allerhand and coworkers[78] investigated, at 15.1 MHz, sucrose in aqueous and D_2O solutions at 42°, using the partially relaxed, Fourier-transform (p.r.F.t. = i.r.F.t.) technique. Under proton-decoupling conditions, each carbon atom had the maximum n.O.e. value of 3 for carbon atoms undergoing pure $^{13}C,H$ dipolar relaxation,[79] so that, with a long pulse-interval, the integrated intensity of each resonance was equal. Spin–lattice relaxation-times for each protonated carbon atom varied from 0.37–0.73 s for a 0.5 M aqueous solution to 0.12–0.17 s for a 2 M solution. These values were interpreted as being due to two sugar rings moving as a single entity. In investigating the structurally related stachyose [O-α-D-galactopyranosyl-(1→6)-O-α-D-galactopyranosyl-(1→6)-O-α-D-glucopyranosyl β-D-fructofuranoside], which gives a more-crowded spectrum, Allerhand and Doddrell[80] did not measure individual T_1 values precisely, but, by using rapid pulses of 0.34 s with a 0.5 M aqueous solution at 65° (15.1 MHz), the lines of the terminal galactosyl group were eliminated, as its segmental mo-

(73) R. Freeman and H. D. W. Hill, *J. Chem. Phys.*, 51 (1969) 3140–3141.
(74) G. C. Levy and I. R. Peat, *J. Magn. Reson.*, 75 (1970) 500–521.
(75) L. J. Burnett and S. B. W. Roeder, *J. Chem. Phys.*, 60 (1974) 2420–2423.
(76) I. M. Armitage, H. Huber, D. H. Live, H. Pearson, and J. D. Roberts, *J. Magn. Reson.*, 15 (1974) 142–149.
(77) K. Bock and L. D. Hall, *Carbohydr. Res.*, 40 (1975) c3–c8.
(78) A. Allerhand, D. Doddrell, and R. Komoroski, *J. Chem. Phys.*, 55 (1971) 189–198.
(79) K. F. Kuhlmann and D. M. Grant, *J. Am. Chem. Soc.*, 90 (1968) 7355–7357.
(80) A. Allerhand and D. Doddrell, *J. Am. Chem. Soc.*, 93 (1971) 2777–2779.

TABLE II

T_1 Values for Monosaccharides Relevant to the Ganglioside Head-Group

Compound	T_1 Value (s)					
	C-1	C-2	C-3	C-4	C-5	C-6
Methyl α-D-glucopyranoside	0.74	0.74	0.76	0.74	0.76	0.40
Methyl β-D-glucopyranoside	0.82	0.81	0.75	0.74	0.73	0.36
Methyl α-D-galactopyranoside	0.86	0.80	0.82	0.83	0.80	0.53
Methyl β-D-galactopyranoside	0.87	0.86	0.84	0.75	0.86	0.53
2-Acetamido-2-deoxy-α-D-galactose	0.54	0.52	0.55	0.53	0.55	0.39
2-Acetamido-2-deoxy-β-D-galactose	0.55	0.52	0.54	0.52	0.54	0.45

tion was more pronounced (having nuclei of $T_1 > 0.5$ s), whereas those of the adjacent galactosyl residue ($T_1 < 0.5$ s) remained.

By using the inversion recovery method described by Canet and coworkers,[81] Czarniecki and Thornton[82] compared the component, carbon T_1 values for a ganglioside that has N-acetylneuraminyllactose (**7**) as the head group with those of α,β-lactose, 2-acetamido-2-deoxygalactose, and the anomers of methyl D-glucopyranoside and methyl D-galactopyranoside (molar solutions in D_2O at 28° and 25.0 MHz). As may be seen from Table II, the T_1 values of the ring-carbon atoms of each of the monosaccharide components of lactose (0.73–0.87 s) are longer than those (0.30–0.40 s) reported for α,β-lactose. (Ring-carbon atoms of molar solutions of methyl α,β-D-glucopyranoside in D_2O at 39° had T_1 values[77] of 1.0–1.2 s.) As in other examples, primary carbon atoms have shorter T_1 values than those of the ring, as would be expected from equation *1*. The spectrum of the N-acetylneuraminyl group of **7** had T_1 0.05 s, shorter than the 0.11–0.26 s for carbon atoms of the galactopyranosyl and glucopyranose residues. Thus, in contrast to stachyose, whose segmental motion radiates from the central units,

N-Acetylneuraminyllactose

7

(81) D. Canet, G. C. Levy, and I. R. Peat, *J. Magn. Reson.*, 18 (1975) 199–204.
(82) M. F. Czarniecki and E. R. Thornton, *J. Am. Chem. Soc.*, 99 (1977) 8279–8282.

the N-acetylneuraminyl group acts as an anchor, a phenomenon that is probably attributable to the highly solvated carboxyl group.

These results indicate that it would be of interest to measure T_1 values, and, thus, segmental motion, for individual units of longer, linear oligosaccharides, for example, those of 5–6 units. However, it is difficult to distinguish resonances of individual units along a chain, although it should be possible through specific ^{13}C-labelling, or by the use of suitable hetero-oligosaccharides whose component units afford distinctive signals. Although such a study has not yet been carried out, T_1 values have been measured for a motionally related glycoside, k-strophanthoside (8), whose trisaccharide moiety is anchored by the

k-Strophanthoside
8

aglycon strophanthidin to a greater extent than it would probably be by a single, adjacent, aldose unit. The T_1 values were determined[83] on a 0.23 M sample in pyridine-^{2}H$_5$ at 90°, observed at 25.2 MHz. They were longest for the ring-carbon atoms of the β-D-glucopyranosyl end-group (0.31–0.36 s), and diminished sequentially for the adjacent β-D-glucopyranosyl (0.22–0.25 s) and cymarosyl residues (0.11–0.19 s), thus indicating decreasing segmental motion going from the end-unit, much the same as occurs with a linear, lipid chain.

At high temperatures, the T_1 values of ^{13}C nuclei of polysaccharides may be sufficiently long to create difficulties in quantitative estimation of integral intensities. Values have been observed, at 25.2 MHz, of 0.27–0.31 s [C-1–C-5 of α-D-(1→6)-linked D-mannopyranan at 90°], and, for side chains of branched-chain mannans, of 0.22 s [C-1 of α-D-mannopyranosyl groups linked (1→2) to a main chain of (1→6)-linked-D-mannopyranosyl residues at 90°] and of 0.31 s [C-1 atoms of 2-O-substituted β-D-mannopyranosyl side-chain units linked (1→2) to a similar main-chain, at 70°]. Under such conditions, the n.O.e. values are[57]

(83) A. Neszmélyi, K. Tori, and G. Lukacs, *J. Chem. Soc. Chem. Commun.*, (1977) 613–614.

TABLE III

^{13}C Spin–Lattice Relaxation-Times, Line Widths, and Nuclear Overhauser Enhancements (n.O.e.) of PS 13140

Carbon atom	Helix								Random coil			
	Resilient gel			0.06 M NaOH			0.19 M NaOH			0.22 M NaOH		Me$_2$SO
	$T_1{}^a$	Line width (T_2)^b	n.O.e.		Line width (T_2)	n.O.e.		Line width (T_2)	n.O.e.	T_1	Line width (T_2)	n.O.e.	T_1	n.O.e.
1	86	172 (1.9)^c	1.0	161 (2.0)	1.2	167 (1.9)	1.3	76	14 (23)	1.5	77 (44)^d	1.4
2	84		1.4		1.3		1.5	78	15 (21)	1.4	73 (48)	1.5
3	76	156 (2.0)	1.2	167 (1.9)	1.3	183 (1.7)	1.3	84	14 (23)	1.4	85 (46)	1.4
4								80	15 (21)	1.5	80 (52)	1.3
5	84		1.4		1.3		1.5	83	14 (23)	1.3	84 (50)	1.4
6	62	50 (6.4)	1.4	50 (6.4)	1.8	53 (6.0)	1.6	54	17 (19)	1.9	57 (40)	1.7

^{a} In milliseconds. ^{b} In hertz. ^{c} In milliseconds calculated from $1/\pi$ (line width). ^{d} T_1 values at 15 MHz.

2.0 to 2.9. At 90° and 25.2 MHz, a T_1 value of 0.25 s was observed for C-1 of a side chain of a dextran.[58]

For other polysaccharides examined, the T_1 values are very short. For example, the β-D-(1→3)-linked D-glucopyranan of *Alcaligenes faecalis* can exist in solution as a random coil, or as a gel in a helical form.[59] Only 20–30% of the total nuclei of the gel could be observed by ^{13}C-n.m.r. spectroscopy, because of immobilization of a majority of the units at junction zones. The signals evidenced restricted motion, as those of C-1, C-3, and C-4 were displaced downfield from the resonances of the solution form by virtue of different ϕ and ψ torsion angles of the glycosidic bond, as with the displacement of C-1 and C-4 signals observed on going from amylose to cycloamyloses.[55] The downfield shifts of C-4 resonances were interpreted as being due to OH-4'⋯O-5 hydrogen-bonds in a region having a single-helix conformation. At 28° and 25.0 MHz, ring-carbon atoms have T_1 values (0.08 s) that are virtually the same for the solution or the gel form. The n.O.e. values are very low, the value of 1.1 being close to the minimum (see Table III).

Low values for T_1 and n.O.e. were also reported for bovine nasal cartilage, which contains collagen (zero spectrum),[84] proteoglycan, and component chondroitin sulfate (see Fig. 4 and Section VI, 7). At 37° and 15.1 MHz, bovine nasal cartilage and chondroitin sulfate[85] have, respectively, T_1 values of 60 ms and 55 ms, and n.O.e. values of 1.35 and 2.0. It is noteworthy that these polymers are linear, and the T_1 values of observed signals are short, and convenient in terms of obtaining spectra, even when the τ_C values are relatively long, at least as evidenced by line width (see Fig. 5). Even with a high-field, n.m.r. spectrometer, low T_1 values of 28–41 ms (D$_2$O, 30°, 68 MHz) were observed for ^{13}C nuclei of inulin of mol. wt. 3060. The exception was C-2, which had[86] $T_1 = 0.35$ s.

The spin–spin relaxation-time (T_2) is related to the natural linewidth according to the expression $\nu_2 = 1/\pi T_2^*$, where $1/T_2^*$ is the sum of $1/T_2$ (the transverse-relaxation rate) and $1/T_2'$ (the dephasing contribution arising from the lack of homogeneity of the magnetic field). Line widths of polysaccharides are greater than those of carbohydrates of low molecular weight, and vary from 5 to 150 Hz (above which width, they become difficult to detect), and these values corre-

(84) J. C. W. Chien and E. B. Wise, *Biochemistry*, 14 (1975) 2786–2792.
(85) D. A. Torchia, M. A. Hasson, and V. C. Hascall, *J. Biol. Chem.*, 252 (1977) 3617–3625.
(86) H. C. Jarrell, T. F. Conway, P. Moyna, and I. C. P. Smith, *Carbohydr. Res.*, 76 (1979) 45–57.

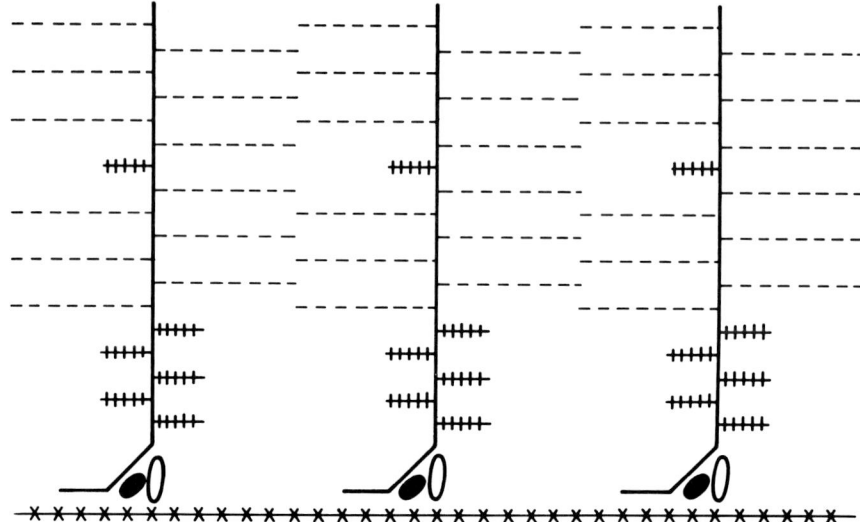

FIG. 4.—Structural Model of Bovine Nasal-Cartilage, Proteoglycan Complex. [———, Core protein; ---------, chondroitin sulfate; +++++++, keratan sulfate; ×××××, hyaluronic acid; and • o, link proteins.]

spond respectively to T_2 values of ~60 and ~2 ms. Such short T_2 values require the use of n.m.r. spectrometers having the lowest possible, ring-down time.[87] The use of the line-width technique for the measurement of T_2 suffers from the disadvantage of preferentially recording, in a system that contains molecules having a range of correlation times (τ_c), those molecules having the longest τ_c values. Also, it

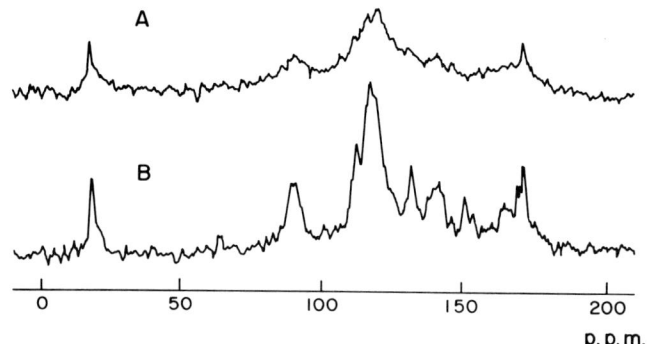

FIG. 5.—^{13}C-N.m.r. Spectra of Bovine Nasal-Cartilage Recorded at 4° (A) and 56° (B). [Solvent, D_2O. The chemical-shift scale is based on external CS_2 ($\delta_c = 0$).]

(87) I. C. P. Smith, personal communication.

might be suspected that, in measuring T_2 values of gel components, if any irregularity of the intermolecular association occurs, broadening of signals could occur through recording of a range of certain resonances, by variation of the ϕ and ψ glycosidic torsion-angles. Several methods based on the spin-echo technique are available for measurement of long and short T_2 values.[88-90]

In terms of spectral resolution, decreased line-widths can be obtained by raising the solution temperature, which decreases τ_c (see equation 2). In most cases where solutions are used, the improvement is modest, but, sometimes, as with bovine nasal cartilage,[85] it is spectacular, as may be seen in Fig. 5 on going from spectrum A to B.

In obtaining, for polysaccharides, ^{13}C-n.m.r. spectra of good quality, the use of elevated temperatures and high magnetic fields, now obtainable by replacement of electromagnets by superconducting solenoids as the spectrometer core, is recommended for increase in spectral dispersion.[63] Both factors increase T_1 values, but, for linear polysaccharides having relatively short values, there is no problem. However, spectral determinations on polysaccharides that contain nonreducing end-groups having relatively high T_1 values should be approached with care, especially in quantitative determinations at elevated temperature.

VI. ^{13}C-N.m.r. Spectra of Polysaccharides

1. α-D-Glucans

(a) **Amylose and Cycloamyloses.**—In the first ^{13}C-n.m.r.-spectroscopic study made on polysaccharides, Dorman and Roberts[47] reported the six-signal spectrum of a soluble starch that contained 10 to 15% of amylopectin. In M sodium hydroxide, signals were displaced upfield substantially, except those of C-5 and C-6, a phenomenon ascribed to disruption of an intramolecularly hydrogen-bonded helix to give a random-coil form.[91,92] However, only small shift-changes were observed in 6.5 M lithium bromide,[47] a compound that alters the specific rotation of amylose. As with many of the early studies, errors were made in the interpretation of spectra because of a lack of accurate information on monosaccharide spectra and on O-substitution

(88) E. L. Hahn, *Phys. Rev.*, 80 (1950) 580–594.
(89) H. Y. Carr and E. M. Purcell, *Phys. Rev.*, 94 (1954) 630–638.
(90) S. Meiboom and M. Gill, *Rev. Sci. Instrum.*, 29 (1958) 688–691.
(91) S. R. Erlander and R. Tobin, *Makromol. Chem.*, 111 (1968) 212–225.
(92) S. R. Erlander, R. M. Purvinas, and H. L. Griffin, *Cereal Chem.*, 45 (1968) 140–153.

shifts. For a solution of structurally related cycloheptaamylose in D_2O, Takeo and coworkers[93] readily assigned C-1, C-4, and C-6 signals, and pinpointed the C-5 signal (δ_c 72.8) by comparison with the C-5 signal of the derived 6-deoxy derivative, which was shifted upfield by 4.4 p.p.m. Assignments of C-2 (72.0) and C-3 (73.0) were incorrectly made on using the shift values available at that time for α-maltose. However, even with correct data, based on the spectra of mono-O-methylglucoses, Usui and coworkers[48] were not successful, and interpreted the signals of cyclohexaamylose in water as C-2 (73.7), C-3 (75.4), and C-5 (73.5). The C-2 and C-5 signal assignments were reversed by Smith and coworkers,[55] who relied on maltose, maltotriose, and methyl 4-O-methyl-α-D-glucopyranoside in D_2O as models and by Perlin and coworkers,[41(b)] who examined cyclohexaamylose labelled specifically with deuterium. Interestingly, the C-2 and C-5 signals of cyclohexaamylose and cycloheptaamylose reversed their relative positions[55] on changing the pD to 14.

The ^{13}C-n.m.r. spectrum of amylose in D_2O at pD 14 was successfully rationalized by Jennings and coworkers,[55,94] using the same models, as C-2 (73.8), C-3 (75.4), and C-5 (72.6). Interpretation was easier than at pD 7, at which the C-2 and C-5 signals are almost superimposed, the shifts being C-2 (72.7), C-3 (74.5), and C-5 (72.6). The assignments corrected those of Dorman and Roberts[47] and Takeo and coworkers,[93] and agree with the assignment of the C-2 signal that was based on a heteronuclear-decoupling experiment.[52(b)]

Small, unexplained differences exist between some shifts of cyclohexa-, -hepta-, and -octa-amyloses, but the C-1 and C-4 resonances are downfield from those of amylose by 2–3 and 3–5 p.p.m., respectively.[48,55,93] These are due to differences in rotational isomers about the C-1–O and O–C-4 bonds arising from the relative restriction of mobility of the cyclic compounds.[55,93] On going from α-maltose to maltooligomers containing ~10 units, no changes in C-1 and C-4 resonances were observed, but, for amylose, they were respectively at 0.4 and 0.5 p.p.m. higher field than for the oligomers, suggesting[95] a different and definite polysaccharide-chain conformation in a 10% solution in Me_2SO-$^2H_6$.

The torsion angles of glycosidic bonds of maltose oligomers and cyclohexaamylose, and their O-acetyl derivatives, have been deter-

(93) K. Takeo, K. Hirose, and T. Kuge, *Chem. Lett.*, (1973) 1233–1236.
(94) H. J. Jennings and I. C. P. Smith, *J. Am. Chem. Soc.*, 95 (1973) 606–608.
(95) H. Freibolin, N. Frank, G. Keilich, and E. Siefert, *Makromol. Chem.*, 177 (1976) 845–858.

mined by way of long-range, ^{13}C,H coupling values, and are discussed elsewhere (see Section II,3).

(b) **β-Limit Dextrins of Amylopectin and Glycogen.**—Usui and co-workers[48] reported difficulties in obtaining ^{13}C-n.m.r. spectra of amylopectin and glycogen, because of low solubilities and excessive linewidths. Investigations were easier with β-limit dextrins of plant and animal glycogens and plant amylopectins, whose signal shifts should be similar. A higher proportion of resonances was observed, at δ_c 67.3, 71.0, and 99.5, on examination of β-dextrins of glycogens, which generally contain twice the degree of branching of amylopectins. The ^{13}C-n.m.r. spectrum of the β-dextrin of rabbit-liver glycogen (see Fig. 6A) was compared with those of isomalto-oligosaccharides and (1→4)-substituted α-D-glucopyranose structures. Signals at δ_c 67.3 and 71.0 were tentatively attributed to O-substituted C-6 and to C-5 of the branch units, and it appeared that C-4 of the nonreducing end-groups also contributed to the latter signal. The resonance at δ_c 99.5 is from C-1 of α-D-

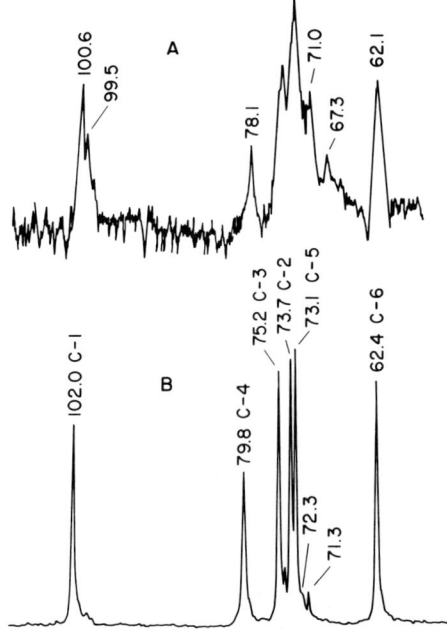

FIG. 6.—^{13}C-N.m.r. Spectra of A, β-Limit Dextrin of Rabbit-Liver Glycogen (aqueous solution; ambient temperature; chemical shifts based on tetramethylsilane) and B, Waxy-barley Amylopectin [in D_2O at 70°; chemical shifts (δ_c) based on external tetramethylsilane].

glucopyranosyl residues attached (1→6) to the branch units. As β-dextrins are pruned amylopectin and glycogen structures containing higher proportions of nonreducing end-groups, it is to be hoped that experimental conditions may be so improved that the extent of branching in parent polymers may be estimated by ^{13}C-n.m.r. spectroscopy. Rabbit-liver glycogen in 5% NaOD in D_2O at 90° gave the usual, main signals (although the assignments were unaccountably incorrect), and minor ones at δ_c 99.5 and 67.5 were barely detectable, so that their quantitation was impossible.[96] In connection with this observation, no difficulty was experienced in obtaining ^{13}C-n.m.r. spectra of amylopectin-containing potato starch and of a 10% solution of waxy-barley amylopectin in 2% NaOD in D_2O at 70°, but signals at δ_c 67.5 were not[97] detected (see Fig. 6,B). In each instance, the spectrum resembled that of amylose, but a narrow signal was present at δ_c 70 to 71, corresponding to C-4 of relatively mobile, nonreducing end-groups, which, like amylose, would give a C-6 signal at δ_c 62. The ^{13}C n.m.r. spectrum of rabbit-liver glycogen (90°), which has a greater degree of branching, contained[98] a comparatively large signal at δ_c 70–71. It is possible that an apparent roll in the baseline at δ_c ~68 is, in fact, a broad signal. Perhaps, the segmental motion of the 4,6-di-O-substituted branch units of amylopectin is comparatively low, resulting in extremely broad signals.

Detection of the C-6 signal of the nonreducing end-group of amylopectin was possible for the 4,6-diphenylboronate derivative, which is associated with nonreducing end-groups and whose C-4 and C-6 signals are displaced[99] upfield and downfield, respectively (see Fig. 7).

(c) **Linear, α-D-(1→6)-Linked Dextran.**—The ^{13}C-n.m.r. spectrum of dextran [NRRL B-512(F)] was markedly improved on using a relatively large amount of substrate (D_2O, 50°) in a 20-mm diameter, n.m.r. tube.[60] Six principal signals were present, arising from α-D-(1→6)-linked units, and, as a result of the high s/n ratio, tiny resonances of C-1, O-substituted carbon atoms, and unsubstituted C-6 were detected that can now be assigned to α-D-glucopyranosyl residues involved in the (1→3)-linkage. Colson and coworkers[55] identified the major signals in the spectrum of a dextran by comparison with the

(96) T. Usui, N. Yamaoka, K. Masuda, K. Tuzimura, H. Sugiyama, and S. Seto, *Agric. Biol. Chem.*, 39 (1975) 1071–1076.
(97) P. A. J. Gorin and M. Mazurek, unpublished results.
(98) F. R. Seymour, R. D. Knapp, T. E. Nelson, and B. Pfannemüller, *Carbohydr. Res.*, 70 (1979) 125–133.
(99) P. A. J. Gorin and M. Mazurek, *Can. J. Chem.*, 31 (1973) 3277–3286.

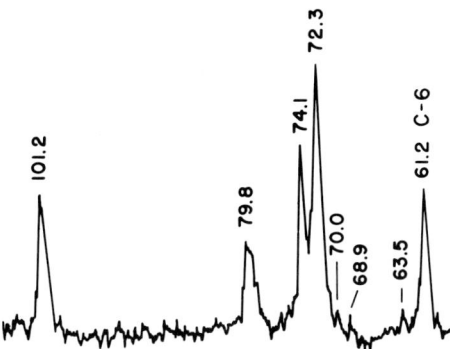

Fig. 7.—[13]C-N.m.r. Spectrum of Diphenylborinate Complex of Potato Amylopectin in D_2O. [Temperature 70°; chemical shifts (δ_c) based on external tetramethylsilane.]

spectra of maltose, maltotriose, isomaltose, panose [O-α-D-glucopyranosyl-(1→6)-O-α-D-glucopyranosyl-(1→4)-α,β-D-glucose], amylose, and methyl 6-O-methyl-α-D-glucopyranoside. Compared with the signals of amylose, △C-1 was −2 p.p.m., △C-6 +4.9 p.p.m., and △C-4 −7.9 p.p.m. Assignments at 32° and pD 7 were C-1 (99.0), C-2 (72.5), C-4 (71.3), C-5 (70.7), and C-6 (66.7), those of C-4 or C-5 being interchangeable. The high-field signal was attributed to C-5 by Gagnaire and Vignon,[26] who unequivocally assigned the ^{13}C-n.m.r. spectrum by selective, proton irradiation, based on the well-resolved n.m.r. spectrum obtained with a high-field spectrometer.

Friebolin and coworkers[95] measured the shifts of C-1 and C-6 signals of dextran and of isomalto-oligosaccharides up to 14 units in length (in $Me_2SO-^2H_6$). No deviations were observed that would be indicative of size-dependent, conformational differences of the glycosidic linkages, such as occur in the malto-oligosaccharide series.

(d) **Branched-Chain Dextrans.**—Quantitative, periodate oxidation of dextran B742 showed it to contain 57% of (1→6)-linked, or nonreducing, end-groups, or both, and 26% of unoxidizable units, having the α-D configuration, that were 3-O-substituted (according to infrared spectroscopy).[100] The ^{13}C-n.m.r. spectrum contained C-1 signals in the high-field region of α-linked D-glucopyranosides, namely, δ_c 100.5 (69%, linked to C-3) and δ_c 99.1 (linked to C-6), which had a small shoulder at δ_c 101.3 corresponding to (1→4)-linkages (see Fig. 8). The

(100) A. Jeanes, W. C. Haynes, C. A. Wilham, J. C. Rankin, E. H. Melvin, M. J. Austin, J. E. Cluskey, B. E. Fisher, H. M. Tsuchiya, and C. E. Rist, *J. Am. Chem. Soc.*, 76 (1954) 5041–5052.

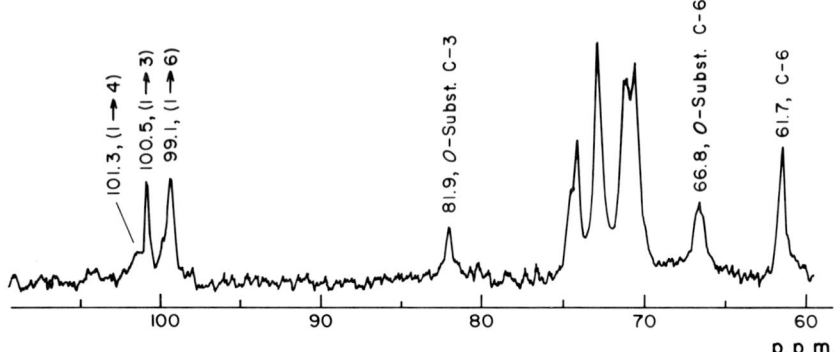

FIG. 8.—^{13}C.N.m.r. Spectrum of Dextran B-742 in D_2O, pD 7, at 32°. [Chemical shifts are based on external tetramethylsilane ($\delta_c = 0$).]

identity of the signals at δ_c 100.5 and 81.9 (D-glucosyloxylated C-3) were confirmed, as they were displaced upfield by ~1 p.p.m. on increase of the pD. These analytical figures for the linkage ratio were close to those obtained by p.m.r. spectroscopy of the free and per-O-methylated derivatives, and by ^{13}C-n.m.r. spectroscopy of the latter. The ^{13}C-n.m.r. spectrum did not distinguish between a linear (1→6), (1→6), (1→3) repeating sequence and the actual branched structure, which was demonstrated by detection of the tetra-O-methylglucose following hydrolysis of the per-O-methylated derivative.[101]

The chemical structures of five dextrans were partially determined by methylation, and found to be branched molecules having the following types of substitution (a) 6-O and 3,6-di-O, (b) 6-O, 3-O, and 3,6-di-O, (c) 6-O, 3,6-di-O, and 2,3-di-O, (d) 6-O, 4-O, and 3,4-di-O, and (e) 6-O and 2,3-di-O. At 27° and pH 7 (external, Me$_4$Si standard), the ^{13}C shifts of O-substituted, non-anomeric carbon atoms were C-2 (76.5), C-3 (81.6), and C-4 (79.5). The C-1 resonances were also recorded, and may be used for reference purposes. Some variation of chemical shifts, relative to each other, was observed with changing temperature. (The work serves to emphasize the importance of accurately measuring the temperature of the solution when determining chemical shifts.[102])

The diagnostic nature of the δ_c 70–75 spectral region, in terms of degree and type of dextran branching, was demonstrated on investigation of dextran fractions from NRCC strains of *L. mesenteroides* B-742,

(101) H. J. Jennings, personal communication.
(102) F. R. Seymour, R. D. Knapp, and S. H. Bishop, *Carbohydr. Res.*, 51 (1976) 179–194.

B-1299, and B-1355, and *Streptobacterium dextranicum* B-1254. Improved spectral resolution was observed[103] at 90°.

Dextran fractions from NRRL strains *L. mesenteroides* B-1299 and B-1399, and dextrans from strains B-640, B-1396, B-1422, and B-1424 were shown by ^{13}C-n.m.r. spectroscopy to be branched exclusively through α-D-(1→2)-linkages.[104] The intensity of the signal at δ_c 77.8 (D_2O at 90°), corresponding to C-2 of 2,6-di-O-substituted α-D-glucopyranosyl residues, was diagnostic of the degree of branching. Significantly absent were resonances at δ_c 82.9 and 80.2 which, respectively, correspond to 3,6-di-O-substituted (C-3 signal) and 4,6-di-O-substituted (C-4 signal)-D-glucopyranosyl residues.

Segmental motion, as representd by T_1 values, is greater in single-unit side-chains of branched-chain dextrans than in the main chains.[58] Dextran B-742 fraction S consists of a main chain of (1→6)-linked α-D-glucopyranosyl residues substituted at O-3 by α-D-glucopyranosyl groups. A T_1 value of 0.25 s was observed for C-1 of the side chains, whereas those of C-3 and C-1 of the main chain were 0.11 s and 0.15 s, respectively. Dextran B-1299 fraction S has O-2 of every other residue of the main chain substituted by α-D-glucopyranosyl groups. T_1 values were 0.14 s (C-1 of 2,6-di-O-substituted, main-chain residues), 0.13 s (C-2 of 2,6-di-O-substituted, main-chain residues), 0.20 s (C-1 of unsubstituted main-chain residues), and 0.24 s (C-1 of side-chain residues). These and other T_1 values demonstrate that such determinations have potential as an aid in assigning ^{13}C signals.

The soluble dextran of *Leuconostoc mesenteroides* NRRL B-1299 was reported[105] to be a branched structure containing nonreducing end-groups and 6-O-, 3-O-, and 2,6-di-O-substituted residues in the ratios of 5:4:1:5. Its ^{13}C-n.m.r. spectrum (see Fig. 9) contains three C-1 signals, at δ_c 99.4, 98.0, and 97.2, corresponding to units B and D, and C and A, respectively, in structure **9**. The C-1 resonance of unit C undergoes a β-shift of −1.4 p.p.m. compared with that of units B and D, as the axial, C-1 substituent is adjacent to the D-glucosyloxy group

```
                    Glcp    (A)
                     1
                     ↓
                     2
     -α-D-Glcp-(1→6)-α-D-Glcp-(1→6)-α-D-Glcp-(1→6)-
         (B)            (C)           (D)
                         9
```

(103) F. R. Seymour, R. D. Knapp, and A. Jeanes, *Carbohydr. Res.*, 68 (1979) 123–140.
(104) F. R. Seymour, R. D. Knapp, E. C. M. Chen, and S. H. Bishop, *Carbohydr. Res.*, 71 (1979) 231–250.
(105) E. J. Bourne, R. L. Sidebotham, and H. Weigel, *Carbohydr. Res.*, 22 (1972) 13–22.

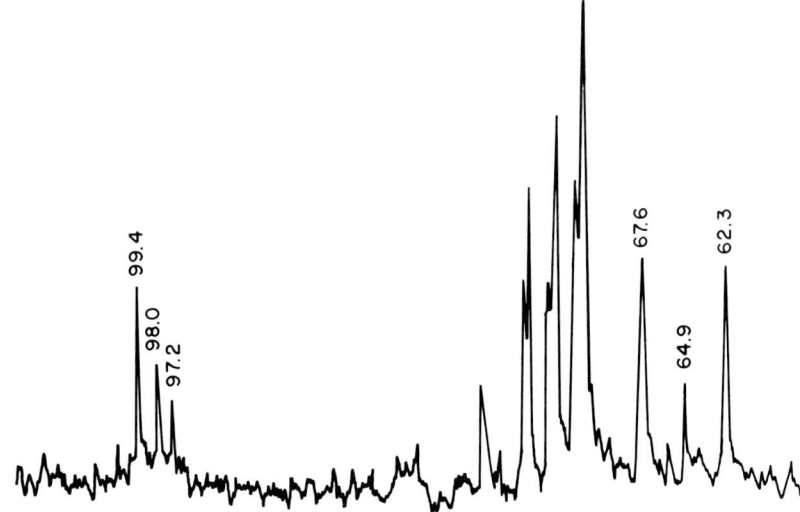

FIG. 9.—^{13}C-N.m.r. Spectrum of D_2O Solution of Water-Soluble Dextran of *Leuconostoc mesenteroides*. [Ambient temperature; chemical shifts (δ_c) based on external tetramethylsilane.]

on C-2. The C-1 signal of the minor (1→3)-linkage was not detected at lower field. Other signals, at δ_c 62.3 and 67.6, arose from free and O-substituted C-6 atoms of group A, and residues B and D, respectively.[96,106] The small, narrow signal at δ_c 64.9 was attributed to C-6 of unit C, but this appears doubtful, because of the small size of the signal. Data on such a model as methyl α-kojibioside are not available, but it is still significant that the OMe resonance of methyl α-D-glucopyranoside at δ_c 56.8 is only displaced upfield by 0.7 p.p.m. by 2-O-substitution in methyl α-sophoroside.[48] Also, there is doubt as to the validity of the described, quantitative estimations of linkage types, based on ^{13}C-n.m.r. spectroscopy, as the total area of the C-6 signals is clearly greater than that of the C-1 signals. The conditions for obtaining spectra were not specified, and, apparently, possible T_1 and n.O.e. effects were ignored. However, C-6 atoms generally have shorter T_1 values than carbon atoms bearing one attached proton, and if, indeed, as appears likely, the interval between pulses was not sufficient, the relative areas of C-1 signals (or C-6 signals) of side-chain units would be disproportionately small, as they have T_1 values longer than those of main-chain units (which undergo less segmental motion).

(106) T. Usui, M. Kobayashi, N. Yamaoka, K. Matsuda, K. Tuzimura, H. Sugiyama, and S. Seto, *Tetrahedron Lett.*, (1973) 3397–3400.

From the cariogenic *Streptococcus mutans* JG2, Usui and coworkers[107] obtained an exocellular α-D-glucan, subjected it to fractional precipitation, and recorded the ^{13}C-n.m.r. spectra of saturated solutions of three fractions in 6 M NaOD in D_2O at 80°. Examined were the starting material, the insoluble product obtained on treatment with alkali followed by neutralization, and the polysaccharide obtained from the resulting mother-liquor. With the aid of suitable glucobiose and glucotriose standards, signals at δ_c 98.9 and 100.6 were identified as resonances of C-1 linked (1→6) and (1→3) to adjacent residues, and those at δ_c 66.9 and 83.7 corresponded to glucosyloxylated C-6 and C-3 atoms, respectively (compare with the ^{13}C-n.m.r. spectrum[55] of dextran-B742). Based on the results of p.m.r. spectroscopy, variance of the relative proportions of each linkage was observed, from 60% of (1→6)-linkages in the water-soluble fraction from the culture to 47% in the water-insoluble fraction. These values corresponded to those determined by other methods.[108]

Colson and coworkers[109] investigated related D-glucans synthesized by enzyme isolates of several strains of *S. mutans*. Methylation results revealed α-(1→3)- and α-(1→6)-linkages, with branching, and comparison of C-1 and O-substituted-carbon resonances with those of known disaccharides and polysaccharides excluded other structures. Fractional precipitation provided D-glucans having high relative proportions of α-(1→6)- or α-(1→3)-linkages, in accord with the finding that the D-glucans have a wide range of structure.[110] The insoluble components were those containing mainly (1→3)-linkages. Soluble, streptococcal D-glucans are more ramified, and it was suggested that soluble D-glucan OMZ-176 contains branch points represented by the minor C-1 signal and δ_c 102.8 in a spectrum obtained at 68 MHz (see Fig. 10). However, the rest of the spectrum was consistent with signals arising from a combination of those of (1→3)- and (1→6)-linked α-D-glucopyranans. The value of the ratio of α-(1→3)- to α-(1→6)-linkages in glucan OMZ-176 was 13:7, as determined both by ^{1}H- and ^{13}C-n.m.r. spectroscopy. The latter technique was used to measure such a linkage ratio in dextranase-degraded glucans. In parallel spectral-de-

(107) T. Usui, H. Sugiyama, S. Seto, S. Araya, T. Nisizawa, S. Imai, and K. Kosaza, *J. Biochem. (Tokyo)*, 78 (1975) 225–227.
(108) T. Nisizawa, S. Imai, H. Akada, M. Hinoide, and S. Araya, *Arch. Oral Biol.*, 21 (1976) 207–213.
(109) P. Colson, H. C. Jarrell, B. L. Lamberts, and I. C. P. Smith, *Carbohydr. Res.*, 71 (1979) 265–272.
(110) M. Ceska, K. Granath, B. Norrman, and B. Guggenheim, *Acta Chem. Scand.*, 26 (1972) 2223–2230.

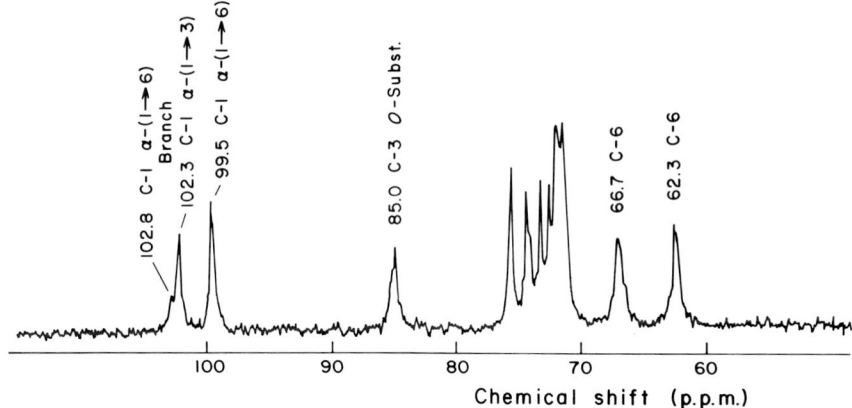

FIG. 10.—^{13}C-N.m.r. Spectrum at 68 MHz of Branched α-D-Glucan from *Streptococcus mutans* OMZ-176. [Solvent, D_2O, pD 14, at 40°; chemical shifts are based on external tetramethylsilane ($\delta_c = 0$).]

terminations on the polysaccharide, increase in the magnetic field by a factor of 2.7 broadened some of the signals to the same extent; this was explained on the basis of dipolar effects, and by small shift-differences of two or more overlapping signals arising from related, but different, chemical structures (in terms of neighboring units). To assess these shift effects, it would be of interest to determine the detailed, chemical structure of the polysaccharide.

(e) **α-D-(1→3)-Linked D-Glucan.**—^{13}C-N.m.r. signals of the α-D-(1→3)-linked D-glucan from *Penicillium patulum* (pD 14) were assigned[55] as C-1 (101.3), C-2 (72.2), C-3 (83.2), C-4 (71.7), C-5 (73.7), and C-6 (62.2). The C-1 to C-5 resonances differed widely from those of β-D-(1→3)-linked laminaran. Signal identification was aided by the ^{13}C-n.m.r. spectrum of 3-O-α-D-glucopyranosyl-α,β-D-glucose (nigerose), which was, in turn, rationalized by comparison with spectra of the methyl glycoside derivatives and 3-O-methyl-D-glucose. The disaccharide assignments agreed with those of Usui and co-workers,[48] except that the C-2 resonance of the α-anomeric form of the reducing end was at δ_c 72.4, not δ_c 71.3. However, it appears that an unequivocal approach is needed in order to identify the polysaccharide signals at δ_c 71.7 or 72.2.

(f) **Pullulan.**—The D-glucan from *Tremella mesenterica* contains (1→4)- and (1→6)-linked α-D-glucopyranosyl residues, and gives three C-1 and three C-6 signals, rationalized by reference to related signals

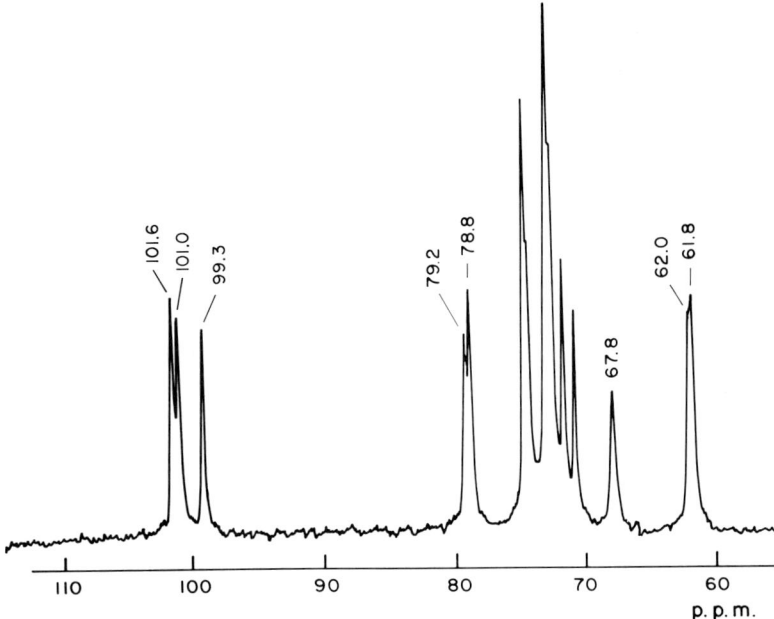

Fig. 11.—¹³C-N.m.r. Spectrum of Pullulan from *Tremella mesenterica*. (Solvent, D₂O; temperature, 32°; chemical shifts expressed as δ_c, relative to external tetramethylsilane.)

of maltose, isomaltose, amylose, and panose.[55,94] C-1 signals at δ_c 101.0 and 101.6 (see Fig. 11) arise from units glycosidically linked to C-4 atoms of adjacent residues, and are sensitive to "long-range" substitution-effects, namely, whether its parent unit is 4-O- or 6-O-substituted. Their combined area, compared with the signal at δ_c 99.3, of C-1 attached (1→6)- to the neighboring unit, was 2.2:1 in agreement with methylation results.[111] Two, almost overlapping, C-6 signals centered at δ_c 61.9 were also present, with an O-substituted C-6 signal at δ_c 67.8. The high-field C-6 and low-field C-1 signals were better resolved, depending on whether the pD was 14 or 7. In accord with the two types of (1→4)-linked structure, the ¹³C-n.m.r. spectrum contained two resolvable signals, at δ_c 78.8 and 79.2, corresponding to deshielded (+8 p.p.m.), C-4 resonances comparable with those of amylose. The similarity of the spectrum to that of pullulan from *Pullularia pullulans* was interpreted as evidence for the same, regular, α-(1→4), α-(1→4), α-(1→6) sequence. A ¹H- and ¹³C-n.m.r. study on an O-methylated pullu-

(111) C. G. Fraser and H. J. Jennings, *Can. J. Chem.*, 49 (1971) 1804–1807.

lan showed that it contained <3% of the α-(1→3)-linkages, rather than the 6% originally estimated by chemical means.[112]

2. β-D-Glucans

(a) **β-D-Glucans Having One Type of Linkage.**—β-D-(1→3)-Linked algal laminaran in D_2O at pD 14 gives six ^{13}C signals, assigned as C-1 (104.7), C-2 (74.9), C-3 (88.0), C-4 (69.9), C-5 (77.8), and C-6 (62.5), the C-1 and C-3 resonances being sensitive[55] to change in pD. The signals were well separated, and identification was readily achieved by use of the ^{13}C-n.m.r. spectrum of laminarabiose, which was, in turn, based on those of its methyl glycosides and of 3-O-methyl-D-glucose. The ^{13}C-n.m.r. spectrum of laminaran was also rationalized by heteronuclear, irradiation experiments following partial analysis of its p.m.r. spectrum (obtained at high magnetic field[51]). As mentioned elsewhere (see Section V,2), a curdlan type of β-D-(1→3)-linked D-glucan from *Alcaligenes faecalis* forms a resilient gel whose ^{13}C signals are broad and are displaced from those of the random-coil form, which gives rise to a ^{13}C-n.m.r. spectrum similar to that of laminaran.[59] Random-coil conformations occur in 0.22 M NaOH or Me_2SO. Fractions having d.p. >25 are capable of forming the gel network.

Gagnaire and Vincendon[51] dissolved cellulose in Me_2SO-$^2H_6$ containing 20% of 4-methylmorpholine N-oxide at 100°, and the solution gave a ^{13}C-n.m.r. spectrum rationalized as C-1 (102.5), C-4 (79.3), C-2, C-3, C-5 (73.1, 74.7, 75.3, not individually assigned), and C-6 (60.5), the C-6 resonance being close to one of those of the N-oxide, at δ_c 61.1. Other resonances rising from the solvent were at δ_c 65.7 and 59.3. Due to difficulties in obtaining the p.m.r. spectrum, specific heteronuclear-decoupling experiments could not be performed, so that the group of signals at δ_c 73.1, 74.7, and 75.3 were left unassigned. Inoue and Chujo[113] found that oligomers of cellulose resemble a homologous series of oligosaccharides based on maltose, isomaltose,[95] and 2-O-α-D-mannopyranosyl-D-mannose,[114] in that chemical shifts of ^{13}C nuclei of internal units are independent of chain length. The gentiobiose series is an exception, but the members thereof were examined by using a higher-field spectrometer having higher resolution.[50] By using the cellobiose assignments of Usui and coworkers[48] as a basis, the following interpretation was suggested for central residues of cellulose oligomers: C-1 (103.4), C-2 (74.3), C-3 (76.1), C-4 (79.9), C-5 (75.4), and C-6 (61.5).

(112) W. Sowa, A. C. Blackwood, and G. A. Adams, *Can. J. Chem.*, 41 (1963) 2314–2319.
(113) Y. Inoue and R. Chujo, *Carbohydr. Res.*, 60 (1978) 367–370.
(114) P. A. J. Gorin, *Can. J. Chem.*, 51 (1973) 2105–2109.

The β-D-(1→2)-linked D-glucan from *Arthrobacter radiobacter* gives six signals; these were given[99] the assignments C-1 (102.7), C-2 (83.1), C-3 (77.0), C-4 (69.3), C-5 (76.1), and C-6 (61.4). As the C-3 and C-5 resonances are very close, their identity should be determined by an unequivocal method, but, even so, there is a strong indication that the C-3 and C-5 assignments should be reversed. C-3 resonances adjacent to a 2-O-substituent should be upfield from that of C-5, as the C-3 and C-5 signals of methyl β-D-glucopyranoside overlap,[35] and 2-O-methylation of β-D-glucopyranose gives a β-upfield shift of 0.5 p.p.m. from the C-3 resonance.[48]

The ^{13}C-n.m.r. spectrum of the β-D-(1→6)-linked glucan of *Gyrophora esculenta* was assigned as follows[56]: C-1 (104.2), C-2 (74.2), C-3 (70.7), C-5 (76.1), and C-6 (70.0), by analogy with that of gentiobiose.[48] Confirmation was effected by selective, heteronuclear-decoupling experiments on the structurally similar pustulan from *Umbilicaria pustulata*, following the assignment of the well-resolved signals of the proton spectrum, obtained at high field.[50] The $^1J_{C-1,H-1}$ value of β-D-linked pustulan is, as would be expected, 160 Hz, as compared with the 171 Hz reported for α-D-linked dextrans.[115]

(b) **β-D-Glucans Having Mixed Linkages.**—Barley glucan and the lichenan of Iceland moss (*Cetraria islandica*) consist of β-D-(1→4)-linked and β-D-(1→3)-linked D-glucopyranosyl residues. For barley glucan, the linkage ratio[116] of 2.5:1 is reflected in its ^{13}C-n.m.r. spectrum[96] (see Fig. 12A). Despite the low s/n ratio, many of the signals could be assigned with the aid of the ^{13}C-n.m.r. spectra of laminarabiose and cellobiose. Only one C-1 signal was detected (δ_c 104.0), and also readily assigned were a C-4 signal of 4-O-substituted units (δ_c 80.0) and smaller C-3 and C-4 resonances of 3-O-substituted residues at δ_c 87.4 and 69.7, respectively. The latter had undergone a β-shift of −1.7 p.p.m.

Lichenan is reported to contain β-D-(1→4)- and β-D-(1→3)-links in the ratios[117,118] of 13:7 and 7:3. Its ^{13}C-n.m.r. spectrum, obtained at 68.2 MHz, with Me$_2$SO-^{2}H$_6$ at 100° as the solvent, was highly resolved, and typical of a (1→4), (1→4), (1→3) repeating-sequence, as it contained 15 signals corresponding to 18 carbon atoms. No splitting of signals associated with (1→3)-linkages was present, suggesting the absence of two, consecutive (1→3)-linkages. Unlike the spectrum of barley D-glucan, the C-1, C-6, and O-substituted C-4 resonances were

(115) D. Gagnaire and M. Vignon, *Carbohydr. Res.*, 51 (1974) 140–144.
(116) O. Igarashi and Y. Sakurai, *Agric. Biol. Chem.*, 29 (1965) 678–686.
(117) R. H. Meyer and P. Gürtler, *Helv. Chim. Acta*, 30 (1947) 751–761.
(118) S. Peat, W. J. Whelan, and J. G. Roberts, *J. Chem. Soc.*, (1957) 3916–3924.

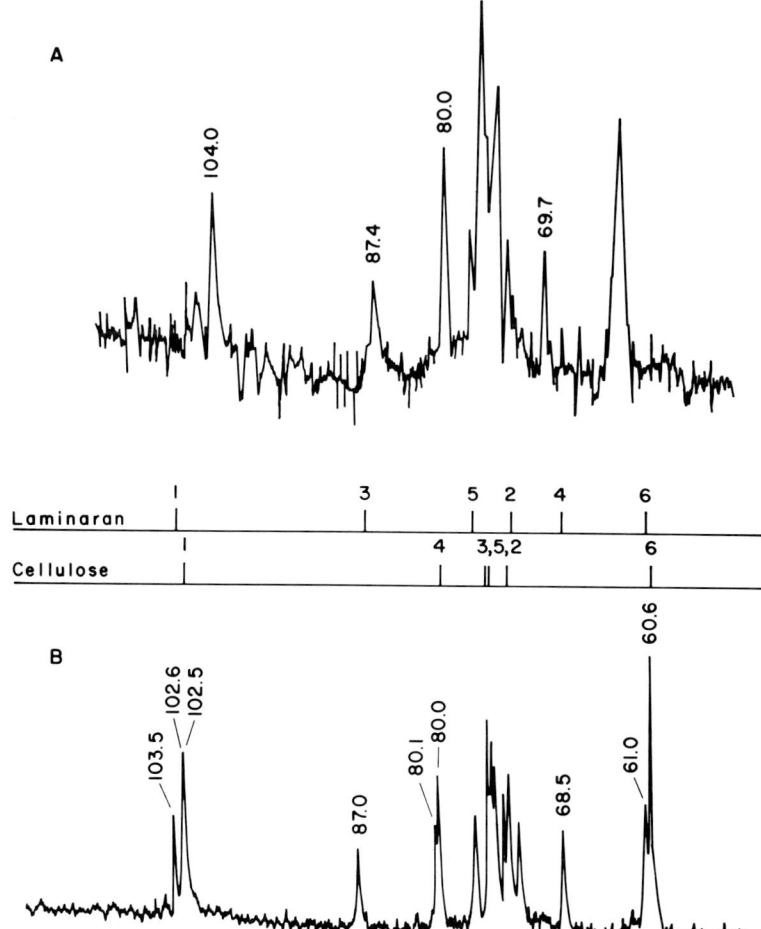

FIG. 12.—¹³C-N.m.r. Spectra of A, β-D-Glucan of Endosperm of Naked Barley (25.2 MHz; D₂O at 90°; chemical shifts expressed as δ_c, relative to internal DSS) and B, Lichenan (62.9 MHz; in Me₂SO-²H₆ at 100°; chemical shifts expressed as δ_c relative to tetramethylsilane; inset lines, corresponding to cellulose and laminaran spectra, are also presented.)

resolved into more than one component. The spectrum of the lichenan consists of a virtual combination of the spectra of laminaran and cellulose in the ratio of 2:1 (see Fig. 12B).

D-Glucans containing β-D-(1→3) and β-D-(1→6) units, often with a β-D-(1→3)-linked main-chain, are common components of fungi. Saito and coworkers[56] showed that the ¹³C-n.m.r. spectra of solutions of D-

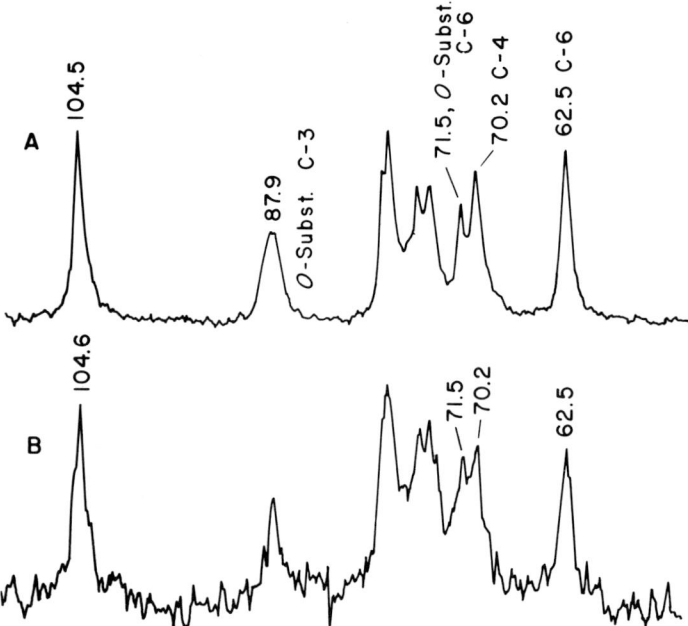

FIG. 13.—¹³C-N.m.r. Spectra of (1→3),(1→6)-Linked β-D-Glucans from *Schizophyllum commune* (A) and *Lentinus edodes* (B). (Solvent, Me₂SO-²H₆; temperature, ambient; chemical shifts expressed as δ_c, downfield from external tetramethylsilane.)

glucans of *Schizophyllum commune* and *Lentinus edodes* (shiitake mushroom) (see Fig. 13, A and B) contain broad signals that are poorly resolved and contain many overlapping resonances. [They, too, correspond to a combination of spectra of D-glucans, namely, of β-D-(1→6)- and β-D-(1→3)-linked structures.] However, C-1, 3-O-substituted C-3, 6-O-substituted C-6, and free C-6 signals were recognizable. These have been useful in ascertaining molecular motion in the gel of the *Lentinus edodes* glucan (lentinan) (see Section V,2), a polysaccharide that contains a β-D-(1→3)-linked D-glucopyranan main-chain having both β-D-(1→3)- and β-D-(1→6)-linked D-glucopyranosyl residues in the side chains. As with many polymers of its type, it has antitumor properties and can, like its acid-degraded products of mol. wt. ~20,000, form a complex with Congo Red by virtue of the ordered conformation of the polymer chains. Both lentinan (mol. wt. ~10⁶) and an acid-degraded product (mol. wt. 1.6 × 10⁴) give soft gels whose ¹³C-n.m.r. spectra are considerably attenuated, showing only resonances of disordered β-D-(1→6)-substituted residues, and lack ¹³C signals of (1→3)-substituted β-D-glucopyranosyl residues of the main chain and side chains. These portions of the gel are immobilized, and appear to exist as single-helix forms that tend to exist as multiple helices at junction zones.

3,6-Di-O-substituted branch-points were detected in lentinan (in Me_2SO-$^2H_6$) and in the glucan of *Schizophyllum commune*, which contains single-unit, β-D-glucopyranosyl side-groups, as a considerably displaced, O-substituted, C-6 signal at $δ_c$ 71.5 (see Fig. 13,A). This signal was also present in the ^{13}C-n.m.r. spectrum of a glucan from *Pleurotus ostretus* that contains β-D-(1→3)-links and a small proportion of 4-O-substituted α-D-glucopyranosyl side-chain units, respectively represented by C-1 and C-4 signals at $δ_c$ 101.9 and 80.7 p.p.m.

Reference has been made by Usui and coworkers[48,96] to a "glycogen" component of *Lentinus edodes*, and an extract was reported to contain (1→6)-, (1→3)-, and (1→4)-linkages in the ratios of 4:1:1, as determined by p.m.r. spectroscopy.[119] However, glycogen was not present, as all units had the β-D configuration, as evidenced by one C-1 signal at $δ_c$ 104.0. A signal for O-substituted C-4 was also observed, at $δ_c$ 81.0. 4-O-Substituted β-D-glucopyranosyl residues were present in linear, soluble glucans extracted by Travassos and coworkers[120] from yeast-like, mycelial, and conidial forms of *Sporothrix schenckii* by means of aqueous alkali. Chemical analysis showed the polysaccharide(s) to contain (1→3)-, (1→6)-, and (1→4)-linkages in the ratios of 11:7:7. All of the D-glucopyranosyl residues had the β-D configuration, as shown by two C-1 signals, at $δ_c$ 104.0 and 103.8 (see Fig. 14). Signal assignments were based on those of cellobiose and laminaran,[48] and the presence of O-substituted C-3, C-4, and C-6 atoms was indicated by signals at $δ_c$ 86.0, 80.3, and 71.2, respectively.

3. Linear D-Mannans

Yeasts may be identified, and classified, according to the H-1 portion of the p.m.r. spectrum of the cell-wall, D-mannose-containing polysaccharide. This portion has served most frequently as a fingerprint for identification of unknown yeasts, but, sometimes, the presence of certain signals of a particular chemical-shift could indicate the presence of certain structures (for example, β-D-linkages).[121] For some

(119) T. Usui, M. Yokoyama, N. Yamaoka, K. Matsuda, K. Tuzimura, H. Sugiyama, and S. Seto, *Carbohydr. Res.*, 33 (1974) 105–116.
(120) J. O. Previato, P. A. J. Gorin, R. H. Haskins, and L. R. Travassos, *Exp. Mycol.*, 3 (1979) 92–105.
(121) P. A. J. Gorin and J. F. T. Spencer, *Adv. Appl. Microbiol.*, 13 (1970) 25–89.

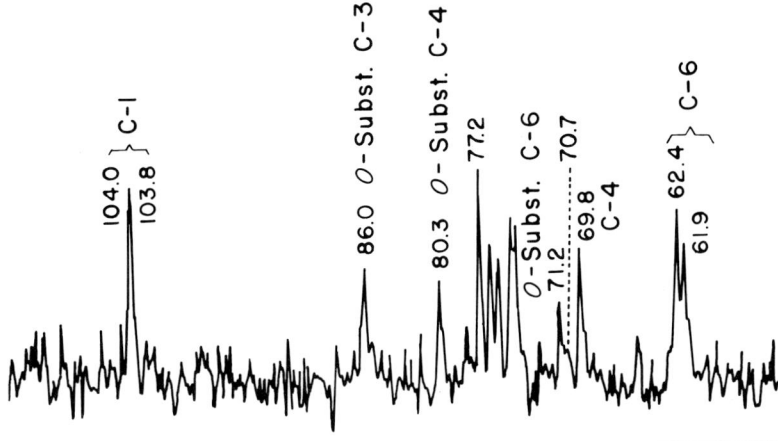

Fig. 14.—¹³C-N.m.r. Spectrum of β-D-Glucan(s) from *Sporothrix schenckii*. (Solvent, D$_2$O; temperature, 70°; chemical shifts expressed as δ$_c$, relative to external tetramethylsilane.)

compounds, ¹³C-n.m.r. spectroscopy provides additional, structural information. For example, *Pichia* and *Hansenula* spp. give D-mannans affording similar p.m.r. spectra, with H-1 signals at δ 5.76 and 5.64, but Gorin and Spencer[45] found that these D-mannans could be distinguished by means of their ¹³C-n.m.r. spectra. *Hansenula capsulata* produces a capsular O-phosphonomannan (see Section VI,8) and an α-D-mannan of a different structural type,[122] mainly linear and having a repeating-sequence of α-D-(1→2)-, α-D-(1→2)-, and α-D-(1→6)-linkages.[45] The ¹³C-n.m.r. spectrum (see Fig. 15) contains three, equal, C-1 resonances, at δ$_c$ 103.7, 102.3, and 100.3, and one signal at δ$_c$ 79.4 corresponding to O-substituted C-2 units. The two, high-field, C-1 signals were assigned to 2-O-substituted α-D-mannopyranosyl residues, as 2-O-methylation of methyl α-D-mannopyranoside results in a β, C-1, upfield shift of 3.8 p.p.m. The C-2 resonances shifted downfield by 8.5 p.p.m. on O-methylation, thus indicating that the prominent signal at δ$_c$ 79.4 is from O-substituted C-2 atoms of the polysaccharide. The more-highly-branched mannans from *Pichia pinus*, *Pichia kluyveri*, and *Hansenula glucozyma* contain an additional signal in this region, most likely from 2,6-di-O-substituted α-D-mannopyranosyl residues (see Section VI,4).

A solution of the β-D-(1→4)-linked D-mannopyranan from ivory

(122) M. E. Slodki, M. J. Safranski, D. E. Hensley, and G. E. Babcock, *Appl. Microbiol.*, 19 (1970) 1019–1020.

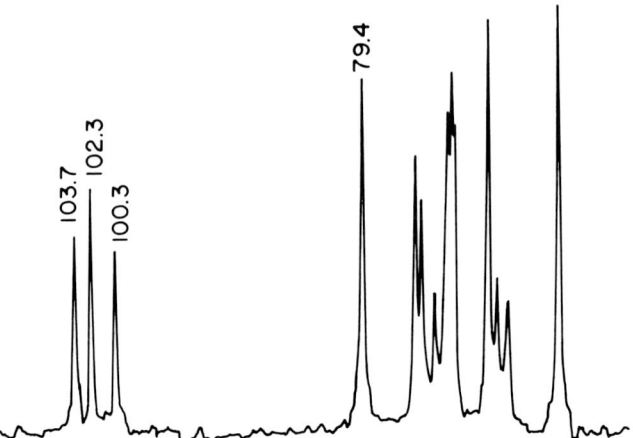

FIG. 15.—^{13}C-N.m.r. Spectrm of Linear α-D-Mannopyranan from *Hansenula capsulata*. (Solvent, D$_2$O; temperature, 70°; chemical shifts expressed as δ$_c$, relative to external tetramethylsilane.)

nuts, in D$_2$O containing 5% of NaOD, at 30°, gives a ^{13}C-n.m.r. spectrum containing five signals, whose assignments were based on those of β-D-mannose[12] in D$_2$O; these were: C-1 (101.7), C-2 (72.2), C-3 (73.8), C-4 and C-5 (78.8), and C-6 (62.1), the signal displacements being C-4 (Δδ$_c$ +9.0 p.p.m.), C-3 (Δδ$_c$ −0.3 p.p.m.) and, unexpectedly, C-5 (Δδ$_c$ +1.6 p.p.m.). The last may be an alkali-induced shift, as signals of a polymer containing alternate (1→3)- and (1→4)-linked β-D-mannopyranosyl residues,[123] also undergo selective displacements on changing[44] the pD.

The aforementioned D-mannan was isolated from *Rhodotorula glutinis*, and the structure having alternate linkages was based on a Smith degradation in which 2-O-β-D-mannopyranosyl-D-erythritol was formed as the sole product. The ^{13}C-n.m.r. spectrum contained 12 signals. Deuterium labelling was conducted by growing the yeast on [6-^{2}H$_2$], [5-^{2}H], and [3-^{2}H] derivatives of D-glucose, which were incorporated without extensive isotope-randomization.[44] Diminution in the size of the signal of the respective ^{13}C nucleus (α-effect) was noted (see Fig. 16), so that C-6, C-5, and C-3 resonances were identifiable. The upfield β-effect[39] of 0.10 p.p.m. was observed only in the spectrum of the D-mannan having C-6 dideuterated. Displacements of −0.06 p.p.m. occurring on monodeuteration were too small (in relation to line width) to be observed. Other shift assignments were based on those of β-D-(1→4)-linked D-mannopyranan (solvent: 5% NaOD in

(123) P. A. J. Gorin, K. Horitsu, and J. F. T. Spencer, *Can. J. Chem.*, 43 (1965) 950–954.

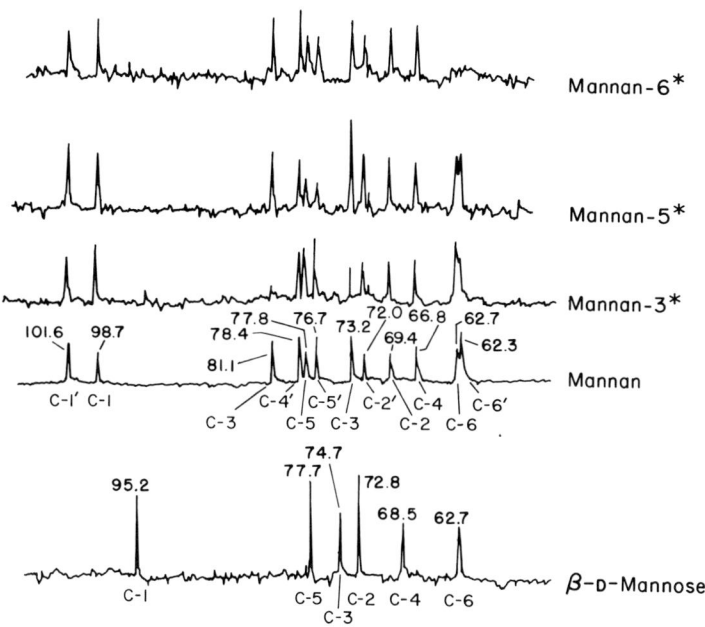

Fig. 16.—^{13}C-N.m.r. Spectra of Polysaccharides (in D_2O) Containing (1→3)- and (1→4)-Linked β-D-Mannopyranosyl Residues, Obtained from [^{2}H]-Labelled D-Glucoses. (Temperature, 70°; chemical shifts expressed as δ_c, relative to external tetramethylsilane; the number marked with a superscript asterisk refers to the position of labelling in each D-glucose derivative.)

D_2O for each polymer) and 4-O-β-D-mannopyranosyl-β-D-mannose and its [6,6'-^{2}H$_4$] derivative, obtained from the deuterated D-mannan by partial hydrolysis. Shifts occurring on 3-O-D-mannosylation could not be determined, as saccharides having two, or more, consecutive β-D-(1→3)-linkages were not available. For purposes of interpretation, it was assumed that only α- and β-shift effects occur on O-D-mannosylation, by analogy with those observed on mono-O-methylation of β-D-mannose at various oxygen atoms. The signal assignments made for the spectrum of the β-D-mannan are presented in Fig. 16, based on the numbering shown in formula **10**.

An α-D-(1→6)-linked D-mannopyranan isolated by Smith degradation of the glucuronorhamnomannan of *Ceratocystis stenoceras* gives a ^{13}C-n.m.r. spectrum[53] (D$_2$O at 70°) whose signals can be assigned as C-1 (101.1), C-2 and C-3 (72.6), C-4 (68.6), C-5 (71.7), and C-6 (67.6) by comparison with that of α-D-mannose.[39] The α- and β-carbon shifts occurring on 6-O-D-mannosylation are +4.9 and −2.3 p.p.m., respectively.

An α-D-(1→3)-linked D-mannopyranan was prepared from a complex heteropolymer of *Candida bogoriensis*,[124] but the quantity was small, and only the C-1 signal (δ_c 103.8) was positively identified.[12]

Alkaline extraction of the insect protozoon, *Crithidia deanei*, provided a β-D-(1→2)-linked D-mannopyranan that could be recognized as a new polysaccharide structure from its six-signal, ^{13}C-n.m.r. spectrum.[61] Assignments for a D$_2$O solution at 70° were C-1 (103.0), C-2 (81.1), C-3 (73.7), C-4 (69.3), C-5 (77.8), and C-6 (62.6), based on those of β-D-mannose (corrected to 70°). Displacements for the C-2 and C-3 signals were +8.2 and −1.0 p.p.m., respectively. A similar D-mannan is present in *Crithidia fasciculata*,[125] but, in *Herpetomonas samuelpessoai*,[11] the β-D-(1→2)-linked components consist of a homologous series of mannose oligosaccharides whose ^{13}C spectra contain a major, C-1 signal at δ_c 103.0, and the low, average molecular weight was indicated by minor signals at δ_c 101.0 (nonreducing end-group), and δ_c 95.2 and 93.8 (β- and α-D anomers of the reducing residue). Extraction of cells with hot, aqueous NaOH–NaBH$_4$ provided a mixture whose ^{13}C-n.m.r. spectrum lacked C-1 signals for the reducing end. The relative shifts of C-1 signals of nonreducing groups and internal residues are unusual, as they represent a 2-O-substitution β-shift of +2.0 p.p.m., different from the +0.4 p.p.m. noted on 2-O-methylation of β-D-mannose.[44] This suggests that the torsion angles of the (1→2)-glycosidic bonds of the nonreducing end-groups differ from those of internal residues, whose rotation is probably more restricted.

4. Branched-Chain D-Mannans

The majority of ascomycetous yeasts form cell-wall mannans or galactomannans having main chains of (1→6)-linked α-D-mannopyranosyl residues, often substituted at O-2 by α-linked glycosyl groups. The linkage types in the rest of the side chains vary widely, and can consist of 2-O-substituted α- or β-D-mannopyranosyl residues, or 3-O-substi-

(124) P. A. J. Gorin and J. F. T. Spencer, *Can. J. Chem.*, 46 (1968) 3407–3411.
(125) P. A. J. Gorin, J. O. Previato, L. Mendonça-Previato, and L. R. Travassos, *J. Protozool.*, 26 (1979) 473–478.

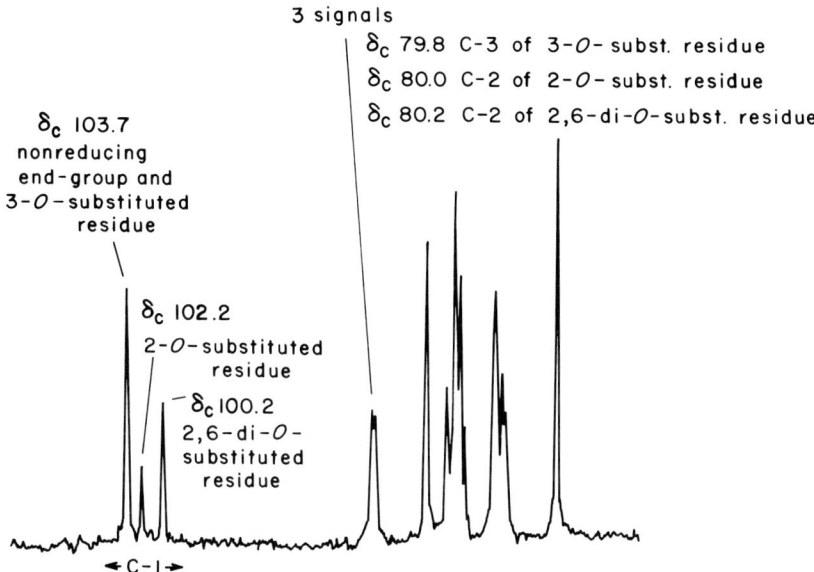

FIG. 17.—13C-N.m.r. Spectrum of Branched-Chain α-D-Mannopyranan from Bakers' Yeast. (Solvent, D₂O; temperature, 70°; chemical shifts expressed as δ_c, relative to external tetramethylsilane.)

tuted α-D-mannopyranosyl residues, which can be arranged in many combinations.[12] The 13C-n.m.r. spectra of D-mannans having α-D-linked side-chains contain, in less-crowded, downfield regions, signals that are readily assigned, as in that, for example, of bakers'-yeast mannan (11) (see Fig. 17). Reference to a homologous series of known

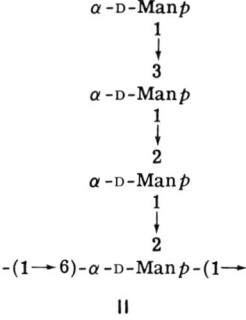

α-D-(1→2)-linked D-mannopyranose oligosaccharides, isolated by partial acetolysis of D-mannans,[12] showed that the C-1 signals of nonreducing end-groups and of internal 2-O-substituted residues are at δ_c

103.7 and 102.2, respectively, the latter signal increasing irregularly with the chain length. The signal at δ_c 103.7 coincides, as shown by examination of oligosaccharides containing α-D-(1→3)-linkages, with the C-1 resonances of 3-O-substituted α-D-mannopyranosyl residues (the H-1 signals of p.m.r. spectra are sensitive to 2-O- or 3-O-substitution.[126] Another common signal, at δ_c 100.2, was assigned to C-1 of 2,6-di-O-substituted α-D-mannopyranosyl residues, as it was the only major, C-1 signal remaining unidentified for several mannans of known structure. Sometimes, a C-1 resonance is present at δ_c 101.2 that corresponds to unsubstituted (1→6)-linked α-D-mannopyranosyl residues of the main chain.[127] Other characteristic signals, at δ_c 79.8, 80.0, and 80.2, correspond to C-3 of 3-O-substituted, C-2 of 2-O-substituted, and C-2 of 2,6-di-O-substituted α-D-mannopyranosyl units, respectively. Although barely resolvable at 25.2 MHz and sweep width 5 kHz, the presence of the signals at δ_c 80.0 and 80.2 distinguishes the ^{13}C-n.m.r. spectrum of the branched-chain D-mannan of *Saccharomyces rouxii*[128] (**12**) from that of the structurally related, linear D-mannan (**13**) of *H.*

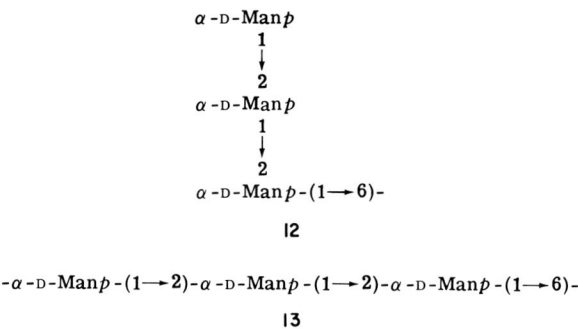

capsulata (see Section VI,3). Differentiation between the spectra can be based on the T_1 values of C-1 resonances, which are uniform for ^{13}C nuclei of the linear polymer. On the other hand, for the branched-chain structure, the T_1 values of C-1 atoms of the side chains are higher than those of the segmentally less-mobile main-chain[57] (see Section IV and Section V,2). Also, differences of T_2 values, as represented by line width, were detected.

(126) P. A. J. Gorin, M. Mazurek, and J. F. T. Spencer, *Can. J. Chem.*, 46 (1968) 2305–2310.
(127) P. A. J. Gorin, J. F. T. Spencer, and D. E. Eveleigh, *Carbohydr. Res.*, 11 (1969) 387–398.
(128) P. A. J. Gorin and A. S. Perlin, *Can. J. Chem.*, 34 (1956) 1796–1803.

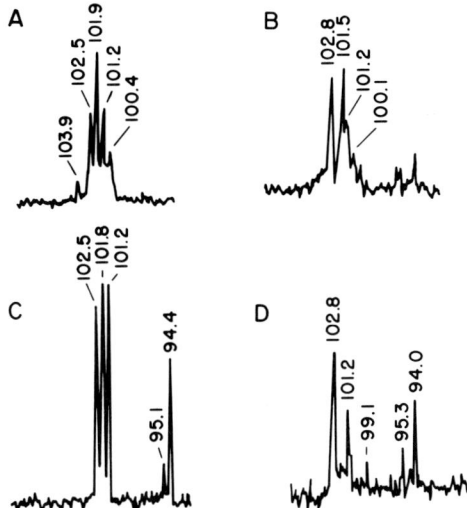

FIG. 18.—C-1 Regions of ^{13}C-N.m.r. Spectra of α,β-D-Mannopyranans from *Pichia pastoris* (A) and *Citeromyces matritensis* (B), and Those of Oligosaccharides Formed by Partial Acetolysis (C and D, Respectively). (Solvent, D$_2$O; temperature, 70°; chemical shifts expressed as δ$_c$, relative to external tetramethylsilane.)

Unlike H-1 shifts,[129] those of C-1 of D-mannopyranosyl residues are not sensitive to anomeric configuration. The values of $^1J_{\text{C-1,H-1}}$ are characteristic of configuration, as is shown by the observation of Bock and coworkers[21,22] that α- and β-D-mannopyranose have splittings of 170 and 160 Hz, respectively. However, D-mannans having α-D-(1→6)-linked main-chains and α-D-mannopyranosyl groups in the side chains (see Fig 17, for example) are distinguishable from those having β-D-mannopyranosyl residues, as there are small differences in C-1 shifts. For example, the D-mannan from *Pichia pastoris* contains α-D-Man*p*-(1→2)-β-D-Man*p*-(1→2)-β-D-Man*p*-(1→2)-α-D-Man*p* side-chains, linked to O-2 atoms of the α-D-(1→6)-linked main-chain, and has C-1 signals at δ$_c$ 102.5, 101.9, and 101.2 (see Fig. 18,A), as shown by comparison with the ^{13}C-n.m.r. spectrum of the pentasaccharide (Fig. 18,C) liberated by partial acetolysis.[12] Also, the C-1 signals of the β-D-(1→2)-linked side-chains of *Citeromyces matritensis* D-mannan, at δ$_c$ 101.2 and 102.8 (Fig. 18,B), can be assigned to nonreducing end-groups and internal residues, respectively, by comparison with ^{13}C-n.m.r. spectra of a β-D-(1→2)-linked tetrasaccharide (Fig. 18,D),

(129) P. A. J. Gorin, J. F. T. Spencer, and S. S. Bhattacharjee, *Can. J. Chem.*, 47 (1969) 1499–1505.

formed by partial acetolysis, and of oligosaccharides having longer chain-lengths.[11] Another distinguishable aspect was the C-2 signal of 2-O-substituted residues; this is at δ_c 80.8 (rather than δ_c 80.0) in the α-D series.[12]

5. Heteropolymers Having D-Mannose-Containing Main-Chains

In dimorphic fungi, the composition of constituent polysaccharides is sensitive to morphology and to the cultural conditions. These effects are accentuated in the case of *Sporothrix schenckii*, and ^{13}C-n.m.r. spectroscopy can be used to detect individual polysaccharides in a qualitative way. The ^{13}C-n.m.r. spectra of mannose-containing polysaccharides of ten *Sporothrix schenckii* and three *Ceratocystis stenoceras* species, grown under various conditions, were distinguishable in terms of the presence of signals,[130] at δ_c 103.3 to 103.7, which arose from O-α-L-rhamnopyranosyl-(1→2)-O-α-L-rhamnopyranosyl-(1→3)-side-chains (**14**; C-1', 103.7; C-1, 96.8), and were not present in

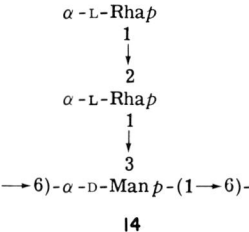

14

polysaccharides of *C. stenoceras*. Also, a signal at δ_c 81.6, arising from C-2 of α-L-rhamnopyranosyl residues substituted at O-2 by β-D-glucopyranosyluronic acid (**15**), is absent from the spectra of polysaccha-

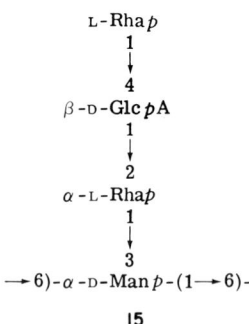

15

rides of *S. schenckii* (see Fig. 19). These differences suggested that *C. stenoceras* is not the imperfect stage of *S. schenckii*.

(130) L. R. Travassos, P. A. J. Gorin, and K. O. Lloyd, *Infect. Immun.*, 9 (1974) 674–680.

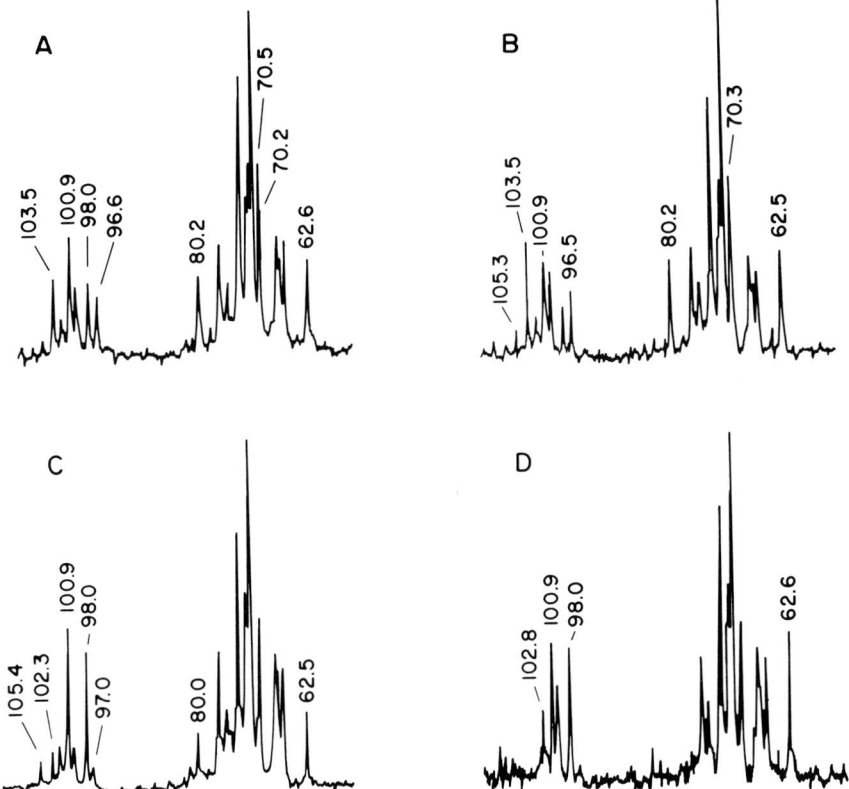

FIG. 19.—¹³C-N.m.r. Spectra (Excluding C-CH₃ Regions) of Polysaccharides (in D₂O) Isolated from Various Strains of *Sporothrix schenckii*. (Polysaccharides A, B, C, and D were isolated from Strains 1099.23, 1099.27, 1099.16, and 1099.18, respectively; solvent, D₂O; temperature, 70°; chemical shifts expressed as δ_c, relative to external tetramethylsilane.)

Yeast forms of *S. schenckii*, grown at 37°, produce a rhamnomannan (**16**) having[16] a main chain of (1→6)-linked α-D-mannopyranosyl resi-

$$\begin{array}{c} \alpha\text{-L-Rha}p \\ 1 \\ \downarrow \\ 3 \\ \rightarrow 6)\text{-}\alpha\text{-D-Man}p\text{-}(1\rightarrow 6)\text{-} \end{array}$$

16

dues substituted at O-3 by single α-L-rhamnopyranosyl groups (C-1, 98.2). At 25°, the yeast form produces both rhamnomannan **16** and a small proportion of a compound that contains 4-*O*- and 2,4-di-*O*-sub-

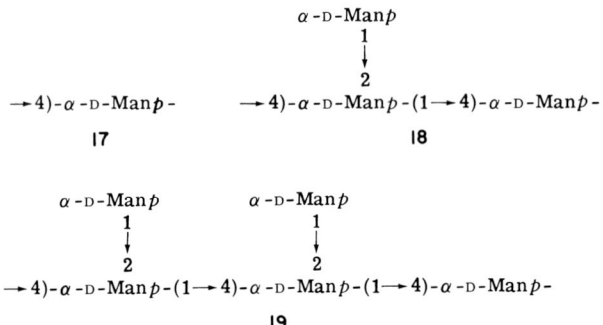

stituted α-D-mannopyranosyl residues (**17, 18**, and **19**) that are components of a larger total structure.

When grown at 25°, mycelia, or conidia mixed with mycelia, produce di-O-rhamnosyl side-chains (see the preceding) that are the main, antigenic determinants, as shown by inhibition experiments with oligosaccharides and rabbit immune serum.[131] The response of human sera to the polysaccharide, and the fact that ^{13}C-n.m.r. spectroscopy showed that conidia of S. *schenckii* form a rhamnomannan similar to the yeast component and different from the di-O-rhamnosylmannan of hyphae,[132] suggest the formation of hyphae *in vivo*.

Sometimes, cells contain a galactomannan[16] whose structure and ^{13}C-n.m.r. spectrum are discussed in Section VI,6. The types of polysaccharides formed under various conditions from different morphologies are summarized in Table IV.

The polysaccharide from *C. stenoceras* mainly contained predominant, single-unit α-L-rhamnopyranosyl side-chains attached (1→3) to an α-D-(1→6)-linked D-mannopyranosyl main-chain (**16**), as is clear from signals at δ_c 101.1 and 98.3. The side-chain structure, which distinguished it from *S. schenckii* polysaccharide, is **15**, which contains 4-O-substituted β-D-glucopyranosyluronic acid residues (C-1, 105.6; C-4, 80.3) linked[53] (1→2) to α-L-rhamnopyranose (C-2, 81.6) residues, as shown in Fig. 20, A. However, in one strain, *S. schenckii* 1099.12, most clones gave rise to spectra similar to that of the original rhamnomannan, but one gave an acidic rhamnomannan identical with that of *C. stenoceras* (Fig. 20,A), suggesting that there were at least two fungal variants present.[133]

(131) K. O. Lloyd and L. R. Travassos, *Carbohydr. Res.*, 40 (1975) 89–97.
(132) L. R. Travassos and L. Mendonça-Previato, *Infect. Immun.*, 19 (1978) 1–4.
(133) L. R. Travassos, L. Mendonça-Previato, and P. A. J. Gorin, *Infect. Immun.*, 19 (1978) 1107–1109.

TABLE IV

Summary of Structures of Mannans Isolated from *S. schenckii* under Different Conditions

Strain	Culture medium	Temp. (°C)	Morphology	Source[a]	Major structure
1099.12	YCV[b]	37	mixed (yeasts predominating)	C	mono-*O*-rhamnosylmannan
1099.12	YCV	37	mixed (yeasts predominating)	S	monorhamnosylmannan having a small proportion of di-rhamnosyl side-chains
1099.12	M[c], BHI[d]	37	100% yeasts	C	monorhamnosylmannan
1099.12	Sab[e]	37	90% yeasts	C	monorhamnosylmannan
1099.12	BHI, Sab	25	mixed (mycelium plus conidia predominating)	C	mainly dirhamnosylmannan
1099.18	M	37	100% yeasts	C	monorhamnosylmannan
1099.18	M	25	100% yeasts	C	monorhamnosylmannan with minor proportions of 4-*O*- and 2,4-di-*O*-substituted α-D-Man*p* residues
1099.12	YCV	25	conidia	C	Monorhamnosylmannan
1099.12	BHI	25	unsporulated mycelium	C	Galactomannan
1099.18	B[f]	25	unsporulated mycelium	C	Dirhamnosylmannan
1099.18	B	25	unsporulated mycelium	S	Mannan and trace of rhamnomannan
1099.18	B	25	mycelium plus a few conidia	S	Mannan and a small proportion of rhamnomannan

[a] Cells (C), or supernatant liquor (S). [b] YCV, Yeast nitrogen base–Casamino acids–Vitamins. [c] M, Synthetic medium. [d] BHI, Brain–heart infusion. [e] Sab, Sabouraud. [f] B, Mariat's medium.

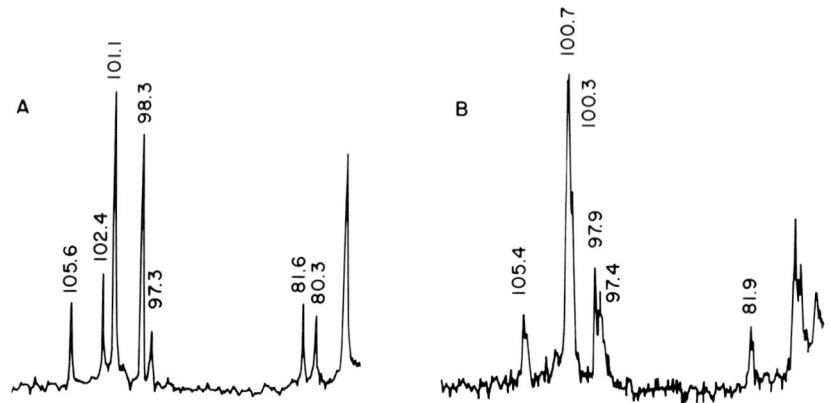

FIG. 20.—[13]C-N.m.r. Spectra of Glucosyluronorhamnomannan of *Ceratocystis stenoceras* (A), and the Acid-Degraded Product Lacking Rhamnosyl End-Units (B). Solvent, D_2O; temperature, 70°; chemical shifts expressed as δ_c, relative to external tetramethylsilane.)

Signals of rhamnomannans and acidic rhamnomannan were assigned as follows. The C-1 resonances of mono- and di-rhamnosyl side-chains were determined by comparison with those of 3-O-α-L-rhamnopyranosyl-α-D-mannose (C-1′, 98.0) and O-α-L-rhamnopyranosyl-(1→2)-O-α-L-rhamnopyranosyl-(1→3)-α-D-mannose (C-1″, 103.7; C-1′, 96.8). C-1 signals of the main chain should be similar in shift to that[12] of an unsubstituted, α-D-(1→6)-linked D-mannopyranan, δ_c 101.2.

Growth of *S. schenckii* at 25° gave a structural component whose C-1 signal at δ_c 102.3 was assigned by comparison with those of 4-O-α-D-mannopyranosyl-α,β-D-mannose (C-1′, 102.8), and O-α-D-mannopyranosyl-(1→4)-O-α-D-mannopyranosyl-(1→4)-α,β-D-mannose (C-1′ and C-1″, 102.5 and 102.8), obtained on partial hydrolysis.[53] As 2-O-α-substitution of an α-D-mannopyranosyl group causes the C-1 signals to move upfield by 1.6 p.p.m., the signal at δ_c 100.3 probably arose from 2,4-di-O-substituted α-D-mannopyranosyl residues, as in **18** and **19**.

The ^{13}C-n.m.r. spectrum of the *C. stenoceras* polysaccharide (see Fig. 20,A), containing structure **15**, contains a C-1 signal at δ_c 105.6 assignable to a D-glucuronic acid residue, as it corresponds to C-1′ of 2-O-(α-D-glucopyranosyluronic acid)-α-L-rhamnose. The acid-degraded polysaccharide lacked signals at δ_c 80.3 and 102.4 (see Fig. 20,B), because of removal of (terminal) L-rhamnopyranosyl groups, indicating that they belonged to C-4 of (1→4)-O-substituted β-D-glucopyranosyluronic acid residues and C-1 of terminal rhamnopyranosyl groups, re-

TABLE V

Assignments of Signals in the ^{13}C-N.m.r. Spectra of Rhamnomannans from *S. schenckii* and *C. stenoceras*

Signal, $\delta_c \pm 0.2$ (p.p.m.)	Assignment
105.6	C-1 of β-D-glucopyranosyluronic acid units
103.7	C-1 of α-L-rhamnopyranosyl (nonreducing) end-group of O-α-L-Rhap-(1 → 2)-O-α-L-Rhap-(1 → 3)-D-Manp
102.3	C-1 of 4-O-substituted α-D-mannopyranosyl residues
	C-1 of L-rhamnopyranosyl (nonreducing) end-group of O-L-Rhap-(1 → 4)-O-β-D-GlcpA-(1 → 2)-α-L-Rhap-
101.1	C-1 of 3,6-di-O-substituted α-D-mannopyranosyl residues
100.3	C-1 of 2,4-di-O-substituted α-D-mannopyranosyl residues
98.2	C-1 of α-L-rhamnopyranosyl (nonreducing) end-group of O-α-L-Rhap-(1 → 3)-D-Manp
97.2	C-1 of 2-O-substituted α-L-rhamnopyranosyl residues of β-D-GlcpA-(1 → 2)-α-L-Rhap-(1 → 3)-D-Manp
96.8	C-1 of 2-O-substituted α-L-rhamnopyranosyl residues of α-L-Rhap-(1 → 2)-α-L-Rhap-
81.6	C-2 of 2-O-substituted α-L-rhamnopyranosyl residues of β-D-GlepA-(1 → 2)-α-L-Rhap-
80.3	C-2 of 2-O-substituted α-L-rhamnopyranosyl residues of α-L-Rhap-(1 → 2)-α-L-Rhap-
	C-4 of 4-O-substituted β-D-glucopyranosyluronic acid residues
62.8	C-6 of unsubstituted α-D-mannopyranosyl units
18.4	C of methyl group of α-L-rhamnopyranosyl units

spectively. The remaining signals, at δ_c 81.6 and 97.3, respectively, arose from C-2 and C-1 of internal (1→2)-O-substituted α-L-rhamnopyranosyl residues. A summary of shift assignments for all rhamnomannan-related structures thus far studied by this technique is presented in Table V.

The C-2 to C-6 signals of monorhamnomannan were rationalized[53] by gauging substitution-shifts (see Fig. 21) and by use of D-glucose precursors labelled with ^{2}H or ^{13}C. D-[6-^{2}H$_2$]Glucose gave a rhamnomannan whose spectrum lacked the signal at δ_c 66.7, characterizing it as from C-6. However, the label of [5-^{2}H] and [3-^{2}H] derivatives was randomized, and the spectrum of the product was indistinguishable from that of unlabelled rhamnomannan. D-[2-^{13}C]Glucose containing twice the natural abundance of carbon-13 was used as the precursor, and increases in the intensities of the C-2 (δ_c 67.0) and C-5 (δ_c 71.8) signals occurred, the latter enhancement being due to fragmentation,

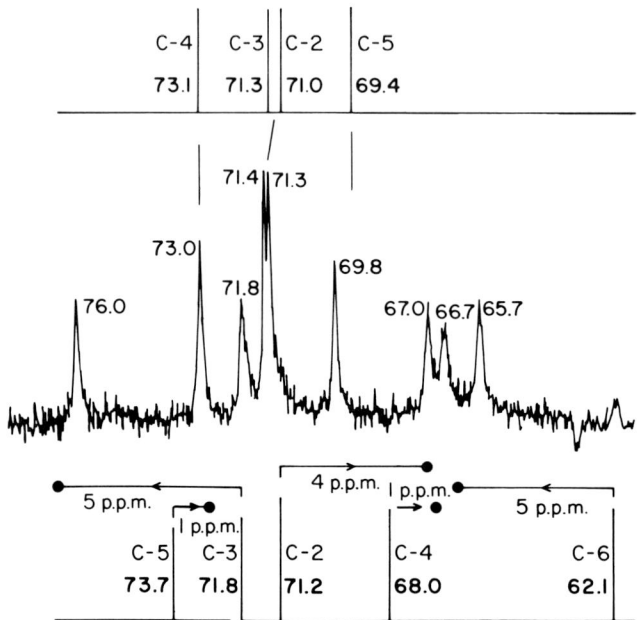

Fig. 21.—Partial ^{13}C-N.m.r. Spectrum of the Rhamnomannan (16) of *Sporothrix schenckii*, with Shift Comparisons with Signals of Methyl α-D-Mannopyranoside (Lower Inset Lines) and Methyl α-L-Rhamnopyranoside (Upper Inset Lines). (Solvent, D_2O; temperature, 33°; chemical shifts expressed as δ_c, relative to external tetramethylsilane.)

and rearrangement of the D-glucose chain. The route of formation of the L-rhamnopyranosyl residues was more circuitous, as none of its ^{13}C signals were enhanced. Use of D-glucose that was 90% labeled at C-1 with ^{13}C was unsuccessful in detecting the ^{13}C-2 signal at δ_c 67.0, which should have been coupled, and present as a doublet ($^1J_{C-1,C-2}$ ~45 Hz),[54] but the C-6 signal at δ_c 65.7 was so large that neighboring signals were obscured.

The glucomannan from *Ceratocystis brunnea*[134] consists mainly of a (1→6)-linked α-D-mannopyranosyl main-chain (C-1, 99.8; C-2, 81.2) substituted at the O-2 atoms by α-D-glucopyranosyl groups (C-1, 102.3). These ^{13}C signals (see Fig. 22,A) were identified by comparison with those of 2-O-α-D-glucopyranosyl-α-D-mannose.[134] Signals having the same chemical-shift were present in the ^{13}C-n.m.r. spectrum of *Ceratocystis paradoxa* 101-C (see Fig. 22,B), which was more complex because of the presence of α-D-mannopyranosyl (nonreduc-

(134) P. A. J. Gorin and J. F. T. Spencer, *Carbohydr. Res.*, 13 (1970) 339–349.

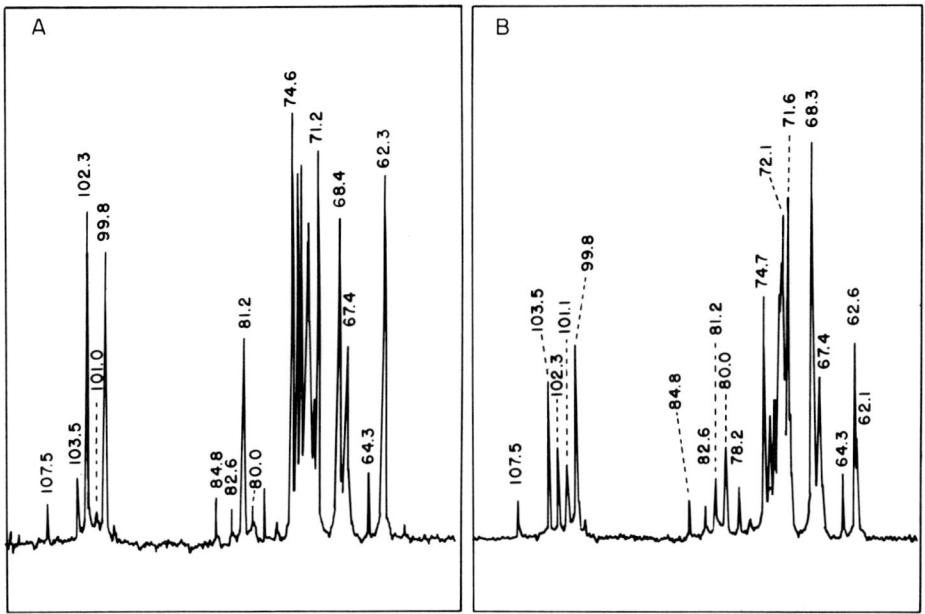

FIG. 22.—¹³C-N.m.r. Spectra of Polysaccharides from *Ceratocystis paradoxa* (A) and the Preponderant Glucomannan from *Ceratocystis brunnea* (B). (Solvent, D_2O; temperature, 70°; chemical shifts expressed as δ_c, relative to external tetramethylsilane.)

ing) end-groups.[135] This polysaccharide was one of a number of mannans, glucomannans, and galactomannans produced by three strains of *Ceratocystis paradoxa* and 2 strains of *Ceratocystis fimbriata* that are phytopathogenic towards certain tropical plants. No correlation was observed between polysaccharide structure and host specialization or host specificity. *C. paradoxa* 102-J from *Crotolaria juncea* was grown at 37°, and gave a polysaccharide affording ¹³C signals at δ_c 175.6 and 23.4, corresponding to carbonyl and methyl carbon atoms, respectively, of acetamido groups, which also gave a signal, at δ 2.6, in the p.m.r. spectrum[136]; these arose from 2-acetamido-2-deoxyglucose units that were not formed at 25°.

The galactomannan from *Trichosporon fermentans* contains a (1→6)-linked α-D-mannopyranose main-chain substituted at the O-2 atoms with *O*-α-D-galactopyranosyl-(1→2)-*O*-α-D-mannopyranosyl

(135) C. S. Alviano, P. A. J. Gorin, and L. R. Travassos, *Exp. Mycol.*, 3 (1979) 174–187.
(136) P. A. J. Gorin, J. F. T. Spencer, and A. J. Finlayson, *Carbohydr. Res.*, 16 (1971) 161–166.

side-chains.[137] The C-1 signal of the α-D-galactopyranosyl group is at δ_c 102.1 or 102.6, at higher field than that of the β-form.[9,125,138,139] Another C-1 signal was present at δ_c 99.9 (main chain), and O-substituted C-2 signals were noted at δ_c 80.8 and 80.2. The anomeric configuration of the D-galactopyranosyl residues in the polysaccharide of *Lipomyces lipofera* was shown to be β from their C-1 shift.[140] Galactomannans containing D-galactofuranosyl residues are dealt with next, in Section VI,6.

6. Furanose-Containing Structures

The β-D-(1→5)-linked D-galactofuranan from *Penicillium citrinum* Thom 1131 contains D-galactosyl and malonic acid residues in the ratio of 3:1, and the resistance of residues to periodate oxidation was interpreted[141] as being due to substitution at O-2 or O-3. Following comparison of the ¹³C-n.m.r. spectrum of the derived galactan with that of the malonogalactan (see Fig. 23,A and B), it was concluded that C-2 substitution was absent, and that the signal at δ_c 77.5, which was assigned to C-3, was smaller in the malonogalactan spectrum because of the esterification. The C-3 resonances of the esterified polysaccharide were then shifted downfield by 5.5 p.p.m., to overlap with the composite[14] C-2,C-4 signal at δ_c 83. Carbonyl signals of malonic acid ester were at δ_c 171.5 and 174, and other, additional signals, at δ_c 76 and 66, were assigned to malonic acid mono- and di-ester; a signal at δ_c ~41, close to the methylene signal of free malonic acid, was absent. The present author considers that the foregoing interpretation is incorrect, and that, in order to explain the two additional signals at δ_c 66 and 76, the malonate group must be situated at C-6, causing an α-shift of +4.5 p.p.m. from δ_c 62.5, and a β-shift of −1.5 p.p.m., analogous to those occurring on O-acetylation. The lesser intensity of the signal at δ_c 77.5 is explained by assigning it to C-5, as with the ¹³C signals of methyl 5-O-methyl-β-D-galactofuranoside, and of a D-galactotetraose obtained from the β-D-(1→5)-linked D-galactofuranan from *Penicillium charlesii*[15] (5-O-glycosylation was accompanied by an α-shift of +5.0 p.p.m., and β-shifts of C-4 and C-6 signals of −1.3 and −1.7 p.p.m.,

(137) P. A. J. Gorin and J. F. T. Spencer, *Can. J. Chem.*, 46 (1968) 2299–2304.
(138) D. D. Cox, E. K. Metzner, L. W. Cary, and E. J. Reist, *Carbohydr. Res.*, 67 (1978) 23–31.
(139) P. Colson and R. R. King, *Carbohyr. Res.*, 47 (1976) 1–13.
(140) A. S. Shashkov, A. F. Svirdlov, S. E. Gorin, O. D. Dzhikiya, O. S. Chizhov, N. Gullyev, and N. K. Kochetkov, *Bioorg. Khim.*, 4 (1978) 752–759.
(141) T. Kohama, M. Fujimoto, A. Kuniaka, and H. Yoshino, *Agric. Biol. Chem.*, 38 (1974) 127–134.

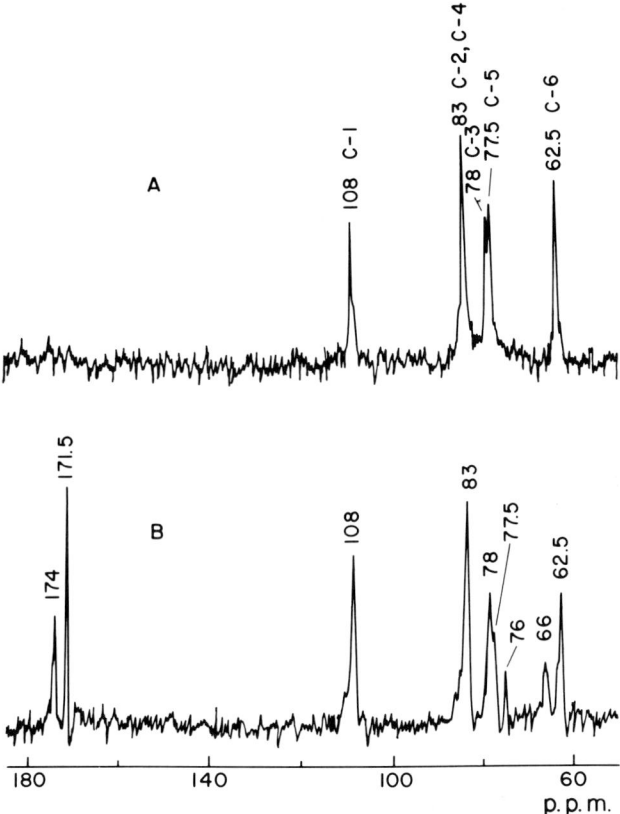

Fig. 23.—^{13}C-N.m.r. Spectra of the β-D-(1→5)-Linked D-Galactofuranan (A) Obtained by Hydrolysis of Its Malonic Ester (B), Produced by *Penicillium citrinum* Thom 1131. (Solvent, D_2O; temperature, ambient; chemical shifts expressed as δ_c, relative to external tetramethylsilane.)

respectively). The resistance of the malonogalactan to periodate oxidation is consistent with the relatively slower attack on *trans*-vicinal hydroxyl groups on a 5-membered ring.

Compared to those of their pyranose analogs, the ring-carbon atoms of furanoses are deshielded,[8] so that ring signals of furanosyl residues of polysaccharides are readily detected in the presence of pyranose structures, as they are in relatively uncrowded portions of the ^{13}C-n.m.r. spectrum. C-1 signals of β-D-galactofuranosyl residues are at δ_c 107–109, and those of unsubstituted C-2, C-3, and C-4 are at δ_c ~82, 78, and 84, respectively. With the aid of such signals, the presence of β-D-galactofuranosyl residues of a minor galactomannan component[16]

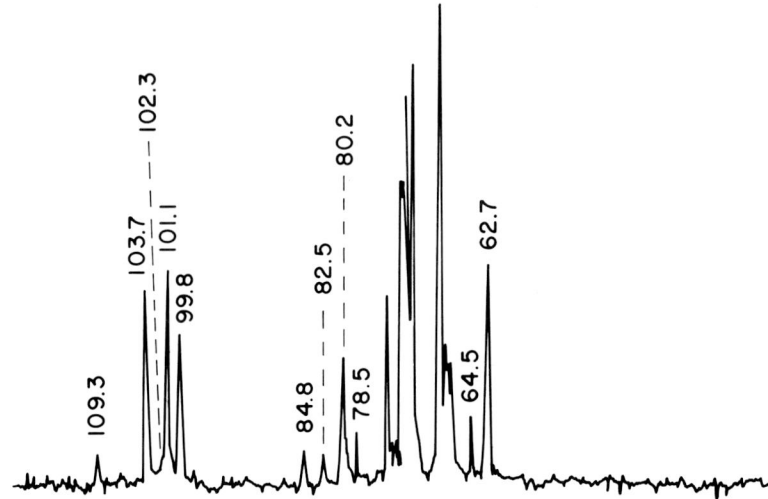

FIG. 24.—^{13}C-N.m.r. Spectrum of a Galactofuranose-Containing Polysaccharide from *Sporothrix schenckii* Grown on a Medium Containing Brain–Heart Infusion. (Solvent, D_2O; temperature, 70°; chemical shifts expressed as δ_c, relative to external tetramethylsilane.)

was detected in polysaccharides of *Sporothrix schenckii* 1099.12 (see Fig. 24). Their single-unit side-chain structure was based on the absence of *O*-substitution at C-2, C-3, C-5, or C-6, as indicated by undisplaced, ^{13}C resonances. Such a C-6 signal was present at δ_c 64.5. This contrasts with the C-6 signal of β-D-(1→5)-linked D-galactofuranose structures, which underwent an upfield β-shift and was[15] at δ_c 62.1. The β-D-galactofuranosyl residues of the galactomannan were linked (1→6) to adjacent mannopyranosyl residues, and their C-1 signal was at δ_c 109.3. Other furanoid structures were found to be present in polysaccharide preparations from *Sporothrix schenckii* 1099.24 and *Ceratocyctis stenoceras* 1099.40, as methylation results showed that (nonreducing) end-groups and a lesser proportion of 6-*O*-substituted galactofuranosyl residues were present.[142]

A signal at δ_c 109.0, displayed[17] by a complex heteropolymer obtained from *Klebsiella* serotype 40, arose from 2,3-di-*O*-substituted β-D-galactofuranosyl residues (see Section VI,14).

Signals having low shifts were also detected on examination of two structurally related polysaccharides (from *Rosa glaucai*) containing α-L-arabinofuranosyl residues.[18] In the study, one of the two arabinans

(142) L. Mendonça-Previato, P. A. J. Gorin, and L. R. Travassos, unpublished results.

contained (1→2)-, (1→3)-, and (1→5)-O-substituted α-L-arabinofuranosyl residues, and its ¹³C-n.m.r. spectrum was complex and difficult to interpret. However, three C-1 signals were present, and units having free and O-substituted C-5 atoms could be distinguished, and shown, under conditions suitable for signal quantitation, to be present, respectively, as 38 and 62% of the total. Three signals were observed in the O-substituted C-5 region, at δ_c 69.8, 69.45, and 69.15, with 21, 32, and 8%, respectively, of the total C-5 area. These underwent an average, α-carbon shift of +5.1 p.p.m. The $^1J_{C-1,H-1}$ and some longer-range coupling values were determined. The former corresponded to those of furanoside structures, but α-D and β-D configurations could not be distinguished from each other. Using the methylation-analysis data as a guide, the ¹³C-n.m.r. spectra of O-methylated arabinans were found to contain assignable OMe-5, OMe-3, and OMe-2 signals. O-Glycosylated, and O-methylated, C-5 resonances could be distinguished, and three, sharp, C-1 signals were present.

The arabinogalactan from green coffee-beans has been tentatively assigned a main chain of β-D-(1→3)-linked D-galactopyranosyl residues, some of which are substituted at O-6 by L-arabinofuranosyl groups or L-arabinofuranosyl-(1→3)-β-D-galactopyranosyl side-chains.[143] C-1 signals, at δ_c 104.0 and 104.4, arose from β-D-galactopyranosyl residues, and the two low-field signals, at δ_c 110.6 and 112.4, corresponded to L-arabinofuranosyl residues having the α-L configuration.[9]

Tomašić and coworkers[144] isolated two D-fructofuranans from rye grass (*Lolium perenne*), and showed the main component to be a β-D-(2→6)-linked levan by ¹³C-n.m.r. spectroscopy. Its configuration was indicated by correspondence with the C-3, C-4, and C-5 resonances of β-D-fructofuranose and methyl β-D-fructofuranoside, rather than with those of the α-D anomers. Signals were assigned as C-1 (61.3), C-2 (105.4), C-3 (77.6), C-4 (76.5), C-5 (81.5), and C-6 (64.4). Compared with resonances of β-D-fructofuranose, small α-shifts were observed for signals of C-2 ($\Delta\delta_c$ +2.8 p.p.m.) and C-6($\Delta\delta_c$ +1.4 p.p.m.), with β-shifts for signals of C-1 ($\Delta\delta_c$ + 2.8 p.p.m.) and C-5 (−0.1 p.p.m.). Based on the relative magnitudes of the main, C-2 signal and the minor one at δ_c 104.9, corresponding to the D-fructofuranosyl residue attached to the terminal D-glucopyranosyl group, an average of ten D-fructofuranosyl residues are linked to the terminal residue of sucrose. Assignments for some carbon nuclei of the α-D-glucopyranosyl residues were corrected by Jarrell and coworkers.[86] Levans were also investigated by

(143) M. L. Wolfrom and D. L. Patin, *J. Org. Chem.*, 30 (1965) 4060–4063.
(144) J. Tomašić, H. J. Jennings, and C. P. J. Glaudemans, *Carbohydr. Res.*, 62 (1978) 127–133.

Seymour and coworkers.[145,146] A preparation obtained by partial hydrolysis of *Streptococcus salivarus* levan gave two minor signals (D_2O, 34°) centered at δ_c 63.4. The signals were better resolved at 90°, being at δ_c 63.97 and 64.23, and were also present in the spectra[146] of levan fractions from *L. mesenteroides* B-512 (F) and *Bacillus* sp. B-1662 (M). An association with the C-1 resonance of 1,6-di-O-substituted β-D-fructofuranosyl residues was suggested. The chemical structure appears more complex than in levan fractions of *Erwinia ananas* B-133, which gave only the signal at δ_c 64.23, and those of *L. mesenteroides* B-523, which gave a spectrum having only the δ_c 63.97 signal.[146] In each case, some type of branching seemed likely, as the ratio of the peak heights for minor and major resonances was in general accord with the degree of branching.

Signal assignments for inulin [β-D-(2→1)-linked D-fructofuranan] were[144] C-1 (62.2), C-2 (104.5), C-3 (78.5), C-4 (76.6), C-5 (82.4), and C-6 (63.4) at pD 13. Comparison of the ^{13}C signals of β-D-fructofuranose with those of inulin, obtained[86] at pD 7, showed remarkable O-glycosylation α-shifts of −1.4 p.p.m. and +1.8 p.p.m. for C-1 and C-2 resonances, respectively. The ^{13}C-n.m.r. spectra for a solution in Me_2SO-$^2H_6$ revealed minor signals that were attributable to the end group of sucrose. The relative magnitudes of the C-2 signals indicated an average of 16 D-fructofuranosyl residues linked to a terminal, sucrose residue. The polysaccharide contained nuclei having T_1 values (68 MHz), in milliseconds, as follows: C-1 (28), C-2 (357), C-3 (49), C-4 (50), C-5 (55), and C-6 (41). There was no correlation between n.O.e. values and line widths and T_1 values.

7. Sulfated Polysaccharides

Beef-lung heparin, classified as belonging to Group B, as its proton-n.m.r. spectrum showed virtually no NHAc groups,[147] was examined by ^{13}C-n.m.r. spectroscopy. Perlin and coworkers[148] found that the spectrum was partly resolved, having 7 of the 12 signals possible (see Fig. 25,A). The signals corresponded to the 2-unit repeating-structure **20**, whose resonances were assigned as follows. The high-field signal at δ_c 59.0 (external standard of tetramethylsilane) arises from C-2 of

---

(145) F. R. Seymour, R. D. Knapp, J. E. Zweig, and S. H. Bishop, *Carbohydr. Res.*, 72 (1979) 57–69.
(146) F. R. Seymour, R. D. Knapp, and A. Jeanes, *Carbohydr. Res.* 72 (1979) 222–228.
(147) A. S. Perlin, M. Mazurek, L. B. Jaques, and L. W. Kavanagh, *Carbohydr. Res.*, 7 (1968) 369–379.
(148) A. S. Perlin, N. M. K. Ng Ying Kin, S. S. Bhattacharjee, and L. F. Johnson, *Can. J. Chem.*, 50 (1972) 2437–2441.

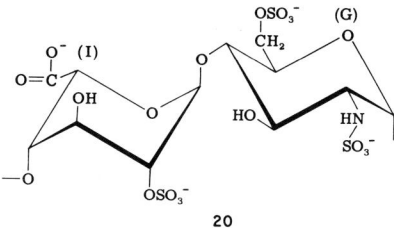

20

residue G, by analogy with the C-2 resonance of 2-amino-2-deoxy-α-D-glucose hydrochloride. The signal at δ_c 67.6 is from sulfated C-6 (G), as, by virtue of its 2 attached protons, it has a lower T_2 value than other nuclei (which resulted in a signal 1.6 times as broad as other signals arising from one nucleus), and because, in an undecoupled spectrum, it appears as a triplet. Compared with the C-6 resonance of 2-amino-2-deoxy-α-D-glucose hydrochloride, it is displaced 6 p.p.m. downfield

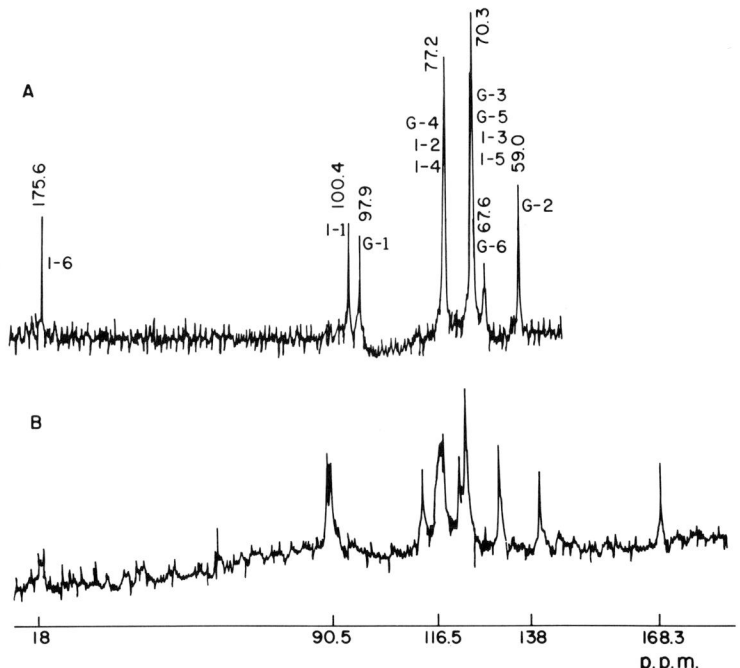

FIG. 25.—(A) ^{13}C-N.m.r. Spectrum of Heparin in D_2O at Ambient Temperature. (I-Signals are ascribed to α-L-idosyluronic acid residues, and G-signals to 2-amino-2-deoxy-α-D-glucosyl residues; chemical shifts are expressed as δ_c, relative to external tetramethylsilane.) (B) ^{13}C-N.m.r. Spectrum of Dermatan Sulfate in D_2O at 70°. (The horizontal axis is based on δ_c, relative to external CS_2.)

by sulfation. Similarly, the C-2 signal of the 2-sulfate of the iduronic acid residue I (δ_c 77.2) is at ~8 p.p.m. lower field than the C-2 resonance of sodium (methyl α-D-idopyranosid)uronate, which is one of a group of signals at δ_c 69–71.0. Of the C-1 signals at δ_c 100.4 and 97.9, the former is assigned to residue I by comparison with the C-1 shift of disaccharide **21**, obtained by deaminative degradation of heparin.

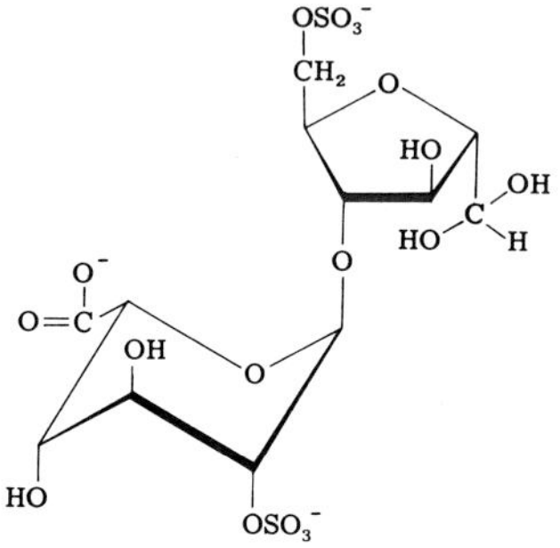

21

The correlation of shifts in the spectrum of heparin with those of disaccharide **21** and sodium (methyl α-D-idopyranosid)uronate indicated that its L-iduronic acid residues may be best described by the $^1C_4(L)$ conformation. Tentative assignments for these signals, made by Lasker and Chiu,[149] proved to be incorrect. The spectrum of bleached, hog-gut heparin contained an additional, minor C-1 signal and two signals at δ_c ~130 p.p.m. (CS_2 as the external standard). The C-1 signal could be due to the presence of β-D-glucosyluronic acid residues of heparitin sulfate, as its shift coincides.[150] Dermatan sulfate ("chondroitin sulfate B"; **22**; see Fig. 25,B) was not an impurity, as its distinc-

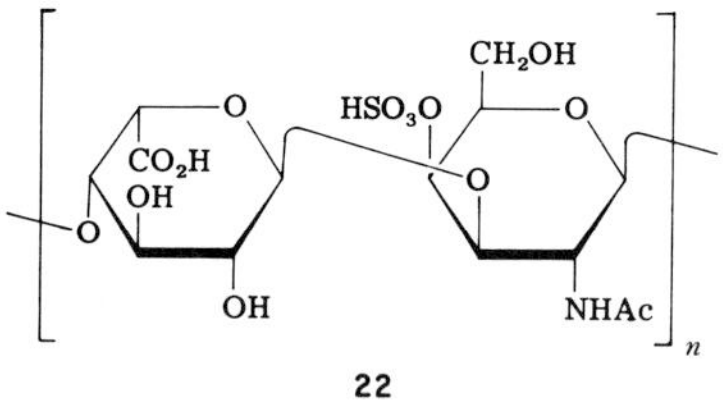

22

tive, ^{13}C signals were absent from the heparin spectrum. From the 13 carbon atoms contained in the two-unit repeating-structure of dermatan sulfate, 11 signals were detected, not including a minor resonance ~21 p.p.m. downfield from those of C-1 and C-1' (see Fig. 25,B). In

(149) S. E. Lasker and M. L. Chiu, *Ann. N. Y. Acad. Sci.*, 221 (1973) 971–977.
(150) A. S. Perlin, personal communication.

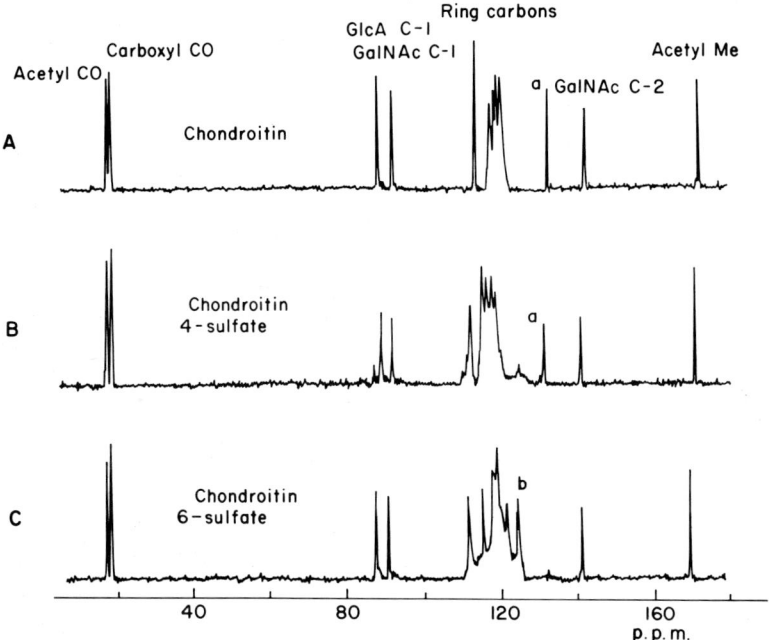

FIG. 26.—^{13}C-N.m.r. Spectra of Chondroitin and its 4- and 6-Sulfates. [In D_2O at 80°; the axis of chemical shifts is based on external CS_2 ($\delta_c = 0$).]

this spectrum, as in that of chondroitin 4-sulfate (4-sulfate of **23**; Fig. 26,B), it is not possible to locate the C-4 resonance,[151] as the downfield shift caused on sulfation (see Fig. 26,A) places it in the middle of a complex group of signals. The ^{13}C-n.m.r. spectrum of chondroitin 6-sulfate (6-sulfate of **23**; Fig. 26,C) lacks signals *a* and contains signal *b*, a useful way of distinguishing it from chondroitin 4-sulfate.

Differences between the ^{13}C-n.m.r. spectra of chondroitins have been used to show that specimens of chondroitin 4- and 6-sulfate contain up to 30% of impurities, and appear to be composites of both

(151) S. Honda, H. Yuki, and K. Takiura, *J. Biochem. (Tokyo)*, 76 (1974) 209–211.

TABLE VI

Displacement of ¹³C Resonances on O-Sulfation of α,β-D-Glucose and
Methyl 3-O-Methyl-α-D-mannopyranoside

Compound	Shift observed on O-sulfation (p.p.m.)							
	C-1	C-2	C-3	C-4	C-5	C-6	OMe-1	OMe-3
D-Glucose derivative								
3-sulfate								
α anomer	0	−1.1	+9.5	−2.2	+0.1	+0.1	—	—
β anomer	+0.2	−1.2	+8.5 or +8.6	−2.2	−0.3 or −0.4	0	—	—
6-sulfate								
α anomer	−0.2	−0.1	0	−0.3	−1.7	+6.5	—	—
β anomer	−0.2	0	−0.1 or −0.2	−0.3	−0.9 or −2.0	+6.6	—	—
Methyl 3-O-methyl-α-D-mannopyranoside derivative								
2-sulfate	−2.2	+7.0	−1.7	−0.2	0	−0.1	+0.3	+0.4
4-sulfate	−0.4	+2.1	−2.7	+8.4	−1.3	−0.2	+0.2	+1.1
6-sulfate	0	+0.2	−0.2	+0.1	−2.0	+6.3	+0.2	+0.1

types of polymer. Dermatan sulfate contains 10–20% of unidentified constituents, including 6-sulfated residues.[25]

The downfield displacement of the C-6 resonance of chondroitin on O-sulfation was 6.5 p.p.m., close to the 6.6 p.p.m. observed on 6-O-substitution of β-D-glucose.[152] As may be seen from Table VI, O-sulfation of α,β-D-glucose and methyl 3-O-methyl-α-D-mannopyranoside[153] causes a strong, downfield shift of the substituted resonance that is accompanied by smaller, upfield displacements of the signals of adjacent carbon atoms. Other, more-distant, substitution effects were not observed, except for the C-2 and OMe-3 signals on 4-O-sulfation of the D-mannopyranoside derivative.

The relaxation of certain ¹³C nuclei in some uronic acids and derivatives is affected by Gd^{3+}. For example, on addition of gadolinium nitrate to solutions of heparin or dermatan sulfate in D_2O, the C-1 and C-6 signals of the uronic acid residues are selectively diminished. This property would be expected from α-L-idopyranosyluronic acid residues, as the C-1 and C-6 signals of sodium α-D-idopyranuronate are similarly lessened in intensity, whereas those of the β anomer are

(152) S. Honda, H. Yuki, and K. Takiura, *Carbohydr. Res.*, 28 (1973) 150–153.
(153) A. I. Usov, S. V. Yarotskii, and L. K. Vasayina, *Bioorg. Khim.*, 1 (1975) 1583–1588.

unaffected. A similar, configurational distinction was observed with the C-1 and C-6 signals of α,β-mixtures of sodium D-glucopyranuronate and sodium D-galactopyranuronate. As anticipated, Gd^{3+} had no effect on the ^{13}C-n.m.r. spectrum of chondroitin 6-sulfate, whose D-glucopyranosyluronic acid residues have the β-D configuration.[25,154]

As indicated later (see Section VI,8), on addition of the chloride of praseodymium, europium, or other lanthanides to mono- or poly-saccharide phosphates in D_2O, the signals of carbon atoms substituted with phosphate groups are recognizable, as they are displaced, relative to the rest of the ^{13}C-n.m.r. spectrum.[155] However, this diagnostic method is not applicable to sulfated polysaccharides, as signal displacements were not observed on addition of praseodymium or europium chloride to a solution of α,β-D-galactose 6-sulfate or its sodium salt.[156]

Red algae can produce many different polysaccharide structures, and four of these were examined, and differentiated, by means of the C-1 region of their ^{13}C-n.m.r. spectra.[157] These were agar (**24**), κ-carrageenan (**25**; R = $SO_3^-Na^+$), partially desulfated κ-carrageenan (**25**; R = $SO_3^-Na^+$ or H^+), and ι-carrageenan (**26**), three of which have predomi-

(154) B. Casu, G. Gatti, N. Cyr, and A. S. Perlin, *Carbohydr. Res.* 41 (1975) c6–c8.
(155) P. A. J. Gorin and M. Mazurek, *Can. J. Chem.*, 52 (1974) 3070–3076.
(156) P. A. J. Gorin and M. Mazurek, unpublished results.
(157) S. S. Bhattacharjee, W. Yaphe, and G. K. Hamer, *Cabohydr. Res.*, 60 (1978) c1–c3.

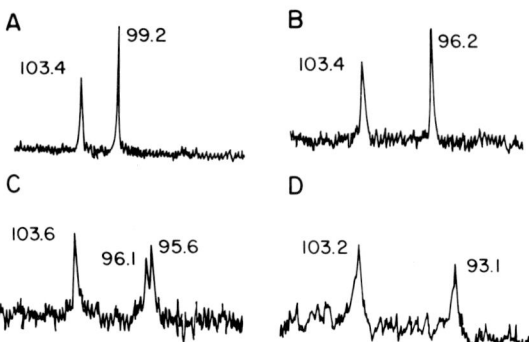

FIG. 27.—C-1 Portions of ^{13}C-N.m.r. Spectra of Agar (A; **24**; from *Ahneltia plicata*), κ-Carrageenan (B; **25**, R = SO$_3^-$Na$^+$; from *Eucheuma striatum*), Partially Desulfated κ-Carrageenan (C; **25**, R = SO$_3^-$Na$^+$ or H), and ι-Carrageenan (D; **26**, R = R' = SO$_3^-$Na$^+$; from *Eucheuma spinosum*). (Samples were examined as sodium salts in D$_2$O at 90°; chemical shifts, δ$_c$, are values relative to external tetramethylsilane.)

nant, two-unit repeating-structures (see Fig. 27) and that, with the exception of agar, contain sulfate. Each partial spectrum contained resonances at δ$_c$ 103.2–103.6, arising from 3-*O*-substituted β-D-galactopyranosyl residues that can be *O*-sulfated or unsulfated. Such a substituent at C-4 influences the shift of the adjacent C-1 of κ-carrageenan, which gives one signal at δ$_c$ 96.2 with its fully sulfated structure, whereas partly sulfated κ-carrageenan (**25**; R = SO$_3^-$Na$^+$ or H) has 2 signals, at δ$_c$ 96.1 and 95.6. A strong β-effect of −3.6 p.p.m. is observed on 2-*O*-sulfation of the 3,6-anhydro-α-D-galactopyranosyl residue; this is apparent in the spectrum of structure (**26**) (see Fig. 27).

Rees and coworkers[158] showed that, at 15°, ι-carrageenan forms a gel whose ^{13}C-n.m.r. signals are so broad that they cannot be detected, in contrast to those given by the solution at 80° (see Fig. 28). At the lower temperature, segmental motion is restricted by frequent, interunit junction-zones in a double-helix structure, in contrast to the gel of a β-D-(1→3)-linked D-glucopyranan, where the intermolecular association is not so complete, and portions of the polymer are sufficiently mobile to provide broad signals.[159]

Brewer and Keiser[160] investigated the ^{13}C-n.m.r. spectra of bovine nasal-cartilage and its component sulfate polysaccharides (see Fig. 4). It consists of 75% of water and equal proportions of proteoglycan com-

(158) T. A. Bryce, A. A. McKinnon, E. R. Morris, D. A. Rees, and D. Thom, *Discuss. Faraday Soc.*, (1974) 221–229.
(159) D. A. Rees, *Adv. Carbohydr. Chem. Biochem.*, 24 (1968) 267–332.
(160) C. F. Brewer and H. Keiser, *Proc. Natl. Acad. Sci. U. S. A.*, 72 (1975) 3421–3423.

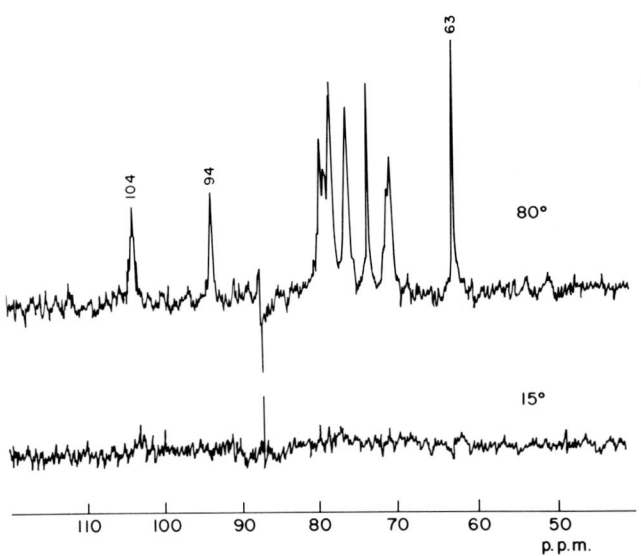

FIG. 28.—Comparison of ^{13}C-N.m.r. Spectra for Aqueous ι-Carrageenan Solution (80°) and Gel(15°). [Shifts are expressed relative to internal DSS ($\delta_c = 0$).]

plex and collagen, the latter existing in a relatively rigid, triple-helix state, so that its ^{13}C resonances are too broad to be detected under the usual conditions.[161] The proteoglycan complex is a large aggregate consisting of subunits (mol. wt. 2.5×10^6) that are non-covalently linked to hyaluronic acid (**23**). The subunit contains 110 chondroitin sulfate chains, mostly the 4-sulfate, each containing 35–40 disaccharide repeating-units, and 50 keratan sulfate chains having 10–12 disaccharide repeating-units each. These polysaccharide chains are each linked covalently at one end to a linear core-protein that is involved in the linkage with hyaluronic acid. The overall composition of the proteoglycan is: chondroitin sulfate, 86%; protein, 8%; keratan sulfate, 6%*; and hyaluronic acid, <1%, so that only the chondroitin 4-sulfate component is detected by ^{13}C-n.m.r. spectroscopy (see Fig. 29). The ^{13}C signals of the preponderant chondroitin 4-sulfate, detected at 25°, are much wider than those obtained[151] at 80°, but the derived τ_c of ~10 ns indicates considerable, segmental flexibility in the molecule. Intermolecular hydrogen-bonding is not significant, as

(161) J. C. W. Chien and W. B. Wise, *Biochemistry*, 12 (1973) 3418–3424.

* Keratan sulfate consists of a (1→3)-*O*-β-D-galactopyranosyl-(1→4)-*O*-2-acetamido-2-deoxy-β-D-glucopyranosyl repeating-unit containing ~40% of galactosyl residues and >70% of 2-acetamido-2-deoxy-D-glucosyl residues esterified at C-6 by sulfate.

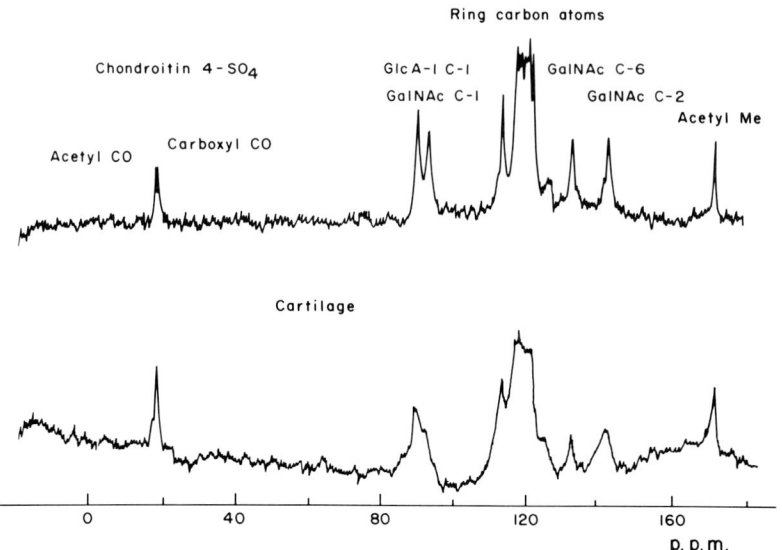

Fig. 29.—^{13}C-N.m.r. Spectra of Chondroitin 4-Sulfate and of Cartilage in D$_2$O at 25°. (Shifts are presented relative to external CS$_2$.)

addition of 6 M guanidine hydrochloride or 6 M urea did not alter the spectrum, in agreement with the observation that similar additions did not affect the sedimentation rate.[162] The spectra of the chondroitin sulfate, the proteoglycan subunit, and the proteoglycan complex were virtually indistinguishable, thus showing that the speed of molecular motions of the polysaccharide components are similar, broadening being observed only with intact, bovine nasal-cartilage. This conclusion contrasts with that from X-ray investigations on oriented films of the proteoglycan complex of pig laryngeal-cartilage, which suggested the adoption of helical structures.[163,164]

In an intriguing study, Torchia and coworkers[85] investigated possible molecular motions of the proteoglycans. It was found that at least 80% of the total, glycosaminoglycan carbon atoms ($\tau < 1$ μs) in bovine, nasal-cartilage tissue contribute to the ^{13}C-n.m.r. spectrum, thus showing greater segmental motion than the 8% protein-component of the proteoglycan, whose signals could not be detected, even on enrichment. Of the glycosaminoglycan carbon atoms of cartilage, 80% had corrected line widths of 100 Hz, the rest being 20 Hz, indicating dy-

(162) V. C. Hascall and S. W. Sajdera, *J. Biol. Chem.*, 245 (1970) 4920–4930.
(163) S. Arnott, J. M. Guss, D. W. L. Hukins, and M. B. Mathews, *Science*, 180 (1973) 743–745.
(164) E. D. T. Atkins, T. E. Hardingham, D. H. Isaac, and H. Muir, *Biochem. J.*, 141 (1974) 919–921.

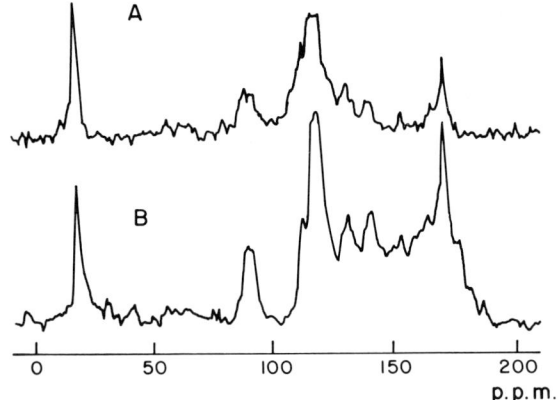

FIG. 30.—Comparison of ^{13}C-N.m.r. Spectra of Bovine, Nasal Cartilage in D_2O at 37°. (Presented are spectrum A, obtained by using scalar decoupling with $\gamma H_2/2\pi$ = 3.5 kHz, and spectrum B, by using dipolar decoupling of 65 kHz.)

namic heterogeneity. The T_1 and n.O.e. values were analyzed according to (a) single correlation times, (b) a log λ^2 distribution of correlation times, and (c) anisotropic motion. A model having a broad distribution of correlation times centering on τ_c 50 ns was preferred. Its two deficiencies were that (a) slow motions of τ_c 500 ns were neglected, although dipolar-decoupling experiments showed them to be present, and (b) all motions were assumed to be isotropic. When dipolar, proton decoupling of $\gamma H_2/2\pi$ = 65 kHz was applied, line-width-broadening contributions from static, dipolar interactions and slow motions were removed from the ^{13}C-n.m.r. spectrum of cartilage. The resulting lines were sharper and of greater intensity, and a background signal was present that was ascribed to the carbon atoms of collagen (see Fig. 30).

8. O-Phosphonohexopyranans

Investigations conducted at the Canadian Government Research Laboratories in Ottawa on the chemical structures of *Neisseria menigitidis* antigens have mainly been based on the interpretation of ^{13}C-n.m.r. spectra. The polysaccharides of serogroups A, B, C, W135, X, Y,[165] 29e,[166] and BO[167] were characterized as complex polysaccharides,

(165) H. J. Jennings, A. K. Bhattacharjee, D. R. Bundle, C. P. Kenny, A. Martin, and I. C. P. Smith, *J. Infect. Dis.*, 136 (1977) 878–883.
(166) A. K. Bhattacharjee, H. J. Jennings, and C. P. Kenny, *Biochemistry*, 17 (1978) 645–651.
(167) A. K. Bhattacharjee, H. J. Jennings, C. P. Kenny, A. Martin, and I. C. P. Smith, *Can. J. Biochem.*, 59 (1976) 1–8.

namely, O-phosphonomannans containing phosphoric diester bridges between aldopyranosyl residues, polysaccharides containing sialic acid, and one that contains 3-deoxy-D-*manno*-octulosylonic acid (KDO) residues. A number of the structures were partly O-acetylated, and some of the position(s) of substitution could be determined.

Such studies, and others on an O-phosphonomannan[155] and a teichoic acid,[168] relied on judicious comparisons (of shift) with signals of model compounds, and these are simpler than conventional, analytical procedures. For example, it is difficult to methylate alkali-labile O-phosphonomannans, and sialic acid and KDO-containing polymers would require difficultly available, O-methylated standards. In addition, periodate-oxidation analyses are restricted to polymers having fortuitously amenable, chemical structures.

Smith and coworkers[169] found ^{13}C-n.m.r. spectroscopy to be of aid in structural studies on O-phosphonoglycans, by virtue of spectral effects occurring on O-phosphorylation. The ^{13}C-n.m.r. spectrum of the known, (1→6)-linked polymer of 2-acetamido-2-deoxy-α-D-glucopyranosyl phosphate from *Staphylococcus lactis*[170] contained 4 signals split by $^2J_{C,^{31}P}$ and $^3J_{C,^{31}P}$ coupling, and a signal at δ_c 65.3, which was from C-6, as it gave a triplet under undecoupled conditions. The polysaccharide of *Neisseria meningitidis* serogroup X, known to consist of (1→3)- or (1→4)-linked residues of 2-acetamido-2-deoxy-α-D-glucopyranosyl phosphate, or both,[171] gave a ^{13}C-n.m.r. spectrum containing 8 signals typical of a linear homopolymer, 5 of which were split. Comparison of each chemical-shift value with those of the S. *lactis* polymer showed that only the C-4 resonance of the O-phosphonoglycan of serogroup X was deshielded, by a value of 4.8 p.p.m. (see Table VII), consistent with phosphorylation at O-4. The α-D configuration[171] was confirmed, as the shifts of C-2, C-4, and C-5 corresponded to those of 2-acetamido-2-deoxy-α-D-glucopyranose or its 1-phosphate, rather than to those of the β-D anomer.[172]

The serotype A polysaccharide of *Neisseria meningitidis* was already known to have a repeating unit consisting of a (1→6)-linked 2-acetamido-2-deoxy-D-mannopyranosyl phosphate partly substituted

(168) W. R. deBoer, F. J. Kruyssen, J. T. M. Wouter, and C. Kruk, *Eur. J. Biochem.*, 62 (1976) 1–6.
(169) D. R. Bundle, I. C. P. Smith, and H. J. Jennings, *J. Biol. Chem.*, 249 (1974) 2275–2281.
(170) A. R. Archibald and G. H. Stafford, *Biochem. J.*, 130 (1972) 681–690.
(171) D. R. Bundle, H. J. Jennings, and C. P. Kenny, *Carbohydr. Res.*, 26 (1973) 268–270.
(172) D. R. Bundle, H. J. Jennings, and I. C. P. Smith, *Can. J. Chem.*, 51 (1973) 3812–3819.

TABLE VII
Carbon-13 Chemical Shifts[a] of O-Phosphonoglycans

Polysaccharide	C-1	C-2	C-3	C-4	C-5	C-6	CH$_3$ (NHCOCH$_3$)	C=O (NHCOCH$_3$)
S. lactis 2102	95.2	55.0	71.5	70.3	73.0	65.3	23.0	175.6
N. meningitidis serogroup X	95.2	54.8	71.1	75.1	73.2	61.3	23.2	175.6
N. meningitidis serogroup A								
acetylated component	96.2	51.9	73.2	64.7			23.0	176.2
non-acetylated component	96.4	54.3	69.7	67.3	73.6	65.6	23.3	175.9
							21.6	174.6
O-deacetylated	96.4	54.3	69.7	67.1	73.5	65.6	23.3	175.8

[a] In parts per million from external tetramethylsilane.

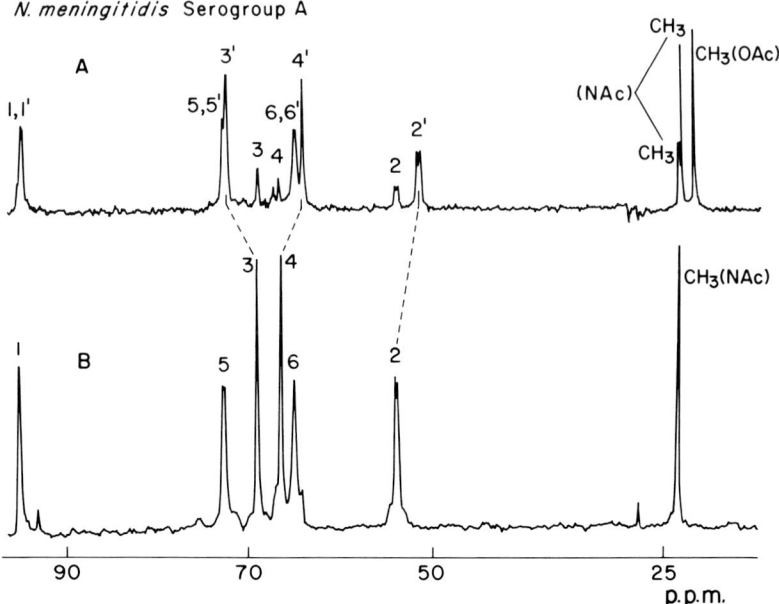

FIG. 31.—^{13}C-N.m.r. Spectra of Native *Neisseria meningitidis* Serogroup A Polysaccharide (A; **27** and **28**) and O-Deacetylated Product (B; **28**). [Solvent, D_2O at 37°; the spectral-shift axis is based on external tetramethylsilane ($\delta_c = 0$).]

with acetyl groups.[173] The ^{13}C-n.m.r. spectrum contained signals (see Fig. 31,A) that corresponded to 70% of mono-O-acetylated (**27**) and 30% of nonacetylated units (see Fig. 31,B; **28**). The 8 signals of the

latter (C=O signal not included) are typical of linear homopolymer, and 4 of the resonances were shortened by reason of small, 2- and 3-bond couplings, some of which were barely discernible. The (1→6)-phosphoric diester linkage was confirmed, as the C-6 resonance was

(173) T. Y. Liu, E. C. Gotschlich, E. K. Jonssen, and J. R. Wysocki, *J. Biol. Chem.*, 246 (1971) 2849–2858.

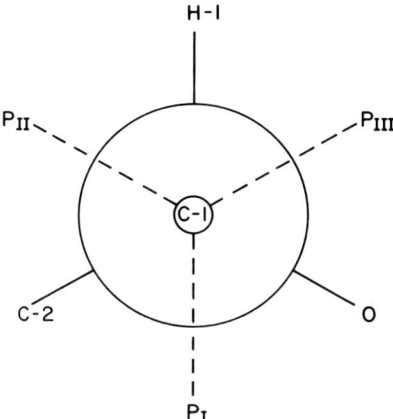

FIG. 32.—Possible Rotamers about the Oxygen to C-1 Bond in a Phosphorylated Polysaccharide When Viewed Along the Bond.

4.3 p.p.m. downfield from that of the (1→4)-linked O-phosphonoglycan from serogroup X (see Table VII). The glycosidic bond was α-D, as the C-2, C-3, and C-5 shifts corresponded to those of 2-acetamido-2-deoxy-α-D-mannopyranose rather than to those of the β-D anomer.[172] The native polymer is acetylated on the O-3 atoms, as the α- and β-carbon signal-displacements on acetylation are close to those occurring on 3-O-acetylation of 2-acetamido-2-deoxy-α,β-D-glucose, namely, +3.0 and −2.0 p.p.m., respectively.[172]

In these studies, it was observed that, when a hydroxyl group is replaced by a phosphate or a phosphoric ester group, the α-carbon resonance was the only one to be displaced. The shift was 2–5 p.p.m. downfield, except in the case of 2-acetamido-2-deoxy-β-D-glucopyranosyl phosphate, when it was only 0.5 p.p.m.

Two-bond, $^{13}C,^{31}P$ splittings[174,175] are 3–6 Hz, but 3-bond couplings vary according to the dihedral angle between $^{31}P-O-C$ and $O-C-^{13}C$ planes. They are 8–10 Hz for the trans, and 2 Hz for the gauche, orientation.[174,175] The ^{13}C-n.m.r. spectra of polysaccharides of serotypes A and X, and S. lactis,[169] contain C-2 resonances exhibiting $^3J_{C,^{31}P}$ values of 8.0–8.6 Hz, consistent with favored adoption by the phosphorus nucleus of the P_{III} position, trans to the (large) 2-acetamido group (see Fig. 32). The C-5–C-6 and P–O bonds are also trans in (1→6)-ortho-

(174) R. D. Lapper, H. H. Mantsch, and I. C. P. Smith, J. Am. Chem. Soc., 94 (1972) 6243–6244.
(175) R. D. Lapper, H. H. Mantsch, and I. C. P. Smith, J. Am. Chem. Soc., 95 (1973) 2878–2880.

phosphorylated *S. lactis* polysaccharide and deacetylated serogroup A polysaccharides, as the $^3J_{C,^{31}P}$ values of C-5 are 7.6 and ~6 Hz, respectively. Thus, these polysaccharides favor an extended conformation in solution. The conformation of the 4-*O*-linked, serotype X polysaccharide is less-well defined, as the 3-bond coupling to C-3 is 2.5 Hz, showing the relative absence of the P_{II} conformation, but the intermediate, C-5 coupling value of 5 Hz indicates no strong preference for either rotamer, P_I or P_{III}.

In addition to the coupled-signal method just described, phosphorylated carbon signals can be detected by use of praseodymium chloride, which displaces α- and β-carbon resonances of α,β-D-mannose 6-phosphate and α-D-mannosyl phosphate downfield, with little effect on other resonances. Europium chloride has analogous properties, except that the displacements are upfield. With certain polysaccharides, such as the *O*-phosphonomannan of *Hansenula capsulata* (**29**), the sig-

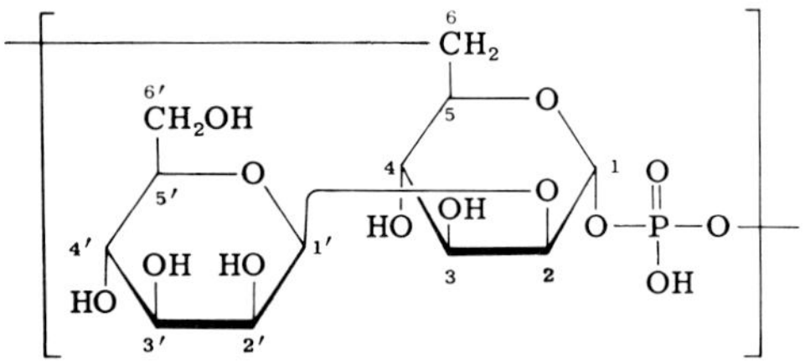

29

nals of a solution in D_2O are broader than 7 Hz (see Fig. 33,A), even after lowering of the viscosity with alkali, so that small $^{13}C,^{31}P$ couplings cannot be detected.[155] The coupled signals of C-1, and of C-2 and C-6 units, are recognizable from their increased line-widths and decreased heights, but the C-5 signal was obscured by others.

Initially, the *O*-phosphonomannan was assigned a repeating unit of 2-*O*-β-D-mannopyranosyl-α-D-mannopyranose joined by (1→6)-phosphoric diester bridges (**30**). With α- and β-D-mannopyranosyl units, it

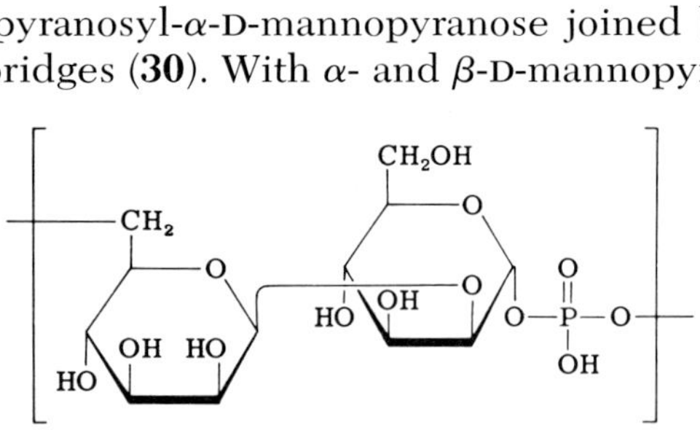

30

would be expected that the lanthanide salts would displace the resonances of C_β-5', C_α-2, and C_α-1, in addition to that of C-6. However, on

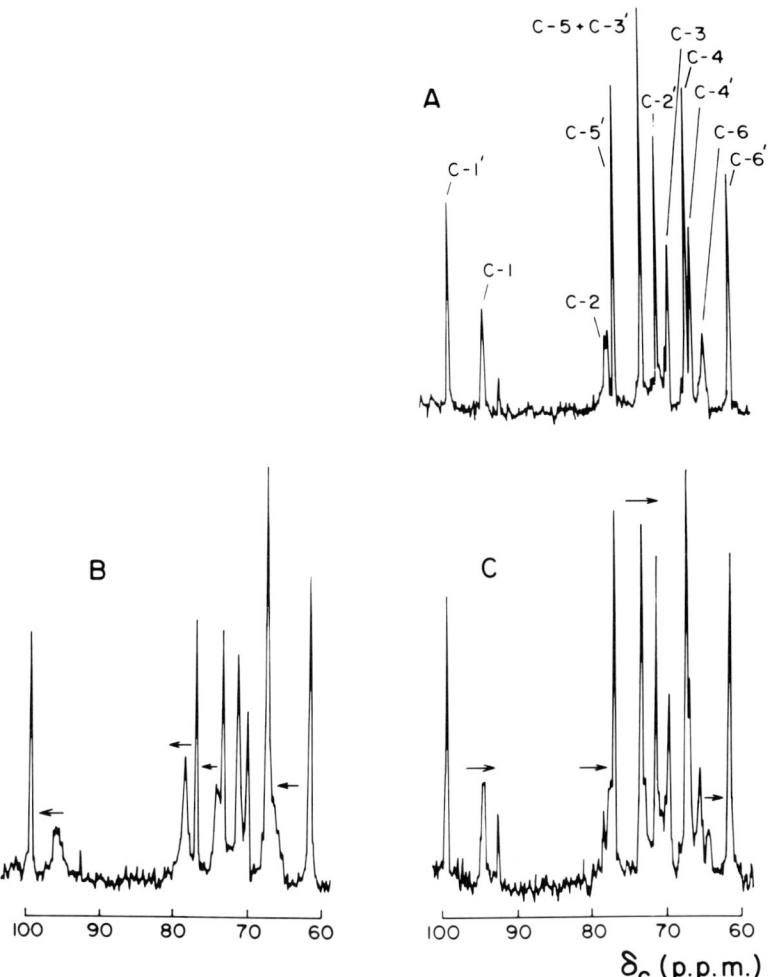

FIG. 33.—^{13}C-N.m.r. Spectra of O-Phosphonomannan (**29**) in D_2O at 33° (A) and the Selective Signal-Displacements Occurring on Addition of $PrCl_3 \cdot 6\ H_2O$ (B) and $EuCl_3 \cdot 6\ H_2O$ (C). [The spectra were recorded using an external standard of tetramethylsilane ($\delta_c = 0$).]

addition of $PrCl_3$ (see Fig. 33,B) and $EuCl_3$ (see Fig. 33,C), spectral shifts were observed for C_α-5 (not previously observed) which emerged from a large signal, and for C_α-2 and C_α-1 signals, consistent with structure **29**. This observation confirmed a separate finding[114] that partial hydrolysis of the O-phosphonomannan gave the 6-phosphate, and not the 6′-phosphate of 2-O-β-D-mannopyranosyl-α,β-D-

TABLE VIII

Carbon-13 Chemical-Shifts of the Glycerol Residue of a
Teichoic Acid and Its Components

Compound	Chemical shift (δ_c)		
	C-1	C-2	C-3
Glycerol	63.8	73.3	63.8
1-O-β-D-Glucopyranosylglycerol	71.4	71.1	62.8
1-O-β-D-Glycopyranosylglycerol 3-phosphate	70.9	69.5	67.5
Teichoic acid	69.5	75.3	66.1

mannose (see structural precursors **29** and **30**) previously claimed.[176] The conclusion was based on its ^{13}C-n.m.r. spectrum, which contained ^{13}C, ^{31}P coupled signals attributable to C-5, and had δ_c values close to those of C_α-5 and C_β-5 of α,β-D-mannose 6-phosphate, consistent with 6-substitution of the reducing residue. The D-mannobiose 6'-phosphate would be expected, by contrast, to give one coupled signal in the C_β-5 region.

9. A Teichoic Acid

The teichoic acid of the walls of *Bacillus subtilis* var. *niger* WM gave, on partial hydrolysis with alkali, a 1-O-β-D-glucopyranosylglycerol 3-phosphate and a little of a glycerol phosphate.[168] Enzymolysis with alkaline phosphatase provided a 1-O-β-D-glucopyranosylglycerol, whose structure was determined by comparison of ^{13}C-signal shifts with those of glycerol and methyl β-D-glucopyranoside. Only the C-1 and C-2 resonances of glycerol differed (see Table VIII), being displaced by +7.6 and −2.2 p.p.m. (no distinction was made between D- and L-glycerol moieties). In turn, the resonances of a D-glucosylglycerol phosphate were similar to those of the D-glucosylglycerol, except for the ^{13}C, ^{31}P coupled C-3 and C-2 signals of the glycerol residue, which were displaced by +4.7 (*J* 5.1 Hz) and −1.6 p.p.m. (*J* 7.9 Hz), respectively, consistent with a 3-phosphate group. As may be seen from Table VIII, on going from the spectrum of the 3-phosphate to that of teichoic acid, C-2 (+5.8 p.p.m.) and C-1 and C-3 (−1.4 p.p.m.) resonances are distinguishable, in accord with a predominant (2→3)- or (1→2)-phosphoric diester link to the D-glycerol residue (see **31**).

(176) M. E. Slodki, *Biochim. Biophys. Acta*, 69 (1963) 96–102.

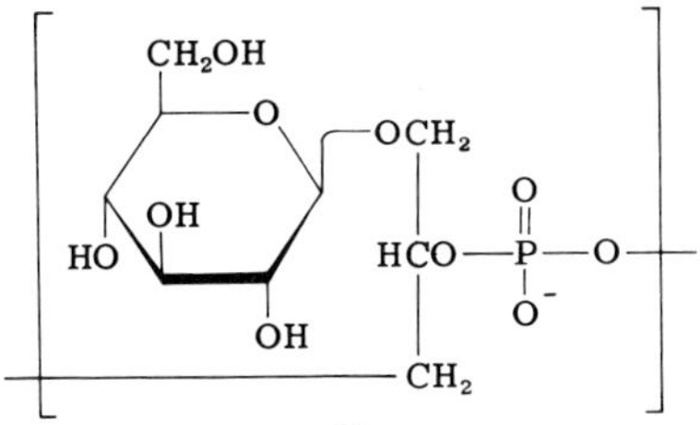

31

10. Polysaccharides Containing Sialic Acid

The partly acetylated, sialic acid polymer of *Neisseria meningitidis* serogroup C was characterized by ^{13}C-n.m.r. spectroscopy.[177] As with D-fructans and polysaccharides containing 3-deoxy-D-*manno*-octulosonic acid residues, p.m.r. spectroscopy cannot be used to determine the configuration directly, because protons are absent from the anomeric centers. The ^{13}C-n.m.r. spectrum of the O-deacetylated polysaccharide contained 10 of the 11 signals possible, typical of a linear homopolymer (see Fig. 34,B). Comparison with previously assigned spectra of methyl α- and β-glycosides of N-acetylneuraminic acid showed a greater correspondence with most of the resonances of the α-D anomer. That of C-9 was displaced downfield by 2.0 p.p.m., indicating an α-D-(2→9) structure (see Fig. 34 and **32**). The linkage posi-

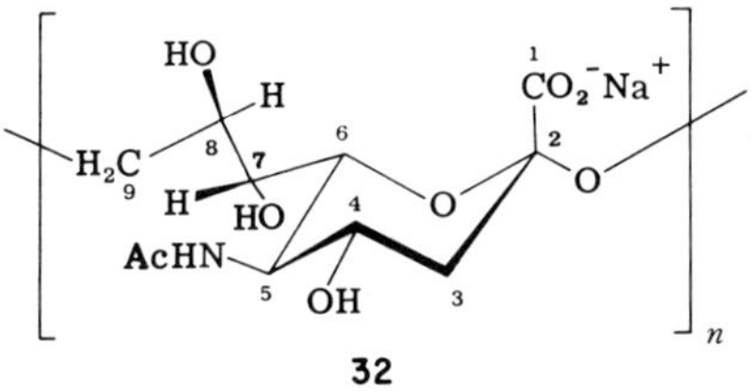

32

tion was confirmed by periodate oxidation. The native polymer contained 1.16 mol of O-acetyl per mol of sialic acid residue, and only limited conclusions could be made as to their position(s). However, it was clear that 4-hydroxyl groups were unacetylated, as the readily identifiable, CH_2 signal of C-3 had identical chemical shifts, whether from native or O-deacetylated polysaccharides (see Fig. 34, A and B), and thus the acetate groups were far removed, and, therefore, restricted to C-7 and C-8.

The 11-signal, ^{13}C-n.m.r. spectrum of the sialic acid polymer from

(177) A. K. Bhattacharjee, H. J. Jennings, C. P. Kenny, A. Martin, and I. C. P. Smith, *J. Biol. Chem.*, 250 (1975) 1926–1932.

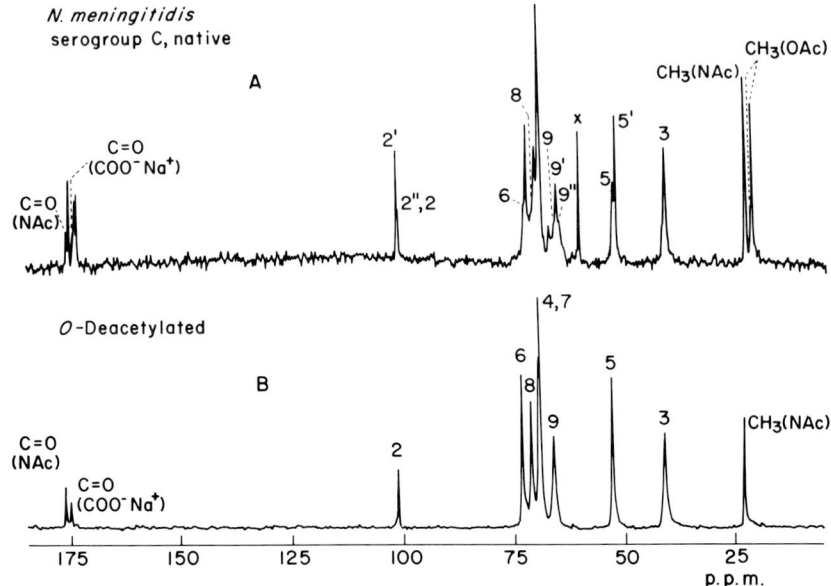

FIG. 34.—^{13}C-N.m.r. Spectra of Native, Partly Acetylated Polysaccharide from *Neisseria meningitidis* Serogroup C (A) and the α-D-(2→9)-Linked N-Acetyl-D-neuraminic acid Polymer (**32**) Obtained on *O*-Deacetylation (B). (Solvent, D$_2$O, pD 7, at 31°; chemical shifts are expressed relative to external tetramethylsilane.)

Neisseria meningitidis, serogroup B, was identical with that of colominic acid, which has α-D-(2→8)-linkages.[178] A comparison similar to the foregoing was conducted, using the methyl α- and β-glycosides of N-acetylneuraminic acid derivatives; this confirmed the α-D configuration, but the *O*-glycosylation effects on the α- and β-carbon resonances were much greater, being +6.8 p.p.m. for the C-8, and −1.9 p.p.m. for the C-9, signal. In such comparisons, highly ionized, carboxylic salts are preferred to the free acids.[179]

The two, somewhat similar, polysaccharide antigens from *Neisseria meningitidis* from serogroups Y and W135 were identified.[165] The partially *O*-acetylated, serogroup Y polymer consisted of glucosyl and N-acetylneuraminic acid residues (**33**), and the complex, ^{13}C-n.m.r. spectrum of the *O*-deacetylated material gave two anomeric signals corresponding to an alternating structure, but only 11 signals of the full complement of 14 could be resolved (see Fig. 35,B). Partial hydrolysis provided a disaccharide identified as 4-*O*-α-D-glucopyranosyl-β-N-

(178) E. J. McGuire and S. B. Binkley, *Biochemistry*, 3 (1964) 247–251.
(179) H. J. Jennings and A. K. Bhattacharjee, *Carbohydr. Res.*, 55 (1977) 105–112.

W-135 X = OH, Z = H
O-Deacetylated Y X = H, Z = OH

33

acetylneuraminic acid, because pertinent signals corresponded to those of α-D-glucose and β-N-acetylneuraminic acid, except for C-3, C-4, and C-5 of the latter, which underwent displacements of −3.3, +5.5, and −2.5 p.p.m. (see Table IX). Similar displacements were observed in the polysaccharide spectrum, together with a deshielding of C-6' (2.0 p.p.m.), consistent with glycosylation at O-6'. The complexity of the ^{13}C-n.m.r. spectrum of the native polymer (see Fig. 35,A), which contains 1.3 mol of O-acetyl groups per mol of repeating unit, precluded location of the O-acetyl groups. The ^{13}C-n.m.r. spectrum of the serogroup BO polysaccharide was identical, except that it indicated that 1.8 mol of O-acetate were present per mol of repeating unit.

Similar reasoning (see Table IX) showed that the structure of the non-O-acetylated polysaccharide of *Neisseria meningitidis* serogroup W135 (for its ^{13}C-n.m.r. spectrum, see Fig. 35,C) is closely related, containing D-galactopyranosyl in place of D-glucopyranosyl residues (in **33**).

11. Polysaccharides Containing 3-Deoxy-D-*manno*-octulosylonic Acid Residues

Observation of displacement effects resulting from O-substitution, and use of standards, led to the assignment of the chemical structure of the polysaccharide of *Neisseria meningitidis* serogroup 29e, which consists of 2-acetamido-2-deoxy-D-galactosyl and partly acetylated 3-deoxy-D-*manno*-octulosylonic acid (KDO) residues.[166] The ^{13}C-n.m.r. spectrum of the O-deacetylated polymer (see Fig. 36,B) contained 15 signals out of the possible 16 expected from an alternating structure. Comparison of these resonances with those of the anomers of 2-aceta-

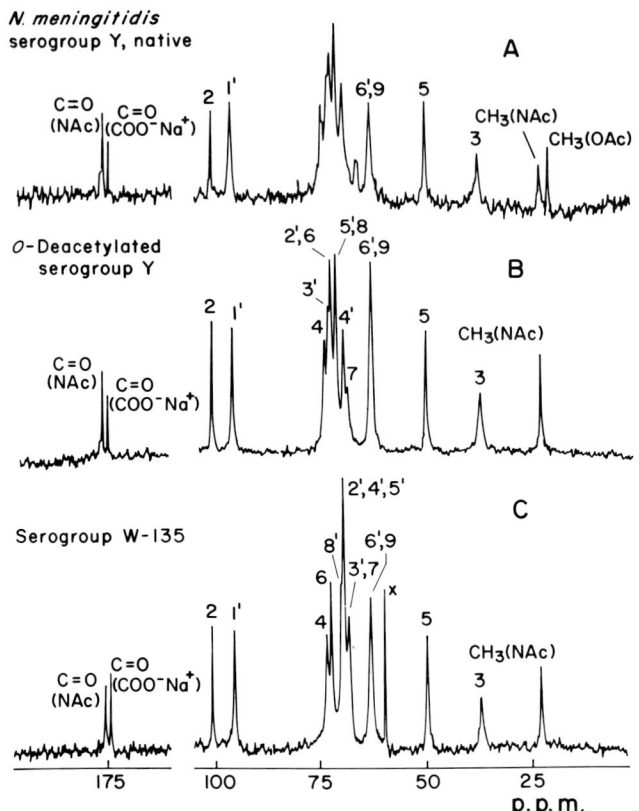

FIG. 35.—^{13}C-N.m.r. Spectra of Acetylated Polymer of *Neisseria meningitidis* Serogroup Y (A), its *O*-Deacetylated Derivative (B; see **33**), and the Polysaccharide of Serogroup W-135 (C; see **33**). (Solvent, D_2O, pD 7, at 31°; chemical shifts are expressed relative to external tetramethylsilane.)

mido-2-deoxy-D-galactopyranose showed that the typical C_β-5 signal at δ_c 76.3 was absent. Also, the C-3 signal of the polymer, compared with that of the α-D anomer, is displaced downfield by 4.7 p.p.m., indicating a 3-*O*-substituted α-D-galactopyranose structure (**34**). The anomers

34

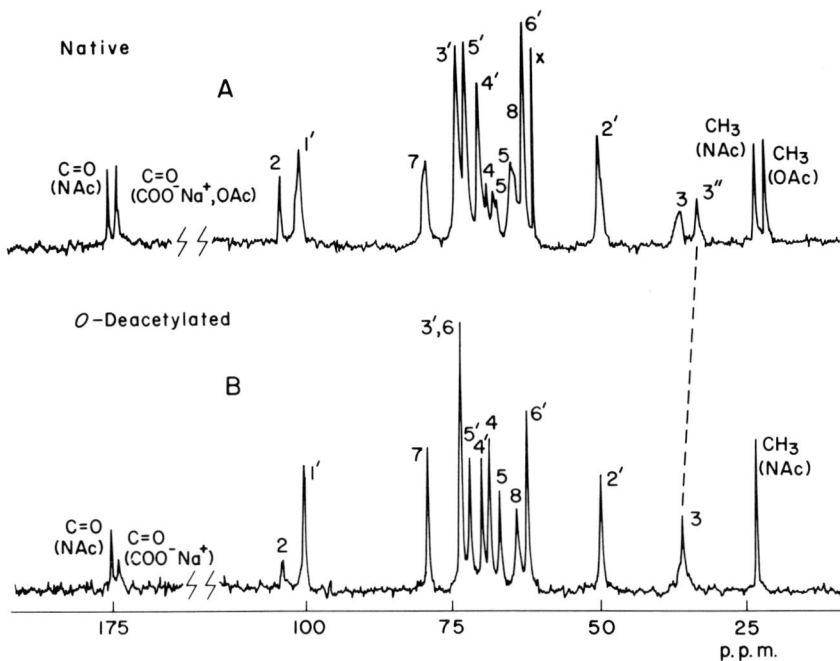

Fig. 36.—¹³C-N.m.r. Spectra of Partly Acetylated, Native (A) and O-Deacetylated Polymer (B; **34**) of *Neisseria meningitidis* Serogroup 29-3. (Solvent, D_2O, pD 7, at 31°; chemical shifts are expressed relative to external tetramethylsilane.)

of methyl 3-deoxy-D-*manno*-octulosonic acid (Na salt) were prepared, and resonances were unambiguously assigned, with the exception of those of C-4 and C-5. The configuration of the KDO residues is β-D, as its C-1 signal had $δ_c$ 174.5 (corresponding to $δ_c$ 174.8 of the β-D form) rather than $δ_c$ 176.5 of the α-D form. The difference in shift was attributed to hydrogen bonding of the carboxylate group with OH-8 of the β-D form in the $^2C_5$(L) conformation (**34**), a property of the structurally related methyl α-N-acetylneuraminidic acid [(methyl 5-acetamido-3,5-dideoxy-D-*glycero*-D-*galacto*-2-nonulopyranosid)onic acid].[177] The specific optical rotation, +90°, is consistent with the β-D form of the KDO residue in combination with the α-D form of the other unit. (1→7)-O-Substitution is present, as the C-7 resonance of the polymer is at 8.6 p.p.m. lower field than that of the corresponding methyl β-D-ketoside, with upfield displacement of 1.3 and 1.4 p.p.m. for the C-6 and C-8 resonances. The acetate groups are present on the KDO residue to the extent of ~0.7 mol per mol of repeating unit, and 40% of the units are substituted at OH-4, as the C-3 signal is displaced upfield by 2.8 p.p.m. on acetylation, as may be seen from the spectra of

TABLE IX

Carbon-13 Chemical-Shifts of Polysaccharides, Disaccharides, and Relevant Monomers[a]

Carbohydrate	Hexose moiety					
	C-1'	C-2'	C-3'	C-4'	C-5'	C-6'
α-D-Glucose	93.1	72.5	73.8	70.7	72.5	61.7
Disaccharide from group Y	96.3	72.6	73.9	70.3	72.1	61.5
O-Deacetylated, group Y polysaccharide	96.4	73.3	74.1	70.2	72.1	63.7
β-N-Acetylneuraminic acid	—	—	—	—	—	—
α-D-Galactose	93.4	70.3	69.4	70.4	71.6	62.3
Disaccharide from group W-135	96.6	70.3	69.4	70.7	72.3	62.0
Group W-135 polysaccharide	96.1	70.2	69.3	70.2	70.2	63.8

[a] In parts per million from external tetramethylsilane.

the O-acetylated and the O-deacetylated polysaccharide (Fig. 36,A and B). A less definite assignment of 5-O-acetyl groups was made on the basis of a downfield, γ-acetylation shift of 0.5 p.p.m. that was observable in the split, C-3 signal of the KDO residue.

12. Alginic Acid

For oligosaccharides derived from alginic acid structures, Boyd and Turvey[180] determined values of C-1 shift that should be applicable to the ^{13}C-n.m.r. spectra of the polymers themselves.[181] The trisaccharide, and an oligosaccharide of high molecular weight consisting of (1→4)-linked β-D-mannopyranosyluronic acid residues, each gave a C-1 signal at δ_c 101.3, corresponding to (nonreducing) end-groups and internal residues. The value for consecutive, (1→4)-linked, α-L-gulopyranosyluronic acid residues was δ_c 102.0. Also characteristic were C-2 to C-5 signals at δ_c 81.2, 69.0, and 65.8 associated with guluronic, but not mannuronic, acid residues. An oligosaccharide fraction containing mannuronic and guluronic acid residues, with some degree of alternation, gave C-1 signals at δ_c 101.1 and 102.2. These were assigned in respective fashion, as the internal residue of a trisaccharide containing 4-deoxy-L-*erythro*-5-hexulosyluronic acid linked (1→4) to β-D-Man*p*A-(1→4)-α,β-L-Gul having a C-1 signal of δ_c 101.2.

(180) J. Boyd and J. R. Turvey, *Carbohydr. Res.*, 61 (1978) 223–226.
(181) D. A. Rees, unpublished results.

TABLE IX (continued)

Carbon-13 Chemical-Shifts of Polysaccharides, Disaccharides, and Relevant Monomers[a]

				Sialic acid moiety						
C-1	C-2	C-3	C-4	C-5	C-6	C-7	C-8	C-9	CH_3 (NHCOCH$_3$)	C=O (NHCOCH$_3$)
174.4	95.4	36.6	73.3	50.7	71.3	69.3	71.3	64.3	23.3	175.4
174.2	101.3	38.0	74.8	50.8	73.3	69.4	72.1	63.7	23.4	175.5
174.3	96.4	39.9	67.8	53.2	71.3	69.4	71.5	64.3	23.2	176.0
—	—	—	—	—	—	—	—	—	—	—
174.4	96.0	36.6	72.5	50.8	71.5	69.1	71.5	64.4	23.4	175.6
174.3	101.4	37.8	74.3	50.7	73.3	69.2	70.9	63.8	23.6	175.6

13. D-Xylose-Containing Polysaccharides

The β-D-(1→4)-linked D-xylopyranan from wood gives a ^{13}C-n.m.r. spectrum containing 5 signals,[11] which, based on the interpretation of the spectrum of methyl β-D-xylopyranoside (with shift correction for different temperatures), arise from C-1 (103.1), C-2 and C-3 (74.3, 75.3), C-4 (78.0, $\Delta\delta_c$ +7.0), and C-5 (64.5, $\Delta\delta_c$ −2.4).

The branched-chain glucuronoxylan from an insect protozoon, *Herpetomonas samuelpessoai*, consists of an acidic, nonreducing endgroup for every seven units, associated with the structures D-GlcpA-(1→2)-α-D-Xylp-(1→2)-D-Xyl, D-GlcpA-(1→2)-β-D-Xylp-(1→4)-D-Xyl, and -(1→4)-α-D-Xyl.[11] Its ^{13}C-n.m.r. spectrum (see Fig. 37) contained C-1 signals at δ_c 103.2 and at δ_c 99.0 and 98.4, which arose from units

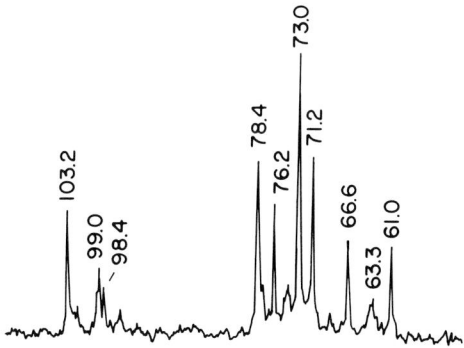

FIG. 37.—^{13}C-N.m.r. Spectrum of Glucuronoxylan of *Herpetomonas samuelpessoai* (D$_2$O, 70°). (Chemical shifts are in δ_c, relative to external tetramethylsilane.)

respectively having β-D and α-D configurations. The signal at δ_c 61.0 was from 4-O-substituted α-D-xylopyranosyl residues, as methyl α-D-xylopyranoside gives[38] a signal at δ_c 62.6 that is modified by a β-shift of −1.6 p.p.m. The β-D-xylopyranosyl residues were substituted at O-2, as no β-shift was observed, the signal at δ_c 66.6 corresponding closely to that[38] of C-5 of methyl β-D-xylopyranoside (δ_c 66.9). Methylation analysis of the glucuronoxylan provided methyl 2,3-di-O-methyl-α,β-D-xylopyranoside, which could have been formed either from 4-O-substituted D-xylopyranose or 5-O-substituted D-xylofuranose structures. Significant proportions of the latter structure were not present, as a series of ^{13}C-n.m.r. spectra obtained with complete proton coupling, and off-resonance decoupling, did not contain a triplet in the region of a signal of O-substituted C-5.

14. Miscellaneous Heteropolymers

The heteropolymers of certain *Klebsiella* species contain three to six sugar residues, some of which carry acetyl or 1-carboxyethylidene acetal substituents. Those of serotypes K36 (ref. 63) and K41 (ref. 17) give complex, ^{13}C-n.m.r. spectra that are well resolved when obtained at high field. The K41 polysaccharide contains rhamnose, galactose, glucose, and glucuronic acid in the ratios of 1:2:3:1, giving, under conditions suitable for quantitation, seven equal, C-1, ^{13}C-n.m.r. signals. Five signals showed $^1J_{C-1,H-1}$ values of 170 Hz, and two signals had splittings of 160 Hz. With the exception of the low-field, C-1 signal at δ_c 109.0 (generally characteristic of β-D-galactofuranosyl or α-L-arabinofuranosyl units, see Section VI,6), which has $^1J_{C-1,H-1}$ 172.5, corresponding to that of methyl β-D-galactofuranoside, but not that of the α-D anomer[28] (~175 Hz), the signals were assigned to α- or β-linked pyranose units, depending on whether they had couplings of ~170 or ~160 Hz, respectively. With the aid of methylation, Smith degradation, and partial-hydrolysis studies, structure **35** was proposed. Two of

→6)-α-D-Glc*p*-(1→3)-α-L-Rha*p*-(1→3)-α-D-Gal*p*-(1→2)-β-D-Gal*f*-(1→
                                                                                       3
                                                                                       ↑
                                                                                       1
β-D-Glc*p*-(1→6)-α-D-Glc*p*-(1→4)-β-D-Glc*p*A

35

the C-1 signals, at δ_c 105.5 and 99.5, were assigned by reference to C-1 signals of gentiobiose (105.5; internal TSP reference) and 3-O-α-D-glucopyranosyl-α-L-rhamnose (98.3). Two other signals were rationalized on the basis of the properties of the O-L-rhamnopyranosyl-(1→3)-

O-D-galactopyranosyl-(1→2)-O-L-arabinofuranosyl-(1→1)-glycerol that was obtained on Smith degradation; this gave anomeric C-1 signals at δ_c 109.15, 105.1, and 101.55 that were attributed to α-L-arabinofuranosyl, α-L-rhamnopyranosyl, and α-D-galactopyranosyl units, respectively. (The spectrum of the polysaccharide contained signals at δ_c 104.45 and 101.55.) The polysaccharide signal at δ_c 104.8 showed ^{13}C,^{1}H coupling corresponding to a β configuration, and should arise from a glucuronic acid unit, as the derived 3-O-(D-glucopyranosyluronic acid)-D-galactose had, according to the p.m.r. spectrum, a β-D-glycosidic linkage. The ^{13}C-n.m.r. spectrum of the polysaccharide also contained well resolved, ^{13}C signals at δ_c 87.90, 86.1, and 85.15; these were assigned to C-2, C-4, and C-3 of 2,3-di-O-substituted β-D-galactofuranosyl residues, by analogy with those of methyl 2,3-di-O-benzyl-β-D-galactofuranoside.[28]

The D-arabino-D-galactan of cells of the insect protozoon *Crithidia fasciculata* contains β-D-(1→3)-linked D-galactopyranosyl main-chain units, some of which are unsubstituted; others are substituted at O-2 by (single-unit) D-arabinopyranosyl groups. The ^{13}C-n.m.r. spectrum shows that these are only partial structures, as the C-1 region contained 8 signals.[125]

15. Synthetic Derivatives of Polysaccharides

(a) **Acetates and O-Benzyl Derivatives.**—The ^{13}C-n.m.r. spectrum of cellulose acetate in dichloromethane was found by Dorman and Roberts[47] to contain 3 single-carbon signals and one broad one, corresponding to the D-glucopyranosyl residues. Of the acetate resonances, 3 resolved C=O signals and only one CH_3 signal were detected. However, at that time, few model systems were available, and interpretation of signals was difficult. In terms of acetylation shifts, it was known that acetylation of simple, primary or secondary alcohols is accompanied by downfield displacements of α-carbon resonances and smaller, upfield shifts of β-carbon resonances.[182] However, these results could not be extrapolated to peracetylated sugars, as, compared to carbon atoms of α- and β-D-glucopyranose, four of those of the pentaacetates were shielded, except C-4, which was unaffected, and C-6, which was deshielded. Presumably, steric effects take precedence over the (smaller) electronic-deshielding mechanism.

Takeo and coworkers[93] examined the ^{13}C-n.m.r. spectra of peracetylated cyclo-hexa-, -hepta-, and -octa-amyloses [consisting of 6, 7,

(182) M. Christl, H. J. Reich, and J. D. Roberts, *J. Am. Chem. Soc.*, 93 (1971) 3463–3468.

and 8 (1→4)-O-substituted α-D-glucopyranosyl residues, respectively], by using C_6D_6 and $CDCl_3$ as solvents. As with unsubstituted polysaccharides (D_2O solvent), there were small, nonsystematic variations of chemical shifts with annular size. Acetylation of the cycloamyloses and of 6-deoxycycloheptaamylose gave rise to large, upfield shifts of 5–7 p.p.m. for C-1 and C-4, whereas smaller, upfield shifts (2–4 p.p.m.) were observed for C-2, C-3, and C-5 resonances (the C-2 and C-3 assignments were incorrectly reversed). The C-6 resonances of cycloamyloses and 6-deoxycycloheptaamylose were respectively deshielded by ~1.5 p.p.m. and shielded by 1.5 p.p.m. Three-bond, $^{13}C,H$ coupling of cyclohexaamylose acetate is discussed in Section II,3.

Gagnaire and Vignon[26] investigated linear, α-D-(1→6)-linked dextran and its per-O-acetyl and per-O-benzyl derivatives. The ^{13}C signals of the spectra (62.84 MHz) given by each compound were unambiguously assigned by selective proton-irradiation, following identification of proton signals by homonuclear irradiation. The dextran assignments agreed with those previously reported.[55,95,102] The $^1J_{^{13}C,H}$ values were determined, where possible, but 2-bond and 3-bond values were not observable, as they were small compared to line widths. As may be seen from Table X, acetylation displacements of C-2 and C-6 resonances were small, but those of C-1, C-3, C-4, and C-5 lay in the range of −1.2 to −3.3 p.p.m. O-Benzylation shifts were analogous to those occurring on O-methylation, only hydroxylated-carbon resonances being affected greatly (+8 to +9 p.p.m.). In terms of mono-O-substitution of polysaccharides, the α- and β-shifts occurring on 3-O-substitution of α,β-D-glucose are of some significance. On replacement of H- by CH_3CO-, CH_2ClCO-, CCl_3CO-, CH_3-, and CH_3SO_2-, the ratio of α- to β-effect is greatest on O-methylation,[183] thus indicating that the ^{13}C-n.m.r. spectra of methyl (and, perhaps, other) ethers of polysaccharides are easier to interpret than those of esters, once the spectrum of the parent polysaccharide has been rationalized.

^{13}C-N.m.r. spectroscopy is useful in determining the position(s) of substitution, as esters of acetic and malonic acid, in polysaccharides. This topic is covered in Section VI,6, 8, and 10.

(b) Methyl, Carboxymethyl, and 2-Hydroxyethyl Ethers.—The spectral characteristics of methyl, carboxymethyl, and 2-hydroxethyl ethers of cellulose have been examined by Parfondry and Perlin,[184] and interpreted in terms of the positions of O-substitution and degree

(183) M. R. Vignon and P. J. A. Vottero, *Tetrahedron Lett.*, (1976) 2445–2448.
(184) A. Parfondry and A. S. Perlin, *Carbohydr. Res.*, 57 (1977) 34–49.

TABLE X

Chemical Shifts and Coupling Constants (in Parentheses) of ^{13}C Signals of Dextran and its Per-O-acetyl and Per-O-benzyl Derivatives

Compound	Solvent	Temperature (degrees C)	C-1	C-2	C-3	C-4	C-5	C-6
Dextran	D_2O	90	97.9 (171)	71.5 (146.5)	73.5 (146.5)	70.0 (146.5)	70.4 (146.5)	66.2 (145)
Peracetate	Cl_2CCCl_2		95.6 (172.5)	70.8	70.2	68.8	68.7	65.8
Perbenzyl ether	$CDCl_3$	55	97.7 (171)	80.8 (143.5)	81.8 (143.5)	77.8 —	71.2 (142)	66.0 (142)

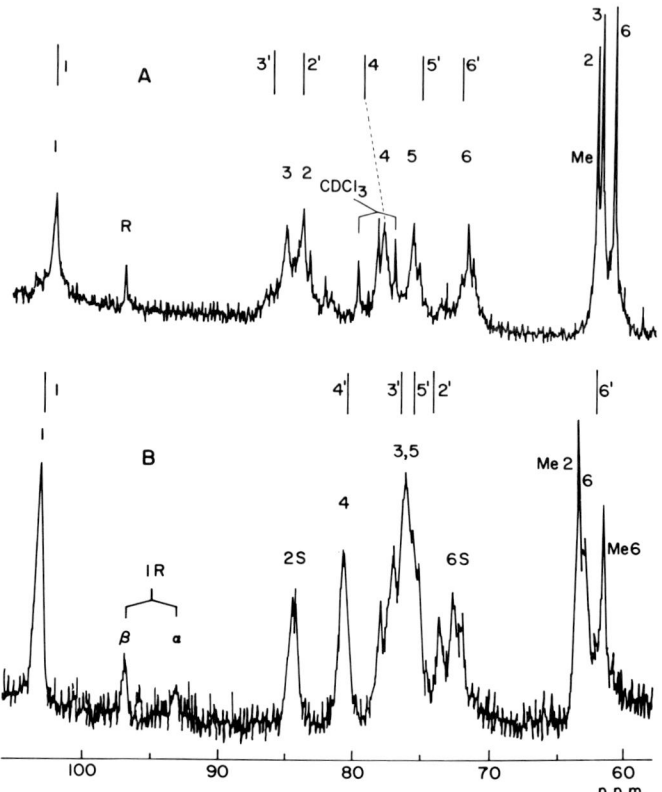

FIG. 38.—¹³C-N.m.r. Spectrum of A, Partly Depolymerized O-Methylcellulose (d.s. ~2.8) in CDCl$_3$ at 30° (R, signal due to reducing-end residue; Me, O-methyl; inset lines represent chemical shifts of corresponding carbon atoms in methyl hepta-O-methyl-β-cellobioside) and of B, O-Methylcellulose (d.s. ~0.7), Partially Degraded by Cellulase, in D$_2$O at 30°. (S represents a ¹³C nucleus bonded to an OMe group; inset lines give the chemical shifts of corresponding carbon atoms in methyl β-cellobioside.)

of substitution (d.s.). Under the conditions used, it was not possible to obtain satisfactory spectra of dilute, viscous solutions, and a partial depolymerization step, involving use of an acid or of cellulase, was necessary. O-Methylcellulose (d.s. ~2.8) that had been degraded slightly with acid in CDCl$_3$ at 30° (as heating can cause gelation) gave a spectrum (see Fig. 38,A) bearing a resemblance to that of methyl hepta-O-methyl-β-cellobioside,[185] the biggest difference being noted for C-4 of the polymer, which resonates ~2 p.p.m. upfield from that of the disac-

(185) J. Haverkamp, M. J. A. de Bie, and J. F. G. Vliegenthart, *Carbohydr. Res.*, 37 (1974) 111–125.

charide. A commercial O-methylcellulose (d.s. ~0.7) gave a somewhat more O-methylated product on treatment with cellulase, which preferentially degraded regions lacking substituents. Its ^{13}C spectrum (see Fig. 38,B) clearly indicated 2-O-substitution, as 2-O-methyl-β-D-glucopyranose gives C-2 and OMe-2 signals corresponding to those at δ_c 84 and 62. Of the two OMe signals, the OMe-6 resonance at δ_c 60 was the weaker, and no evidence of 3-O-substitution was visible at δ_c 87 p.p.m. However, 3-O-substitution was present to a small extent in an O-methylcellulose of d.s. 1.5–2.0. These data were consistent with those in a previous report that stated[186] that, in terms of ease of methylation, O-2 > O-6 >> O-3.

The ^{13}C-n.m.r. spectra of per-O-methylated arabinans are discussed elsewhere (see Section VI,6).

A pullulan reported to contain 6% of α-D-(1→3)-linkages[112] was reinvestigated by ^{13}C-n.m.r. spectroscopy of its per-O-methylated derivative,[55] which was shown to contain <3% of this type of unit. Similarly, the permethyl ether of a dextran containing α-D-(1→4)-, α-D-(1→6)-, and α-D-(1→3)-linkages was used for quantitation purposes, but the H-1 resonances in the proton-n.m.r. spectrum were better resolved than those of the C-1 atom in the ^{13}C-n.m.r. spectrum and, furthermore, the p.m.r. method was superior, as T_1 and n.O.e. effects were not involved.

An enzyme-degraded sample of O-(carboxymethyl)cellulose having d.s. 0.7 had the same pattern of substitution.[184] In D$_2$O at 30°, it gave a ^{13}C-n.m.r. spectrum (see Fig. 39,A) containing a prominent, substituted C-2 signal at δ_c 82, corresponding to that of 2-O-(carboxymethyl)-β-D-glucose. Derivatization with the carboxymethyl group causes an α-shift of 8–9 p.p.m., with little β-shift. Also observed were a strong signal at δ_c 62, due to unsubstituted C-6 residues, and a relatively weak signal in the region 69–72 p.p.m. that corresponded to 6-O-(carboxymethyl)ated units, as it was larger for an O-(carboxymethyl)cellulose having a higher d.s. value.

The ^{13}C-n.m.r. spectrum of O-(2-hydroxyethyl)cellulose partly degraded enzymically (d.s. 0.8) in D$_2$O at 30° (see Fig. 39,B) was similar, except for the region arising from the substituent, which is [-O-(CH$_2$-CH$_2$-O)$_n$-CH$_2$-CH$_2$O] with n = 2 or 3. Because the spectral shifts occurring on O-(2-hydroxyethyl)ation are close to those caused by O-(carboxymethyl)ation, a pattern of O-substitution similar to those already discussed was present.

Examination of undegraded O-(2-hydroxyethyl)cellulose, d.s. 2.5, in

(186) I. Croon, *Sven. Papperstidn.*, 63 (1960) 247–257.

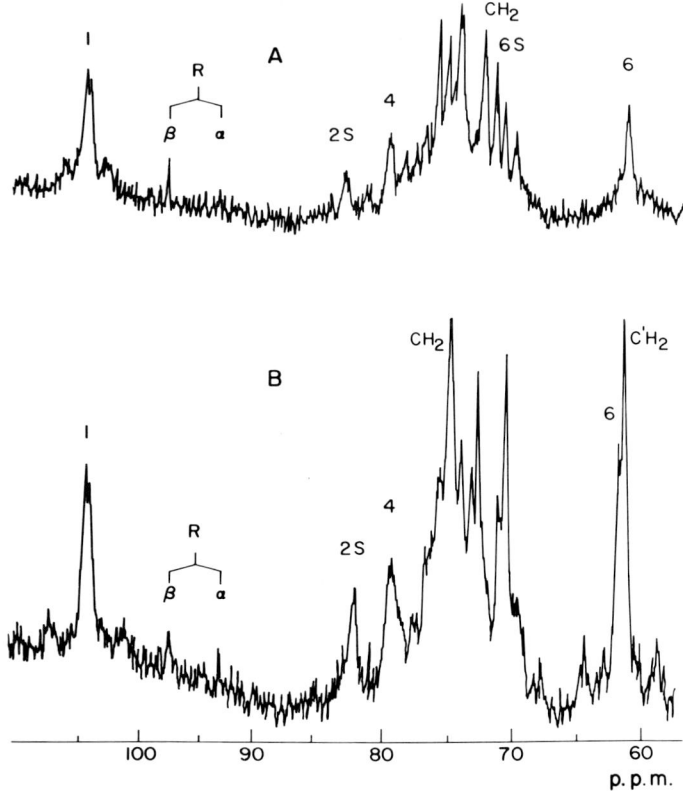

Fig. 39.—^{13}C-N.m.r. Spectra of A, O-(Carboxymethyl)cellulose (d.s. 0.7), Partially Degraded by Cellulase, in D$_2$O at 30° (R, signal of reducing-end residue; S represents a ^{13}C nucleus bonded to an alkoxyl group) and of B, O-(2-Hydroxyethyl)cellulose (d.s. 0.8), Partly Degraded by Cellulase, in D$_2$O at 30°. (R, signal due to reducing-end residue; S represents a ^{13}C nucleus bonded to an alkoxyl group.)

D$_2$O at 75°, by use of an n.m.r. system having a weaker magnetic field, was less successful, as the signals of the D-glucosyl residues were broad, and poorly resolved.[187] However, comparison of the signals with those reported by Parfondry and Perlin,[184] using the C-1 signals as the reference basis, showed that it contained both 2-O- and 3-O-substituted residues. The proportion of the latter, as represented by a signal at δ_c 82, was actually higher than had been realized by the authors, as they incorrectly assigned it to C-4 also. Significantly, a high-field, C-6 signal was present, corresponding to unsubstituted C-6

(187) J. R. DeMember, L. D. Taylor, S. Trummer, L. E. Rubin, and C. K. Chiklis, *J. Appl. Polym. Sci.*, 21 (1977) 621–627.

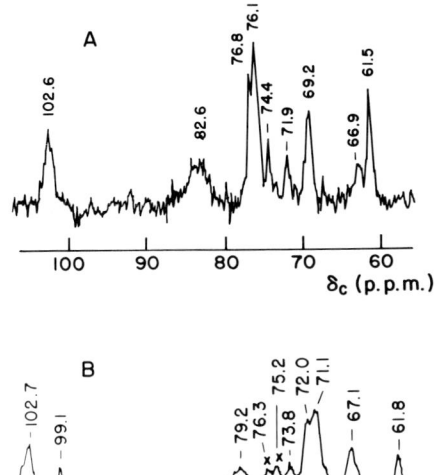

FIG. 40.—^{13}C-N.m.r. Spectra of Diphenylborinate Derivatives of β-D-(1→2)-Linked D-Glucopyranan (A), and Branched α-D-Mannopyanan of *Endomycopsis fibuliger* (B). [Solvent, D$_2$O at ambient temperature; chemical shifts are based on external tetramethylsilane; on derivatization, additional signals appear at δ_c 66.9, 71.9, and 74.4 (A) and at δ_c 75.2 and 76.3 (B)].

atoms of D-glucosyl residues. Cohen and Haas[188] reported that, in the reaction of ethylene oxide with cellulose, the primary hydroxyl group is less reactive than had theretofore been supposed.

(c) **Diphenylborinates.**—The ability of sodium diphenylborinate to complex with polysaccharides may be assessed by using ^{13}C-n.m.r. spectroscopy.[99] For example, in D$_2$O solution, the D-mannan of *Endomycopsis fibuliger* [which consists mainly of a (1→6)-O-linked α-D-mannopyranose main-chain substituted on O-2 by a single α-D-mannopyranosyl group] and a β-D-(1→2)-linked D-glucopyranan have been shown to exist in equilibrium with their diphenylborinate complexes, as evidenced by their ^{13}C signals, some of which, as yet unidentified, differ from those of the parent polysaccharide (see Fig. 40,A and B, respectively). Broadening of signals occurred, probably because of an increased correlation-time on complexing. In the case of potato starch, the (nonreducing) end-groups of the amylopectin com-

(188) S. G. Cohen and C. Haas, *J. Am. Chem. Soc.*, 72 (1950) 3954–3958.

ponent were shown to complex with diphenylborinate (see Fig. 7), as 2 additional signals were evident in the ^{13}C-n.m.r. spectrum.

(d) **O-D-Glucopyranosyl Derivatives of Amylose.**—Comb-like derivatives of D-glucopyranosyl residues attached to amylose were synthesized[98] by orthoester[189] and Helferich[190] condensations. Both products gave ^{13}C-n.m.r. resonances characteristic of 4,6-di-O-substituted α-D-glucopyranosyl residues, although the Helferich conditions gave rise to a greater degree of branching. Based on the observation of configuration-dependent, C-1 signals of D-glucopyranosyl residues, it was clear that, as expected, β-D-(1→6)-linkages were formed under the conditions of orthoester condensation. Unexpectedly, however, α-D-glucosylation could not be detected following use of the Helferich conditions; this was attributed to occurrence of the condensation with a primary, rather than a secondary, alcohol group.

ACKNOWLEDGMENT

The author thanks Dr. Lucia Mendonça-Previato of the Department of General Microbiology, Federal University of Rio de Janeiro, for valuable help in the preparation of this article.

(189) N. K. Kochetkov, A. F. Bochkov, and T. A. Sokolovskaya, *Carbohydr. Res.*, 19 (1971) 1–4.
(190) B. Helferich and J. Zirner, *Chem. Ber.*, 95 (1962) 2604–2611.

PHOTOCHEMICAL REACTIONS OF CARBOHYDRATES

By Roger W. Binkley

Department of Chemistry, Cleveland State University, Cleveland, Ohio 44115

I. Introduction	105
II. Alkenes	106
III. Carbonyl Compounds	122
1. Aldehydes and Ketones	122
2. Esters	129
IV. Acetals	142
V. Unprotected Alditols, Aldohexoses, and Related Compounds	147
VI. Sulfur-Containing Compounds	150
1. Dithioacetals and Sulfides	151
2. Sulfoxides, Sulfones, and Sulfonates	153
3. Thiocarbamates and Xanthides	157
VII. Nitrogen-Containing Compounds	160
1. Nitro Compounds	160
2. Azides	176
3. Diazo Compounds	178
4. Oximes	179
5. Azines	186
VIII. Iodo Compounds	186
1. Deoxyiodo Sugars	186
2. o-Iodobiphenylyl Ethers	187
IX. Organometallic Compounds	190
X. Phosphates	192

I. Introduction

Study of the photochemical reactions of carbohydrates is a topic of longstanding interest among chemists; in fact, publications on this subject first appeared shortly after the beginning of the present century. Much of the early work in this area was concerned with changes, in the physical properties of reaction mixtures, arising from irradiations conducted under various conditions. Relatively little identification of the photoproducts was undertaken prior to the introduction of modern instrumentation.

In recent years, interest in the photochemistry of carbohydrates has intensified, and the major orientation of research in this field has changed. Synthetic chemists have become increasingly aware of the advantages offered by photochemical reactions. The possibilities raised by these advantages (for example, conducting reactions under mild conditions, synthesizing molecules not obtainable by other routes, and selectively reacting particular functional groups in multifunctional molecules) have stimulated study of the photolysis of a wide variety of carbohydrates and their derivatives. Several photochemical processes already have emerged as important, synthetic reactions, and others are certain to follow.

The purpose of the present article is to review comprehensively those photochemical reactions of carbohydrates for which product structures have been established, and to apply current mechanistic reasoning to the understanding of these reactions. Two reviews of carbohydrate photochemistry already exist. The first[1] was published in this Series in 1963, prior to the considerable recent growth in this field, and the second[2] was a brief summary published in the Japanese literature in 1973.

An organizational plan based upon the type of functional group has been adopted here in presenting and discussing the photochemical reactions of carbohydrates.

II. ALKENES

The most important photochemical reaction of carbon to carbon unsaturated carbohydrates is addition to the unsaturated system. Two types of addition reaction are readily recognized. The first consists of those in which the molecule adding to the carbohydrate does so by involving a π-bond of its own. Processes of this type, listed in Table I, are those which lead to formation of a new ring-system (cycloaddition). The second class of addition reaction is one in which a σ-bond is broken in the molecule adding to the unsaturated carbohydrate. The reactions that belong to the latter category (see Tables II and III) follow two basic patterns, and comprise the majority of the addition processes reported.

Cycloaddition reactions (see Table I) involving unsaturated carbohydrates are regio- and stereo-selective. These selectivities can be understood by assuming that the photochemical interaction between the two π-systems results in formation of the more stable 1,4-diradical. The reaction between 3,4,6-tri-O-acetyl-D-glucal (1) and acetone pro-

(1) G. O. Phillips, *Adv. Carbohydr. Chem.*, 18 (1963) 9–59.
(2) K. Matsuura, *Kagaku No Ryoiki*, 27 (1973) 35–42.

TABLE I

Photochemical Cycloaddition Reactions between Unsaturated Carbohydrate Derivatives and Carbonyl Compounds

Reactants	Products	Yields (%)	Ref.
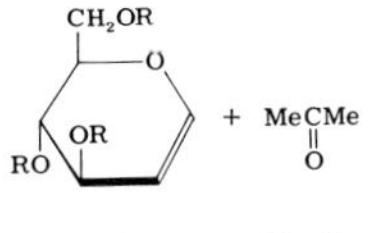 **1** (CH$_2$OR / OR / RO, R = Ac) + MeCMe(=O)	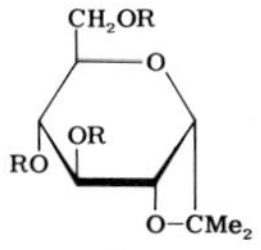 **2** (bicyclic oxetane product)	27[a] 99 99 33	3 4 5 6
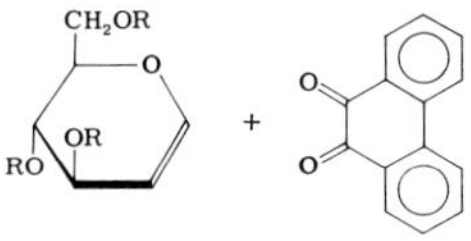 (CH$_2$OR / OR / RO, R = Ac) + phenanthrenequinone	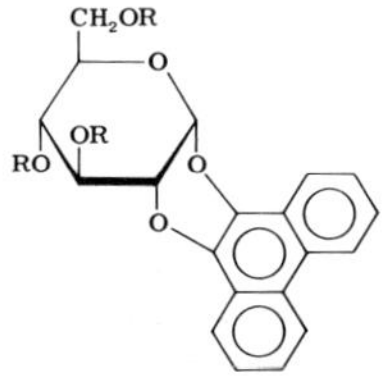 fused tricyclic product	33	7
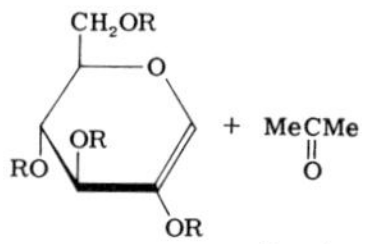 (CH$_2$OR / OR / RO / OR, R = Ac) + MeCMe(=O)	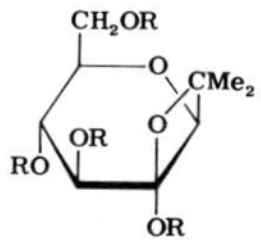 oxetane product	14[b]	8
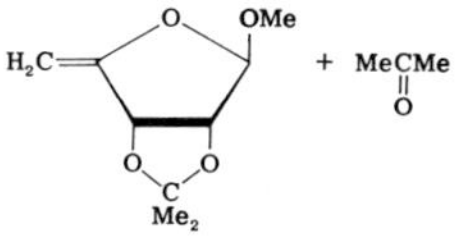 (H$_2$C= furanose with OMe, Me$_2$ acetal) + MeCMe(=O)	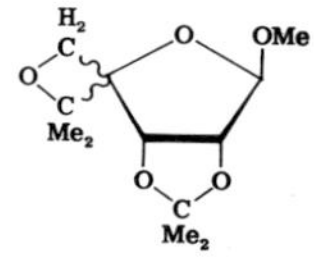 cyclobutane adduct (both isomers)	44 (combined yield)	9
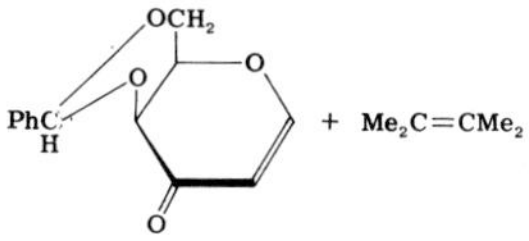 (PhCH, OCH$_2$, pyranone) + Me$_2$C=CMe$_2$	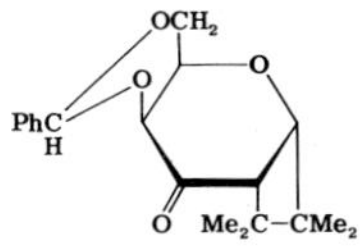 cyclobutane adduct Me$_2$C–CMe$_2$	11[c]	10, 11
	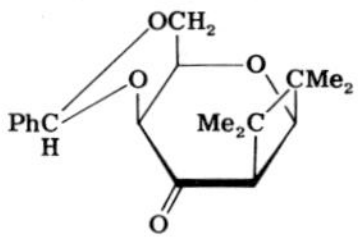 oxetane adduct Me$_2$C	32	

(Continued)

TABLE I (Continued)

Photochemical Cycloaddition Reactions between Unsaturated
Carbohydrate Derivatives and Carbonyl Compounds

Reactants	Products	Yields (%)	Ref.
	[structure with PhCH, OCH$_2$, O, CMe$_2$]	12	
[structure with PhCH, OCH$_2$, O] + Me$_2$C=CMe$_2$	[structure with PhCH, OCH$_2$, O, Me$_2$C–CMe$_2$]	d	10, 11
	[structure with PhCH, OCH$_2$, O, CMe$_2$]	d	

[a] Oxetane hydrolysis product formed. [b] Non-oxetane products formed. [c] Dimer formed. [d] Not determined.

vides an illustration (see Scheme 1). Diradical stability is maximized by attack at C-2 (regiochemistry determined) and by allowing the larger group attached to C-2 to become the equatorial substituent (stereochemistry determined).[4–6]

For addition reactions other than cycloaddition, unsaturated carbohydrates follow one of two fundamental pathways. The more common of the two begins with radical formation arising from bond homolysis in the noncarbohydrate reactant. Radicals are either produced directly, from absorption of light, or indirectly, from hydrogen abstrac-

(3) Y. Araki, K. Senna, K. Matsuura, and Y. Ishido, *Carbohydr. Res.*, 60 (1978) 389–395.
(4) K. Matsuura, Y. Araki, and Y. Ishido, *Carbohydr. Res.*, 29 (1973) 459–468.
(5) K. Matsuura, Y. Araki, and Y. Ishido, *Bull. Chem. Soc. Jpn.*, 45 (1972) 3496–3498.
(6) K.-S. Ong and R. L. Whistler, *J. Org. Chem.*, 37 (1972) 572–574.
(7) B. Helferich and E. von Gross, *Chem. Ber.*, 85 (1952) 531–535.
(8) Y. Araki, K. Senna, K. Matsuura, and Y. Ishido, *Carbohydr. Res.*, 64 (1978) 109–117.
(9) Y. Araki, K. Senna, K. Matsuura, and Y. Ishido, *Carbohydr. Res.*, 65 (1978) 159–165.
(10) P. M. Collins and B. H. Whitton, *J. Chem. Soc. Perkin Trans. 1*, (1973) 1470–1476.
(11) P. M. Collins and B. Whitton, *Carbohydr. Res.*, 21 (1972) 487–489.

TABLE II

Photochemically Initiated, Radical-Addition Reactions to Unsaturated Carbohydrate Derivatives

Reactants	Products and Yields (%)	Ref.
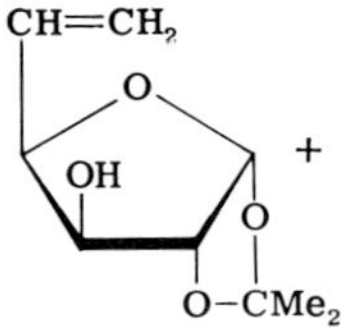 **3** (CH=CH₂ furanose with OH and O–CMe₂) + RH	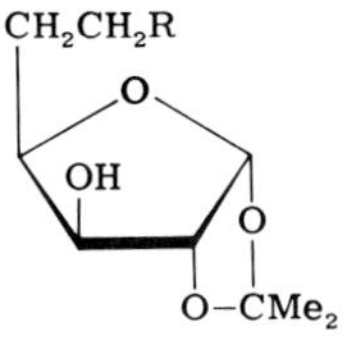 CH₂CH₂R furanose analog	
R = —SCMe (‖O)	54	12,13
R 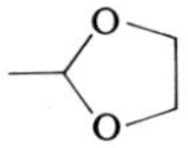 = (1,3-dioxolan-2-yl)	45 (**4**)	14
R = —PHPh	75	15
R = —PH₂	24	15[a]
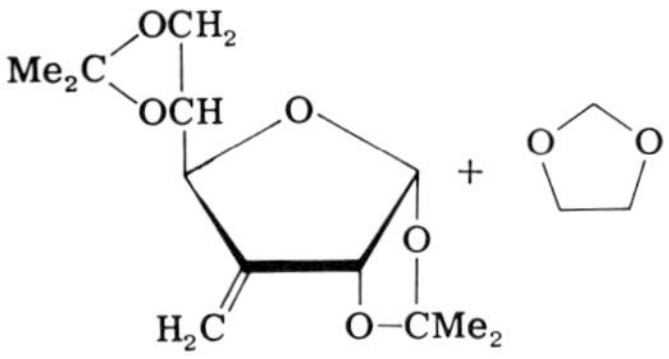 **5** (Me₂C–OCH₂–OCH furanose with =CH₂ and O–CMe₂) + dioxolane	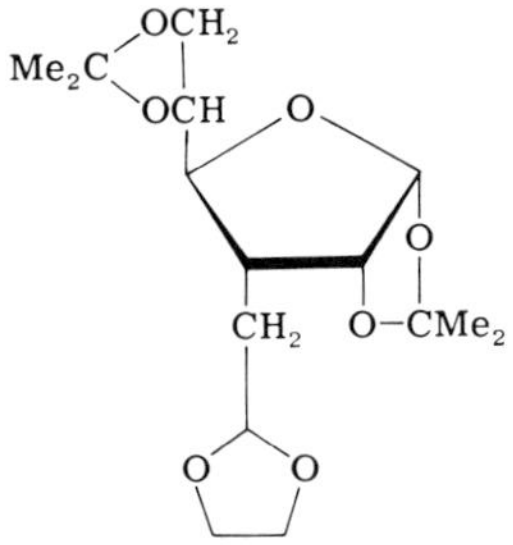 **6** (Me₂C–OCH₂–OCH furanose with CH₂–dioxolane and O–CMe₂) 55	14
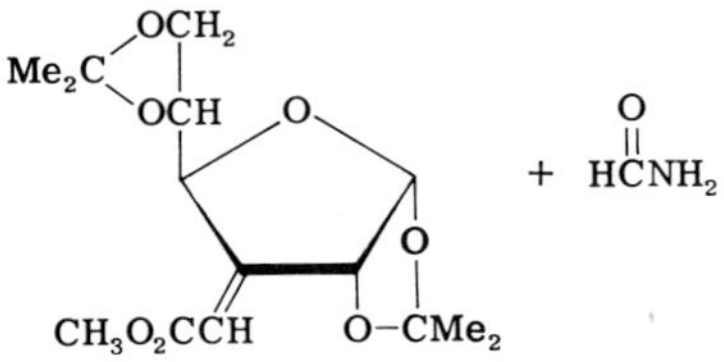 (Me₂C–OCH₂–OCH furanose with CH₃O₂CCH= and O–CMe₂) + HC(=O)NH₂	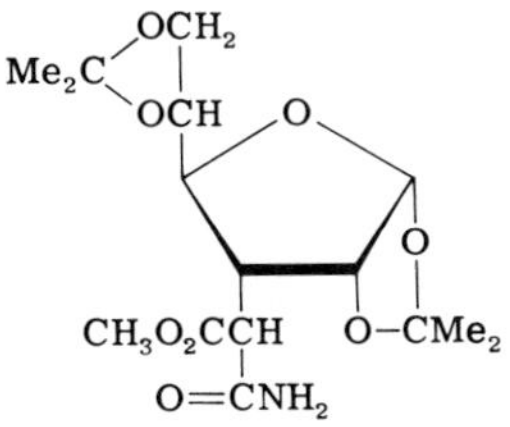 (Me₂C–OCH₂–OCH furanose with CH₃O₂CCH(–O=CNH₂) and O–CMe₂) 45	16

(*Continued*)

Table II (Continued)

Photochemically Initiated, Radical-Addition Reactions to Unsaturated Carbohydrate Derivatives

Reactants	Products and Yields (%)	Ref.
[structure: hex-enose with XO groups, R substituent] R = SEt R = OEt X = Ac	[cyclized product with R^1, R^2, XO] $R^1 = R^2 = $ SEt 60 $R^1 = $ H, $R^2 = $ OEt ⎫ $R^1 = $ OEt, $R^2 = $ H ⎬ 47	17
[furanose structure with H$_2$C=, OMe, Me$_2$ acetal] + RH R = —SCMe (‖O) R = —SCH$_2$Ph R = —P(OEt)$_2$ (‖S) R = [1,3-dioxolane]	[product with CH$_2$R, OMe, Me$_2$] 61 69 95 10	18 18 19 20[b]
[bicyclic furanose with H$_2$C—CH, Me$_2$, O—CMe$_2$] + RH R = —CNH$_2$ (‖O) R = —CMe$_2$OH R = —P(OEt)$_2$ (‖S)	[two products shown with Me$_2$C, OCH$_2$, OCH, R, O—CMe$_2$ and second isomer with HCO, H$_2$CO, CMe$_2$] 16 15 31 33 9	21 19 19

TABLE II (Continued)

Photochemically Initiated, Radical-Addition Reactions to Unsaturated Carbohydrate Derivatives

Reactants	Products and Yields (%)	Ref.
(structure) + RH; X = Ac, R = –C(O)NH$_2$	(two products) 26, 65	22
(structure with CH=CHOBz) + HP(S)(OEt)$_2$	(product) 55	19
(structure) + RH; X = Ac, R = –(1,3-dioxolan-2-yl)	44, 35	23
(structure) + RH	46, 13 (X = Ac, R = –C(O)NH$_2$)	24
	38, 21, 31 (X = Ac, R = –(1,3-dioxolan-2-yl))	23[c]
	76 (X = Ac, R = –CMe$_2$OH)	23

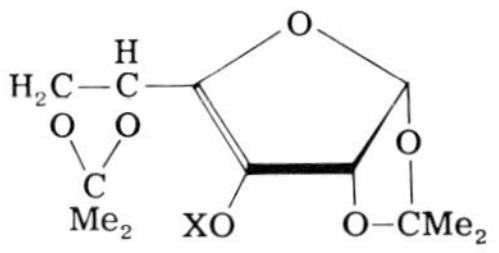

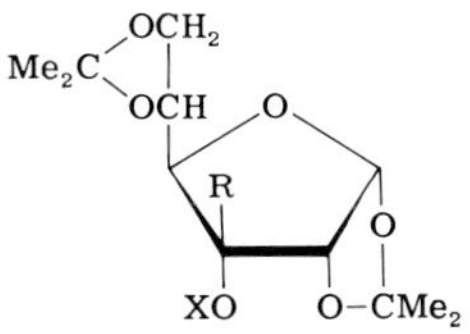

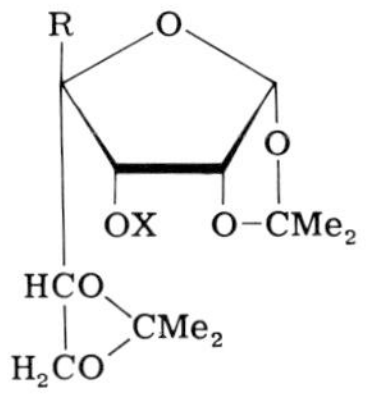

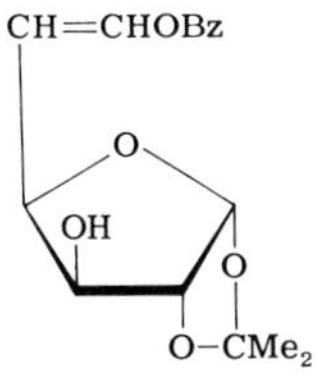

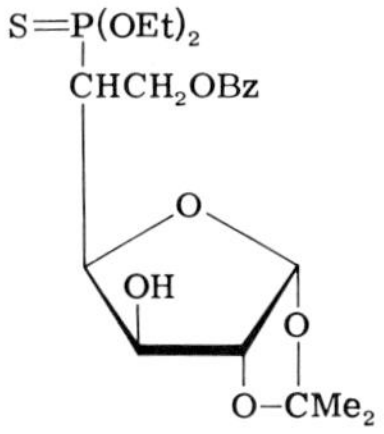

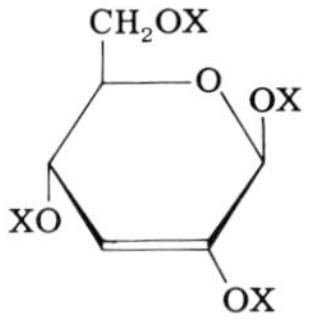

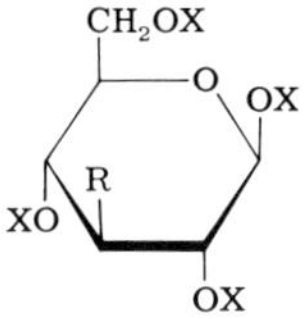

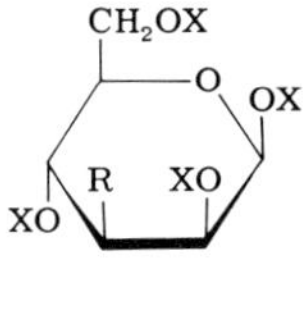

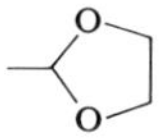

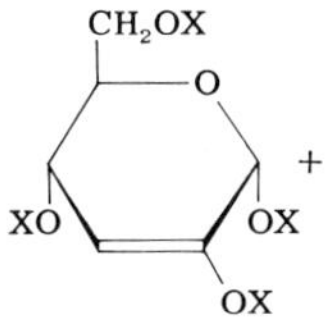

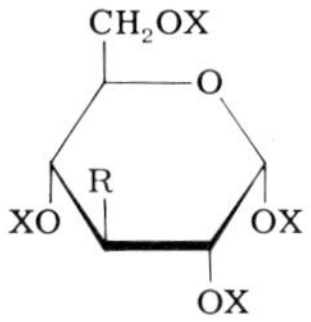

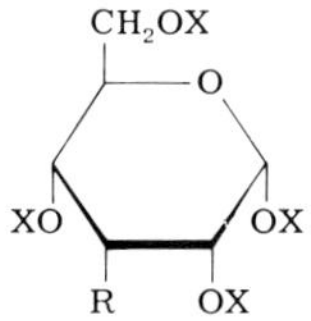

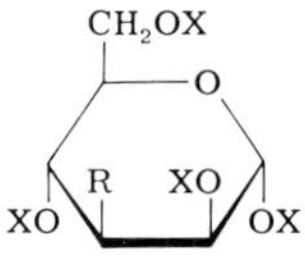

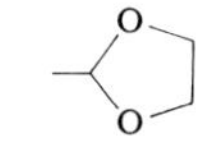

(Continued)

Table II (Continued)

Photochemically Initiated, Radical-Addition Reactions to Unsaturated Carbohydrate Derivatives

Reactants	Products and Yields (%)			Ref.
(glycal with CH$_2$OX, XO, OMe) + RH	(product with R axial down)	(product with R axial up)	(product with R at other position)	
X = Ac, R = —SEt	87			25
X = Ac, R = —SPr	94			25
X = Ac, R = (1,3-dioxolan-2-yl)		39	39	20
X = Ac, R = —CMe$_2$OH		66		26
X = Ac, R = (tetrahydrofuran-2-yl)		32	15	20
(glycal with CH$_2$OX, OX, XO, OX) + RH	(product 1)	(product 2)	(product 3)	
X = Ac, R = —SEt	77			25
X = Ac, R = —SPr	79			25
X = Ac, R = —CNH$_2$ (C=O)	55		7	16
X = Ac, R = (1,3-dioxolan-2-yl)		42	17	20
X = Ac, R = —CMe$_2$OH	16	51		8^c
(glycal with CH$_2$OX, OX, XO) + RH	(product 1) + (product 2)			
X = H, R = —P(OEt)$_2$ (‖S)	90			19
X = Ac, R = —SEt	44	44		25
X = Ac, R = —SPr	47	47		25

TABLE II (*Continued*)

Photochemically Initiated, Radical-Addition Reactions to
Unsaturated Carbohydrate Derivatives

Reactants	Products and Yields (%)	Ref.
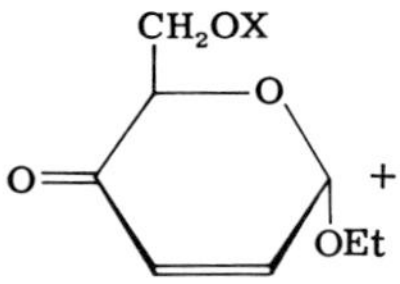 [enone with CH$_2$OX, OEt] + RH	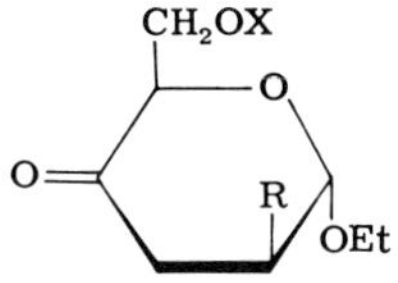 [product with R added]	
X = H, R = —CH$_2$OH	66	27, 28
X = H, R = —CMe$_2$OH	65–85	28
X = Ac, R = —CH$_2$OH	75	27, 28, 29
X = Ac, R = —CHCH$_3$ (OH)	65–85	28
X = Ac, R 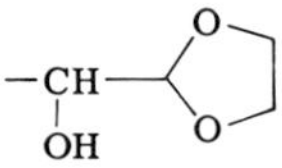 = —CH(OH)—[dioxolane]	42	29
X = Tr, R = —CH$_2$OH	75	27
X = Tr, R = —CMe$_2$OH	65–85	28
X = Tr, R = —CHCH$_2$OH (OH)	75–79	30, 31
X = Tr, R = —CHCH$_2$CH$_2$OH (OH)	49	30
X = Tr, R = —CHCH$_2$CO$_2$Me (OH)	32	30
X = Tr, R = Ac	67	30, 32
X = Tr, R 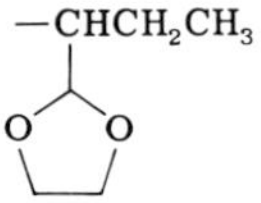 = —CHCH$_2$CH$_3$ [dioxolane]	62	32
X = Tr, R 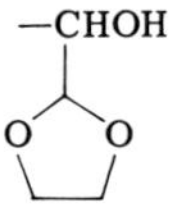 = —CHOH [dioxolane]	42	32
X = Tr, R = Bz	58	30, 32

(*Continued*)

TABLE II (*Continued*)

Photochemically Initiated, Radical-Addition Reactions to Unsaturated Carbohydrate Derivatives

Reactants	Products and Yields (%)	Ref.
[enone sugar structure] + RH	[adduct structure]	
X = Ac, Y = Et, R = —CH(OH)(1,3-dioxolan-2-yl)	30	29
X = Ac, Y = Et, R = —CH$_2$OH	60	29
X = Tr, Y = Et, R = —CH$_2$OH	61	27
X = Tr, Y = Me, R = —CH(OH)CH$_2$OH	71	30
X = Tr, Y = Me, R = —CH(OH)(1,3-dioxolan-2-yl)	46	30
[enone structure with OEt, CH$_3$] + EtOH	[adduct with HOCHCH$_3$ group] — 56	33
[enone sugar structure, X = Tr] + MeOH	[two products: 65 and 24]	28

TABLE II (Continued)

Photochemically Initiated, Radical-Addition Reactions to Unsaturated Carbohydrate Derivatives

Reactants	Products and Yields (%)	Ref.
[Structure: CH$_2$OX, AcO, pyranone with enone] + MeOH X = Tr X = Bz	[Structure: CH$_2$OX, AcO, pyranone with CH$_2$OH substituent] 60	34

[a] Minor product formed. [b] Five other products formed. [c] Acetone addition-product formed.

(12) D. Horton and W. N. Turner, *Carbohydr. Res.*, 1 (1966) 444–454.
(13) D. Horton and W. N. Turner, *Chem. Ind. (London)*, (1964) 76.
(14) J. S. Jewell and W. A. Szarek, *Tetrahedron Lett.*, (1969) 43–46.
(15) R. L. Whistler, C.-C. Wang, and S. Inokawa, *J. Org. Chem.*, 33 (1968) 2495–2497.
(16) A. Rosenthal and M. Ratcliffe, *Can. J. Chem.*, 54 (1976) 91–96.
(17) A. A. Othman, N. A. Al-Masudi, and U. S. Al-Timari, *J. Antibiot.*, 31 (1978) 1007–1012.
(18) K. Matsuura, S. Maeda, Y. Araki, and Y. Ishido, *Tetrahedron Lett.*, (1970) 2869–2872.
(19) K. Kumamoto, H. Yoshida, T. Ogata, and S. Inokawa, *Bull. Chem. Soc. Jpn.*, 42 (1969) 3245–3248.
(20) K. Matsuura, K. Nishiyama, K. Yamada, Y. Araki, and Y. Ishido, *Bull. Chem. Soc. Jpn.*, 46 (1973) 2538–2542.
(21) A. Rosenthal and K. Shudo, *J. Org. Chem.*, 37 (1972) 1608–1612.
(22) A. Rosenthal and M. Ratcliffe, *Carbohydr. Res.*, 54 (1977) 61–73.
(23) Y. Araki, K. Nishiyama, K. Senna, K. Matsuura, and Y. Ishido, *Carbohydr. Res.*, 64 (1978) 119–126.
(24) A. Rosenthal and M. Ratcliffe, *Carbohydr. Res.*, 39 (1975) 79–86.
(25) Y. Araki, K. Matsuura, Y. Ishido, and K. Kushida, *Chem. Lett.*, (1973) 383–386.
(26) K. Matsuura, Y. Araki, Y. Ishido, and S. Satoh, *Chem. Lett.*, (1972) 849–852.
(27) B. Fraser-Reid, N. L. Holder, D. R. Hicks, and D. L. Walker, *Can. J. Chem.*, 55 (1977) 3978–3985.
(28) B. Fraser-Reid, N. L. Holder, and M. B. Yunker, *J. Chem. Soc. Chem. Commun.*, (1972) 1286–1287.
(29) B. Fraser-Reid, D. R. Hicks, D. L. Walker, D. E. Iley, M. B. Yunker, S. Y.-K. Tam, and R. C. Anderson, *Tetrahedron Lett.*, (1975) 297–300.
(30) B. Fraser-Reid, R. C. Anderson, D. R. Hicks, and D. L. Walker, *Can J. Chem.*, 55 (1977) 3986–3995.
(31) D. L. Walker and B. Fraser-Reid, *J. Am. Chem. Soc.*, 97 (1975) 6251–6253.
(32) D. R. Hicks, R. C. Anderson, and B. Fraser-Reid, *Synth. Commun.*, 6 (1976) 417–421.
(33) H. Paulsen and W. Koebernick, *Carbohydr. Res.*, 56 (1977) 53–66.
(34) D. L. Walker, B. Fraser-Reid, and J. K. Saunders, *J. Chem. Soc. Chem. Commun.*, (1974) 319–320.

TABLE III

Addition Reactions between Solvent Molecules and Electronically Excited, Unsaturated Carbohydrate Derivatives

Reactants	Products and Yields (%)			Ref.
X = Ac, R = —CMe$_2$(OH)	74	8 (**9**)		3
X = Ac, R = —CMe$_2$(OH)	86			4
X = Ac, R = —C(=O)NH$_2$	22	30	22	35
X = Ac, R = (1,3-dioxolan-2-yl)	37		37	20
X = Ac, R = (1,3-dioxolan-2-yl)	29	10		36
(methylenated furanose + CH$_3$CHCH$_3$(OH) + MeCMe(=O))	6	19	5	9

TABLE III (*Continued*)

Addition Reactions between Solvent Molecules and Electronically Excited, Unsaturated Carbohydrate Derivatives

Reactants	Products and Yields (%)	Ref.
[structure] + ROH R = Me R = H	[structure] 30 (unstable)	37

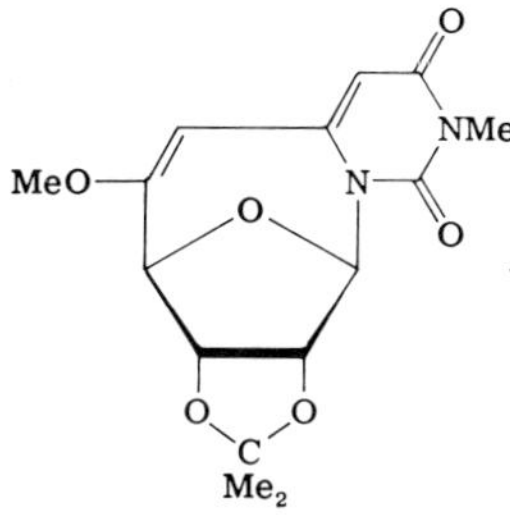

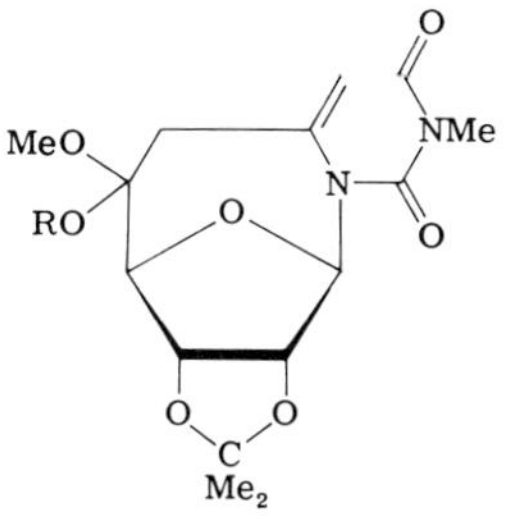

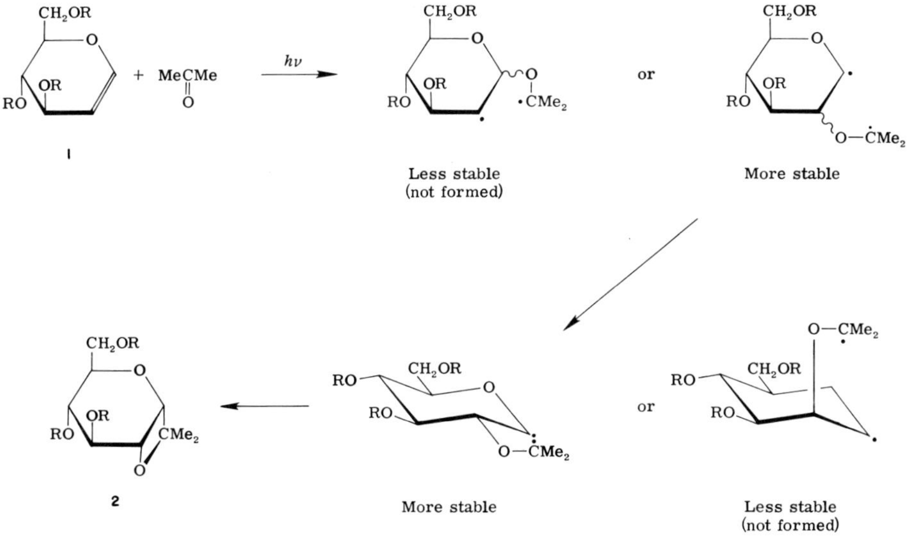

Scheme 1.—Proposed Mechanism for Cycloaddition between 3,4,6-Tri-*O*-acetyl-D-glucal (**1**) and Acetone.

(35) A. Rosenthal and A. Zanlungo, *Can. J. Chem.*, 50 (1972) 1192–1198.
(36) Y. Araki, K. Nishiyama, K. Matsuura, and Y. Ishido, *Carbohydr. Res.*, 63 (1978) 288–292.
(37) B. A. Otter and E. A. Falco, *Tetrahedron Lett.*, (1978) 4383–4386.

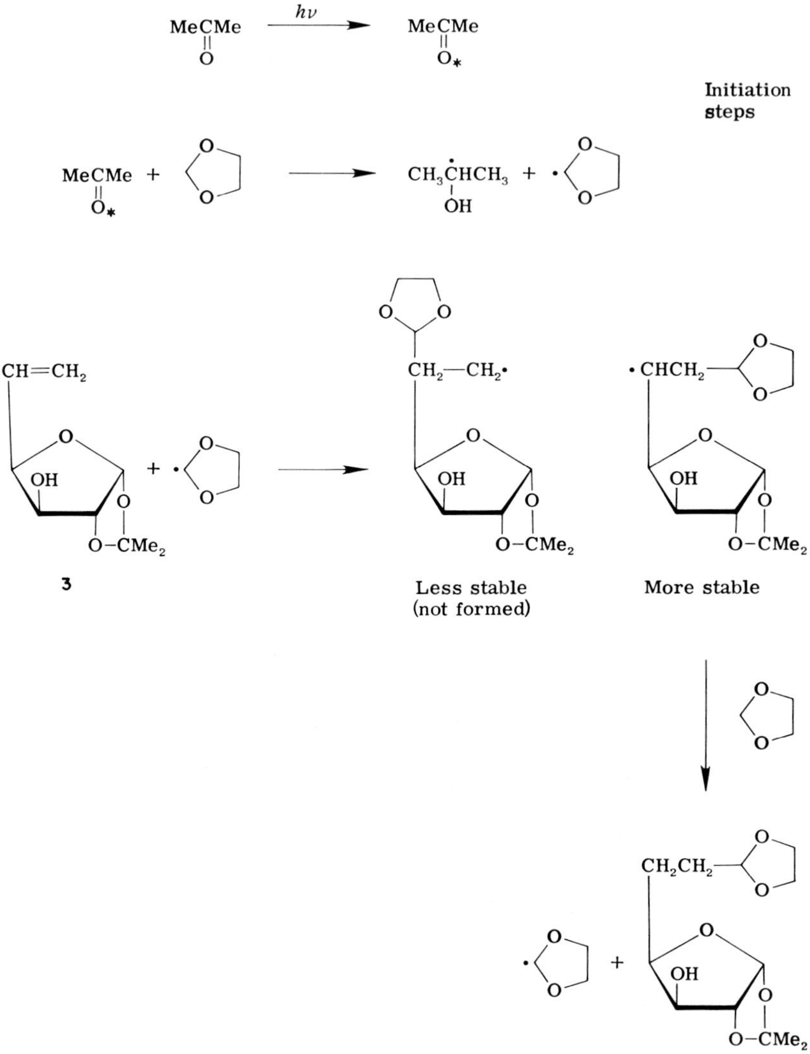

Scheme 2.—Proposed Mechanism for Photochemically Initiated, Radical Addition of 1,3-Dioxolane to 5,6-Dideoxy-1,2-O-isopropylidene-α-D-*xylo*-hex-5-enofuranose (3).

tion by excited acetone (see Scheme 2). Regioselectivity in these reactions (see Table II) is determined by addition (anti-Markovnikov) to the double bond in order to maximize the stability of the new radical produced. An example of this selectivity is the photochemical addition of 1,3-dioxolane to 5,6-dideoxy-1,2-O-isopropylidene-α-D-*xylo*-

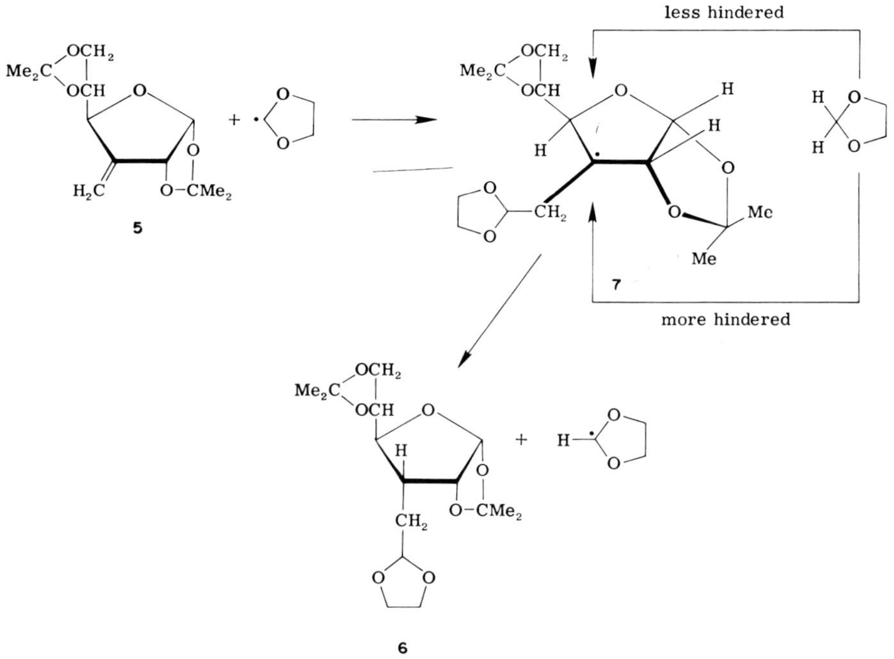

Scheme 3.—Proposed Mechanism for Photochemically Initiated, Radical Addition of 1,3-Dioxolane to 3-Deoxy-1,2:5,6-di-O-isopropylidene-3-C-methylene-α-D-*ribo*-hexofuranose (**5**).

hex-5-enofuranose (**3**) to give exclusively[14] 2-(5,6-dideoxy-1,2-O-isopropylidene-α-D-*xylo*-hexofuranose-6-yl)-1,3-dioxolane (**4**) (see Scheme 2). The stereochemistry in these reactions can usually be predicted by assuming that the radical resulting from initial reaction (for example, radical **7** in Scheme 3) will be approached from its less-hindered side.

In contrast to the radical addition just described, the less-common of the two reaction pathways mentioned in the preceding paragraph requires that the unsaturated carbohydrate be electronically excited (see Scheme 4). A significant characteristic of alkenes experiencing this type of reaction (see Table III) is that the double bond is conjugated with an oxygen atom. The triplet energy of such a molecule is sufficiently lowered by conjugation to allow transfer of energy from acetone to the carbohydrate (see Scheme 4). Once transfer of energy has occurred, the excited carbohydrate can abstract a hydrogen atom from the solvent in the way shown in Scheme 4 for the reaction of 3,4,6-tri-O-acetyl-D-glucal (**1**) with 2-propanol.

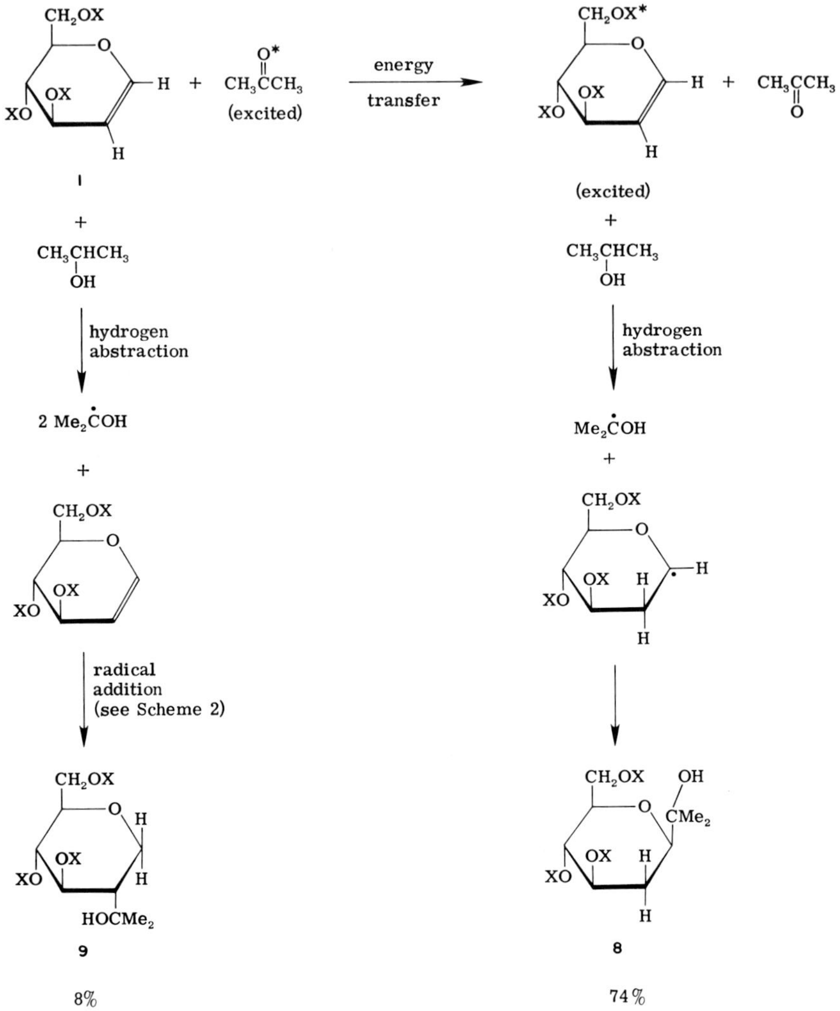

X = Ac

Scheme 4.—Proposed Mechanism for Reaction between 2-Propanol and Excited 3,4,6-Tri-*O*-acetyl-D-glucal (**1**).

The relative proportions of unsaturated carbohydrate, sensitizer (usually acetone), and solvent may have a decided effect upon a photochemical addition reaction, as at least three competing processes (cycloaddition, radical addition, and energy transfer) are possible. The irradiation of **1** in the presence of 2-propanol and acetone provides an illustration (see Scheme 4). When a small proportion of sensitizer

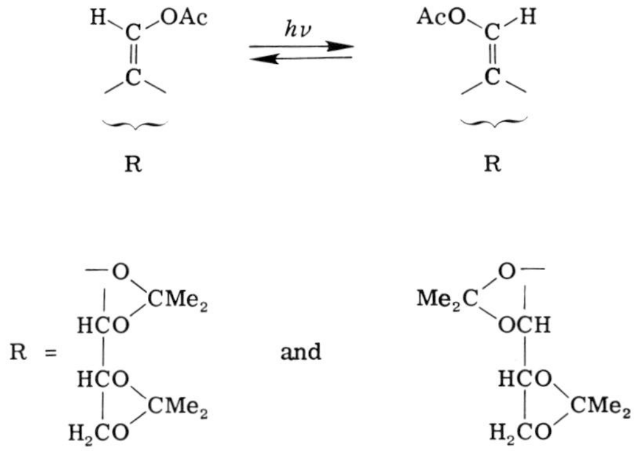

Scheme 5.—Photochemical, E–Z Isomerization of Groups Attached to a Double Bond.

(acetone) is present, energy transfer and radical addition occur, although energy transfer dominates the reaction. (A detailed study of the competition between these two processes would be interesting.) If the concentration of acetone is raised sufficiently, the third process, cycloaddition (see Scheme 1), becomes the major reaction-pathway.[4]

Several other types of photochemical reactions involving unsaturated carbohydrates have been reported. One of these is[38] photochemical, E–Z isomerization of the groups attached to a double bond (see Scheme 5). A second is the internal cycloaddition between two double bonds connected by a carbohydrate chain.[39–41] Although the carbohydrate portion of the molecule is not directly involved in this cycloaddition, its presence induces optical activity in the cyclobutane derivatives produced photochemically. Finally, a group of acid-catalyzed addition-reactions has been observed for which the catalyst appears to arise from photochemical decomposition of a noncarbohydrate reactant.[42–44]

(38) A. Ducruix, C. Pascard-Billy, S. J. Eitelman, and D. Horton, *J. Org. Chem.*, 41 (1976) 2652–2653.
(39) B. S. Green, Y. Rabinsohn, and M. Rejtö, *J. Chem. Soc. Chem. Commun.*, (1975) 313–314.
(40) B. S. Green, Y. Rabinsohn, and M. Rejtö, *Carbohydr. Res.*, 45 (1975) 115–126.
(41) B. S. Green, A. T. Hagler, Y. Rabinsohn, and M. Rejtö, *Isr. J. Chem.*, 15 (1976/77) 124–130.
(42) K. Matsuura, Y. Araki, Y. Ishido, and M. Kainosho, *Chem. Lett.*, (1972) 853–856.
(43) K. Matsuura, K. Senna, Y. Araki, and Y. Ishido, *Bull. Chem. Soc. Jpn.*, 47 (1974) 1197–1200.
(44) J. Csaszar and V. Bruckner, *Ann. Univ. Sci. Budapest. Rolando Eotvos Nominatae, Sect. Chim.*, (1973) 87.

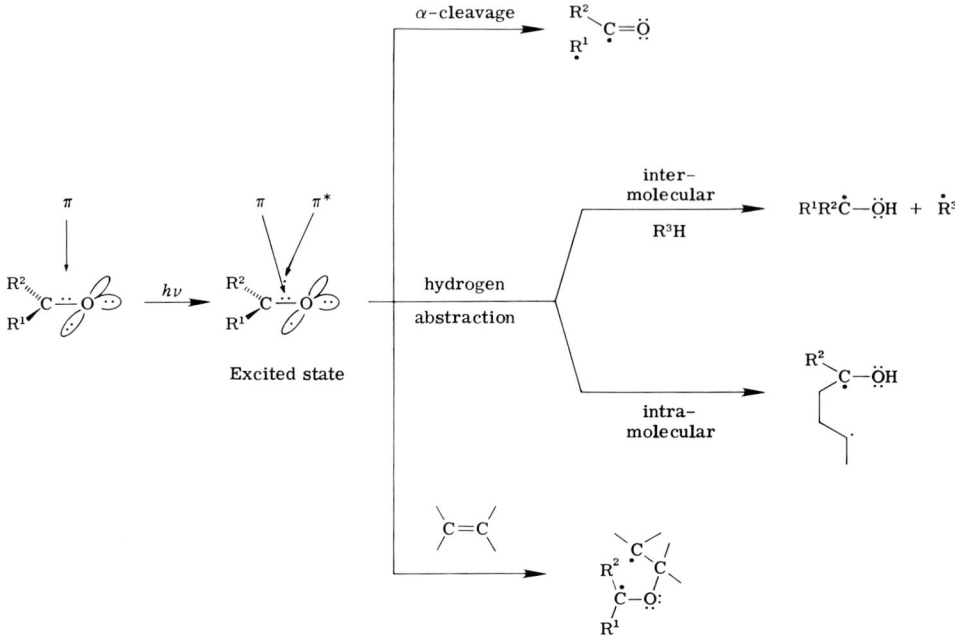

π = Electron in bonding π-orbital
π* = Electron in antibonding π-orbital

Scheme 6.—Potential Reactions of a Carbonyl, n → π*, Excited State.

III. Carbonyl Compounds

Carbonyl compounds have, thus far, more frequently been the subject of photochemical study than any other group of organic molecules. The reactions of these compounds may be understood by assuming that absorption of light leads to an n → π* excited state, that is, an excited state in which an electron has been promoted from the highest, nonbonding orbital (on oxygen) to the lowest, antibonding (π*) orbital (see Scheme 6). The reactions of such an excited system are similar to those of an alkoxyl radical and, thus, may include α-cleavage, hydrogen abstraction (intra- and inter-molecular), and addition to a π-system (see Scheme 6).

1. Aldehydes and Ketones

A photochemical reaction initiated by α-cleavage in a carbonyl compound is called a Type I process.[45] Photolysis of *tert*-butyl-3,4-O-iso-

(45) C. H. Bamford and R. G. W. Norrish, *J. Chem. Soc.*, (1938) 1521–1525.

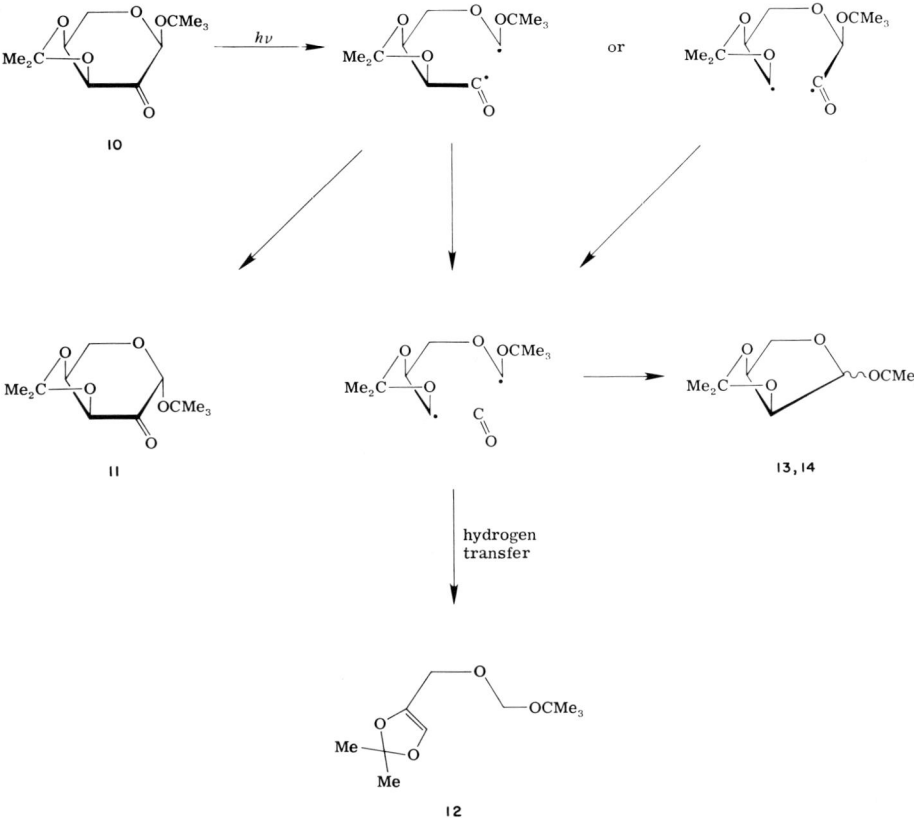

Scheme 7.—Proposed Mechanism for Type I Reaction of *tert*-Butyl 3,4-*O*-Isopropylidene-α-L-*erythro*-pentopyranosid-2-ulose (**10**).

propylidene-α-L-*erythro*-pentopyranosid-2-ulose (**10**) (see Scheme 7) produces three such reactions (decarbonylation, disproportionation, and stereoisomerization).[46,47] Irradiation of 5-deoxy-1,2-*O*-isopropylidene-β-L-*threo*-pentofuranos-3-ulose (**15**) (see Scheme 8) provides an example of a fourth (carbene formation).[48] Type I reactions are more common for carbohydrate systems than for most other types of molecules, because the likelihood of occurrence of a Type I process is related to the stabilization, during formation, of the radicals produced

(46) P. M. Collins, P. Gupta, and R. Iyer, *Chem. Commun.*, (1970) 1261–1262.
(47) P. M. Collins, R. Iyer, and A. S. Travis, *J. Chem. Res. (S)*, (1978) 446–447 and *J. Chem. Res. (M)*, (1978) 5344–5360.
(48) P. M. Collins, N. N. Oparaeche, and B. R. Whitton, *J. Chem. Soc. Chem. Commun.*, (1974) 292–293.

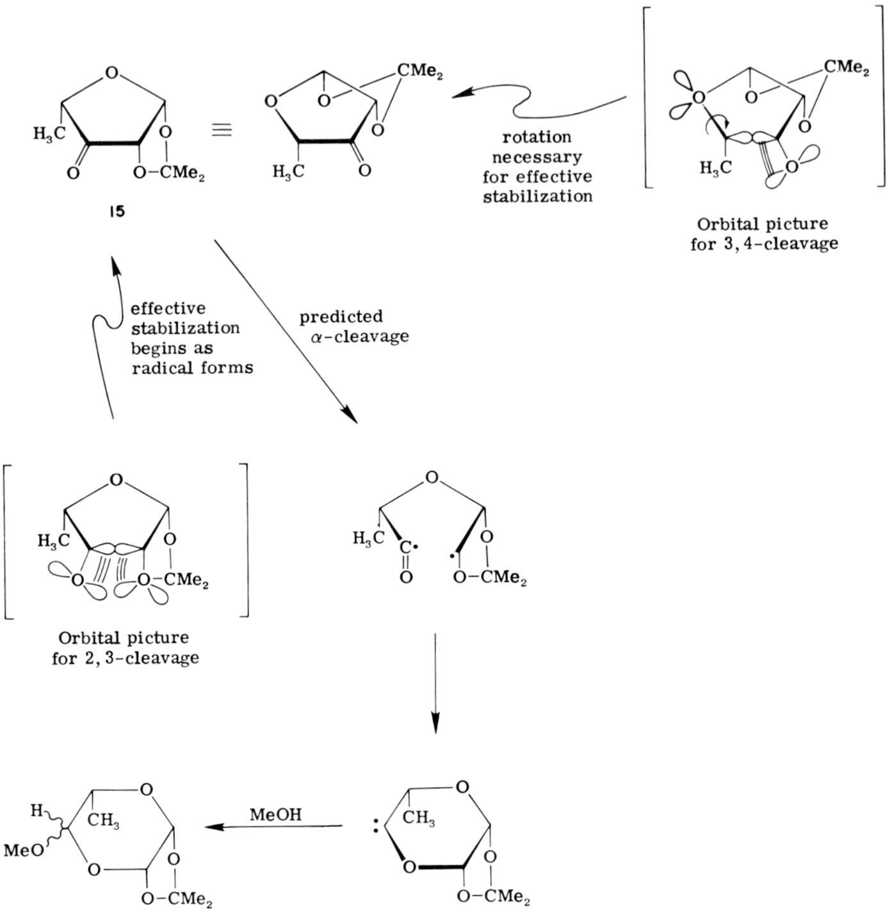

Scheme 8.—Proposed Mechanism for Carbene Formation from Photolysis of 5-Deoxy-1,2-O-isopropylidene-β-L-*erythro*-pentofuranos-3-ulose (**15**). Rationalization for the Direction of α-Cleavage.

by α-cleavage. In carbohydrate systems, where a radical center on carbon will usually have an adjacent, stabilizing, oxygen atom, α-cleavage is particularly favored.[49,50] The Type I reactions of carbohydrates reported are listed in Table IV.

Although an excited carbonyl can fragment at either of the two carbon to α-carbon bonds thereof, the favored direction of fragmentation

(49) K.-G. Seifert, *Tetrahedron Lett.*, (1974) 4513–4516.
(50) S. Steenken, W. Jaenicke-Zauner, and D. Schulte-Frohlinde, *Photochem. Photobiol.*, 21 (1975) 21–26.

TABLE IV

Type I Reactions of Excited Carbohydrate Derivatives

Reactants	Products and Yields (%)				Ref.
[structure]	[structure] 54	+	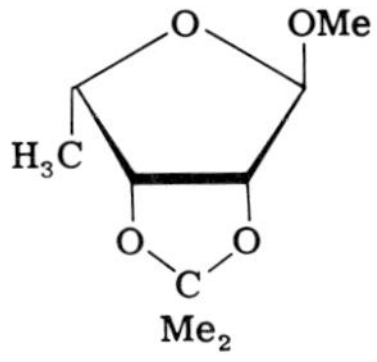 [structure] 6		51-53
[structure]	[structure] 77	+	[structure] 8		53
[structure]	[structure] 17				53
[structure] **10**	[structure] **13** 13	+	[structure] **14** 11	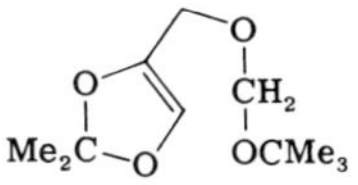 + [structure] **12** 40	46,47
[structure]	[structure] **13** 13	+	[structure] **14** 11	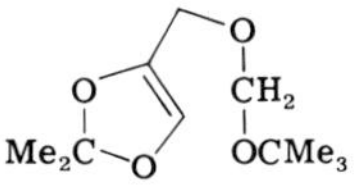 + [structure] **12** 40	46,47

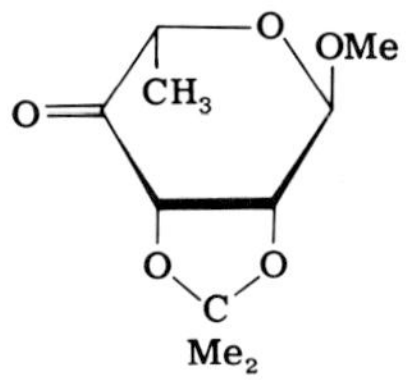

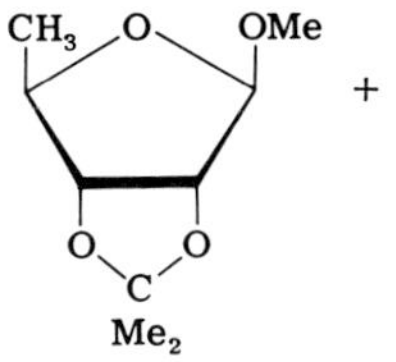

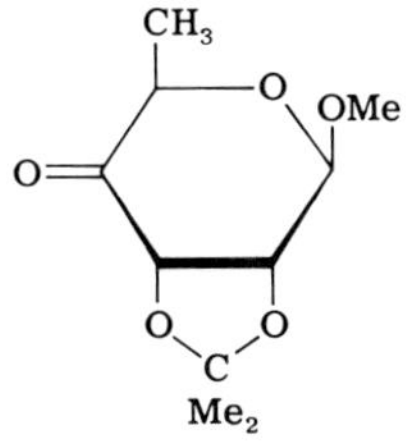

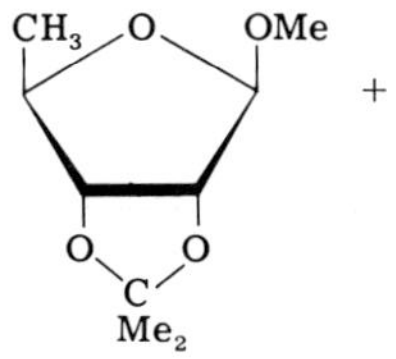

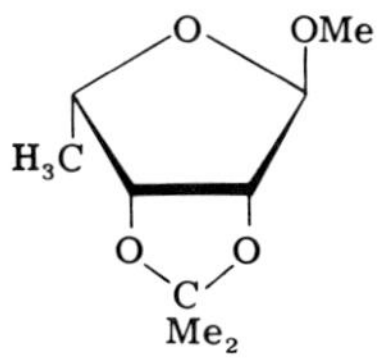

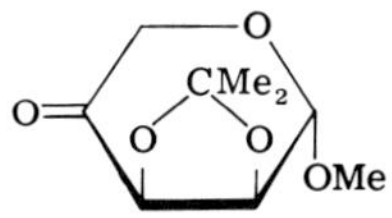

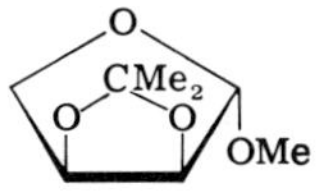

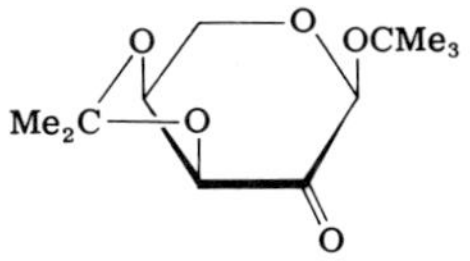

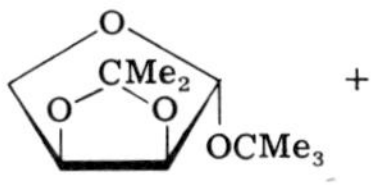

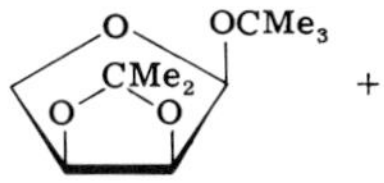

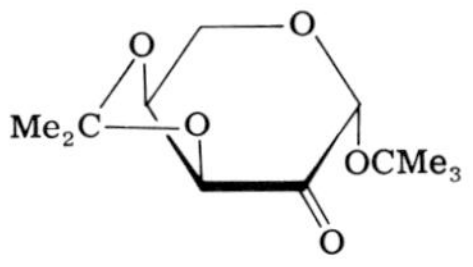

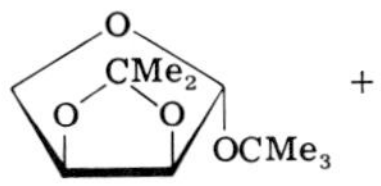

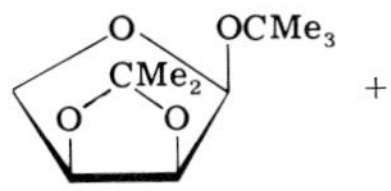

(Continued)

TABLE IV (*Continued*)

Type I Reactions of Excited Carbohydrate Derivatives

Reactants	Products and Yields (%)	Ref.
[structure]	[structure] 60	47
[structure]	[structure] + [structure] 60	54
[structure]	[structure] 10	55
[structure]	[structure] 10	55
[structure] + EtOH	[structure] 65	48

TABLE IV (Continued)
Type I Reactions of Excited Carbohydrate Derivatives

Reactants	Products and Yields (%)	Ref.
(furanone with H₃C, O, O–CMe₂ substituents)	(pyranone with MeO, CH₃, O–CMe₂ substituents)	48
CHO | HCOAc | AcOCH | HCOAc | HCOAc | CH₂OAc	CH₂OAc | AcOCH | HCOAc | HCOAc | CH₂OAc 16	56

may often be predicted by examining the structure of the reacting molecule. The critical factor in determining which α-cleavage will result is stabilization of the developing radicals (or diradicals, in cyclic ketones) during formation. This stabilization is not only dependent upon the presence of radical-stabilizing groups on the α-carbon atoms but also on the positioning of these groups. The photolysis of **15** (see Scheme 8) provides an informative illustration, because, as the C-2–C-3 bond in excited **15** begins to break, a nonbonding orbital on the oxygen atom on C-2 is properly aligned to stabilize the developing radical-center on C-2. In contrast, the radical center on C-4 that is being formed by fragmentation of the C-3–C-4 bond will not experience stabilization from a nonbonding orbital on the oxygen atom on C-4 until the C-3–C-4 bond is broken and rotation can occur. As a result, cleavage of the C-2–C-3 bond is favored.[53]

Intramolecular hydrogen-abstraction to produce a 1,4-diradical is the major, photochemical pathway for molecules in which the oxygen

---

(51) P. M. Collins, *Chem. Commun.*, (1968) 403–405.
(52) P. M. Collins and P. Gupta, *J. Chem. Soc., C*, (1971) 1965–1968.
(53) P. M. Collins, *J. Chem. Soc., C*, (1971) 1960–1965.
(54) P. M. Collins and P. Gupta, *Chem. Commun.*, (1969) 1288–1289.
(55) K. Heyns, R. Neste, and M. Paal, *Tetrahedron Lett.*, (1978) 4011–4014.
(56) R. L. Whistler and K.-S. Ong, *J. Org. Chem.*, 36 (1971) 2575–2576.

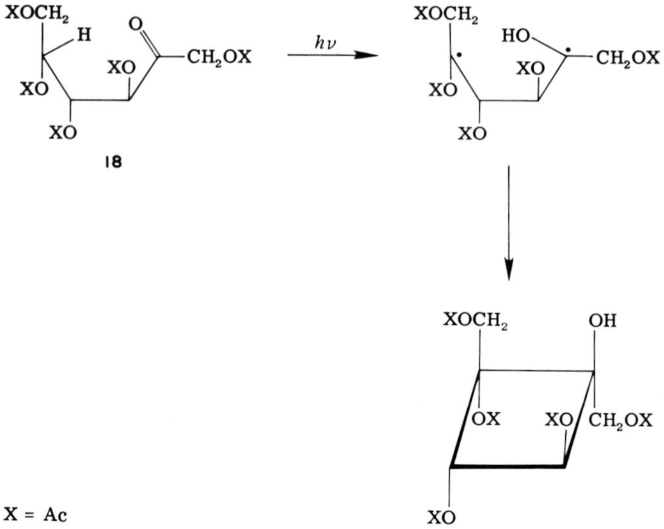

18

19

X = Ac

Scheme 9.—Proposed Mechanism for Formation of Cyclobutanol from the Photolysis of 1,3,4,5,6-Penta-*O*-acetyl-*keto*-L-sorbose (**18**).

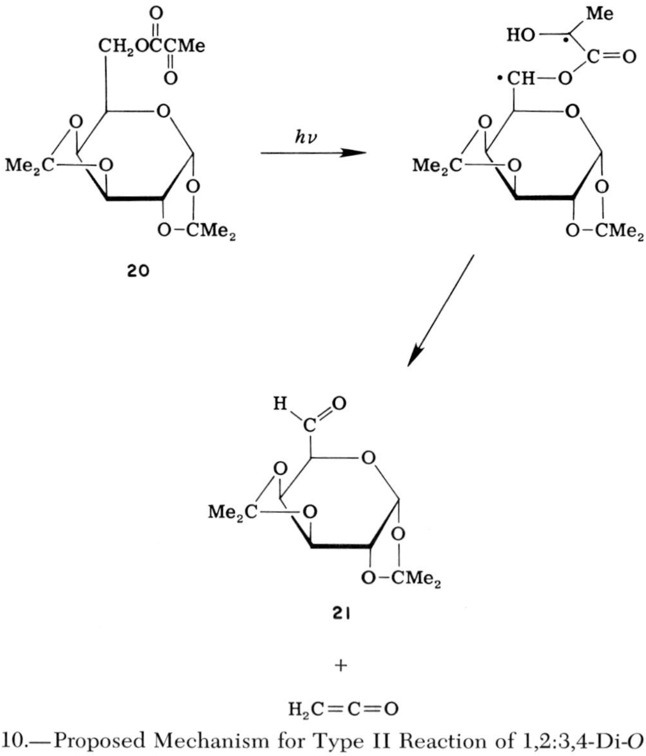

20

21

Scheme 10.—Proposed Mechanism for Type II Reaction of 1,2:3,4-Di-*O*-isopropylidene-6-*O*-pyruvoyl-α-D-galactopyranose (**20**).

atom of an excited carbonyl can come within bonding distance of a hydrogen atom attached to a γ-related carbon atom. (Such a process is possible for all of the compounds in Tables V and VI.) The reactions of the 1,4-diradical produced by hydrogen abstraction are called Type II processes, and may include closure to form a cyclobutanol (see Scheme 9), fragmentation of the 2,3-bond to produce a new carbonyl compound (see Scheme 10), and reversal of the hydrogen-abstraction process to regenerate the starting material (or an epimer). Photolysis of α-ketoesters provides an example of the synthetic potential of the Type II reaction, as synthesis and irradiation of these compounds accomplish oxidation of alcohols to carbonyl compounds under quite mild conditions (see Table VI).

For molecules in which intramolecular hydrogen-abstraction (Type II reaction) is structurally prevented, intermolecular abstraction from a hydrogen-donating solvent may take place (see Scheme 6). This type of reaction usually results in reduction of the carbonyl group (see Table VII).

Cycloaddition reactions between alkenes and noncarbohydrate, carbonyl compounds have been described in discussing the reactions of alkenes (see Table I and Scheme 1). The depiction of the excited carbonyl given in Scheme 6 is useful in understanding the regiochemistry of the cycloaddition process, as it suggests that the electron-deficient oxygen atom in the excited carbonyl will react with the alkene to produce the (more-stable) 1,4-diradical. Table VIII lists cycloaddition reactions in which the excited carbonyl is part of a carbohydrate.

2. Esters

When considered as a part of the photochemistry of carbonyl compounds, irradiations of esters constitute a minor component. The more frequent photolyses of other carbonyl compounds, in particular ketones, is not surprising, as, even though parallels exist between ester and ketone photochemistry (for example, both experience α-cleavage and hydrogen abstraction-reactions), esters require radiation of higher energy for reaction, and typically produce more-complex mixtures of products. In addition to their similarity to other carbonyl compounds in their reactivity, esters also experience reactions that are uniquely their own.

When the esters listed in Table IX are irradiated, typical carbonyl reactions result; that is, each of these compounds (**22–27**) experiences both α-cleavage and hydrogen-abstraction reactions. For the esters of 1,2:3,4-di-O-isopropylidene-α-D-galactopyranose (**22–26**), the hydrogen-abstraction reaction is internal, and leads[69] to a Type II reaction

Table V
Type II Reactions of Excited Carbohydrate Derivatives[a]

Reactants	Products and Yields (%)	Ref.
(structure) X = Ac	12	57
(structure) 18	26 (19)	57
(structure)	(structure)	58,59
(structure) R = H	(No yield reported)	58,59
R = Me	50	58
(structure) R = Ph	3 / 46 / (No yields reported)	58
R = Me		58

TABLE V (Continued)

Type II Reactions of Excited Carbohydrate Derivatives[a]

Reactants	Products and Yields (%)	Ref.
[structure] R = Ph	65 + 4	58
R = Me	(No yields reported)	58

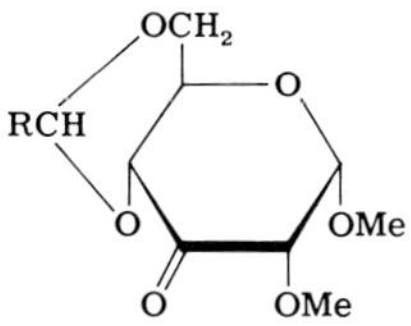

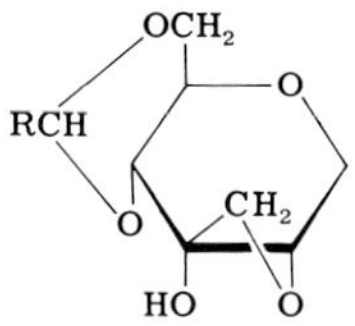

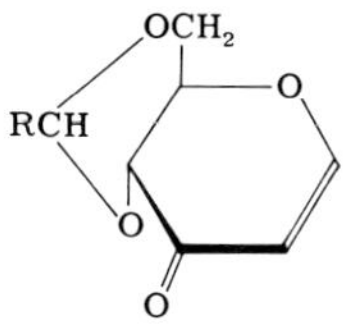

[a] Additional examples of Type II reactions are probable, but product structures not established.[60]

TABLE VI. **Type II Reactions of α-Ketoesters of Carbohydrates**

Reactants	Products and Yields (%)	References
RCH_2OCCMe with $\overset{\parallel\parallel}{OO}$ → $RCH\overset{\parallel}{O}$		
R = [pyranose with Me₂C and O-CMe₂ groups] 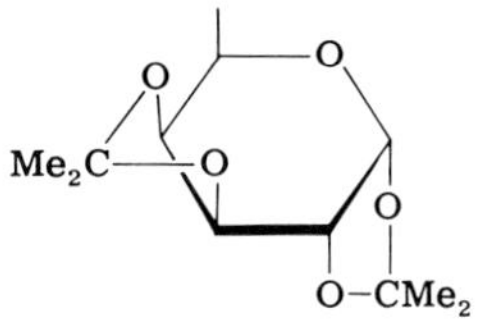	70	61,62
R 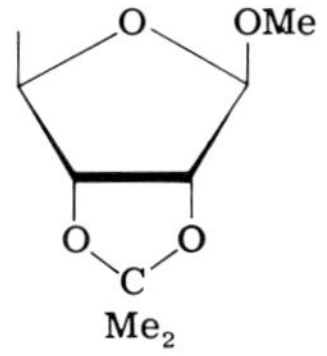 = [furanose with OMe and CMe₂ groups]	65	61,62
R 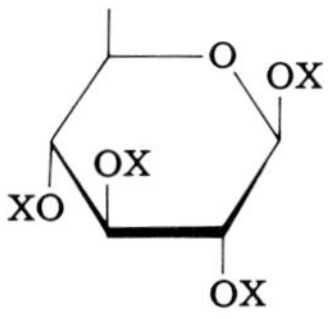 = [pyranose with OX groups], X = Ac	85	62

(Continued)

TABLE VI (Continued)
Type II Reactions of α-Ketoesters of Carbohydrates

Reactants	Products and Yields (%)	References
R = (sugar with Me₂C–O acetals, CMe₂ group)	62	61,62
–CH–OCCMe with OO, R	>C=O	
R = (furanose with OCH₂, Me₂C, OCH, O–CMe₂)	74	61,62
R = (pyranose with O–CMe₂, CH₂O, Me₂C–O)	57	61,62
R = (pyranose with XOCH₂, OX, XO, OX); X = Ac	100	62
R = (pyranose with XOCH₂, OX, XO, OX); X = Ac	70	62

TABLE VI (*Continued*)

Type II Reactions of α-Ketoesters of Carbohydrates

Reactants	Products and Yields (%)	References
R 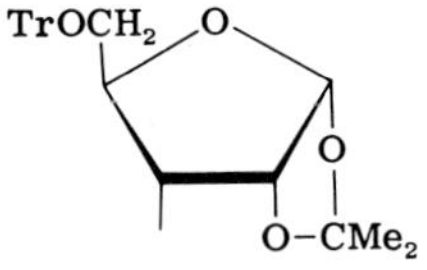 = [TrOCH$_2$ furanose with O–CMe$_2$]	38	63
R 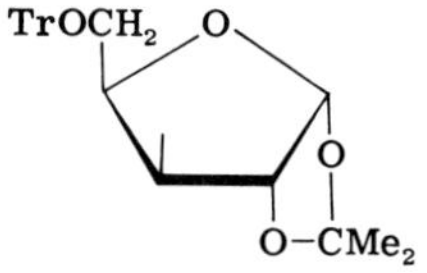 = [TrOCH$_2$ furanose with O–CMe$_2$]	59	63
R 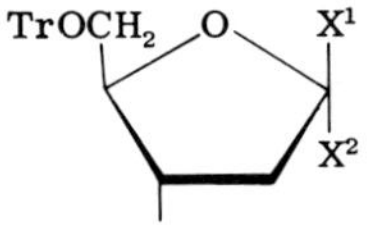 = [TrOCH$_2$ furanose with X^1, X^2]		
X^1 = H, X^2 = Me	80	63
X^1 = Me, X^2 = H	79	63
R 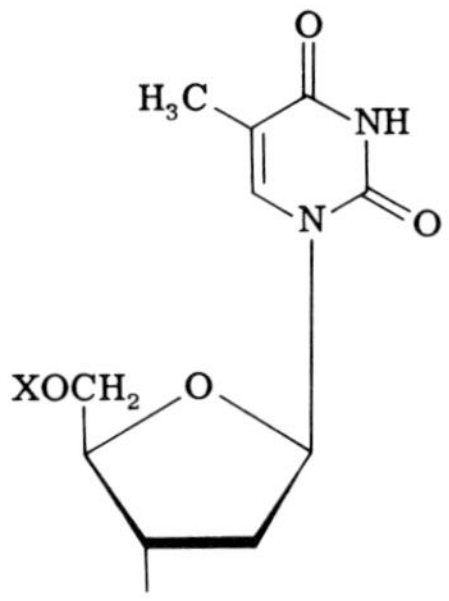 = [thymidine derivative with XOCH$_2$]		
X = Tr	61	63,64
X = Bz	68	63,64

(*Continued*)

TABLE VI (Continued)
Type II Reactions of α-Ketoesters of Carbohydrates

Reactants	Products and Yields (%)	References
R = (thymidine derivative with XOCH₂, X = Tr)	57	63,64
X = Bz	57	63,64

TABLE VII
Reactions of Excited Carbonyl Groups in Carbohydrate Derivatives with Solvent Molecules

Reactants	Products and Yields (%)			Ref.
(MeCH, OCH₂, O–CMe₂ pyranose ketone)	30	20	22	65
(PhCH, OCH₂, O–CHEt pyranose ketone)				65
(PhCH, OCH₂, OMe, OSO₂C₆H₄Me ketone)	(PhCH, OCH₂, OMe product)			66

TABLE VIII

Cycloaddition Reactions between Excited Carbonyl Groups in Carbohydrate Derivatives and Unsaturated, Non-Carbohydrates

Reactants	Products and Yields (%)	References
R = —CH$_2$OAc	57	67
R = —COEt (C=O)	23	67
R = (diacetone sugar fragment with OCH, HCO, CMe$_2$ groups)	19	68
R = (furanose with OX, O–CMe$_2$); X = CH$_2$Ph	14 10	68
X = Me	19	68
(Me$_2$C/OCH$_2$/OCH furanose ketone) + furan	(adduct shown)	68

Table IX
Photochemical Reactions of Esterified Carbohydrates[a]

Reactants	Products and Yields (%)		References
(structure with CH₂OCR, Me₂C, O-CMe₂)	**28** (CH₃ structure)	**29** (H₂C= structure)	
R = Ph (**22**)	3	12	69
R = CH₂Ph (**23**)	26	50	69
R = CH₂CH₂Ph (**24**)	3	8	69
R = —CHPh₂ (**25**)	10	29	69
R = —C(Me)Ph₂ (**26**)	5	27	69
27 (COMe pyranose)	**30**: R¹ = R² = H **31**: R¹ = H, R² = —CH₂OH **32**: R¹ = —CH₂OH, R² = H (No yields reported)	**33** + **34**	70

[a] Additional examples of this type of reaction are probable, but product structures not established.[71–73]

Table X
Photochemical Reactions of Esterified Carbohydrates with Hexamethylphosphoric Triamide

Reactants ROAc	Products and Yields (%)		Ref.
	RH	ROH	
R = (sugar structure with Me_2C, $-CH_2$, $O-CMe_2$)	85		74
	65		75
R = (furanose with Me_2C, OCH_2, OCH, $O-CMe_2$)	65	30	74
	65		75
R = (pyranose with OMe, CH_3, $O-C-O$ with Me_2)	81		74
	70		75
R = (furanose with OMe, H_3C, $O-C-O$ with Me_2)	78		74
R = (pyranose with OMe, CH_3, Me_2C-O)	60		74

(Continued)

TABLE X (Continued)
Photochemical Reactions of Esterified Carbohydrates with Hexamethylphosphoric Triamide

Reactants ROAc	Products and Yields (%)		Ref.
	RH	ROH	
R = (structure: 1,2:3,5-di-O-isopropylidene furanose derivative)	76		74
R = (structure: methyl 6-deoxy-3,4-O-isopropylidene pyranoside)	(No yields reported)		75
R = (structure: methyl 4,6-O-isopropylidene pyranoside with OMe)	(No yields reported)		75
R = (structure: di-O-isopropylidene pyranose derivative)	55		75
R = (structure: di-O-isopropylidene pyranose derivative)	70		75
(structure with OX, X = Ac)	(structure with H) 51	(structure with CH$_3$) 9	75

TABLE X (Continued)

Photochemical Reactions of Esterified Carbohydrates with Hexamethylphosphoric Triamide

Reactants ROAc	Products and Yields (%)		Ref.
	RH	ROH	
(structure) X = Ac	50		75
(structure) X = Ac	10	20	75
X = —CCMe₃ (‖O)	10	20	75
(structure) X = —CCMe₃ (‖O)	75		75

(57) R. L. Whistler and L. W. Doner, *J. Org. Chem.*, 38 (1973) 2900–2904.
(58) P. M. Collins, P. Gupta, and R. Iyer, *J. Chem. Soc. Perkin Trans. 1*, (1972) 1670–1677.
(59) P. M. Collins and P. Gupta, *Chem. Commun.*, (1969) 90.
(60) I. Kitagawa, K. S. Im, and Y. Fujimoto, *Chem. Pharm. Bull.*, 25 (1977) 800–808.
(61) R. W. Binkley, *Carbohydr. Res.*, 48 (1976) c1–c4.
(62) R. W. Binkley, *J. Org. Chem.*, 42 (1977) 1216–1221.
(63) R. W. Binkley, D. G. Hehemann, and W. W. Binkley, *J. Org. Chem.*, 43 (1978) 2573–2576.
(64) R. W. Binkley, D. G. Hehemann, and W. W. Binkley, *Carbohydr. Res.*, 58 (1977) c10–c12.

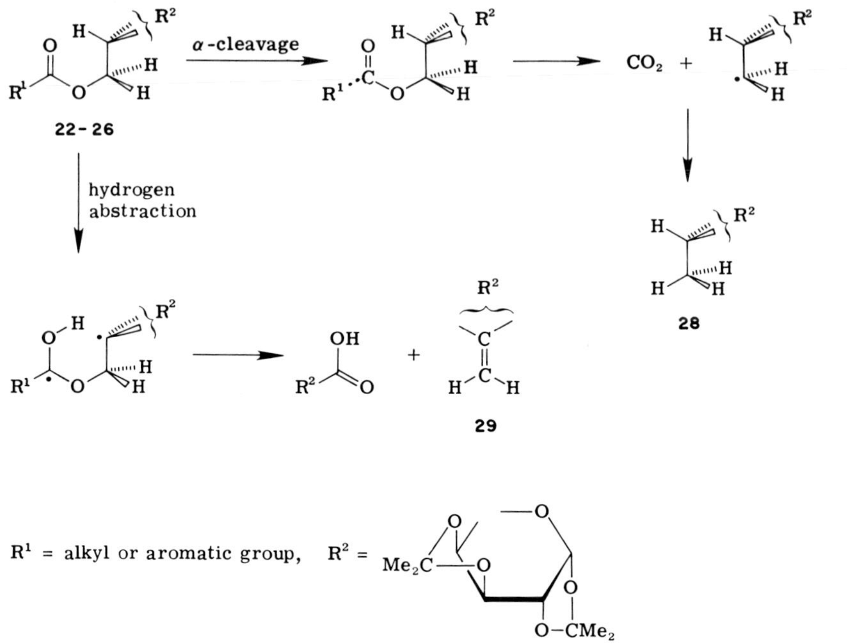

Scheme 11.—Proposed Mechanisms for Type I and Type II Reactions of Esters **22–26**.

(Scheme 11). For methyl (methyl D-glucopyranosid)uronate (**27a,b**), α-cleavage is accompanied by a rearrangement[70] initiated by hydrogen abstraction from the solvent (Scheme 12).

A reaction unknown for other carbonyl compounds occurs when es-

---

(65) P. M. Collins, V. R. N. Munasinghe, and N. N. Oparaeche, *J. Chem. Soc. Perkin Trans. 1*, (1977) 2423–2428.
(66) W. A. Szarek and A. Dmytraczenko, *Synthesis*, (1974) 579–580.
(67) Y. Araki, J.-I. Nagasawa, and Y. Ishido, *Carbohydr. Res.*, 58 (1977) c4–c6.
(68) J. M. J. Tronchet and B. Baehler, *J. Carbohydr. Nucleos. Nucleot.*, 1 (1974) 449–459.
(69) R. W. Binkley and J. L. Meinzer, *J. Carbohydr. Nucleos. Nucleot.*, 2 (1975) 465–469.
(70) A. G. W. Bradbury and C. von Sonntag, *Carbohydr. Res.*, 60 (1978) 183–186.
(71) I. Kitagawa, M. Yoshikawa, and I. Yosioka, *Tetrahedron Lett.*, (1973) 3997–3998.
(72) I. Kitagawa, M. Yoshikawa, Y. Imakura, and I. Yosioka, *Chem. Pharm. Bull.*, 22 (1974) 1339–1347.
(73) I. Kitagawa, M. Yoshikawa, Y. Imakura, and I. Yosioka, *Chem. Ind. (London)*, (1973) 276–277.
(74) J.-P. Pete and C. Portella, *Synthesis*, (1977) 774–775.
(75) P. M. Collins and V. R. Z. Munasinghe, *J. Chem. Soc. Chem. Commun.*, (1977) 927–928.

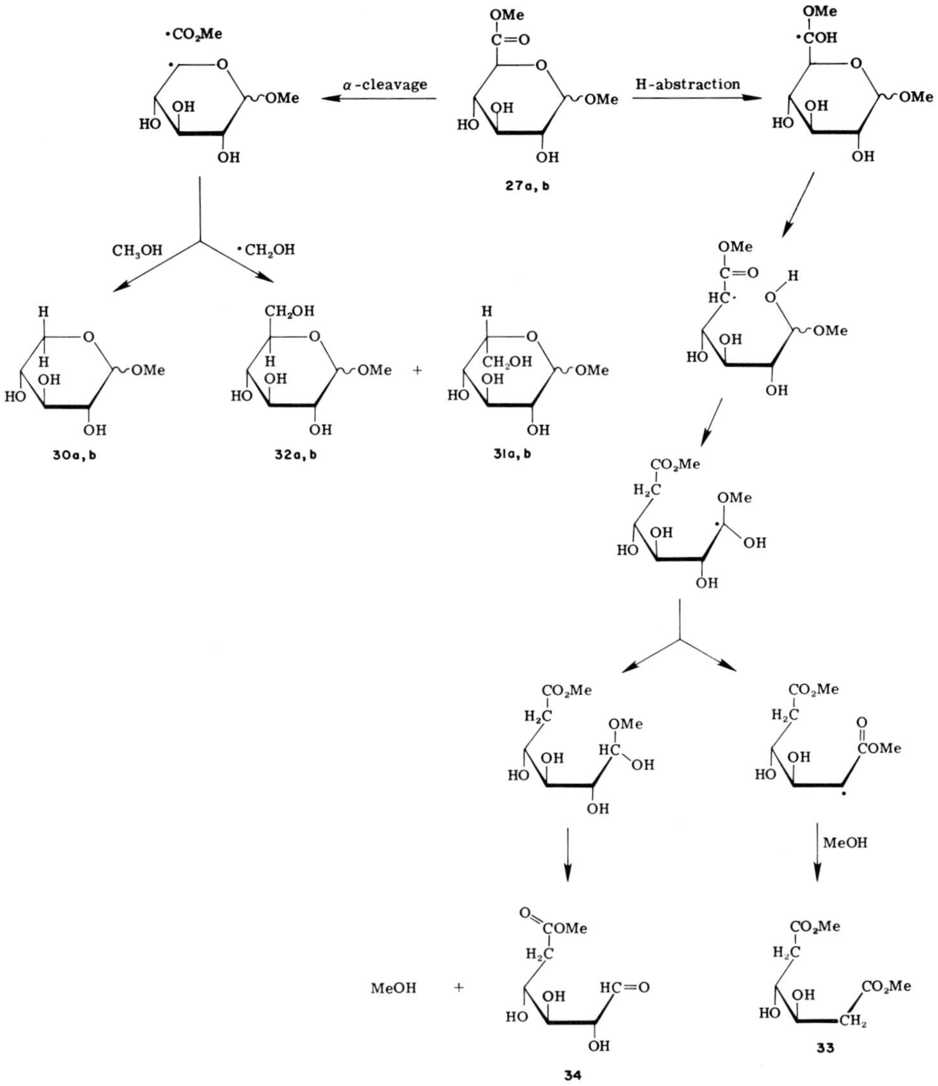

Scheme 12.—Proposed Mechanism for Photochemical Reactions of Methyl (Methyl D-Glucopyranosid)uronates (**27a,b**).

terified carbohydrates are photolyzed in hexamethylphosphoric triamide. Such irradiations result in replacement of the ester group by a hydrogen atom and, thus, are useful[74,75] for synthesizing deoxy sugars (see Table X). Although the yields of deoxy sugars are normally good, formation of an alcohol competes in some instances. Depicted in

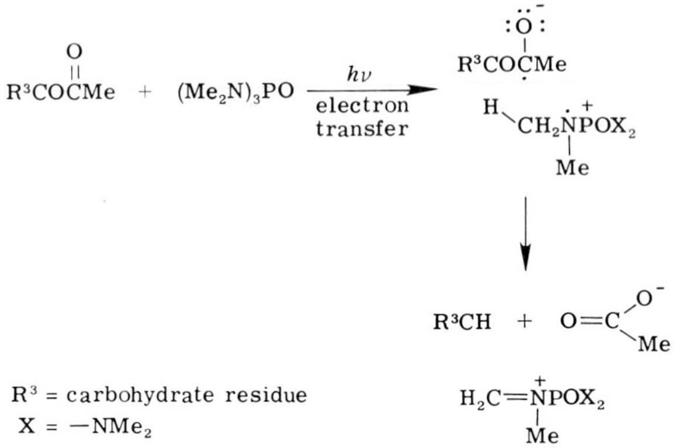

$$R^3COCMe + (Me_2N)_3PO \xrightarrow[\text{electron transfer}]{h\nu} \begin{array}{c} :\ddot{O}:^- \\ | \\ R^3CO\dot{C}Me \\ H \diagdown \\ CH_2\overset{+}{N}POX_2 \\ | \\ Me \end{array}$$

$$\downarrow$$

$$R^3CH + O=C\begin{array}{c}O^- \\ \diagup \\ \diagdown Me\end{array}$$

R^3 = carbohydrate residue
X = $-NMe_2$

$$H_2C=\overset{+}{N}POX_2 \\ | \\ Me$$

Scheme 13.—Proposed Mechanism for Photochemical Reaction between Esters and Hexamethylphosphoric Triamide.

Scheme 13 is a mechanism for this reaction that is based upon an electron-transfer process. An electron-transfer mechanism has been proposed in order to explain similar reactions in noncarbohydrate systems.[76]

IV. ACETALS

Three types of photochemical reaction of carbohydrate acetals have been investigated. Early studies centered on the photochemical fragmentation of phenyl glycosides, and the photolysis of o-nitrobenzylidene acetals. (The latter reactions will be discussed with the photolysis of other nitro compounds; see Sect. VII,1.) Later experiments were concerned with hydrogen-abstraction reactions from acetal carbon atoms by excited carbonyl compounds.

The reactions that follow initial hydrogen-abstraction from an acetal carbon atom are those characteristic of radicals, and include coupling, ring opening, and reaction with molecular oxygen (see Table XI). The presence of molecular oxygen, even in low concentration, appears to exert a controlling influence on the course of acetal photochemistry by typically producing hydroxy esters as the major photoproducts. A mechanism proposed[81] for the formation of these hydroxy esters (and other acetal photoproducts), using methyl 3,4-O-ethylidene-β-L-arabinopyranoside (**35**) as an example, is given in Scheme 14. (Scheme 16 contains a mechanism proposed for a related reaction.)

Study of model systems has revealed a tendency for photochemical abstraction of axial, rather than equatorial, acetal hydrogen atoms in

(76) H. Deshayes, J. P. Pete, and C. Portella, *Tetrahedron Lett.*, (1976) 2019–2022.

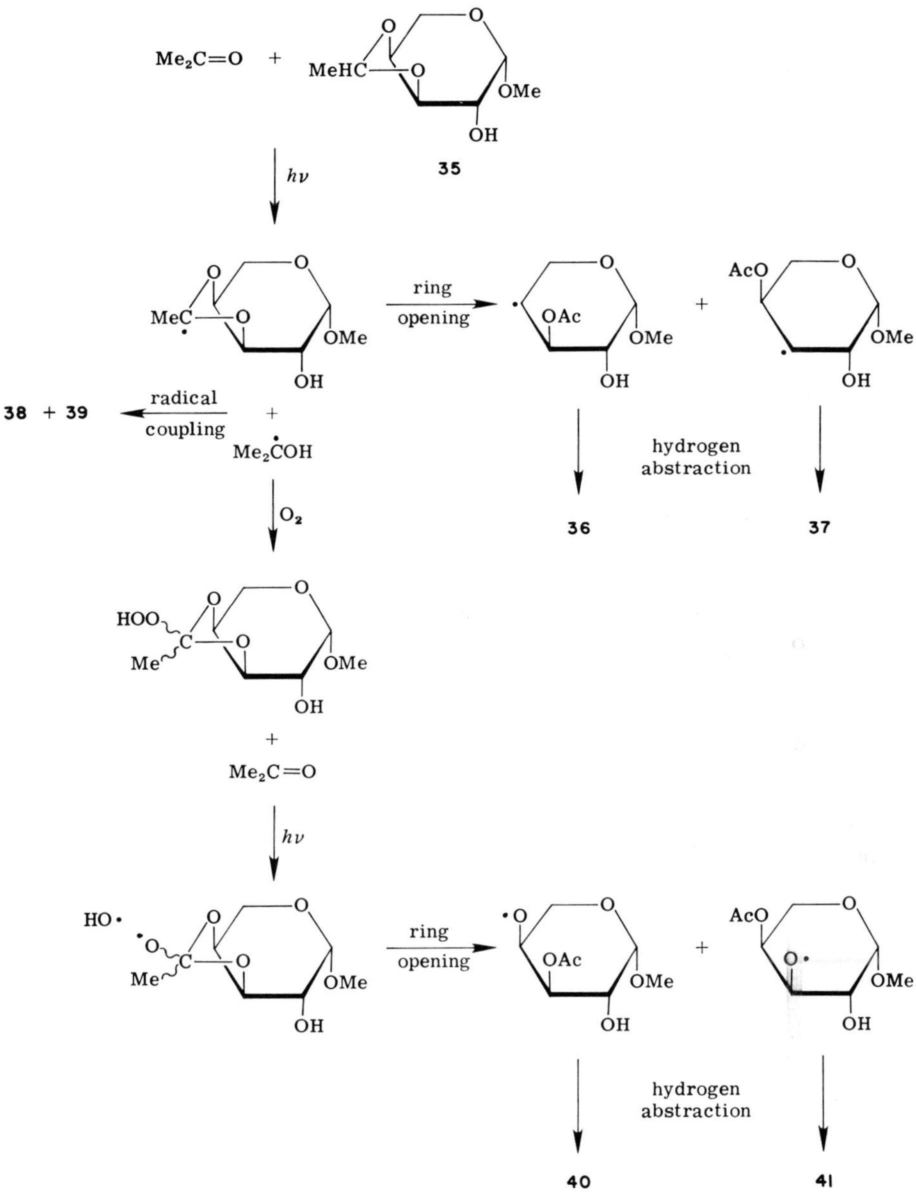

Scheme 14.—Proposed Mechanism for Photochemical Reaction of Methyl 3,4-*O*-Ethylidene-β-L-arabinopyranoside (**35**) with Excited Acetone.

TABLE XI

Photochemical Reactions of Benzylidene and Ethylidene Acetals of Carbohydrates

Reactants	Products and Yields (%)	Ref.

(First reaction: PhCH acetal of sugar + Me$_2$C=O)

X = Ac
R^1 = R^2 = Ph — 38 — — — 77,78
R^1 = R^2 = Me — 23 — 9 — 77,78
R^1 = Me, R^2 = Ph — 33 — 7 — 77,78

(Second reaction: PhCH acetal + Me$_2$C=O + O$_2$)

X = Ac
R^1 = R^2 = Ph — 19 — 27 — 77,78
R^1 = R^2 = Me — 23 — 41 — 77,78

(Third reaction: furanoside benzylidene acetal + Me$_2$C=O) — 58 — 79

(Fourth reaction: furanoside with PhCH and O-CMe$_2$ + Me$_2$C=O) — 79

TABLE XI (Continued)
Photochemical Reactions of Benzylidene and Ethylidene Acetals of Carbohydrates

Reactants	Products and Yields (%)	Ref.
PhCH benzylidene acetal + Me$_2$C=O, X = Bz	HOCH$_2$ product (BzO, OX, OMe, OX) + BzOCH$_2$ product (HO, OX, OMe, OX)	80
PhHC benzylidene acetal + Me$_2$C=O, X = Bz	CH$_2$OBz product (HO, OX, OMe, OX) + CH$_2$OH product (BzO, OX, OMe, OX)	80
35 ethylidene acetal + Me$_2$C=O	**36** (H, OAc, OMe, OH) + **37** (AcO, H, OMe, OH) 7	81
	38 Me$_2$COH–C–Me bridged (OMe, OH) 17 + **39** Me, OH–C–CMe$_2$ bridged (OMe, OH) 7	
	40 (HO, OAc, OMe, OH) + **41** (AcO, OH, OMe, OH)	

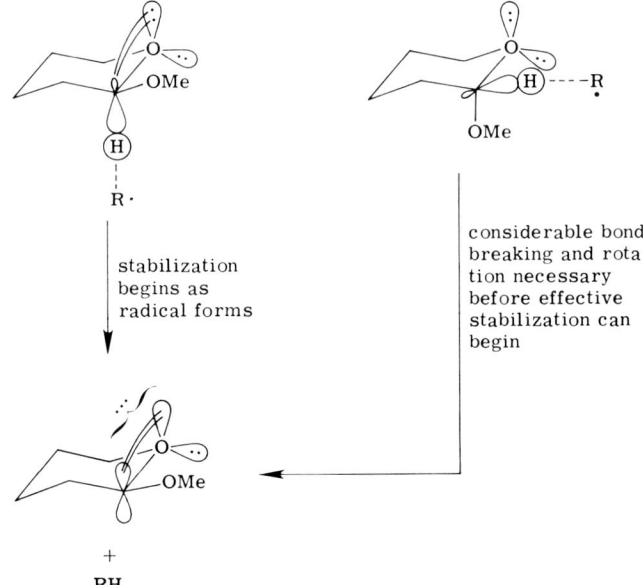

Scheme 15.—Rationalization for the Importance of Orbital Overlap to Radical Stabilization During Hydrogen Abstraction from an Acetal Carbon Atom.

pyranoid systems.[82-87] This tendency has been rationalized[83,86] as arising from more-effective stabilization of the developing radical-center on the carbon atom by the adjacent oxygen atom during axial hydrogen-abstraction (see Scheme 15). Interestingly, there is a parallel between the importance of orbital overlap (a) to radical stabilization dur-

---

(77) M. Suzuki, T. Inai, and R. Matsushima, *Bull. Chem. Soc. Jpn.*, 49 (1976) 1585–1589.
(78) S. Morio, R. Matsushima, T. Inai, and S. Tsujimoto, *Asahi Garasu Kogyo Gijutsu Shoreikai Kenkyu Hokoku*, (1977) 235–247; *Chem. Abstr.*, 89 (1978) 180,240.
(79) K. Matsuura, S. Maeda, Y. Araki, and Y. Ishido, *Bull. Chem. Soc. Jpn.*, 44 (1971) 292.
(80) W. Szeja and M. Lapokowski, *Pol. J. Chem.*, 52 (1978) 673–675.
(81) W. A. Szarek, R. J. Beveridge, and K. S. Kim, *J. Carbohydr. Nucleos. Nucleot.*, 5 (1978) 273–284.
(82) R. D. McKelvey, *Carbohydr. Res.*, 42 (1975) 187–191.
(83) K. Hayday and R. D. McKelvey, *J. Org. Chem.*, 41 (1976) 2222–2223.
(84) C. Bernasconi and G. Descotes, *C. R. Acad. Sci. Ser. C*, 280 (1975) 469–472.
(85) C. Bernasconi, L. Cottier, and G. Descotes, *Bull. Soc. Chim. Fr.*, (1977) 107–112.
(86) C. Bernasconi, L. Cottier, and G. Descotes, *Bull. Soc. Chim. Fr.*, (1977) 101–106.
(87) C. Bernasconi, L. Cottier, G. Descotes, M. F. Grenier, and F. Metras, *Nouv. J. Chim.*, 2 (1978) 79–84.

ing hydrogen abstraction from an acetal carbon atom (see Scheme 15) and (b) in α-cleavage of excited carbonyl compounds (see Scheme 8).

Photolysis of aryl glycosides was not only one of the first photochemical reactions of carbohydrates to be studied[88-91] but has continued to be an area of active investigation. Intriguing observations have been made concerning the effect of aromatic substituents upon glycosidic hydrolysis.[92,93] These observations suggested that the photochemical, aromatic substitution observed in noncarbohydrate systems[94] may be responsible for the hydrolysis of aryl glycosides. Complete understanding of these effects must await further studies on product structures, chemical yields, and reaction mechanisms.

V. Unprotected Alditols, Aldohexoses, and Related Compounds

Irradiation of unprotected carbohydrates was the subject of greatest photochemical interest to the early carbohydrate photochemists.[1] The investigations of these early workers were concerned primarily with the effect of reaction conditions upon the photochemical process, rather than with the identity of the reaction products. Between 1960 and 1969, a comprehensive series of papers was published on the photochemical reactions of D-glucose and D-glucitol.[95-99] These studies contributed greatly to understanding of the photochemistry of unprotected carbohydrates, as not only was the result of variation in the reaction conditions studied but also the structures of the products were determined.

Extended irradiation of D-glucose and D-glucitol in the presence of oxygen results[95-99] in considerable degradation of the starting materials (see Table XII). Such degradation is consistent with the radical

(88) L. J. Heidt, *J. Am. Chem. Soc.*, 61 (1939) 2981–2982.
(89) L. J. Heidt, *J. Franklin Inst.*, (1942) 473–486.
(90) G. Tanret, *C. R. Acad. Sci.*, 202 (1936) 881–883.
(91) G. Tanret, *Bull. Soc. Chim. Biol.*, 18 (1936) 1344–1351; *Chem. Abstr.*, 30 (1936) 7119.
(92) T. Yamada, M. Sawada, and M. Taki, *Agric. Biol. Chem.*, 39 (1975) 909–910.
(93) W. G. Filby, G. O. Phillips, and M. G. Webber, *Carbohydr. Res.*, 51 (1976) 269–271.
(94) N. J. Turro, *Modern Molecular Photochemistry*, Benjamin/Cummings, Menlo Park, CA, 1978, pp. 404–408.
(95) G. O. Phillips and G. J. Moody, *J. Chem. Soc.*, (1960) 3398–3404.
(96) G. O. Phillips and W. J. Criddle, *J. Chem. Soc.*, (1963) 3984–3989.
(97) G. O. Phillips and P. Barber, *J. Chem. Soc.*, (1963) 3990–3997.
(98) G. O. Phillips, P. Barber, and T. Rickards, *J. Chem. Soc.*, (1964) 3443–3450.
(99) G. O. Phillips and T. Rickards, *J. Chem. Soc.*, B, (1969) 455–461.

TABLE XII

Photochemical Reactions of Unprotected Alditols and Aldohexoses

	Major Products and Percent Yields				
Reactants	D-Arabinose	D-Gluconic Acid	D-Glucuronic Acid	Formal-dehyde	References
D-Glucose (low conversion)	23	47	6		99
D-Glucose (high conversion)	13	0.5	0.6	22	95
D-Glucitol (high conversion)	4	4	5		96

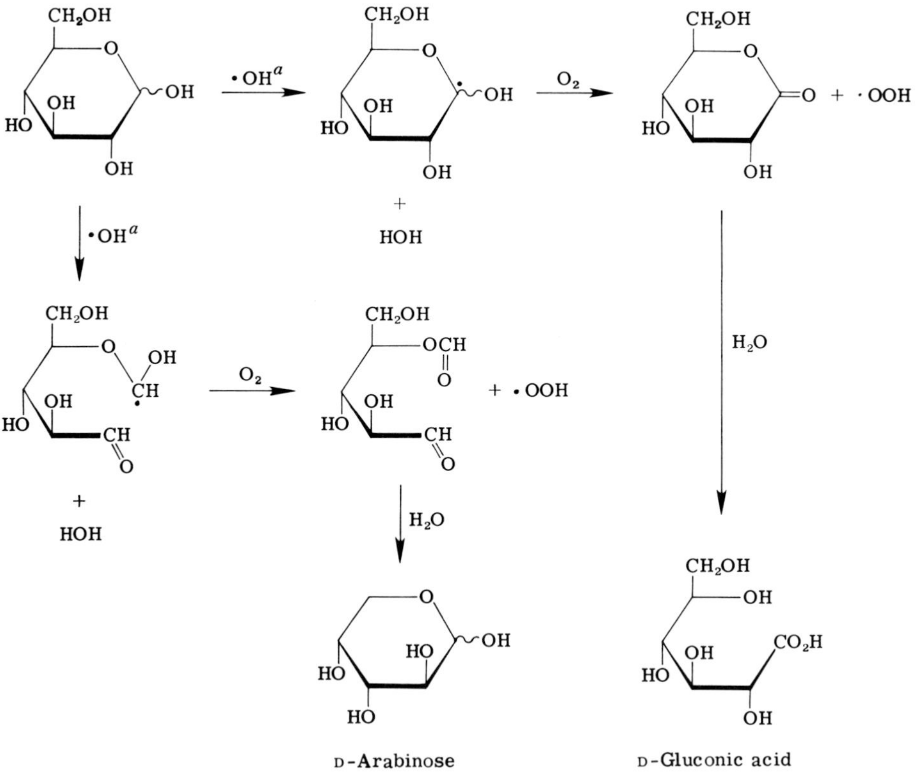

D-Arabinose D-Gluconic acid

[a]Hydroxyl radicals derived from photolysis of water[100]

Scheme 16.—Proposed Mechanism for Photochemical Reaction between D-Glucose and the Oxygen in Water.

reactions expected under these photolytic conditions. When reaction of D-glucose is conducted under conditions that maximize the percentage yield of primary products [that is, low (10%) conversion of D-glucose], D-arabinose and D-gluconic acid are the major products. These two products are those expected from a photochemically initiated, radical decomposition of an acetal (see Scheme 16).

Several investigations have since been conducted on the photolysis of unprotected sugars.[100–105] Spectroscopic and chemical evidence has been offered for the formation of malonaldehyde from the photolysis of D-glucose in neutral or alkaline solution.[100] The formation of this compound has been suggested as the basis for the use of D-glucose as a chemical actinometer.[101] Also, photolysis of D-glucose in the presence of L-lysine or glycine results in low yields of D-glucopyranosylamine; however, this photochemical reaction is probably not of D-glucose, but rather, of the amino acids to form ammonia, which then reacts with the carbohydrate.[102] Finally, synthesis of a complex mixture of monosaccharides results when aqueous formaldehyde is irradiated in basic solutions.[103]

Two derivatives of unprotected carbohydrates, riboflavine[106,107] (**42**)

$$CH_2OH$$
$$HOCH$$
$$HOCH$$
$$HOCH$$
$$CH_2$$

42

and 2'-deoxyuridine[108] (**43**), experience intramolecular, photochemical reactions. The excited portion of each of these molecules (**42** and **43**)

(100) K. Scherz, *Carbohydr. Res.*, 14 (1970) 417–419.
(101) R. K. Datta and K. N. Rao, *Indian J. Chem., Sect. A*, 14 (1976) 122–123.
(102) L. W. Doner, R. Balicki, and R. L. Whistler, *Carbohydr. Res.*, 47 (1976) 342–344.
(103) Y. Shigemasa, Y. Matsuda, C. Sakazawa, and T. Matsuura, *Bull. Chem. Soc. Jpn.*, 50 (1977) 222–226.
(104) W. J. Criddle, B. Jones, and E. Ward, *Chem. Ind. (London)*, (1967) 1833–1834.
(105) J.-Y. Peng, K. Minami, and T. Yoshimoto, *Mokuzai Gakkaishi*, 22 (1976) 511–517.
(106) M. S. Jorns and P. Hemmerich, *Z. Naturforsch. Teil B*, 27 (1972) 1040–1044.
(107) M. S. Jorns, G. Schöllnhammer, and P. Hemmerich, *Eur. J. Biochem.*, 57 (1975) 35–48.
(108) J. Cadet, L.-S. Kan, and S. Y. Wang, *J. Am. Chem. Soc.*, 100 (1978) 6715–6720.

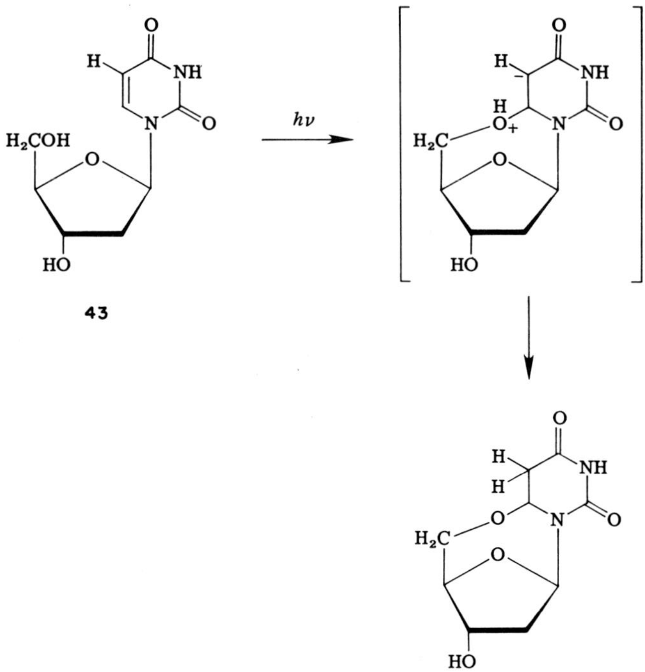

Scheme 17.—Proposed Mechanism for Photochemical Reaction of 2′-Deoxyuridine (**43**).

does not involve the carbohydrate part of the structure; consequently, the carbohydrate portion of each molecule is involved in reaction by adding to the excited system. This addition process, which is believed to involve zwitterionic intermediates, is illustrated for 2′-deoxyuridine (**43**) in Scheme 17.

VI. Sulfur-Containing Compounds

The types of sulfur compounds irradiated include dimethyl thiocarbamates, disulfides, dithioacetals, sulfides, sulfonates, sulfones, sulfoxides, and tetrasulfides. Also, such sulfur-containing compounds as simple alkanethiols have been added to unsaturated carbohydrates (see Table II). Several of the photochemical reactions of sulfur-containing carbohydrates clearly have synthetic value (for example, photolytic removal of *p*-toluenesulfonate protecting groups), and study of such other reactions as sulfone transformations has contributed significantly to mechanistic understanding of the photochemistry of organic sulfur compounds.

$$\underset{\mathbf{44}}{\underset{|}{\text{RCHSEt}}}\overset{h\nu}{\longrightarrow}\left[\underset{\overset{\bullet}{\text{SEt}}}{\overset{\bullet}{\text{RCHSEt}}}+\right]\overset{\text{MeOH}}{\longrightarrow}\underset{\mathbf{45}}{\text{RCH}_2\text{SEt}}+{}^{\bullet}\text{CH}_2\text{OH}$$

$$\downarrow h\nu$$

$$[\text{RCH}_2{}^{\bullet}+{}^{\bullet}\text{SEt}]$$

$$\downarrow \text{MeOH}$$

$$R = \begin{array}{c}\text{HOCH}\\|\\\text{HOCH}\\|\\\text{HCOH}\\|\\\text{HCOH}\\|\\\text{CH}_2\text{OH}\end{array} \qquad \underset{\mathbf{46}}{\text{RCH}_3}+{}^{\bullet}\text{CH}_2\text{OH}$$

Scheme 18.—Proposed Mechanism for Photochemical Reaction of D-Galactose Diethyl Dithioacetal (**44**).

1. Dithioacetals and Sulfides

Photolysis of D-galactose diethyl dithioacetal (**44**) results in carbon–sulfur bond-fragmentation, leading to 1-S-ethyl-1-thio-D-galactitol (**45**) in 55–60% yield; in addition, minor proportion of L-fucitol (**46**) and 1-deoxy-1-ethylsulfinyl-D-galactitol (**47**) are formed.[109] Similar irradiation of acetates (**48–51**) of the diethyl dithioacetals produces the corresponding 1-S-ethyl-1-thioalditol acetates in 79, 85, 87, and 84% yields, respectively.[110] The excellent yields from these reactions suggest that the photochemical transformation of dialkyl dithioacetals is an attractive means for accomplishing a net replacement of an S-alkyl group by hydrogen. A mechanism based upon homolytic carbon–sulfur bond-fragmentation well rationalizes product formation[109] (see Scheme 18).

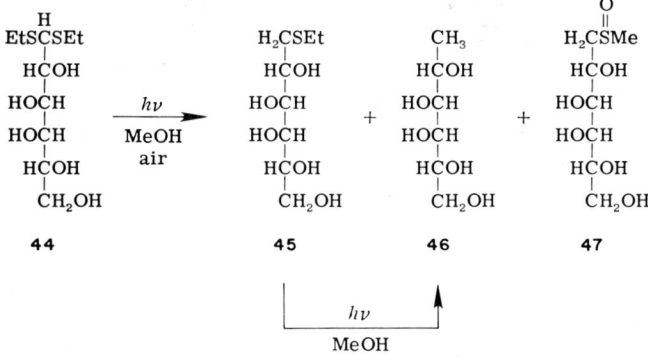

(109) D. Horton and J. S. Jewell, *J. Org. Chem.*, 31 (1966) 509–513.
(110) K. Matsuura, Y. Araki, and Y. Ishido, *Bull. Chem. Soc. Jpn.*, 46 (1973) 2261–2262.

```
      H                    H
   EtSCSEt              EtSCSEt
      |                    |
     ROCH                 HCOR
      |                    |
     HCOR                 HCOR
      |                    |
     HCOR                 HCOR
      |                    |
    CH₂OR                CH₂OR
      48                   49


      H                    H
   EtSCSEt              EtSCSEt
      |                    |
     HCOR                 HCOR
      |                    |
     ROCH                 ROCH
      |                    |
     HCOR                 HCOR
      |                    |
    CH₂OR                 HCOR
                           |
                         CH₂OR
      50                   51

            R = Ac

          CH₂OH
           |
          HCOH
           |
          HOCH
           |
          HOCH
           |
          HCOH
           |
          CH₂OH
            52
```

Extended irradiation of the diethyl dithioacetal **44** increases the yield of L-fucitol (**46**) at the expense of 1-S-ethyl-1-thio-D-galactitol (**45**); also galactitol (**52**) was isolated from the photolysis mixture.[109] The intermediacy of **45** in the formation of **46**, a possibility that was suggested by the extended photolysis of **44**, is supported by the observation that irradiation of **45** in methanol produces L-fucitol (**46**) in 44% yield.[109] (Compounds **47** and **52** also are formed during irradiation of **45**, but in low yield.) The mechanism for sulfide photoreaction parallels that[109] for the alkyl dithioacetals (see Scheme 18).

Other sulfide photoreactions result when appropriate nucleoside derivatives are photolyzed. Upon irradiation in acetonitrile, 9-[5-deoxy-2,3-O-isopropylidene-5-(phenylthio)-β-D-ribofuranosyl]-adenine (**53**) is converted into the anhydronucleoside 8,5'-anhydro-(5'-deoxy-2',3'-O-isopropylidene-adenosine) (**55**) in 66% yield.[111] Similar reactions were observed when other sulfur-containing nucleosides were irradiated (see Table XIII). The reaction of the sulfide **53**

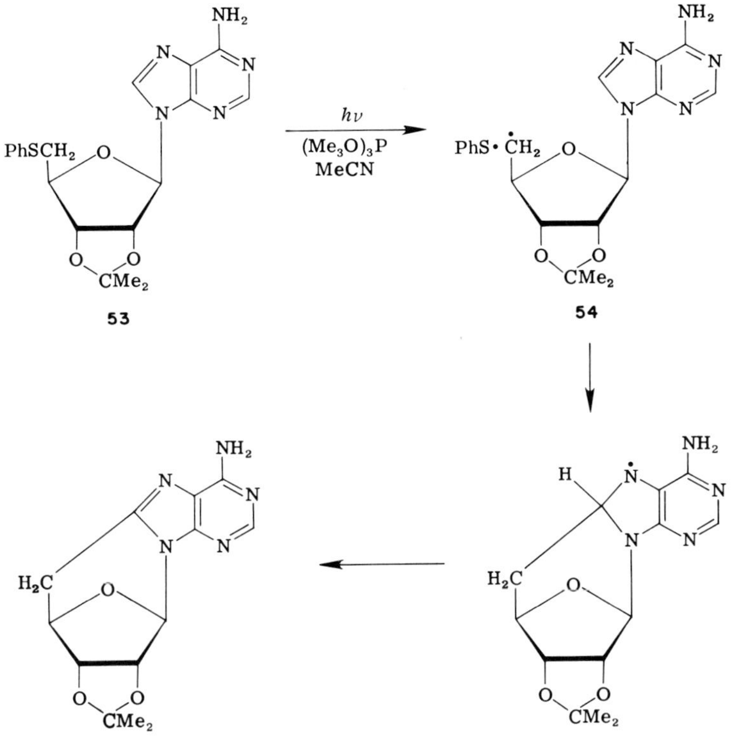

Scheme 19.—Proposed Mechanism for Photochemical Reaction of 2′,3′-O-Isopropylidene-5′-S-phenyl-5′-thioadenosine (**53**).

differs from the photoreaction of the sulfide **45**, because a radical such as **54** (presumably an intermediate in the reaction of **53**) can add internally to the nitrogenous-base moiety of the molecule. Such an intramolecular addition (see Scheme 19) should be favored over a bimolecular reaction (see Scheme 18).

2. Sulfoxides, Sulfones, and Sulfonates

Photochemical, carbon–sulfur bond cleavage is also observed in compounds containing sulfur in oxidation states higher than that which exists in sulfides and in dialkyl dithioacetals. For example, the irradiation of the sulfoxide **47** in methanol produces[109] a 58% yield of galactitol (**52**). Even though homolysis of the carbon–sulfur bond does occur in **47**, it is unlikely that **52** results from a simple, carbon–sulfur bond-cleavage, as such a reaction predicts products that were not observed

TABLE XIII

Photochemical Reactions of Sulfur-Containing, Nucleoside Derivatives

Reactants	Products and Yields (%)	Ref.
	66	111
	(Yield not reported)	111
	(Yield not reported)	111
	R^1 = OH, R^2 = H 19 R^1 = H, R^2 = OH 10 6	112

TABLE XIII (Continued)

Photochemical Reactions of Sulfur-Containing, Nucleoside Derivatives

Reactants	Products and Yields (%)	Ref.
[structure: XOCH$_2$ furanose with S-pyridyl, XO OX, X = Ac]	[structure 5] + [structure 9] + [structure 5 with SH]	113

(see Scheme 20). A rationalization for formation of **52** on photolysis of **47** may be based on the suggestion that **47** experiences prior photochemical rearrangement to **56**, a compound that, on photolysis, should be converted into **52** (see Scheme 21). This type of rearrangement (**47** to **56**) has been observed during photolysis of sulfoxides in noncarbohydrate systems.[114,115]

The photochemical reaction of 2,3,4,6-tetra-O-acetyl-β-D-glucopyranosyl sulfone (**57**) in benzene under nitrogen has been carefully studied, and a number of products identified[116] (see Scheme 22). A mechanism that involves a photochemically initiated series of free-radical processes has been proposed that is consistent not only with product formation but also with the extent of incorporation of deuterium found in the various products following irradiation of **57** in benzene-d_6. The mechanism shown in Scheme 22 is compatible with proposals offered to explain sulfone photochemistry in noncarbohy-

(111) A. Matsuda, M. Tezuka, and T. Ueda, *Nucleic Acids Res.*, (1976) s13–s16.
(112) A. Matsuda, M. Tezuka, and T. Ueda, *Tetrahedron*, 34 (1978) 2449–2452.
(113) J.-L. Fourrey and P. Jouin, *J. Org. Chem.*, 44 (1979) 1892–1894.
(114) A. G. Schultz and R. H. Schlessinger, *Chem. Commun.*, (1970) 1294–1295.
(115) I. W. J. Still, M. S. Chauhau, and M. T. Thomas, *Tetrahedron Lett.*, (1973) 1311–1315.
(116) P. M. Collins and B. R. Whitton, *J. Chem. Soc. Perkin Trans. 1*, (1974) 1069–1075.

O
RCH2SEt
47
hν
path A
path B
path C
RCH2 •SEt
MeOH
RCH3
46
(not formed)
RCH2 SEt
RCH2OMe
(not formed)
RCH3
46
(not formed)
HCOH
HOCH
R = HOCH
HCOH
CH2OH

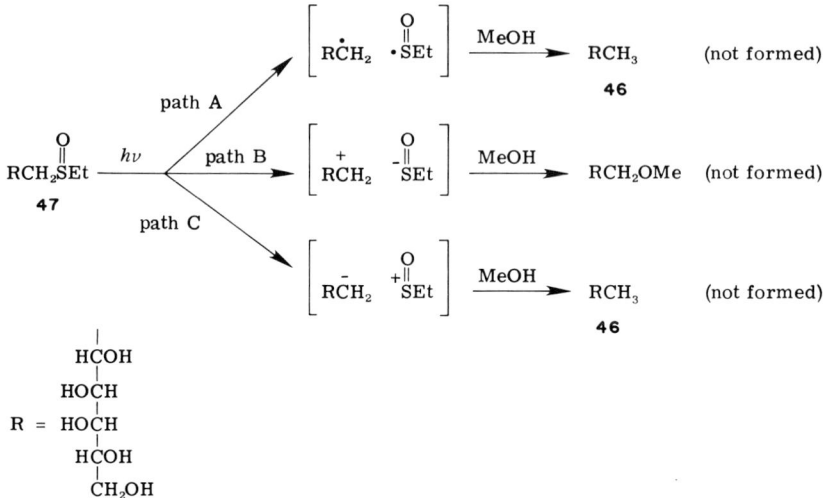

Scheme 20.—Potential, Photochemical-Reaction Pathways for Reaction of 1-Deoxy-1-(ethysulfinyl)-D-galactitol (**47**).

drate systems,[117] and is also in accordance with the absence of methyl glycosides (the products expected from heterolytic, carbon–sulfur bond-cleavage) from irradiation of **57** in benzene–methanol. The carbohydrate sulfones that have been irradiated are listed in Table XIV.

Quartz-filtered irradiation of carbohydrate *p*-toluenesulfonates in

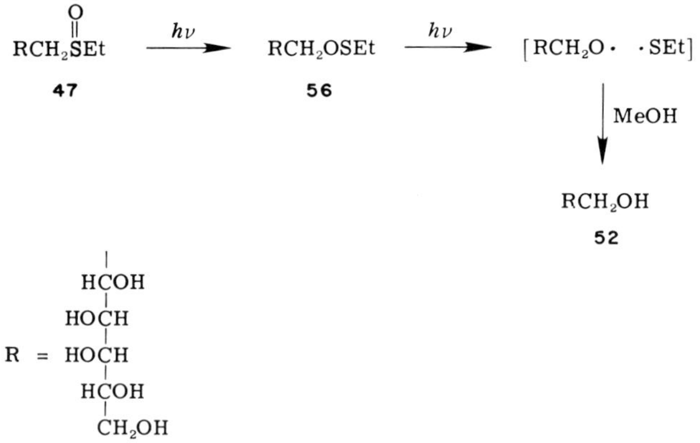

O
RCH2SEt
47
hν
RCH2OSEt
56
hν
[RCH2O• •SEt]
MeOH
RCH2OH
52
HCOH
HOCH
R = HOCH
HCOH
CH2OH

Scheme 21.—Proposed Mechanism for Photochemical Reaction of 1-Deoxy-1-(ethylsulfinyl)-D-galactitol (**47**).

(117) N. Kharasch and A. I. A. Khodair, *Chem. Commun.*, (1967) 98–99.

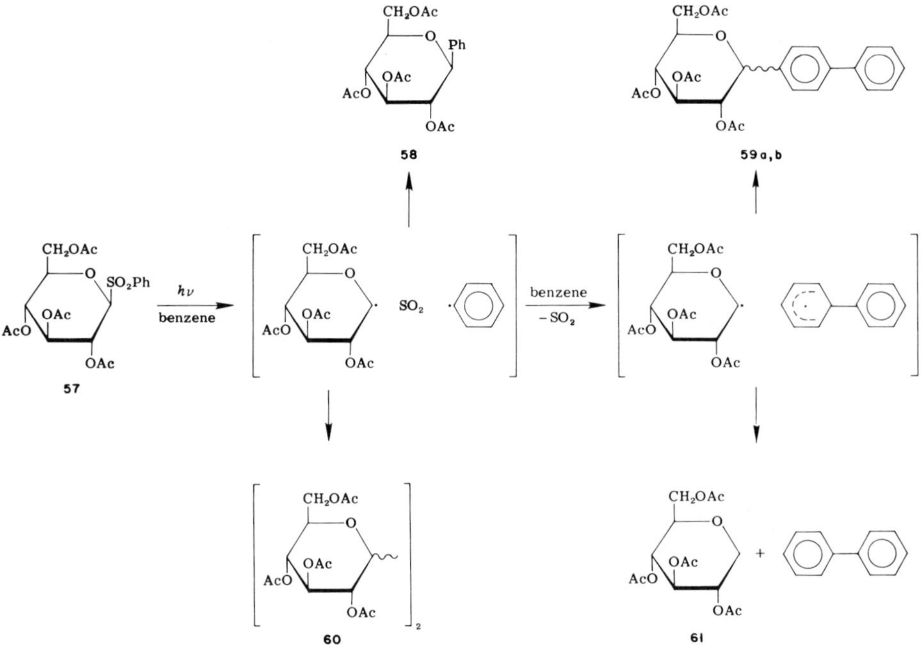

Scheme 22.—Proposed Mechanism for Photochemical Reaction of Phenyl 2,3,4,6-Tetra-O-acetyl-β-D-glucopyranosyl Sulfone (**57**).

methanol in the presence of a base has been found to lead to desulfonylation. The results from various such photolyses are summarized in Table XV. The retention of configuration at C-3 during reaction of **62** and **63**, for example, and the absence of deoxy compounds and methyl ethers, argue against any carbon–oxygen bond-cleavage and in favor of a sulfur–oxygen fission.

3. Thiocarbamates and Xanthides

Studies have also been conducted on compounds containing a thiocarbonyl group. Photolysis of dimethylthiocarbamates produces[128,129] mixtures of deoxy sugars and alcohols (see Table XVI). As the alcohols can be reconverted into dimethylthiocarbamates, and these re-irradiated, the synthesis and irradiation of these compounds offers a useful pathway to deoxy sugars. A simple mechanism for this reaction, based on the two possible α-cleavages resulting from thiocarbonyl excitation, is shown in Scheme 23. In addition, the possibility has been

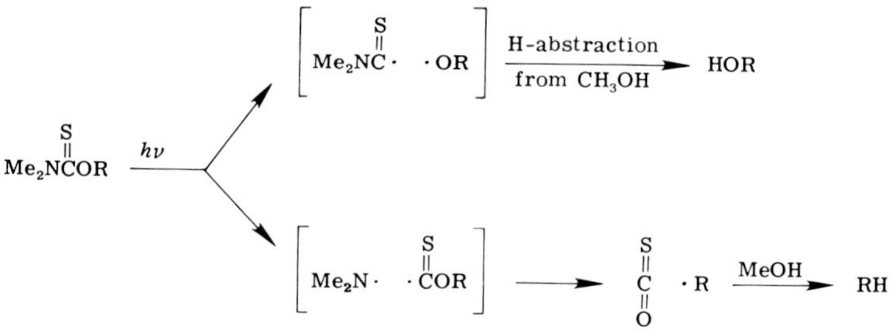

$$\underset{\text{Me}_2\text{N}\overset{\text{S}}{\overset{\|}{\text{C}}}\text{OR}}{} \xrightarrow{h\nu} \begin{cases} \left[\text{Me}_2\text{N}\overset{\text{S}}{\overset{\|}{\text{C}}}\cdot \quad \cdot\text{OR}\right] \xrightarrow[\text{from CH}_3\text{OH}]{\text{H-abstraction}} \text{HOR} \\[2ex] \left[\text{Me}_2\text{N}\cdot \quad \cdot\overset{\text{S}}{\overset{\|}{\text{C}}}\text{OR}\right] \longrightarrow \overset{\text{S}}{\underset{\text{O}}{\overset{\|}{\underset{\|}{\text{C}}}}}\cdot\text{R} \xrightarrow{\text{MeOH}} \text{RH} \end{cases}$$

R = carbohydrate moiety

Scheme 23.—Proposed Mechanism for Photochemical Reaction of Dimethylthiocarbamates.

raised that a photochemical, "Freudenberg type" of rearrangement (see Scheme 24) prior to further photochemical reaction could explain the formation of the deoxy compounds from the photolysis of dimethylthiocarbamates.[128]

Several, oxidatively coupled xanthates (**64–66**), compounds (also called xanthides) containing the photochemically reactive, sulfur–sulfur bond, have been studied.[130] Homolytic cleavage of this reactive bond is the primary reaction for these compounds, although this process is normally masked by recombination of the radicals produced. This primary, light-initiated process becomes apparent when a mixture of the xanthide **64** and ethyl xanthide (**67**) is irradiated in cyclohexane, because an equilibrium between **64**, **67**, and the mixed xanthide **68** is rapidly established.

The photochemical reactions of xanthides are quite complex. They are solvent-, concentration-, temperature-, wavelength-, and time-dependent.[130] The most thoroughly studied of these compounds is compound **64**, whose irradiation (through a Corex filter) in cyclohexane under nitrogen produces tetrasulfide **69** (37% yield), xanthate **70** (35%), 1,2:3,4-di-O-isopropylidene-α-D-galactopyranose (**71**, 13%), sulfur, and carbonyl sulfide. Irradiation of a dilute solution of **64** produced only **70** (in 74% yield). The most intriguing finding from irradiation of the xanthides **64–66** is the fact that **66** produces a xanthate

$$\text{RO}\overset{\text{S}}{\overset{\|}{\text{C}}}\text{NMe}_2 \xrightarrow{h\nu} \text{RS}\overset{\text{O}}{\overset{\|}{\text{C}}}\text{NMe}_2 \xrightarrow{h\nu} \text{RH}$$

Scheme 24.—Proposed, Photochemical, "Freudenberg" Rearrangement of Dimethylthiocarbamates.

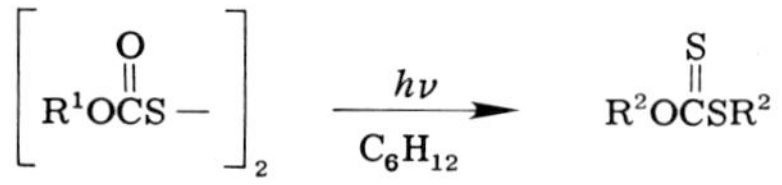

$$\left[\text{R}^1\text{O}\overset{\text{O}}{\underset{\|}{\text{C}}}\text{S}- \right]_2 \xrightarrow[\text{C}_6\text{H}_{12}]{h\nu} \text{R}^2\text{O}\overset{\text{S}}{\underset{\|}{\text{C}}}\text{SR}^2$$

65 R¹ =
70 R² =

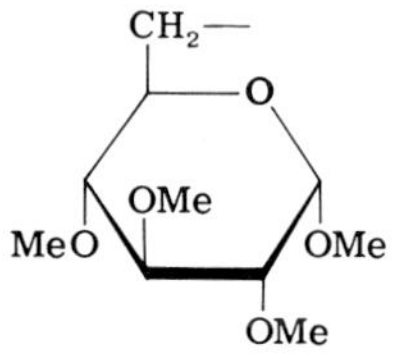

[structure: methyl 2,3,4-tri-O-methyl-hexopyranoside derivative]

66 R¹ =
72 R² =

[structure: 1,2:5,6-di-O-isopropylidene furanose derivative]

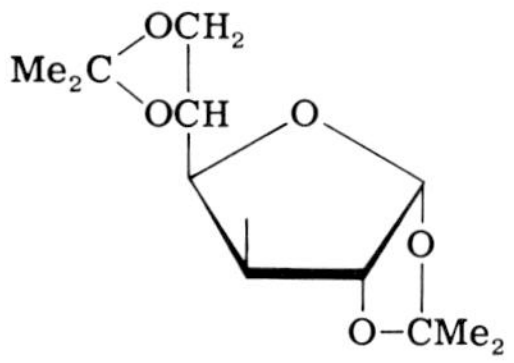

(**72**) in which sulfur has replaced oxygen with at least 92% retention of configuration.

VII. NITROGEN-CONTAINING COMPOUNDS

1. Nitro Compounds

A number of carbohydrates protected by groups containing *o*-nitrophenyl substituents has been studied photochemically. The primary reason for interest in these compounds is that they can be deprotected by irradiation. For example, *o*-nitrobenzyl glycosides of carbohydrates,[131,132] and ethers of nucleosides[133,134] and oligoribonucleotides,[135–137] experience loss of the *o*-nitrobenzyl group upon photolysis. The reported examples of this reaction are listed in Table XVII; in addition, there is preliminary evidence that glycosides formed from polymer-bound, *o*-nitrobenzyl-containing groups, and partially protected carbohydrates, release the carbohydrate upon photolysis.[138] A detailed explanation for the photochemical reactions of the *o*-nitrobenzyl

TABLE XIV
Photochemical Reactions of Carbohydrate Sulfones

Reactants	Products	Ref.
57 $R^1 = H, R^2 = SO_2Ph$; $X = Ac$	**61** + **58**	
	59a { $R^3 = H$; $R^4 = $ –C$_6$H$_4$–Ph } + [**60**]$_2$	116
	59b { $R^3 = $ –C$_6$H$_4$–Ph; $R^4 = H$ }	116
$R^1 = SO_2Ph, R^2 = H, X = Ac$	(Same products as above)	116
$R^1 = H, R^2 = SO_2Ph$; $X = Ac$	+ $R^3 = H$, $R^4 = $ –C$_6$H$_4$–Ph and $R^3 = $ –C$_6$H$_4$–Ph, $R^4 = H$ + [ ]$_2$	118
$R^1 = SO_2Ph, R^2 = H, X = Ac$	(Same products as above)	118

TABLE XV

Photochemical Reactions of p-Toluenesulfonic Esters of Carbohydrates

p-Toluenesulfonate	Product	Yield (%)	References
(structure)	(structure)	90	119
(structure)	(structure)	100	119
(structure)	(structure)	80	119
(structure)	(structure)	100	119
(structure)	(structure)	80	120, 121, 122
(structure)	(structure)	> 50	123

TABLE XV (Continued)
Photochemical Reactions of p-Toluenesulfonic Esters of Carbohydrates

p-Toluenesulfonate	Product	Yield (%)	References
(structure with OTs, OMe, CH₂OH)	(structure with OH, OMe, CH₂OH)	63	123
(pyranose with CH₂OX, OR, TsO, R = Me, X = Me)	(pyranose with CH₂OX, OR, HO)	87	124
R = Me, X = H		62	124
(pyranose with CH₃, OMe, OTs, epoxide)	(pyranose with CH₃, OMe, OH, epoxide)	60	125
(pyranose with CH₂OMe, OX, XO, MeO, OMe; X = Ts)	(pyranose with CH₂OMe, OH, HO, MeO, OMe)	61	126
(furanose with HOCH₂, dioxolane, HO, OTs)	(furanose with HOCH₂, dioxolane, HO, OH)		127

TABLE XVI

Photochemical Reactions of Carbohydrate-Containing Dimethylthiocarbamates[a]

Reactant	Products and Yields (%)	
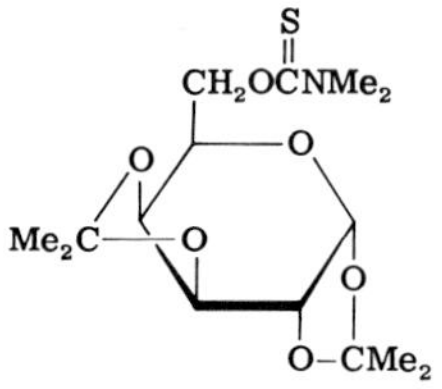	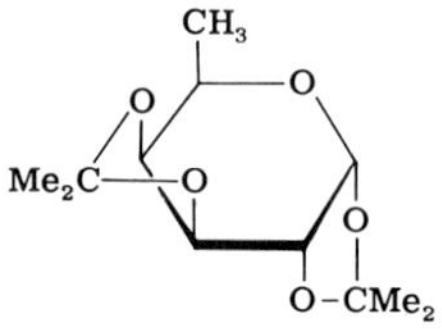 15	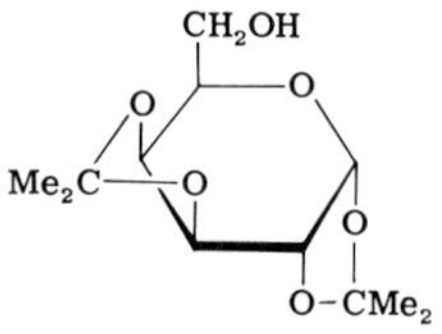 36
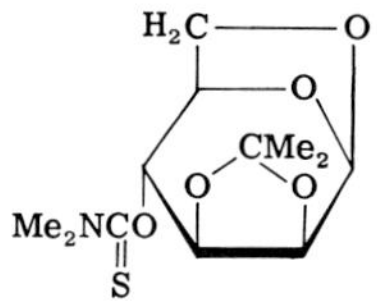	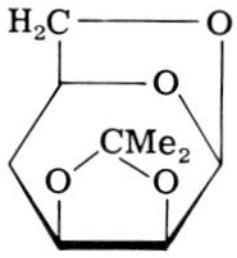 19	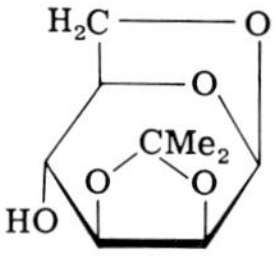 17
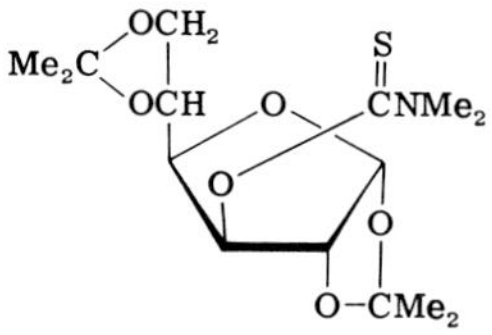	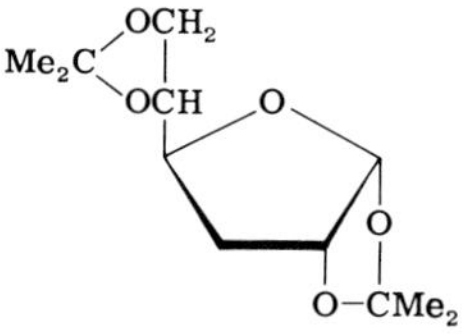 17–23	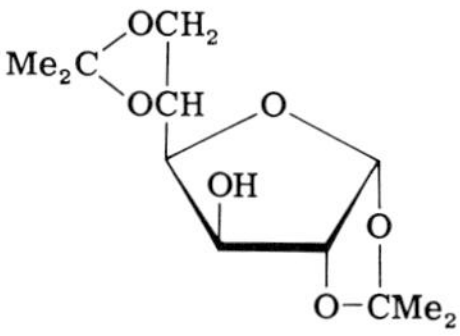 26–32
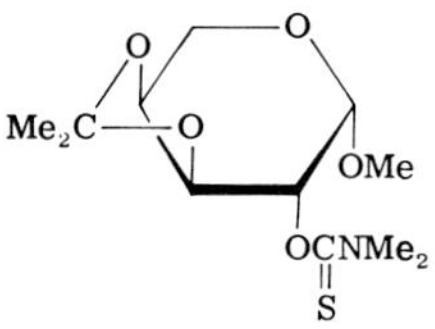	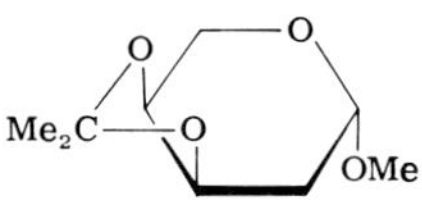 11	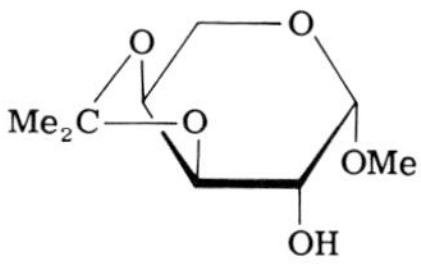 36
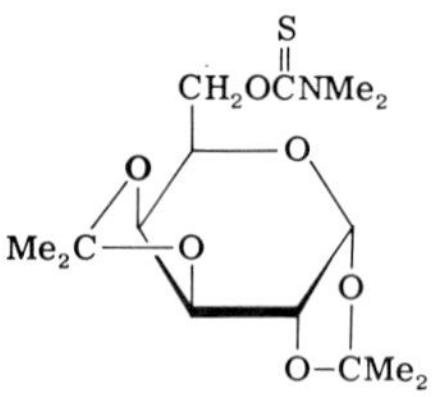	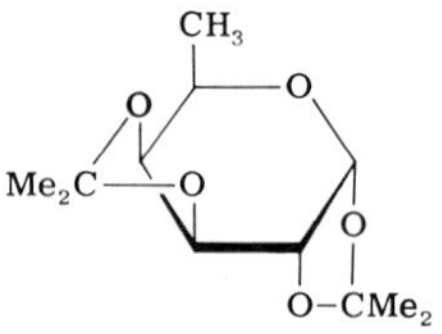 25	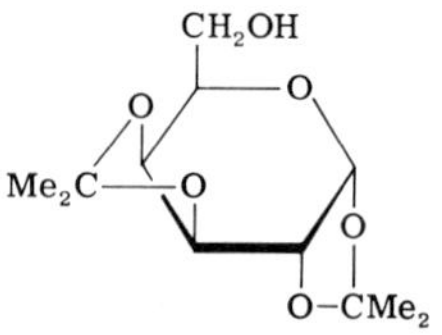 35

[a] Refs. 128 and 129.

group is proposed in Scheme 25. This mechanism is consistent with the suggestion that photolysis of o-nitrobenzyl compounds causes an intramolecular, oxidation–reduction reaction,[131,139] and also with the finding[140] that excited nitro groups, such as those in nitrophenyl β-D-glycosides, are effective hydrogen-atom abstractors. A process similar to that proposed for o-nitrobenzyl ethers (see Scheme 25, path a), but including loss of carbon dioxide in a final step (see Scheme 25, path b), accounts for the photochemical regeneration of an amino function protected by an (o-nitrobenzyl)oxycarbonyl group.[141] Table XVIII contains examples of this reaction involving carbohydrate systems.

Cyclic acetals derived from diols and o-nitrobenzaldehyde produce hydroxynitrosobenzoates upon irradiation (see Scheme 26). Early studies of this reaction demonstrated that acetals involving carbohydrates and o-nitrobenzaldehyde experience facile, photochemical

---

(118) P. M. Collins and B. R. Whitton, *Carbohydr. Res.*, 36 (1974) 293–301.
(119) S. Zen, S. Tashima, and S. Kotō, *Bull. Chem. Soc. Jpn.*, 41 (1968) 3025.
(120) A. D. Barford, A. B. Foster, and J. H. Westwood, *Carbohydr. Res.*, 13 (1970) 189–190.
(121) A. D. Barford, A. B. Foster, J. H. Westwood, L. D. Hall, and R. N. Johnson, *Carbohydr. Res.*, 19 (1971) 49–61.
(122) L. Vegh and E. Hardegger, *Helv. Chim. Acta*, 56 (1973) 2020–2025.
(123) W. A. Szarek, R. G. S. Ritchie, and D. M. Vyas, *Carbohydr. Res.*, 62 (1978) 89–103.
(124) F. R. Seymour, *Carbohydr. Res.*, 34 (1974) 65–70.
(125) R. A. Borgegrain and B. Gross, *Carbohydr. Res.*, 41 (1975) 135–140.
(126) F. R. Seymour, M. E. Slodki, R. D. Plattner, and L. W. Tjarks, *Carbohydr. Res.*, 46 (1976) 189–193.
(127) C. D. Chang and T. L. Hullar, *Carbohydr. Res.*, 54 (1977) 217–230.
(128) R. H. Bell, D. Horton, D. M. Williams, and E. Winter-Mihaly, *Carbohydr. Res.*, 58 (1977) 109–124.
(129) R. H. Bell, D. Horton, and D. M. Williams, *Chem. Commun.*, (1968) 323–324.
(130) E. I. Stout, W. M. Doane, C. R. Russell, and L. B. Jones, *J. Org. Chem.*, 40 (1975) 1331–1336.
(131) U. Zehavi and A. Patchornik, *J. Org. Chem.*, 37 (1972) 2285–2289.
(132) U. Zehavi, B. Amit, and A. Patchornik, *J. Org. Chem.*, 37 (1972) 2281–2285.
(133) D. G. Bartholomew and A. D. Broom, *J. Chem. Soc. Chem. Commun.*, (1975) 38.
(134) E. Ohtsuka, S. Tanaka, and M. Ikehara, *Nucleic Acid Chem.*, 1 (1978) 410–414.
(135) E. Ohtsuka, S. Tanaka, and M. Ikehara, *Synthesis*, (1977) 453–454.
(136) E. Ohtsuka, S. Tanaka, and M. Ikehara, *Chem. Pharm. Bull.*, 25 (1977) 949–959.
(137) E. Ohtsuka, S. Tanaka, and M. Ikehara, *Nucleic Acids Res.*, 1 (1974) 1351–1357.
(138) U. Zehavi and A. Patchornik, *J. Am. Chem. Soc.*, 95 (1973) 5673–5677.
(139) A. Patchornik, B. Amit, and R. B. Woodward, *J. Am. Chem. Soc.*, 92 (1970) 6333–6334.
(140) P. J. Baugh, K. Kershaw, and G. O. Phillips, *Carbohydr. Res.*, 22 (1972) 233–243.
(141) B. Amit, U. Zehavi, and A. Patchornik, *J. Org. Chem.*, 39 (1974) 192–196.

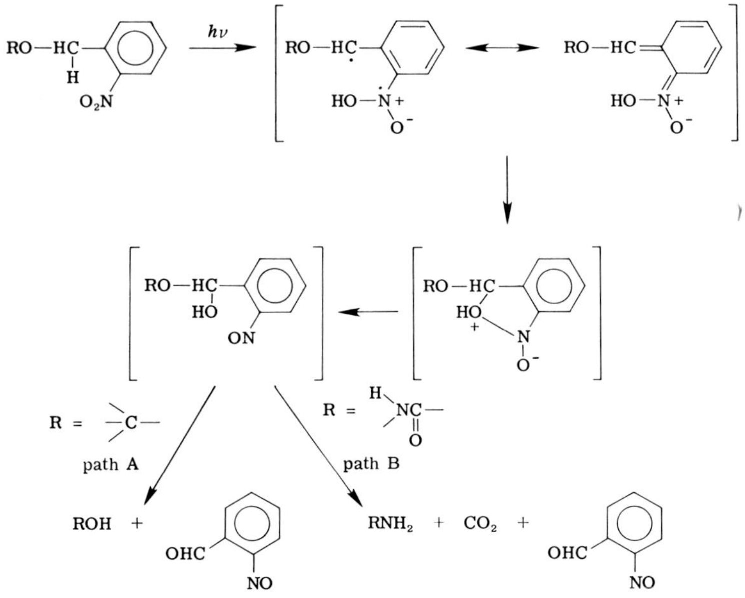

R = carbohydrate, nucleoside, or nucleotide residue

Scheme 25.—Proposed Mechanism for Photochemical Reaction of *o*-Nitrobenzyl Compounds.

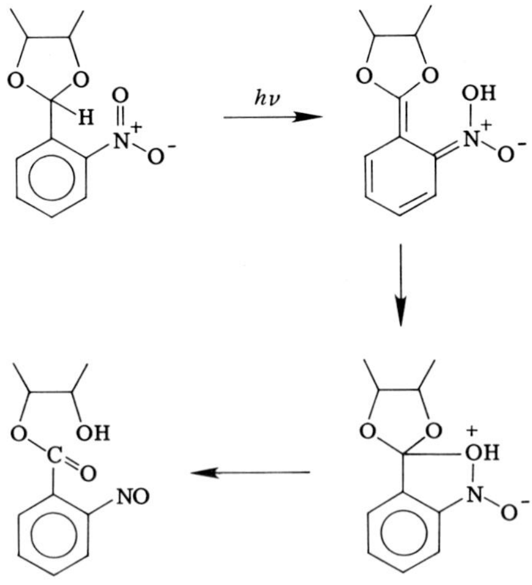

Scheme 26.—Proposed Mechanism for Photochemical Reaction of *o*-Nitrobenzylidene Acetals.

TABLE XVII

Photochemical Reactions of *o*-Nitrophenyl Ethers of Carbohydrates and Nucleosides

Reactants	Products	Yields (%)	Ref.
[rhamnopyranose with O-CH₂-(o-nitrophenyl) ether]	[rhamnopyranose diol]	100	131
[isomeric rhamnopyranose o-nitrobenzyl ether]	[rhamnopyranose triol]	100	131
[glucopyranose O-CH₂-aryl ether, aryl = 2-nitro, 4,5-R₂] R = H R = OMe	[glucopyranose]	80 100	132 132
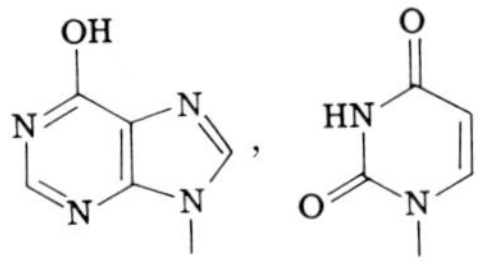 [nucleoside with R¹OCH₂, R²O, and O-CH₂-(o-nitrophenyl); B = adenine, cytosine, hypoxanthine, uracil], R¹ = R² = H	[nucleoside with R¹OCH₂, R²O, OH]	100	133, 134

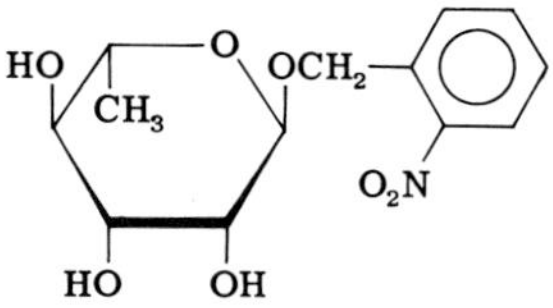

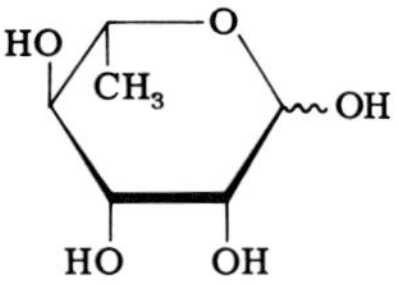

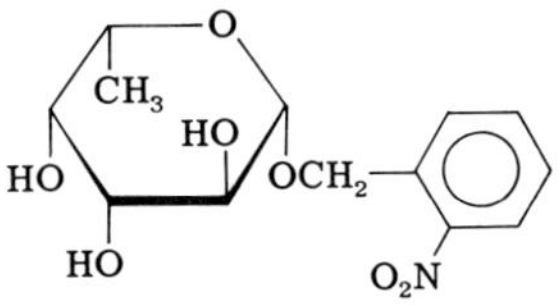

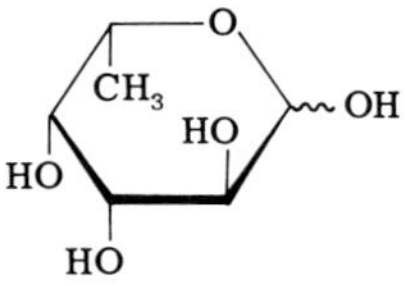

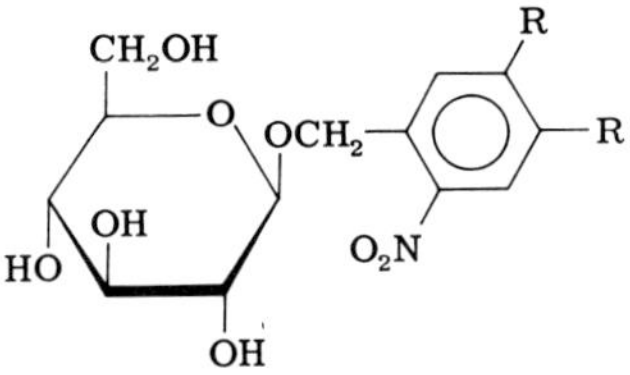

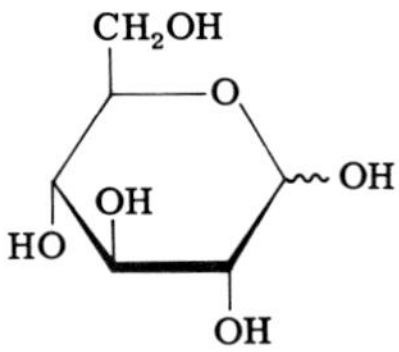

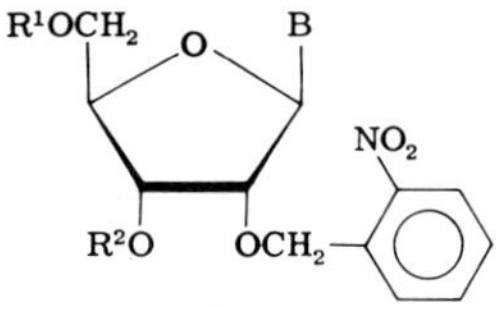

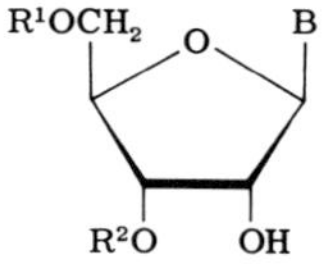

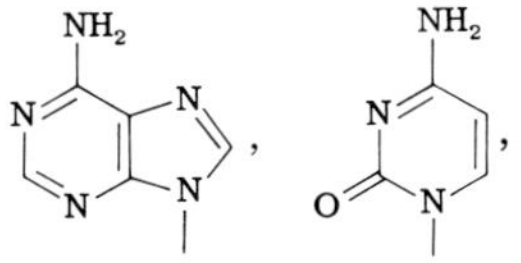

(*Continued*)

TABLE XVII (Continued)
Photochemical Reactions of o-Nitrophenyl Ethers of Carbohydrates and Nucleosides

Reactants	Products	Yields (%)	Ref.
B = guanine, $R^1 = R^2 = H$		(Not reported)	135
B = adenine, R^1 = Cytidylyl-(3'—5')-cytidylyl-(3'—5'), $R^2 = H$		95	136
B = adenine, R^1 = Cytidylyl-(3'—5'), $R^2 = H$		(Not reported)	136
B = uracil, R^1 = Uridylyl(3'—5'), $R^2 = H$		(Not reported)	137

reaction[142–146]; however, the products formed were complex and not fully characterized. Investigations of o-nitrobenzylidene derivatives of carbohydrates have since established the structure of the photoproducts by oxidation of these unstable (and often dimeric) nitroso compounds to their corresponding monomeric nitro derivatives, compounds that can readily be characterized.[147–149] Because the 1,3-dioxo-

(142) I. Tanasescu and E. Craciunescu, *Bull. Soc. Chim. Fr.*, 3 (1936) 581–598.
(143) I. Tanasescu and E. Craciunescu, *Bull. Soc. Chim. Fr.*, 5 (1936) 1517–1527.
(144) I. Tanasescu and M. Ionescu, *Bull. Soc. Chim. Fr.*, 7 (1940) 84–90.
(145) I. Tanasescu and M. Ionescu, *Bull. Soc. Chim. Fr.*, 7 (1940) 77–83.
(146) I. Tanasescu and M. Ionescu, *Bull Soc. Chim. Fr.*, 5 (1936) 1511–1517.
(147) P. M. Collins and N. N. Oparaeche, *J. Chem. Soc. Chem. Commun.*, (1972) 532–533.
(148) P. M. Collins and N. N. Oparaeche, *J. Chem. Soc. Perkin Trans. 1*, (1975) 1695–1700.
(149) P. M. Collins, N. N. Oparaeche, and V. R. N. Munasinghe, *J. Chem. Soc. Perkin Trans. 1*, (1975) 1700–1706.

TABLE XVIII

Photochemical Reactions of *o*-Nitrobenzyloxycarbonyl Derivatives of Carbohydrates

Reactants	Products and Yields (%)	References
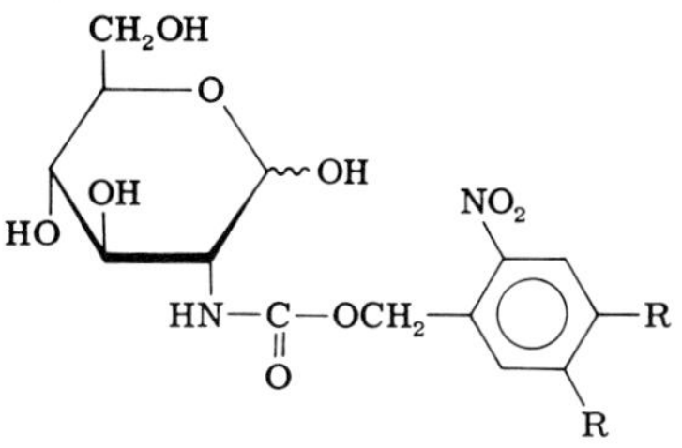	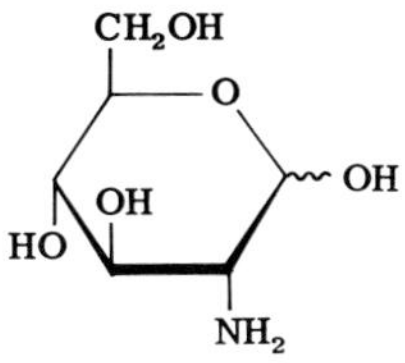	141
R = H	82	
R = OMe	89	
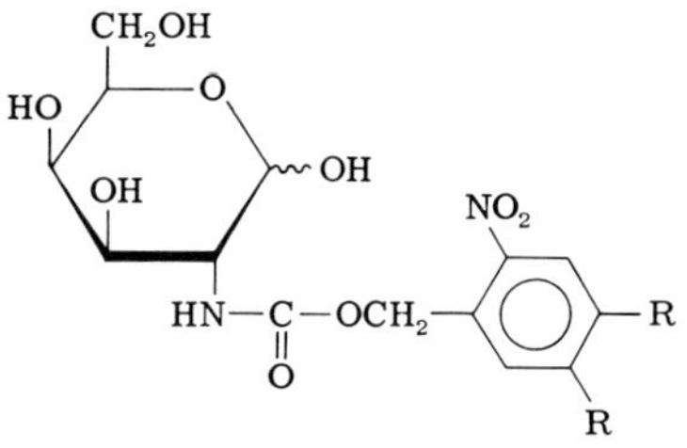	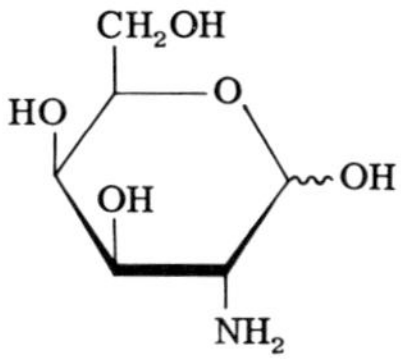	141
R = H	80	
R = OMe	91	
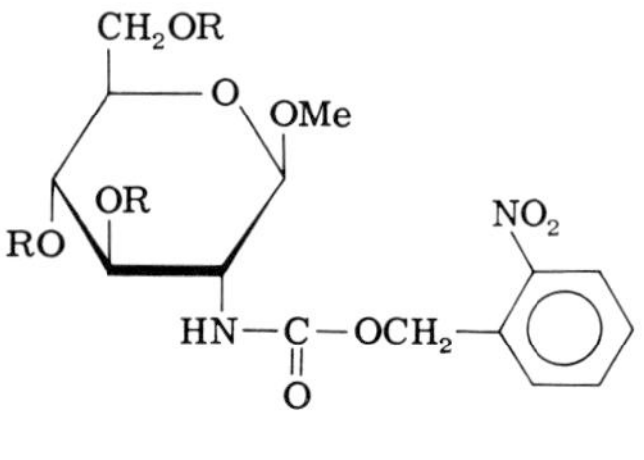	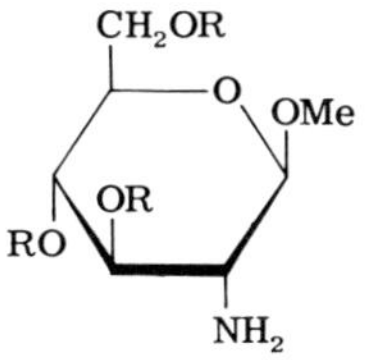	141
R = Ac		

lane and 1,3-dioxane rings of benzylidene acetals can open in two ways, photolysis of compounds containing *o*-nitrobenzylidene acetal groups invariably produces mixtures of products. Table XIX contains a tabulation of acetals irradiated, and the products formed.[147–149]

TABLE XIX

Photochemical Reactions of *o*-Nitrobenzylidene Acetals of Carbohydrates

Reactants	Products and Yields (%)		Ref.
[structure with AcO, CH₃, OMe, CH, NO₂]	[structure, 95]	[structure, 5]	147,148
[structure with CH₂OMe, OMe, MeO, CH, NO₂]	[structure, < 5]	[structure, 93]	148
[structure with CH₃, OMe, AcO, HC–O, NO₂]	[structure, < 5]	[structure, 93]	147,148
[structure with HC, OMe, OAc, O₂N]	[structure, 5]	[structure, 95]	148

TABLE XIX (*Continued*)

Photochemical Reactions of *o*-Nitrobenzylidene Acetals of Carbohydrates

Reactants	Products and Yields (%)		Ref.
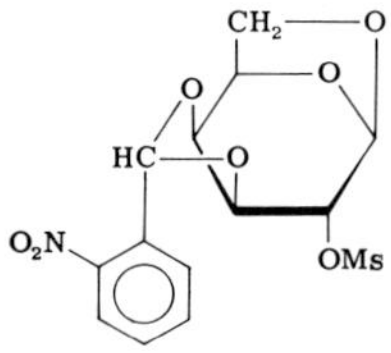	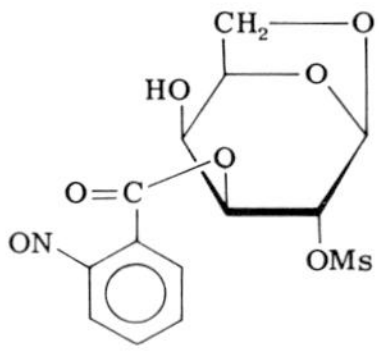 89	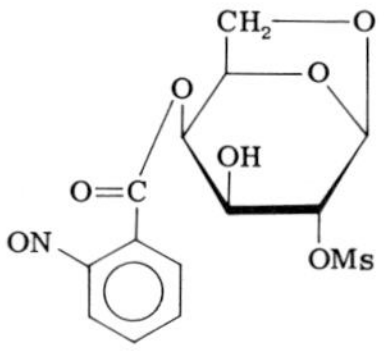 < 5	148
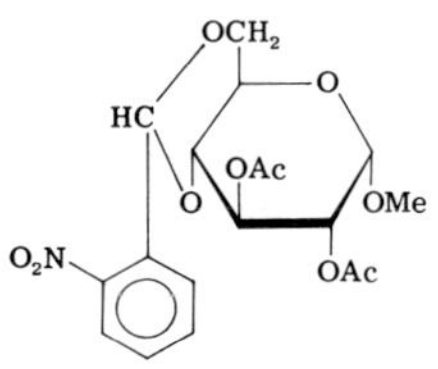	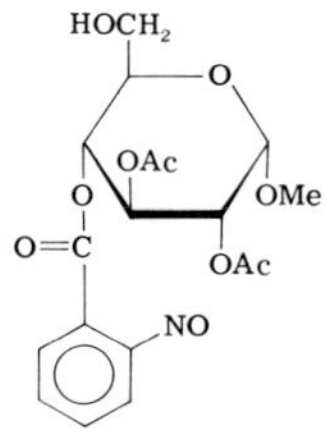 30	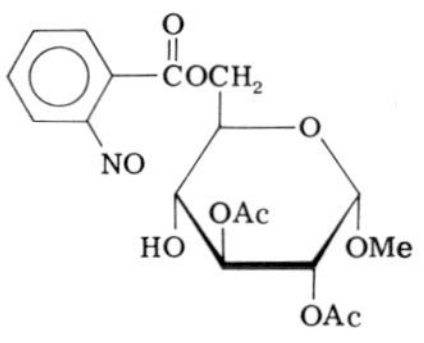 59	147,149
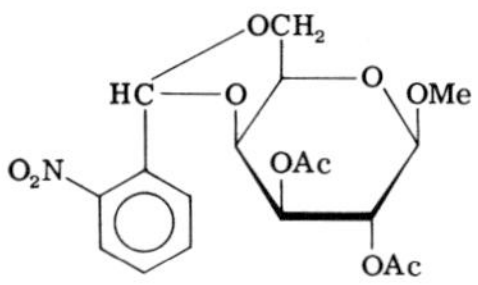	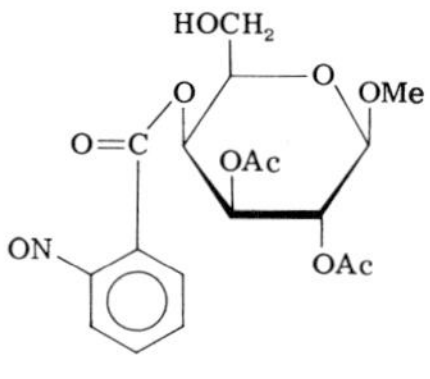 50	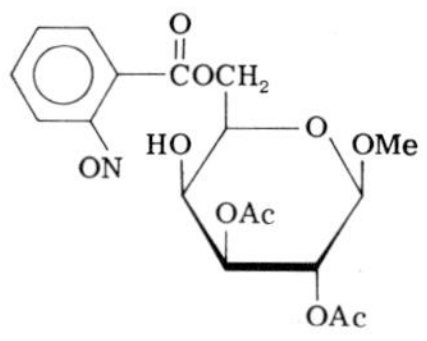 23	147,149
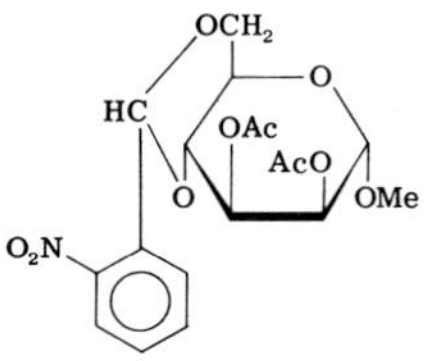	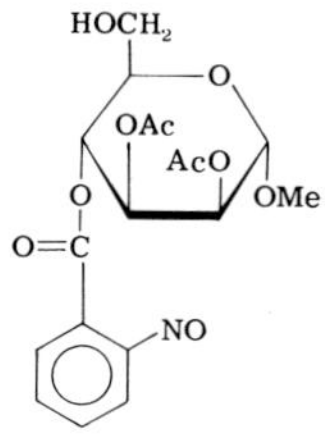 32	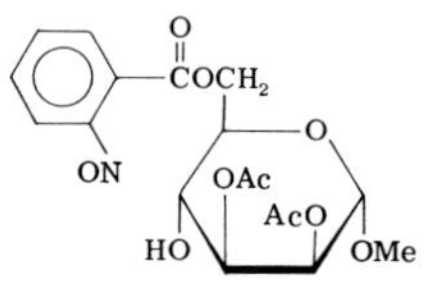 59	149

(*Continued*)

TABLE XIX (*Continued*)

Photochemical Reactions of *o*-Nitrobenzylidene Acetals of Carbohydrates

Reactants	Products and Yields (%)		Ref.
	40	60	149
	50	20	149
	48	17	149
	96 (total)		147, 149

TABLE XIX (Continued)
Photochemical Reactions of o-Nitrobenzylidene Acetals of Carbohydrates

Reactants	Products and Yields (%)	Ref.
	(No yields reported)	149
	45 32	149

Photolysis of three 2,4-dinitroanilino-substituted carbohydrates, compounds that differ considerably from each other in photochemical reactivity, has been reported.[150,151] 1-Deoxy-1-(2,4-dinitroanilino)-D-glucitol (**73**) is photochemically unreactive; in contrast, sodium 2-deoxy-2-(2,4-dinitroanilino)-D-gluconate (**74**) produces D-arabinose in 52% yield upon irradiation.[150] The behavior of compounds **73** and **74** indicates that oxidative loss of the 2,4-dinitroanilino group during photolysis is only possible when it is accompanied by simultaneous decarboxylation. The evidence gathered from the considerable study of this reaction for noncarbohydrate systems suggested that this process is quite complex. Although useful, mechanistic proposals have

(150) A. E. El Ashmawy, D. Horton, and K. D. Philips, *Carbohydr. Res.*, 9 (1969) 350–355.
(151) T. L. Nagabhushan, J. J. Wright, A. B. Cooper, W. N. Turner, and G. H. Miller, *J. Antibiot.*, 31 (1978) 43–48.

[Structures of compounds 73 and 74 with photochemical reaction conditions shown]

73: H₂CNH—(2,4-dinitrophenyl)—... polyol chain; hv, H₂O/MeOH → no reaction

74: sodium carboxylate derivative with 2,4-dinitrophenyl group; hv, H₂O → pyranose sugar + 2-amino-4-nitro-nitrosobenzene

been advanced,[152] a completely satisfactory pathway for this reaction has not yet been described.

The nitroalkene (E)-6,7,8-trideoxy-1,2:3,4-di-O-isopropylidene-7-C-nitro-α-D-*galacto*-oct-6-enose (**75**) is the single nitro sugar thus far photochemically investigated in which the nitro group is not attached to an aromatic ring.[153,154] Irradiation of **75** results in isomerization of the double bond, producing the Z isomer **76** (27% yield), and in a rearrangement–fragmentation process yielding (E)- and (Z)-6,8-dideoxy-1,2:3,4-di-O-isopropylidene-α-D-*galacto*-oct-5-enos-7-ulose (**77** and **78**) in 8 and 6% yields, respectively. A reasonable pathway for formation of compounds **77** and **78** begins with rearrangement of **75** to the intermediate nitrous ester **79**, a process similar to that proposed for excited β-methyl-β-nitrostyrene.[155] The reaction is complete when the nitrous ester experiences light-initiated, nitrogen–oxygen-bond homolysis and subsequent disproportionation (see Scheme 27).

Photolysis of the nitric esters **80–82** causes nitrogen–oxygen bond-

(152) O. Meth-Cohn, *Tetrahedron Lett.*, (1970) 1235–1236.
(153) G. B. Howarth, D. G. Lance, W. A. Szarek, and J. K. N. Jones, *Chem. Commun.*, (1968) 1349.
(154) G. B. Howarth, D. G. Lance, W. A. Szarek, and J. K. N. Jones, *Can. J. Chem.*, 47 (1969) 81–87.
(155) O. L. Chapman, P. G. Cleveland, and E. D. Hoganson, *Chem. Commun.*, (1966) 101–102.

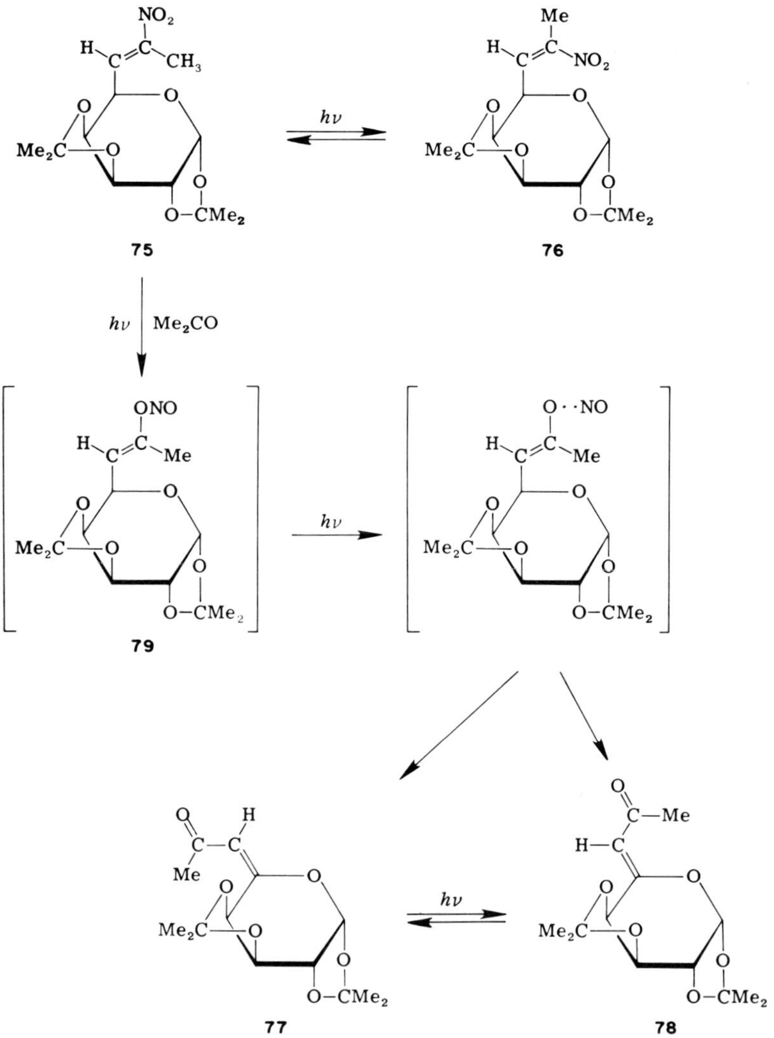

Scheme 27.—Proposed Mechanism for Photochemical Reaction of (E)-6,7,8-Trideoxy-1,2:3,4-di-O-isopropylidene-7-C-nitro-α-D-galacto-oct-6-enose (75).

homolysis, to produce nitrogen dioxide and an alkoxyl radical.[156] The alkoxyl radical abstracts a hydrogen atom from the solvent, to form an alcohol (see Scheme 28). This photochemical removal of a nitro group occurs in essentially quantitative yield (see Table XX). If an unstable

(156) R. W. Binkley and D. J. Koholic, *J. Org. Chem.*, 44 (1979) 2047–2048.

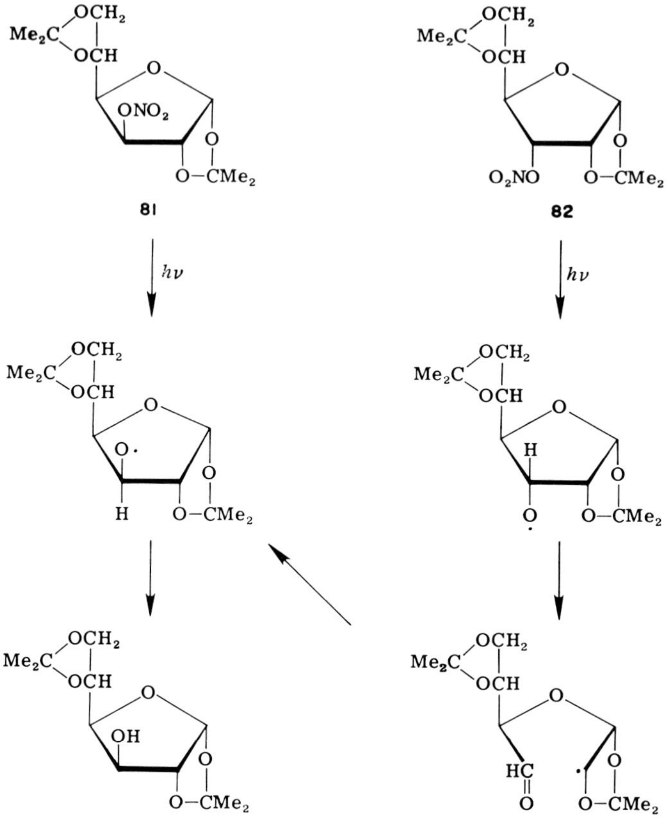

Scheme 28.—Proposed Mechanism for Photochemical Reaction of 1,2:3,4-Di-O-isopropylidene-3-O-nitro-α-D-allofuranose (**81**) and 1,2:3,4-Di-O-isopropylidene-3-O-nitro-α-D-glucofuranose (**82**).

alkoxyl radical is formed by photolysis, rearrangement occurs (see Scheme 28).

2. Azides

Irradiation of a carbohydrate azide results in loss of nitrogen, and rearrangement (probably by way of a nitrene intermediate) to produce an imine (see Scheme 29). The imine, which, in many instances, exists

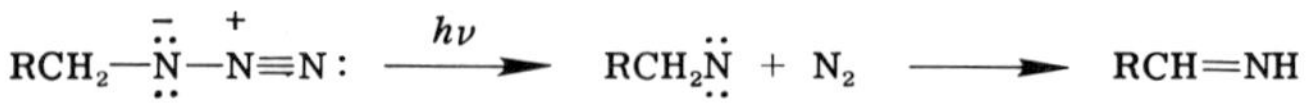

Scheme 29.—Proposed Mechanism for Photochemical Reaction of Carbohydrate Azides.

TABLE XX

Photochemical Reactions of Nitrous Esters of Carbohydrates

Reactants	Products and Yields[156] (%)
[structure with CH$_2$ONO$_2$, Me$_2$C, O-CMe$_2$] **80**	[structure with CH$_2$OH, Me$_2$C, O-CMe$_2$] 100
[structure with OCH$_2$, Me$_2$C, OCH, R^1, R^2, O-CMe$_2$] R^1 = H, R^2 = ONO$_2$ (**81**) R^1 = ONO$_2$, R^2 = H (**82**)	[structure with OCH$_2$, Me$_2$C, OCH, OH, O-CMe$_2$] 100 92

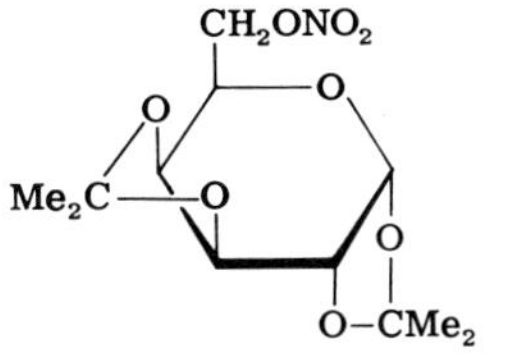

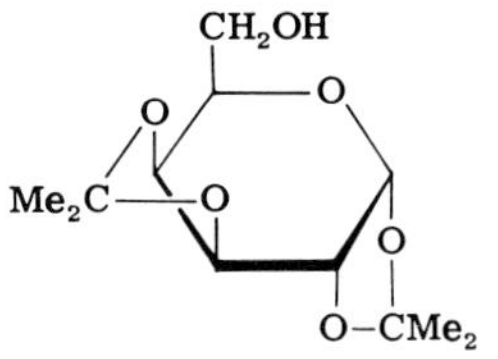

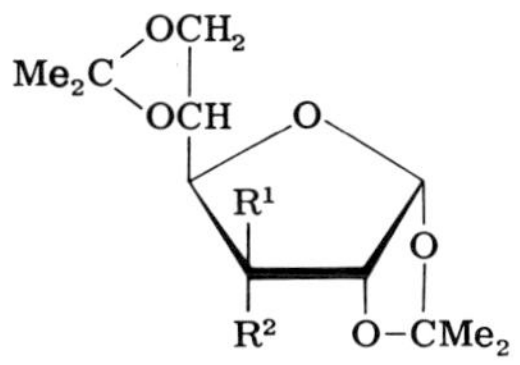

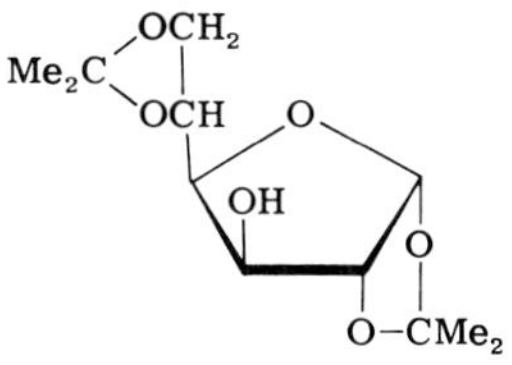

in a polymeric form,[164] is readily hydrolyzed to the corresponding carbonyl compound. This photolysis–hydrolysis sequence is effective in converting primary azides into aldehydes, but it produces low yields of ketones from secondary azides (see Table XXI).

Photolysis of glycosyl azides affords products (see Table XXI) that differ significantly from those formed by photolysis of other azides.[175,176] A probable explanation for this difference in reactivity is that, unlike other imines, the ring in iminolactones produced by irradiation of glycosyl azides can open spontaneously, to give unstable cyanohydrins[176] (see Scheme 30). The cyanohydrins can then lose hydrogen cyanide, to offord the next-lower aldoses. During photolysis of some glycosyl azides, an additional product is formed, one for which the general structure **83** has been proposed. This additional product,

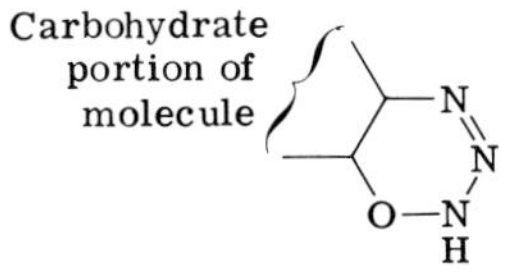

[structure labeled: Carbohydrate portion of molecule, with N=N, O—N—H]

83

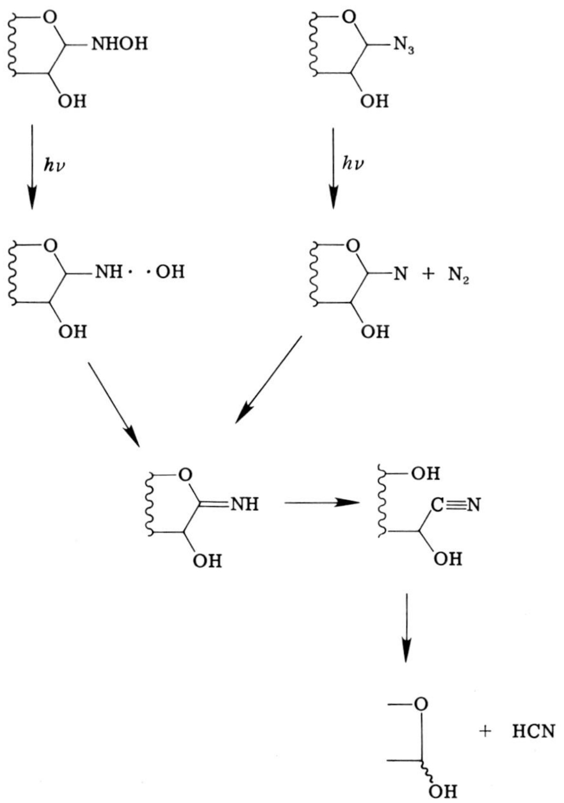

Scheme 30.—Proposed Mechanism for Photochemical Reactions of Glycosyl Azides and of Sugar Oximes.

which may be photochemically reactive,[176] reverts to the starting material on standing.

3. Diazo Compounds

Methyl 3,4,5,6-tetra-O-acetyl-2-deoxy-2-diazo-D-*arabino*-hexonate (**84**) has been irradiated in methanol and in 2-propanol.[177] In methanol, the only photoproduct is the enol acetate **85**; however, irradiation in 2-propanol results in formation of minor proportions (6%) of **85** and the alkene **86** (7%), but the major product is the deoxy sugar **87** (61%). The difference in reactivity of **84** in these two solvents is probably a reflection of the difference in the ability of methanol and 2-propanol to function as hydrogen donors when reacting with a carbene (see Scheme 31). In methanol, a 1,2-hydrogen atom shift to the divalent

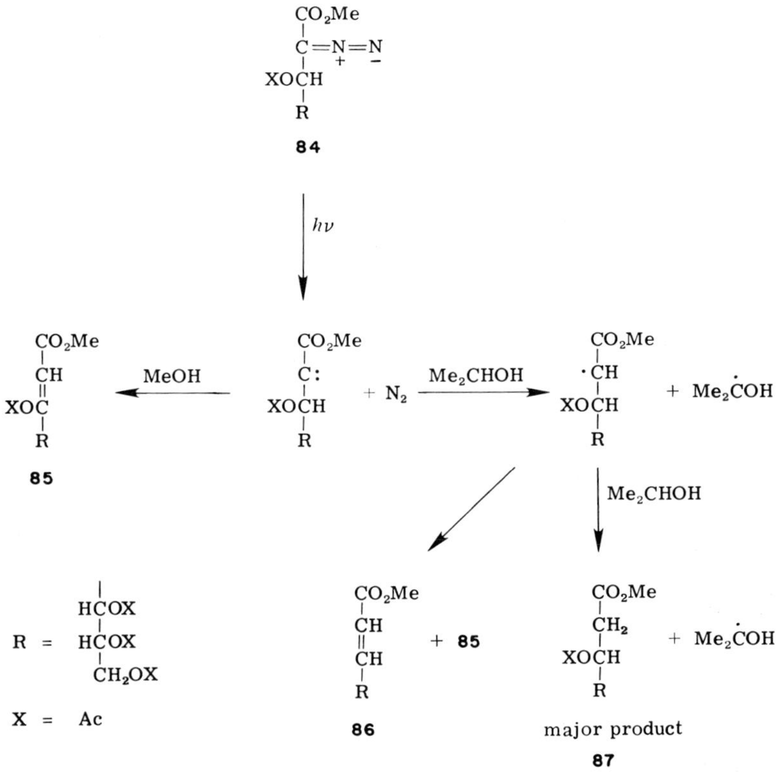

Scheme 31.—Proposed Mechanism for Photochemical Reaction of Methyl 3,4,5,6-Tetra-O-acetyl-2-deoxy-2-diazo-D-*arabino*-hexonate (**84**).

carbon atom occurs faster than reaction with the solvent. In 2-propanol, a more effective hydrogen-donor than methanol, hydrogen abstraction from the solvent competes effectively with internal rearrangement.

4. Oximes

Photolysis of sugar oximes produces iminolactones.[178,179] These are the compounds proposed to arise from irradiation of glycosyl azides; however, the mechanism leading to formation of iminolactones from these two starting-materials (azides and oximes) must be quite different (see Scheme 30). Photolysis of azides is considered to generate a nitrene, whereas photolysis of oximes produces an iminolactone that

Table XXI
Photochemical Reactions of Carbohydrate Azides

Reactants RCH_2N_3	Products and Yields[a] (%) $RCH{=}NH$	Ref.
R = [pyranose with XO, OMe, NHAc]		
X = Bz	(Yield not reported)	157
X = H	38	157
X = Ac	62	157
R = [pyranose with XO, OX, OCH₃, OX] X = Ac	62 50	158 159
R = [bicyclic sugar with Me₂C and O–CMe₂ acetals]	43	160
R = [adenosine derivative with F and Me₂ acetal]	(Yield not reported)	161

TABLE XXI (Continued)
Photochemical Reactions of Carbohydrate Azides

Reactants RCH_2N_3	Products and Yields[a] (%) $RCH=NH$	Ref.
R = (ribofuranose with O-C(Me)₂-O and base B)		
B = adenine	54	162
B = uracil	58	162
R = AcOCH(–)–CH₂–CH₂–C(CH₃)₂(OH)–CH₃	51	163
(diacetone azido sugar)	(imine product) 18	164
	34	165

(Continued)

TABLE XXI (*Continued*)

Photochemical Reactions of Carbohydrate Azides

Reactants RCH_2N_3	Products and Yields[a] (%) $RCH=NH$	Ref.
R = sugar (HCOH, HCOH, ring with OX, HO, OMe, OX; X = CH$_2$Ph)	33	166
R = sugar (OCH$_2$Ph, O–CMe$_2$, HOCH, CH$_2$, X^{1}CX^2) X^1 = OH, X^2 = H X^1 = H, X^2 = OH	16 25	167 167
R = disaccharide (OMe, OH, OH, OH, HO, OH)		
[CH$_2$N$_3$ sugar unit]$_n$	[HC=NH sugar unit]$_n$	169, 170, 171

TABLE XXI (Continued)

Photochemical Reactions of Carbohydrate Azides

Reactants RCH_2N_3	Products and Yields[a] (%) $RCH{=}NH$		Ref.
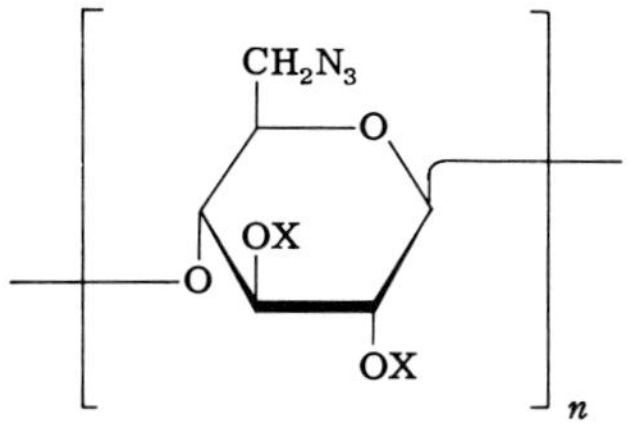 [polysaccharide with CH₂N₃, OX, OX substituents; X = H; X = Ac]	[polysaccharide with HC=NH, OX, OX substituents]		172, 173, 174
[pyranose with N₃, R = H]	HCN + N₂ + [pyranose with HO, HO, OH]	47	175, 176
R 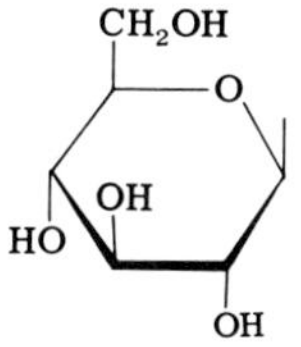 = [pyranose with CH₂OH, OH, HO, OH]		53	176
R 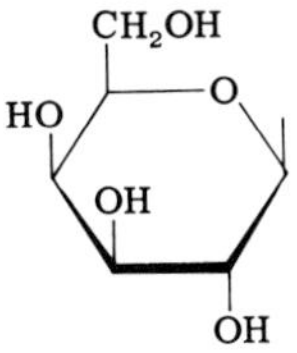 = [pyranose with CH₂OH, HO, OH, OH]		51	176
R 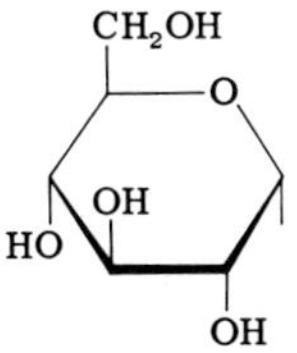 = [pyranose with CH₂OH, HO, OH, OH]		0[b]	175, 176

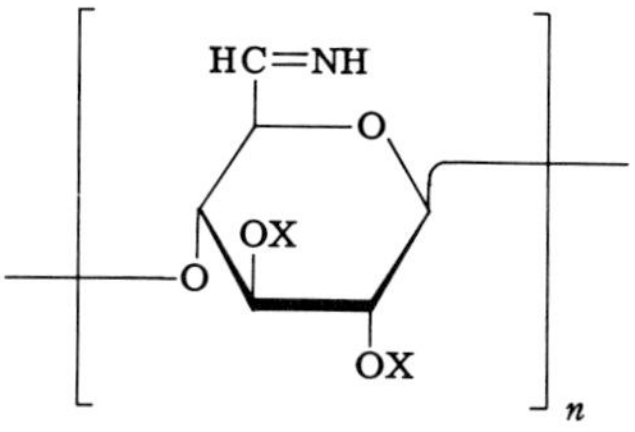

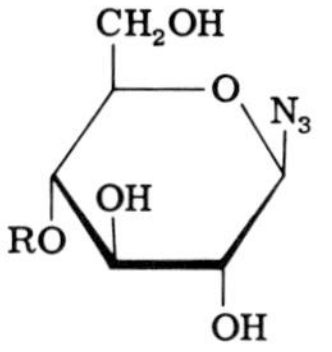

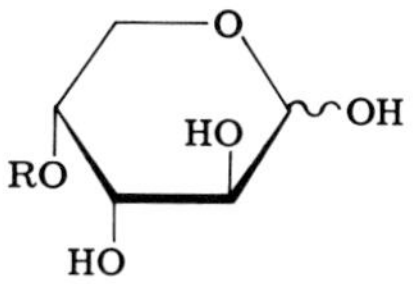

(Continued)

TABLE XXI (*Continued*)

Photochemical Reactions of Carbohydrate Azides

Reactants RCH$_2$N$_3$	Products and Yields[a] (%) RCH=NH	Ref.
[structure: pyranose with CH$_2$OH, OH, HO, OH, N$_3$]	HCN + N$_2$ + [structure: pyranose with O, HO, HO, HO, OH] 60	175, 176
[structure: pyranose with CH$_2$OH, HO, OH, OH, N$_3$]	HCN + N$_2$ + [structure: pyranose with O, HO, OH, HO, OH] 65	175, 176
[structure: pyranose with HO, OH, OH, N$_3$]	b	175, 176
[structure: furanose with HOCH$_2$, HO, OH, N$_3$]	b	175, 176
[structure: pyranose with CH$_2$OX, XO, OX, OX, N$_3$] X = Ac	[structure: pyranose with CH$_2$OX, XO, OX, OX, =NH] 37	175, 176

[a] Product usually isolated as aldehyde or its derivative. [b] Unstable product formed that reverts to the starting azide.

TABLE XXII

Photochemical Reactions of Carbohydrate Oximes

Reactant	Products and Percent Yields	
	Iminolactone	Aldose
D-Galactose oxime	81	D-Lyxose (14)
D-Mannose oxime	33[a]	
D-Ribose oxime	23	
D-Arabinose oxime	28	
D-Xylose oxime	15	D-Threose (3)
D-Lyxose oxime		D-Erythrose (18)

[a] D-Arabinose is formed on standing.

may arise by way of nitrogen–oxygen-bond homolysis followed by hydrogen-atom transfer (see Scheme 30). The yield of the next-lower aldose from an oxime irradiation is quite dependent upon the particular oxime photolyzed (see Table XXII).

(157) D. Horton, W. Weckerle, and B. Winter, *Carbohydr. Res.*, 70 (1979) 59–73.
(158) R. L. Whistler and A. K. M. Anisuzzaman, *J. Org. Chem.*, 34 (1969) 3823–3824.
(159) D. Horton, A. E. Luetzow, and J. C. Wease, *Carbohydr. Res.*, 8 (1968) 366–367.
(160) A. R. Gibson, L. D. Melton, and K. N. Slessor, *Can. J. Chem.*, 52 (1974) 3905–3912.
(161) I. D. Jenkins, J. P. H. Verheyden, and J. G. Moffatt, *J. Am. Chem. Soc.*, 93 (1971) 4323–4324.
(162) D. C. Baker and D. Horton, *Carbohydr. Res.*, 21 (1972) 393–405.
(163) W. A. Szarek, D. M. Vyas, and L.-Y. Chen, *Carbohydr. Res.*, 53 (1977) c1–c4.
(164) D. M. Clode and D. Horton, *Carbohydr. Res.*, 14 (1970) 405–408.
(165) R. L. Whistler and L. W. Doner, *J. Org. Chem.*, 35 (1970) 3562–3563.
(166) H. C. Jarrell and W. A. Szarek, *Can. J. Chem.*, 56 (1978) 144–146.
(167) H. C. Jarrell and W. A. Szarek, *Carbohydr. Res.*, 67 (1978) 43–54.
(168) H. F. G. Beving, A. E. Luetzow, and O. Theander, *Carbohydr. Res.*, 41 (1975) 105–115.
(169) D. M. Clode, D. Horton, M. H. Meshreki, and H. Shoji, *Chem. Commun.*, (1969) 694–695.
(170) D. M. Clode and D. Horton, *Carbohydr. Res.*, 17 (1971) 365–373.
(171) D. Horton, A. E. Luetzow, and O. Theander, *Carbohydr. Res.*, 27 (1973) 268–272.
(172) D. Horton, A. E. Luetzow, and O. Theander, *Carbohydr. Res.*, 26 (1973) 1–19.
(173) D. M. Clode and D. Horton, *Carbohydr. Res.*, 19 (1971) 329–337.
(174) D. M. Clode and D. Horton, *Carbohydr. Res.*, 12 (1970) 477–479.
(175) J. Plenkiewicz, G. W. Hay, and W. A. Szarek, *Can. J. Chem.*, 52 (1974) 183–185.
(176) W. A. Szarek, O. Achmatowicz, J. Plenkiewicz, and B. K. Radatus, *Tetrahedron*, 34 (1978) 1427–1433.
(177) D. Horton and K. D. Philips, *Carbohydr. Res.*, 22 (1972) 151–162.
(178) W. W. Binkley and R. W. Binkley, *Tetrahedron Lett.*, (1970) 3439–3442.
(179) R. W. Binkley and W. W. Binkley, *Carbohydr. Res.*, 23 (1972) 283–288.

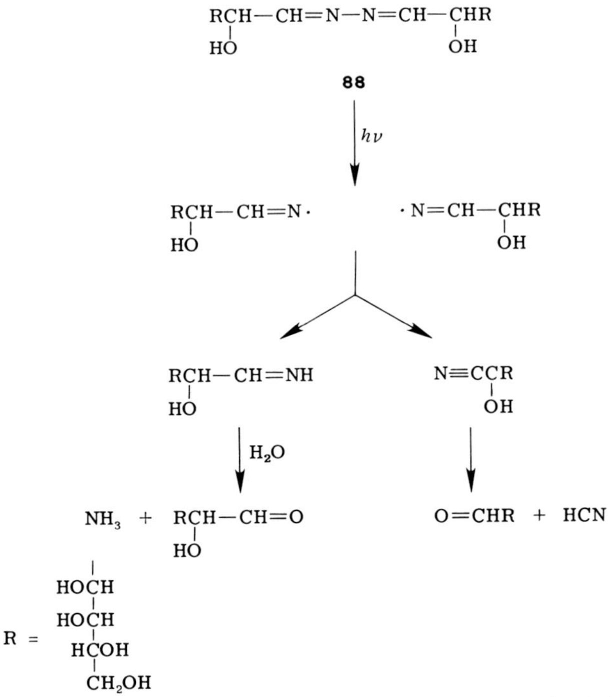

Scheme 32

$$RCH(OH)-CH=N-N=CH-CHR(OH)$$
88

$$\downarrow h\nu$$

$$RCH(OH)-CH=N\cdot \quad\quad \cdot N=CH-CHR(OH)$$

$$\swarrow \quad\quad \searrow$$

$$RCH(OH)-CH=NH \quad\quad N\equiv C-CR(OH)$$

$$\downarrow H_2O \quad\quad\quad\quad\quad\quad \downarrow$$

$$NH_3 + RCH(OH)-CH=O \quad\quad O=CHR + HCN$$

$$R = \begin{array}{c} HOCH \\ | \\ HOCH \\ | \\ HCOH \\ | \\ CH_2OH \end{array}$$

Scheme 32.—Proposed Mechanism for Photochemical Reaction of D-Galactose Azine (**88**).

5. Azines

An iminolactone also has been suggested as a key intermediate in the photolysis of D-galactose azine (**88**). Irradiation of **88** is thought to fragment the weak nitrogen–nitrogen bond. Subsequent disproportionation of the radical pair produces[180] a cyanohydrin and an iminolactone (see Scheme 32). The cyanohydrogen loses hydrogen cyanide to form D-lyxose (23% yield), and the acyclic iminolactone is hydrolyzed to D-galactose (37% yield).

VIII. Iodo Compounds

1. Deoxyiodo Sugars

Photolysis of deoxyiodo sugars is an effective method for forming the corresponding deoxy compounds (see Scheme 33); however, both

(180) R. W. Binkley and W. W. Binkley, *Carbohydr. Res.*, 13 (1970) 163–166.

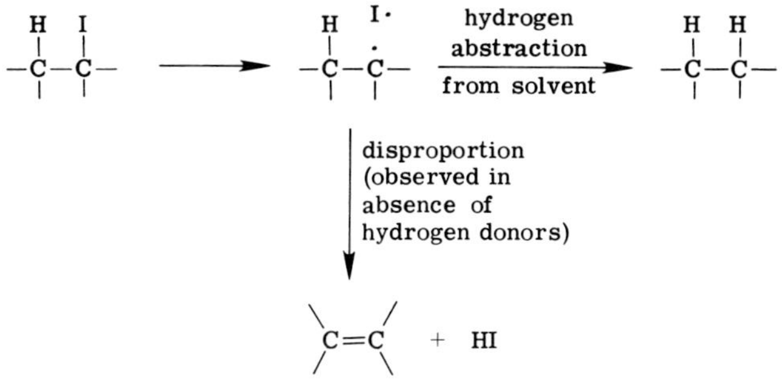

Scheme 33.—Proposed Mechanism for Photochemical Reaction of Deoxyiodo Sugars.

the type of solvent used, and the energy of the incident light, have a decided effect upon the yield of products and the complexity of the reaction mixture. For example, irradiation of 6-deoxy-6-iodo-1,2:3,4-di-O-isopropylidene-α-D-galactopyranose (**89**) in methanol[181,182] or 2-propanol[183] produces 6-deoxy-1,2:3,4-di-O-isopropylidene-α-D-galactopyranose (**90**) in much higher yield than irradiation in 2-methyl-2-propanol (see Table XXIII). Also, the yield of **90** obtained from photolysis of **89** with unfiltered light improves when the higher-energy light from the radiation source is removed by use of an appropriate filter (see Table XXIII). Irradiation of 3-deoxy-3-iodo-1,2:5,6-di-O-isopropylidene-α-D-glucofuranose (**91**) led to a similar observation concerning the dependence on wavelength.[128,183]

The effect of the reaction solvent on iodide photolysis is similar to that observed in irradiation of the diazo compound **84**. Among the various solvents used in iodide reactions, 2-propanol produces the highest yields of deoxy sugars (see Table XXIII). Irradiations in methanol, a good, but less effective, hydrogen donor than 2-propanol, afford deoxy sugars in fair to good yields (see Table XXIII). Hydrogen-atom transfer from 2-methyl-2-propanol is sufficiently difficult that other reactions, such as alkene formation (see Scheme 33), can become significant reaction-pathways in this solvent.

2. o-Iodobiphenylyl Ethers

The o-iodobiphenylyl ethers, **92** and **93**, of 1,2:3,4-di-O-isopropylidene-α-D-galactopyranose and 1,2:5,6-di-O-isopropylidene-α-D-glucofuranose, respectively, have been irradiated in 2-propanol and in 2-

(181) W. W. Binkley and R. W. Binkley, *Carbohydr. Res.*, 8 (1968) 370–371.
(182) W. W. Binkley and R. W. Binkley, *Carbohydr. Res.*, 11 (1969) 1–8.
(183) R. W. Binkley and D. G. Hehemann, *Carbohydr. Res.*, 74 (1979) 337–340.

TABLE XXIII

Photochemical Reactions of Deoxyiodo Sugars

Reactants	Solvent	Filter	Yield (%) of Deoxy Sugar	Ref.
89	Me$_2$CHOH	Corex[a]	95	183
	MeOH	Pyrex[b]	97	181,182
	MeOH	none	83	181,182
	Me$_3$COH	none[c]	36	181,182
91	Me$_2$CHOH	Corex	99	183
	MeOH	none	32	128
	Me$_2$CHOH	Corex	100	183
	Me$_2$CHOH	Corex	93	183
	Me$_2$CHOH	Corex	80	183

TABLE XXIII (*Continued*)

Photochemical Reactions of Deoxyiodo Sugars

Reactants	Solvent	Filter	Yield (%) of Deoxy Sugar	Ref.
(CH₂I, OMe furanose with isopropylidene)	Me₂CHOH	Corex	72	183
(iodo furanose with di-isopropylidene)	Me₂CHOH	Corex	90	183
(CH₂I pyranose disaccharide, R = Ac)	MeOH	none	57	184
(CH₂I pyranose with OMe, HO groups)	H₂O, HCHO	none	9	185

[a] Transmittance at wavelengths <260 nm, 0%. [b] Transmittance at wavelengths <280 nm, 0%. [c] 6-Deoxy-1,2:3,4-di-*O*-isopropylidene-β-L-*arabino*-hex-5-enopyranose (**96**) formed in 32% yield.

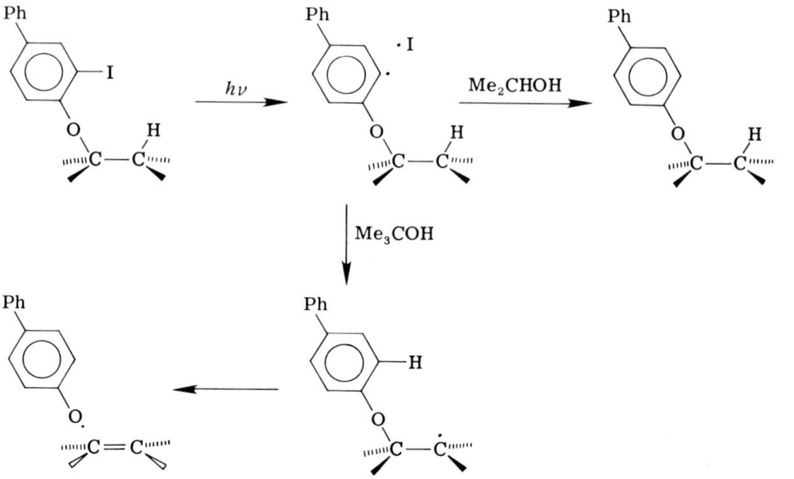

Scheme 34.—Proposed Mechanism for Photochemical Reaction of o-Iodobiphenylyl Ethers.

methyl-2-propanol.[186] Irradiation of these ethers (**92** and **93**) in 2-propanol resulted in essentially quantitative replacement of iodine by hydrogen (see Scheme 34) to give **94** and **95**, respectively. In contrast, irradiation of **92** in 2-methyl-2-propanol produced the enol ether **96** as the major product. No alkene was formed from irradiation of **93** in 2-methyl-2-propanol. Intramolecular hydrogen-abstraction, a process possible only when hydrogen-donating solvents are absent, has been proposed for explaining alkene formation during irradiation in 2-methyl-2-propanol (see Scheme 34). The failure of **93** to form an enol ether upon photolysis demonstrated that other factors, such as ring strain arising from introduction of unsaturation, can determine the course of the reaction.

IX. Organometallic Compounds

The single, organometallic, carbohydrate derivative whose irradiation has thus far been reported[187] is methyl 2-(acetoxymercuri)-3,4,6-tri-O-acetyl-2-deoxy-β-D-glucopyranoside (**97**). Irradiation of **97** in

---

(184) E. R. Guilloux, J. Defaye, R. H. Bell, and D. Horton, *Carbohydr. Res.*, 20 (1971) 421–426.
(185) P. J. Garegg, J. Hoffman, B. Lindberg, and B. Samuelsson, *Carbohydr. Res.*, 67 (1978) 263–269.
(186) R. W. Binkley and D. J. Koholic, *J. Org. Chem.*, 44 (1979) 3357–3360.
(187) D. Horton, J. M. Tarelli, and J. D. Wander, *Carbohydr. Res.*, 23 (1972) 440–446.

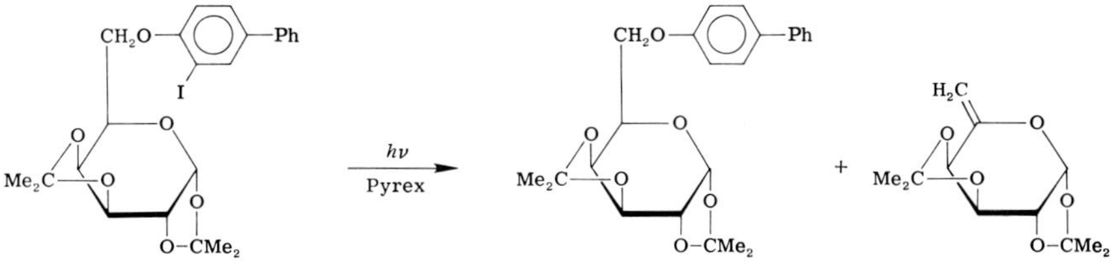

	94	96
solvent		
Me₃COH	14%	80%
Me₂CHOH	94%	0%

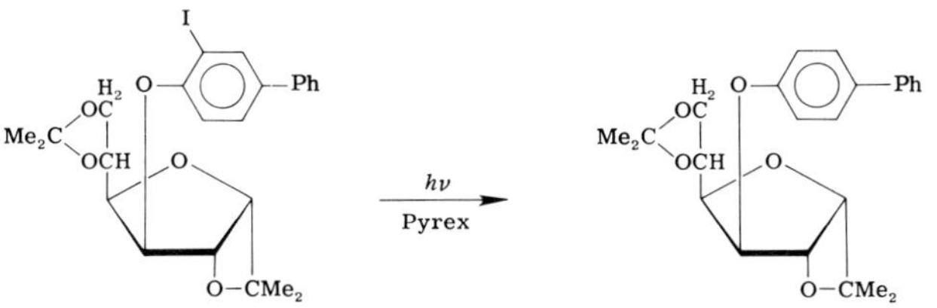

	95	
solvent		
Me₃COH	51%	(no alkene formed)
Me₂CHOH	96%	

methanol yields the corresponding deoxy compound, 3,4,6-tri-*O*-acetyl-2-deoxy-D-*arabino*-hexopyranoside (**99**) and elemental mercury. The radical **98** is a probable intermediate in the photochemical process (see Scheme 35).

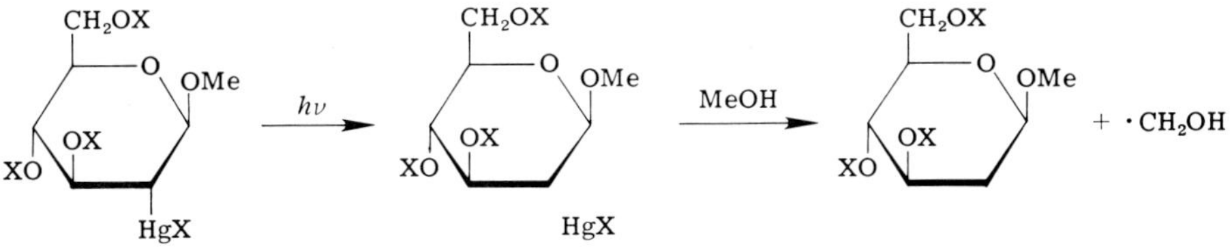

X = Ac

Scheme 35.—Proposed Mechanism for Photochemical Reaction of Methyl 2-(Acetoxymercuri)-3,4,6-tri-*O*-acetyl-2-deoxy-β-D-glucopyranoside (**97**).

Table XXIV
Photochemical Reactions of Carbohydrate Phosphates

Reactants	Products	References
CHO–HCOH–HOCH–HCOH–HCOH–CH$_2$OPO$_3^{2-}$ + O$_2$ **100**	CO$_2$H–HCOH–HOCH–HCOH–HCOH–CH$_2$OPO$_3^{2-}$ + CHO–HOCH–HCOH–HCOH–CH$_2$OPO$_3^{2-}$	188
(pyranose with CH$_2$OH, OH, OH, OH, OPO$_3^{2-}$) + O$_2$ **101**	(lactone pyranose) + (pyranose with HO, OH, OH, OH)	189
CH$_2$OH–C=O–HOCH–HCOH–HCOH–CH$_2$OPO$_3^{2-}$ + O$_2$ **102**	CHO–CH$_2$–HOCH–CH$_2$OPO$_3^{2-}$ + (cyclobutane with O$_3^{2-}$PO, OH, HO, H)	190
	+ OCHCH$_2$OPO$_3^{2-}$ + CHO–HCOH–CH$_2$OPO$_3^{2-}$	

X. Phosphates

Compounds **100** and **101**, unprotected carbohydrates containing a phosphate group, exhibit in photochemical reactivity a marked similarity to other unprotected, carbohydrate systems[188–191] (see Scheme

(188) C. Triantaphylides and M. Halmann, *J. Chem. Soc. Perkin Trans. 1*, (1975) 34–40.
(189) M. Trachtman and M. Halmann, *J. Chem. Soc. Perkin Trans. 2*, (1977) 132–137.
(190) C. Triantaphylides and R. Gerster, *J. Chem. Soc. Perkin Trans. 2*, (1977) 1719–1724.
(191) J. Greenwald and M. Halmann, *J. Chem. Soc. Perkin Trans. 2*, (1972) 1095–1101.

16). Irradiation of α-D-glucose 6-(disodium phosphate)[188] (**100**) yields products that retain the phosphate group (see Table XXIV) but are otherwise the same as those formed from irradiation of D-glucose itself (see Table XII). Photolysis of α-D-glucosyl (dipotassium phosphate) (**101**) causes rapid release of orthophosphate, followed by formation of the same carbohydrate photoproducts[189] as are produced by irradiation of D-glucose (see Table XXIV). D-Fructose 6-(disodium phosphate)[190] (**102**), a compound that exists to a significant extent in the *keto* form, experiences the α-cleavage reaction characteristic of carbonyl compounds (see Scheme 6 and Table XXIV).

FLUORINATED CARBOHYDRATES

By Anna A. E. Penglis*

Department of Chemistry, Queen Elizabeth College, Campden Hill Road, London W8 7AH, United Kingdom

I. Introduction	195
II. Synthesis of Fluorinated Carbohydrates	199
1. Glycosyl Fluorides	199
2. Displacement of Primary Sulfonyloxy Groups by Fluorine	204
3. Epoxide Cleavage by Fluoride Ion	212
4. Displacement of Secondary Sulfonyloxy Groups by Fluorine	218
5. Direct Displacement of Oxygen Functions by Fluorine	225
6. Addition to Glycals	229
7. Total Syntheses	237
8. Isotopically Modified, Fluorinated Carbohydrates	238
9. Miscellaneous	240
III. Physical Methods	253
1. Mass Spectrometry	253
2. ^{19}F-Nuclear Magnetic Resonance Spectroscopy	256
3. Other Methods	280
IV. Some Biological Applications of Fluorinated Carbohydrates	281
V. Addendum	284

I. Introduction

Interest in organic fluorine chemistry was very limited until the second quarter of this century, when the application of fluorocarbons in refrigeration initiated a wave of research in their synthesis and properties. The importance of organic fluorine compounds in general was becoming gradually apparent, for example, in their use as insecticides, and there developed a continuing interest in their preparation and manufacture and in the study of their properties. During World War II, a number of events contributed to the widening of this inter-

---

* Present address: The Dyson Perrins Laboratory, University of Oxford, South Parks Road, Oxford OX1 3QY, U. K. The author thanks the University of London for the award of a University Postgraduate Studentship (1973–1975).

est. In the United Kingdom, for instance, precautions had to be taken against the possibility of use of fluorinated compounds against the Allies, but mainly, fluorinated compounds were needed in the United States in connection with the fractionation of the isotopes of uranium as the volatile hexafluorides (in the "Manhattan project" for the development of the atomic bomb). Additional impetus was given to research in organic fluorine chemistry by the realization that replacement by fluorine of hydrogen in naturally occurring compounds altered their biological properties dramatically. For example, fluoroacetate causes "lethal synthesis" and 9α-fluorohydrocortisone exhibits enhanced corticoid activity. Such observations undoubtedly had a profound effect towards increasing interest in the synthesis of fluorinated carbohydrates.

The selective introduction of fluorine into naturally occurring compounds and organic compounds having functional groups has, in general, presented the synthetic chemist with a continuous challenge. Mild methods are required. In carbohydrates, it is usually the hydroxyl groups that have been replaced by fluorine, although fluorine has replaced hydrogen in a few instances, such as in nucleocidin[1] (**1**), in 2-deoxy-2,2-difluoro-D-arabinopyranose[2] (**2**), and in 2,5-anhydro-1-deoxy-1,1-difluoro-D-ribitol[3,4] (**3**). Fluorine and hydrogen have similar

Van der Waals radii, but differ dramatically in the polarization of their bonds to carbon. Even greater comparison can be made between the fluoro and the hydroxyl groups. The C–F and C–OH bonds are more similar physicochemically (for instance, in their bond lengths and polarizations) than C–F and C–H bonds. Comparative X-ray studies

(1) G. O. Morton, J. E. Lancaster, J. E. Van Lear, G. E. Fulmor, and W. E. Meyer, *J. Am. Chem. Soc.*, 91 (1969) 1535–1537.
(2) J. Adamson, A. B. Foster, and J. H. Westwood, *Carbohydr. Res.*, 18 (1971) 345–347.
(3) P. W. Kent, J. E. G. Barnett, and K. R. Wood, *Tetrahedron Lett.*, (1963) 1345–1348.
(4) P. W. Kent and J. E. G. Barnett, *Tetrahedron, Suppl.*, 7 (1966) 69–74.

of erythriol and 2-deoxy-2-fluoro-DL-erythritol[5] have illustrated this similarity in behavior; more support was obtained from other X-ray studies, for example, that of methyl 4-deoxy-4-fluoro-α-D-glucopyranoside[6] (**4**), which showed the molecule to exist in the $^4C_1(D)$ conformation, with all bond lengths identical to those of methyl α-D-glucopyranoside (**5**), except that the C-6–O-6 bond was found to be

shorter in the former. In **4**, O-6 is antiperiplanar to H-5, whereas in **5**, O-6 is antiperiplanar to C-4. The hydrogen-bonding pattern is different in the two molecules. This difference in hydrogen-bonding capability, where the hydroxyl group can be both a donor and an acceptor, whereas the fluoro group can act only as an acceptor, has been used to advantage in biochemical studies, for example, in studies of yeast galactokinase with deoxyfluoro-D-galactoses,[7] and will be discussed in more detail later (see Section IV). Other characteristics of the C–F bond have also played a part in biological work. For example, the strong inductive effect of the $\rangle CF_2$ group in **2**, which increases the acidity of the vicinal HO-3, was invoked as a possible contributing factor to the observation that **2** is a good substrate for yeast hexokinase[8] (see Section IV).

The application of fluorine-19 nuclear magnetic resonance (^{19}F-n.m.r.) spectroscopy to fluorinated carbohydrates made investigations directed towards their synthesis even more attractive. As the fluoro group does not impose any gross conformational changes on the molecule (see, however, Section III,2e for the consequences of the strong anomeric effect exhibited by fluorine replacing an anomeric hydroxyl group), it is a very sensitive probe for the conformations and configurations of specifically fluorinated analogs. Specifically fluorinated carbohydrates have been found useful for the study of various dependencies of ^{19}F-n.m.r. parameters (see Section III,2). ^{19}F-N.m.r. spectroscopy of carbohydrates has also been used for the study of biologi-

(5) A. Bekoe and H. M. Powell, *Proc. R. Soc., Ser. A*, 250 (1959) 301–315.
(6) W. Choong, N. C. Stephenson, and J. D. Stevens, *Cryst. Struct. Commun.*, 4 (1975) 491–496; *Chem. Abstr.*, 83 (1975) 171,188.
(7) P. Thomas, E. M. Bessell, and J. H. Westwood, *Biochem. J.*, 139 (1974) 661–664.
(8) E. M. Bessell, A. B. Foster, and J. H. Westwood, *Biochem. J.*, 128 (1972) 199–204.

cal problems (see Section IV). It is important, however, to realise that use of the fluorine nucleus as a nuclear magnetic resonance probe, and application of the knowledge so acquired, have been strongly challenged by studies of other nuclei. For example, carbon-13 at natural-abundance levels gives simple spectra (proton-decoupled) relative to hydrogen. This is mainly due to advanced technology that has led to the possibility of a variety of n.m.r. experiments that can be performed routinely, for example, the application of Fourier-transform, n.m.r. techniques, which greatly complement knowledge that can be obtained by ^{19}F-n.m.r. studies, where the fluorinated analogs would usually have to be specifically prepared.

The number of naturally occurring, organic fluorine compounds is very small indeed. Compound 1 is[1] a nucleoside antibiotic that contains fluorine attached to the carbohydrate ring, replacing H-4 in the D-ribosyl moiety. The presence of fluorine was shown by ^{1}H-n.m.r.-, ^{19}F-n.m.r.-, and mass-spectral data, and the structure of this compound has been confirmed by independent synthesis.[9,10]

In the past 25 years, a number of specifically fluorinated carbohydrates has been prepared, and they are reviewed here (to the end of 1978), with the exception of nucleosides containing fluorinated carbohydrate moieties (other than that in 1) and carbohydrate derivatives having groups bearing fluorine, other than mention of general synthetic procedures for sugars bearing the fluoroacetamido and trifluoroacetamido groups, and examples of biological applications of such sugar derivatives involving ^{19}F-n.m.r. spectroscopy. Applications of trifluoroacetic anhydride in carbohydrate chemistry have been reviewed in this Series.[11] The extensive, essentially mechanistic studies of Pedersen and his group on the action of hydrogen fluoride on acylated aldoses, aldosides, glycals, and some selected derivatives are not reviewed here. A brief description of the nature of these reactions, and a selected number of reactions that afford synthetically significant yields of glycosyl fluorides are given (see Sections II,1 and II,6); however, the studies on aldoses and aldosides have been partly reviewed.[12–15] The

(9) I. D. Jenkins, J. P. H. Verheyden, and J. G. Moffatt, *J. Am. Chem. Soc.*, 93 (1971) 4323–4324.
(10) I. D. Jenkins, J. P. H. Verheyden, and J. G. Moffatt, *J. Am. Chem. Soc.*, 98 (1976) 3346–3357.
(11) T. G. Bonner, *Adv. Carbohydr. Chem.*, 16 (1961) 59–84.
(12) J. Lenard, *Chem. Rev.*, 69 (1969) 625–638.
(13) H. Paulsen, *Fortsch. Chem. Forsch.*, 14 (1970) 472–525.
(14) H. Paulsen, *Adv. Carbohydr. Chem. Biochem.*, 26 (1971) 127–195; *Pure Appl. Chem.*, 41 (1975) 69–92.
(15) C. U. Pittman, Jr., S. P. McManus, and J. W. Larsen, *Chem. Rev.*, 72 (1972) 357–438.

reactivity of fluorinated carbohydrates is also not covered in this article; it has been reviewed.[16-19]

II. SYNTHESIS OF FLUORINATED CARBOHYDRATES

Methods for the chemical synthesis of fluorinated carbohydrates were reviewed by Micheel and Klemer[16] in 1961, partly by Barnett[17] in 1967, partly by Hanessian[20] in 1968, by Kent[18] in 1969 and[19] 1972, partly by Wolfrom and Szarek[21] in 1972, by Foster and Westwood[22] in 1973, and by Podešva and Pacák[23] in 1973. Kent,[18,19] and Foster and Westwood,[22] emphasized the replacement by fluorine of hydroxyl groups other than the anomeric hydroxyl group; so did Podešva and Pacák,[23] but with special emphasis on the use of anhydro sugars for the synthesis of fluoro carbohydrates and their derivatives. On the other hand, Micheel and Klemer[16] emphasized the glycosyl fluorides. In 1955, Haynes and Newth[24] partly reviewed the glycosyl fluorides, although very little was known about them at that time. All this is a reflection of the development of interest in fluoro analogs of carbohydrates.

1. Glycosyl Fluorides

This Section deals with the methods of preparation of glycosyl fluorides generally (for the synthesis of glycosyl fluorides by addition reactions to glycals, see Section II,6), and gives a number of examples from the types of glycosyl fluorides that have been prepared.

Glycosyl fluorides were first prepared by brief treatment of acylated aldoses with hydrogen fluoride[25] (it being soon recognized that prolonged treatment could affect other parts of the molecule[26]) as exemplified by the synthesis of tetra-O-acetyl-α-D-glucopyranosyl fluoride

(16) F. Micheel and A. Klemer, Adv. Carbohydr. Chem., 16 (1961) 85–103.
(17) J. E. G. Barnett, Adv. Carbohydr. Chem., 22 (1967) 177–227.
(18) P. W. Kent, Chem. Ind. (London), (1969) 1128–1132.
(19) P. W. Kent, Ciba Found. Symp., (1972) 169–213.
(20) S. Hanessian, Adv. Chem. Ser., 74 (1968) 159–201.
(21) M. L. Wolfrom and W. A. Szarek, in W. Pigman and D. Horton (Eds.), The Carbohydrates: Chemistry and Biochemistry, 2nd edn., Vol IA, Academic Press, New York, 1972, pp. 239–251.
(22) A. B. Foster and J. H. Westwood, Pure Appl. Chem., 35 (1973) 147–168.
(23) J. Podešva and J. Pacák, Chem. Listy, 67 (1973) 785–807; Chem. Abstr., 79 (1973) 115,787.
(24) L. J. Haynes and F. H. Newth, Adv. Carbohydr. Chem., 10 (1955) 207–255.
(25) D. H. Brauns, J. Am. Chem. Soc., 45 (1923) 833–835.
(26) D. H. Brauns, J. Am. Chem. Soc., 48 (1926) 2776–2788.

from peracetylated α-[16] and β-D-glucopyranose.[16,25] This method generally affords the thermodynamically stable isomer, although this generalization does not always hold. For example, treatment of perbenzoylated β-D-glucopyranose with hydrogen fluoride afforded a large proportion of the corresponding β-fluoride, together with some of the α anomer.[27]

As has been elegantly demonstrated[27–37] (for reviews of earlier work, see Refs. 12–15), reaction with hydrogen fluoride of acylated (or partially acylated) aldofuranoses, aldopyranoses, aldofuranosides, and aldopyranosides, together with some selected, methylated or halogenated derivatives, immediately leads to an equilibrium between the fluoride and the 1,2-acyloxonium ion, when formation of such an ion is possible (the glycosyl fluoride is then obtained upon processing); the equilibrium depends on the nature of the reacting carbohydrate and that of the acyl group. On prolonged treatment with hydrogen fluoride, under various conditions, and depending upon the nature of the carbohydrate derivative studied, acyloxonium ions at other parts of the molecule may arise (either following the 1,2-acyloxonium ion initially formed, with simultaneous formation of the glycosyl fluoride, or induced independently by the hydrogen fluoride); these ions may depend on the nature of the acyl group, and their formation is sometimes accompanied by 1,6-anhydro ring-formation, in the case of pyranoses. Acyloxonium ions involving inversion at C-3, where such an inversion was not expected, have also been observed. Unsaturated intermediates also occur in the case of acylated 2-deoxyaldopyranoses. Ring contractions (and an instance of ring expansion) have been observed in these reactions. If the group attached at C-2 in pyranoses is not an acyloxy but a methoxyl group or halogen (Cl or Br), ring contraction usually predominates over other courses. It has been suggested that pyranose–furanose interconversion in hydrogen fluoride is a reversible reaction, and that the equilibrium is determined by the

(27) K. Bock and C. Pedersen, *Acta Chem. Scand.*, 27 (1973) 2701–2709.
(28) C. Pedersen, *Angew. Chem. Int. Ed. Engl.*, 11 (1972) 241.
(29) K. Bock and C. Pedersen, *Acta Chem. Scand.*, 25 (1971) 2757–2764.
(30) I. Lundt and C. Pedersen, *Acta Chem. Scand.*, 25 (1971) 2749–2756.
(31) K. Bock and C. Pedersen, *Acta Chem. Scand.*, 26 (1972) 2360–2366.
(32) S. Jacobsen, S. R. Jensen, and C. Pedersen, *Acta Chem. Scand.*, 26 (1972) 1561–1568.
(33) K. Bock, C. Pedersen, and L. Weibe, *Acta Chem. Scand.*, 27 (1973) 3586–3590.
(34) K. Bock and C. Pedersen, *Acta Chem. Scand.*, Sect. B, 29 (1975) 181–184.
(35) K. Bock and C. Pedersen, *Acta Chem. Scand.*, Sect. B, 30 (1976) 727–732.
(36) K. Bock and C. Pedersen, *Acta Chem. Scand.*, Sec. B, 30 (1976) 777–780.
(37) C. Pedersen and S. Refn, *Acta Chem. Scand.*, Sect. B, 32 (1978) 687–689.

stability of the products. Some selectively halogenated derivatives have also been studied. In all of these reactions, the relative stabilities of the various intermediates in hydrogen fluoride play a dominant role in determining their outcome. Upon processing, glycosyl fluorides different from those expected from the starting compounds are thus usually obtained, often in synthetically significant yields. In addition to glycosyl fluorides and 1,6-anhydrohexopyranoses, which have often been obtained in good yield (on prolonged treatment with hydrogen fluoride, they themselves furnish small proportions of glycosyl fluorides), dimeric species, products arising from the hydrolysis of the glycosyl fluorides formed, and decomposition products, have been observed. The products also depend upon the method of processing.

Those fluorides arising by fluorination accompanied by Walden inversion or ring contraction that had been prepared up to 1969 have been listed.[12] Those that arise from reactions not accompanied by such inversions and contractions include perbenzoylated α-D-mannopyranosyl fluoride,[38] perbenzoylated α-D-lyxopyranosyl fluoride,[39] and perbenzoylated α- and β-D-xylopyranosyl fluorides.[39] Subsequent work arising from these studies, which has furnished glycosyl fluorides in synthetically significant yields, includes the synthesis of perbenzoylated β-D-glucofuranosyl fluoride from the corresponding anomeric pentabenzoate,[31] perbenzoylated α-D-mannofuranosyl fluoride from the corresponding β-D-galactofuranosyl fluoride,[31] perbenzoylated β-D-galactofuranosyl fluoride from perbenzoylated ethyl β-D-galactofuranoside,[33] perbenzoylated α-D-talofuranosyl fluoride from the corresponding β-D-galactofuranosyl fluoride,[33] peracetylated α- and β-D-xylofuranosyl fluoride from peracetylated 1,2-O-isopropylidene-α-D-xylofuranose,[35] and peracetylated 2-bromo-2-deoxy-β-D-glucopyranosyl fluoride from peracetylated 2-bromo-2-deoxy-β-D-glucopyranose.[36] Ring expansion occurred when 1,3,5-tri-O-benzoyl-2-O-methyl-β-D-arabinofuranose was treated with hydrogen fluoride at room temperature, an anomeric mixture of 3,4-di-O-benzoyl-2-O-methyl-D-arabinopyranosyl fluorides resulting.[32] Perbenzoylated 2-deoxy-α-D-*lyxo*-hexopyranosyl fluoride was obtained from the anomeric tetrabenzoates.[30]

Treatment of the acylated glycosyl halide (bromide or chloride) of the opposite anomeric configuration with silver fluoride in acetonitrile was the method initially developed[40] for the synthesis of the less stable anomers, as exemplified by the synthesis of peracylated β-D-glucopyranosyl fluoride[40] from the corresponding α-bromide. It has

(38) C. Pedersen, *Acta Chem. Scand.*, 17 (1963) 673–677.
(39) C. Pedersen, *Acta Chem. Scand.*, 17 (1963) 1269–1275.
(40) B. Helferich and R. Gootz, *Ber.*, 62 (1929) 2505–2507.

been observed,[41] however, that the presence of a participating group at C-2 always leads to the anomer having the fluorine atom *trans*-related to that group. Among others, this was observed in the preparation of peracetylated α-D-mannopyranosyl fluoride from the corresponding α-bromide,[42] and of peracetylated 2-bromo-2-deoxy-α-D-mannopyranosyl fluoride from the corresponding α-bromide.[43] However, treatment of perbenzoylated (α-D-glucosyl fluoride)urono-6,3-lactone with hydrogen fluoride for ~10 min led to a mixture of the anomeric fluorides.[41] The stable α-fluoride has been obtained by reaction of perbenzoylated 2-deoxy-α-D-*arabino*-hexopyranosyl bromide with silver fluoride in acetonitrile containing 10% of dichloromethane.[44] Peracetylated β-D-xylopyranosyl fluoride[45] has been synthesized from the corresponding α-bromide by the action of silver fluoride in acetonitrile. A very minor product subsequently isolated from this reaction was shown by X-ray analysis to be, surprisingly, peracetylated 2-deoxy-2-fluoro-α-D-xylopyranose.[46]

Peracetylated β-D-glucopyranosyl fluoride has also been obtained, in unspecified yield, by treatment of peracetylated 1,6-anhydro-β-D-glucopyranose with hydrogen fluoride in acetic anhydride.[47] Treatment of peracetylated β-D-glucopyranose with 50% hydrogen fluoride in acetic anhydride afforded peracetylated α-D-glucopyranosyl fluoride, together with a small proportion of peracetylated α-D-glucopyranose.[48] Treatment of peracetylated α-lactose with hydrogen fluoride in acetic acid afforded peracetylated α-lactosyl fluoride.[49]

Silver tetrafluoroborate in ether or toluene has also been used for the synthesis of glycosyl fluorides. Peracetylated 2-chloro-2-deoxy-D-gluco- and -mannopyranosyl fluorides have been prepared by treatment of the corresponding chlorides with the aforementioned reagent.[50,51] Products of kinetic control were obtained when diethyl ether was used as the solvent, whereas products of thermodynamic control were obtained when toluene was used instead. Peracetylated

(41) L. D. Hall and P. R. Steiner, *Can. J. Chem.*, 48 (1970) 2439–2443.
(42) L. D. Hall and J. F. Manville, *Can. J. Chem.*, 47 (1969) 1–17.
(43) L. D. Hall and J. F. Manville, *Can. J. Chem.*, 47 (1969) 361–377.
(44) I. Lundt and C. Pedersen, *Acta Chem. Scand.*, 21 (1967) 1239–1243.
(45) I. Lundt and C. Pedersen, *Mikrochim. Acta*, 1 (1966) 126–132.
(46) V. G. Kothe, P. Luger, and H. Paulsen, *Acta Crystallogr., Sect. B*, (1976) 2710–2712; *Chem. Abstr.*, 85 (1976) 170,047.
(47) F. Micheel, A. Klemer, M. Nolte, H. Nordick, L. Tork, and H. Westermann, *Chem. Ber.*, 90 (1957) 1612–1616.
(48) C. Pedersen, *Acta Chem. Scand.*, 16 (1962) 1831–1836.
(49) P. W. Kent and S. D. Dimitrijevich, *J. Fluorine Chem.*, 10 (1977) 455–478.
(50) K. Igarashi, T. Honma, and J. Irisawa, *Carbohydr. Res.*, 11 (1969) 577–578.
(51) K. Igarashi, T. Honma, and J. Irisawa, *Carbohydr. Res.*, 13 (1970) 49–55.

α-D-glucopyranosyl chloride in diethyl ether afforded a mixture of the α- and β-fluorides, with no detectable formation of the 1,2-acetoxonium ion; with a prolonged reaction-time, the products of thermodynamic control were obtained. The same behavior was observed when the 1,2-acetoxonium ion was used as the starting material.

Among others, and in addition to the glycosyl fluorides already described, a number of acetylated aldohexo- and -pentopyranosyl fluorides,[45] perbenzoylated and peracetylated (β-D-glucosyl fluoride)urono-6,3-lactone,[41] and 2-O-acetyl-3,6-anhydro-5-O-benzoyl-β-L-idofuranosyl fluoride[41] have been prepared, together with peracetylated 6-bromo-6-deoxy-α- and -β-D-glucopyranosyl fluoride,[52] peracetylated 2-bromo-2-deoxy-α-D-mannopyranosyl fluoride (by the foregoing method[43] and by addition of the elements of "BrF" to tri-O-acetyl-D-glucal[43,53–57]), perbenzoylated 2-bromo-2-deoxy-α-D-arabino-furanosyl fluoride from the corresponding methyl β-D-furanoside,[58] methylated derivatives of α- and β-D-glucopyranosyl fluoride,[59] 3,4,5-tri-O-acetyl-1-O-methyl-α-D-fructopyranosyl fluoride[60] and the corresponding β anomer,[60,61] peracetylated α- and β-D-fructofuranosyl fluoride,[62] 2-acetamido-2-deoxy-α-[63] and -β-D-glucopyranosyl fluoride,[64] 2-deoxy-2-(p-toluenesulfonamido)-α-D-glucopyranosyl fluoride[65] and its β anomer,[63] perbenzoylated 2-deoxy-α-D-*arabino*-hexopyranosyl fluoride,[44,66] and di-N-acetyl-β-chitobiosyl fluoride.[67]

A number of fluorinated glycosyl fluorides has also been obtained;

(52) F. Micheel, *Chem. Ber.*, 90 (1957) 1612–1616.
(53) L. D. Hall and J. F. Manville, *Chem. Commun.*, (1968) 35–36.
(54) P. W. Kent, F. O. Robson, and V. A. Welch, *Proc. Chem. Soc. (London)*, (1963) 24–25.
(55) P. W. Kent, F. O. Robson, and V. A. Welch, *J. Chem. Soc.*, (1963) 3273–3276.
(56) J. C. Campbell, R. A. Dwek, P. W. Kent, and C. K. Prout, *Chem. Commun.*, (1968) 34–35.
(57) J. C. Campbell, R. A. Dwek, P. W. Kent, and C. K. Prout, *Carbohydr. Res.*, 10 (1969) 71–77.
(58) K. Bock, C. Pedersen, and P. Rasmussen, *Acta Chem. Scand., Sect. B*, 29 (1975) 185–190.
(59) A. Klemer, I. Ridder, and G. Drolshagen, *Chem. Ber.*, 96 (1963) 1976–1985.
(60) F. Micheel and L. Tork, *Chem. Ber.*, 93 (1960) 1013–1020.
(61) D. H. Brauns and H. L. Frush, *Bur. Stand. J. Res.*, 6 (1931) 449–456.
(62) B. Erbing and B. Lindberg, *Acta Chem. Scand., Sect. B*, 30 (1976) 12–14.
(63) F. Micheel and H. Wulff, *Chem. Ber.*, 89 (1956) 1521–1530.
(64) F. W. Ballardie, B. Capon, W. M. Dearie, and R. L. Foster, *Carbohydr. Res.*, 49 (1976) 79–92.
(65) F. Micheel and E. Michaelis, *Chem. Ber.*, 91 (1958) 188–194.
(66) L. D. Hall and J. F. Manville, *Can. J. Chem.*, 45 (1967) 1299–1303.
(67) F. W. Ballardie, B. Capon, M. W. Cuthbert, and W. M. Dearie, *Bioorg. Chem.*, 6 (1977) 483–509.

for example, the peracetylated 2-deoxy-2-fluoro-α-D-glucopyranosyl and -β-D-mannopyranosyl fluorides,[68,69] peracetylated 2-deoxy-2-fluoro-β-D-gluco- and -α-D-mannopyranosyl fluoride,[70,71] peracetylated 2-deoxy-2-fluoro-α-D-xylo- and -β-D-lyxo-pyranosyl fluoride,[72,73] peracetylated 2-O-acetyl-2-deoxy-2-fluoro-β-D-ribopyranosyl fluoride,[74] peracetylated 3-deoxy-3-fluoro-α- and -β-D-glucopyranosyl fluoride and 3-deoxy-3-fluoro-β-D-xylopyranosyl fluoride,[75] peracetylated 4-deoxy-4-fluoro-α- and -β-D-galactopyranosyl fluoride,[76] 6-deoxy-6-fluoro-α- and -β-D-glucopyranosyl fluoride,[77] peracetylated 2-deoxy-2-fluoro-4-O-β-D-galactopyranosyl-α-D-glucopyranosyl fluoride, and peracetylated 2-deoxy-2-fluoro-4-O-β-D-galactopyranosyl-β-D-mannopyranosyl fluoride.[49]

The cis-glycosyl fluorides of the peracetylated 2-deoxy-2-fluoro saccharides have been obtained by addition of trifluoro(fluoroxy)methane to the corresponding peracetylated glycals, whereas the preparation of the other derivatives was achieved by treatment of a suitable precursor with hydrogen fluoride or silver fluoride in acetonitrile; peracetylated 2-deoxy-2-fluoro-α-D-mannopyranosyl fluoride[70,71] was prepared by equilibration of the β anomer in hydrogen fluoride.

2. Displacement of Primary Sulfonyloxy Groups by Fluorine

With certain exceptions, primary fluorination of carbohydrate derivatives is very facile. Fluoride displacement of sulfonyloxy groups by potassium, or cesium, fluoride in 1,2-ethanediol and a variety of other solvents has frequently been used. Tetrabutylammonium (sometimes tetraethylammonium) fluoride in dipolar, aprotic solvents (mainly ace-

(68) J. Adamson, A. B. Foster, L. D. Hall, and R. H. Hesse, Chem. Commun., (1969) 309–310.
(69) J. Adamson, A. B. Foster, L. D. Hall, R. N. Johnson, and R. H. Hesse, Carbohydr. Res., 15 (1970) 351–359.
(70) L. D. Hall, R. N. Johnson, J. Adamson, and A. B. Foster, J. Chem. Soc., D, (1970) 463–464.
(71) L. D. Hall, R. N. Johnson, J. Adamson, and A. B. Foster, Can. J. Chem., 49 (1971) 118–123.
(72) R. A. Dwek, P. W. Kent, P. T. Kirby, and A. S. Harrison, Tetrahedron Lett., (1970) 2987–2990.
(73) C. G. Butchard and P. W. Kent, Tetrahedron, 27 (1971) 3457–3463.
(74) E. L. Albano, R. L. Tolman, and R. V. Robins, Carbohydr. Res., 19 (1971) 63–70.
(75) L. D. Hall, R. N. Johnson, A. B. Foster, and J. H. Westwood, Can. J. Chem., 49 (1971) 236–240.
(76) A. B. Foster, J. H. Westwood, B. Donaldson, and L. D. Hall, Carbohydr. Res., 25 (1972) 228–231.
(77) E. M. Bessell, A. B. Foster, J. H. Westwood, L. D. Hall, and R. N. Johnson, Carbohydr. Res., 19 (1971) 39–48.

tonitrile) has also been employed extensively; the importance of the latter reagent, however, lies in the fact that it effects ready displacement of secondary sulfonyloxy groups by fluorine, whereas the former has been used, often with limited success, for such displacements. The sulfonic ester precursors are usually protected with such normally nonparticipating groups as cyclic acetal or ether (methyl or benzyl) groups that are stable to the conditions of fluorination; exceptions have, however, been encountered; for example, participation by the benzyloxy group (see Section II,4). See also, later in this Section, for reactions involving acetal groups. Ester groups have also been used; however, they provide precursors which rarely lead to high yields of the desired products; also, sometimes, no fluorinated products have been obtained.

A small number of fluorinated analogs has been prepared without prior protection of the hydroxyl groups, but, owing to the strongly basic conditions prevailing, such reactions as concurrent formation of the 3,6-anhydro derivative (7) may occur, as in the fluorination of methyl 6-O-p-tolylsulfonyl-α-D-glucopyranoside (6) with potassium fluoride in 1,2-ethanediol.[77] A previous attempt to use this reaction, in

both 1,2-ethanediol and methanol, did not give rise to fluorinated products.[78] When treated with potassium fluoride in 2-methoxyethanol, phenyl 2,3-di-O-acetyl-6-O-(methylsulfonyl)-4-O-(2,3,4,6-tetra-O-acetyl-α-D-glucopyranosyl)-α-D-glucopyranoside[79] gave a mixture of the 6-deoxy-6-fluoro and the 3,6-anhydro derivatives. When methyl 2-benzamido-2-deoxy-3-O-[D-1-(methoxycarbonyl)ethyl]-6-O-p-tolylsulfonyl-β-D-glucopyranoside (8) was treated with tetrabutylammonium fluoride in butanone, the 6-deoxy-6-fluoro derivative[80] (9) was obtained. The same treatment of the 4-O-acetyl-6-O-(methylsulfonyl) derivative 10 resulted in solvolysis, and a mixture of the 3,6-diol mono-

(78) J. B. Lee and M. M. El Sawi, *Tetrahedron*, 12 (1961) 226–235.
(79) H. Arita and Y. Matsushima, *J. Biochem. (Tokyo)*, 69 (1971) 409–413; *Chem. Abstr.*, 74 (1971) 120,501.
(80) G. D. Diana, *J. Org. Chem.*, 35 (1970) 1910–1912.

acetates (**11** and **12**) was obtained, indicating competing attack by the neighboring ester group. Treatment of 3,4-di-O-acetyl- or -di-O-benzoyl-6-O-p-tolylsulfonyl-D-glucal with fluoride ion in 1,2-ethanediol afforded products of solvolysis.[81] Treatment of the latter compound with cesium fluoride in N,N-dimethylformamide led to the corresponding unsaturated product,[81] whereas treatment of the former with the same reagents led, additionally, to the corresponding, fluorinated product, albeit in low yield.[82] A low yield of the corresponding fluorinated product was also obtained by treatment of ethyl 4-O-acetyl-2,3-dideoxy-6-O-p-tolylsulfonyl-α-D-*erythro*-hex-2-enopyranoside with cesium fluoride in N,N-dimethylformamide.[82] In contrast, 2-acetamido-1,3,4-tri-O-acetyl-2,6-dideoxy-6-fluoro-β-D-glucopyranose has been prepared in 55% yield by treatment of 2-acetamido-1,3,4-tri-O-acetyl-2-deoxy-6-O-(p-nitrophenylsulfonyl)-β-D-glucopyranose with tetrabutylammonium fluoride in acetonitrile for[67] 6 h at 70°. The corresponding 6-p-toluenesulfonate, however, afforded products of decomposition.

(81) I. D. Blackburne, P. M. Fredericks, and R. D. Guthrie, *Aust. J. Chem.*, 29 (1976) 381–391.
(82) G. Descotes, J.-C. Martin, and Tachi-Dung, *Carbohydr. Res.*, 62 (1978) 61–71.

The original preparation of 6-deoxy-6-fluoro-D-glucose[83] was accomplished by treatment of 3,5-O-benzylidene-1,2-O-isopropylidene-6-O-(methylsulfonyl)-α-D-glucofuranose with potassium fluoride dihydrate in methanol. This synthesis was later repeated, using an improved synthesis of the precursor.[77] 3,5-O-Benzylidene-6-deoxy-6-fluoro-1,2-O-isopropylidene-α-D-glucofuranose has also been prepared by treatment of the corresponding ethanesulfonic ester with potassium fluoride dihydrate in methanol.[84] The formation of 6-deoxy-6-fluoro-D-galactose from 1,2:3,4-di-O-isopropylidene-6-O-(methylsulfonyl)-α-D-galactopyranose was accompanied by some O-methylation.[85,86] Use of 1,2-ethanediol as the solvent was introduced[87] to minimize ether formation, although it has also sometimes led to ether formation (for example, see Refs. 81 and 88–90). Anhydrous potassium fluoride in refluxing N,N-dimethylformamide was found inferior to potassium fluoride dihydrate in methanol for the fluorination of methyl 2,3-O-isopropylidene-5-O-(methylsulfonyl)-D-ribofuranoside[91] (see also, Section II,4). On treatment of methyl 2-deoxy-5-O-(methylsulfonyl)- or -p-tolylsulfonyl-D-*erythro*-pentofuranoside with potassium or cesium fluoride in 1,2-ethanediol, the corresponding 5-deoxy-5-fluoro analog was obtained in moderate yield, whereas a better yield (35%) of the fluoride was obtained with tetrabutylammonium fluoride in acetonitrile (together with several unidentified products). Fluorination of methyl 2-deoxy-5-O-p-tolylsulfonyl-3-O-trityl-D-*erythro*-pentofuranoside afforded a higher yield (53%) of the desired fluoro product[92] (compare Section II,4); this route, however, was deprived of convenience owing to the low yields obtained in the preparation of the starting material.

Potassium fluoride in 1,2-ethanediol has further been used, as, for example, in the preparation of 5-deoxy-5-fluoro-D-ribose from methyl 2,3-O-isopropylidene-5-O-(methylsulfonyl)-α-D-ribofuranoside,[87] and of 6-deoxy-6-fluoro-D-glucose from methyl 6-O-p-tolylsulfonyl-α-D-glucopyranoside.[77] In one study of primary fluorinations with potas-

(83) B. Helferich and A. Gnüchtel, *Ber.*, 74 (1941) 1035–1039.
(84) B. Helferich and M. Vock, *Ber.*, 74 (1941) 1807–1811.
(85) N. F. Taylor and P. W. Kent, *Research*, 9 (1956) S28–S29.
(86) P. W. Kent, A. Morris, and N. F. Taylor, *J. Chem. Soc.*, (1960) 298–303.
(87) N. F. Taylor and P. W. Kent, *J. Chem. Soc.*, (1958) 872–875.
(88) A. B. Foster and R. Hems, *Carbohydr. Res.*, 10 (1969) 168–171.
(89) L. Evelyn and L. D. Hall, *Carbohydr. Res.*, 47 (1976) 285–297.
(90) H. S. Srivastava and V. K. Srivastava, *Carbohydr. Res.*, 60 (1978) 210–218.
(91) H. M. Kissman and M. J. Weiss, *J. Am. Chem. Soc.*, 80 (1958) 5559–5564.
(92) M. von Janta-Lipinski, G. Kowollik, K. Gaertner, and P. Langen, *Nucleic Acid Res.-Special Publ.*, 1 (1975) 45–47.

sium fluoride in 1,2-ethanediol, the yields were found to be variable.[89] Reaction[90] of 1,2:3,5-di-O-methylidene-6-O-p-tolylsulfonyl-α-D-glucofuranose with anhydrous potassium fluoride in dry 1,2-ethanediol was shown to afford the corresponding 6-deoxy-6-fluoride, alkene, and 6-(2-hydroxyethyl) ether. Conditions were optimized (short reaction-times) for fluorination; longer reaction-times resulted in higher yields of the alkene, whereas ether formation was fairly constant with the different reaction-times. Cesium fluoride in 1,2-ethanediol was used for the preparation of 2-acetamido-2,6-dideoxy-6-fluoro-D-glucose.[93] 3-O-Benzyl(or p-nitrobenzyl)-1,2-O-isopropylidene-5-O-(methylsulfonyl)-(or p-tolylsulfonyl)-α-D-xylofuranose was resistant to fluoride displacement when potassium fluoride in methanol or 1,2-ethanediol was the fluoride source, whereas treatment of 3-O-benzyl-1,2-O-isopropylidene-5-O-p-tolylsulfonyl-α-D-xylofuranose with tetrabutylammonium fluoride in acetonitrile afforded the 5-deoxy-5-fluoro analog, accompanied by some 3-O-benzyl-5-deoxy-1,2-O-isopropylidene-β-L-*threo*-pent-4-enofuranose.[94] Methyl 2-acetamido-3,4-di-O-acetyl-2,6-dideoxy-6-fluoro-α-D-gluco- (**15**) and -galacto-pyranoside (**16**) have been prepared from methyl 2-benzamido-3-O-benzyl-2-deoxy-4,6-di-O-(methylsulfonyl)-α-D-glucopyranoside (**13**) by selective displacement of the primary mesyloxy group with tetrabutylammonium fluoride in acetonitrile, followed by O-debenzylation, O-acetylation, N-acetyl–N-benzoyl exchange, and O-deacetylation.[95] The *galacto*

(93) M. L. Shulman and A. Ya. Khorlin, *Carbohydr. Res.*, 27 (1973) 141–147.
(94) P. W. Kent and R. C. Young, *Tetrahedron*, 27 (1971) 4057–4064.
(95) A. A. E. Penglis, Ph.D. Thesis (London), 1977.

configuration of **16** was produced by benzoate displacement of the intermediate 4-methanesulfonate (**14**). Among other fluorinated, unsaturated hexoses (see also, earlier in this Section, and Section II,9), 3,4-di-O-benzyl-6-deoxy-6-fluoro-D-glucal has been prepared by treatment of the corresponding 6-p-toluenesulfonate with tetrabutylammonium fluoride in N,N-dimethylformamide, 4-O-benzyl-3,6-dideoxy-6-fluoro-D-glucal for the corresponding 6-p-toluenesulfonate by treatment with cesium fluoride in N,N-dimethylformamide, and ethyl 4-O-benzyl-2,3,6-trideoxy-6-fluoro-α-D-*erythro*-hex-2-enopyranoside by treatment of the corresponding 6-p-toluenesulfonate with cesium fluoride in 1,2-ethanediol.[81]

Several fluoro analogs of ketoses have been reported: 1,6-dideoxy-1,6-difluoro-D-fructose was readily obtained from 2,3-O-isopropylidene-1,6-di-O-p-tolylsulfonyl-β-D-fructofuranose by treatment with potassium fluoride in 1,2-ethanediol under a stream of carbon dioxide.[96] Surprisingly, although the 6-sulfonyloxy group would be expected to be more reactive than the 1-sulfonyloxy group,[97–99] no selectivity was observed. The failure to obtain 1-deoxy-1-fluoro-D-fructopyranose[96] from 2,3:4,5-di-O-isopropylidene-1-O-(methylsulfonyl) (or p-nitrophenylsulfonyl)-β-D-fructopyranose or phenyl 3,4,5-tri-O-acetyl-1-O-(methylsulfonyl)-β-D-fructopyranoside by treatment with potassium or sodium fluoride in 1,2-ethanediol, N,N-dimethylformamide, or formamide at elevated temperatures may be attributed to the fact that nu-

(96) J. Pacák, J. Halásková, V. Štěpán, and M. Černý, *Collect. Czech. Chem. Commun.*, 37 (1972) 3646–3651.
(97) P. A. Levene and R. S. Tipson, *J. Biol. Chem.*, 120 (1937) 607–619.
(98) H. Müller and T. Reichstein, *Helv. Chim. Acta*, 21 (1938) 263–272.
(99) M. S. Feather and R. L. Whistler, *J. Org. Chem.*, 28 (1963) 1567–1569.

cleophilic displacement with charged nucleophiles at C-1 of 2-hexulopyranose 1-sulfonates is known to be difficult.[100,101] 1-Deoxy-1-fluoro-β-D-fructopyranose has, however, been synthesized, albeit in low yield, by treatment of 2,3:4,5-di-O-isopropylidene-1-O-(methylsulfonyl)-β-D-fructopyranose at 144° with tetrabutylammonium fluoride in N,N-dimethylformamide containing a small proportion of pyridine to minimize acid-catalyzed decomposition.[102]

Another example of the fact that primary displacement of sulfonic esters is not always facile[101,103–105] is the discovery that vigorous conditions were required for fluorination of 1,2:3,5-di-O-isopropylidene-6-O-(methylsulfonyl)-α-D-galactofuranose with potassium fluoride dihydrate in methanol,[85,86] and a comparatively long reaction-time was necessary for the fluorination of the 6-O-p-tolylsulfonyl derivative with tetrabutylammonium fluoride in acetonitrile.[106]

1-Deoxy-1-fluoro-L-glycerol (18) has been prepared by, among other methods, the treatment of 3,4-O-benzylidene-2,5-O-methylene-1,6-di-O-p-tolylsulfonyl-D-mannitol[107] (17) with tetrabutylammonium fluoride in acetonitrile, followed by removal of the benzylidene group, periodate oxidation, reduction with borohydride, and hydrolysis. 1,6-Dideoxy-1,6-difluorogalactitol[108] was obtained by treatment of 2,3:4,5-di-O-isopropylidene-1,6-di-O-(methylsulfonyl)galactitol with tetra-

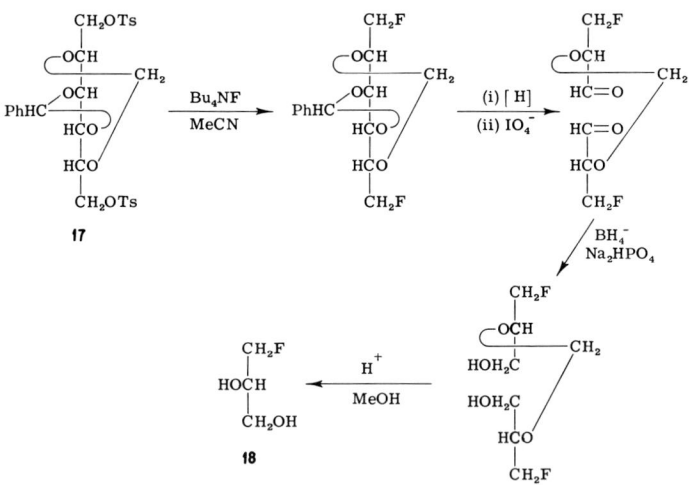

(100) R. S. Tipson, Adv. Carbohydr. Chem., 8 (1953) 107–215, see p. 190.
(101) A. C. Richardson, Carbohydr. Res., 10 (1969) 395–402.
(102) J. E. G. Barnett and G. R. S. Atkins, Carbohydr. Res., 25 (1972) 511–515.
(103) R. S. Tipson, Adv. Carbohydr. Chem., 8 (1953) 107–215.
(104) D. H. Ball and F. W. Parrish, Adv. Carbohydr. Chem. Biochem., 24 (1969) 139–197.

ethylammonium fluoride in N,N-dimethylformamide, and deprotection. Attempts to synthesize 1,6-dideoxy-1,6-difluoro-D-mannitol[108] by treatment of the 2,4:3,5-di-O-benzylidene-1,6-di-O-(methylsulfonyl) derivative with tetraethylammonium fluoride in N,N-dimethylformamide, butanone, or 1,4-dioxane gave very low yields of the desired 1,6-dideoxy-1,6-difluoro-D-mannitol, elimination products preponderating. Treatment of the corresponding di-O-methylene derivative afforded a somewhat better yield of the corresponding 1,6-dideoxy-1,6-difluoride, together with (presumably) elimination products, but the acetal groups were resistant to hydrolysis, even with 80% trifluoroacetic acid. Good yields of substitution products were obtained in both cases with other nucleophiles. The elimination reaction was partly attributed to the nature of the fluoride ion, but it was considered that these results, compared with those in the galactitol series, also suggested a generally increased steric strain in the diacetals of mannitol.

Some fluoro analogs of disaccharides that have been prepared include 6,6'-dideoxy-6,6'-difluoro-α,α-trehalose and its *galacto* analog. 2,3,2',3'-Tetra-O-benzyl-6,6'-di-O-p-tolylsulfonyl-α,α-trehalose was fluorinated with tetrabutylammonium fluoride in N,N-dimethylformamide: alternatively, 2,3,2',3'-tetra-O-benzyl-4,6,4',6'-tetra-O-(methylsulfonyl)-α,α-trehalose was selectively fluorinated with tetrabutylammonium fluoride in acetonitrile to give 2,3,2',3'-tetra-O-benzyl-6,6'-dideoxy-6,6'-difluoro-4,4'-di-O-(methylsulfonyl)-α,α-trehalose. The latter compound was treated with sodium benzoate in hexamethylphosphoric triamide to afford the 6,6'-dideoxy-6,6'-difluoro-*galacto* analog. The fluorinated derivatives were subsequently deprotected by conventional procedures.[109] 6-Deoxy-6-fluoro-α-D-glucopyranosyl α-D-glucopyranoside has been synthesized from 2,3,4,6-tetra-O-benzyl-α-D-glucopyranosyl 2,3-di-O-benzyl-6-O-p-tolylsulfonyl-α-D-glucopyranoside by treatment with tetrabutylammonium fluoride in acetonitrile, followed by hydrogenolysis. The intermediate was synthesized by standard procedures from 2,3-di-O-acetyl-4,6-O-isopropylidene-α-D-glucopyranosyl 2,3,4,6-tetra-O-acetyl-α-D-gluco-

---

(105) N. F. Taylor, *Nature*, 182 (1958) 660–661.
(106) A. B. Foster, R. Hems, and J. M. Webber, *Carbohydr. Res.*, 5 (1967) 292–301.
(107) W. J. Lloyd and R. Harrison, *Carbohydr. Res.*, 26 (1973) 91–98.
(108) E. M. Acton, M. Keyanpour-Rad, J. E. Christensen, H. H. Tong, R. P. Kwok, and L. Goodman, *Carbohydr. Res.*, 22 (1972) 477–486.
(109) L. Hough, A. K. Palmer, and A. C. Richardson, *J. Chem. Soc., Perkin Trans. 1*, (1972) 2513–2517.

pyranoside,[110] obtained as the major product from the selective acetalation of α,α-trehalose with methyl or ethyl isopropenyl ether in the presence of p-toluenesulfonic acid in N,N-dimethylformamide. Controlled hydrolysis of 2,3,2′,3′-tetra-O-benzyl-4,6,4′,6′-di-O-benzylidene-α,α-trehalose has been used to obtain 2,3-di-O-benzyl-4,6-O-benzylidene-α-D-glucopyranosyl 2,3-di-O-benzyl-α-D-glucopyranoside, which was per-O-(methanesulfonyl)ated, and the product selectively fluorinated, to give the corresponding 6-deoxy-6-fluoro-4-O-(methylsulfonyl) derivative, which, in turn, afforded the 6-deoxy-6-fluoride[110] by standard saponification and hydrogenolysis.[111] Per-O-acetylated 4-acetamido-4,6-dideoxy-6-fluoro-α-D-galactopyranosyl α-D-glucopyranoside was obtained[112] from 2,3-di-O-benzyl-4,6-O-benzylidene-α-D-glucopyranosyl 2,3-di-O-benzyl-4,6-di-O-(methylsulfonyl)-α-D-glucopyranoside by fluoride displacement (tetrabutylammonium fluoride in acetonitrile) at C-6, followed by azide displacement at C-4, reduction of the azide, acetylation of the amine, deprotection, and O-acetylation.

3. Epoxide Cleavage by Fluoride Ion

One of the methods by which replacement of secondary hydroxyl groups by fluorine in pentofuranoses and pento- and hexo-pyranoses has been achieved is by cleavage of epoxide rings with fluoride ion. The epoxide rings are always cleaved *trans*-diaxially, according to the Fürst–Plattner rule.

Treatment with hydrogen fluoride in 1,4-dioxane, followed by hydrogenolysis and acid hydrolysis, has been used for the preparation of methyl 3-deoxy-3-fluoro-β-L-xylopyranoside (**20**) from methyl 2,3-anhydro-4-O-benzyl-β-L-ribopyranoside[113] (**19**). An improved preparation of the fluoride of the D enantiomer (**22**) (subsequently hydrolyzed to the free sugar) was achieved[114] by treatment of methyl 2,3-anhydro-4-O-benzyl-β-D-ribopyranoside (**21**) with potassium hydrogenfluoride, and 3-deoxy-3-fluoro-D-xylose[115] (**24**) was successfully prepared by treatment of benzyl 2,3-anhydro-β-D-ribopyranoside (**23**) with potassium hydrogenfluoride in 1,2-ethanediol or potassium fluo-

---

(110) J. Defaye, H. Driguez, B. Henrissat, J. Gelas, and E. Bar-Guilloux, *Carbohydr. Res.*, 63 (1978) 41–49.
(111) A. F. Hadfield, L. Hough, and A. C. Richardson, *Carbohydr. Res.*, 63 (1978) 51–60.
(112) A. F. Hadfield, Ph.D. Thesis (London), 1974.
(113) N. F. Taylor, R. F. Childs, and R. V. Brunt, *Chem. Ind. (London)*, (1964) 928–929.
(114) J. A. Wright and N. F. Taylor, *Carbohydr. Res.*, 3 (1967) 333–339.
(115) S. Cohen, D. Levy, and E. D. Bergmann, *Chem. Ind. (London)*, (1964) 1802–1803.

ride in molten acetamide, followed by deprotection of the hydroxyl group. 3-Deoxy-3-fluoro-D-arabinose (**26**) has been obtained by treatment of methyl 2,3-anhydro-5-O-benzyl-α-D-lyxofuranoside (**25**) with potassium hydrogenfluoride in 1,2-ethanediol, followed by removal of the protecting groups.[114] 3-Deoxy-3-fluoro-β-D-arabinose (**28**) was obtained, together with 2-deoxy-2-fluoro-α-D-xylose (**29**), by the action of potassium hydrogenfluoride and sodium fluoride in 1,2-ethanediol on methyl 2,3-anhydro-5-O-benzyl-β-D-lyxofuranoside[116] (**27**), followed by removal of the protecting groups. Potassium hydrogenfluoride–1,2-ethanediol treatment of methyl 2,3-anhydro-5-O-benzyl-α-[117] and -β-

(116) J. A. Wright and J. J. Fox, *Carbohydr. Res.*, 13 (1970) 297–300.
(117) J. A. Wright, N. F. Taylor, and J. J. Fox, *J. Org. Chem.*, 34 (1969) 2632–2636.

[Scheme showing conversion of 27 to 28 and 29 via KHF₂–NaF/MeCONH₂, then [H], giving 3-deoxy-3-fluoro-β-D-arabinose (28) and 2-deoxy-2-fluoro-α-D-xylose (29)]

D-ribofuranoside[118] (**30** and **32**) (or potassium hydrogenfluoride and sodium fluoride in the latter case[119]) yielded 2-deoxy-2-fluoro-D-arabinose (**31**) and 3-deoxy-3-fluoro-D-xylose (**33**), respectively, after deprotection.

Absence of attack at C-2 in the case of **32** is, presumably, due to the combined effect of two factors. Firstly, two oxygen substituents at C-1 would be expected to destabilize a transition state involving a positive charge on C-2, and, secondly, steric hindrance by the methoxyl group must hinder attack at C-2. This steric hindrance by the methoxyl group would be expected to be larger than that of the substituent at

[Scheme showing conversion of 30 via KHF₂/1,2-ethanediol, then Pd–C/H₂, then H₂SO₄, giving open-chain aldehyde 31]

(118) J. A. Wright and N. F. Taylor, *Carbohydr. Res.*, 6 (1968) 347–354.
(119) J. A. Wright, *Methods Carbohydr. Chem.*, 6 (1972) 201–205.

C-4, because of an interaction similar to that which leads to the anomeric effect in pyranoses.[120] For **30**, steric hindrance by the substituent at C-4 evidently inhibits the polar tendency for attack at C-3.

In the aldohexopyranose series, treatment of methyl 2,3-anhydro-4,6-di-*O*-methyl-α-D-allopyranoside (**34**), or of the corresponding 4,6-diacetate (**39**), with hydrogen tetrafluoroborate in hydrogen fluoride at 70° afforded[121] the corresponding derivatives of 2-deoxy-2-fluoro-D-altropyranosyl fluoride (**35** and **40**) and 3-deoxy-3-fluoro-D-glucopyranosyl fluoride (**36** and **41**). Compounds **35** and **36** were hydrolyzed, and

(120) E. J. Reist and S. L. Holton, *Carbohydr. Res.*, 9 (1969) 71–77.
(121) I. Johansson and B. Lindberg, *Carbohydr. Res.*, 1 (1966) 467–473.

the products O-demethylated, to afford the free sugars **37** and **38**, respectively. Compounds **40** and **41** were acetolyzed, and the products treated with aqueous, methanolic triethylamine for removal of the ester groups. At the latter stage, epimerization of **37** occurred, to give 2-deoxy-2-fluoro-D-allopyranose (**42**).

1,6-Anhydroaldohexopyranoses have been extensively used for the preparation of epoxide precursors having rigid conformations. Rigid conformations are needed in order to ensure the almost exclusive formation of one of the possible positional isomers expected from *trans*-diaxial ring-opening of the epoxide ring by nucleophiles. For example, 2-deoxy-2-fluoro-D-glucose,[122,123] 2,4-dideoxy-2,4-difluoro-D-glucose,[123,124] 2,6-dideoxy-2-fluoro-D-glucose,[125] 2-amino-1,6-anhydro-2,4-dideoxy-4-fluoro-β-D-glucopyranose[126] (**43**), and 4-acetamido-2,4-dideoxy-2-fluoro-β-D-glucose[126] have been prepared, the introduction of fluorine in all cases being achieved by treatment of a suitable epoxide with potassium hydrogenfluoride in 1,2-ethanediol. 4-Deoxy-4-fluoro-D-glucose has been prepared by the action of potassium hydrogenfluoride in 1,2-ethanediol on 1,6-anhydro-3-O-p-tolylsulfonyl-β-D-glu-

---

(122) J. Pacák, F. Točík, and M. Černý, *Chem. Commun.*, (1969) 77.
(123) J. Pacák, J. Podešva, F. Točík, and M. Černý, *Collect. Czech. Chem. Commun.*, 37 (1972) 2589–2599.
(124) J. Pacák, J. Podešva, and M. Černý, *Chem. Ind. (London)*, (1970) 929.
(125) M. Černý, V. Přiklylová, and J. Pacák, *Collect. Czech. Chem. Commun.*, 37 (1972) 2978–2984.
(126) J. Pacák, P. Drašar, D. Štropová, M. Černý, and M. Budesinsky, *Collect. Czech. Chem. Commun.*, 38 (1973) 3936–3939.

copyranose,[127,128] followed by hydrolysis of the anhydro ring. The intermediate 1,6-anhydro-4-deoxy-4-fluoro-β-D-glucopyranose was then selectively p-toluenesulfonylated at OH-2, and the product treated with sodium methoxide, to afford 1,6:2,3-dianhydro-4-deoxy-4-fluoro-β-D-mannopyranose. On treatment with potassium hydrogenfluoride in 1,2-ethanediol, the latter afforded 1,6-anhydro-2,4-dideoxy-2,4-di-

(127) A. D. Barford, A. B. Foster, J. H. Westwood, and L. D. Hall, *Carbohydr. Res.*, 11 (1969) 287–288.
(128) A. D. Barford, A. B. Foster, J. H. Westwood, L. D. Hall, and R. N. Johnson, *Carbohydr. Res.*, 19 (1971) 49–61.

fluoro-β-D-glucopyranose.[128] A more efficient synthesis of 4-deoxy-4-fluoro-D-glucose (**45**) has, however, been achieved by photolytic removal of TsO-2 from 1,6:3,4-dianhydro-2-O-p-tolylsulfonyl-β-D-galactopyranose (**44**), treatment of the product with potassium hydrogenfluoride in 1,2-ethanediol, and hydrolysis of the anhydro ring.[128,129]

Tetrabutylammonium fluoride in acetonitrile has been successfully used for epoxide-ring cleavage of 3,4-anhydro-1,2-O-isopropylidene-β-D-*ribo*-hexulopyranose (**46**), to give 4-deoxy-4-fluoro-1,2-O-isopropylidene-β-D-*xylo*-hexulopyranose[130] (**47**). The 3,4-anhydride has been found resistant to attack by potassium fluoride in N,N-dimeth-

ylacetamide (1,2-O-isopropylidene-β-D-*xylo*-hexulopyranose was obtained, probably owing to the presence of water), and potassium hydrogenfluoride under similar conditions gave at least eight, unidentified products.

4-Deoxy-4-fluoro-α-D-sorbose was prepared by the action of potassium hydrogenfluoride on 3,4-anhydro-1,2-O-isopropylidene-β-D-psicopyranose in 2-methoxyethanol, followed by O-deisopropylidenation.[131] When 3,4-anhydro-1,2-O-isopropylidene-β-D-tagatopyranose was treated with potassium hydrogenfluoride in 2-methoxyethanol, epoxide migration occurred, and 4,5-anhydro-1,2-O-isopropylidene-β-D-fructopyranose was obtained. Subsequent attack by fluoride on this epoxide gave, after O-deisopropylidenation, 4-deoxy-4-fluoro-D-tagatose and 5-deoxy-5-fluoro-α-L-sorbopyranose.[131]

4. Displacement of Secondary Sulfonyloxy Groups by Fluorine

An alternative procedure for the introduction of the fluorine substituent in secondary positions of carbohydrates consists in nucleophilic displacement of sulfonic esters. Walden inversion always accompanies such displacements, and the success of this method may be attributed to two factors.

(129) A. D. Barford, A. B. Foster, and J. H. Westwood, *Carbohydr. Res.*, 13 (1970) 189–190.
(130) M. Sarel-Imber and E. D. Bergmann, *Carbohydr. Res.*, 27 (1973) 73–77.
(131) G. V. Rao, L. Que, Jr., L. D. Hall, and T. P. Fondy, *Carbohydr. Res.*, 40 (1975) 311–321.

Firstly, displacements occur under relatively mild conditions when the fluoride source is tetrabutylammonium fluoride used in conjunction with a dipolar, aprotic solvent (tetraethylammonium fluoride has also been used). This reagent was first used in the steroid field[132] for successful displacement of sulfonyloxy groups attached to five-membered carbocyclic rings (testosterone p-toluenesulfonate) and axial or equatorial sulfonyloxy groups on six-membered carbocyclic rings (cholestan-3-α- and -3-β-yl p-toluenesulfonates). Its successful, wide application to the carbohydrate field was initiated with the preparation of 3-deoxy-3-fluoro-1,2:5,6-di-O-isopropylidene-α-D-glucofuranose from 1,2:5,6-di-O-isopropylidene-3-O-p-tolylsulfonyl-α-D-allofuranose.[106,133] The course of such reactions with tetrabutylammonium fluoride was found to depend upon the quality of the reagent (it is thermally unstable, and extremely hygroscopic), and an alternative preparation was published.[94] It is now commercially available. When initially used in the carbohydrate field, the fluoride was obtained[134,135] by neutralization of tetrabutylammonium hydroxide with aqueous hydrogen fluoride, and evaporation of the water, followed by drying *in vacuo* over phosphorus pentaoxide. A salt of better quality was obtained by neutralizing the base and then cooling the solution, thus allowing the precipitation of the clathrate ($Bu_4NF \cdot 32.8 \, H_2O$), and dehydration of this clathrate, initially by evaporation of the water in a rotary evaporator, and then by drying over phosphorus pentaoxide in a desiccator *in vacuo*.[94] After a certain amount of predrying of the clathrate over phosphorus pentaoxide *in vacuo*, the last traces of water have also been removed by drying a solution of the salt in acetonitrile with molecular sieves. Fluorinations have been successfully conducted with this dry solution.[112]

In a comparative study of fluorination of 1,2:3,4-di-O-isopropylidene-6-O-p-tolylsulfonyl-α-D-galactopyranose with tetrabutylammonium fluoride in a variety of dipolar, aprotic solvents (as well as 1,2-ethanediol, in which no reaction was observed), acetonitrile was found to give the highest proportion of substitution of the sulfonic esters relative to their elimination.[106] Elimination is the major, competing reaction in these nucleophilic-substitution reactions, because of the high basicity and low nucleophilicity of the fluoride ion or, in terms of the

(132) H. B. Henbest and W. R. Jackson, *J. Chem. Soc.*, (1962) 954–959.
(133) K. W. Buck, A. B. Foster, R. Hems, and J. M. Webber, *Carbohydr. Res.*, 3 (1966) 137–138.
(134) A. B. Foster and R. Hems, *Methods Carbohyd. Chem.*, 6 (1972) 197–200.
(135) D. L. Fowler, W. V. Loebenstein, D. B. Pall, and C. A. Kraus, *J. Am. Chem. Soc.*, 62 (1940) 1140–1142.

HSAB principle, because it is a "hard" base. Acetonitrile was also shown to give the highest ratio of substitution to elimination in comparative displacements of a secondary sulfonyloxy group of the alkyl chain of stearic acid in a number of solvents.[136]

Secondly, the factors controlling nucleophilic displacements of sulfonic esters had already been determined to a considerable extent, and have been described in reviews in this Series[103,104] and, in addition, the steric and polar factors governing such displacements have been summarized qualitatively,[101] and will therefore not be discussed here in any detail. This accumulated knowledge made possible the prediction of the course of such fluoride displacements.

4-Deoxy-4-fluoro-D-glucose[137] and -galactose[138] were obtained by treatment of methyl 2,3-di-O-methyl-4-O-(methylsulfonyl)-6-O-trityl-α-D-galactopyranoside and methyl 2,3-di-O-benzyl-4-O-(methylsulfonyl)-6-O-trityl-α-D-glucopyranoside, respectively, with tetrabutylammonium fluoride in acetonitrile, followed by deprotection of the hydroxyl groups. Some elimination of the sulfonyloxy group accompanied the preparation of 4-deoxy-4-fluoro-D-glucose, which is not surprising in view of the favorable, *trans*-diaxial arrangement existing between the 4-(methylsulfonyl)oxy group, H-5, and H-3. Similarly, methyl 2-acetamido-3,4-di-O-acetyl-2,4-dideoxy-4-fluoro-α-D-glucopyranoside has been prepared from methyl 2-benzamido-3,4-di-O-benzyl-2-deoxy-4-O-(methylsulfonyl)-6-O-trityl-α-D-galactopyranoside by fluoride displacement followed by removal of the benzyl and trityl groups, acetylation, and N-acetyl–N-benzoyl exchange.[95]

When methyl 2,3-di-O-methyl-4-O-(methylsulfonyl)-6-O-trityl-α-D-galactopyranoside was treated with cesium fluoride in 1,2-ethanediol, no fluorination was observed; however, fluorination did occur, albeit to a limited extent, after removal of the trityl group[137] (compare Section II,2). Use of potassium fluoride for displacement of secondary sulfonyloxy groups has often been unsuccessful, as in attempted displacements with potassium fluoride in methanol in 5,6-O-benzylidene-1,2-O-isopropylidene-3-O-(methylsulfonyl)-α-D-glucofuranose (which afforded starting material) and 1,2-O-isopropylidene-3,5,6-tri-O-(methylsulfonyl)-α-D-glucofuranose (which afforded the primary fluoride only),[84] 1,6-anhydro-3,4-O-isopropylidene-2-O-(methylsulfonyl)-β-D-galactopyranose (which afforded 1,6-anhydro-3,4-O-isopropylidene-2-O-methyl-β-D-talopyranose)[139] and 1,6-anhydro-2-O-

(136) N. J. M. Birdsall, *Tetrahedron Lett.*, (1971) 2675–2678.
(137) A. B. Foster, R. Hems, and J. H. Westwood, *Carbohydr. Res.*, 15 (1970) 41–49.
(138) D. M. Marcus and J. H. Westwood, *Carbohydr. Res.*, 17 (1971) 269–274.
(139) P. W. Kent, D. W. A. Farmer, and N. F. Taylor, *Proc. Chem. Soc. London*, (1959) 187–188.

(methylsulfonyl)-β-D-altropyranose (which did not undergo exchange),[139] as well as attempted displacements with potassium fluoride in 1,2-ethanediol in methyl 2,3-di-O-(methylsulfonyl)-4,6-O-benzylidene-α-D-glucopyranoside (which afforded methyl 2,3-anhydro-4,6-O-benzylidene-α-D-alloside, together with an unidentified product containing fluorine, and starting material), methyl 4,6-O-benzylidene-2,3-di-O-p-tolylsulfonyl- and -3-O-(methylsulfonyl)-α-D-glucopyranoside (which afforded methyl 2,3-anhydro-4,6-O-benzylidene-α-D-allopyranoside).[78] Treatment of 3-O-benzoyl-1,2:5,6-di-O-isopropylidene-4-O-(methylsulfonyl)-D-mannitol with potassium fluoride in N,N-dimethylformamide gave 3(4)-O-benzoyl-1,2:5,6-di-O-isopropylidene-D-talitol.[140] However, treatment of 3-O-benzyl-1,2-O-isopropylidene-5,6-di-O-(methylsulfonyl)-α-D-glucofuranose with potassium fluoride in 1,2-ethanediol gave, by benzyloxy-group participation, 3,6-anhydro-5-deoxy-5-fluoro-1,2-O-isopropylidene-β-L-idofuranose, whereas treatment with tetrabutylammonium fluoride in acetonitrile afforded 3,6-anhydro-1,2-O-isopropylidene-5-O-(methylsulfonyl)-α-D-glucofuranose. The latter compound afforded 3,6-anhydro-5-deoxy-5-fluoro-1,2-O-isopropylidene-β-L-idofuranoside upon treatment with potassium fluoride in 1,2-ethanediol.[141] Fluorination with potassium fluoride in 1,2-ethanediol proceeds satisfactorily[95] with methyl 2-benzamido-3-O-benzyl-2-deoxy-4,6-di-O-(methylsulfonyl)-α-D-gluco- (**48**) and -galacto-pyranoside (**50**), to yield the corre-

(140) B. R. Baker and A. H. Haines, *J. Org. Chem.*, 28 (1963) 438–441.
(141) L. D. Hall and P. R. Steiner, *Can. J. Chem.*, 48 (1970) 451–458.

sponding 4,6-difluoro-D-galacto- (49) (77%) and -D-gluco-pyranoside (51) (40%). In the latter reaction, methyl 2-benzamido-3-O-benzyl-2,4,5,6-tetradeoxy-6-fluoro-β-L-*threo*-hex-4-enopyranoside (52) (25%) was also obtained. (For further successful displacements with cesium and potassium fluoride, see also, later in this Section; compare also, Section II,2.)

Several deoxyfluoro disaccharide analogs have also been synthesized by fluoride displacement (with tetrabutylammonium fluoride in acetonitrile) of suitably protected sulfonic esters. These include 4,6,4′,6′-tetradeoxy-4,6,4′,6′-tetrafluoro-α,α-trehalose and its *galacto* analog from the corresponding perbenzylated methanesulfonic esters,[142] and 4,4′-dideoxy-4,4′-difluoro-α,α-trehalose and its *galacto* analog from 2,3,2′,3′-tetra-O-benzyl-4,4′-di-O-(methylsulfonyl)-6,6′-di-O-trityl-α,α-"*galacto*-trehalose" and -trehalose,[142] followed by deprotection; the unsymmetrically substituted 4,6-dideoxy-4,6-difluoro-α-D-galactopyranosyl α-D-glucopyranoside from 2,3-di-O-benzyl-4,6-di-O-(methylsulfonyl)-α-D-glucopyranosyl 2,3-di-O-benzyl-4,6-O-benzylidene-α-D-glucopyranoside; and 4-deoxy-4-fluoro-α-D-galactopyranosyl α-D-glucopyranoside from 2,3-di-O-benzyl-4-O-(methylsulfonyl)-6-O-trityl-α-D-glucopyranosyl 2,3-di-O-benzyl-4,6-O-benzylidene-α-D-glucopyranoside.[112]

As might be expected, no displacement of the sulfonyloxy group was observed when 1,2:4,5-di-O-isopropylidene-3-O-*p*-tolylsulfonyl-β-D-fructo- or -psico-pyranose was treated with fluoride under a variety of conditions.[130,131]

Reaction of methyl 2,3-di-O-benzyl-4-O-(*p*-bromophenylsulfonyl)-6-O-trityl-β-D-glucopyranoside with tetrabutylammonium fluoride in acetonitrile afforded the corresponding 4-deoxy-4-fluoro *galacto* derivative[143] in 30% yield. However, treatment of methyl 2,3,6-tri-O-benzoyl-4-O-(*p*-bromophenylsulfonyl)-β-D-glucopyranoside with the reagent resulted in the loss of the benzoyl, as well as the bromophenylsulfonyl, group, in contrast to the successful application of esters as protecting groups in displacement of a primary *p*-nitrophenylsulfonyloxy group. Fluoride displacement of primary sulfonates of sugars protected with ester groups has occurred, to a limited extent, in other cases as well; see Section II,2.

3-Deoxy-3-fluoro-D-glucose[106,133,144] (54), 3-deoxy-3-fluoro-D-galac-

(142) L. Hough, A. K. Palmer, and A. C. Richardson, *J. Chem. Soc., Perkin Trans. 1*, (1973) 784–788.
(143) A. Maradufu and A. S. Perlin, *Carbohydr. Res.*, 32 (1974) 261–277.
(144) A. B. Foster, R. Hems, and L. D. Hall, *Can. J. Chem.*, 48 (1970) 3937–3945.

tose[145,146] (**56**), and 3-deoxy-3-fluoro-1,2:5,6-di-*O*-isopropylidene-β-L-idofuranose[147,148] (**58**) have been prepared from the 3-sulfonic esters of 1,2:5,6-di-*O*-isopropylidene-α-D-allo- (**53**), -D-gulo- (**55**), and -β-L-talo-furanose (**57**), respectively. Derivatives of 3-deoxy-3-fluoro-L-idose have also been prepared, more conveniently, from 3-deoxy-3-fluoro-1,2:5,6-di-*O*-isopropylidene-α-D-glucofuranose by mild, partial hydrolysis with acid to 3-deoxy-3-fluoro-1,2-*O*-isopropylidene-α-D-glucofuranose, per(methanesulfonyl)ation of the resulting diol, and benzoate displacement of the disulfonate, to afford 5,6-di-*O*-benzoyl-

(145) J. S. Brimacombe, A. B. Foster, R. Hems, and L. D. Hall, *Carbohydr. Res.*, 8 (1968) 249–250.
(146) J. S. Brimacombe, A. B. Foster, R. Hems, J. H. Westwood, and L. D. Hall, *Can. J. Chem.*, 48 (1970) 3946–3952.
(147) J. S. Brimacombe, P. A. Gent, and J. H. Westwood, *Carbohydr. Res.*, 12 (1970) 475–478.
(148) J. S. Brimacombe, P. A. Gent, and J. H. Westwood, *J. Chem. Soc., C*, (1970) 1632–1635.

3-deoxy-3-fluoro-1,2-O-isopropylidene-β-L-idofuranose and 6-O-benzoyl-3-deoxy-3-fluoro-1,2-O-isopropylidene-β-L-idofuranose. O-Debenzoylation of the mono- and di-benzoates afforded 3-deoxy-3-fluoro-1,2-O-isopropylidene-β-L-idofuranose.[149] 3-Deoxy-3-fluoro-1,2-O-isopropylidene-β-L-idofuranose was hydrolyzed to the free sugar by mild treatment with acid.[150] In these precursors, the departing sulfonyloxy groups are *endo* to the two *cis*-fused, five-membered rings. It was noted[145,146] that elimination accompanies substitution only in the case of 1,2:5,6-di-O-isopropylidene-3-O-p-tolylsulfonyl-α-D-gulofuranose (**55**), where the 3-p-tolylsulfonyloxy group and H-4 exist in the *trans* arrangement. Elimination by attack at H-2 is probably not favored in 1,2:5,6-di-O-isopropylidene-3-O-p-tolylsulfonyl-α-D-allofuranose (**53**), as the resulting double bond would impose increased strain on the molecule. Potassium fluoride in acetamide has also been used for the direct displacement of the 3-p-tolylsulfonyloxy group in **53**, to yield the corresponding 3-deoxy-3-fluoro-D-glucofuranose in 62% yield.[151] This same 3-deoxy-3-fluoride has been obtained by treatment of 1,2:5,6-di-O-isopropylidene-3-O-(trifluoromethylsulfonyl)-α-D-allofuranose with cesium fluoride in N,N-dimethylformamide at reflux temperature[152] (see also, Section II,2).

Nucleophilic displacements at carbon atoms adjacent to the anomeric center of aldopyranosides do not normally proceed satisfactorily, but, surprisingly, treatment of methyl 3,4-O-isopropylidene-2-O-(p-nitrophenylsulfonyl)-α-D-glucoseptanoside with tetrabutylammonium fluoride in acetonitrile afforded the corresponding mannoside, from which 2-deoxy-2-fluoro-β-D-mannopyranose was obtained by subsequent deprotection.[153] Whether this is a general property of sep-

---

(149) J. S. Brimacombe, A. M. Mofti, and J. H. Westwood, *Carbohydr. Res.*, 21 (1972) 297–300.
(150) A. B. Foster, R. Hems, J. H. Westwood, and J. S. Brimacombe, *Carbohydr. Res.*, 25 (1972) 217–227.
(151) U. Reichman, K. A. Watanabe, and J. J. Fox, *Carbohydr. Res.*, 42 (1975) 233–240.
(152) T. J. Tewson and M. J. Welch, *J. Org. Chem.*, 43 (1978) 1090–1092.
(153) C. J. Ng and J. D. Stevens, unpublished results.

tanosides is not yet known, because of the present lack of availability of common sugars in the septanoid form.

5. Direct Displacement of Oxygen Functions by Fluorine

Sulfur tetrafluoride is a well-known reagent for the displacement of oxygen functions by fluorine[154]; it has been applied to cellulose and cellulose derivatives (for examples, see Refs. 155–158). The attempted reaction of phenylsulfur trifluoride with the hydrate of 1,6-anhydro-2,4-dideoxy-2,4-difluoro-β-D-*ribo*-hexopyranos-3-ulose[159] afforded unchanged starting-material; the expected exchange of the two geminal hydroxyl groups for two fluorine atoms did not occur. Phenylsulfur trifluoride[160–162] effects the conversion of carbonyl and carboxyl groups into difluoromethylene and trifluoromethylene groups, respectively, and has a number of advantages over sulfur tetrafluoride, including the possibility of carrying out the reaction at atmospheric pressure and in glass apparatus.[161] Dialkylaminosulfur trifluorides have been shown to achieve the following transformations: $-CO_2H \rightarrow -CO(F)$,[163] $-CHO$ and $\rangle CO$ to $\rangle CHF$ and $\rangle CF_2$,[163,164] and $C-OH \rightarrow CF$.[164] These reagents have a number of advantages over sulfur tetrafluoride, again including the possibility of carrying out the reaction at atmospheric pressure in glass apparatus[163]; also, they are suitable for use with acid-sensitive compounds.[164] A number of 6-deoxy-6-fluoro aldohexoses (including 2-acetamido-2-deoxy-D-gluco, -galacto, and -manno derivatives) has been obtained from pyranoses containing a free primary hydroxyl group and otherwise protected with acetyl groups, by treatment with diethylaminosulfur trifluoride in diglyme [1,1'-oxybis(2-methoxyethane)]. This reagent was also found to be

(154) For a review, see G. A. Boswell, Jr., W. C. Ripka, R. M. Scribner, and C. W. Tullock, *Org. React.*, 21 (1974) 1–124.
(155) B. N. Gorbunov, S. S. Radchenko, and A. P. Khardin, USSR Pat. 232,228 (1968); *Chem. Abstr.*, 70 (1969) 98,102.
(156) V. P. Bezsolitsen, B. N. Gorbunov, S. S. Radchenko, and A. P. Khardin, USSR Pat. 254,767 (1969); *Chem. Abstr.*, 72 (1970) 133,578.
(157) B. N. Gorbunov, P. A. Protopopov, and A. P. Khardin, *Khim. Tekhnol. Proizvod. Tsellyul.*, (1971) 341–344; *Chem. Abstr.*, 78 (1973) 112,881.
(158) B. N. Gorbunov, P. A. Protopopov, and A. P. Khardin, *Funkts. Org. Soedin, Polim.*, (1973) 89–93; *Chem. Abstr.*, 81 (1974) 154,840.
(159) J. Pacák, M. Braunová, D. Štropová, and M. Černý, *Collect. Czech. Chem. Commun.*, 42 (1977) 120–131.
(160) W. A. Sheppard, *J. Am. Chem. Soc.*, 84 (1962) 4751–4752.
(161) W. A. Sheppard, *J. Am. Chem. Soc.*, 84 (1962) 3058–3063.
(162) W. A. Sheppard, *Org. Synth.*, 44 (1964) 82–85.
(163) L. N. Markovskii, V. G. Pashinnik, and A. V. Kirsanov, *Synthesis*, (1973) 787–788.
(164) W. J. Middleton, *J. Org. Chem.*, 40 (1975) 574–578.

compatible with the benzyl protecting-group.[165] Diethylaminosulfur trifluoride was also used in dichloromethane, or benzene, for the reaction of a carbonyl oxygen atom in sugars and glycosides protected with isopropylidene groups, to give geminal difluoro saccharides.[166] For example, 6-deoxy-6,6-difluoro-1,2:3,4-di-O-isopropylidene-α-D-galactopyranose (**60**) has been prepared from 1,2:3,4-di-O-isopropylidene-α-D-*galacto*-hexodialdo-1,5-pyranose (**59**), and methyl 2-deoxy-2,2-di-

fluoro-3,4-O-isopropylidene-β-L-*erythro*-pentopyranoside (**62**) from methyl 3,4-O-isopropylidene-β-L-*erythro*-pentopyranosid-2-ulose (**61**). However, treatment of 1,2:5,6-di-O-isopropylidene-α-D-*ribo*-hexulofuranose with the same reagent led to decomposition. Treatment of 1,2:5,6-di-O-isopropylidene-α-D-glucofuranose with diethyl-

aminosulfur trifluoride in dichloromethane at 0°, followed by processing with water, afforded the starting material in excellent yield, whereas the inclusion of two equivalents of pyridine in the mixture,

(165) M. Sharma and W. Korytnyk, *Tetrahedron Lett.*, (1977) 573–576.
(166) R. A. Sharma, I. Kavai, Y. L. Fu, and M. Bobek, *Tetrahedron Lett.*, (1977) 3433–3436.

followed by direct distillation, afforded the alkene **63** (see also, Section II,8). ^{19}F-N.m.r. evidence was obtained that suggested that an intermediate (such as **64**) was formed that did not undergo spontaneous decomposition to the fluorinated product, but that appeared to be readily hydrolyzed to the starting material, indicating that solvolysis occurred at the sulfur rather than at the carbon atom.[152] The reaction between 1,2:5,6-di-O-isopropylidene-α-D-allofuranose with diethylaminosulfur trifluoride in the presence of pyridine, followed by direct distillation, afforded 3-deoxy-3-fluoro-1,2:5,6-di-O-isopropylidene-α-D-glucofuranose, whereas, as before, processing with water afforded starting material. The suggested intermediate **65** in the allose series was found to be more labile than **63** in the glucose series. When the intermediate **65** was added to a solution of H^{18}F in pyridine and heated to 150°, the 3-deoxy-3-[^{18}F]fluoro-1,2:5,6-di-O-isopropylidene-α-D-glucofuranose was isolated in >90% radiochemical yield, indicating that the reaction proceeds *via* an S$_N$2 displacement, rather than an ion pair or a cyclic transition-state[152] (see also, Refs. 164, 167, and 168). The observed difference in reaction between the gluco- and allofuranose derivatives was attributed[152] either to the difference in stability of the two alkenes that would be formed by *trans*-diaxial elimination, or to the steric effects of the 1,2-O-isopropylidene ring blocking the approach of the nucleophile to the "lower" face of the furanose ring.

Treatment of 1,2:3,5-di-O-methylene-α-D-glucofuranose with (2-chloro-1,1,2-trifluoroethyl)diethylamine, the Yarovenko–Rashka reagent,[169] yielded the corresponding 6-deoxy-6-fluoro derivative, but attempts with other monosaccharide precursors proved unsuccessful, the method being limited by steric hindrance of the hydroxyl group that it had been intended should react.[170] For example, treatment of 1,2:3,4-di-O-isopropylidene-α-D-galactopyranose in dichloromethane afforded the corresponding chlorofluoroacetate,[89,171] the formation of which is usually attributed[172,173] to the presence of a trace of water in the mixture; however, as care was taken to exclude moisture,[171] the

---

(167) M. von Biollaz and J. Kalvoda, *Helv. Chim. Acta*, 60 (1977) 2703–2710.
(168) M. J. Green, H.-J. Shue, M. Tanabe, D. M. Yassuda, A. T. McPhail, and K. D. Onan, *J. Chem. Soc., Chem. Commun.*, (1977) 611–612.
(169) N. N. Yarovenko and M. A. Rashka, *Zh. Obshch. Khim.*, 29 (1959) 2159–2163; *J. Gen Chem. USSR*, 29 (1959) 2125–2128; *Chem. Abstr.*, 54 (1960) 9724.
(170) L. D. Hall and L. Evelyn, *Chem. Ind. (London)*, (1968), 183–184.
(171) K. R. Wood, D. Fisher, and P. W. Kent, *J. Chem. Soc., C*, (1966) 1994–1997.
(172) D. Ayer, *Tetrahedron Lett.*, (1962) 1065–1069.
(173) L. H. Knox, E. Velarde, S. Berger, D. Cuadriello, and A. D. Cross, *J. Org. Chem.*, 29 (1964) 2187–2195.

suggestion has been made[174] that this product results by decomposition of an intermediate complex by the water added in the processing. It was subsequently shown,[89] by ^{19}F-n.m.r. spectroscopy, that, in the reaction of 1,2:3,4-di-O-isopropylidene-α-D-galactopyranose (**66**) with reagent **67** in dichloromethane, the intermediate best described by **69** (rather than by **68**) is formed (see also, Refs. 172 and 173) and that it is not attacked by fluoride ion, but is stable indefinitely; addition of water causes immediate hydrolysis to the chlorofluoroacetate **70**. Support for such a reaction course was also provided by the observation[89]

(174) J. E. G. Barnett, *Adv. Carbohydr. Chem.*, Ref. 17, see p. 186.

that the cyclic acetal[89,170] **72** is obtained on treatment of 1,2-O-isopropylidene-α-D-xylofuranose (**71**) with **67**. It was suggested[89] that the first stage is attack of the reagent by OH-5, followed by attack by OH-3. When the reaction was conducted in N,N-dimethylformamide in the presence of anhydrous potassium fluoride, 6-chloro-6-deoxy-1,2:3,4-di-O-isopropylidene-α-D-galactopyranose was obtained, in addition to the chlorofluoroacetate.[171] This was interpreted as either[174] being due to intramolecular attack by chlorine at the stage of the intermediate, alcohol–reagent complex, or[174] as being due to attack by chloride ion formed by exchange with **67**. The one-step conversion of hydroxyl derivatives into the corresponding fluoro compounds by means of the Yarovenko–Rashka reagent has been reviewed, with inclusion of examples from the carbohydrate field.[175]

6. Addition to Glycals

Reactions of hydrogen fluoride with acylated glycals have been studied.[29,30,176–178] The reactions essentially proceed by protonation, and loss of the group at C-3, with formation of the corresponding allylic carbonium ion, which is in equilibrium with the corresponding 1,2-unsaturated, 3,4-acyloxonium ion, followed by fluoride attack at C-1 either before or after processing. Formation of the 1,2-unsaturated, 3,4-acyloxonium ion appears to be the driving force in the reaction, as the ion is stable in solution in hydrogen fluoride. The products depend on the reaction conditions and the nature of the glycal. 2,3-Unsaturated glycosides may be formed by attack at C-1 of the allylic carbonium ion,[30] or attack at C-1 of the 1,2-unsaturated, 3,4-acyloxonium ion which is accompanied by an allylic shift and opening of the acyloxonium ring during processing.[177]

2-Deoxyglycosyl fluorides are obtained by fluoride attack at C-1 of the 1,2-unsaturated, 3,4-acyloxonium ion, followed by opening of the acyloxonium ring during processing[177] (see also Ref. 30); also, probably, by fluoride attack at C-1 of the allylic carbonium ion accompanied by 3,4-acyloxonium-ion formation, the latter then opening up on processing.[30] In cases where the groups at C-3 and C-4 were cis-related, and loss of the group at C-3 was considered slow, products of direct addition of hydrogen fluoride to the glycal were observed.[29,30] In one instance,[178] n.m.r. evidence pointed to the presence of an ion-pair between the 1,2-unsaturated, 3,4-acyloxonium ion and hydrogen fluo-

(175) F. Liska, Chem. Listy, 66 (1972) 189–197; Chem. Abstr., 76 (1972) 99,936.
(176) I. Lundt and C. Pedersen, Acta Chem. Scand., 20 (1966) 1369–1375.
(177) I. Lundt and C. Pedersen, Acta Chem. Scand., 24 (1970) 240–246.
(178) I. Lundt and C. Pedersen, Acta Chem. Scand., 25 (1971) 2320–2326.

ride (the bond between C-1 and F-1 being ionic). In certain instances,[27,28,164] disaccharides have been found in the product mixtures. In the case of peracylated 2-acyloxyglycals studied,[179,180] treatment with hydrogen fluoride led to immediate formation of the 2,3-unsaturated glycosyl fluoride which, on further treatment, was converted into the corresponding 3,4-dideoxyhex-3-enopyranosylulose fluoride.

Glycosyl fluorides prepared by addition of hydrogen fluoride to glycals in synthetically significant yields include the unstable 4,6-di-O-benzoyl-2,3-dideoxy-D-*erythro*-hexopyranosyl fluoride, obtained in good yield by treatment of tri-O-benzoyl-D-glucal with hydrogen fluoride for 30 min at −70°, followed by an aqueous processing[177] (see also, Ref. 176). Prolonged treatment afforded 3,6-di-O-benzoyl-2-deoxy-α-D-*ribo*-hexopyranosyl fluoride.[177] This is a convenient method for the preparation of 2-deoxy-D-*ribo*-hexose derivatives from tri-O-benzoyl-D-glucal. When tetra-O-acetyl-1,5-anhydro-2-deoxy-D-*arabino*-hex-1-enitol was treated with hydrogen fluoride in benzene for 3 h at 0°, 2,4,6-tri-O-acetyl-3-deoxy-D-*ribo*-hex-2-enopyranosyl fluorides were obtained in 93% yield, in an $\alpha:\beta$ ratio of 17:3, whereas, when a solution of the same glycal in hydrogen fluoride at −70° was kept for 3 h at −30°, 6-O-acetyl-3,4-dideoxy-D-*glycero*-hex-3-enopyranosylulose fluorides were obtained[179,180] in 98% yield in the $\alpha:\beta$ ratio of 7:1.

An alternative, synthetic procedure for the introduction of fluorine at C-2 of aldopentoses and aldohexoses is addition of electrophilic fluorine to glycals. Polar additions to glycals usually occur with complete electrospecificity, and any reaction with fluoride ion results in the formation of glycosyl fluorides by attack of fluoride ion at C-1, the more electropositive center. However, when the fluorine atom is attached to a highly electronegative, good leaving-group, as, for example, in trifluoro(fluoroxy)methane ($CF_3OF \rightarrow CF_3O^- + F^+$) or perchloryl fluoride ($ClO_3F \rightarrow ClO_3^- + F^+$), electrophilic fluorine becomes available, yielding 2-deoxy-2-fluoro derivatives. Trifluoro(fluoroxy)methane, a hypofluorite or fluoroxy[181] compound, is a mild, selective, highly active reagent. Hypofluorites are more powerful than perchloryl fluoride, both previously used in steroid chemistry, and, provided that certain compounds are absent from the reaction mixtures, they cause no explosions.[181] The mechanism of addition of this reagent has been discussed in the course of studies in the steroid field.[182,183] *cis*-Addition in

(179) K. Bock and C. Pedersen, *Tetrahedron Lett.*, (1969) 2983–2986.
(180) K. Bock and C. Pedersen, *Acta Chem. Scand.*, 25 (1971) 1021–1030.
(181) For a review, see R. H. Hesse, *Isr. J. Chem.*, 17 (1978) 60–70.
(182) D. H. R. Barton, L. S. Godinho, R. H. Hesse, and M. M. Pechet, *Chem. Commun.*, (1968) 804–806.
(183) D. H. R. Barton, L. J. Danks, A. K. Ganguly, R. H. Hesse, G. Tarzia, and M. M. Pechet, *J. Chem. Soc.*, (1969) 227–228.

the Markovnikoff manner occurs, and has been explained by the fact that the intermediate α-fluorocarbonium ion initially formed is not bridged (as it is with other common electrophiles). The carbonium ion rapidly combines with a counter-ion [CF_3O^- or F^- (obtained from $CF_3O^- \rightarrow COF_2 + F^-$, or by virtue of the ambivalent, nucleophilic character of CF_3O^-)], and subsequently behaves as would be expected for an unstabilized, carbonium ion. With glycals, this exclusive cis-addition is also operative; the ion-pairs collapse to form trifluoromethyl glycosides and glycosyl fluorides. Attack occurs mainly from the less-hindered side of the double bond. Details of the handling of trifluoro(fluoroxy)methane as used in the carbohydrate field have been given.[69] Treatment[68,69] of tri-O-acetyl-D-glucal with trifluoro(fluoroxy)methane at −80° afforded trifluoromethyl 3,4,6-tri-O-acetyl-2-deoxy-2-fluoro-α-D-glucopyranoside (26%), 3,4,6-tri-O-acetyl-2-deoxy-2-fluoro-α-D-glucopyranosyl fluoride (34%), trifluoromethyl 3,4,6-tri-O-acetyl-2-deoxy-2-fluoro-β-D-mannopyranoside (6%), and 3,4,6-tri-O-acetyl-2-deoxy-2-fluoro-β-D-mannopyranosyl fluoride (8%).

The ease of hydrolysis of both the fluorine atom on C-1 and the trifluoromethoxyl group makes this an excellent route for the preparation of monosaccharides and related compounds fluorinated at C-2. For example, 2-deoxy-2-fluoro-D-glucose and -mannose[68,69] have been prepared from tri-O-acetyl-D-glucal, and 2-deoxy-2-fluoro-D-galactose from tri-O-acetyl-D-galactal.[184,185] The *talo* adducts were also isolated in low yield in the latter reaction. The *galacto*:*talo* ratio[184,185] was larger than the *gluco*:*manno* ratio,[68,69] and this was interpreted[185] as probably being due to the steric influence exerted by the quasi-axial AcO-4 group "above" the ring in tri-O-acetyl-D-galactal, which further hinders attack by the double bond from the "upper" side of the ring. Such an influence of AcO-4 in reactions of peracetylated D-galactal (compared to those of peracetylated D-glucal) was first observed, and discussed in stereoelectronic terms, in halogenomethoxylation reactions of these glycals.[186] In addition, 2-deoxy-2-fluoro-L-fucose has been prepared from 3,4-di-O-acetyl-L-fucal by treatment with trifluoro(fluoroxy)methane.[187] 2-Deoxy-2-fluoro-D-glucose (**73**), by way of its conversion into 3,4,6-tri-O-acetyl-D-glucal (**74**), has been used[2] for the preparation of **2**. It was noted[2] that the trifluoromethyl α-D-glycoside is the preponderant product from tri-O-acetyl-D-glucal, whereas the α-D-glycosyl fluoride is the major product from **74**, and this result was explained as probably being due to the greater stability of the ini-

(184) J. Adamson and D. M. Marcus, *Carbohydr. Res.*, 13 (1970) 314–316.
(185) J. Adamson and D. M. Marcus, *Carbohydr. Res.*, 22 (1972) 257–264.
(186) R. U. Lemieux and B. Fraser-Reid, *Can. J. Chem.*, 43 (1965) 1460–1475.
(187) C. G. Butchard, D. J. Winterbourne, and P. W. Kent, in preparation.

tially formed 2-deoxy-2,2-difluoro-oxonium ion (compared with the 2-deoxy-2-fluoro-oxonium ion), which facilitates loss of COF_2 from the intermediate ion-pair, with the result that there is a higher proportion of glycosyl fluorides in the mixture.

Hexa-*O*-acetyl-lactal required relatively vigorous conditions for reaction to occur.[49] At 0°, the following products were obtained: trifluoromethyl 3,6-di-*O*-acetyl-2-deoxy-2-fluoro-4-*O*-(2,3,4,6-tetra-*O*-acetyl-β-D-galactopyranosyl)-α-D-glucopyranoside (6%), 3,6-di-*O*-acetyl-2-deoxy-2-fluoro-4-*O*-(2,3,4,6-tetra-*O*-acetyl-β-D-galactopyranosyl)-α-D-glucopyranosyl fluoride (17%), trifluoromethyl 3,6-di-*O*-acetyl-2-deoxy-2-fluoro-4-*O*-(2,3,4,6-tetra-*O*-acetyl-β-D-galactopyranosyl)-β-D-mannopyranoside (29%), and 3,6-di-*O*-acetyl-2-deoxy-2-fluoro-4-*O*-(2,3,4,6-tetra-*O*-acetyl-β-D-galactopyranosyl)-β-D-mannopyranosyl fluoride (23%). The differences in both the rate and the products of the addition observed for hexa-*O*-acetyl-lactal and comparable reactions in the monosaccharide series (compare, however, additions to peracetylated D-xylal[72,73] and peracetylated 2-acetoxy-D-arabinal,[74] later in this Section) were discussed in relation to the steric influence exerted by the presence of the nonreducing ring of hexa-*O*-acetyl-lactal. It was suggested that the substituent on C-6 of the nonreducing ring is brought into close proximity to the double bond in the other ring (because of the relative positions of the two rings), in such a way that it considerably hinders the approach of the reagent from the α-D-*gluco* side, thus facilitating approach from the β-D-*manno* side. It was also considered possible that, at lower temperatures, the rotational mobility about the glycosidic bond between the two rings is so lessened that it hinders the formation of the F-2–C-1 carbonium ion, causing an overall diminution in the reaction rate. Furthermore, the ring attached at C-4 of the D-glucal ring could somewhat lessen the conformational mobility of the latter ring. Similarly, a relatively high temperature (0°) was needed for reaction to occur between 2-acetoxy-3,4-di-O-acetyl-D-arabinal and trifluoro(fluoroxy)methane.[74] Trifluoromethyl 2,3,4-tri-O-acetyl-2-fluoro-β-D-ribopyranoside and 2,3,4-tri-O-acetyl-2-fluoro-β-D-ribopyranosyl fluoride were obtained.

In the pentose series, attack has also been observed to occur mainly from the less-hindered side of the double bond, as in the addition of trifluoro(fluoroxy)methane to di-O-acetyl-D-arabinal. Trifluoromethyl 3,4-di-O-acetyl-2-deoxy-2-fluoro-β-D-arabinopyranoside and 3,4-di-O-acetyl-2-deoxy-2-fluoro-β-D-arabinopyranosyl fluoride were obtained,[72,74] together with a very small proportion of 3,4-di-O-acetyl-2-deoxy-2-fluoro-α-D-ribopyranosyl fluoride.[74] With di-O-acetyl-D-xylal, however, the lyxose derivatives were mainly obtained.[72,73] The suggestion has been made[73] that the greater conformational mobility of pentopyranoses results in a preponderance of the $^1C_4(D)$ conformer of the D-lyxose derivatives,[72,73] such that F-1 and CF_3O-1 are axially disposed. For further reactions of unsaturated carbohydrates with trifluoro(fluoroxy)methane, see Section II,9.

Compound **67** and 2-deoxy-2-fluoro-D-mannose have also been obtained by direct fluorination of 3,4,6-tri-O-acetyl-D-glucal with anhydrous fluorine diluted with an inert gas, followed by hydrolysis of the acetylated glycosyl fluorides resulting.[188] Fluorination of peracetylated D- and L-arabinal with lead tetraacetate and hydrogen fluoride led to monoacetates of the corresponding 2,5-anhydro-1,1-difluororibitols, instead of the vicinal difluorides expected[3,4]; it was suggested[4] that rearrangement occurs after the initial formation of a vicinal difluoro adduct (compare also, reaction with peracetylated D-glucal, following in this Section); an intermediate, 1,3-dioxolan-2-ylium ion has also been invoked for the explanation of this reaction.[15] Similar treatment of peracetylated D-glucal[3,189] and peracetylated ethyl 2,3-dideoxy-α-D-*erythro*-hexopyranoside led to the formation (after deacetylation) of 2,5-anhydro-1-deoxy-1,1-difluoro-D-mannitol, instead of the vicinal difluorides expected.[189] The *cis*-1,1-difluorides cannot have been intermediates, as originally suggested,[189] as, in another study,[69] 3,4,6-tri-O-acetyl-2-deoxy-2-fluoro-α-D-glucopyranosyl fluoride and 3,4,6-tri-O-acetyl-2-deoxy-2-fluoro-β-D-mannopyranosyl fluoride (both *cis*-1,2-difluorides) were recovered unchanged after treatment with the foregoing reagent.

(188) T. Ido, C.-N. Wan, J. S. Fowler, and A. P. Wolf, *J. Org. Chem.*, 42 (1977) 2431–2432.
(189) K. R. Wood and P. W. Kent, *J. Chem. Soc.*, C, (1967) 2422–2425.

Electrophilic additions of the elements of "ClF", "BrF," and "IF" to glycals, always furnishing 2-deoxy-2-halogenoglycosyl fluorides, have been described. One method whereby bromofluorinations and iodofluorinations (see also Section II,9) have been achieved involves treatment of the peracylated glycal with N-bromo- or N-iodo-succinimide and liquid hydrogen fluoride in a solvent exhibiting Lewis basicity.[190-192] When tri-O-acetyl-D-glucal was treated with N-bromosuccinimide and liquid hydrogen fluoride in diethyl ether,[54,55] tri-O-acetyl-2-bromo-2-deoxy-α-D-mannopyranosyl fluoride (27%) (this compound was initially assigned the β-D configuration,[54,55] but this attribution was later revised[56,57]) and tri-O-acetyl-2-bromo-2-deoxy-α-D-glucopyranosyl fluoride (14%) were the products obtained in the crystalline state. In the same reaction, other workers[43,53] obtained tri-O-acetyl-2-bromo-2-deoxy-α-D-mannopyranosyl fluoride (55%), tri-O-acetyl-2-bromo-2-deoxy-α-D-glucopyranosyl fluoride (9%), tri-O-acetyl-2-bromo-2-deoxy-β-D-glucopyranosyl fluoride (30%), and tri-O-acetyl-2-deoxy-α-D-*arabino*-hexopyranosyl fluoride (8%). Only the major product was isolated crystalline, but all four products were identified by comparison of the ^{19}F-n.m.r. spectra with those of compounds obtained by independent synthesis.

When the same glycal was treated with N-iodosuccinimide and liquid hydrogen fluoride in diethyl ether,[193] tri-O-acetyl-2-deoxy-2-iodo-α-D-mannopyranosyl fluoride (22%) (a compound also initially assigned the β-D configuration,[193] but this was later revised[56,57]) and tri-O-acetyl-2-deoxy-2-iodo-α-D-glucopyranosyl fluoride (1.3%) were obtained crystalline. Again, in the same reaction, other workers[43,53] obtained tri-O-acetyl-2-deoxy-2-iodo-α-D-mannopyranosyl fluoride (71%), tri-O-acetyl-2-deoxy-2-iodo-β-D-glucopyranosyl fluoride (3%), tri-O-acetyl-2-deoxy-2-iodo-α-D-glucopyranosyl fluoride (23%), and an unidentified, fluorine-containing compound (3%). The first three were identified in the product mixture by comparison of their ^{19}F-n.m.r. spectra with those of compounds prepared by independent synthesis. The revision by X-ray analysis[56,57] and ^{19}F-n.m.r. spectroscopy[43,56,57] of the original assignments for the major products obtained from these reactions, independent synthesis,[43] the possibility of examining all of the products (which cannot always be isolated in a pure state), and the product ratios (by ^{19}F-n.m.r. spectroscopy) of the product mixtures

(190) A. Bowers, *J. Am. Chem. Soc.*, 81 (1959) 4107–4108.
(191) A. Bowers, E. Denot, L. C. Ibanes, and R. Becerra, *J. Am. Chem. Soc.*, 82 (1960) 4001–4007.
(192) A. Bowers, E. Denot, and R. Becerra, *J. Am. Chem. Soc.*, 82 (1960) 4007–4012.
(193) K. R. Wood, P. W. Kent, and D. Fisher, *J. Chem. Soc.*, C, (1966) 912–915.

(sometimes confirmed by independent synthesis[43]) established the fact that *trans*-adducts preponderate (not exclusively *cis*, as was initially postulated[54,55]). ^{19}F-N.m.r. spectroscopy confirmed[56] that the product isolated crystalline from the reaction[194] of di-O-acetyl-D-arabinal with N-bromosuccinimide and hydrogen fluoride is di-O-acetyl-2-bromo-2-deoxy-β-D-arabinopyranosyl fluoride, and that that isolated crystalline, after deacetylation, from the reaction[195] with tri-O-acetyl-D-galactal is 2-bromo-2-deoxy-α-D-galactopyranosyl fluoride. Treatment of tri-O-acetyl-D-glucal with N-chlorosuccinimide and liquid hydrogen fluoride did not result in fluorinated products.[196]

A milder method of achieving such electrophilic additions to glycals is by reaction with halogen in the presence of silver monofluoride.[197] It has been observed that the success of the reaction depends upon the physical state of the silver monofluoride used and the rate of addition of the halogen during the initial stages of the reaction.[197] The addition of "BrF" in benzene, according to this method, to tri-O-acetyl-D-glucal in acetonitrile has been described.[43,53] Tri-O-acetyl-2-bromo-2-deoxy-α-D-mannopyranosyl fluoride (70%), tri-O-acetyl-2-bromo-2-deoxy-α- (9%) and -β-D-glucopyranosyl fluoride (21%) were obtained. Similar products and product ratios were obtained by the addition of "IF" to tri-O-acetyl-D-glucal by this method.[43,53] Aspects of the mechanisms of this reaction were discussed.[43,53,197] In addition, it was suggested[43] that the similarities between the products and product ratios obtained by the two different methods of addition of "BrF" and "IF" to peracetylated D-glucal indicate that the mechanisms of addition must be very similar. When the preceding method of bromofluorination and iodofluorination to tri-O-acetyl-D-glucal was conducted in benzene alone, however, different product ratios were obtained.[43]

Tentative assignments have been made for the products obtained by bromofluorination and iodofluorination of tri-O-acetyl-D-galactal[43] by the halogen–silver halide method; there is a low percentage of the α-*talo* isomer, presumably[43] owing to the presence of the quasi-axial substituent[186] at C-4. Tentative assignments have also been made for the products obtained by similar bromofluorination and iodofluorination of di-O-acetyl-D-xylal and -D-arabinal.[43] For the former glycal, all four possible products were obtained; the suggestion was made that this may be due to the fact that, upon reaction with halogen, the result-

(194) P. W. Kent and J. E. G. Barnett, *J. Chem. Soc.*, (1964) 6196–6204.
(195) P. W. Kent and M. R. Freeman, *J. Chem. Soc., C*, (1966) 910–912.
(196) L. D. Hall and J. F. Manville, *Can. J. Chem.*, 47 (1969) 379–385.
(197) L. D. Hall, D. L. Jones, and J. F. Manville, *Chem. Ind. (London)*, (1967) 1787–1788.

ing ion, having at least two substituents axial, may be conformationally labile [$^4C_1(D) \rightleftharpoons {}^1C_4(D)$], thus allowing the formation of adducts from both conformers; this would, of course, only be true were fluoride attack slower than conformational inversion.[43]

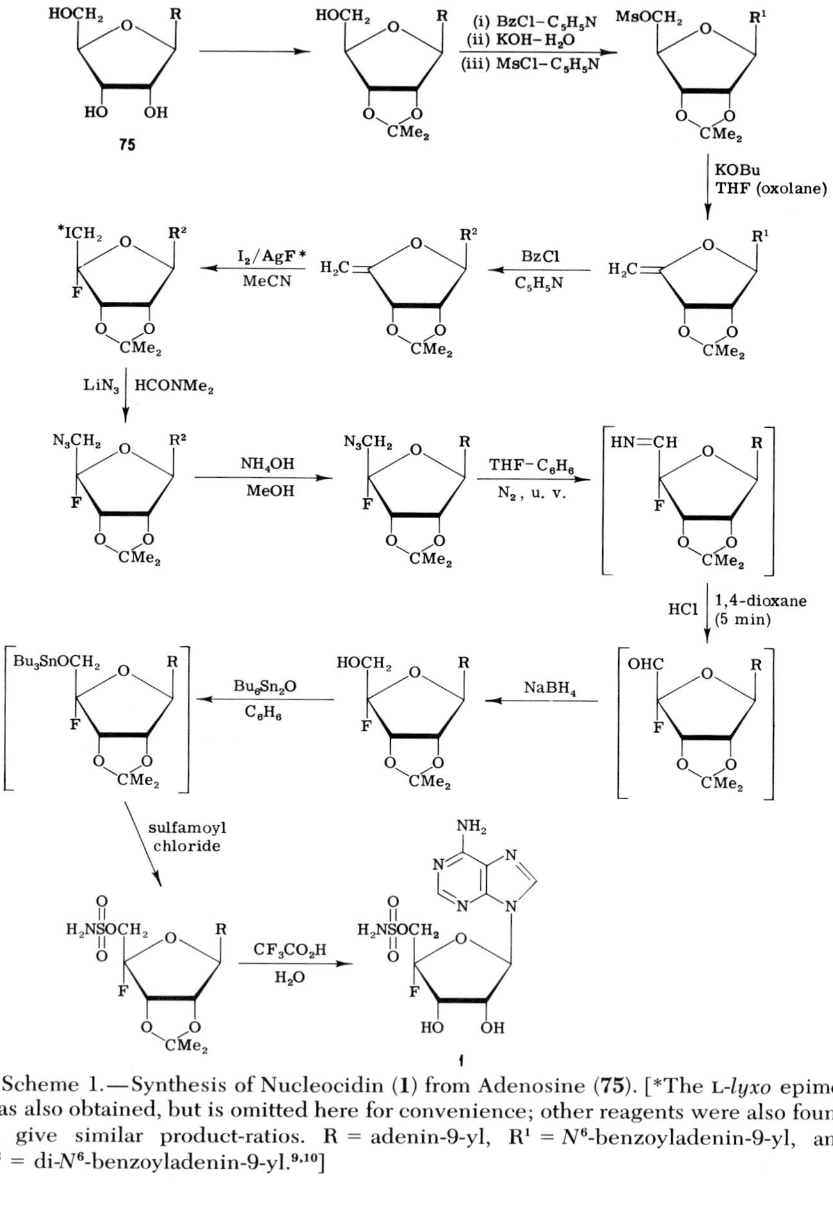

Scheme 1.—Synthesis of Nucleocidin (1) from Adenosine (75). [*The L-*lyxo* epimer was also obtained, but is omitted here for convenience; other reagents were also found to give similar product-ratios. R = adenin-9-yl, R^1 = N^6-benzoyladenin-9-yl, and R^2 = di-N^6-benzoyladenin-9-yl.[9,10]]

On treatment of tri-O-acetyl-D-glucal with chlorine and silver fluoride,[196] all four possible isomers were formed; tri-O-acetyl-2-chloro-2-deoxy-α-D-mannopyranosyl fluoride (16%), tri-O-acetyl-2-chloro-2-deoxy-β-D-mannopyranosyl fluoride (16%), tri-O-acetyl-2-chloro-2-deoxy-α-D-glucopyranosyl fluoride (6%), and tri-O-acetyl-2-chloro-2-deoxy-β-D-glucopyranosyl fluoride (62%). These product ratios differ significantly from those of the corresponding bromofluorination and iodofluorination reactions of tri-O-acetyl-D-glucal,[43,53] and this behavior has been discussed[196] in terms of the differences observed between the addition of bromine (or iodine) and chlorine to tri-O-acetyl-D-glucal, and the nature of the chlorination reaction itself.

This method of addition of interhalogens to double bonds has been applied to the synthesis of nucleocidin (1) from adenosine [9,10] (75) (see Scheme 1).

7. Total Syntheses

Total synthesis was the method by which carbohydrates fluorinated at secondary positions were first obtained. It is, however, a method of limited usefulness, owing to the mixtures of isomers that are inevitably obtained. The fluoro analogs were obtained by Claisen condensation[198-200] (fluorine being introduced as esters of fluoroacetic acid derivatives), followed by selective reduction.[201-203] For example, 2-deoxy-2-fluoro-DL-erythritol and 2-deoxy-2-fluoro-DL-threitol,[202,203] the first carbohydrates fluorinated at secondary positions to be synthesized, were obtained by vigorous reduction of ethyl ethoxalylfluoroacetate (prepared by condensation of ethyl fluoroacetate with ethyl oxalate[198]) with lithium aluminum hydride. Chemical manipulation of these fluoro analogs led to further fluorinated carbohydrates, as in the synthesis of 2-deoxy-2-fluoroglycerol from 2-deoxy-2-fluoro-DL-erythritol by periodate oxidation, followed by reduction with lithium aluminum hydride.[202,203] Mild treatment of ethyl 2,4-difluoro-3-oxobutanoate[199] (obtained by Claisen self-condensation of ethyl fluoroacetate) with potassium borohydride afforded ethyl (±)-2,4-difluoro-3-hydroxybutanoate which, on further treatment with potassium borohydride under more vigorous conditions, yielded (±)-2,4-difluoro-1,3-

(198) D. E. Rivett, *J. Chem. Soc.*, (1953) 3710–3711.
(199) E. T. McBee, O. R. Pierce, H. W. Kilbourne, and E. R. Wilson, *J. Am. Chem. Soc.*, 75 (1953) 3152–3153.
(200) P. W. Kent and J. E. G. Barnett, *J. Chem. Soc.*, (1964) 2497–2500.
(201) N. F. Taylor and P. W. Kent, *Nature (London)*, 174 (1954) 401.
(202) P. W. Kent and N. F. Taylor, *Research*, 8 (1955) S66.
(203) N. F. Taylor and P. W. Kent, *J. Chem. Soc.*, (1956) 2150–2154.

butanediols.[204] 2-Deoxy-2-fluoro-DL-ribitol (**78**) was obtained by condensation of ethyl fluoroacetate (**76**) with methyl 2,3-O-isopropylidene-DL-glycerate (**77**), followed by reduction, deprotection, and chromatographic separation of the isomers.[200] 3-Fluoropyruvic acid

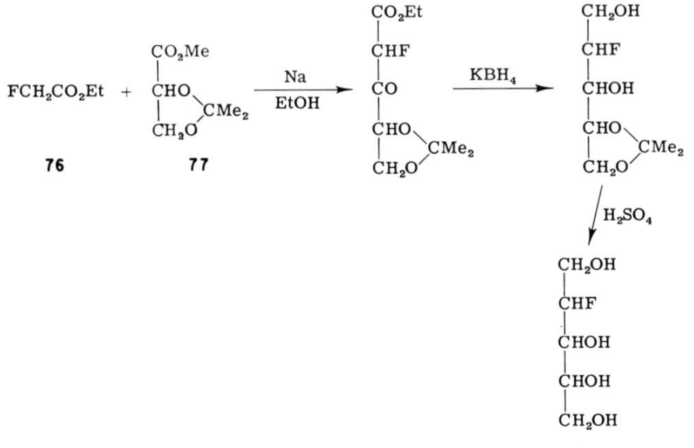

FCH$_2$CO$_2$Et +

76

CO$_2$Me
|
CHO
| \
     CMe$_2$
CH$_2$O /

77

$\xrightarrow{\text{Na, EtOH}}$

CO$_2$Et
|
CHF
|
CO
|
CHO
| \
     CMe$_2$
CH$_2$O /

$\xrightarrow{\text{KBH}_4}$

CH$_2$OH
|
CHF
|
CHOH
|
CHO
| \
     CMe$_2$
CH$_2$O /

$\downarrow$ H$_2$SO$_4$

CH$_2$OH
|
CHF
|
CHOH
|
CHOH
|
CH$_2$OH

78

(**79**) was condensed with 2-acetamido-2-deoxy-D-glucose (**80**) or -mannose in alkaline solution, to give 5-acetamido-3,5-dideoxy-3-fluoro-D-*glycero*-D-*galacto*-nonulosonic acid (**81**) (5-acetamido-3,5-dideoxy-3-fluoro-neuraminic acid) in 1.5–3% yield.[205]

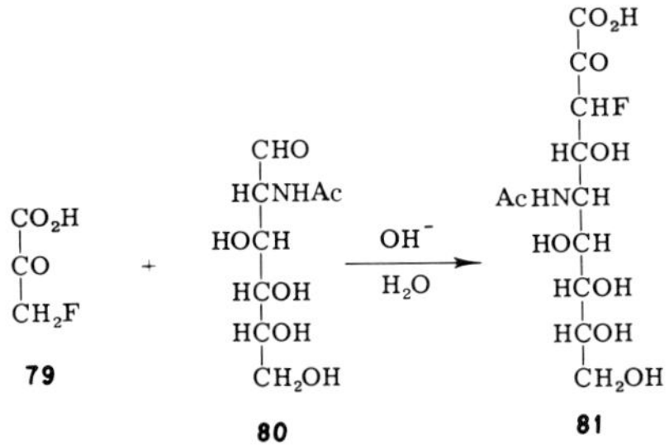

CO$_2$H
|
CO
|
CH$_2$F

79

+

CHO
|
HCNHAc
|
HOCH
|
HCOH
|
HCOH
|
CH$_2$OH

80

$\xrightarrow[\text{H}_2\text{O}]{\text{OH}^-}$

CO$_2$H
|
CO
|
CHF
|
HCOH
|
AcHNCH
|
HOCH
|
HCOH
|
HCOH
|
CH$_2$OH

81

8. Isotopically Modified, Fluorinated Carbohydrates

The radioactive isotope ^{18}F has been used for the preparation of substituted compounds of potential biomedical interest.[206,207] Among

(204) J. E. G. Barnett and P. W. Kent, *J. Chem. Soc.*, (1963) 2743–2747.
(205) R. Gantt, S. Millner, and S. B. Binkley, *Biochemistry*, 3 (1964) 1952–1960.
(206) J. P. de Kleijn, *J. Fluorine Chem.*, 10 (1977) 341–350.
(207) A. J. Palmer, J. C. Clark, and R. W. Goulding, *Int. J. Appl. Radiat. Isot.*, 28 (1977) 53–65.

others, tetraethylammonium [^{18}F]fluoride,[208] diethylaminosulfur [^{18}F]trifluoride,[152,209,210] ^{18}F$_2$ (Refs. 211 and 212), and silver [^{18}F]fluoride[213] have been used as fluorinating agents. In the sugar series, 6-deoxy-6-[^{18}F]fluoro-α-D-galactopyranose has been synthesized by treatment of 1,2:3,4-di-O-isopropylidene-6-O-p-tolylsulfonyl-α-D-galactopyranose with tetraethylammonium [^{18}F]fluoride in acetonitrile, and subsequent deprotection of the hydroxyl groups.[214,215] The preparation of 3-deoxy-3-[^{18}F]fluoro-D-glucose from 1,2:5,6-di-O-isopropylidene-α-D-glucofuranose by treatment with diethylaminosulfur [^{18}F]trifluoride, followed by deprotection, has been reported,[210] but no conditions were given. [However, these reactions should be compared with that of 1,2:5,6-di-O-isopropylidene-α-D-glucofuranose with (isotopically unmodified) diethylaminosulfur difluoride, which did not afford the corresponding 3-deoxy-3-fluoride[152] (see Section II,5).] Treatment of the mixture obtained by reaction of 1,2:5,6-di-O-isopropylidene-α-D-allofuranose with diethylaminosulfur trifluoride and pyridine in dichloromethane (prior to processing by direct distillation) with H^{18}F in pyridine, afforded 3-deoxy-3-[^{18}F]fluoro-1,2:5,6-di-O-isopropylidene-α-D-glucofuranose in >90% radiochemical yield.[152] 2-Deoxy-2-[^{18}F]fluoro-D-glucose[211,212] (see also, Ref. 216) and 2-deoxy-2-[^{18}F]fluoro-D-mannose[212] have been prepared from ^{18}F$_2$ and tri-O-acetyl-D-glucal, followed by hydrolysis; β-D-glucopyranosyl [^{18}F]fluoride has been obtained by treatment of tetra-O-acetyl-α-D-glucopyranosyl bromide with silver [^{18}F]fluoride in acetonitrile, followed by de-esterification with sodium methoxide.[213]

^{18}F is not the only isotope with which fluorinated carbohydrates

(208) J. P. de Kleijn, R. F. Ariaansz, and B. van Zanten, *Radiochem. Radioanal. Lett.*, 28 (1977) 257–262.
(209) M. G. Straatman and M. J. Welch, *J. Nucl. Med.*, 18 (1977) 151–158.
(210) M. G. Straatman and M. J. Welch, *J. Labelled Compd. Radiopharm.*, 13 (1977) 210.
(211) V. R. Casella, T. Ido, and A. P. Wolf, *J. Labelled Compd. Radiopharm.*, 13 (1977) 209.
(212) T. Ido, C.-N. Wan, V. Casella, J. S. Fowler, A. P. Wolf, M. Reivich, and D. E. Kuhl, *J. Labelled Compd. Radiopharm.*, 14 (1978) 175–183.
(213) A. E. Lemire and M. F. Reed, *J. Labelled Compd. Radiopharm.*, 15 (1978) 105–109.
(214) D. R. Christman, Z. Orhanovic, and A. P. Wolf, *J. Labelled Compd. Radiopharm.*, 13 (1977) 283.
(215) D. R. Christman, Z. Orhanovic, W. W. Shreeve, and A. P. Wolf, *J. Labelled Compd. Radiopharm.*, 13 (1977) 555–559.
(216) B. M. Gallagher, A. Ansari, H. Atkins, V. Casella, D. R. Christman, J. S. Fowler, T. Ido, R. R. MacGregor, P. Som, C.-N. Wan, A. P. Wolf, D. E. Kuhl, and M. Reivich, *J. Nucl. Med.*, 18 (1977) 990–996.

have been modified. For instance, 2-deoxy-2-fluoro-D-[^{14}C]glucose has been prepared by addition either of fluorine or trifluoro(fluoroxy)-methane[212] to 3,4,6-tri-O-acetyl-D-[^{14}C]glucal, followed by hydrolysis. 2-Deoxy-2-fluoro-D-[^{3}H]glucose and -mannose have been prepared by catalytically tritiating 2-deoxy-2-fluoro-D-glucose and -mannose, respectively.[217]

9. Miscellaneous

2-Deoxy-2-fluoro-D-ribose (**83**) has been prepared by fluorination,[218–220] hydrogenation, and hydrolysis of 2,2'-anhydro-(1-β-D-arabinofuranosyluracil) (**82**), followed, for purification purposes, by benzoylation of the resulting product, and hydrolysis[221] (see also, Ref. 222).

Methyl 3-benzamido-4,6-di-O-benzoyl-2,3-dideoxy-2-fluoro-α-D-altropyranoside (**87**) was obtained by treatment of methyl 2-benzamido-4,6-O-benzylidene-3-O-p-tolylsulfonyl-α-D-glucopyranoside (**84**) with tetrabutylammonium fluoride in hexamethylphosphoric triamide, followed by O-debenzylidenation and benzoylation.[95] The reaction is

(217) M. F. G. Schmidt, P. Biely, Z. Krátký, and R. T. Schwarz, *Eur. J. Biochem.*, 87 (1978) 55–68.
(218) J. F. Codington, I. L. Doerr, D. Van Pragg, A. Bendich, and J. J. Fox, *J. Am. Chem. Soc.*, 83 (1961) 5030.
(219) J. F. Codington, I. L. Doerr, L. Kaplan, and J. J. Fox, *Fed. Proc.*, 22 (1963) 532.
(220) J. F. Codington, I. L. Doerr, and J. J. Fox, *J. Org. Chem.*, 29 (1964) 558–564; see also, J. F. Codington, I. L. Doerr, and J. J. Fox, *ibid.*, 29 (1964) 564–560.
(221) J. F. Codington, I. L. Doerr, and J. J. Fox, *Carbohydr. Res.*, 1 (1966) 455–466.
(222) R. J. Cushley, J. F. Codington, and J. J. Fox, *Carbohydr. Res.*, 5 (1967) 31–35.

presumed to proceed by ring-opening of methyl 2,3-(benzoylepimino)-4,6-O-benzylidene-2,3-dideoxy-α-D-allopyranoside (**86**) by fluoride ion, as **86** could be isolated at the initial stages of the reaction, prior to formation of any fluorinated product. Together with the 2-deoxy-2-fluoro-D-altropyranoside (**85**) and the benzoylepimine **86**,

two more products were formed; one of them is, presumably, the free epimine formed by loss of the benzoyl group under the strongly basic conditions used.

Treatment of 6-deoxy-6-iodo-D-glucose with silver fluoride in pyridine led to unsaturated products,[223] whereas ethyl 2,3,6-trideoxy-6-

(223) B. Helferich and E. Himmen, *Ber.*, 61 (1928) 1825–1835.

fluoro-α-D-*erythro*-hex-2-enopyranoside (40%) was obtained by treatment of the corresponding 6-deoxy-6-iodide with silver fluoride in pyridine, and[82] ethyl 4-*O*-acetyl-2,3,6-trideoxy-6-fluoro-α-D-*erythro*-hex-2-enopyranoside (10%) by treatment of the corresponding 6-iodide with the same reagents. Fluorination was also not achieved when methyl 4-*O*-acetyl-6-deoxy-6-iodo-2,3-di-*O*-*p*-tolylsulfonyl-β-D-glucopyranoside (not well characterized) was treated with silver fluoride in ether.[224] Treatment of methyl 3,4-di-*O*-acetyl-6-bromo-2,6-dideoxy-2-fluoro-β-D-glucopyranoside with potassium fluoride in 1,2-ethanediol yielded the 3,6-anhydride. Alternatively, treatment of the bromide with potassium fluoride in *N*,*N*-dimethylformamide, or tetrabutylammonium fluoride in acetonitrile, afforded the corresponding unsaturated derivative.[225] However, 1,4-anhydro-5-deoxy-5-fluoroxylitol was obtained[226] in 60–65% yield by reaction of 1,4-anhydro-5-chloro-5-deoxyxylitol with anhydrous potassium fluoride in 1,2-ethanediol at 150–160°.

Treatment of 3-deoxy-1,2:5,6-di-*O*-isopropylidene-α-D-*erythro*-hex-3-enofuranose with iodine and thallous fluoride in anhydrous ether afforded[227] 3-deoxy-1-fluoro-3-iodo-1,2:5,6-di-*O*-isopropylidene-D-*xylo*-4-hexulose in 80% yield, together with small proportions of (tentatively identified) 3-deoxy-4-fluoro-3-iodo-1,2:5,6-di-*O*-isopropylidene-α-D-allofuranose and two unidentified products. A mechanism proposed for the furanose ring-opening involves the formation of a 3,4-iodonium ion, and attack by fluoride at C-1.

Reactions that have led to other deoxyhalogeno sugars do not necessarily lead to deoxyfluoro sugars, as, for example, in the attempted decomposition of fluoroformates, the treatment of diazoketones and of 2-deoxy-2-diazohexonates with hydrogen fluoride, and the reaction of benzoxonium ions with halide ions. The reaction[228,229] by which fluoroformates are thermally or catalytically decarbonylated to give alkyl fluorides has been applied to carbohydrates. Both thermal and catalytic treatment of 6-*O*-(fluoroformyl)-1,2:3,4-di-*O*-isopropylidene-

(224) K. Hess, O. Littmann, and R. Pfleger, *Ann.*, 507 (1933) 55–61.
(225) J. Pacák, J. Hrinak, and M. Černý, *Collect. Czech. Chem. Commun.*, 39 (1974) 3332–3337.
(226) S. N. Danilov and E. Ya. Afanas'eva, *Zh. Obshch. Khim.*, 36 (1966) 1406–1407; *J. Gen. Chem. USSR*, 36 (1966) 1413–1414; *Chem. Abstr.*, 66 (1967) 18,827.
(227) A. A. Akhrem, N. B. Kripach, and I. A. Mikhailopoulo, *Carbohydr. Res.*, 50 (1976) C6–C8.
(228) S. Nakanishi, T. C. Myers, and E. V. Jensen, *J. Am. Chem. Soc.*, 77 (1955) 3099–3100.
(229) S. Nakanishi, T. C. Myers, and E. V. Jensen, *J. Am. Chem. Soc.*, 77 (1955) 5033–5034.

α-D-galactopyranose (**88**) (prepared from the chloroformate by fluoride exchange with thallous fluoride), for instance, did not result in fluorination.[230] The dialkyl carbonate **89** was obtained on heating **88** in pyridine or in pyridine–N,N-dimethylformamide containing potassium fluoride.[231]

The reaction[232–235] (see also Ref. 236) of 1-diazoketones with hydrogen fluoride that affords 1-fluoro ketones has also been applied to carbohydrates, with, however, no success. Treatment of 3,4-di-O-benzoyl-1-deoxy-1-diazo-D-*glycero*-tetrulose with hydrogen fluoride in ether mainly afforded 3,4-di-O-benzoyl-D-*glycero*-tetrulose, whereas, with hydrogen bromide and chloride, 3,4-di-O-benzoyl-1-bromo-1-deoxy- and 3,4-di-O-benzoyl-1-chloro-1-deoxy-D-*glycero*-tetrulose, respec-

(230) V. A. Welch and P. W. Kent, *J. Chem. Soc.*, (1962) 2266–2270.
(231) N. Baggett, K. W. Buck, A. B. Foster, R. Jefferis, and J. M. Webber, *Carbohydr. Res.*, 4 (1967) 343–351.
(232) I. L. Knunyants, Ya. M. Kisel, and E. G. Bykhovskaya, *Izv. Akad. Nauk SSSR, Otdel. Khim. Nauk*, (1956) 377–378; *Bull. Acad. Sci. USSR, Div. Chem. Sci.*, (1956) 363–364; *Chem. Abstr.*, 50 (1956) 15,454.
(233) G. Olah and S. Kuhn, *Chem. Ber.*, 89 (1956) 864–865.
(234) E. D. Bergmann and R. Ikan, *Chem. Ind. (London)*, (1957) 394.
(235) R. R. Fraser, J. E. Millington, and F. L. M. Pattison, *J. Am. Chem. Soc.*, 79 (1957) 1959–1961.
(236) G. A. Olah and J. Welch, *Synthesis*, (1974) 896–898.

tively, were obtained.[237] Treatment of 2-deoxy-2-diazohexonates with hydrogen fluoride also did not give fluorinated products,[237] whereas 2-bromo and 2-chloro derivatives (not well characterized) have been obtained.[237–239] Treatment of the 3,4-benzoxonium ion of methyl 2-O-benzoyl-β-D-arabinopyranose with sodium fluoride did not afford any deoxyfluorobenzoates (as was the case with chloride and bromide ions); ^{19}F-n.m.r. spectroscopy indicated that an orthoacid fluoride was formed, and that this was thermally unstable and was readily hydrolyzed.[240]

The Arbuzov reaction has been applied for the synthesis of fluorinated carbohydrates. 1,2:3,4-Di-O-isopropylidene-α-D-galactopyranose 6-(N,N-diethyl-P-methylphosphonamidite) was treated with ethyl fluoroacetate, to afford 6-deoxy-6-fluoro-1,2:3,4-di-O-isopropylidene-α-D-galactopyranose in 19% yield.[241] The corresponding 6-deoxy-6-fluoride was obtained in 60% yield by treatment of 1,2:3,4-di-O-isopropylidene-α-D-galactopyranose 6-(dipropylphosphinite) with hexafluoropropene.[242] The mechanism of this reaction has been discussed.[243,244] In contrast, treatment of 1,2:3,4-di-O-isopropylidene-α-D-galactopyranose 6-(tetraethylphosphorodiamidite) with benzoyl fluoride yielded the corresponding 6-benzoate, not the 6-deoxy-6-fluoride expected.[245]

The availability, in gram quantities, of the fluorinated carbohydrates thus far synthesized has also made possible their use as starting materials for the synthesis of other analogs and intermediates; for example, the preparation of 3,5-dideoxy-3,5-difluoro-D-xylose from 3-deoxy-3-fluoro-D-xylose,[88] which is itself preparable from 3-deoxy-3-fluoro-1,2-O-isopropylidene-α-D-glucofuranose. Electrolytic oxidation of 3-deoxy-3-fluoro-D-glucose afforded, after treatment with an acidic

(237) P. W. Kent, K. R. Wood, and V. A. Welch, *J. Chem. Soc.*, (1964) 2493–2497.
(238) P. A. Levene and F. B. LaForge, *J. Biol. Chem.*, 21 (1915) 345.
(239) P. A. Levene, *J. Biol. Chem.*, 53 (1922) 449.
(240) S. Jacobsen, B. Nielsen, and C. Pedersen, *Acta Chem. Scand.*, 31 (1977) 359–364.
(241) E. E. Nifant'ev, I. N. Sorochkin, and A. P. Tuseev, *Zh. Obshch. Khim.*, 35 (1965) 2256; *J. Gen. Chem. USSR*, 35 (1965) 2248; *Chem. Abstr.*, 64 (1966) 11,295.
(242) K. A. Petrov, E. E. Nifant'ev, A. A. Shchegolev, and V. Terekhov, *Zh. Obshch. Khim.*, 34 (1964) 1459–1462; *J. Gen. Chem. USSR*, 34 (1964) 1463–1466; *Chem. Abstr.*, 61 (1964) 5738.
(243) I. L. Knunyants, E. Ya. Pervova, and V. V. Tyuleneva, *Dokl. Akad. Nauk SSSR*, 129 (1959) 576–577; *Dokl. Chem.*, 129 (1959) 1025–1026; *Chem. Abstr.*, 54 (1960) 7536.
(244) J. E. G. Barnett, Ref. 17, see p. 182.
(245) E. E. Nifant'ev, M. P. Koroteev, and N. S. Raboskaya, *Zh. Obshch. Khim.*, 43 (1973) 1806–1811; *J. Gen. Chem. USSR*, 43 (1973) 1790–1794; *Chem. Abstr.*, 79 (1973) 137,414.

resin, 3-deoxy-3-fluoro-D-gluconic acid.[246] Reduction of 6-deoxy-6-fluoro-D-galactose with sodium borohydride afforded 1-deoxy-1-fluoro-L-galactitol, and phosphorylation of methyl 2,3,4-tri-O-acetyl-6-deoxy-6-fluoro-β-D-galactopyranoside with silver diphenyl phosphate afforded, after hydrogenolysis and deacetylation, 6-deoxy-6-fluoro-α-D-galactopyranosyl (dipotassium phosphate).[247] 2-Deoxy-2-[^{18}F]fluoro-D-glucose 6-phosphate was synthesized enzymically from 2-deoxy-2-[^{18}F]fluoro-D-glucose.[248] 5-O-Benzoyl-2-deoxy-2-fluoro-3-O-formyl-D-arabinofuranose (**92**) has been obtained from 6-O-benzoyl-3-deoxy-3-fluoro-D-glucofuranose (**91**) by periodate oxidation, thus providing the key step in a practical synthesis of 2-deoxy-2-fluoro-D-arabinose derivatives (**93–95**) from the readily available 3-deoxy-3-fluoro-1,2:5,6-di-O-isopropylidene-α-D-glucofuranose (**90**), without the concurrent formation of any fluorinated isomers.[151] Similarly, 3-C-(acetoxymethyl)-6-

---

(246) N. F. Taylor, B. Hunt, P. W. Kent, and R. A. Dwek, *Carbohydr. Res.*, 22 (1972) 467–469.
(247) P. W. Kent and J. R. Wright, *Carbohydr. Res.*, 22 (1972) 193–200.
(248) B. M. Gallagher, J. S. Fowler, N. I. Gutterson, R. R. MacGregor, C.-N. Wan, and A. P. Wolf, *J. Nucl. Med.*, 19 (1978) 1154–1161.

O-benzoyl-3-deoxy-3-fluoro-D-glucofuranose has been converted into 1,2-di-O-acetyl-5-O-benzoyl-2-deoxy-2-fluoro-3-O-formyl-D-arabino-furanose.[249]

1,6-Anhydro derivatives of 2,4-dideoxy-2-fluoro-β-D-*xylo*-hexopyranose and 2,3-dideoxy-2-fluoro-β-D-*ribo*-hexopyranose, and the corresponding, reducing sugars, have been synthesized from 1,6-anhydro-4-O-benzyl-2-deoxy-2-fluoro-β-D-glucopyranose by standard procedures.[250] Oxidation of 1,6-anhydro-4-O-benzyl-2-deoxy-2-fluoro-β-D-glucopyranose with ruthenium tetraoxide, and of 1,6-anhydro-2,4-dideoxy-2,4-difluoro-β-D-glucopyranose with chromium trioxide, gave 1,6-anhydro-4-O-benzyl-2-deoxy-2-fluoro-β-D-*ribo*-3-hexulopyranose and the hydrate of 1,6-anhydro-2,4-dideoxy-2,4-difluoro-β-D-*ribo*-3-hexulopyranose, respectively. Reduction of these products gave 1,6-anhydro-2-deoxy-2-fluoro-β-D-allopyranose and 1,6-anhydro-2,4-dideoxy-2,4-difluoro-β-D-allopyranose, from which 2,4-dideoxy-2,4-difluoro-D-allopyranose was obtained. Derivatives of 1,6-anhydro-2-deoxy-2-fluoro-3-C-methyl-β-D-allopyranose were also obtained from 1,6-anhydro-4-O-benzyl-2-deoxy-2-fluoro-β-D-*ribo*-3-hexulopyranose by conventional procedures.[159] Methyl 3,4-di-O-acetyl-6-bromo-2,6-dideoxy-2-fluoro-β-D-glucopyranoside was treated with sodium azide in N,N-dimethylformamide, to give the azide derivative which, after deacetylation, catalytic reduction, and hydrolysis, afforded 6-amino-2,6-dideoxy-2-fluoro-D-glucose hydrochloride.[225] For further examples, see later in this Section.

Unsaturated derivatives of saccharides in which fluorine is attached to sp^2-hybridized carbon atoms have been synthesized by application of the Wittig reaction to sugars bearing a carbonyl group. For example, 1,2-O-isopropylidene-3-O-methyl-α-D-*xylo*-pentodialdo-1,4-furanose (**96**) was treated with (difluoromethylene)tris(dimethylamino)phosphorane, to yield an unsaturated, geminal difluoro saccharide, namely, 5,6-dideoxy-6,6-difluoro-1,2-O-isopropylidene-3-O-methyl-α-D-*xylo*-hex-5-enofuranose (**97**) which, on treatment with 1.2 to 1.5 equivalents of lithium methylphenylamide gave 6-deoxy-6-fluoro-1,2-O-isopropylidene-3-O-methyl-6-(methylphenylamino)-α-D-*xylo*-hex-5-enofuranose (**98**), together with a small quantity of the ynamine (**99**), in contrast to the dehydrohalogenated product obtained from the dibromo and dichloro analogs.[251] (Larger proportions of lithium methyl-

(249) A. J. Brink, O. G. De Villiers, and A. Jordaan, *Carbohydr. Res.*, 54 (1977) 285–291.
(250) J. Doležalová, M. Černý, T. Trnka, and J. Pacák, *Collect. Czech. Chem. Commun.*, 41 (1976) 1944–1953.
(251) J. M. J. Tronchet, B. Baehler, and A. Bonenfant, *Helv. Chim. Acta*, 59 (1976) 941–944.

phenylamide led to larger proportions of **99**.) Such compounds as **97** are versatile synthons; for example, treatment of **97** with lithium thiophenoxide or a lithium methyl(thiophenoxide) afforded the corresponding Z and E germinal (fluoroaryl)thioalkenes, with the former preponderating.[252] Treatment of **96** with (chlorocyanofluoromethyl)-phenylmercury (prepared by treatment of phenylmercuric chloride and 2-chloro-2-fluoropropionitrile with potassium *tert*-butoxide in 1,4-dioxane) and triphenylphosphine afforded) E- and Z-6-fluoro-1,2-O-isopropylidene-3-O-methyl-α-D-*xylo*-hept-5-enofuranuronontrile.[253]

1,2:5,6-Di-O-isopropylidene-α-D-*ribo*-hexofuranos-3-ulose (**100**) was treated with (chlorofluoromethylene)triphenylphosphorane (prepared by reaction of triphenylphosphine on difluorocarbene generated *in situ* by reaction of potassium *tert*-butoxide with dichlorofluoromethane), to give* *cis*- and *trans*-3-C-(chlorofluoromethylene)-3-deoxy-1,2:5,6-di-O-isopropylidene-α-D-*ribo*- (**101** and **103**) and -*xylo*-hexofuranoses (**105** and **107**), which, on treatment with lithium aluminum hydride, gave *cis*- and *trans*-3-deoxy-3-C-(fluoromethyl)-1,2:5,6-di-O-isopropylidene-α-D-*ribo*- (**102** and **104**) and -*xylo*-hexofura-

---

* The terms *cis* and *trans* (relative to C-2) are preferable to the alternative Z and E designations, as n.m.r. observations can be rationalized much more readily if the compounds are distinguished by the former, rather than by the latter, terminology; see Sections III,2a, III,2b(*v*), and III,2c(*i*).

(252) J. M. J. Tronchet and A. Bonenfant, *Helv. Chim. Acta*, 60 (1977) 892–895.
(253) J. M. J. Tronchet and O. R. Martin, *Helv. Chim. Acta*, 60 (1977) 585–589.

noses[254] (**106** and **108**); treatment of the ketose with (fluoromethylene)triphenylphosphorane did not give good results. In contrast, treatment of **100** with (chlorofluoromethylene)triphenylphosphorane [prepared *in situ* from (dichlorofluoromethyl)phenylmercury and triphenylphosphine] afforded **101** and **103** in 73% yield. Similar treatment of 1,2:5,6-di-*O*-isopropylidene-α-D-*xylo*-hexofuranos-3-ulose afforded **105** and **107** in 60% yield. Treatment of **100** with (chlorodifluoromethyl)phosphonium bromide (generated *in situ* from dichloro-

(254) J. M. J. Tronchet, D. Schwarzenbach, and F. Barbalat-Rey, *Carbohydr. Res.*, 46 (1976) 9–17.

fluoromethane and triphenylphosphine) and potassium fluoride (that is, effectively, with difluorocarbene and triphenylphosphine) afforded 3-deoxy-3-C-(difluoromethylene)-1,2:5,6-di-O-isopropylidene-α-D-*ribo*-hexofuranose (**109**), and similar treatment of the corresponding *xylo*-hexos-3-ulose afforded 3-deoxy-3-C-(difluoromethylene)-1,2:5,6-di-O-isopropylidene-α-D-*xylo*-hexofuranose. The *ribo* epimer (**109**) is very sensitive to moisture, and gives 3-C-(difluoromethylene)-1,2:5,6-di-O-isopropylidene-α-D-gluco(or allo)furanose[254] (**110**).

$$100 \xrightarrow{Ph_3P, :CF_2} 109$$

$$109 \xrightarrow{H_2O} 110$$

Treatment[255] of (*E*)-3-deoxy-3-C-[ethoxycarbonyl(formylamino)-methylene]-1,2:5,6-di-O-isopropylidene-α-D-glucofuranose (**111**) in moist dichloromethane containing ethanol with trifluoro(fluoroxy)-methane[181] yielded 3-deoxy-3-C-ethoxalyl-3-fluoro-1,2:5,6-di-O-isopropylidene-α-D-glucofuranose (**112**) and 3-deoxy-3-fluoro-3-C-[ethoxy(ethoxycarbonyl)-(formylimino)methyl]-1,2:5,6-di-O-isopropylidene-α-D-glucofuranose (**113**), whereas, in ethanol-free, dry dichloromethane, 3-deoxy-3-C-[(ethoxycarbonyl)-(formylimino)methyl]-3-fluoro-1,2:5,6-di-O-isopropylidene-α-D-glucofuranose (**114**) was obtained. (For reactions of trifluoro(fluoroxy)methane with glycals, see Section II,6.) The oxo ester (**112**) was reduced with lithium aluminum

(255) K. Bischofberger, A. J. Brink, and A. Jordaan, *J. Chem. Soc. Perkin Trans. 1*, (1975) 2457–2460.

hydride, to give an epimeric mixture of glycols (**115**). Oxidative cleavage of **115** gave the hydrated aldehyde **116**, which was reduced with sodium borohydride to 3-deoxy-3-fluoro-3-*C*-(hydroxymethyl)-1,2:5,6-di-*O*-isopropylidene-α-D-glucofuranose (**117**), which was, in turn, hydrolyzed to the 5,6-diol **118**, and this converted into the D-*xylo* compound **119** by glycol cleavage and reduction; **119** was then acetylated, to give 3-*C*-(acetoxymethyl)-5-*O*-acetyl-3-deoxy-3-fluoro-1,2-*O*-isopropylidene-α-D-xylofuranose[255] (**120**). Compound **120** was subsequently suitably modified, and coupled with uracil, to yield fluorinated nucleosides.[249] The α-oxoester (**112**) has also been treated with carbamoylmethylenephosphorane to afford (*E*)-3-*C*-[2-(carboxamido)-1-(ethoxycarbonyl)ethylene]-3-deoxy-3-fluoro-1,2:5,6-di-*O*-isopropylidene-α-D-glucofuranose (**121**) and [5-(*S*)-(3-deoxy-3-fluoro-1,2:5,6-di-*O*-isopropylidene-α-D-glucofuranose-3-yl)-5-hydroxy-2,4-dioxopyrrolidin-3-ylidene]triphenylphosphorane (**122**). Similarly treated, 3-deoxy-3-C-ethoxalyl-3-fluoro-1,2-O-isopropylidene-5-O-trityl-α-D-xylofuranose (**123**) afforded (E)-3-C-[2-(carboxamido)-1-(ethoxycarbonyl)ethylene]-3-deoxy-3-fluoro-1,2-O-isopropylidene-5-O-trityl-α-D-xylofuranose (**124**) and [5-(R)-(3-deoxy-3-fluoro-1,2-O-isopropylidene-5-O-trityl-α-D-xylofuranos-3-yl)-5-hydroxy-2,4-dioxopyrrolidin-3-ylidene]triphenylphosphorane (**125**). The course of these reactions has been investigated in detail.[256]

(Fluoroacetyl)ation and (trifluoroacetyl)ation of the amino groups provides a simple means of labelling amino sugars with the ^{19}F nucleus. N-(Fluoroacetyl)ation occurs readily by treatment of, for example, 1,3,4,6-tetra-O-acetyl-2-amino-2-deoxy-β-D-glucopyranose with

(256) J. C. A. Boeyens, A. J. Brink, and A. Jordaan, S. Afr. J. Chem., 31 (1978) 7–13.

N,N-dicyclohexylcarbodiimide and fluoroacetic acid in chloroform–pyridine.[257] 2-Deoxy-2-(fluoroacetamido)-D-glucopyranose has also been obtained directly from 2-amino-2-deoxy-D-glucose hydrochloride by treatment with sodium fluoroacetate and N,N-dicyclohexylcarbodiimide in pyridine.[258] N-(Fluoroacetyl)neuraminic acid has been prepared by treatment of the methyl β-D-glycoside of neuraminic acid with fluoroacetic anhydride in triethylamine at 0°, followed by hydrolysis[259] of the N-(fluoroacetyl)ated glycoside with Dowex-50 ion-exchange resin at 80°. On the other hand, N-(trifluoroacetyl)ation is readily achieved by treatment of the amine with S-ethyl trifluorothioacetate in methanol,[260] for example, in the preparation of 2-deoxy-2-(trifluoroacetamido)-D-glucopyranose, without the need for prior protection of the hydroxyl groups. For carbohydrates having pro-

(257) C. G. Grieg, D. H. Leaback, and P. G. Walker, J. Chem. Soc., (1961) 879–883; see also, C. G. Grieg and D. H. Leaback, ibid., (1963) 2644–2647.
(258) P. F. Daniel, cited in R. A. Dwek, P. W. Kent, and A. V. Xavier, Eur. J. Biochem., 23 (1971) 343–348.
(259) R. Schauer, F. Wirtz-Peitz, and H. Faillard, Hoppe-Seyler's Z. Physiol. Chem., 351 (1970) 359–364.
(260) M. L. Wolfrom and P. J. Conigliaro, Carbohydr. Res., 11 (1969) 63–76.

tected hydroxyl groups, N-(trifluoroacetyl)ation may be achieved with trifluoroacetic anhydride, in the presence of sodium carbonate,[261] or in pyridine,[262] or alone, as in the preparation of 3,4,6-tri-O-acetyl-2-deoxy-2-(trifluoroacetamido)-α-D-glucopyranosyl bromide from 3,4,6-tri-O-acetyl-2-amino-2-deoxy-α-D-glucopyranosyl bromide hydrobromide.[263]

III. Physical Methods

1. Mass Spectrometry

The derivatives most commonly used for electron-bombardment, mass-spectrometric study of fluorinated carbohydrates have been acetates, although other derivatives have also been examined, for example, methyl 6-O-acetyl-4-deoxy-4-fluoro-2,3-di-O-methyl-α-D-glucopyranoside.[137] Among others, the mass spectra of some per-O-(trimethylsilyl)ated and benzylated derivatives of certain unsymmetrically fluorinated α,α-trehalose analogs[111,112] have been described; those of perbenzylated benzyl 2-acetamido-2,6-dideoxy-6-fluoro-α-D-glucopyranoside and of permethylated methyl 2-acetamido-2,6-dideoxy-6-fluoro-α-D-glucopyranoside[93] and those of some geminal difluoro saccharides bearing isopropylidene groups[166] and of unsaturated, fluorinated carbohydrates[254,264] have also been recorded.

In more detail, the mass spectra of peracetylated 6-deoxy-6-halogeno-α-D-glucopyranoses have been examined,[265] and it was found that the ease with which atomic halogen, or hydrogen halide, is cleaved from the molecular ion, or any daughter ions, increases with decreasing electronegativity of the halogen substituent. In the case of peracetylated 6-deoxy-6-fluoro-α-D-glucopyranose, loss of a fluorine radical or of hydrogen fluoride was not observed (except from one daughter ion of a particular series), and this has been found to be generally true for primary fluorides in deoxyfluoroglycopyranoses, although loss of hydrogen fluoride has been observed from 4-acetamido-2,3,2′,3′,4′,6′-hexa-O-acetyl-4,6-dideoxy-6-fluoro-α,α-trehalose,[112] and loss of hydrogen fluoride (together with acetic acid and ketene) has

(261) M. L. Wolfrom and H. B. Bhat, *Chem. Commun.*, (1966) 146.
(262) M. L. Wolfrom and H. B. Bhat, *J. Org. Chem.*, 31 (1966) 1821–1823.
(263) R. G. Strachan, W. V. Ruyle, T. Y. Shen, and R. Hirschmann, *J. Org. Chem.*, 31 (1966) 507–509.
(264) J. M. J. Tronchet, B. Gentile, J. Ojha-Poncet, G. Moret, D. Schwarzenbach, F. Barbalat-Rey, and J. Tronchet, *Carbohydr. Res.*, 59 (1977) 87–93.
(265) O. S. Chizhov, A. B. Foster, M. Jarman, and J. H. Westwood, *Carbohydr. Res.*, 22 (1972) 37–42.

been observed from 3,4-di-O-acetyl-1,5-anhydro-1,2,6-trideoxy-6-fluoro-D-*arabino*-hex-1-enitol.[82] Cleavage of the C-5–C-6 bond of peracetylated 6-deoxy-6-fluoro-α-D-glucopyranose was not generally observed; this appears to be a common feature of deoxyfluoro saccharides, that is, the cleavage of a C–C bond, where one of the carbon atoms is substituted by fluorine, is an unfavorable process,[265,266] and the strength of such a bond is not surprising in view of the high electron-attracting power of fluorine. This feature may be used diagnostically for the determination of the position of substitution by fluorine; for example, in the configurational assignment of methyl 6-O-acetyl-4-deoxy-4-fluoro-2,3-di-O-methyl-α-D-glucopyranoside,[137] although some exceptions to this have been observed. For example, in a study of the mass spectra of acetylated deoxyfluoro-D-glucitols, 2-deoxy-2-fluoro-D-glucitol was found[267] to give an abundant fragment containing C-3–C-4–C-5–C-6 arising from a C-2–C-3 bond-fragmentation, and, surprisingly, for 3,4-di-O-acetyl-1,5-anhydro-1,2,6-trideoxy-6-fluoro-D-*arabino*-hex-1-enitol, where two abundant fragments involving the loss of the CH_2F group were observed.[82]

In a study of peracetylated deoxyfluoroaldo-pento- and -hexo-pyranose derivatives,[266] it was established that the fragmentation pathways are essentially the same as those for the corresponding, non-fluorinated saccharide derivatives, with the exception that the fluorine substituent hinders cleavage of the C–C fragment to which it is attached. Loss of hydrogen fluoride has been observed,[266] for instance, from peracetylated 3-deoxy-3-fluoro-β-D-gluco- and -xylo-pyranose following loss of the substituent at C-1; such a loss of the C-3 substituent is a common feature of mass spectra of aldopyranose derivatives. Similarly, loss of hydrogen fluoride has been observed when the fluorine substituent is attached to C-4 of pyranoses; for example, in methyl 6-O-acetyl-4-deoxy-4-fluoro-2,3-di-O-methyl-α-D-glucopyranoside[137] and in methyl 2,3,6-tri-O-acetyl-4-deoxy-4-fluoro-α-D-glucopyranoside.[138] The aldopyranoses substituted at C-2 with fluorine exhibited a new fragmentation-pathway involving loss of two molecules of acetic acid, one of ketene, and one of carbon monoxide. Some parallelism was also observed between the spectra of such compounds and that of 1,3,4,6-tetra-O-acetyl-2-deoxy-β-D-glucopyranose.[266]

It is worth noting that, for low-resolution, electron-bombardment,

(266) O. S. Chizhov, V. I. Kadentsev, B. M. Zolotarev, A. B. Foster, M. Jarman, and J. H. Westwood, *Org. Mass Spectrom.*, 5 (1971) 437–445.
(267) J. Adamson, A. D. Barford, E. M. Bessell, A. B. Foster, M. Jarman, and J. H. Westwood, *Org. Mass Spectrom.*, 5 (1971) 865–875.

mass-spectrometric examination, trimethylsilyl ethers of fluorinated carbohydrates[112] are not particularly suitable. It is well known that they give a number of fragments and daughter ions that typically arise from migration of trimethylsilyl groups and, also, a parallel series initiated by alternative loss of a methyl radical, a trimethylsilanol molecule, or a trimethylsilyloxy radical.[268] Under these circumstances, where the absence of a fragmentation pathway indicates the position of fluorination in the carbohydrate molecule, O-trimethylsilyl derivatives take second place to peracetates, for instance, as suitable derivatives for low-resolution, mass-spectrometric studies.[95] Despite the successful application of mass spectrometry to structural studies of fluorinated carbohydrates, it is less useful than the extremely sensitive technique of ^{19}F-n.m.r. spectroscopy.

However, high-resolution, field-desorption, mass spectrometry of disodium deoxyfluoro-D-glucose 6-phosphates has been studied.[269,270] This technique can be usefully applied to molecules of low volatility or thermal instability. It could, for example, be used in the study of drug and carcinogen metabolism, where, frequently, only small amounts of material are available, and derivatization is often inconvenient. At both low and high resolution, (M + 1), [(M − Na + H) + H]$^+$ were present as strong peaks; moderate peaks corresponding to [(M − 2 Na + 2 H) + H]$^+$ were also observed. On the basis of the few observations made, it was not possible to determine the effect of fluorine substitution on the cleavage of the C–C fragment to which it was attached. At a low emitter-temperature (20°), the molecular ion and daughter fragments were observed, but, as the temperature was raised, pyrolysis fragments became abundant and interfered with the detection of the ions from field-desorption-induced fragmentation. It was further suggested that application of the field-desorption method might result in new fragmentation-pathways. However, owing to the limited data then available on (a) fragmentation pathways of carbohydrates under field-desorption conditions, and (b) the relative contribution of pyrolytic processes, further investigations will be needed in order to provide the information necessary for defining the positions of the functional groups; tentative identification of two-carbon fragments was made.

(268) For example, see D. C. DeJongh, T. Radford, J. D. Hribar, S. Hanessian, M. Bieber, G. Dawson, and C. C. Sweeley, *J. Am. Chem. Soc.*, 91 (1969) 1728–1740.
(269) H.-R. Schulten, H. D. Beckey, E. M. Bessell, A. B. Foster, M. Jarman, and J. H. Westwood, *J. Chem. Soc. Chem. Commun.*, (1973) 416–418.
(270) H.-R. Schulten, *Methods Biochem. Anal.*, 24 (1977) 313–448.

2. ^{19}F-Nuclear Magnetic Resonance Spectroscopy

Specifically fluorinated carbohydrates having rigid conformations provide suitable systems for the study of the various dependencies of the ^{19}F-n.m.r. parameters, especially as complementary information on conformations and configurations is usually available from ^{1}H-n.m.r. spectroscopy. The application of ^{19}F-n.m.r. spectroscopy to specifically fluorinated carbohydrates added a new dimension to structural studies of carbohydrates in solution (including aqueous solution), the fluorine nucleus acting as an additional n.m.r. probe. In certain instances, the stereospecific dependencies of ^{19}F-n.m.r. parameters resemble those of the corresponding ^{1}H-n.m.r. parameters; ^{19}F-n.m.r. parameters, compared to the corresponding ^{1}H values, are much more sensitive to the steric environment. The chemical shifts of fluorine nuclei occur over a wider range than ^{1}H values, and the coupling constants involving fluorine nuclei are much larger than the corresponding proton–proton values. The ^{19}F isotope is of 100% natural abundance, has a spin of $\frac{1}{2}$, and a sensitivity of $\sim 10^{-1}$, compared to that of ^{1}H. Specifically fluorinated carbohydrates having rigid conformations are also very useful for the study of ^{19}F–^{13}C coupling-constants; very few such studies have so far been reported.

Reviews of ^{19}F-n.m.r. spectroscopy of specifically fluorinated carbohydrates have been included in a number of articles, namely, by Inch[271] in 1969 and 1972 (Ref. 272), simultaneously by Fields[273] and Bentley[274] in 1972, by Kotowycz and Lemieux[275] in 1973, by Emsley and coworkers,[276] who dealt exclusively with ^{19}F coupling constants, in 1976, and by Wasylishen[277] in 1977 (see also, Refs. 278–280).

a. Chemical Shifts.—Chemical shifts that are at least one order of magnitude greater than the corresponding ^{1}H values are generally in-

(271) T. D. Inch, *Annu. Rev. NMR Spectrosc.*, 2 (1969) 35–82.
(272) T. D. Inch, *Annu. Rev. NMR Spectrosc.*, 5A (1972) 305–352.
(273) R. Fields, *Annu. Rev. NMR Spectrosc.*, 5A (1972) 99–513.
(274) R. Bentley, *Annu. Rev. Biochem.*, 41 (1972) 812–996.
(275) G. Kotowycz and R. U. Lemieux, *Chem. Rev.*, 73 (1973) 669–698.
(276) J. W. Emsley, L. Phillips, and V. Wray, *Prog. Nucl. Magn. Reson. Spectrosc.*, 10 (1976) 83–756.
(277) R. E. Wasylishen, *Annu. Rev. NMR Spectrosc.*, 7 (1977) 245–291.
(278) K. Bock and C. Pedersen, *Acta Chem. Scand.*, 29 (1975) 682–686; it has been pointed out[279] that study of ^{13}C, observed at only one field, may lead to erroneous assignments. As a consequence, it is possible that some of the assignments made here may need revision.
(279) V. Wray, *J. Chem. Soc., Perkin Trans. 2*, (1976) 1598–1605.
(280) V. Wray, personal communication; see also, H. J. Schneider, W. Gschwendter, D. Heiske, V. Hoppen, and F. Thomas, *Tetrahedron*, 33 (1977) 1769–1773.

dicative of the position of substitution by fluorine on the carbohydrate ring.[72,144,170,281,282] Primary fluorides generally resonate at higher fields than secondary fluorides, which, in turn, generally resonate at higher fields than glycosyl fluorides. Tertiary fluorides, including instances where the carbon atom bears an additional fluorine atom[2,166] and secondary fluorine atoms (where the carbon atom bears an additional fluorine atom)[166] also resonate at low fields. An extreme exception to this rule is the high F-2 chemical-shift observed for 3,4,6-tri-O-acetyl-2-deoxy-2-fluoro-β-D-mannopyranosyl fluoride.[71] Few examples are available for fluorine attached to sp²-hybridized carbon atoms; in these, the chemical shifts occur over a wide range, approximately over that of primary and secondary fluorides. The observation has been made that, for a number of 3-C-(fluoromethylene)-2-D-ribo- and -xylo-furanoses[254,264] bearing hydrogen, deuterium, fluorine, or chlorine on the C-methylene group, F_{trans} resonates to higher field of F_{cis} (*cis* and *trans* relative to C-2).

In a study of aldohexopyranosyl fluoride derivatives, it was observed that, for H-2 equatorial, axial fluorine resonates at higher fields than equatorial fluorine[283] [in this series, β-D-xylo- and β-D-ribo-pyranosyl fluoride were included; this comparison is no longer valid, in view of the fact that they were later shown to exist in the $^1C_4(D)$ conformation, where F-1 is axial[284,285]]. It was also suggested that the chemical-shift values depend upon the orientation of substituents at adjacent centers in the molecule. In a later study of hexopyranosyl fluoride derivatives,[42] it was observed that, for α- and β-D-glucopyranosyl fluoride, where the substituents on C-2, C-3, C-4, and C-6 are equatorially oriented, the ^{19}F chemical-shifts provide a convenient means of unequivocally identifying the orientation of the fluorine substituent with respect to the pyranose ring (in all of the derivatives, equatorial F-1 resonated at lower field than its axial counterpart; the same has been found true for tri-O-benzoyl-α- and -β-D-*arabino*-hexopyranosyl fluoride[66]). The shifts of the β anomers of the aldohexopyranosyl fluo-

---

(281) P. W. Kent, R. A. Dwek, and N. F. Taylor, *Tetrahedron*, 27 (1971) 3887–3891. Subsequent determinations for additional fluorinated pyranoses showed that $^3J_{a,e}$ values are not always larger than $^3J_{e,e}$ values; as described in this Section, the values of *gauche*-vicinal coupling-constants have been shown to depend on a number of complex factors.
(282) P. W. Kent, R. A. Dwek, and N. F. Taylor, *Biochem. J.*, 121 (1971) 10P–11P.
(283) L. D. Hall and J. F. Manville, *Chem. Ind. (London)*, (1965) 991–993. For conformations of β-D-arabino-, ribo- and -xylo-furanosyl fluorides, see Ref. 285.
(284) L. D. Hall and J. F. Manville, *Carbohydr. Res.*, 4 (1967) 512–513.
(285) L. D. Hall and J. F. Manville, *Can. J. Chem.*, 47 (1969) 19–30.

rides were found to be temperature-dependent, in contrast to those of the α anomers, which were not.[42,43] It was considered probable that, in the former situation, the O-5–C-1–C-2 part of the ring is distorted into a shape that may undergo inversion with a very small activation-energy barrier.[42] In addition, it was found that alteration of configuration at adjacent and remote centers of the molecule (in the case of F-1 axial) induced chemical-shift changes that were additive. Hence, after comparing the ^{19}F-induced shifts for peracetylated α-D-allo-, -galacto-, and -manno-pyranosyl fluorides with the ^{19}F chemical-shifts for α-D-glucopyranosyl fluoride (in deuteriochloroform), it was possible to predict that of the corresponding α-D-altropyranosyl fluoride. The value predicted for peracetylated α-D-idopyranosyl fluoride[42] was later shown to be correct.[27] In a parallel investigation,[285] determination of chemical shifts of pentopyranosyl fluorides generally confirmed the conclusions made for the hexopyranosyl fluoride series. In fact, the dependence of the chemical-shift value of an axial, anomeric fluorine substituent upon the configuration of substituents at adjacent and remote centers of the molecule was elegantly applied to confirm that the same series of pentopyranosyl fluorides studied exist in that conformation which has the fluorine atom axially oriented. In addition, comparison between the chemical-shift values of pento- and hexo-pyranosyl fluorides indicated that an acetoxymethyl group equatorially oriented at C-5 has a deshielding effect upon an axially oriented, anomeric fluorine substituent.

An angular dependence of the ^{19}F chemical-shifts was considered to be operative in the case of peracetylated 2-deoxy-2-halogenohexopyranosyl fluoride derivatives of the α-D-*manno*, α-D-*gluco*, and β-D-*gluco* configurations.[286] On comparing the shifts, which were found to increase regularly through the series X = Cl, Br, I, with those for 2-deoxy derivatives (X = H), it became evident that, when a halogen substituent is placed in a *gauche* relationship with respect to the fluorine substituent (α- and β-D-*gluco* series), the ^{19}F resonance (ϕ_c value) is shifted to high field, whereas a halogen substituent in an anti-planar orientation (α-D-*manno* series) results in a shift to low field.

No simple rationalization for the observed values of the ^{19}F resonances could be made for the peracetylated 2-deoxy-2-fluoro-D-gluco- and -manno-pyranosyl fluorides, apart from the fact that the anomeric fluorine substituent always gives the lower field-resonance of the two[196]; this should, however, be compared with the explanation described[287] (see later in this Section) for the observed differences in ^{19}F

(286) L. D. Hall and J. F. Manville, *Chem. Commun.*, (1968) 37–38.
(287) L. Phillips and V. Wray, *J. Chem. Soc., B*, (1971) 1618–1624.

resonances for 2-deoxy-2-fluoro-α- and -β-D-gluco- and -manno-pyranose. No explanation could be given for the abnormally high shift of F-2 for the β-D-*manno* derivative, already mentioned. In addition to the foregoing, peracetylated 2-deoxy-2-halogeno-α- and -β-D-gluco- and -α-D-mannopyranosyl fluorides (halogen = Br, Cl, F, and I) have been used in a theoretical study[288] (see also, Refs. 289 and 290) of the effects of vicinal substituents on the shielding of fluorine nuclei, where the observations made with these compounds (and related, noncarbohydrate systems) indicated that the changes in chemical shift observed for fluorine upon changing the vicinal substituent may well arise from structurally dependent, inductive effects. It was noted that, for 2-, 3-, and 6-deoxyfluoro-D-glucopyranose[128,287] and the D-glucopyranosyl fluorides, the ^{19}F chemical-shifts of the α anomers occur to higher field than those of the β anomers; for the 4-deoxyfluoro analog, the reverse is true.[128,287] Some of these observations have been rationalized[287] (see later in this Section).

In addition to the foregoing, a systematic study of ^{19}F-n.m.r. parameters of the anomeric pairs of all five monodeoxyfluoro-D-glucopyranoses and 2-deoxy-2-fluoro-D-mannopyranose[287] revealed that chemical shifts and chemical-shift differences between pairs of anomeric isomers depend on the stereochemistry of electronegative substituents in the molecule. In a qualitative fashion, chemical shifts are subject to the same types of interactions that influence the variation in ^{19}F–^{1}H coupling-constants (although chemical shifts depend on more factors than do coupling constants). *trans*-Diaxial interactions of the ^{19}F substituent with a hydroxyl group (an electronegative group) were found to be larger than *gauche* interactions. For example, a comparison was made between the small difference in the shielding of the F-2 nucleus in 2-deoxy-2-fluoro-α- and -β-D-glucopyranose (−0.18 p.p.m.), where, in both, F-2 bears a *gauche* relationship to HO-1, and the large difference (+18.48 p.p.m.) observed between 2-deoxy-2-fluoro-α- and -β-D-mannopyranose, where the corresponding relationship changes from *trans* in the former to *gauche* in the latter. This accords with the fact that a hydroxyl group or ring-oxygen atom that is *trans* to a fluorine substituent by way of the coupling pathway leads to a large diminution in the coupling constant with a *gauche*-vicinal, hydrogen nucleus.

Stereospecificity was also encountered in long-range interactions. For 3-deoxy-3-fluoro-D-glucopyranose, the α anomer resonated to

(288) L. Phillips and V. Wray, *J. Chem. Soc., Perkin Trans.* 2, (1972) 223–228.
(289) H. Stahl-Larivière, *Org. Magn. Reson.*, 6 (1974) 170–177.
(290) J. B. Lambert and J. E. Goldstein, *J. Am. Chem. Soc.*, 99 (1977) 5689–5693.

higher field of the β anomer by 4.91 p.p.m.; this pointed to an influence of the anomeric hydroxyl group on the shielding of F-3. Thus, it was concluded that a considerable difference must exist between a 1,3-diequatorial interaction of the ^{19}F nucleus with an electronegative substituent and a corresponding 1-axial, 3-equatorial interaction. In 4-deoxy-4-fluoro-D-glucopyranose, the α anomer resonates to lower field of the β anomer by 2.01 p.p.m. It was suggested that through-bond, electronic interactions (which may also be made larger by the presence of the ring-oxygen atom) must be favored between the equatorial substituents at C-4 and C-1, especially as such an arrangement also leads to relatively large ^{19}F–^{1}H coupling-constants. The same distance through bonds, and, approximately, through space, exists in the 6-deoxy-6-fluoro-D-glucopyranose anomers as for the 4-deoxy-4-fluoro analogs; however, the α anomer resonates to higher field than the β anomer by a smaller amount (0.73 p.p.m.) than the difference observed for the 4-deoxy-4-fluoro analogs (and in the opposite sense). This was considered probably due to the fact that, in the 4-deoxy-4-fluoro analogs, the fluorine atom is equatorial, whereas, in the 6-deoxy-6-fluoro analogs, it is pseudo-axial. The same rationalization was propounded[287] for the observation that the fluorine atom in 4-deoxy-4-fluoro-α-D-galactopyranose resonates to high field of that in the corresponding β anomer. Interactions between the fluorine nucleus and the OH groups were presumed to occur through bonds.

b. ^{19}F–^{1}H Coupling-Constants.—(i) $^{2}J_{F,H}$. Measurements of the geminal, ^{19}F–^{1}H coupling-constants of the anomeric, peracetylated D-gluco-, -galacto-, -manno-, and L-rhamno-pyranosyl fluorides[42] revealed that, irrespective of anomeric configuration, those derivatives that possess an axial substituent at C-2 exhibit a geminal, ^{19}F–^{1}H coupling-constant of ~49 Hz, whereas those having the substituent on C-2 equatorial give coupling constants of ~53 Hz. This stereospecific dependence was also noted in a study of the peracylated pentopyranosyl fluorides, where it was also found that this dependence parallels that found for ^{1}H–^{1}H, geminal coupling-constants.[285] Geminal coupling-constants in glycopyranosyl fluorides occur in the approximate range of 48–54 Hz, and, although it is generally true that the value of the coupling is larger when the substituent on C-2 is equatorial rather than axial, the difference is sometimes insignificant. For example, this occurs[43] in the case of peracetylated 2-deoxy-2-iodo-α-D-galactopyranosyl fluoride (50 Hz) and 2-deoxy-2-iodo-α-D-talopyranosyl fluoride (50.5 Hz). The reverse has also been found to be true; that is, the glycopyranosyl fluoride having the substituent on C-2 axial exhibits a larger coupling-constant than that having it equatorial; for peracetylated 2-iodo-2-deoxy-α-D-glucopyranosyl fluoride, $^2J_{F,H}$ = 50.5 Hz, and,[43] for peracetylated 2-iodo-2-deoxy-α-D-mannopyranosyl fluoride, $^2J_{F,H}$ = 51.7 Hz.

This dependence of the geminal, $^{19}F-^1H$ coupling-constant on the orientation of a neighboring, electronegative substituent (AcO-4, in this case) has been suggested as being the cause of the difference in the observed values of this coupling for peracetylated 3-deoxy-3-fluoro-β-D-gluco- and -galacto-pyranose; the value for the *gluco* is somewhat larger than that for the *galacto* epimer.[146] The values of these coupling constants for secondary fluorine substituents in pyranoses are in the approximate range of 43–59 Hz. Values of geminal, $^{19}F-^1H$ coupling-constants, for furanoses, generally occur in the approximate range of 49–66 Hz. A few geminal, $^{19}F-^1H$ coupling-constants have been observed[254] in cases where fluorine is attached to sp²-hybridized carbon, $^2J_{F,H}$ = ~82 Hz.

It has also been observed[281] that, for a number of fluorinated monosaccharides, geminal $^{19}F-^1H$ coupling-constants for the pyranose series are larger when the fluorine substituent is axially rather than equatorially disposed. Exceptions to this generalization were subsequently found, as in the virtually identical value of $J_{F-2, H-2}$ for peracetylated 2-deoxy-2-fluoro-α-D-gluco- and -manno-pyranosyl fluorides.[71]

In addition to the foregoing, an empirical approach to the "substituent contributions" to geminal, $^{19}F-^1H$ coupling-constants have been described[291] for fluorinated D-gluco- and D-manno-pyranose derivatives, and some noncarbohydrate compounds of defined stereochemistry and conformation. These contributions behave additively, and are defined by the effect they have, upon a $^{19}F-^1H$, geminal coupling-constant, of replacing a hydrogen atom, vicinal to the coupled nuclei, by a substituent X (X = OR, R = alkyl, Ac, Bz, or H; or X = F). The substituent contributions have effectively similar magnitude (although it does depend on the electron-withdrawing power of the group), and the only factor that seems to cause a variation is the geometrical disposition of X relative to the coupled nuclei. Application of these empirically derived contributions (taking the average, observed value of $^2J_{HF}$ = 50 ± 1 Hz of a number of 3α- and 3β-fluoro steroids in which C-2 and C-4 are unsubstituted as the unperturbed value of $^2J_{HF}$) to the calculation of geminal, $^{19}F-^1H$ coupling-constants for compounds that have been reported in the literature furnished calculated values in good agreement with values observed; for example, for 2-deoxy-2-fluoro-β-D-mannopyranose, $^2J_{H,F}$ (calc.) = 53 Hz, and $^2J_{H,F}$ (obs.)[69,287] = 52.0 Hz. It could be seen from these comparisons that, ir-

(291) L. Phillips and V. Wray, *J. Chem. Soc. Perkin Trans. 2*, (1974) 928–933.

respective of whether fluorine was axial or equatorial, change in the vicinal substituent from an axial to an equatorial orientation results in an increase (~3 Hz) in $^2J_{H,F}$, whereas (except for the cases where the substituent is a ring-oxygen atom or sulfur) changing the fluorine atom from an axial to an equatorial orientation has little or no effect upon $^2J_{H,F}$. For the series of D-gluco- and -manno-pyranosyl fluorides having the substituent X = F, OBz, Cl, Br, I, and H, it was observed[291] that these general trends are true, and, in addition, for F-1 axial, X causes $^2J_{H,F}$ to decrease in the order: I > Br > Cl > OBz > F, whereas, with X equatorial, the reverse is true. For F-1 equatorial, axial X causes $^2J_{H,F}$ to decrease in the order Cl > F, whereas, with X equatorial, $^2J_{H,F}$ increases in the order I < Br < Cl < OR < F. It was, therefore, concluded from these observations that it is possible to distinguish between isomers in which the substituent is equatorial or axial, when the value of $^2J_{H,F}$ is available for both compounds.

The signs of geminal, $^{19}F-^1H$ coupling-constants (in contrast to those of geminal, $^1H-^1H$ coupling-constants) are absolutely positive. They were determined for derivatives of glycopyranosyl fluorides,[75,292,293] 2-deoxy-2-fluoro-D-gluco- and -manno-pyranosyl fluorides,[71] 3-deoxy-3-fluoro-β-D-gluco- and -xylo-pyranosyl fluorides,[75] and 4-deoxy-4-fluoro-D-glucopyranose.[128]

(ii) $^3J_{F,H}$. The angular dependence of vicinal, $^{19}F-^1H$ coupling-constants parallels that of $^1H-^1H$ coupling-constants[42,66,281,283,286,287] and $J_{trans} > J_{gauche}$.

That vicinal, $^{19}F-^1H$ coupling-constants depend on the electronegativity of neighboring substituents is well illustrated for the 2-deoxy-2-halogeno-aldohexopyranosyl fluorides,[286] where it was additionally shown that a linear relationship exists between the $^3J_{F,H}$ values observed and the substituent electronegativity at C-1 and C-2. For a series of peracylated glycopyranosyl fluorides,[42] a "configurational" electronegativity-dependence, similar to that of $^1H-^1H$ vicinal coupling-constants, was observed; namely, for compounds having F-1a and H-2e, the values of the coupling constants ranged between 1.0 and 1.5 Hz, whereas, for F-1e and H-2a, the coupling constants ranged between 7.5 and 12.6 Hz. As with $^1H-^1H$ coupling-constants, the low values observed for the former coupling-constants were explained in terms of the antiperiplanar orientations of the C-2–O-2, C-1–F-1, C-1–O-5, and C-2–H-2 bonds; such relationships are absent in the last example. These dependencies have been observed in a number of cases; for instance, the small $J_{F-2e, H-1e}$ value (<0.5 Hz) observed for

(292) L. D. Hall and J. F. Manville, *Chem. Ind. (London)*, (1967) 468–469.
(293) L. D. Hall and J. F. Manville, *Carbohydr. Res.*, 9 (1969) 11–19.

peracetylated trifluoromethyl 2-deoxy-2-fluoro-α-D-glucopyranoside and 2-deoxy-2-fluoro-α-D-glucopyranosyl fluoride was attributed[68,69] to the high, total electronegativity of substituents at C-1 and C-2, and, in particular, the antiplanar relationship between C-2–F-2 and C-1–O-5. Comparison between the F-3–H-4 coupling constants of peracetylated 3-deoxy-3-fluoro-α- and -β-D-galactopyranose (~6 Hz) and the corresponding *gluco* epimers (~13 Hz) suggested[146] that the antiplanar relationship between C-4–H-4 and C-5–O-5 is responsible for the lower value of the coupling constants of the *galacto* epimer. As the total electronegativity of the substituents at C-1 and C-2 in peracylated 2-deoxy-α-D-*arabino*-hexopyranosyl fluoride[66] is the same as that of the substituents at C-5 and C-6 in derivatives of 3,6-anhydro-5-deoxy-5-fluoro-L-idofuranose,[141] it is surprising that only one set of $^{19}F-^{1}H$ coupling-constants has similar values. The other is grossly different[141]; it was considered that this is due to a configurational dependence or that comparisons between coupling constants of cyclic systems of different size are impermissible. The latter possibility was also invoked to explain the "anomaly" observed in the F-3–H-2 coupling-constants (~10.5 Hz for a dihedral angle of $\phi = \sim 30°$) for derivatives of 3-deoxy-3-fluoro-D-glucofuranose, compared with a value of ~12.5 Hz, $\phi = \sim 60°$ for derivatives of 3-deoxy-3-fluoro-D-glucopyranose.[144] A larger coupling would be expected for $\phi = \sim 30°$ than for $\phi = \sim 60°$ were a $\cos^2 \phi$ relationship operative for the furanose derivatives.

An alternative explanation[144] is the possibility of a configurational dependence. The distinction between the ^{19}F resonances of 3-deoxy-3-fluoro-α- and -β-D-idofuranose was made[150] on the basis of the $^{19}F-^{1}H$ coupling-constants, which were interpreted on the basis of the Karplus curve[294] for vicinal, $^{19}F-^{1}H$ coupling-constants. Various dependencies of ^{19}F-n.m.r. parameters for furanoses are usually difficult to determine, because of the conformational inhomogeneity of the five-membered ring (see also, Section III,2,e). However, for a series of peracylated arabino-, ribo-, and xylo-furanosyl fluorides, it has been observed[32,295] that the vicinal, $^{19}F-^{1}H$ coupling-constants for *trans*-related nuclei are ~20 Hz, whereas those for *cis*-related nuclei are ~6 Hz.

In addition to the foregoing (with respect to deoxyfluoropyranoses), vicinal $^{19}F-^{1}H$ coupling-constants were examined in a study of the

(294) K. L. Williamson, Y.-F. Li, F. H. Hall, and S. Swager, *J. Am. Chem. Soc.*, 88 (1966) 5678–5680.
(295) L. D. Hall, P. R. Steiner, and C. D. Pedersen, *Can. J. Chem.*, 48 (1970) 1155–1165.

^{19}F-n.m.r. spectra of deoxyfluoro-D-glucopyranoses and 2-deoxy-2-fluoro-D-mannopyranose,[287] and it was found that the factors that affect the ^{1}H–^{1}H coupling-constants also affect ^{19}F–^{1}H coupling-constants. trans-Vicinal coupling-constants were found to depend on the nature of the substituents on the carbon atoms to which the coupled nuclei were themselves bonded. The value of the coupling constant decreased as the number of oxygen atoms increased. This was well illustrated in, for example, a comparison of cyclohexyl fluoride ($J_{F-1, H-3}$ = 43.5 Hz) with 2-deoxy-2-fluoro-α-D-glucopyranose ($J_{F-2, H-3}$ = 30.0 Hz) and peracetylated α-D-glucopyranosyl fluoride ($J_{F-1, H-2}$ = 27.2 Hz). The factors that control gauche-vicinal coupling-constants were found to be more complex than those controlling trans-vicinal coupling-constants. The former depend on (i) the number of substituent oxygen atoms on the carbon atoms that bear the coupled nuclei, (ii) the presence of an oxygen atom that is trans to the fluorine atom by way of the coupling pathway, and (iii) the presence of a hydroxyl group trans to the C–C bond bearing the coupled nuclei. The latter dependence indicated a stereospecificity in the coupling mechanism, because it was only observed when the electronegative substituent was trans to the C–C bond bearing the coupled nuclei (it was also found to depend on the stereochemical relationships between the hydroxyl group and the coupled nuclei). Empirical contributions towards the unperturbed value of gauche $^{3}J_{F,H}$ (taken to be that of fluoroethane, and equal to 16.0 Hz) were assigned to these factors. Calculated and observed values for a number of deoxyfluoropyranoses were in good agreement. For example, for methyl 4-deoxy-4-fluoro-2,3-di-O-methyl-α-D-glucopyranoside, $J_{F-4, H-3}$ (calc.)[287] = 15.5 Hz, and $J_{F-4, H-3}$ (obs.)[137] = 15.0 Hz.

Coupling constants for vicinal, trans-diaxially related, coupled nuclei in pyranoses and pyranosides are generally in the range of 21–30 Hz. Lower values have been observed, as, for example, $^{3}J_{F-2, H-1}$ = ~13 Hz for peracetylated 2-deoxy-2-fluoro-β-D-mannopyranosyl fluoride[68,69,71]; this was explained in terms of the anomeric effect of fluorine, which affects the $^{4}C_{1}$(D) conformation expected (see Section III,2,e). In contrast, a high value ($J_{F-1a, H-2a}$ = 38.0 Hz) has been observed for perbenzoylated 2-deoxy-α-D-arabino-hexopyranosyl fluoride.[66] Coupling constants for vicinal, gauche-related, coupled nuclei in pyranoses and pyranosides are generally in the approximate range of 0–18 Hz. With few exceptions, coupling constants for trans-related, vicinal nuclei in furanoses and furanosides are in the approximate range of 14–30 Hz, and, for the corresponding cis-related nuclei, in the approximate range of 2–15 Hz. $^{3}J_{F,H}$ Coupling-constants have been observed in a few instances where fluorine is attached to sp²-hybridized carbon, as, for example, for E- and Z-5,6-dideoxy-6-fluoro-3-O-methyl-α-D-*xylo*-hept-5-enofuranurononitrile,[252] $J_{F6,H5}$ = 13.5 Hz (E), and 33.3 Hz (Z), and for 5,6-dideoxy-6,6-difluoro-1,2-O-isopropylidene-3-O-methyl-α-D-*xylo*-hex-5-enofuranose,[251] $^3J_{Ftrans,H5}$ = 24 Hz and $^3J_{Fcis,H5}$ = 2.0 Hz (*cis* and *trans* relative to H-5).

The signs of vicinal, $^{19}F-^1H$ coupling-constants have been determined for dihedral angles of ~60° and ~180° in glycosyl fluorides, and these were found to be absolutely positive in sign,[292,293] like the $^1H-^1H$ values. Further determinations for 2-deoxy-2-fluoro-α- and -β-D-gluco- and -manno-pyranosyl fluoride,[71] 3-deoxy-3-fluoro-α- and -β-D-glucopyranosyl fluoride, and 3-deoxy-3-fluoro-β-D-xylopyranosyl fluoride,[75] and for 4-deoxy-4-fluoro-D-gluco-[128] and -galacto-pyranose[76] derivatives and their glycopyranosyl fluorides showed that they are absolutely positive.

(iii) $^4J_{F,H}$. Long-range, $^{19}F-^1H$ couplings are a common feature of ^{19}F-n.m.r. signals. Four-bond, $^{19}F-^1H$ coupling-constants show the same stereospecificity as the corresponding $^1H-^1H$ coupling-constants, and generally achieve their largest magnitudes when ^1H and ^{19}F have the planar, "M" relationship (or occur at the termini of a W-coplanar conformation).[296] This is illustrated by, for example, the coupling constants observed for 3-deoxy-3-fluoro-α- and -β-D-glucopyranose; for the α anomer, $J_{F-3e,H-1e}$ = 4.9 Hz, and for the β anomer, $J_{F-3e,H-1a}$ = 0.5 Hz. This difference was later reproduced in calculations for model systems related to carbohydrates.[279] Other examples illustrating this dependence were afforded by 1,2,4-tri-O-acetyl-3-deoxy-3-fluoro-β-D-xylopyranose, which exhibits $J_{F-3,H-5e}$ = +4.2 Hz and $J_{F-3,H-5a}$ = 0.5 Hz, and by 1,2,4,6-tetra-O-acetyl-β-D-allopyranosyl fluoride, for which $J_{F-1e,H-3e}$ = +3.6 Hz, whereas 1,3,4,6-tetra-O-acetyl-β-D-glucopyranosyl fluoride exhibits[296] no resolvable coupling between F-1e and H-3a. Resolvable coupling-constants have, however, been observed for a—e relationships, as, for example, $J_{F-3e,H-5a}$ = 1.1 Hz for peracetylated 3-deoxy-3-fluoro-β-D-glucopyranose,[144,296] and $J_{F-4e,H-2a}$ = −0.9 Hz for methyl 6-O-acetyl-4-deoxy-4-fluoro-2,3-di-O-methyl-α-D-glucopyranoside.[128] Peracetylated 2-deoxy-2-fluoro-α-D-glucopyranosyl fluoride shows[71] $J_{F-1a,H-3a}$ = 0.5 Hz, whereas the corresponding α-D-*manno* derivative shows $J_{F-1a,H-3a}$ = +2.5 Hz, and, although possible explanations were adduced, no unequivocal rationalization was made. For the β-D-*manno*

(296) A. B. Foster, R. Hems, L. D. Hall, and J. F. Manville, *Chem. Commun.*, (1968) 158–159.

derivative, $J_{F-2a,H-4a} = +2.0$ Hz was observed, and, although it was suggested that this might reflect an abnormality in the system, some caution should be exercised before such couplings are used for making configurational assignments. Further examples of sizable coupling constants of axially related nuclei have been observed, as, for example, $J_{F-4a,H-2a} = 2.8$ Hz for peracetylated 4-deoxy-4-fluoro-α-D-galactopyranosyl fluoride.[76]

A configurational dependence is exhibited by the signs of the $^{19}F-^{1}H$ coupling-constants over four bonds in pyranoses, as observed for the corresponding $^{1}H-^{1}H$ values. The values $^4J_{e,e} = \sim 4.0$ Hz were found to be absolutely positive, and $^4J_{e,a} = \sim 1.0$ Hz, absolutely negative; these were mainly determined for peracetylated 3-deoxy-3-fluoro-β-D-xylo- and peracetylated -α-D-gluco-pyranose,[144] peracetylated 3-deoxy-3-fluoro-α-D-glucopyranosyl fluoride,[75] peracetylated 4-deoxy-4-fluoro-α-D-glucopyranosyl fluoride, and methyl 6-O-acetyl-4-deoxy-4-fluoro-2,3-di-O-methyl-α-D-glucopyranoside.[128] The value $^4J_{Fa,Ha} = 2.5$ Hz observed for peracetylated 2-deoxy-2-fluoro-α-D-mannopyranosyl fluoride[71] and peracetylated 3-deoxy-3-fluoro-β-D-xylopyranosyl fluoride[75] was found to be absolutely positive.

Stereospecificity has also been observed for furanoses. Derivatives of D-arabino-,[32,295] D-ribo-,[295] and D-xylo-[32,295] furanosyl fluorides exhibited large $^4J_{F-1,H-4}$ values (5.5 − 7.9 Hz) for *trans*-related nuclei, and $^4J_{F-1,H-4} = 1.0$–1.8 Hz for *cis*-related nuclei. Conversely, $^4J_{F-1,H-3}$ <0.7 Hz for *trans*-related nuclei and $^4J_{F-1,H-3} = 1.7$–2.4 Hz for *cis*-related nuclei,[295] a stereospecifity that parallels that for pyranoses.[296] (β-D-Glucofuranosyl fluoride)urono-6,3-lactone and 2-O-acetyl-3,6-anhydro-5-O-benzoyl-α-L-idofuranosyl fluoride (probably favoring the $^3T_2$ or ^3E conformation) exhibited $^4J_{F-1,H-4} = \sim 4.5$ Hz and $^4J_{F-1,H-3} = <0.5$ Hz; as a *trans* relationship exists between the coupled nuclei, it was suggested[41] that the ring-oxygen atom must have a significant influence on the magnitude of $^4J_{F-1,H-4}$.

Coupling constants ($^4J_{F,H}$) have been observed in a few instances where fluorine is attached to sp²-hybridized carbon, as, for example, in derivatives of 3-deoxy-3-C-(fluoromethylene)-1,2:5,6-di-O-isopropylidene-α-D-ribo- and -xylo-furanose, where $^4J_{F,H-2}$ and $^4J_{F,H-4}$ occur[254,264] in the range of 0.5–5.5 Hz, and in 5,6-dideoxy-6,6-difluoro-1,2-O-isopropylidene-3-O-methyl-α-D-*xylo*-hex-5-enofuranose,[251] where $^4J_{Fcis,H-4} = 2.0$ Hz and $^4J_{Ftrans,H-4} = 1.5$ Hz (*cis* and *trans*, relative to H-5).

(iv) $^5J_{F,H}$. Long-range, $^{19}F-^{1}H$ coupling-constants over five bonds have also been observed, but they have been studied to a lesser extent than those already described. Partial, stereospecific dependencies for

$^5J_{F,H}$ were manifested by 2,3,6-tri-*O*-acetyl-4-deoxy-4-fluoro-α-D-glucopyranosyl fluoride[128] and methyl 6-*O*-acetyl-4-deoxy-4-fluoro-2,3-di-*O*-methyl-α-D-glucopyranoside,[128,137] which exhibit $J_{F-4e, H-1e}$ = +3.0 and +3.3 Hz, respectively, values larger than those ($J_{F-1e, H-4a}$ = 0.8 Hz and $J_{F-4e, H-1a}$ = +0.5 Hz) found for 2,3,6-tri-*O*-acetyl-4-deoxy-4-fluoro-β-D-glucopyranosyl fluoride (not allowing for any slight conformational distortion of the latter compound).[128] It was also remarked[128] that, assuming a through-bond coupling to be operative, it would be of interest to establish the individual contributions to the operating, coupling pathway of the two possible alternatives, namely, F-4, C-4, C-5, O-5, C-1, and H-1, and F-4, C-4, C-3, C-2, C-1, and H-1. Similarly, methyl 2,3,6-tri-*O*-acetyl-4-deoxy-4-fluoro-α-D-galactopyranoside exhibits[138] $J_{F-4a, H-1e}$ = 0 Hz, in contrast to derivatives of 4-deoxy-4-fluoro-α-D-glucopyranose, for example, methyl 6-*O*-acetyl-4-deoxy-4-fluoro-2,3-di-*O*-methyl-α-D-glucopyranoside, which exhibits[128,137] $J_{F-4e, H-1e}$ = +3.3 Hz.

A coupling constant (1.5 Hz) is observed between F-3*e* and that H-6 atom which resides for most of the time in the orientation antiparallel to the C-5–O-5 bond in 1,2,4,6-tetra-*O*-acetyl-3-deoxy-3-fluoro-β-D-glucopyranose (**126**), and it was suggested[296] that the favored pathway for $^{19}F-^1H$ coupling to occur over five bonds would approximate such

126

a stereochemical arrangement, essentially[287] a linear extension of the favored, planar, "M" arrangement[296] for the corresponding coupling over four bonds.

In addition to the foregoing, however, it was suggested[287] that maximal $^{19}F-^1H$ coupling over five bonds most probably occurs when the coupled nuclei are in a *trans*-coplanar relationship to the bond that is the "mid-point" of the coupling pathway, and this was illustrated with the example of 4-deoxy-4-fluoro-D-glucopyranose, for which a coupling of F-4 with the anomeric proton is observed for the α but not for the β anomer. Again, it was suggested[287] that, as this $^5J_{F-4e, H-1e}$ value for the α anomer has a magnitude comparable to that of the $^4J_{F-3e, H-1e}$ value for 3-deoxy-3-fluoro-α-D-glucopyranose, the ring-oxygen atom may participate in the coupling. It was later shown,[279] by calculations for model systems related to carbohydrates, that such a presence of an oxygen atom in the coupling pathway (compared to the all-carbon pathway) leads to an increased value of $^5J_{F,H}$; the calculations also reproduce the orientational dependence of this coupling.

$^{19}F-^1H$ coupling over five bonds has also been observed for carbohydrate derivatives containing the C-(OMe)-C-F fragment. For example, the following coupling constants were observed[297] between F-1 and MeO-2 of peracetylated 2-O-methyl-α-D-glucopyranosyl fluoride (1.0 Hz), 2-O-methyl-β-D-glucopyranosyl fluoride (1.4 Hz), and 2-O-methyl-α-D-mannopyranosyl fluoride (0.55 Hz). It was considered[297] that such couplings could be of diagnostic value in structural determinations. The variation in value of ^5J between these e-a, e-e, and a-a arrangements was, however, too small for assignment of the geometrical dependence with any accuracy; more examples are needed in order to make this possible.

A stereochemical dependence has been observed for the $^5J_{F,H-1}$ values observed for a number of 3-deoxy-3-C-(fluoromethylene)-1,2:5,6-di-O-isopropylidene-α-D-ribofuranose derivatives: $J_{Ftrans, H-1}$ = 1–1.5 Hz, $J_{Fcis, H-1}$ = 0 Hz (cis and trans relative to C-2). Also, a through-space coupling-constant was considered to be operative between Ftrans and H-5 in these compounds[254,264] and a number of corresponding xylofuranoses.[254]

(v) $^6J_{F,H}$. A long-range coupling (1.5 Hz) has been observed between one of the protons on C-6 and F-1 of 2-O-acetyl-3,6-anhydro-5-O-benzoyl-α-L-idofuranosyl fluoride,[41] as well as between H-6 (exo) and F-2 (1.5 Hz) of 4-acetamido-3-O-acetyl-1,6-anhydro-2,4-dideoxy-2-fluoro-β-D-glucopyranose.[126] A coupling constant is observed between Ftrans and the protons on C-6 in a number of derivatives of 3-deoxy-3-C-(fluoromethylene)-1,2:5,6-di-O-isopropylidene-α-D-ribo- and -xylo-furanoses, which is considered to occur through space.[254]

c. **$^{19}F-^{19}F$ Coupling-Constants.**—(i) $^2J_{F,F}$. Germinal, $^{19}F-^{19}F$ coupling-constants have been reported in the carbohydrate field; for example, that of **60** (Ref.166) ($J_{F-6,F-6}$ = 298.2 Hz), that of the α anomer of **2** (Ref. 2) ($J_{F-2a,F-2e}$ = 246 Hz), and that of **109** (Ref. 254) ($J_{Ftrans, Fcis}$ = 38.0 Hz).

(ii) $^3J_{F,F}$. Vicinal, $^{19}F-^{19}F$ coupling-constants have been observed in a number of instances, as, for example, for peracetylated 2-deoxy-2-fluoro-α-D-galactopyranosyl fluoride[72,185] and peracetylated 2-deoxy-2-fluoro-β-D-talopyranosyl fluoride.[185] In addition, the $^{19}F-^{19}F$ coupling-

(297) A. B. Foster, R. Hems, J. H. Westwood, and L. D. Hall, *Carbohydr. Res.*, 23 (1972) 316–318.

constants for the four steric relationships possible for the pyranose form have been studied in detail for peracetylated 2-deoxy-2-fluoro-α- and -β-D-gluco- and -manno-pyranosyl fluoride[70,71] ($J_{F-1a, F-2e}$ = −18.8, $J_{F-1e, F-2e}$ = −15.8, $J_{F-1a, F-2a}$ = −20.0, and $J_{F-1e, F-2a}$ = −13.5 Hz, respectively). These observations, together with comparison of such coupling-constants of related, noncarbohydrate systems, revealed that the angular dependence of these coupling constants is more complex than that of the corresponding $^1H-^1H$ and $^{19}F-^1H$ coupling-constants, and that at least one inversion of sign occurs. It also became obvious that *trans*-vicinal coupling-constants are more sensitive to changes in neighboring-group electronegativity than *gauche*-vicinal coupling-constants, and it was therefore suggested that they can only be used with caution for stereochemical assignments.[70] The signs of vicinal, $^{19}F-^{19}F$ coupling-constants were determined for peracetylated 2-deoxy-2-fluoro-α- and -β-D-gluco- and -manno-pyranosyl fluorides, and were found to be absolutely negative.[70,71]

(iii) $^4J_{F,F}$. $^{19}F-^{19}F$ coupling-constants over four bonds have been observed for peracetylated 3-deoxy-3-fluoro-α- and -β-D-gluco- and -β-D-xylo-pyranosyl fluoride.[75] By comparison with $^4J_{F,F}$ couplings for perfluorinated cyclohexanes, it was seen that the relative magnitudes for the two series are identical, and, especially, that the 1,3-diaxially oriented nuclei gave the largest couplings. The coupling is not favored when the ^{19}F nuclei have the planar, "M" relationship, as is the case for the corresponding $^1H-^1H$ and $^{19}F-^1H$ coupling-constants.[296] Their magnitude does, however, depend upon the relative orientations of the fluorine substituents and the nature of nearby substituents.[75] A through-space contribution to the coupling has been considered possible when the fluorine substituents are in *syn*-axial positions, because the distance of separation then approximates the sum of the Van der Waals radii.[76] As $^4J_{a,a}$ and $^4J_{a,e}$ values were found to be absolutely positive in sign, and $^4J_{e,e}$ values were found to be absolutely negative in sign, from measurements made with 2,4,6-tri-O-acetyl-3-deoxy-3-fluoro-α- (+1.0 Hz) and -β-D-glucopyranosyl fluoride (−3.0 Hz) and 2,4-di-O-acetyl-3-deoxy-3-fluoro-β-D-xylopyranosyl fluoride (+10.4 Hz), it was suggested[75] that certain stereochemical arrangements would lead to very small values; until further information becomes available about the dependencies of these coupling constants, stereochemical assignments based on $^4J_{F,F}$ coupling-constants cannot therefore be made with any confidence. The suggestion has, however, been made, by consideration of these values (and other parameters), that the two fluorine atoms in 3-O-acetyl-1,6-anhydro-2,4-dideoxy-2,4-difluoro-β-D-glucopyranose are in almost diequatorial orientation, as they exhibit no coupling.[128]

(iv) $^5J_{F,F}$. Long-range, $^{19}F-^{19}F$ coupling-constants over five bonds have also been observed. In the peracetylated 6-deoxy-6-fluoro-α- and -β-D-glucopyranosyl fluorides, the F-6–F-1e arrangement is apparently more suitable[77] for coupling than the F-6–F-1a. $^5J_{F,F}$ coupling-constants have also been observed for the anomeric pairs of peracetylated 4-deoxy-4-fluoro-D-gluco- and -galacto-pyranosyl fluorides,[76] where relatively rigid geometries, related to those of the 6-deoxy-6-fluoro-D-glucopyranosyl fluorides, exist. These couplings supposedly occur to a large extent by way of through-bond interactions, rather than through-space interactions, for which the two fluorine nuclei are too far apart. It has, however, been presumed that the condition most probably needed for production of the maximal value of $^5J_{F,H}$ (namely, a *trans*-coplanar arrangement of the coupled nuclei to the bond that is the mid-point of the coupling pathway[287]) may have no bearing on the $^5J_{F,F}$ value, and that the stereospecific dependencies are more complex in the latter than in the former. Magnitudes of the coupling constants are often similar, but the signs are opposite. For example, for peracetylated 4-deoxy-4-fluoro-β-D-glucopyranosyl fluoride, the two coupled, ^{19}F nuclei are *trans* to the two mid-point bonds C-2–C-3 and C-5–O-5, and exhibit $^5J_{F-4e,F-1e} = +3.1$ Hz. Alternatively, for the β-D-*galacto* derivative, for which the two coupled, ^{19}F nuclei are *gauche* to the two mid-point bonds, a $^5J_{F-4a,F-1a}$ value of -3.7 Hz was observed.[76] Further observations have been made that again suggest a stereospecificity, but do not define it in an unequivocal way. The $^5J_{F-4e,F-1e}$ value for 2,3,6-tri-O-acetyl-4-deoxy-4-fluoro-β-D-glucopyranosyl fluoride was found to be larger than the $^5J_{F4e,F-1a}$ value for 2,3,6-tri-O-acetyl-4-deoxy-4-fluoro-α-D-glucopyranosyl fluoride.[94] Again, caution was advised[94] before assigning any stereochemical arrangements on the sole basis of $^5J_{F,F}$ coupling constants, unless their source and sign was known, and the following example was given. The $^4J_{F-3e,F-1e}$ value for peracetylated 3-deoxy-3-fluoro-β-D-glucopyranosyl fluoride is similar in magnitude to, but opposite in sign from, the $^5J_{F-4e,F-1e}$ value for peracetylated 4-deoxy-4-fluoro-β-D-glucopyranosyl fluoride.

d. $^{19}F-^{13}C$ Coupling Constants.—(i) $^1J_{F,C}$. In a study of derivatives of glycopyranosyl and glycofuranosyl fluorides,[278] it was found that, for pyranosyl fluorides (α- and β-D-gluco, α- and β-D-galacto, α- and -β-D-manno, α- and -β-D-altro, α-D-ido, 6-deoxy-α-D-ido, 6-deoxy-β-D-gulo, α- and β-D-xylo, and β-D-arabino), the $^1J_{F,C}$ values are numerically higher for those pyranosyl fluorides that have the fluorine atom axial than for those having it equatorial. In almost all of the cases examined, the α-fluorides (F-axial) exhibited $^1J_{F,C}$ values numerically higher, by ~10 Hz, than those of the corresponding β anomers (F-equatorial). It was suggested that the lone-pair electrons of the ring-oxygen atom affect the coupling constant. In addition, it was noted that an axial substituent at C-2 causes a numerical decrease in the coupling constant, as exemplified by peracetylated α-D-glucopyranosyl fluoride, $^1J_{F,C}$ = 228.9 Hz, and peracetylated α-D-mannopyranosyl fluoride, $^1J_{F,C}$ = 223.1 Hz. This was invoked as an explanation for the virtually identical values observed for the anomeric tetra-O-acetyl-D-mannopyranosyl fluorides. The tri-O-acetyl-D-xylopyranosyl fluorides, which have different conformations, with F-1 axial in both, exhibit a numerically lower coupling-constant for the β anomer (having the substituent at C-2 equatorial) than for the α anomer (having the substituent at C-2 axial).

In addition, $^1J_{F,C}$ coupling-constants have been studied for the anomeric pairs of the deoxyfluoro-D-glucopyranoses, 2-deoxy-2-fluoro-D-mannopyranose, and 4-deoxy-4-fluoro-D-galactopyranose,[279] for which it was shown that the $^1J_{F,C}$ value depends on the electronegativity of the attached atoms on the coupled carbon atom. A comparison[279] of the average values for the coupling constants for glycopyranoses[278] (224.6 ±3.8 Hz), the deoxyfluoroglycopyranoses fluorinated at secondary positions other than the anomeric (179.5 ±3.6 Hz), and the 6-deoxy-6-fluoro-D-glucopyranoses (167.4 ±0.3 Hz) was made, and it was found that the magnitude of the coupling constant increases as the magnitude of the electronegativity of the atoms attached to the coupled carbon atom increases. Furthermore, the $^1J_{F,C}$ values depend on the orientation, with respect to the fluorine atom under consideration, of substituents on adjacent carbon atoms. $^1J_{F,C}$ values also depend upon the orientation of β (and longer-range) substituents. Comparisons were made in situations where the electronegativity of the α-atoms was kept constant, namely, between 2-deoxy-2-fluoro-β-D-manno- and -gluco-pyranose, 4-deoxy-4-fluoro-α-D-galacto- and -gluco-pyranose, and 4-deoxy-4-fluoro-β-D-galacto- and -gluco-pyranose, where the orientation of the fluorine atom changes from *gauche* to *trans* with respect to the ring-oxygen atom, and 2-deoxy-2-fluoro-β- and -α-D-mannopyranose, where the orientation of the hydroxyl group at C-1 (a β substituent) changes from *gauche* to *trans* with respect to F-2. In the former cases, $^1J_{F,C}$ values increase by 2.8 ±0.3 Hz, whereas, in the latter case, the value decreases by 7.8 Hz. It was therefore concluded that this dependence is not simple, possibly due to the complex nature of the mutual interactions between the polar groups.

For derivatives of glycofuranosyl fluorides (α- and β-D-gluco, α- and β-D-manno, β-D-xylo, α- and β-D-lyxo, α- and β-D-ribo, and α-D-arabino)[278] on the other hand, it was observed that, when the fluorine atom and the substituent at C-2 are *cis*-oriented, the $^1J_{F,C}$ value is numerically higher (231.6–236.8 Hz) than when they are *trans*-oriented (224.7–228.46 Hz).

Values for $^1J_{F,C}$ have been recorded[251] for fluorine attached to sp²-hybridized carbon, as in 5,6-dideoxy-6,6-difluoro-1,2,O-isopropylidene-3-O-methyl-α-D-*xylo*-hex-5-enofuranose, where $^1J_{F,C} = 285$ Hz. The signs of the $^1J_{F,C}$ values both for the pyranosyl and the furanosyl fluorides were found to be negative.[278]

(ii) $^2J_{F,C}$. In the study of glycopyranosyl and glycofuranosyl fluorides[278] [listed in Section III,2d(i)], it was observed, for glycopyranosyl fluorides, that, irrespective of anomeric configuration, $^2J_{F-1,C-2} = $ ~22.1–27.8 Hz when the substituent at C-2 is equatorial, and $^2J_{F-1,C-2} = 35.2$–41.3 Hz when axial, with the exception of peracetylated β-D-manno- and -altro-pyranosyl fluorides. In addition, in a study with deoxyfluoropyranoses[279] [listed in Section III,d(i)], the geminal, $^{19}F–^{13}C$ coupling-constants were found to depend to a much smaller extent on the electronegativity of the atoms attached to the coupled fragment than the $^1J_{F,C}$ values, whereas they are much more sensitive to the orientation of these electronegative substituents with respect to fluorine. The values observed for the following fragments were examined: (*i*) C–CHF–*C*HOH–O, (*ii*) C–CHF–*C*HOH–C, and (*iii*) CFH₂–*C*H(C)–O, where it was observed that, for (*i*), as illustrated by 2-deoxy-2-fluoro-β-D-manno- ($^2J_{F-2,C-1} = 15.8$ Hz) and -gluco-pyranose ($^2J_{F-2,C-1} = 23.0$ Hz), an increase occurs on going from a *gauche* orientation between F-2 and O-5 to a corresponding *trans* orientation. In addition, by further comparison with 2-deoxy-2-fluoro-α-D-mannopyranose ($^2J_{F-2,C-1} = 29.6$ Hz), it was found that a *trans*-hydroxyl group (OH-1) has a larger effect than the *trans* ring-oxygen atom. For (*ii*), for a *gauche* relationship between fluorine and the hydroxyl group, values of 17.5 ± 0.3 Hz were observed, whereas, for a *trans* relationship, values of 24.2 ± 0.4 Hz were found; for example, for 4-deoxy-4-fluoro-α-D-galactopyranose, $^2J_{F-4,C-5} = 17.6$ Hz, and for 4-deoxy-4-fluoro-α-D-glucopyranose, $^2J_{F-4,C-5} = 23.9$ Hz (the oxygen atom attached to the coupled carbon atom being the ring-oxygen atom, instead, in this example).

The glycofuranosyl fluorides[278] [listed in Section III,2d(i)] exhibit numerically larger values when the substituents at C-1 and C-2 are *trans*- (31.4–40.3 Hz) than *cis*-related (18.9–20.5 Hz).

Values of $^2J_{F,C}$ have been observed[253] for fluorine attached to sp²-hybridized carbon, as in *E*- and *Z*-5,6-dideoxy-6-fluoro-3-O-methyl-α-D-

xylo-hept-5-enofuranurononitrile, where $^2J_{F-6,C-5} = 19.8$ Hz (E) and $J_{F-6,C-5} = 9.2$ Hz (Z).

The signs of the geminal, $^{19}F-^{13}C$ coupling-constants both for glycopyranosyl and glycofuranosyl fluorides were found to be positive,[278] whereas, in another investigation, the spectra were considered to be too complex for determination of the signs, but, by comparison with noncarbohydrate systems, they were assumed to be positive.[279]

(iii) $^3J_{F,C}$. In a study on the glycopyranosyl and glycofuranosyl fluorides[278] [listed in Section III,2d(i)], it was found that, for the β-pyranosyl fluorides for which F-1 and C-3 are *trans*-oriented, $^3J_{F,C} = 4-10$ Hz, whereas, for the α-pyranosyl fluorides for which F-1 and C-3 are *gauche*-oriented, $^3J_{F,C} = 0-6$ Hz. The suggestion of a Karplus type of relationship was therefore made. Irrespective of anomeric configuration, coupling constants between F-1 and C-5 were found to be 2-5 Hz. In a study with the deoxyfluoropyranoses[279] [listed in Section III,2d(i)], on the other hand, it was found that (*i*) an increase in the electronegativity of the substituents attached to the coupled carbon atom, on the carbon atom bearing the fluorine atom, causes a small decrease in $^3J_{F,C}$, for example, as between β-D-glucopyranosyl fluoride ($^3J_{F-1,C-3} = 11.7$ Hz) and 2-deoxy-2-fluoro-α-D-glucopyranose ($^3J_{F-2,C-4} = 8.0$ Hz), (*ii*) a change from a *trans* to a *gauche* orientation of the coupled nuclei leads to a large decrease in the magnitude of the coupling constant, as, for example, the difference observed between 2-deoxy-2-fluoro-α-D-glucopyranose ($^3J_{F-2,C-4} = 8.0$ Hz) and 2-deoxy-2-fluoro-α-D-mannopyranose ($^3J_{F-2,C-4} = 1.4$ Hz), as well as that observed between 3-deoxy-3-fluoro-α-D-glucopyranose ($^3J_{F-3,C-1} = 10.9$ Hz) and 4-deoxy-4-fluoro-D-glucopyranose ($^3J_{F-4,C-6} = 0$ Hz) or 4-deoxy-4-fluoro-D-galactopyranose ($^3J_{F-4,C-6} = 5.5$ Hz). The latter two compounds illustrate a third dependence, namely, (*iii*) that, if the relative positions of the coupled nuclei are held constant, but the orientation of the electronegative substituent (O-5) attached to the middle carbon atom of the coupled fragment changes from *trans* orientation (*gluco*) to a *gauche* orientation (*galacto*) relative to the fluorine atom, a large increase in the coupling constant is observed. Calculations[279] with model compounds qualitatively rationalized all of these observations, and predicted $^3J_{F,C}$ to be positive and to show an angular dependence similar to that found for $^3J_{H,H}$ and $^3J_{F,H}$, namely, $^3J_{F,Ctrans} > {}^3J_{F,Cgauche}$. A Karplus type of relationship was later confirmed[280] by a comparison of values obtained for carbohydrates and other systems. In addition, the smaller $^3J_{F-1,C-5}$ value (4.4 Hz) compared to the $^3J_{F-1,C-3}$ value (8.8 Hz) found for the coupling across the ring-oxygen atom in β-D-glucopyranosyl fluoride was also reproduced by such calculations.[279]

Very few of the glycofuranosyl fluorides studied[278] [listed in Section III,2d(i)] showed any coupling between F-1 and C-3, as, for example, perbenzoylated β-D-ribofuranosyl fluoride (1.5 Hz).

(iv) $^4J_{F,C}$. In a study with the glycopyranosyl and glycofuranosyl fluorides[278] [listed in Section III,2d(i)], very few four-bond coupling-constants were observed; for example, $^4J_{F-1,C-4} = 1.20$ Hz for peracetylated β-D-galactopyranosyl fluoride and $^4J_{F-1,C-OMe} = 1.83$ Hz for peracetylated 2-O-methyl-β-D-mannopyranosyl fluoride. In a study with deoxyfluoropyranoses[279] [listed in Section III,2d(i)], it was observed that the values of $^{19}F-^{13}C$ coupling-constants over four bonds depend on orientation; equatorial fluorine on any of the carbon atoms (except the anomeric) had observable values, whereas their axial counterparts did not, as, for example, for 4-deoxy-4-fluoro-α- or -β-D-glucopyranose, $^4J_{F-4,C-1} = 1.44$ Hz, and for the corresponding *galacto* epimers, $^4J_{F-4,C-1} = 0$ Hz. Calculations made with simpler, related systems, where two alternative, coupling pathways were possible, namely, that including an oxygen atom, and the all-carbon pathway, were considered, showed that only for the former is a significant $^4J_{F,C}$ value observed that reproduces the experimental trend observed. The presence of the oxygen atom is necessary, both in order to increase the magnitude of the coupling constant and to reproduce its orientational dependence.

Small coupling-constants, in the range of 1.17–2.9 Hz, were observed for the glycofuranosyl fluorides[279] [listed in Section III,2d(i)], and $^4J_{F-1,C-OMe} = 0.9$ Hz was found for perbenzoylated 2-O-methyl-α-D-glucofuranosyl fluoride.

e. Applications and Miscellaneous.—As with other halogens (see, for example, Refs. 298–301), a fluorine substituent attached to C-1 of a pyranose ring exhibits a strong anomeric effect to the extent that, in certain instances, it may cause gross conformational changes. In the latter instance, an extreme example is provided by 2,3,4-tri-O-acetyl-β-D-xylopyranosyl fluoride.[284,285] When 2,3,4-tri-O-acetyl-α-D-xylo-

(298) C. V. Holland, D. Horton, and J. S. Jewell, *J. Org. Chem.*, 32 (1967) 1818–1821.
(299) P. L. Durette and D. Horton, *Carbohydr. Res.*, 18 (1971) 57–80; *Adv. Carbohydr. Chem. Biochem.*, 26 (1971) 49–125.
(300) B. Coxon, *Tetrahedron*, 22 (1966) 2281–2302; H. J. Jennings, *Can. J. Chem.*, 47 (1969) 1157–1162.
(301) "Anomeric Effect: Origin and Consequences," W. A. Szarek and D. Horton (Eds.), *ACS Symp. Ser.*, 87 (1979).

pyranosyl bromide in the $^4C_1(\text{D})$ conformation is converted into the corresponding β-D-xylopyranosyl fluoride,[45,285] the latter is found to exist in the $^1C_4(\text{D})$ conformation (with all of the substituents axial), as suggested by ^{1}H-n.m.r. and ^{19}F-n.m.r. observations.[284,285] There is, however, evidence, that discussion in terms of the two chair conformations could be an oversimplification, as some of the ^{1}H-n.m.r. parameters are not entirely consistent with an exclusive population of the $^1C_4(\text{D})$ conformation.[284] Also, ^{19}F-n.m.r., chemical-shift changes were observed as the temperature was varied. Such observations could imply a rapid equilibrium between $^1C_4(\text{D})$ and $^4C_1(\text{D})$ conformers, or the existence of the molecule in a nonchair conformation, or both.[284,285] In addition to the foregoing, however, it has been observed,[285] in a comparison of the calculated and observed values of the ^{19}F chemical-shifts for a series of pentopyranosyl fluorides, that the β-D-*xylo* fluoride is one of those that show the largest deviations. The suggestion was consequently made that classical, nonbonded, repulsive interactions (attractive forces could not be considered, owing to insufficiency of data) between substituents on the ring are capable of distorting the conformation favored by the predominant effect of anomeric fluorine. Comparison of ^{19}F–^{1}H and ^{1}H–^{1}H coupling-constants of peracetylated and perbenzoylated β-D-xylopyranosyl fluoride (as well as the β-D-*arabino* fluoride) with those expected for the $^4C_1(\text{D})$ conformation showed that better agreement is obtained for the tribenzoates than for the triacetates.

In contrast to β-D-pentopyranosyl fluorides, β-D-hexopyranosyl fluorides[42] do not suffer conformational inversion due to anomeric fluorine. It was found that, for a series of β-D-glucopyranosyl fluorides, the ^{1}H–^{1}H coupling-constants are smaller than those of the corresponding α anomers (including a 4,6-O-benzylidene derivative); this was attributed either to time-averaging between the two forms, or to some local, static distortion for the $^4C_1(\text{D})$ conformer. The anomeric effect of fluorine apparently influences even the $^4C_1(\text{D})$ conformation of the β-D-*gluco* configuration, wherein all of the substituents are equatorial. In addition, the ^{19}F chemical-shifts of the β anomers (including a 4,6-O-benzylidene derivative) were temperature-dependent; the most likely explanation was, therefore, the presence of a distortion about the O-5,C-1,C-2 region of the pyranose ring, which could lead to a conformation capable of undergoing inversion with a very small activation-energy. Numerous examples of such effects of anomeric fluorine upon conformation have been observed for hexopyranosyl fluorides, as, for instance, with 2-deoxy-2-chloro-, -bromo-, and -iodo-D-glucopyranosyl fluorides, where such an explanation was considered probable for the systematic difference found between the induced shifts of the α- and β-D-*gluco* series,[286] and for peracetylated 2-deoxy-2-fluoro-α- and -β-D-mannopyranosyl fluorides, where $J_{H-3,H-4}$, $J_{H-4,H-5}$ and $J_{F-2,H-3}$ were found to be smaller for the β than for the α anomer.[71] Attempts[71] at recording the spectra at −90°, supposedly to prevent any low-barrier inversion processes from occurring, were unsuccessful.

Allowing for the uncertainty with which the favored conformations of derivatives of β-D-ribo- and -xylo-furanosyl fluorides were evaluated, it has been observed that the fluorine substituent is pseudo-axially oriented.[295]

Vicinal, $^{19}F-^{1}H$ coupling-constants can be diagnostic of anomeric configuration[283] (for further examples, see Refs. 42, 43, 71, 75–77, 128, 196, 283, 285, and 292). This fact has also been confirmed in one instance by the results of X-ray studies.[56,57]

Vicinal, $^{19}F-^{1}H$ coupling-constants have been used to determine the favored rotamer of 6-deoxy-6-fluorohexopyranoses.[77,89,93,165,170,287] A number of 6-deoxy-6-fluoro-D-glucopyranose derivatives[89,170] and 2-acetamido-2,6-dideoxy-6-fluoro-D-glucopyranose derivatives[93] exhibit $J_{F-6,H-5}$ = 27–29 Hz, indicating that the favored rotamer about C-5–C-6 has F-6 antiperiplanar[89,170] to H-5 (see Fig. 1a). Further sets of 6-deoxy-6-fluoro-D-glucopyranose derivatives[77,165] and 2-acetamido-2,6-dideoxy-6-fluoro-D-glucopyranose derivatives[165] exhibit $J_{F-6,H-5}$ = 20.6–23.9 Hz, and it was considered that these coupling constants were sufficiently large to be indicative of the preponderance of the same conformation (see Fig. 1a). Similar coupling constants (26.0 and 23.5 Hz, not assigned to the individual anomers) were also observed for peracetylated 2-acetamido-2,6-dideoxy-6-fluoro-D-mannopyranoses.[165] The corresponding D-galactopyranose derivatives exhibit[89,165,170] $J_{F-6,H-5}$ = 12–18.5 Hz, and this was interpreted as indicating that the favored rotamer has F-6 antiperiplanar to O-5 (see Fig. 1b). These arguments were used as supporting evidence in a determination of the favored rotamers of analogous 6-O-acetyl-D-*gluco* and

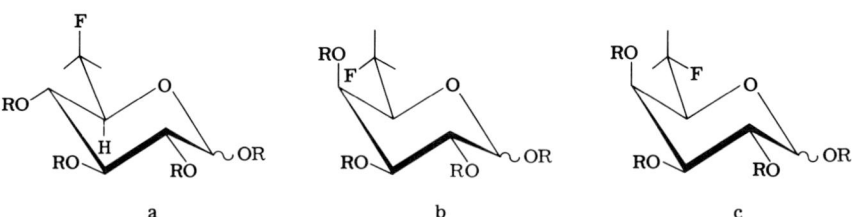

Fig. 1.—Favored Rotamers About the C-5–C-6 Bond in 6-Deoxy-6-fluoro-D-gluco- (a) and -galacto-pyranoses (b,c).

-*galacto* derivatives.[42] For the 6-deoxy-6-fluoro-D-galactopyranoses, however, two further suggestions have been made[287]: either the favored rotamer has F-6 antiperiplanar to C-4 (see Fig. 1c), or there is very little conformational bias, and what is observed is a fully averaged value of coupling constants.

6-Deoxy-6-fluoro-1,2:3,5-di-*O*-methylene-α-D-glucofuranose (**127**) and 3,5-*O*-benzylidene-6-deoxy-6-fluoro-1,2-*O*-isopropylidene-α-D-glucofuranose (**128**) have $^3J_{F-6,H-5} = \sim 36$ Hz, indicating that F-6 is antiperiplanar to H-5 and oriented towards the axial hydrogen atom of the acetal group. When, however, this hydrogen atom is replaced by a methyl group, in 1,2:3,5-di-*O*-isopropylidene-α-D-glucofuranose (**129**), a small coupling-constant ($^3J_{F-6,H-5} = 22.2$ Hz) is observed, indicating that, as expected, such a structure as that described is destabilized.[89] A large difference in $^3J_{F-5,H-4}$ values was observed for the α (**130**) ($^3J_{F-5,H-4} = 30.0$ Hz) and β (**131**) ($^3J_{F-5,H-4} = 10.5$ Hz) anomers of methyl 2,3-*O*-isopropylidene-D-ribofuranoside, and this was attributed to the methoxyl group on C-1.

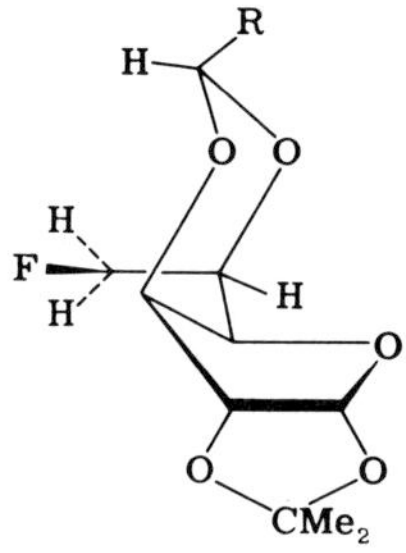

127 R = H
128 R = Ph

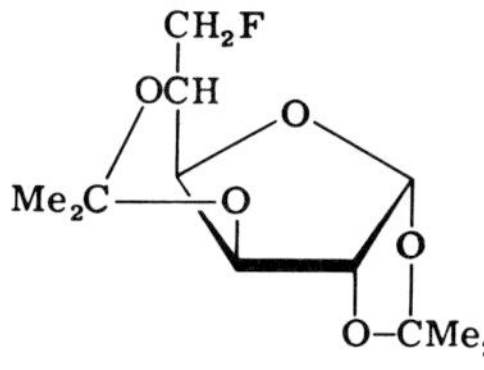

129

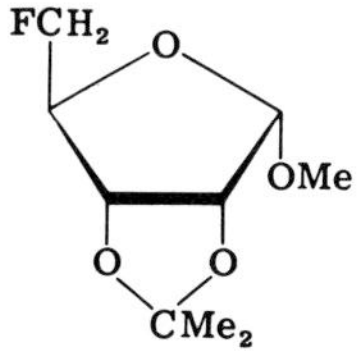

130

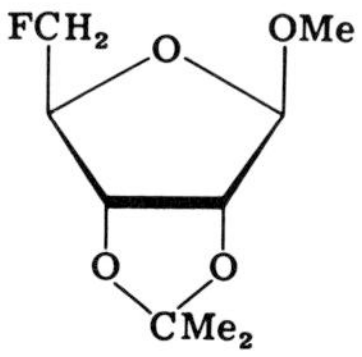

131

As is apparent from the discussion thus far, fluoro analogs in the furanose form have been studied to a lesser extent than those in the pyranose form, fundamentally because the data they provide are more difficult to interpret, because the conformational mobility of the furanose forms is high (on the n.m.r. scale). A study of 3,6-anhydro-5-deoxy-5-fluoro-1,2-O-isopropylidene-α-L-idofuranose, a molecule having a relatively rigid conformation, is of particular interest, because fluorine couplings were observed[141] with all of the ring protons except H-2. The ^{19}F-n.m.r. spectrum observed was in excellent agreement with a predicted one based on the observed values of the ^{19}F–^{1}H coupling-constants. Among others, derivatives of 3-deoxy-3-fluoro-D-gluco-[144] and -galacto-furanose[145] have been studied. In a study of D-arabino-, -ribo-, and -xylo-furanosyl fluorides, an approach to the investigation of such conformationally mobile compounds was described.[295] The various possible conformations of the furanose ring were related in a pseudorotational cycle, and it was found that discussion of conformation for substituted furanose derivatives could be centered around a particular segment of the pseudorotational cycle. This approach may afford chemically significant conclusions; see, for example, Ref. 146. This approach was also applied [295] to the conformation of **1**.

The possibility of making configurational and conformational assignments from the ^{19}F-n.m.r. data already described was excellently illustrated in a mutarotational-equilibrium study of an aqueous solution of 3-deoxy-3-fluoro-L-idose.[150] Two pyranose and two furanose forms were observed. The chemical shifts of the furanose forms were assigned on the basis of the vicinal, ^{19}F–^{1}H coupling-constants expected by comparison with 3-deoxy-3-fluoro-1,2-O-isopropylidene-α-D-glucofuranose. The chemical shifts of the pyranose forms were assigned on the basis of ^{19}F–^{1}H coupling-constants over four bonds, where the $^{4}C_{1}$(L) conformation of the α-L form was considered to be favored over the $^{1}C_{4}$(L) conformation; this was also considered to be true for the β-L form. On the basis of observed and calculated, vicinal, ^{19}F–^{1}H coupling-constants, however, the $^{4}C_{1}$(L) conformation for the α-L form was considered to be favored over the $^{1}C_{4}$(L) conformation to a greater extent than predicted from the long-range coupling-constants, and it was concluded that there is no conformational bias in the case of the β-L form. These conclusions were further supported by values calculated for the chemical shifts and by comparisons with 3-deoxy-3-fluoro-α- and -β-D-glucopyranose, and they indicated that the α-L form in the $^{4}C_{1}$(L) conformation preponderates over the β-L form. It was also pointed out that these conclusions concerning the favored conformations of the pyranose forms were tentative, especially because the methods used for calculating vicinal, ^{19}F–^{1}H coupling-constants and chemical shifts are firmly established[287] only for conformationally homogeneous systems. The preponderance of the $^{4}C_{1}$(L) forms is in contrast to that known for α- and β-L-idopyranose, and it was partially explained on the basis of *gauche*, F–OH interactions present in the fluoro analogs. Two further points were made; this study is an example of the fact that useful, ^{19}F-n.m.r. spectra can be readily recorded for aqueous solutions, whereas this is not always so for ^{1}H-n.m.r. spectra. Also, assuming that the chemical-shift assignments were correct, the percentage composition of the fluorinated α- and β-pyranose and -furanose forms at equilibrium parallels that known for the parent sugar. This is an example of the fact that, in the aldohexose series, replacement of OH by F does not significantly influence mutarotational equilibria. Other examples include the anomeric-equilibrium mixtures of 2-deoxy-2-fluoro-D-glucose and -mannose,[69,287] 4-deoxy-4-fluoro-D-glucose,[128] and 6-deoxy-6-fluoro-D-glucose.[77,287] Somewhat larger differences have also been observed, as, for example, in the case of 4-deoxy-4-fluoro-D-galactose.[138] That fluorine can take the place of a hydroxyl group without much disturbance of the overall structure has also been illustrated by the fact that 3-deoxy-3-fluoro-D-arabinose, which failed to crystallize spontaneously, crystallized immediately upon addition of a crystal of β-D-arabinose[114] (see also, Section I). Conformational equivalence between C–F and C–OAc has been observed, for example, for derivatives of 3-deoxy-3-fluoro-D-glucose.[144]

In addition to those mentioned thus far, ^{19}F-n.m.r. spectra of a large number of fluorinated carbohydrate derivatives have been recorded for the purpose of structural elucidation, or routinely. They include those of acyclic derivatives, for example, those of derivatives of 1,6-dideoxy-1,6-difluoro-galactitol and -D-mannitol,[108] and that of 2,3-di-O-acetyl-2,5-anhydro-1-deoxy-1,1-difluoro-D-ribitol.[4] 5-Deoxy-5-fluoro-D-xylofuranose,[94] benzyl 3-deoxy-3-fluoro-β-D-xylopyranoside,[302] and 1,3,4-tri-O-benzoyl-2-deoxy-2-fluoro-β-D-ribopyranose[222] have also been studied. In addition, the spectra of fluorinated 1,6-anhydrohexopyranoses have been described; for example, that of 1,6-anhydro-2-O-acetyl-4-O-benzyl-3-deoxy-3-fluoro-β-D-altropyranose[122,123]; those of fluorinated, branched-chain sugars, for example, that[255] of **114**; those of unsaturated fluorinated carbohydrates, for example, that of 5,6-dideoxy-6-fluoro-1,2-O-isopropylidene-3-O-methyl-6-(phenylthio)-α-D-xylo-hex-5-enofuranose[252]; and those of geminal difluoro saccharides, for example, those[164] of **60** and **62**; as well as those of fluorinated ketoses, for example, 4-deoxy-4-fluoro-α-D-sorbopyranose, 5-deoxy-5-fluoro-α-L-sorbopyranose, and 4-deoxy-4-fluoro-1,2-O-isopropylidene-β-D-tagatopyranose.[131] Coupling between fluorine and deuterium has been routinely recorded.[254]

(302) A. DeBruyn, M. J. Anteunis, G. Aerts, and E. Saman, *Carbohydr. Res.*, 41 (1975) 290–294.

It has been shown[43] that ^{19}F-n.m.r. spectroscopy is a valuable tool for studying certain electrophilic addition-reactions of unsaturated carbohydrates, by providing a simple method for identification of the resulting products. It has also been of great value in studies of the reactions of sugar esters and glycals with hydrogen fluoride, as it provides a ready means of identification, and characterization, of any glycosyl fluorides formed; see, for example Refs. 30 and 35. ^{19}F-N.m.r. spectroscopy (including measurement of relaxation times) has been used for the study of biological problems; two examples are discussed in Section IV.

3. Other Methods

In a study of the infrared spectra of acetylated and benzoylated glycopyranosyl fluorides, it was found that, with few exceptions, such compounds exhibit an absorption band at 802–748 cm^{-1} if the fluorine atom is axial, whereas no such absorption band is observed if the fluorine atom is equatorial; therefore, it is necessary to know the conformation of the molecule before the anomeric configuration can thus be determined.[45] It is possible that reinterpretation of some of these observed absorptions may be necessary, because some of the conformations of some of the glycopyranosyl fluorides are now known to exist in the conformation opposite to that originally anticipated without proof, because of the strong anomeric effect of a fluorine substituent (see Section III,2e).

Apart from the examples already mentioned (see Section I), a small number of fluorinated carbohydrates have been studied by X-ray crystallography. For example, X-ray analysis indicated that compound **68** obtained by total synthesis had a structure very similar to that of ribitol and very different from that of xylitol and arabinitol.[200] The crystal structure of 2-deoxy-2-fluoro-β-D-mannopyranose has been reported,[303] and compared with that of α-DL-mannopyranose. The presence of fluorine has no significant influence on the conformation, although the orientation of the hydroxyl group of the hydroxymethyl group is different in the two molecules. In α-DL-mannopyranose, HO-6 is antiperiplanar to C-4 and forms a hydrogen bond (it is the acceptor) with HO-2 (the donor) of an adjacent molecule, whereas F-2 in the fluoro analog is not involved in any hydrogen bonding, and HO-6 is antiperiplanar to H-5. X-Ray crystallographic analysis was used to identify tri-O-acetyl-2-deoxy-2-fluoro-α-D-xylopyranose,[46] an unexpected by-product of the reaction of tri-O-acetyl-α-D-xylopyrano-

(303) W. Choong, D. C. Craig, N. C. Stephenson, and J. D. Stevens, *Cryst. Struct. Commun.*, 4 (1975) 111–115; *Chem. Abstr.*, 83 (1975) 19,572.

syl bromide with silver fluoride in acetronitrile. The compound has the expected $^4C_1(\text{D})$ conformation, with AcO-1 axial and F-2 equatorial. X-Ray crystallographic information was also used[56,57] in an elucidation of the anomeric configuration of 3,4,6-tri-O-acetyl-2-bromo-2-deoxy-α-D-mannopyranosyl fluoride obtained from the reaction of tri-O-acetyl-D-glucal with N-bromosuccinimide in diethyl ether.[54,55] The structure of **122** has been studied by X-ray crystallography.[256]

Optical-rotation studies that have been made on fluorinated carbohydrates are not included in this Section.

IV. SOME BIOLOGICAL APPLICATIONS OF FLUORINATED CARBOHYDRATES

This Section draws attention to a selected number of review articles, and discusses some biological uses to which fluorinated carbohydrates have been put, so that an indication of their biological activity and utility may be given. The examples described are not necessarily representative of the types of uses to which fluorinated carbohydrates have, or can be, put. In 1967, Barnett[12] partly reviewed the biological activity of fluorinated carbohydrates, and, in 1972, he discussed[304] the use of fluorinated carbohydrates in studying the binding of sugars to membrane carriers, and as substrates for glycosidases. Taylor[305] simultaneously discussed the metabolism and enzymology of fluorinated carbohydrates and related compounds.

3-Deoxy-3-fluoro-D-glucose affects the metabolism of D-glucose in *Saccharomyces cerevisiae;* it inhibits polysaccharide synthesis and the uptake of D-glucose. For the metabolism of D-galactose, however, a significant respiratory inhibition, a significant increase in total polysaccharide synthesis, and no significant inhibition of the uptake of D-galactose were observed[306] in the presence of 3-deoxy-3-fluoro-D-glucose.

Deoxyfluoro-D-glucopyranoses and related compounds were used in a study of the specificity of yeast hexokinase.[8] The study was undertaken for the determination of the precise definition of the binding requirements of the hexokinase isoenzymes, so that selective inhibitors could be designed, and evaluated as antitumor agents. It was confirmed by kinetic studies that only modifications at C-2 of the D-glucose molecule could be made without loss of binding to the hexokinase. It was suggested from other evidence that the group on C-2 should be a hydrogen-bond donor, but this supposition ignores the

(304) J. E. G. Barnett, *Ciba Found. Symp.*, (1972) 95–115.
(305) N. F. Taylor, *Ciba Found. Symp.*, (1972) 215–238.
(306) B. Woodward, N. F. Taylor, and R. V. Brunt, *Biochem., J.*, 114 (1969) 445–447.

fact that 2-deoxy-2-fluoro-D-glucose, 2-deoxy-2-fluoro-D-mannose, and 2,2-difluoro-D-*arabino*-hexose are all good substrates. One explanation proffered for this behavior is that it would be possible for HO-2 to displace a water molecule on the receptor site of the enzyme, and compensate by providing a hydrogen bond, whereas a fluorine or a hydrogen substituent might be too small to effect such a displacement. The 2-deoxy-2,2-difluoro-D-*arabino*-hexose could also be regarded as a special case, in the sense that the strong inductive effect of the $\rangle CF_2$ group would increase the acidity of HO-3 and, hence, its hydrogen-bonding ability as a donor. Decreased binding to the hexokinase was observed on replacement of a hydroxyl group by fluorine at C-3, C-4, and C-6 of the D-glucose molecule. Again, kinetic data suggested that HO-3 and HO-4 could play similar parts in the binding, although, OH-3, OH-4, and OH-6 may not necessarily function only as hydrogen-bond donors in the binding. Neither of the anomeric D-glucopyranosyl fluorides is a substrate or an inhibitor, and it was concluded either that D-glucose does not bind in the pyranose form, or that HO-1 indeed plays an important role in the binding.

The specificity of yeast galactokinase has been studied with the aid of 2-, 3-, 4-, and 6-deoxyfluoro-D-galactoses and of 2-deoxy-D-*lyxo*-hexose, which were all found to be substrates.[7] Comparison of kinetic results obtained with 2-deoxy-2-fluoro-D-galactose and 2-deoxy-D-*lyxo*-hexose indicated that HO-2 behaves as a hydrogen-bond acceptor in the binding of the sugar substrate to the enzyme, whereas comparative kinetic results between the 3-, 4-, and 6-deoxyfluoro-D-galactoses and D-galactose indicated that HO-3, HO-4, and HO-6 behave as hydrogen-bond donors in the binding of D-galactokinase. The specificity of yeast D-galactokinase for the D-*galacto* configuration was manifested by the fact that 4-deoxy-4-fluoro-D-glucose is neither an inhibitor nor a substrate. Again, the ability of the enzyme to distinguish between epimeric hydroxyl and epimeric fluoro groups at C-4 pointed to its high specificity.

Di-N-acetyl-β-chitobiosyl fluoride has been used as a substrate for hen's egg-white lyzozyme in place of chitohexaose and 3,4-dinitrophenyl chitotetraoside, in order to avoid the problem of rotational isomerism about the C–O bonds of the leaving groups [the terminal $(GlcNAc)_2$ and the 3,4-dinitrophenoxy group, respectively].[67] The use of fluorine provides a cylindrically symmetrical leaving-group. The rate of hydrolysis of the fluoride was comparable to that of the 3,4-dinitrophenyl chitotetraoside; this relatively fast reaction of the fluoride could either be due to neighboring-group participation by the acetamido group, or be a result of enzymic action.[67] 3,4-Dinitrophenyl

2-acetamido-2,6-dideoxy-6-fluoro-β-D-glucopyranoside has been used, together with related compounds, in kinetic studies aimed at the acquisition of further information relating to the structural requirements for lysozyme catalysis.[67]

6-Deoxy-6-fluoro-α-D-glucopyranosyl α-D-glucopyranoside has been found to be a substrate for cockchafer trehalase,[110] and a competitive, reversible inhibitor of trehalase isolated from the flight muscle of a green butterfly (*Lucilia sericata*),[111] whereas 6-deoxy-6-fluoro-4-*O*-(methylsulfonyl)-α-D-glucopyranosyl α-D-glucopyranoside is a weak inhibitor of the latter trehalase.[111]

Investigations, with the aid of ^{19}F-n.m.r. spectroscopy, directed towards the study of the binding of the substrate to lysozyme include a preliminary communication wherein some important points relating to this technique were also made.[307] 2-Deoxy-2-(trifluoroacetamido)-α- and -β-D-glucose and methyl 2-deoxy-2-(trifluoroacetamido)-α- and -β-D-glucopyranoside were used as substrates. Detectable, chemical-shift changes were observed (at various concentrations in the presence of lysozyme) only for 2-deoxy-2-(trifluoroacetamido)-α-D-glucopyranoside. These gave values for the chemical shift of the bound substrate and for the binding constant. It was found, however, by separate competition-experiments, that both analogs of methyl D-glucoside compete less strongly for the same binding site as the known inhibitor, namely, methyl 2-acetamido-α-D-glucopyranoside. The rate of association of 2-deoxy-2-(trifluoroacetamido)-α-D-glucose with lysozyme was found to be greater than that for 2-acetamido-2-deoxy-α-D-glucose. From these observations, the following conclusions became evident: (*i*) ^{19}F chemical-shift changes are less sensitive to intermolecular changes in environment (but more than the corresponding, ^{1}H changes) than to intramolecular changes, (*ii*) molecular association of a small, reversible, specifically fluorinated inhibitor of a macromolecule may be readily detected by ^{19}F chemical-shift changes, but absence of such shift-changes does not necessarily imply absence of molecular association, and (*iii*) as introduction of a fluorine substituent can cause changes in the nature of the molecular association, specifically fluorinated substrates must be used with caution when their behavior is to be compared with that of the parent compounds.

^{19}F-N.m.r. spectroscopy has been used in structural studies of the binding of sugars to lysozyme in a solution containing paramagnetic ions.[308] The substrates used were *N*-(fluoroacetyl)ated and *N*-(tri-

(307) L. D. Hall and C. W. M. Grant, *Carbohydr. Res.*, 24 (1972) 218–220.
(308) C. G. Butchard, R. A. Dwek, P. W. Kent, R. J. P. Williams, and A. V. Xavier, *Eur. J. Biochem.*, 27 (1972) 548–553.

fluoracetyl)ated derivatives of 2-amino-2-deoxy-D-glucose. The paramagnetic ion Gd^{3+} causes an increase in the relaxation time of the fluorine nucleus. Under conditions of fast chemical exchange of the substrate between the free and the bound state, this increase is a function of (a) the distance of the fluorine nucleus from the metal ion, (b) the correlation time for the n.m.r. relaxation-behavior of the complex, and (c) the relative concentrations of the free and bound sugar. Thus, information concerning the molecular conformation of the enzyme–substrate complex with respect to the metal site can be obtained. The observed distances between the fluorine in 2-deoxy-2-(fluoroacetamido)-α- and -β-D-glucose and in methyl 2-deoxy-2-(fluoroacetamido)-α- and -β-D-glucopyranoside and Gd^{3+} in the complexes were found to be shorter than those expected from a study of molecular models of lysozyme with the fluoro substrates, positioned as 2-acetamido-2-deoxy-β-D-glucose in its known complex with the enzyme. (The binding site of Gd^{3+} was already known from X-ray studies.) This indicated that there exists hindered rotation of the fluoromethyl group, and that the fluorine is oriented towards the Gd^{3+} in the complex. Support for this interpretation was obtained from the distances observed between fluorine and Gd^{3+} in 2-deoxy-2-(trifluoroacetamido)-α- and -β-D-glucose, which are the average of the distances of the three fluorine atoms in the trifluoromethyl group. The values observed for these distances were in good agreement with those known from X-ray studies and with those of the three protons of the acetamido methyl group in methyl 2-deoxy-2-acetamido-β-D-glucopyranose. The ^{19}F chemical shifts of the 2-deoxy-2-(trifluoroacetamido)-D-glucose–lysozyme complexes could best be explained as being due to electrostatic effects of the tryptophan-108 and tryptophan-63 residues, by reference to a molecular model of lysozyme. The geometrical information that could be obtained from chemical shifts from through-space interactions, together with the distance information just described, suggested that the position of the trifluoromethyl group of the substrate in the complex in solution is almost the same as that expected in the crystalline state. The latter position was assumed to be the same as that of the acetamido methyl group in molecular models of methyl 2-acetamido-2-deoxy-α- and -β-D-glucopyranoside–lysozyme complexes.

V. ADDENDUM

Articles describing work on fluorinated carbohydrates that have appeared since about the end of 1978 are listed here, with the exceptions already mentioned (see Section I) and excepting those describing biological work. Syntheses of fluorinated carbohydrates by established methods[309-316] (and an attempted synthesis[317]), including those of aminocyclitol antibiotics,[318,319] of derivatives of fluorinated carbohydrates,[320,321] and of ^{18}F-labelled carbohydrates,[322,323] and by new methods,[324,325] have been reported, as well as ^{19}F.-n.m.r. work,[309,311-313,326] including measurement of $J_{^1H-^{19}F}$ by ^{1}H, 2-d-J experiments,[327] and mass-spectrometric[311,328,329] and crystallographic studies.[330,331]

(309) J. M. J. Tonchet, A. Bonenfant, and F. Barbalat-Rey, *Carbohydr. Res.*, 67 (1978) 564–573; see also, J. M. J. Tronchet and O. R. Martin, *Helv. Chim. Acta*, 62 (1979) 1401–1405.
(310) A. F. Hadfield, L. Hough, and A. C. Richardson, *Carbohydr. Res.*, 71 (1979) 95–102.
(311) D. P. Lopes and N. F. Taylor, *Carbohydr. Res.*, 73 (1979) 125–134.
(312) G. C. Butchard and P. W. Kent, *Tetrahedron*, 35 (1979) 2439–2443.
(313) G. C. Butchard and P. W. Kent, *Tetrahedron*, 35 (1979) 2551–2554.
(314) M. Sharma and W. Korytnyk, *Carbohydr. Res.*, 79 (1980) 39–51.
(315) A. F. Hadfield, L. Hough, and A. C. Richardson, *Carbohydr. Res.*, 80 (1980) 123–180.
(316) J. R. Surfin, R. J. Bernacki, M. J. Morin, and W. Korytnyk, *J. Med. Chem.*, 23 (1980) 143–149.
(317) A. A. Akhrem, E. I. Krasyuk, I. A. Mikhailopulo, G. Kovollik, P. Langen, and A. Otto, *Zh. Obshch. Khim.*, 49 (1979) 1175–1176; *J. Gen. Chem. USSR*, 49 (1979) 1028–1029.
(318) P. J. L. Daniels, and D. F. Rane, S. Afr. Pat. (1978) 78 06,385; *Chem. Abstr.*, 90 (1979) 104,301.
(319) G. Vass, A. Rolland, J. Cleophax, D. Mercier, B. Quichet, and S. D. Gero, *J. Antibiot.*, 32 (1979) 670–672.
(320) J. Pačak, Z. Köllnerová, and M. Černý, *Collect. Czech. Chem. Cmmun.*, 44 (1979) 933–941.
(321) J. M. J. Tronchet, B. Gentile, A. Bonenfant, and O. R. Martin, *Helv. Chim. Acta*, 62 (1979) 696–699.
(322) J. S. Fowler, R. E. Lade, R. R. MacGregor, C. Shiue, C.-N. Wan, and A. P. Wolf, *J. Labelled Compd. Radiopharm.*, 16 (1979) 7–9.
(323) T. J. Tewson, M. J. Welch, and M. E. Raichle, *J. Labelled Compd. Radiopharm.*, 16 (1979) 10–11.
(324) S. Colonna, A. Re, G. Gelbard, and E. Cesarotti, *J. Chem. Soc., Perkin Trans. 1*, (1979) 2248–2252.
(325) W. Korytnyk and S. Valentekovic-Horvat, *Tetrahedron Lett.*, (1980) 1493–1496.
(326) A. DeBruyn, M. Anteunis, G. Aerts, and C. K. DeBruyn, *Acta Cienc. Ind.*, 4 (1978) 103–108.
(327) L. D. Hall and S. Sukumar, *J. Am. Chem. Soc.*, 101 (1979) 3120–3121.
(328) G. Descotes, P. Boullanger, T. Dung, and J.-C. Martin, *Carbohydr. Res.*, 71 (1979) 305–314.
(329) See also, J. Kiburis, A. B. Foster, M. Jarman, and J. H. Westwood, *Chim. Chron., New Ser.*, 8 (1979) 77–82; *Chem. Abstr.*, 91 (1979) 91,885.
(330) G. Kothe, P. Luger, and H. Paulsen, *Acta Crystallogr., Sect. B*, 35 (1979) 2079–2087.
(331) W. Choong, *Cryst. Struct. Commun.*, 8 (1979) 27–32.

THE GULONO-1,4-LACTONES
A Review of Their Synthesis, Reactions, and Related Derivatives

By Thomas C. Crawford

Central Research, Chemical Process Research, Pfizer Inc., Groton, Connecticut 06340

I. Introduction 287
II. Synthesis 288
III. Crystal and Solution Structure, and Spectroscopic Properties
    of the Gulono-1,4-lactones and Derivatives 296
IV. Reactions of the Lactone Moiety in the Gulono-1,4-lactones 298
    1. Hydrolysis 298
    2. Reactions with Ammonia, Amines, and Hydrazines 299
    3. Alcoholysis 301
V. Reactions of the Hydroxyl Groups in the Gulono-1,4-lactones 302
    1. With Ketones 302
    2. With Aldehydes 303
    3. Miscellaneous 304
VI. Deoxy and Anhydro Derivatives of the Gulono-1,4-lactones 305
VII. Amino Derivatives of the Gulono-1,4-lactones 308
VIII. Miscellaneous Derivatives of the Gulono-1,4-lactones 310
IX. Oxidation of the Gulono-1,4-lactones and Derivatives 314
X. Reduction of the Gulono-1,4-lactones and Derivatives 315
XI. Nucleophilic Additions to the Carbonyl Group of Derivatives
    of the Gulono-1,4-lactones 317
XII. Analytical Procedures 318
    1. Paper Chromatography 318
    2. Thin-Layer Chromatography 318
    3. Gas–Liquid Chromatography and Mass Spectrometry 318
    4. Ion-Exchange Chromatography 319
    5. High-Performance, Liquid Chromatography 319
XIII. Biological Role of the Gulono-1,4-lactones and Gulonic Acids 320
XIV. Addendum 321

I. Introduction

The following article summarizes the methods of synthesis for L-gulono-1,4-lactone (**1**), D-gulono-1,4-lactone (**2**), and the corresponding sugar acids (L-**3** and D-**4**), as well as the chemistry related to the lactone ring and the hydroxyl groups, and to a variety of related derivatives. An attempt has been made to cover the literature comprehensively, but the methods by which L-gulono-1,4-lactone (**1**) and L-*xylo*-2-hexulosonic acid (**5**) ("2-keto-L-gulonic acid") have been converted into L-ascorbic acid (**6**) have not been included. For a summary of these and related transformations, see an article in the preceding volume in this Series[1] that covers the methods for the synthesis of L-ascorbic acid.

II. Synthesis

The most efficient synthesis of L-gulono-1,4-lactone (**1**) entails the reduction of D-glucofuranurono-6,3-lactone (**7**), which can be obtained from D-glucose[2] (see Scheme 1). Catalytic hydrogenation[3,4] of **7** in the presence of Raney nickel afforded **1** in 81% yield. Alternatively, D-

(1) T. C. Crawford and S. A. Crawford, *Adv. Carbohydr. Chem. Biochem.*, 37 (1980) 79–155.
(2) The methods by which D-glucofuranurono-6,3-lactone (**7**) can be obtained from D-glucose are described in more detail in Ref. 1.
(3) M. Ishidate, Y. Imai, Y. Hirasaka, and K. Umemoto, *Chem. Pharm. Bull.*, 11 (1965) 173–176.
(4) W. Berends and J. Konings, *Recl. Trav. Chim. Pays-Bas*, 74 (1955) 1365–1370.

glucuronic acid (**8**) has been reduced catalytically[5] to L-gulonic acid (**3**) which, on lactonization, affords **1**. Compound **8** has also been reduced to **3** with sodium borohydride[6,7] and with sodium amalgam.[8,9] The yields reported in the latter case were low (15–30%). To the best of the author's knowledge, there are no literature reports of the preparation of D-gulono-1,4-lactone from L-glucofuranurono-6,3-lactone (L-**7**, see Ref. 10) or the corresponding acid (L-**8**). As L-**8** is not readily available,[11] it is not the starting material of choice for the preparation of D-gulono-1,4-lactone (**2**) or the corresponding acid (**4**).

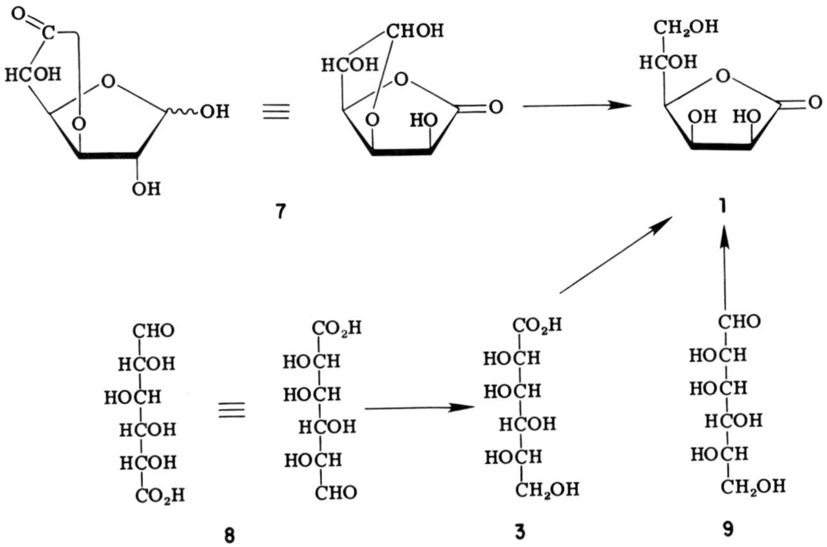

Scheme 1

(5) F. Hoffmann-La Roche and Co., Ger. Pat. 618,907 (1935); *Chem. Abstr.*, 30 (1936) 1070[7].
(6) R. L. Taylor and H. E. Conrad, *Biochemistry*, 11 (1972) 1383–1388.
(7) M. L. Wolfrom and K. Anno, *J. Am. Chem. Soc.*, 74 (1952) 5583–5584.
(8) E. Fischer and O. Piloty, *Ber.*, 24 (1891) 521–528.
(9) H. Thierfelder, *Hoppe-Seyler's Z. Physiol. Chem.*, 15 (1891) 71–76.
(10) Throughout this article, whenever a compound number is preceded by either D or L, it signifies that the compound being discussed is the enantiomer of that actually depicted by the structure as drawn.
(11) L-Glucuronic acid has been prepared by the direct, periodate oxidation of D-*glycero*-D-*gulo*-heptono-1,4-lactone to L-**7**, or by oxidizing 3,5:6,7-di-*O*-isopropylidene-D-*glycero*-D-*gulo*-heptono-1,4-lactone with periodic acid, which resulted in the hydrolysis of the 6,7-isopropylidene group, and cleavage of the C-6–C-7 bond to afford 2,4-*O*-isopropylidene-L-glucurono-6,3-lactone which, on hydrolysis, gave L-**7** in 83% yield; see W. Sowa, *Can. J. Chem.*, 47 (1969) 3931–3934.

Both L-gulose[12,13] (**9**) and D-gulose (D-**9**, see Ref. 10) have been oxidized with bromine to L- or D-gulonic acid, respectively (see Scheme 1). L-Gulose (**9**) has been prepared by a number of different procedures. Schemes 2 and 3 show two early procedures that afforded L-gulose, but the overall yields were low (<30%). In the first procedure[12] (see Scheme 2), D-glucitol (**10**) was converted into 2,4-*O*-benzylidene-D-glucitol (**11**), which was then oxidized with lead tetraacetate to 2,4-*O*-benzylidene-L-xylose (**12**). Nitromethane was added to **12** to afford crystalline **13** in 50% yield. Hydrolysis of **13** provided **14**, which was treated with sodium hydroxide, followed by sulfuric acid, to afford **9**, which was isolated in 52% yield as the 2-benzyl-2-phenylhydrazone.

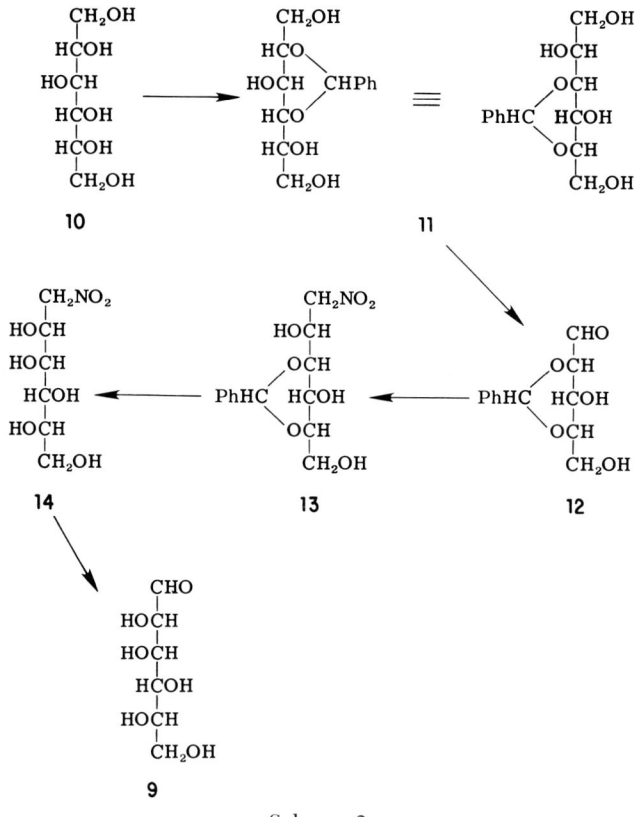

Scheme 2

(12) J. C. Sowden and H. O. L. Fischer, *J. Am. Chem. Soc.*, 67 (1945) 1713–1715; see also, R. L. Whistler and J. N. BeMiller, *Methods Carbohydr. Chem.*, 1 (1962) 137–139; H. S. Isbell, *Bur. Stand. J. Res.*, 5 (1930) 741–755.
(13) K. Heyns and W. Stein, *Ann.*, 558 (1947) 194–201.

In the second procedure[13] (see Scheme 3), D-glucitol (10) was converted into 5-O-benzoyl-1,3:2,4-di-O-ethylidene-6-O-trityl-D-glucitol (17) by successive treatment with acetaldehyde to produce 15, with chlorotriphenylmethane to provide 16, and with benzoyl chloride to afford 17. Selective removal of the trityl protecting group from 17, to give 18, followed by oxidation of 18, provided 19 which, on hydrolysis, afforded L-gulose (9).

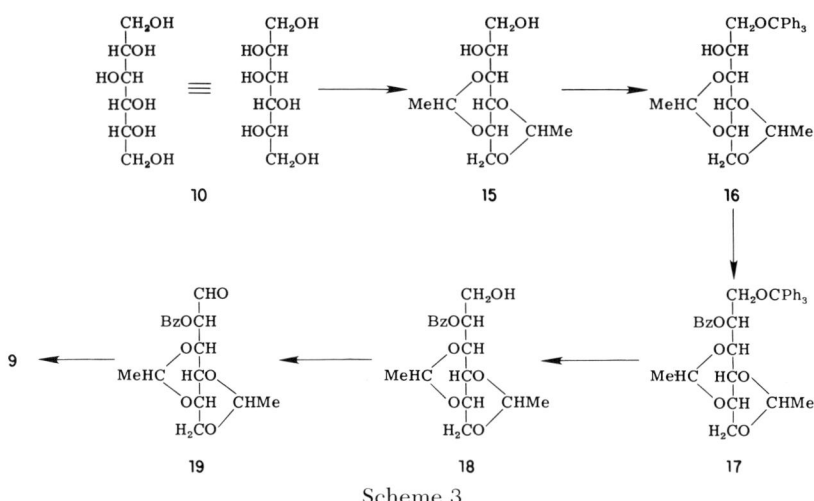

Scheme 3

In later work, Heyns and Beck[14] obtained L-gulose (9), along with D-glucose, D-fructose, and L-sorbose, and various glycuronic acids, by the oxidation of D-glucitol (10) with oxygen in the presence of platinum-on-carbon. L-Gulose was isolated in 18–20% yield as the corresponding 2-benzyl-2-phenylhydrazone.

The conversion of D-mannose (20) into L-gulose (9) was reported by Evans and Parrish,[15] and is shown in Scheme 4. D-Mannose (20) was converted into 21 by condensation with acetone, methanol, and 2,2-dimethoxypropane in the presence of an acid, and mild hydrolysis of 21 afforded 22. Methanesulfonylation of 22 provided 23, which was transformed into 24 with sodium acetate in refluxing N,N-dimethylformamide. The overall yield of 24 from D-mannose was >50%. Base hydrolysis, followed by acid hydrolysis, afforded L-gulose (9).

Minster and Hecht[16] reported a synthesis of a derivative of L-gulose (9) from D-glucose (25) that takes advantage of the ready availability of

(14) K. Heyns and M. Beck, Chem. Ber., 91 (1958) 1720–1724.
(15) M. E. Evans and F. W. Parrish, Carbohydr. Res., 28 (1973) 359–364.
(16) D. K. Minster and S. H. Hecht, J. Org. Chem., 43 (1978) 3987–3988.

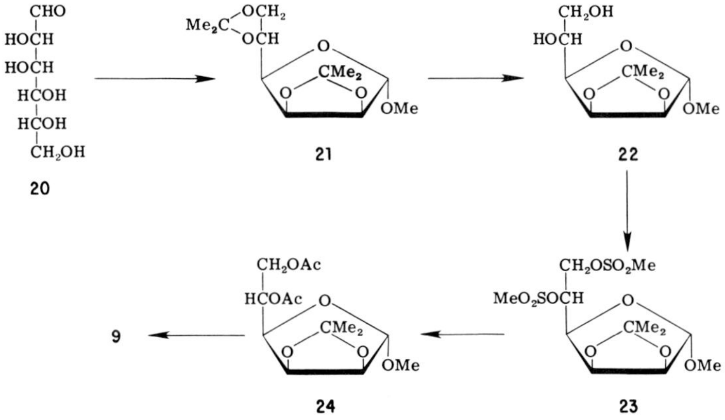

Scheme 4

D-glucose and of the fact that **9** and **25** differ only in the oxidation state at C-1 and C-6 [the relationship between **9** and **25** was originally recognized by Fischer and Piloty[8], and it subsequently played a major role[1] in the development of practical syntheses of L-ascorbic acid (**6**) from D-glucose (**25**) by carbon-chain inversion: that is, reduction at C-1 in **25** and oxidation at C-5 and C-6]. This synthesis of L-gulose resulted in the preparation of 3,4-di-O-benzyl-*aldehydo*-L-gulose 1-(2,2-dimethylhydrazone) (**27**) by oxidation of 1,2-di-O-acetyl-3,4-di-O-ben-

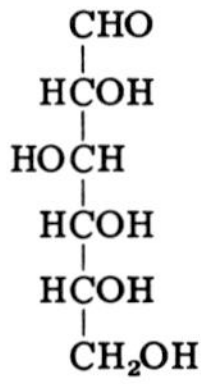

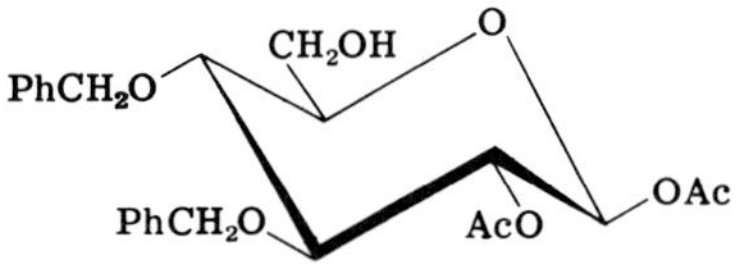

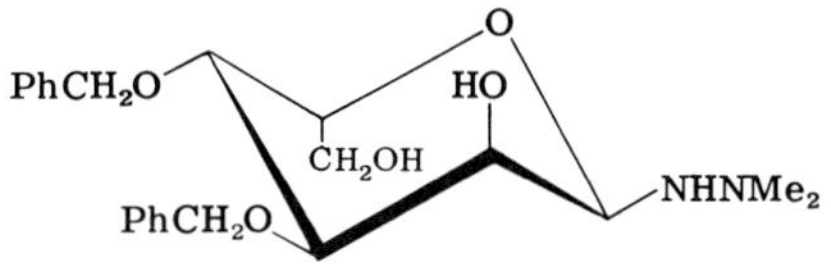

zyl-D-glucopyranose (**26**) at C-6, followed by trapping of the aldehyde with N,N-dimethylhydrazine, hydrolysis of the acetyl groups, and reduction of the resulting free 1-aldehyde with sodium borohydride, to provide **27**. Clearly, **26** could be directly oxidized to the corresponding derivative of D-glucuronic acid which, by steps similar to those used for the conversion of **26** into **27**, could potentially provide 3,4-di-O-benzyl-L-gulonic acid.

As already noted, D-gulose (D-**9**) has been oxidized with bromine[17] to D-gulono-1,4-lactone (**2**). Although D-gulose was originally prepared[18,19] by the reduction of D-gulono-1,4-lactone (**2**), a number of alternative syntheses of D-**9** have since been developed, thus qualifying the oxidation of D-gulose to **2** as a method for preparing **2**. Meyer zu Reckendorf[20-22] was able to prepare D-gulose from D-glucose (**25**) by the procedure shown in Scheme 5. Formation of 1,2:5,6-di-O-isopropylidene-α-D-glucofuranose (**28**) from **25** was followed by oxidation of **28** to **29**, and acetylation–enolization (77%) of **29** to enol acetate **30**. Reduction of **30** with sodium borohydride to **31** (34%), followed by hydrolysis, provided D-**9** (100%). This synthesis of D-**9** was improved when it was found[23] that catalytic hydrogenation of **30** affords acetate

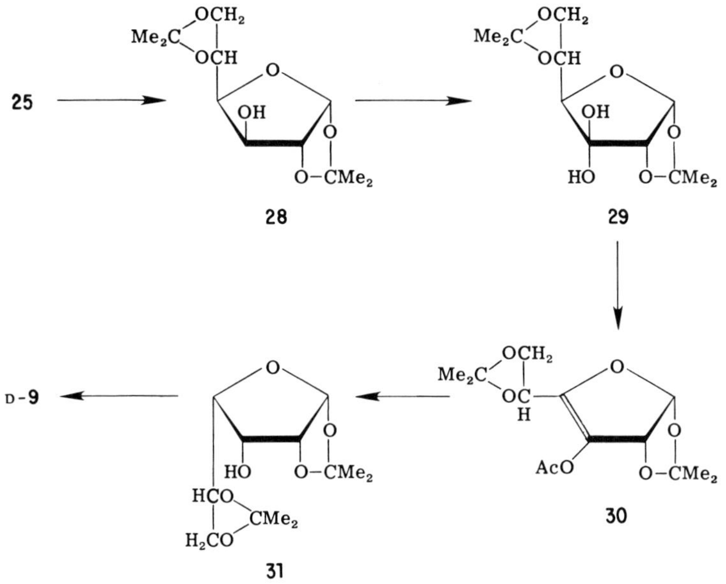

Scheme 5

(17) L. C. Stewart and N. K. Richtmyer, *J. Am. Chem. Soc.*, 77 (1955) 1021–1024.
(18) E. Fischer and R. Stahel, *Ber.*, 24 (1891) 528–539.
(19) H. S. Isbell, *Methods Carbohydr. Chem.*, 1 (1962) 135–136.
(20) W. Meyer zu Reckendorf, *Angew. Chem. Int. Ed. Engl.*, 6 (1967) 177.
(21) W. Meyer zu Reckendorf, *Chem. Ber.*, 102 (1969) 1071–1075.
(22) W. Meyer zu Reckendorf, *Methods Carbohydr. Chem.*, 6 (1972) 129–131.
(23) K. N. Slessor and A. S. Tracey, *Can. J. Chem.*, 47 (1969) 3989–3995.

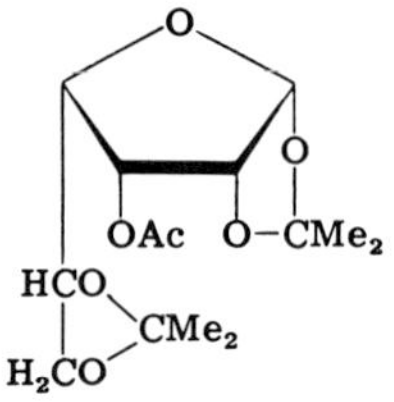

32

32 (75% yield), which may be converted by base hydrolysis into **31** (95%).

D-Gulose (D-**9**) has also been prepared[24] by the oxidation of 4,6-O-ethylidene-1,2-O-isopropylidene-α-D-galactopyranose (**33**), followed by reduction (resulting in overall inversion of the configuration of C-3) and hydrolysis.

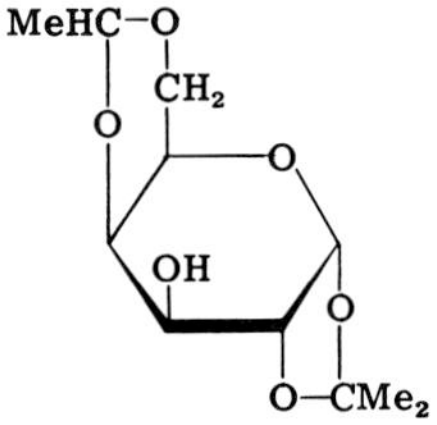

33

A short synthesis of D-gulono-1,4-lactone (**2**) from the inexpensive and readily available D-xylose (**34**) was first reported by Fischer and Stahel,[18] and subsequently by others,[12,25–31] and is shown in Scheme 6. The addition of hydrogen cyanide to D-xylose (**34**) resulted in the formation of cyanohydrins **35** and **36** which, on hydrolysis, afforded a mixture of D-gulonic acid (**4**) and D-idonic acid (**37**). D-Gulono-1,4-lactone may be obtained in 30–33% yield by recrystallization of the reaction products. In a similar way, L-xylose (L-**34**) has been converted[32,33] in high yield into a mixture of L-gulonic and L-idonic acids.

(24) G. J. F. Chittenden, *Chem. Commun.*, (1968) 779–780.
(25) J. V. Karabinos, *Org. Synth. Coll. Vol.*, 4 (1963) 506–508.
(26) R. K. Hulyalkar and J. K. N. Jones, *Can. J. Chem.*, 41 (1963) 1898–1904.
(27) E. Fischer and I. W. Fay, *Ber.*, 28 (1895) 1975–1983.
(28) F. B. LaForge, *J. Biol. Chem.*, 36 (1918) 347–349.
(29) F. B. LaForge, U. S. Pat. 1,285,248 (1918); *Chem. Abstr.*, 13 (1919) 230.
(30) P. A. Levene and G. M. Meyer, *J. Biol. Chem.*, 26 (1916) 355–365.
(31) J. U. Nef, *Ann.*, 403 (1914) 204–383.
(32) S. L. Ruskin and R. C. Hockett, U. S. Pat. 2,853,495 (1958); *Chem. Abstr.*, 53 (1959) 5150d; S. L. Ruskin, U. S. Pat. 2,934,481 (1960); *Chem. Abstr.*, 54 (1960) 16,696h.
(33) Methods for the preparation of L-idonic acid and L-xylose were discussed in Ref. 1.

It has also been reported that D-galactose can be isomerized with molybdic acid to a mixture containing D-talose and D-gulose, but the yields are low.[34,35] A number of derivatives of gulono-1,4-lactone have been prepared that, by reduction, would provide selectively protected derivatives of gulose. These derivatives will be discussed in subsequent Sections of this article.

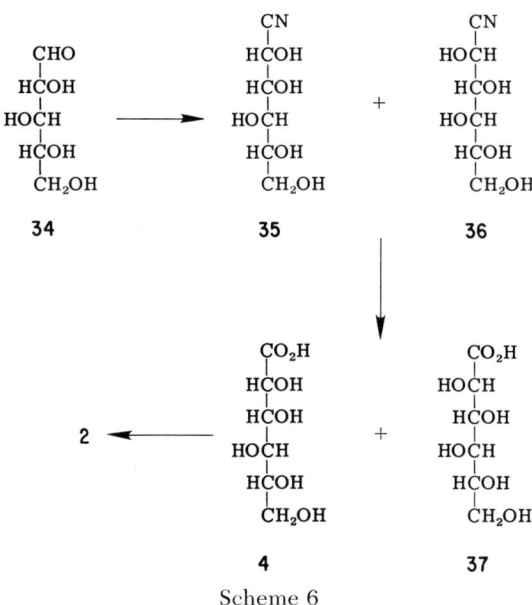

Scheme 6

L-Gulono-1,4-lactone has been prepared[36] by the oxidation of 1,3:2,4-di-O-ethylidene-D-glucitol (**15**) with oxygen in the presence of platinum-on-carbon to 3,5:4,6-di-O-ethylidene-L-gulonic acid (50–70%) which, after hydrolysis, afforded L-gulono-1,4-lactone (**1**). L-Idonic acid[33] has been converted[37] into a mixture of L-**37** and L-gulonic acid (**3**) by heating in aqueous pyridine at 140° under pressure. Compound **3** was isolated by forming the 2,4:3,5-di-O-benzylidene derivative of L-**37**, which is insoluble and can be removed by filtration, leaving **3** in solution. L-Gulonic acid has been obtained by the reduction of L-*xylo*-2-hexulosonic acid (**5**) with a *Lactobacillus* organism,[38]

(34) V. Bilik, W. Voelter, and E. Bayer, *Ann.*, (1974) 1162–1166.
(35) V. Bilik, Czech. Pat. 157,429 (1975); *Chem. Abstr.*, 83 (1975) 164,510f.
(36) A. A. D'Addieco, U. S. Pat. 2,847,421 (1958); *Chem. Abstr.*, 53 (1959) 3084g.
(37) B. E. Gray, U. S. Pat. 2,421,612 (1947); *Chem. Abstr.*, 41 (1947) 5684d.
(38) S. A. Barker, E. J. Bourne, E. Salt, and M. Stacey, *J. Chem. Soc.*, (1959) 593–598.

or with sodium borohydride.[39] The latter reduction affords an approximately equal mixture of L-gulonic (**3**) and L-idonic acid (L-**37**). Finally, L-*xylo*-3-hexulosonic acid was reduced with sodium borohydride to a mixture of L-gulonic acid and L-galactonic acid.[40,41]

III. Crystal and Solution Structure, and Spectroscopic Properties of the Gulono-1,4-lactones and Derivatives

The crystal structure of D-gulono-1,4-lactone (**2**) has been determined,[42] and is closely approximated by formula **38**. Interestingly, the conformation of the side chain in **38** differs from that in crystalline L-ascorbic acid,[1] very probably because of different steric interactions between the 3- and 5-hydroxyl groups in **2** (that is, **38**) and **6**. Bridgman[43] measured the compressibility of gulono-1,4-lactone crystals.

38

In principle, dehydration of the gulonic acids can afford a 5-, a 6-, or a 7-membered lactone ring. By monitoring the change in rotation of an aqueous solution of **3**, Rehorst and Naumann[44] were able to show that L-gulonic acid (**3**) and L-gulonolactone (**1**) are in equilibrium. In an earlier study (in 1925), Levene and Simms[45] had concluded, from the change in rotation of an aqueous solution, that **1** and **3** are present at equilibrium in the ratio of 4–6:1, with possibly some (<10%) L-gulono-1,5-lactone present, but no direct evidence was obtained and, therefore, these conclusions should be considered tentative.

Matsunaga and Tamura[46] showed that the rate of hydrolysis of L-gulono-1,4-lactone (**1**) at pH 5.5 (0.2 M pyridine and 0.2 M acetic acid)

(39) E. R. Nelson, P. F. Nelson, and O. Samuelson, *Acta Chem. Scand.*, 22 (1968) 691–693.
(40) P. Dworsky and O. Hoffmann-Ostenhof, *Acta Biochim. Pol.*, 11 (1964) 269–277.
(41) J. D. Smiley and G. Ashwell, *J. Biol. Chem.*, 236 (1961) 357–364.
(42) H. M. Berman, R. D. Rosenstein, and J. Southwick, *Acta Crystallogr., Sect. B*, 27 (1971) 7–10; see *Adv. Carbohydr. Chem. Biochem.*, 30 (1974) 448.
(43) P. W. Bridgman, *Proc. Am. Acad. Arts Sci.*, 77 (1949) 187–234; *Chem. Abstr.*, 43 (1949) 4917b; *ibid.*, 68 (1933) 27–93; *Chem. Abstr.*, 27 (1933) 3128.
(44) K. Rehorst and A. Naumann, *Ber.*, 77 (1944) 24–29.
(45) P. A. Levene and H. S. Simms, *J. Biol. Chem.*, 65 (1925) 31–47.
(46) I. Matsunaga and Z. Tamura, *Chem. Pharm. Bull.*, 20 (1972) 284–286.

and 30° is reasonably low. After 24 h, >70% of **1** remained, and, after 96 h, >30% of **1** remained. The components of the mixture were determined by freeze-drying an aliquot, per(trimethylsilyl)ating the residue, and assaying by gas–liquid chromatography. No components but **1** and L-gulonic acid (**3**) were reported. Taylor and Conrad[6] showed that L-gulonic acid (**3**) may be converted into lactone **1** in high yield in aqueous solution at pH 4.75 by the use of water-soluble carbodiimides. This procedure is, in general, unnecessary, as, on removal of the water from an aqueous solution of **3**, the 1,4-lactone forms spontaneously. At the present time, the 1,5-lactone of gulonic acid has not been isolated from aqueous solution, or characterized by spectroscopic methods.

S. A. Barker and coworkers[47] reported the carbonyl-stretching frequencies of a number of 1,4- and 1,5-lactones of aldonic acids, including D- and L-gulono-1,4-lactone (1780 cm^{-1}). The ^{1}H-n.m.r. spectrum of L-gulono-1,4-lactone (**1**) in dimethyl sulfoxide-d_6 shows[48] the four hydroxyl protons at δ 5.80 (d), 5.30 (d), 4.95 (d), and 4.65 (t), and the protons on carbon atoms as two broad peaks at δ 4.07–4.45 and 3.35–4.00. The ^{13}C-n.m.r. spectrum of **1** shows[48] C-1 at 177.7, C-4 at 81.3, C-5 at 70.1, C-6 at 61.6, and C-2 and -3 at either 70.8 or 69.6 p.p.m. The ^{13}C-n.m.r. spectrum of 3,4,5-tri-O-acetyl-2,6-anhydro-D-gulononitrile has been reported,[49] along with those of a number of glycosyl cyanides. The mass spectrum of D-gulono-1,4-lactone was obtained[50] by using field-desorption techniques; it showed an (M + 1) peak at m/e 179 (base peak), a peak at m/e 147 (6%) corresponding to loss of -CH$_2$OH, and a peak at m/e 104 (60%).

A number of papers have been published in which the configuration of the carbon atoms in aldonic acids and related 1,4-lactones has been correlated to their circular dichroism (c.d.) curves. Included in this work have been derivatives and complexes of D- and L-gulono-1,4-lactone and D-gulonic acid. Meguro and coworkers[51] measured the c.d. spectrum of 2,3,5,6-tetra-O-(trimethylsilyl)-D-gulono-1,4-lactone (**55**, R = SiMe$_3$). Later, they[52] measured the c.d. curves of borate complexes of aldono-1,4-lactones, including **2**, and found that the rate of

(47) S. A. Barker, E. J. Bourne, R. M. Pinkard, and D. H. Whiffen, *Chem. Ind. (London)*, (1958) 658–659.
(48) T. C. Crawford, G. C. Andrews, and G. Chmurny, (Pfizer, Inc.), unpublished results.
(49) B. Coxon, *Ann. N. Y. Acad. Sci.*, 222 (1973) 952–970.
(50) M. M. Bursey and M. C. Sammons, *Carbohydr. Res.*, 37 (1974) 355–358.
(51) H. Meguro, K. Hachiya, A. Tagiri, and K. Tuzimura, *Agric. Biol. Chem.*, 36 (1972) 2075–2079.
(52) H. Meguro, A. Tagiri, and K. Tuzimura, *Agric. Biol. Chem.*, 38 (1974) 595–598.

hydrolysis of **2** is higher than that of its borate complex at pH 9. Beecham[53] and Bystricky and coworkers[54] reported the c.d. curves for a number of 1,4-lactones of aldonic acids, including that of **2**, and related the c.d. curves to the conformation and the substitution pattern. The c.d. curves of complexes of praseodymium ions,[55] nickel(II) complexes,[56] and molybdenum(V) complexes[57] of gulonic acid, as well as of other hydroxy acids, were reported, as were those of complexes between amino acids and D-gulonic acid and other aldonic acids.[58] Complexes between germanic acid and **1** have also been examined by conductimetric, polarimetric, pH, and viscosity measurements.[59] Similar complexes were not observed with silicic acid. Finally, the instability constants of 1:1 iron complexes of gulonic acid and gulono-1,4-lactone have been determined.[60]

IV. REACTIONS OF THE LACTONE MOIETY IN THE GULONO-1,4-LACTONES

1. Hydrolysis

Ishidate and coworkers[3] prepared L-gulonic acid (**3**) by passing L-gulono-1,4-lactone (**1**) down a strongly basic, anion-exchange resin [Amberlite IRA-410 (OH$^-$)] on which the lactone ring is cleaved, and the resulting, free acid is neutralized by the hydroxide, forming the anion of **3**, and thus being retained on the resin. Elution of the resin with 62.5 mM sulfuric acid resulted in the isolation of **3** as a syrup.

CO_2R
AcOCH
AcOCH
HCOAc
AcOCH
CH_2OAc

39

(53) A. F. Beecham, *Tetrahedron Lett.*, (1968) 2355–2358.
(54) S. Bystricky, T. Sticzay, S. Kucár, and C. Peciar, *Collect. Czech. Chem. Commun.*, 41 (1976) 2749–2754.
(55) L. I. Katzin, *Inorg. Chem.*, 7 (1968) 1183–1191.
(56) L. I. Katzin and E. Gulyas, *J. Am. Chem. Soc.*, 92 (1970) 1211–1214.
(57) D. H. Brown and J. MacPherson, *J. Inorg. Nucl. Chem.*, 33 (1971) 4203–4207.
(58) L. I. Katzin, *Inorg. Chem.*, 12 (1973) 649–655.
(59) E. R. Clark and J. A. Waddams, *Nature*, 180 (1957) 904–905.
(60) V. A. Chernyshev and M. E. Shishniashvili, *Khelaty Met. Prir. Soedin. Ikh Primen.*, 1 (1974) 34–44; *Chem. Abstr.*, 83 (1975) 184,351h.

Acid **3** was characterized by esterification with diazomethane, and acetylation of the methyl ester, to afford crystalline methyl penta-*O*-acetyl-L-gulonate (**39**, R = Me). Acid **3** has also been obtained by adding one equivalent of hydrochloric acid to an aqueous solution of sodium L-gulonate.[6,44,45] A number of salts of gulonic acid have been prepared, including the sodium,[30,31,44,61] ferric,[62] brucine,[30,31] quinine,[31] strychnine,[31] and cinchonine[31] salts. Ferric L-gulonate was obtained from sodium L-gulonate by passing a solution of the sodium salt through Amberlite IR-120 (H^+) resin, and neutralizing the resulting **3** with ferric carbonate.

2. Reactions with Ammonia, Amines, and Hydrazines

The lactone rings in **1** and **2** have been cleaved by a number of amines.[63] D-[64,65] and L[66]-Gulonamide (**40** and L-**40**, R = H) have been prepared by treating **1**, or **2**, respectively, with liquid ammonia[64] or with an ammonia–methanol solution.[65,66] Robbins and Upson[64,67] acetylated **40** (R = H), to afford penta-*O*-acetyl-D-gulonamide (**40**, R = Ac), which was oxidatively hydrolyzed to penta-*O*-acetyl-D-gulonic acid (D-**39**, R = H). The methyl ester (D-**39**, R = Me) was formed by treatment with diazomethane. The phenylhydrazides of D- and L-gulonic acid (**41** and L-**41**, R = H) have been prepared by a number of groups.[31,67,68] By a procedure similar to that used for the preparation of **40** (R = Ac), penta-*O*-acetyl-D-gulonic phenylhydrazide (**41**, R = Ac) was prepared.[67] The unsubstituted hydrazide (**42**) of **3** was prepared[69] by treating **1** with hydrazine. Treatment of **42** with benzaldehyde[69] or 4-methoxybenzaldehyde[69] afforded the corresponding hydrazones.

(61) P. A. Levene, *J. Biol. Chem.*, 59 (1924) 123–127.
(62) Y. Nitta, Y. Nakajima, and Y. Hoski, Jpn. Pat. 15,665 (1964); *Chem. Abstr.*, 62 (1965) 7855d.
(63) In a number of the original references, either no indication was given as to which enantiomer was prepared, or it was designated by using *d* or *l* instead of D or L. An attempt has been made to correlate these compounds with compounds of known configuration, or, by using the rules developed by C. S. Hudson for the rotation of amides of the aldonic acids and related compounds, correlate the reported rotation with the correct absolute configuration (see C. S. Hudson, *Conf. Union Int. Chim.*, *10th*, Liége, September 14–20, (1930) 59–78; *The Collected Papers of C. S. Hudson*, Academic Press, New York, 1946, pp. 804–825.
(64) G. B. Robbins and F. W. Upson, *J. Am. Chem. Soc.*, 60 (1938) 1788–1789.
(65) C. S. Hudson and S. Komatsu, *J. Am. Chem. Soc.*, 41 (1919) 1141–1147.
(66) R. A. Weerman, *Recl. Trav. Chim. Pays-Bas*, 37 (1917) 16–51.
(67) G. B. Robbins and F. W. Upson, *J. Am. Chem. Soc.*, 62 (1940) 1074–1076.
(68) P. A. Levene and G. M. Meyer, *J. Biol. Chem.*, 31 (1917) 623–626.
(69) T. W. J. van Marle, *Recl. Trav. Chim. Pays-Bas*, 39 (1920) 549–572.

```
        CONH₂                    CONHNHPh
        |                        |
        HCOR                     HCOR
        |                        |
        HCOR                     HCOR
        |                        |
        ROCH                     ROCH
        |                        |
        HCOR                     HCOR
        |                        |
        CH₂OR                    CH₂OR
         40                       41
```

The hydrazide **42** has been used[70] for preparing the hydrazone of the cephalosporin **43**. Grinsteins and coworkers[71] prepared the benzylhydrazide and (2-hydroxyethyl)hydrazide of L-gulonic acid, and found that they have biological activity as antidepressants.

```
     CONHNH₂
     |
     HOCH
     |
     HOCH
     |
     HCOH
     |
     HOCH
     |
     CH₂OH
       42
```

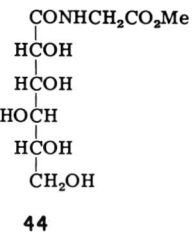

```
         43
```

Several substituted amides of D- or L-gulonic acid, or both, have been prepared, including N-allyl-[72], N,N-dimethyl-D-[73], and N-(4-ethoxyphenyl)- and N-(4-methoxyphenyl)-L-gulonamide.[74] The latter two amides were claimed to be analgesics and antipyretics. The triethylamine salt of D-gulonic acid was coupled[75] with methyl glycinate by the use of dicyclohexylcarbodiimide, to afford N-(methoxycarbonylmethyl)-D-gulonamide (**44**).

```
     CONHCH₂CO₂Me
     |
     HCOH
     |
     HCOH
     |
     HOCH
     |
     HCOH
     |
     CH₂OH
        44
```

(70) M. Yoshioka, Y. Sendo, K. Ishikura, M. Murakami, and S. Miyazaki, Jpn. Kokai 75 131,982 (1975); *Chem. Abstr.*, 85 (1976) 46, 704r.
(71) V. Grinsteins, L. Serina, R. Bluma, and N. Ratenbergs, *Latv. PSR Zinat. Akad. Vestis, Kim. Ser.*, (1967) 705–716; *Chem. Abstr.*, 68 (1968) 76,830v.
(72) D. L. Tabern, U. S. Pat. 2,084,626 (1937); *Chem. Abstr.*, 31 (1937) 5949⁸.
(73) K. Kefurt, Z. Kefurtova, J. Nemec, J. Jarý, I. Fric, and K. Blaha, *Collect. Czech. Chem. Commun.*, 36 (1971) 124–131.
(74) Y. Nakajima and Y. Hoshi, Jpn. Pat. 21,240 (1964); *Chem. Abstr.*, 62 (1965) 11,900d.

Several amides of derivatives of L-gulonic acid (3) have been prepared. On treatment of 2,3:5,6-di-O-isopropylidene-L-gulono-1,4-lactone (the preparation of which is discussed in Section V) with ammonia in methanol, 2,3:5,6-di-O-isopropylidene-L-gulonamide (45) is formed.[76] Similarly, 3,5-O-benzylidene-L-gulonamide (46) was prepared[76] from the corresponding lactone.

$$
\begin{array}{cc}
\text{CONH}_2 & \text{CONH}_2 \\
\text{Me}_2\text{C}\diagup\text{OCH} & \text{HOCH} \\
\phantom{\text{Me}_2\text{C}}\diagdown\text{OCH} & \phantom{\text{PhHC}}\diagup\text{OCH} \\
\text{HCOH} & \text{PhHC}\text{HCOH} \\
\text{Me}_2\text{C}\diagup\text{OCH} & \phantom{\text{PhHC}}\diagdown\text{OCH} \\
\phantom{\text{Me}_2\text{C}}\diagdown\text{OCH}_2 & \text{CH}_2\text{OH} \\
\mathbf{45} & \mathbf{46}
\end{array}
$$

3. Alcoholysis

The simple cleavage of lactones **1** or **2** with alcohol and acid has not been reported. However, when **1** is treated with benzaldehyde diethyl acetal and hydrochloric acid, ethyl 3,5:4,6-di-O-benzylidene-L-gulonate (**47**) is formed in >90% yield.[77,78] No other isomers were observed, and other acetals of benzaldehyde, as well as aliphatic aldehydes, afford similar products in good yield.[77] D'Addieco prepared[36] similarly protected derivatives of L-gulonic acid by oxidation of 1,3:2,4-di-O-ethylidene-D-glucitol (**15**), followed by esterification of the resulting acid with diazomethane.

$$
\begin{array}{c}
\text{CO}_2\text{Et} \\
\text{HOCH} \\
\diagup\text{OCH} \\
\text{PhHC}\text{HCO}\diagdown \\
\phantom{\text{PhHC}}\diagdown\text{OCH}\text{CHPh} \\
\text{H}_2\text{CO}\diagup \\
\mathbf{47}
\end{array}
$$

---

(75) M. E. Shishniashvili, L. D. Kvaratskheliya, I. M. Privalova, and A. Ya. Khorlin, *Soobshch. Akad. Nauk Gruz. SSR*, 72 (1973) 581–583; *Chem. Abstr.*, 80 (1974) 146,441*t*; A. Ya. Khorlin, I. M. Privalova, L. D. Kvaratskheliya, and M. E. Shishniashvili, *Izv. Akad. Nauk SSSR, Ser. Khim.*, (1973) 1619–1624; *Chem. Abstr.*, 79 (1973) 126,723*n*.

(76) M. Matsui, M. Okada, and M. Ishidate, *Yakugaku Zasshi*, 86 (1966) 110–113.

(77) T. C. Crawford and R. Breitenbach, *J. Chem. Soc., Chem. Commun.*, (1979) 388–389.

(78) T. C. Crawford, Eur. Pat. 0,000,243 (1977).

Although not derived by lactone cleavage, it is appropriate at this point to include a number of derivatives of gulonic acid that have been prepared. Haworth and Wiggins[79] obtained methyl 2,3:4,5-di-O-methylene-L-gulonate[80] (**48**) in low yield by the oxidation of 2,3:4,5-di-O-methylene-D-glucitol[79] followed by methyl esterification. Matsui and coworkers[81] prepared 3-O-benzoyl-2,4,5,6-tetra-O-benzyl-L-gulonic acid (**49**, R = benzoyl) by oxidation of 4-O-benzoyl-1,2,3,5-tetra-O-benzyl-D-glucitol. Hydrolysis of **49** with base afforded 2,4,5,6-tetra-O-benzyl-L-gulonic acid (**49**, R = H). Esterification with benzyl alcohol then gave benzyl 2,4,5,6-tetra-O-benzyl-L-gulonate. The corresponding methyl ester was also prepared. Catalytic hydrogenation of the benzyl ester afforded L-gulono-1,4-lactone.

CO_2Me	CO_2H
OCH	BzlOCH
H$_2$C⟨		ROCH
OCH	HCOBzl
HCO	BzlOCH
H$_2$COCH	CH_2OBzl
CH_2OH	
48	**49**

V. REACTIONS OF THE HYDROXYL GROUPS IN THE GULONO-1,4-LACTONES

1. With Ketones

When **1**, or **2**, is treated with acetone and a strong acid (sulfuric or hydrochloric acid), 2,3:5,6-di-O-isopropylidene-L-gulono-1,4-lactone[76,82] (**50**), or the D enantiomer,[26,83,84] is formed in 50–70% yield. Compound D-**50** has been prepared by the oxidation of 2,3:5,6-di-O-isopropylidene-D-gulose.[84] When D-gulono-1,4-lactone was stirred in acetone in the presence of anhydrous copper sulfate, 5,6-O-isopropylidene-D-gulono-1,4-lactone (**51**) was formed[26,84] in 10% yield, along

---

(79) W. N. Haworth and L. F. Wiggins, *J. Chem. Soc.*, (1944) 58–61.
(80) The structure considered to be 2,3:4,5-di-O-methylene-D-glucitol could also be 2,4:3,5-di-O-methylene-D-glucitol. Haworth and Wiggins[79] did not determine which of these structures the methyl L-gulonate derivative actually possesses.
(81) M. Matsui, M. Saito, M. Okada, and M. Ishidate, *Chem. Pharm. Bull.*, 16 (1968) 1294–1299.
(82) H. Ogura, H. Takakashi, and T. Itoh, *J. Org. Chem.*, 37 (1972) 72–75.
(83) L. M. Lerner, B. D. Kohn, and P. Kohn, *J. Org. Chem.*, 33 (1968) 1780–1783.
(84) K. Iwadare, *Bull. Chem. Soc. Jpn.*, 18 (1943) 226–231; *Chem. Abstr.*, 41 (1947) 4457i.

with D-**50** (25%). When D-**50** is mixed with aqueous acetic acid, the 5,6-acetal ring is selectively hydrolyzed, to afford 2,3-O-isopropylidene D-gulonolactone (**52**) in 89% yield.[26,85] It has been reported[83] that **52** was isolated as a by-product during the preparation of D-**50**.

2. With Aldehydes

3,5-O-Benzylidene-L-gulono-1,4-lactone (**53**, R = Ph) was first prepared in 43% yield by Matsui and coworkers[76] by treating L-gulono-1,4-lactone (**1**) with benzaldehyde and concentrated hydrochloric acid. Subsequently, it was found[77,86] that the yield could be improved to >65% by using hydrogen chloride instead of aqueous hydrochloric acid. The formation of **53** with other aldehydes (both aromatic and aliphatic) has been reported,[86] and the acetals include the following: R = 2-chlorophenyl (80%), ethyl (25%), methyl (25%), 2-methoxyphenyl (91%), 2-methylphenyl (83%), 3-methylphenyl (52%), and 2-propyl (24%).

When L-gulono-1,4-lactone was treated with acetaldehyde diethyl acetal in N,N-dimethylformamide–benzene in the presence of a catalytic amount of p-toluenesulfonic acid, 5,6-O-ethylidene-L-gulono-1,4-lactone (**54**) was formed[86] in 61% yield (40%, recrystallized). It is interesting that compound **54** was clearly a mixture of diastereo-

(85) L. M. Lerner, *Carbohydr. Res.*, 9 (1969) 1–4.
(86) T. C. Crawford, U. S. Pat. 4,111,958 (1978).

isomers, whereas **53** (R = Me, Ph, or other substituents) appeared to be a single diastereoisomer. When **54** was treated with a catalytic amount of *p*-toluenesulfonic acid in acetonitrile, it was converted into **53** (R = Me) in 85% yield.

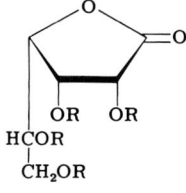

54

It was reported that, when L-gulono-1,4-lactone was treated with benzaldehyde and concentrated sulfuric acid, 2,6:3,5-di-O-benzylidene-L-gulono-1,4-lactone was formed.[76,87]

3. Miscellaneous

Several, fully protected derivatives of gulonolactone have been prepared; these include 2,3,5,6-tetra-O-(trimethylsilyl)-D-gulono-1,4-lactone (**55**, R = SiMe$_3$),[51] 2,3,5,6-tetra-O-acetyl-D-gulono-1,4-lactone (**55**, R = acetyl),[88,89] 2,3,5,6-tetra-O-benzoyl-D-gulono-1,4-lactone[90,91] (**55**, R = benzoyl), and L-**55** (R = benzoyl).[91,92]

55

---

(87) Because of the novel structure assigned to this compound, additional confirmatory data could prove useful in order to ensure that the original assignment of structure was correct.

(88) F. W. Upson, J. M. Brackenburry, and C. Linn, *J. Am. Chem. Soc.*, 58 (1936) 2549–2552.

(89) R. K. Ness, H. G. Fletcher, Jr., and C. S. Hudson, *J. Am. Chem. Soc.*, 73 (1951) 4759–4761.

(90) P. Kohn, R. H. Samaritano, and L. M. Lerner, *J. Am. Chem. Soc.*, 86 (1964) 1457–1458.

(91) P. Kohn, R. H. Samaritano, and L. M. Lerner, *J. Am. Chem. Soc.*, 87 (1965) 5475–5480.

(92) M. Matsui, M. Okada, and M. Ishidate, *Chem. Pharm. Bull.*, 16 (1968) 1288–1293.

A number of triply protected derivatives of L-gulono-1,4-lactone (**1**) have been prepared. When **1** was treated with 3 equivalents of *tert*-butylchlorodimethylsilane, 2,5,6- and 3,5,6-tri-*O*-(*tert*-butyldimethylsilyl)-L-gulono-1,4-lactone were formed[86] in >90% yield. When 3,5-*O*-benzylidene-L-gulono-1,4-lactone was treated with chlorotriphenylmethane, 3,5-*O*-benzylidene-6-*O*-trityl-L-gulono-1,4-lactone was isolated in 38% yield.[86]

Finally, the reduction of 4-*O*-methyl-D-glucuronic acid afforded 3-*O*-methyl-L-gulono-1,4-lactone,[39] which was not further characterized, but was hydrolyzed with hydrobromic acid to L-gulonic acid. By reducing 4-*O*-β-D-glucopyranosyl-D-glucuronic acid[93] with potassium borohydride, 3-*O*-β-D-glucopyranosyl-L-gulonic acid was obtained[94] (but not characterized).

VI. DEOXY AND ANHYDRO DERIVATIVES OF THE GULONO-1,4-LACTONES

Galkowski and coworkers[95] prepared 6-deoxy-L-gulono-1,4-lactone (**56**) by treating D-glucuronamide, which was obtained from D-glucofuranurono-6,3-lactone (**7**), with *p*-tolylsulfonylhydrazine, and reducing the resulting hydrazone (78%) with potassium borohydride, to afford **56** in 8.5% yield from the hydrazone. In a longer synthesis,[96] L-xylose was converted into 5-deoxy-L-xylose (**57**), which was converted into a mixture of 6-deoxy-L-idonic acid and 6-deoxy-L-gulonic acid, from which **56** was isolated. This procedure was first used by Levene and Compton[97] to prepare 6-deoxy-D-gulono-1,4-lactone from 5-deoxy-D-xylose.

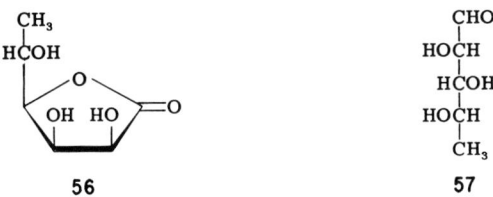

56   57

(93) I. Johansson, B. Lindberg, and O. Theander, *Acta Chem. Scand.*, 17 (1963) 2019–2024.
(94) O. Smidsrød, A. Haug, and B. Larsen, *Acta Chem. Scand.*, 20 (1966) 1026–1034.
(95) T. T. Galkowski, R. W. Kocon, and W. C. Griffiths, *Carbohydr. Res.*, 13 (1970) 187–188.
(96) H. Müller and T. Reichstein, *Helv. Chim. Acta*, 21 (1938) 251–262.
(97) P. A. Levene and J. Compton, *J. Biol. Chem.*, 111 (1935) 335–346; 112 (1936) 775–783; *J. Am. Chem. Soc.*, 57 (1935) 777–778.

Several, alternative syntheses of 6-deoxy-L-gulono-1,4-lactone (**56**) have been developed by Ireland and Wilcox.[98] In the first, 1-deoxy-D-glucitol was converted into 1-deoxy-2,4:3,5-di-O-methylene-D-glucitol, which was oxidized (with chromium trioxide) at C-6 to the acid, and this esterified with diazomethane. This ester was deprotected with boron trichloride, to afford **56** in 33% overall yield. In a much shorter sequence, D-glucofuranurono-6,3-lactone (**7**) was converted into D-glucofuranurono-6,3-lactone 1,2-ethanediyl dithioacetal (**58**) with 1,2-ethanedithiol and acid, and **58** was desulfurized with Raney nickel in ethanol, to afford **56** in 71% overall yield. Thioacetal **58** was transformed into 2,4-O-isopropylidene-D-glucofuranurono-6,3-lactone 1,2-ethanediyl dithioacetal (**59**) (77%) with 2,2-dimethoxypropane and acid, and then **59** was desulfurized, and the product treated with *tert*-butylchlorodimethylsilane to afford **60** (R = *tert*-BuMe$_2$Si) in 77% yield. Earlier, Zinner and Dässler[99] had prepared a number of derivatives of D-glucofuranurono-6,3-lactone (**7**) that were closely related to

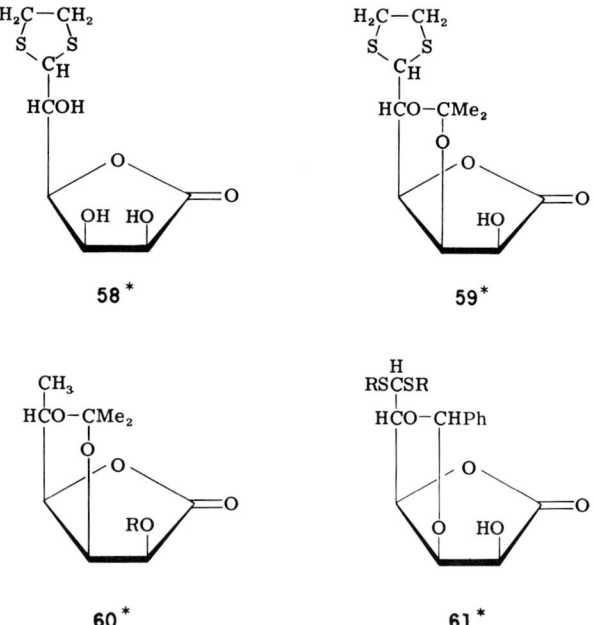

* These structures are drawn to show their relationship to L-gulono-1,4-lactone, rather than in the conventional fashion with C-6 at the left.

(98) R. E. Ireland and C. S. Wilcox, *J. Org. Chem.*, 45 (1980) 197–203, and personal communication.
(99) H. Zinner and C.-G. Dässler, *Chem. Ber.*, 93 (1960) 1597–1608.

59 by treating a dialkyl dithioacetal of **7** with benzaldehyde, to afford the 2,4-*O*-benzylidene-D-glucofuranurono-6,3-lactone dialkyl dithioacetal (**61**). These derivatives could clearly also be used for preparing 6-deoxy-L-gulono-1,4-lactone by desulfurization.

3,6-Anhydro-L-gulono-1,4-lactone (**62**) was prepared[100] in 62% yield by the platinum-catalyzed oxidation of 1,4-anhydro-D-glucitol. Methyl 3,4,5-tri-*O*-acetyl-2,6-anhydro-L-gulonate (**63**, R = H) was obtained from D-glucuronic acid by way of the tetraacetate (**63**, R = OAc) and the thioglycoside[101,102] (**63**, R = SPh) (76%), followed by reduction in the presence of Raney nickel to afford **63** (R= H) (68%).

When 1,5-anhydro-1-phenyl-D-glucitol (**64**, R = CH_2OH) was stirred in nitric acid, 2,6-anhydro-6-deoxy-6-phenyl-L-gulonic acid (**64**, R = CO_2H) was formed in 22% yield.[103] On treating 3,4,5-tri-*O*-acetyl-2,6-

(100) K. Heyns, E. Alpers, and J. Weyer, *Chem. Ber.*, 101 (1968) 4199–4208.
(101) R. J. Ferrier and R. H. Furneaux, *Carbohydr. Res.*, 52 (1976) 63–68; *J. Chem. Soc., Chem. Commun.*, (1977) 332–333; *J. Chem. Soc. Perkin Trans. 1*, (1978) 1996–2000.
(102) B. Helferich, D. Türk, and F. Stoeber, *Chem. Ber.*, 89 (1956) 2220–2224.
(103) Yu. A. Zhdanov, G. V. Bogdanova, and A. I. Chuvileva, *Dokl. Akad. Nauk SSSR*, 128 (1959) 953–955; *Chem. Abstr.*, 54 (1960) 8645e.

anhydro-D-gulononitrile (**65**, R = CN) with 2-aminobenzenethiol, or sodium azide and ammonium chloride, the corresponding benzothiazole (**65a**) or tetrazole (**65b**) was formed in ~50% yield.[104] Deacetylation of **65a** and **65b** afforded the corresponding, anhydro sugar derivatives.

VII. AMINO DERIVATIVES OF THE GULONO-1,4-LACTONES

A variety of amino derivatives of L-gulonic acid has been reported. Weidmann and Fauland[105] prepared 6-amino-6-deoxy-L-gulono-1,6-lactam (**66**) (and the tetraacetate derivative) by treating D-glucofuranurono-6,3-lactone (**7**) with hydroxylamine, to afford the corresponding oxime in 50–70% yield. This oxime was hydrogenated in the presence of palladium-on-carbon, to give 6-amino-6-deoxy-L-gulonic acid hydrochloride (51% yield). Stirring of the hydrochloride with a base in methanol resulted in the formation of lactam **66** in 23% yield. The reduction of 2-acetamido-2-deoxy-D-glucofuranurono-6,3-lactone[106] with sodium borohydride resulted in the formation of 5-acetamido-5-deoxy-L-gulono-1,4-lactone (**67**).

66   **67**

Several 2-amino-2-deoxy-D-gulononitrile derivatives (**68**) have been obtained by treating D-xylose with hydrogen cyanide and an amine. Among the amines that were used in preparing these derivatives were aniline,[107–109] 9-aminofluorene,[110] p-methylphenylamine,[109] and benzylamine.[109] In all cases, as expected, both the D- and L-*glycero* configurations at C-2 are formed. When a solution of **68** (R = Ph)

---

(104) I. Farkas, I. F. Szabó, and R. Bognár, *Carbohydr. Res.*, 56 (1977) 404–406.
(105) H. Weidmann and E. Fauland, *Ann.*, 679 (1964) 192–194.
(106) H. Weidmann, E. Fauland, R. Helbig, and H. K. Zimmerman, *Ann.*, 694 (1966) 183–189.
(107) R. Kuhn, D. Weiser, and H. Fischer, *Ann.*, 628 (1959) 207–239.
(108) R. Kuhn, W. Kirschenlohr, and W. Bister, *Angew. Chem.*, 69 (1957) 60–61.
(109) R. Kuhn and W. Bister, *Ann.*, 617 (1958) 92–108.
(110) R. Kuhn and J. C. Jochims, *Ann.*, 641 (1961) 143–152.

in a base was heated, the corresponding ammonium 2-anilino-2-deoxy-D-gulonate was formed.[107] On hydrogenation of **68** (R = Ph or p-MeC$_6$H$_4$) in hydrochloric acid in the presence of palladium oxide, 2-amino-2-deoxy-D-gulose hydrochloride was formed.[109] By conducting the initial reaction with ammonia, hydrogen cyanide, and D-[5-^{14}C]xylose, a mixture of ^{14}C-labelled 2-amino-2-deoxy-D-gulononitrile and -D-idononitrile was obtained, the *ido* epimer preponderating.[111]

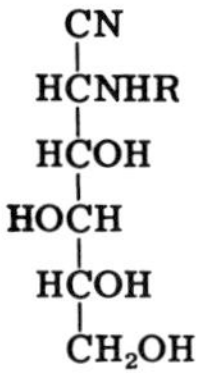

CN
|
HCNHR
|
HCOH
|
HOCH
|
HCOH
|
CH$_2$OH

68

On treatment of 2,3:4,5-di-*O*-isopropylidene-D-xylose with cyclohexyl isonitrile and ammonium acetate, 2-acetamido-*N*-cyclohexyl-2-deoxy-3,4:5,6-di-*O*-isopropylidene-D-gulonamide was formed[112]; it was reported that only one isomer was produced in this reaction.

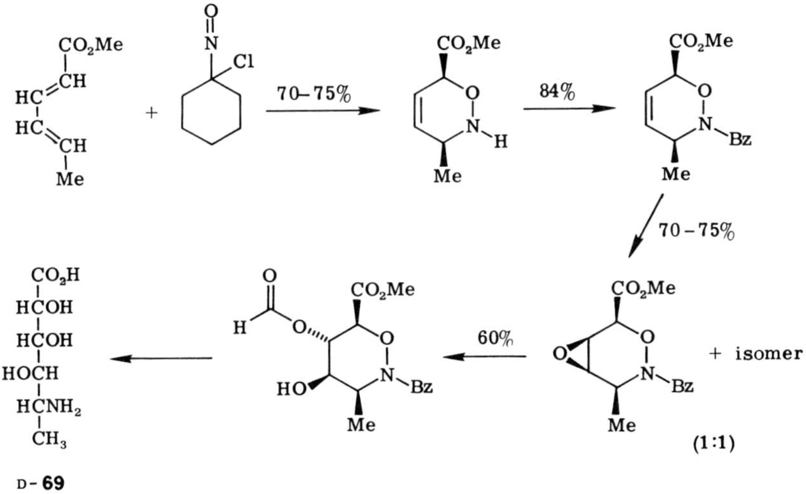

Scheme 7

(111) K. K. De and J. Rodia, *J. Carbohydr. Nucleos. Nucleot.*, 3 (1976) 359–363.
(112) M. F. Shostakovskii, K. F. Lavrova, N. N. Aseeva, and A. I. Polyakov, *Izv. Akad. Nauk SSSR., Ser. Khim.*, (1969) 1168–1169; *Chem. Abstr.*, 71 (1969) 70,856y.

By the route shown in Scheme 7, 5-amino-5,6-dideoxy-DL-gulonic acid (DL-**69**) was prepared.[113] When D-glucuronic acid reacted with L-glutamic acid under reductive conditions, 6-deoxy-6-[1,3-di(carboxypropyl)amino]-L-gulonic acid was formed.[114]

VIII. Miscellaneous Derivatives of the Gulono-1,4-lactones

A variety of miscellaneous derivatives of gulonolactone has been reported. Linn treated D-glucofuranurono-6,3-lactone (**7**) with chlorobenzene and aluminum trichloride in order to obtain[115] 6-deoxy-6,6-diphenyl-L-gulonic acid (**70**). Various related compounds were also prepared. When **7** was treated with sulfur dioxide and 2-(isonicotinoyl)hydrazine in water, 6-deoxy-6-(2-isonicotinoylhydrazino)-6-sulfo-D-gulono-1,4-lactone (**71**) was formed in 81% yield.[116,117] Kawabata

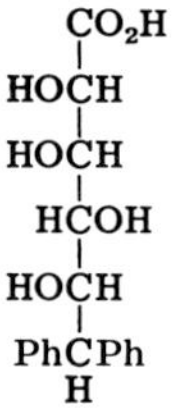

CO$_2$H
HOCH
HOCH
HCOH
HOCH
PhCPh
H

70

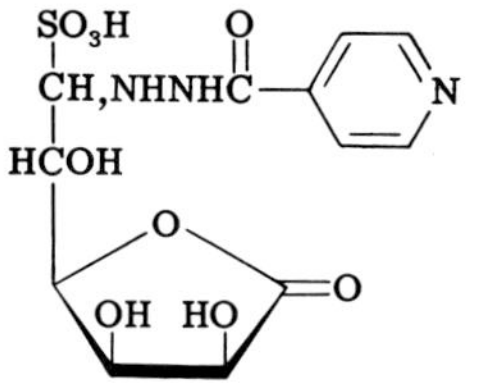

71

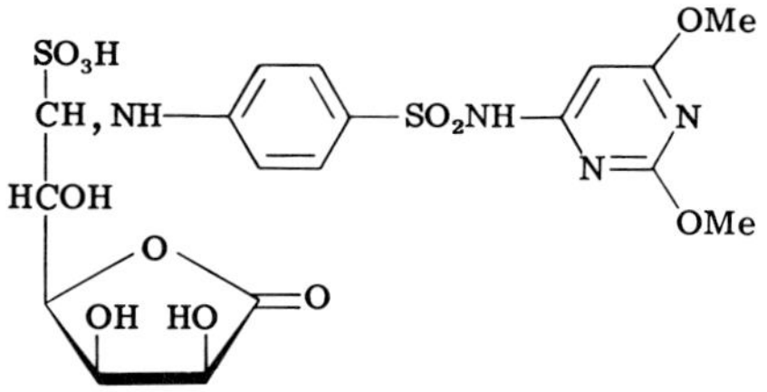

72

(113) B. Belleau and Y. Au-Young, *J. Am. Chem. Soc.*, 85 (1963) 64–71.
(114) A. Heins, H. Moeller, and R. Osberghaus, Ger. Offen. 2,364,525 (1975); *Chem. Abstr.*, 83 (1975) 114,927b; U. S. Pat. 4,032,676 (1977).
(115) C. B. Linn, Br. Pat. 788,200 (1957); *Chem. Abstr.*, 52 (1958) 6400b; U. S. Pat. 2,798,079 (1957); *Chem. Abstr.*, 52 (1958) 2072a; U. S. Pat. 2,938,911 (1960); *Chem. Abstr.*, 55 (1961) 468c.
(116) M. Kawada, Jpn. Pat. 10,674 (1957); *Chem. Abstr.*, 52 (1958) 17,293a.
(117) M. Kawada, Jpn. Pat. 8080 (1957); *Chem. Abstr.*, 52 (1958) 13,805a.

and coworkers[118] prepared 6-deoxy-6{p-[(2,6-dimethoxypyrimidin-4-yl)sulfamoyl]anilino}-6-sulfo-L-gulonic acid (**72**) from **7**.

Levene[119] isolated a complex derivative of L-gulonic acid ("methyl N-acetylchondrosaminidogulonate") during determination of the structure of chondrosine. Later, Wolfrom and coworkers,[120] while also attempting to elucidate the structure of chondrosine, isolated, as a degradation product, 4-O-(2-acetamido-2-deoxy-D-galactopyranosyl)-L-gulonamide.

When 2-4-O-ethylidene-D-glucofuranurono-6,3-lactone [obtained by the hydrolysis of thioacetal **61** (Ph = Et)] was treated with 5,5-dimethyl-1,3-cyclohexanedione (dimedone), 6-deoxy-6,6-bis(4,4-dimethyl-2,6-dioxocyclohexyl)-3,5-O-ethylidene-L-gulono-1,4-lactone (**73**) was formed in 71% yield.[99]

Methyl 2-O-methyl-L-gulonate has been erroneously reported.[121] Treatment of 2,3,5-tri-O-(p-nitrobenzoyl)-D-xylofuranosyl bromide with mercuric cyanide gave 2,5-anhydro-3,4,6-tri-O-(p-nitrobenzoyl)-D-gulononitrile (**74**) in 22% yield.[122]

(118) H. Kawabata, M. Kuranari, and M. Tanino, Jpn. Pat. 2633 (1962); *Chem. Abstr.*, 58 (1963) 7952b.
(119) P. A. Levene, *J. Biol. Chem.*, 140 (1941) 267–277.
(120) M. L. Wolfrom, R. K. Madison, and M. J. Cron, *J. Am. Chem. Soc.*, 74 (1952) 1491–1494.
(121) In the 9th Collective Index (1972–1976) of *Chemical Abstracts*, methyl 2-O-methyl-L-gulonate is reported (Registry number 61275-33-0). However, neither the original reference [P. Kovać, R. Brezny, V. Mihalov, and R. Palovcik, *J. Carbohydr. Nucleos. Nucleot.*, 2 (1975) 445–458] nor the abstract [*Chem. Abstr.*, 85 (1975) 160,444f] gives reference to methyl 2-O-methyl-L-gulonate. Through the kind help of J. T. Dickman of Chemical Abstracts, it was determined that an error in abstracting was made and the compound in question was not reported in the article; registry number 61275-33-0 is to be reassigned.
(122) H. S. El Khadem, T. D. Audichya, D. A. Niemeyer, and J. Kloss, *Carbohydr. Res.*, 47 (1976) 233–240.

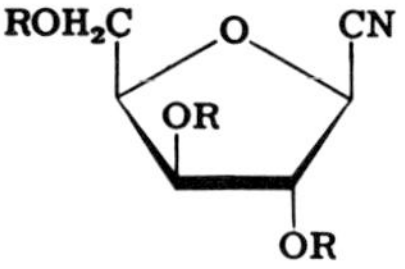

$R = COC_6H_4NO_2\text{-}p$

74

Polymerization of vinylene carbonate in the presence of benzoyl peroxide and carbon tetrachloride gave[123] a mixture of products that included compound **75** (R = CCl_3; low yield). Irradiation of **75** in oxolane (tetrahydrofuran) with the light from a high-pressure, mercury lamp afforded the monoreduced compound **75** (R = $CHCl_2$) in 80% yield. Compound **75** (R = $CHCl_2$) has been converted into DL-xylose by reduction with sodium borohydride, followed by hydrolysis with silver nitrate. When **75** (R = $CHCl_2$) was treated with sodium cyanide in dry N,N-dimethylformamide, a mixture of products was obtained, from which 2,3:4,5-di-O-carbonyl-6,6-dichloro-6-deoxy-DL-gulononitrile (**76**, R = CN) was isolated in 31% yield. Hydrolysis of **76** (R = CN) in anhydrous, methanolic HCl afforded[123] a mixture of the methyl ester (**76**, R = CO_2Me; 36%) and the amide (**76**, R = $CONH_2$; 27%).

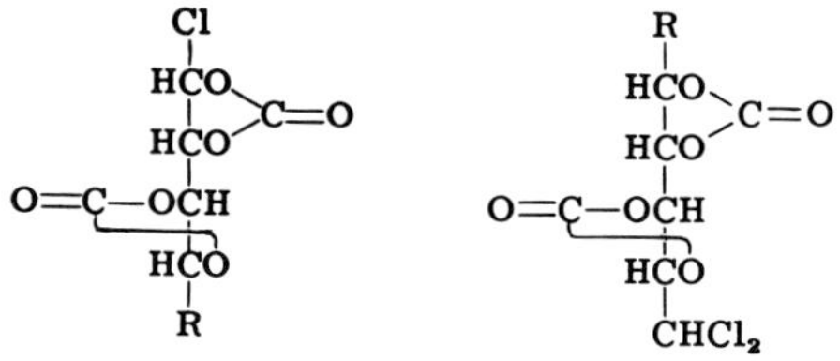

75   **76**

The branched-chain sugar 2-deoxy-3,5:4,6-di-O-ethylidene-2-C-methyl-L-gulononitrile (**77**) has been prepared by oxidizing **78** (R = $CHOHCH_2OH$) with sodium periodate[124] to 2,4:3,5-di-O-ethylidene-L-xylose (**78**, R = CHO) which, on treatment with nitromethane, gave[125] a mixture of 1-deoxy-3,5:4,6-di-O-ethylidene-1-C-nitro-L-iditol and -gulitol (**78**, R = $CHOHCH_2NO_2$) in 70% yield. In the presence of sodium hydrogencarbonate in benzene, **78** was converted into 1,2-dideoxy-3,5:4,6-di-O-ethylidene-1-C-nitro-L-*xylo*-hex-1-enitol (**79**) in 80% yield.[125] When **79** was treated with hydrogen cyanide and trieth-

(123) H. Takahata, T. Kunieda, and T. Takizawa, *Chem. Pharm. Bull.*, 23 (1975) 3017–3026.
(124) R. C. Hockett and F. C. Schaefer, *J. Am. Chem. Soc.*, 69 (1947) 849–851.
(125) H. Paulsen and W. Greve, *Chem. Ber.*, 106 (1973) 2114–2123.

ylamine, 2-cyano-1,2-dideoxy-3,5:4,6-di-*O*-ethylidene-L-*xylo*-hex-1-enitol (**80**, 76%) was formed; on reduction with sodium borohydride, **80** afforded an approximately equal mixture of **77** and the *ido* isomer.[126]

```
        CN                              R
        |                               |
      MeCH                             OCH
        |                               |
       OCH                      EtHC  HCO
       /   \                        \     \
    EtHC   HCO                     OCH   CHEt
       \   /   \                    |   /
       OCH   CHEt                  H₂CO
        |   /
       H₂CO

        77                            78


        NO₂                         H₂C    CN
        |                              \  /
        HC                              C
          \CH                           |
           |                           OCH
          OCH                           |
         /   \                   EtHC  HCO
      EtHC   HCO                     \     \
         \   /   \                  OCH   CHEt
         OCH   CHEt                  |   /
          |   /                     H₂CO
         H₂CO

        79                            80
```

Finally, when L-sorbose (**81**) was treated with hydrogen cyanide, a branched-chain, sugar lactone was formed which was characterized by converting it into a diacetal.[127] An X-ray structure determination of this material revealed it to be 2,2¹:5,6-di-*O*-isopropylidene-[2-*C*-(hydroxymethyl)-L-gulono-1,4-lactone] (**82**). However, all subsequent efforts to prepare **82** resulted in the formation of 2,3:5,6-di-*O*-isopropylidene-2-*C*-(hydroxymethyl)-L-gulono-1,4-lactone (**83**).

```
       CH₂OH
        |
        C=O
        |
       HOCH
        |
       HCOH
        |
       HOCH
        |
       CH₂OH

        81
```

(126) H. Paulsen and W. Greve, *Chem. Ber.*, 107 (1974) 3013–3019.
(127) R. A. Anderson, F. W. B. Einstein, R. Hoge, and K. N. Slessor, *Acta Crystallogr.*, Sect. B, 33 (1977) 2780–2783.

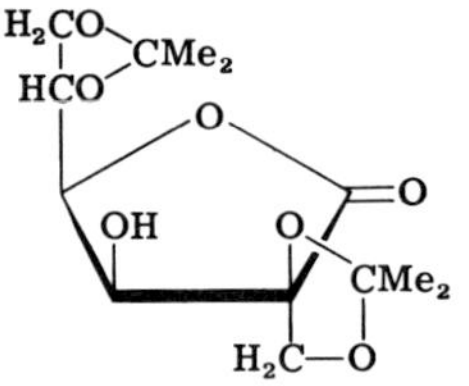

82

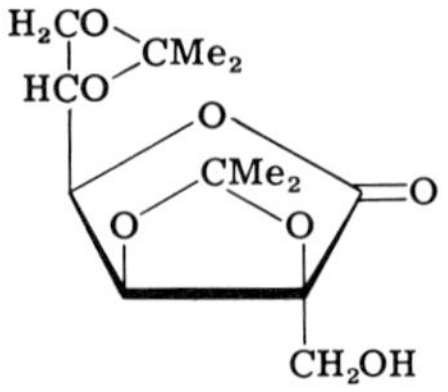

83

IX. Oxidation of the Gulono-1,4-lactones and Derivatives

The most important oxidation product of L-gulono-1,4-lactone (**1**) is, without a doubt, L-ascorbic acid (**6**; vitamin C), and the most important oxidation product of L-gulonic acid (**3**) is L-*xylo*-2-hexulosonic acid (**5**), which serves as a key intermediate in the commercial production of L-ascorbic acid. The literature covering the methods by which **1** or **3** (or derivatives thereof) has been converted into **6** or **5**, as well as other methods for the preparation of **6**, has been reviewed,[1] and will not be discussed here.

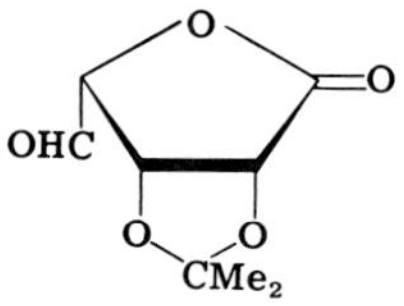

84

When 2,3-*O*-isopropylidene-D-gulono-1,4-lactone (**52**) was treated with sodium periodate, 3,4-*O*-isopropylidene-L-lyxurono-1,4-lactone (**84**) was formed[26]; this was converted in several steps into D-arabinose. Benzyl 2,4,5,6-tetra-*O*-benzyl-L-gulonate was oxidized to benzyl 2,4,5,6-tetra-*O*-benzyl-L-*xylo*-3-hexulosonate (**85**) which, on debenzylation, underwent[81] decarboxylation, and formation of L-*threo*-2-pentulose (**86**). Heyns and coworkers[128] oxidized 3,6-anhydro-L-gulono-1,4-lactone (**62**) with air in the presence of platinum oxide, to afford 3,6-anhydro-L-*xylo*-2-hexulosono-1,4-lactone in 75% yield. On oxidation of D-gulonic acid with nitric acid and sodium metavanadate, L-threaric (*dextro*-tartaric) acid (13.6%), oxalic acid, and D-glucaric acid are formed.[129]

(128) K. Heyns, E. Alpers, and J. Weyer, *Chem. Ber.*, 101 (1968) 4209–4213.
(129) J. K. Dale and W. F. Rice, *J. Am. Chem. Soc.*, 55 (1933) 4984–4985.

```
        CO₂Bzl                    CH₂OH
BzlOCH                      HCOH
    C=O                         C=O
 HCOBzl                      HCOH
BzlOCH                       HOCH
    CH₂OBzl                  CH₂OH
       85                       86
```

X. Reduction of the Gulono-1,4-lactones and Derivatives

The gulonolactones or their derivatives have been reduced to L-gulitol, or derivatives of D- or L-gulitol. Bragg and Hough[130] reported that the reduction of gulonolactone with potassium borohydride was incomplete, primarily because of the formation of potassium gulonate under the conditions used. However, Frush and Isbell[131] had found that D-gulono-1,4-lactone (2) is reduced to D-gulitol (L-glucitol, D-**10**) in 95% yield with sodium borohydride when the reduction is conducted in aqueous acid or in anhydrous methanol. Sixteen years later, similar results were reported by Taylor and Conrad.[132] Hudson and coworkers[89] reduced 2,3,5,6-tetra-*O*-acetyl-D-gulono-1,4-lactone to D-gulitol in 47% yield with lithium aluminum hydride. When 2,3-*O*-isopropylidene-L-gulono-1,4-lactone[85] (L-**52**) was reduced with an excess of sodium borohydride, 2,3-*O*-isopropylidene-L-gulitol (**87**) was formed. In a similar reaction, 3,5-*O*-benzylidene-L-gulono-1,4-lactone (**53**, R = Ph) was reduced with sodium borohydride to 3,5-*O*-benzylidene-L-gulitol (**88**).

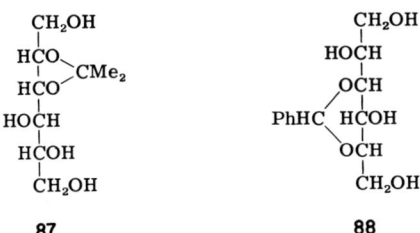

```
    CH₂OH               CH₂OH
    HCO                 HOCH
        CMe₂              OCH
    HCO                      CHPh
    HOCH          PhHC  HCOH
    HCOH                 OCH
    CH₂OH               CH₂OH
      87                   88
```

A variety of methods has been developed for the reduction of the gulonolactones and derivatives to D- and L-gulose and the corresponding derivatives. D-Gulono-1,4-lactone (2) was reduced to D-gulose (D-

(130) P. D. Bragg and L. Hough, *J. Chem. Soc.*, (1957) 4347–4352.
(131) H. L. Frush and H. S. Isbell, *J. Am. Chem. Soc.*, 78 (1956) 2844–2846.
(132) R. L. Taylor and H. E. Conrad, *Biochemistry*, 11 (1972) 1383–1388.

9) with sodium amalgam in ~50–60% yield.[133,134] The same reduction was accomplished in ~75% yield[135,136] with sodium borohydride at pH 3–4 (to prevent the formation of sodium D-gulonate). By using pyridine–oxolane as the solvent, lithium aluminum hydride was successfully used in the preparation of D-gulose (59%) from D-gulono-1,4-lactone,[137] even in the presence of an excess of the hydride.

Reduction of the tetra-(tetrahydropyran-2-yl) ether of D-gulono-1,4-lactone with lithium aluminum hydride–aluminum chloride afforded a 3:1 mixture of D-gulose and D-gulitol derivatives.[138] When 2,3:5,6-di-O-isopropylidene-D-gulono-1,4-lactone (D-**50**) was reduced with sodium borohydride in ether to which acetic acid was slowly added, 2,3:5,6-di-O-isopropylidene-D-gulose (**89**) was formed in 70% yield.[83,139] By the same procedure, 2,3-O-isopropylidene-D-gulose was prepared[139] from 2,3-O-isopropylidene-D-gulono-1,4-lactone (**52**).

89

Kohn and coworkers[90,91,140] found that 2,3,5,6-tetra-O-benzoyl-D-gulono-1,4-lactone (**55**, R = Bz) is reduced to 2,3,5,6-tetra-O-benzoyl-D-gulofuranose (88–97%) by bis(3-methyl-2-butyl)borane in oxolane. Finally, Ireland and coworkers prepared[98] 6-deoxy-L-gulose by two methods; in the first, 6-deoxy-L-gulono-1,4-lactone (**56**) was reduced with sodium amalgam to 6-deoxy-L-gulose in 40% yield (isolated as the methyl pyranoside). A higher yield was obtained when **60** (R = tert-BuMe₂Si) was reduced with diisobutylaluminum hydride in oxo-

---

(133) N. Sperber, H. E. Zaugg, and W. M. Sandstrom, *J. Am. Chem. Soc.*, 69 (1947) 915–920.
(134) H. S. Isbell, *Methods Carbohydr. Chem.*, 1 (1962) 135–136.
(135) L. M. Lerner, *Carbohydr. Res.*, 44 (1975) 116–120; *J. Org. Chem.*, 40 (1975) 2400–2402.
(136) M. L. Wolfrom and A. Thompson, *Methods Carbohydr. Chem.*, 2 (1963) 65–68.
(137) J. Némec and J. Jarý, *Collect. Czech. Chem. Commun.*, 34 (1969) 1611–1614.
(138) S. S. Bhattacharjee, J. A. Schwarcz, and A. S. Perlin, *Carbohydr. Res.*, 42 (1975) 259–266.
(139) R. K. Hulyalkar, *Can. J. Chem.*, 44 (1966) 1594–1596.
(140) P. Kohn, L. M. Lerner, A. Chan, S. D. Ginocchio, and C. A. Zitrin, *Carbohydr. Res.*, 7 (1968) 21–26.

lane, followed by acid hydrolysis, to provide 6-deoxy-L-gulose in 96% yield.

XI. Nucleophilic Additions to the Carbonyl Group of Derivatives of the Gulono-1,4-lactones

In 1971, Ogura and coworkers[82] demonstrated that, when lithium phenylacetylide reacts with lactone **90**, the mono-adduct **91** is formed, in contrast to the fact that, when magnesium phenylacetylide is used, the diadduct **92** is formed. This selectivity for the mono-addition of lithium acetylides to lactones was confirmed by Chabala and Vincent[141] for 4-hydroxybutanoic 1,4-lactone, 5-hydroxypentanoic 1,5-lactone, and 6-hydroxyhexanoic 1,6-lactone. Ogura and coworkers[82] utilized this selective mono-addition of lithium salts to lactones in preparing a variety of 1-(ethynyl)-2,3:5,6-di-O-isopropylidene-L-gulofuranose derivatives (**93**) from 2,3:5,6-di-O-isopropylidene-L-gulono-1,4-lactone (**50**). The following compounds (**93**) are representative of those prepared: R = —C≡CPh (50%), R = C≡C—Cl (45%), R = —C≡CCH(OEt)$_2$ (45%), and R = —C≡C—CH$_2$O-tetrahydropyran-2-yl (40%). Similar results were obtained for D-ribono-1,4-lactone derivatives. In subsequent publications, Ogura and Takahashi[142,143] utilized the selective mono-addition of lithiated heterocycles to **50** in obtaining a number of C-nucleoside analogs. Included in the com-

(141) J. C. Chabala and J. E. Vincent, *Tetrahedron Lett.*, (1978) 937–940.
(142) H. Ogura and H. Takahashi, *J. Org. Chem.*, 39 (1974) 1374–1379.
(143) H. Ogura and H. Takahashi, *Synth. Commun.*, 3 (1973) 135–143.

pounds prepared were **93**, where R = pyridin-2-yl (74%), 1-benzothiazol-2-yl (56%), or 1-benzylbenzimidazol-2-yl (40%).

XII. ANALYTICAL PROCEDURES

A number of methods has been used to assay for aldonic acids and aldonolactones. In this Section, these methods will be summarized. Comprehensive coverage of the literature related to analytical techniques has not been attempted in this article.

1. Paper Chromatography

Articles involving the paper chromatography of aldonic acids, aldono-1,4-lactones, glucuronic acids, and related compounds, under a variety of conditions, are given in References 144–151. Although this method of analysis has proved to be a very useful tool, it has been replaced to a major degree by other analytical procedures, primarily because of the length of time (4–8 h) needed to complete paper-chromatographic assays, although this method of assay is still valuable when large numbers of assays are to be performed.

2. Thin-Layer Chromatography

Smith and coworkers[152] published a relatively complete paper on the thin-layer chromatography, on silica gel, of carbohydrates of low molecular weight. Bancher and coworkers[153] reported the thin-layer chromatography of degradation products of carbohydrates, including aldonic acids and aldonolactones.

3. Gas–Liquid Chromatography and Mass Spectrometry

The preparation of volatile derivatives of carbohydrates has permitted the development of useful gas–liquid chromatographic analyses

---

(144) G. W. Oertel, *J. Chromatogr.*, 8 (1962) 486–490.
(145) R. J. Ferrier, *J. Chromatogr.*, 9 (1962) 251–252.
(146) D. V. Myhre and F. Smith, *J. Org. Chem.*, 23 (1958) 1229–1231.
(147) H. T. Gordon, W. Thornburg, and L. N. Werum, *Anal. Chem.*, 28 (1956) 849–855.
(148) L. Hough, *Methods Biochem. Anal.*, 1 (1954) 205–242.
(149) E. Puhakainen and O. Hanninen, *Acta Chem. Scand.*, 26 (1972) 3599–3604.
(150) G. Cerutti, A. Garzia, and I. Trabucchi, *Chimica (Milan)*, 36 (1960) 362–366.
(151) C. A. Knutson, *Carbohydr. Res.*, 43 (1975) 225–231.
(152) G. W. Hay, B. A. Lewis, and F. Smith, *J. Chromatogr.*, 11 (1963) 479–486.
(153) E. Bancher, H. Scherz, and K. Kaindl, *Mikrochim. Ichnoanal. Acta*, (1964) 1043–1051.

of this class of compound. Sweeley and coworkers[154] described the use of trimethylsilyl derivatives of sugars and related substances for analyses by g.l.c. In subsequent years, trimethylsilyl derivatives of aldonic acids, aldonolactones, and other sugar derivatives have been analyzed by g.l.c.[46,155–160] By using the trimethylsilyl derivatives, the aldonic acid, the 1,4-lactone, and the 1,5-lactone can, in most cases, be readily separated.

The volatility of the trimethylsilyl derivatives of the aldonolactones and related carbohydrates has made these derivatives suitable for use in mass spectrometry.[161,162] Petersson and coworkers[161,162] reported the mass spectra of a variety of trimethylsilyl derivatives of aldonolactones, including the spectrum of **54**.

Other derivatives of aldonolactones and alditols that have found some use in g.l.c. are trifluoroacetates[156] and methyl ethers.[163]

4. Ion-Exchange Chromatography

Aldonic acids and glucuronic acids have been separated by use of anion-exchange chromatography on Dowex-1 X-8, a strongly basic resin.[164–166]

5. High-Performance, Liquid Chromatography

The most promising method for rapid, direct analysis of aldonic acids and related compounds is high-performance, liquid chromatography. A number of different supports are likely to prove useful, in-

(154) C. C. Sweeley, R. Bentley, M. Makita, and W. W. Wells, *J. Am. Chem. Soc.*, 85 (1963) 2497–2507.
(155) G. Petersson, *Carbohydr. Res.*, 33 (1974) 47–61.
(156) M. Matsui, M. Okada, T. Imanari, and Z. Tamura, *Chem. Pharm. Bull.*, 16 (1968) 1383–1387.
(157) F. Loewus, *Carbohydr. Res.*, 3 (1966) 130–133.
(158) O. Raunhardt, H. W. H. Schmidt, and H. Neukom, *Helv. Chim. Acta*, 50 (1967) 1267–1274.
(159) G. Petersson, H. Riedl, and O. Samuelson, *Sven. Papperstidn.*, 70 (1967) 371–375; *Chem. Abstr.*, 68 (1968) 18,477z.
(160) J. Szafranek, C. D. Pfaffenberger, and E. C. Horning, *J. Chromatogr.*, 88 (1974) 149–156.
(161) G. Petersson, O. Samuelson, K. Anjou, and E. von Sydow, *Acta Chem. Scand.*, 21 (1967) 1251–1256.
(162) G. Petersson, *Arch. Mass Spectral Data*, 1 (1970) 584–623; *Chem. Abstr.*, 74 (1971) 142,260m.
(163) J. N. C. Whyte, *J. Chromatogr.*, 87 (1973) 163–168.
(164) O. Samuelson and L. Thede, *J. Chromatogr.*, 30 (1967) 556–565.
(165) B. Carlsson and O. Samuelson, *Anal. Chim. Acta*, 49 (1970) 247–254.
(166) K. Larsson, L. Olsson, and O. Samuelson, *Carbohydr. Res.*, 38 (1974) 1–11.

cluding anion and cation-exchange resins, reverse-phase supports, or silica gel. Although no descriptions have been published of work specifically related to the separation of aldonic acids or lactones, it is clear that this method of analysis will become quite useful in the future. It has been found in our laboratories that L-gulono-1,4-lactone (**1**), L-*xylo*-2-hexulosonic acid (**5**), and L-ascorbic acid (**6**) can be separated from each other on a column of Aminex A-25 anion-exchange resin by using M ammonium formate at pH 5.5 as the eluant.[167] In addition L-gulonic, L-idonic, D-mannonic, and D-gluconic acids are separated on a column of Dowex-50 X-8 (Ca^{2+}) resin by using an aqueous solution of calcium chloride as the eluant.[167]

XIII. BIOLOGICAL ROLE OF THE GULONO-1,4-LACTONES AND GULONIC ACIDS

No effort has been made here to cover the literature on the biological importance of L-gulonolactone and L-gulonic acid comprehensively. It has been found[168–175] that, in both plants and animals, L-gulonic acid (**3**) can be converted into L-gulono-1,4-lactone (**1**) *in vivo*, and subsequently oxidized enzymically to L-*xylo*-2-hexulosono-1,4-lactone (**94**) which, by tautomerization, affords L-ascorbic acid (**6**). As the enzyme needed for the conversion of **1** into **94** or **6** is absent from human beings, L-ascorbic acid (vitamin C) is a dietary requirement for them in order that good health may be maintained. In addition, it has been demonstrated that L-gulonic acid (**3**) may be enzymically oxidized at C-3 to L-*xylo*-3-hexulosonic acid (**95**), which is subsequently decarboxylated to afford[40,41] L-*threo*-2-pentulose (**86**). That **95** is, in fact, produced enzymically was additionally supported by the formation[40,41] of L-gulonic acid and L-galactonic acid on reduction of **95** with sodium borohydride.

(167) T. C. Crawford, C. V. Scanio, and R. Breitenbach (Pfizer, Inc.), unpublished results; G. C. Andrews, B. Bacon, T. C. Crawford, and R. Breitenbach, *Chem. Commun.*, (1979) 740–741; G. C. Andrews, U. S. Pat. 4,159,990 (1979).
(168) I. B. Chatterjee, *Methods Enzymol.*, 18 (1970) 28–34.
(169) O. Hanninen, R. Raunio, and J. Marniemi, *Carbohydr. Res.*, 16 (1971) 343–351.
(170) F. A. Loewus, *Phytochemistry*, 2 (1963) 109–128.
(171) F. Loewus and M. M. Baig, *Methods Enzymol.*, 18 (1970) 22–28.
(172) M. M. Baig, S. Kelly, and F. Loewus, *Plant Physiol.*, 46 (1970) 277–280.
(173) F. Loewus, *Annu. Rev. Plant Physiol.*, 22 (1971) 337–364.
(174) J. J. Burns, *Kirk-Othmer Encyclopedia of Chemical Technology*, 2nd edn., Vol. 2, Wiley, New York, 1963, pp. 747–762.
(175) C. G. King and J. J. Burns (Eds.), *Ann. N. Y. Acad. Sci.*, 258 (1975) 1–552; this is a series of papers by various workers on ascorbic acid, including its biosynthesis and metabolism.

XIV. Addendum

When D-gulono-1,4-lactone was treated with isopropenyl methyl ether and a catalytic amount of p-toluenesulfonic acid in N,N-dimethylformamide, 5,6-O-isopropylidene-D-gulono-1,4-lactone (**51**) was formed in 95% yield.[176] Previously, this compound was available in only low yield as a mixture of products. In addition, **51** was converted into 2,3-di-O-acetyl-5,6-O-isopropylidene-D-gulono-1,4-lactone by treatment with acetic anhydride.

Hall and Bischofberger[177] found that, when 2,3:5,6-di-O-isopropylidene-D-gulono-1,4-lactone was oxidized with ruthenium(VIII) oxide and an excess of sodium periodate, it gave 2,3:5,6-di-O-isopropylidene-D-*ribo*-4-hexulosono-1,4-[(R) or (S)]-lactol. Similar results were observed with 2,3:5,6-di-O-isopropylidene-D-mannono-1,4-lactone and 2,3:5,6-di-O-isopropylidene-D-allono-1,4-lactone. This oxidation presumably proceeds by way of lactone cleavage and oxidation of the free 4-hydroxyl group followed, on acidification by relactonization, and formation of the new lactol.

Ogura and coworkers[178,179] continued their study of the addition of lithiated heterocycles to 2,3:5,6-di-O-isopropylidene-L-gulono-1,4-lactone (**50**) and related compounds. In the case of the addition of lithiated 1,3-dithiane to **50**, it was shown[178] by X-ray crystal-structure analysis that 1-(1,3-dithian-2-yl)-2,3:5,6-di-O-isopropylidene-β-L-gulofuranose is R at the anomeric carbon atom. This demonstrates that the product of this reaction is, surprisingly, the more sterically hindered of the two products possible. This is the opposite of that predicted for addition of the lithiated dithiane from the less hindered side of **50** if no equilibration of the initial adduct is involved.

(176) C. Copeland and R. V. Stick, *Aust. J. Chem.*, 31 (1978) 1371–1374.
(177) R. H. Hall and K. Bischofberger, *Carbohydr. Res.*, 65 (1978) 139–143.
(178) H. Ogura, K. Furuhata, H. Takahashi, and Y. Iitaka, *Chem. Pharm. Bull.*, 26 (1978) 2782–2787.
(179) H. Ogura, K. Furuhata, and H. Takahashi, *Nucleic Acids Res., Spec. Publ.*, 3 (1977) 23–26; *Chem. Abstr.*, 89 (1978) 75,429t.

THE CHEMISTRY AND BIOLOGICAL SIGNIFICANCE OF 3-DEOXY-D-*manno*-2-OCTULOSONIC ACID (KDO)*

By Frank M. Unger

Sandoz Forschungsinstitut Wien, A-1235 Vienna, Austria

I. Introduction 324
II. Occurrence, Location, and Linkages of KDO Residues
    in Bacterial Polysaccharides 326
    1. The Periodate–Thiobarbituric Acid Assay 326
    2. Occurrence and Linkages of KDO in Lipopolysaccharides
       from Gram-Negative Bacteria 334
    3. KDO in Acidic Exopolysaccharides (K-Antigens) 356
III. Elucidation of the Structure of KDO 357
    1. The KDO Molecule 357
    2. Spectroscopic Analysis of KDO and of Its Natural Derivatives .. 359
IV. Chemical Synthesis and Monosaccharide Chemistry of KDO 365
    1. Synthesis of KDO by the Cornforth Reaction 365
    2. Synthesis of KDO by the Kuhn Reaction 369
    3. Cyanohydrin Synthesis of KDO 371
    4. Wittig Syntheses of KDO 371
    5. Monosaccharide Chemistry of KDO 373
V. Enzymes of KDO Metabolism 378
    1. D-Arabinose 5-Phosphate Isomerase (EC 5.3.1.13) 378
    2. 3-Deoxyoctulosonate 8-Phosphate Synthetase
       (KDO 8-Phosphate Synthetase; EC 4.1.2.16) 379
    3. 3-Deoxyoctulosonate 8-Phosphate Phosphatase
       (KDO 8-Phosphate Phosphatase) 380
    4. CTP:CMP-3-deoxyoctulosonate Cytidylyltransferase
       (CMP-KDO-synthetase; EC 2.7.7.38) 381
    5. CMP-3-deoxyoctulosonate:3-deoxyoctulosylono-lipid
       3-Deoxyoctulosylonotransferase (KDO-transferase) 384
    6. 3-Deoxyoctulosonate Aldolase (KDO Aldolase; EC 4.1.2.23) ... 386
VI. The Design of Inhibitors of KDO Metabolism 387

* The author thanks Professor Stephen Hanessian, at whose suggestion this article was written, for friendly and informative discussions on the chemical modification of KDO.

I. INTRODUCTION

3-Deoxy-D-*manno*-2-octulosonic acid (KDO, **1**) occurs as a ketosidic component in all lipopolysaccharides (LPS) of Gram-negative bacteria so far investigated, and it has also been identified in several acidic exopolysaccharides (K-antigens). The location of LPS and of K-antigens (capsular materials) at the cell surface of Gram-negative bacteria is shown diagrammatically in Fig. 1.

KDO appears to be unique to Gram-negative bacteria. In the LPS that have been studied, KDO residues are situated at the reducing ends of the polysaccharide domains, linking them, by ketosidic bonds, to the fatty-acid-substituted 2-amino-2-deoxy-D-glucosyl disaccharides referred to as lipid A. Fig. 2 is a block diagram indicating the location of KDO in the LPS from *Salmonella*.

The incorporation of KDO appears to be a vital step in LPS biosynthesis, and, indeed, in growth of the Gram-negative bacteria. Therefore, the immunochemistry, biochemistry, and synthetic carbohydrate chemistry involving KDO has become of increasing interest in pharmaceutical research. As the significant body of literature published

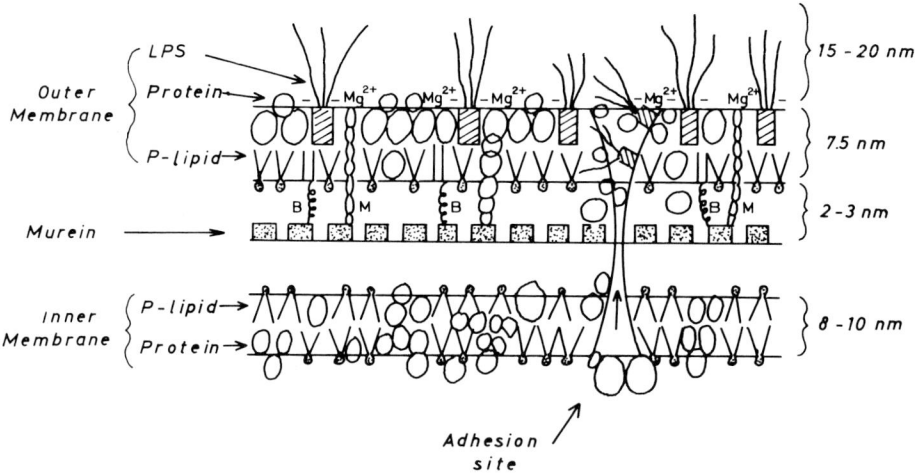

FIG. 1.—Schematic Representation of the Cell Envelope of Gram-Negative Bacteria. [The envelope is seen to consist of an inner (cytoplasmic) membrane, a murein (peptidoglycan) layer, and an outer membrane. The region between the inner and outer membranes is referred to as the periplasmic space. The outer membrane consists of phospholipid, protein, and lipopolysaccharide (LPS). The polysaccharide chains of LPS are seen to extend into the surrounding medium. Approximate figures are given for the thickness of each layer. An adhesion site, through which LPS is exported from the cytoplasm to the outer membrane, is also shown. Drawing by Paul Ray, with permission (compare, Refs. 1–8).]

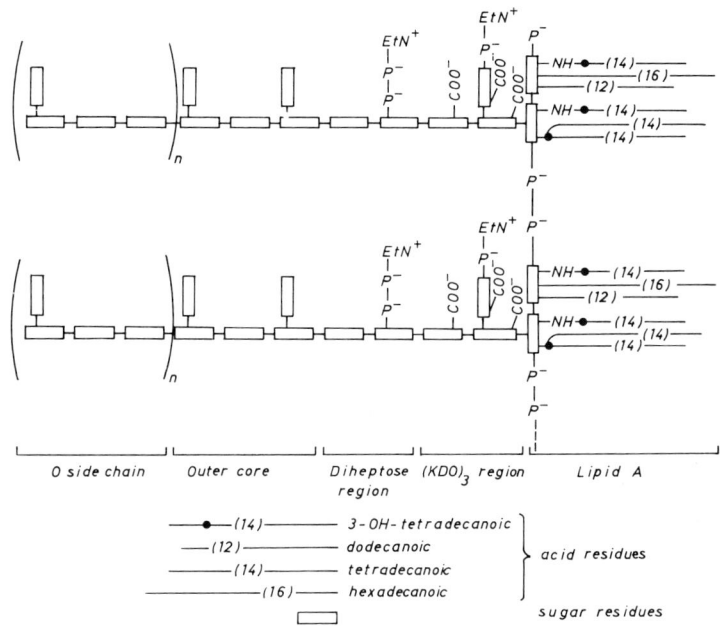

Fig. 2.—Block Diagram Representing the Constituents of the Gram-Negative-Bacterial LPS. [The inside of the bacterial cell is at the right, and the surroundings of the cell are at the left of the drawing. The diheptose–KDO region is sometimes referred to as the "inner core." P^- represents phosphate groups (from H. Nikaido, Ref. 6, with permission).]

on the subject is widely distributed among analytical, biochemical, carbohydrate, general organic-chemical, and microbiological journals, it seemed desirable to discuss within the present Chapter the more significant articles on KDO that have appeared between the time of its discovery (in 1959) and mid-1979.

A brief overview of the biosynthesis of KDO, and its incorporation into LPS, is found in a paper by Heath's group.[1] Some aspects of KDO are discussed in chapters on somatic and capsular antigens of Gram-negative bacteria by Lüderitz and coworkers.[2,3] The early work on KDO has been reviewed in a chapter on the carbohydrates of bacterial LPS by Ashwell and Hickman.[4] Selected papers concerning KDO

(1) E. C. Heath, R. M. Mayer, R. D. Edstrom, and C. Beaudreau, *Ann. N. Y. Acad. Sci.*, 133 (1966) 315–333.
(2) O. Lüderitz, A. M. Staub, and O. Westphal, *Bacteriol. Rev.*, 30 (1966) 192–255.
(3) O. Lüderitz, O. Westphal, A. M. Staub, and H. Nikaido, in G. Weinbaum, S. Kadis, and S. J. Ajl (Eds.), *Microbiol Toxins*, Vol. IV, Academic Press, New York, 1971, pp. 145–233.
(4) G. Ashwell and J. Hickman, in Ref. 3, pp. 262–263.

have been discussed in volumes of wider scope, such as those edited by Landy and Braun,[5] Leive,[6] Kass and Wolf,[7] and Sutherland.[8]

II. OCCURRENCE, LOCATION, AND LINKAGES OF KDO RESIDUES IN BACTERIAL POLYSACCHARIDES

1. The Periodate–Thiobarbituric Acid Assay

Most of the structural and biochemical work related to KDO is based on the estimation of the compound or its derivatives by the periodate–thiobarbituric acid (TBA) assay in its various modifications. Indeed, KDO (see Fig. 3) was discovered[9] through the formation of a characteristic, purple, TBA chromophore (λ_{max} 549 nm) from its 8-phosphate (**2**), which is the product of the condensation of D-arabinose 5-phosphate with enolpyruvate phosphate, catalyzed by 3-deoxy-8-O-phosphonooctulosonate synthetase (EC 4.1.2.16) (see Scheme 1 and Section V,2).

Originally, the periodate–TBA assay was recommended by Waravdekar and Saslaw[10] for the determination of 2-deoxy sugars. In this ap-

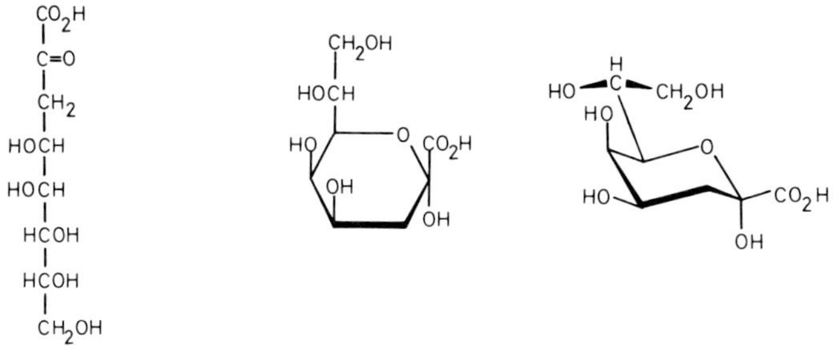

FIG. 3.—Formulas Representing KDO (**1**). [a, Fischer projection-formula (acyclic form); b, Haworth formula (ketopyranose); c, conformational drawing (ketopyranose).]

(5) M. Landy and W. Braun (Eds.), *Bacterial Endotoxins*, Rutgers University Press, New Brunswick, N.J., 1964.
(6) L. Leive (Ed.), *Bacterial Membranes and Walls*, Dekker, New York, 1973.
(7) E. H. Kass and S. M. Wolff (Eds.), *Bacterial Lipopolysaccharides. The Chemistry, Biology, and Clinical Significance of Endotoxins*, University of Chicago Press, Chicago, Ill., 1973.
(8) I. Sutherland (Ed.), *Surface Carbohydrates of the Prokaryotic Cell*, Academic Press, New York, 1977.
(9) D. H. Levin and E. Racker, *J. Biol. Chem.*, 234 (1959) 2532–2539.
(10) V. S. Waravdekar and L. D. Saslaw, *Biochim. Biophys. Acta*, 24 (1957) 439; *J. Biol. Chem.*, 234 (1959) 1945–1950.

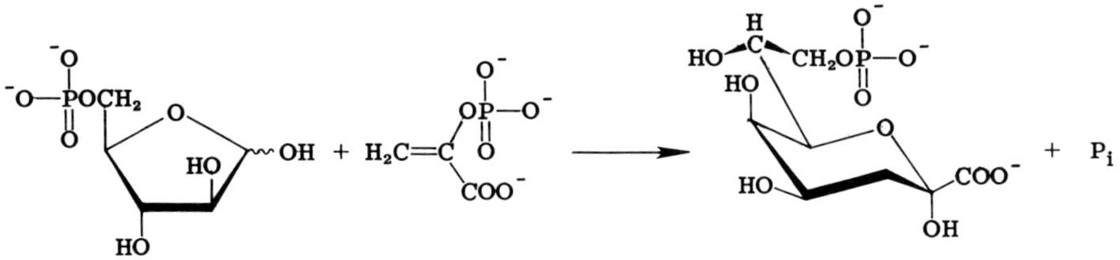

Scheme 1

plication, periodate cleavage of a diol grouping at C-3–C-4 liberates malonaldehyde, which reacts with two molecules of TBA to give chromophore **3** (λ_{max} 532 nm; see Scheme 2a).

Weissbach and Hurwitz[11] found that 3-deoxyglyc-2-ulosonic acids can undergo a similar reaction, whereby 3-formylpyruvic acid is formed through periodate cleavage of a diol grouping located at C-4–C-5. 3-Formylpyruvic acid may be considered to be a derivative of malonaldehyde, containing a carboxyl group in the place of one aldehyde hydrogen atom. Its reaction with two molecules of TBA (see Scheme 2b) yields the chromophore **4** (λ_{max} 549 nm). Kuhn and Lutz[12] synthesized the chromophores **3** and **4**, and reported their molar ex-

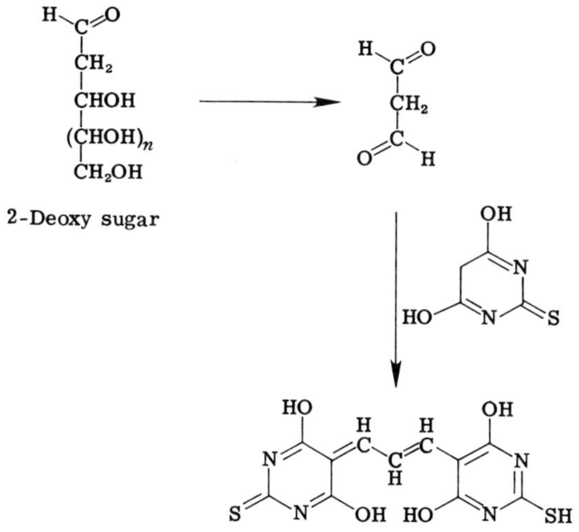

Scheme 2a

(11) A. Weissbach and J. Hurwitz, *J. Biol. Chem.*, 234 (1959) 705–709.
(12) R. Kuhn and P. Lutz, *Biochem. Z.*, 338 (1963) 554–560.

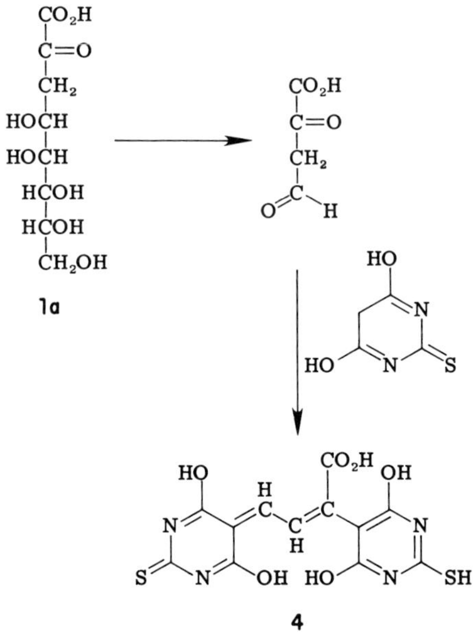

Scheme 2b

tinction coefficients. Table I contains absorption maxima and molar extinction coefficients of TBA chromophores determined under various conditions. The color due to chromophore **3** is stable to alkali, whereas that due to chromophore **4** is not.[11]

The TBA assay for 3-deoxyglyc-2-ulosonic acids, used by Weissbach and Hurwitz[11] to determine 3-deoxy-D-*arabino*-2-heptulosonic acid, was subsequently applied by Warren[13] and by Aminoff[14] to the analysis of N-acetylneuraminic acid (NeuAc, **5**; see Fig. 4). The formation of the 3-formylpyruvate-derived chromophore **4** in the periodate

TABLE I

Molar Extinction Coefficients of the TBA Chromophore (4) in Different Solvents

$\epsilon \times 10^{-4}$	Solvent	References
9.2 ±0.5	aqueous acid	96
8.1	neutral, aqueous solution	12
13.3	cyclohexanone	12
10.8	acidified 1-butanol	96

(13) L. Warren, *Fed. Proc. Fed. Am. Soc. Exp. Biol.*, 18 (1959) 347; *J. Biol. Chem.*, 234 (1959) 1971–1975.
(14) D. Aminoff, *Virology*, 7 (1959) 355–357; *Biochem. J.*, 81 (1961) 384–392.

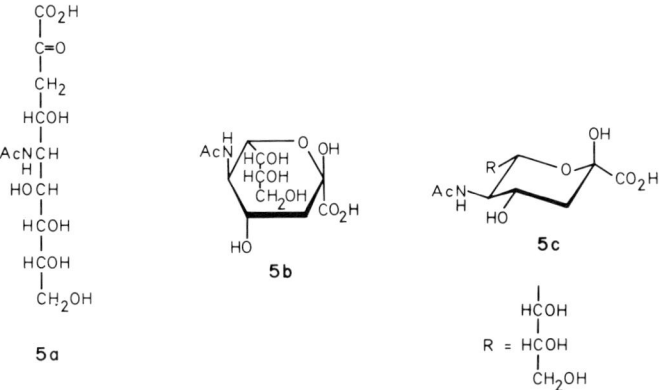

FIG. 4.—Formulas Representing NeuAc (**5**). [a, Fischer projection-formula (acyclic form); b, Haworth formula (ketopyranose); c, conformational drawing (ketopyranose).]

–TBA reaction of NeuAc (not to be expected from the structure of this compound) had initially been explained by a presumed deacetylation at N-5 under the strongly acidic conditions of the assay.[13] Later, Paerels and Schut[15] demonstrated that deacetylation of NeuAc is unlikely to occur under the conditions of the assay, and that free 3-formylpyruvic acid is not a product of the periodate oxidation of NeuAc. They presented evidence for the existence of an intermediate hexos-5-ulosuronic acid, **6** (see Scheme 3), which, they suggested, would give rise to chromophore **4** by condensation, at the ketone group, with one molecule of TBA, followed by a fragmentation reaction[16] with loss of the original C-5–C-6, and by another condensation with TBA through the aldehyde group presumably formed at the original C-4 in the fragmentation step (see Scheme 3).

Charon and Szabó[17] demonstrated that chromophore **4** may be produced to a significant extent from a 5-O-substituted derivative of KDO (which does not contain a free diol grouping at C-4–C-5). These authors synthesized 3-deoxy-5-O-methyl-2-octulosonic acid (**7**; configuration at C-5–C-7, *arabino;* at C-4, unknown) by the Cornforth reaction[18,19] from 4-O-formyl-2-O-methyl-D-arabinose (see Scheme 4 and Section IV,1). Compound **7** gave a millimolar extinction coefficient of 13 in the TBA assay (as compared to 92 ± 5 for KDO). Based on this result, Charon and Szabó[17] formulated for the TBA reaction of 5-O-

(15) G. B. Paerels and J. Schut, *Biochem. J.*, 96 (1965) 787–792.
(16) C. A. Grob and P. W. Schiess, *Angew. Chem. Int. Ed. Engl.*, 6 (1967) 1–15.
(17) D. Charon and L. Szabó, *Eur. J. Biochem.*, 29 (1972) 184–187.
(18) J. W. Cornforth, M. E. Firth, and A. Gottschalk, *Biochem. J.*, 68 (1958) 57–61.
(19) P. M. Carroll and J. W. Cornforth, *Biochim. Biophys. Acta*, 39 (1960) 161–162.

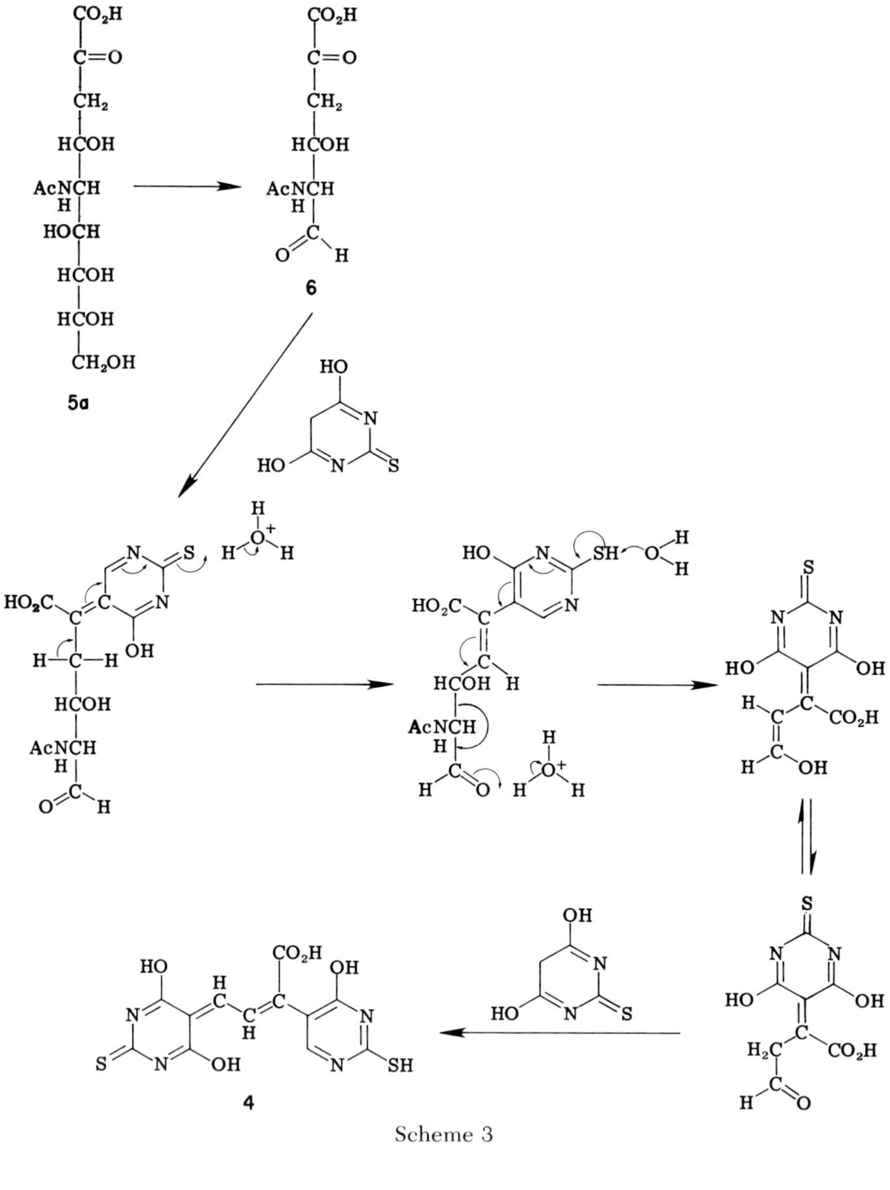

Scheme 3

substituted KDO derivatives a "fragmentation" mechanism analogous to that suggested by Paerels and Schut[15] for the TBA reaction of NeuAc (see Scheme 3). The 4-deoxyhexos-5-ulosuronic acid, **8**, is assumed to arise from periodate cleavage of compound **7**, and is presumably converted into **4** by a sequence of reactions similar to those sug-

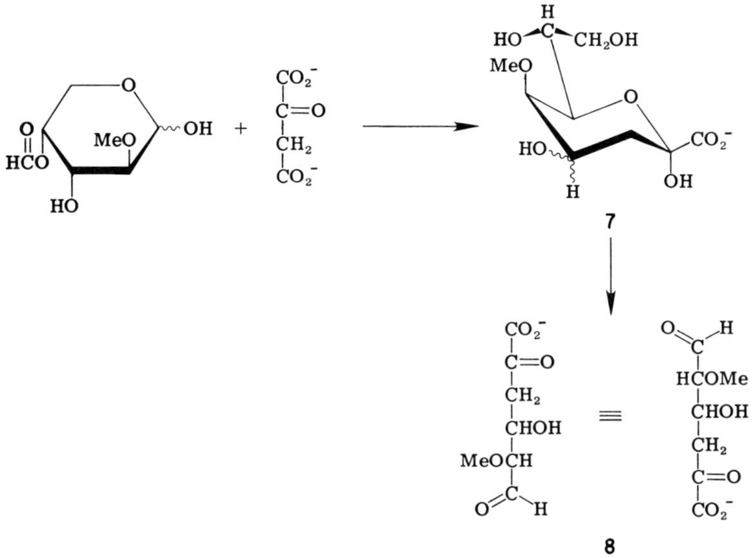

Scheme 4

gested for **6** in Scheme 3. As the molar extinction coefficient of the pure dye **4** is about seven times that obtained in the periodate–TBA reaction of **7**, it follows that over 85% of **7** is degraded by unknown processes that do not give rise to the chromophore. Similarly, ~20–40% of NeuAc may be assumed to escape the TBA assay under the conditions used by Aminoff,[14] Kuhn and Lutz,[12] and Preiss and Ashwell.[20]

Using 3-deoxy-4-*O*-methyl-D-*arabino*-2-heptulosonic acid (**9**) as a model compound, Charon and Szabó[21] showed that 4-*O*-substituted 3-deoxyglyc-2-ulosonic acids can also give rise to chromophore **4** in the TBA assay (ϵ_{mM} 3.7–5). The following two mechanisms were discussed by the authors to account for this observation (see Scheme 5). First, the furanose form of **9** could react with periodate to give **10**, which would presumably be converted into **4** by the "fragmentation" mechanism[15,17] (compare Scheme 3). Alternatively, periodate cleavage of the acyclic form of **9** would give rise to 3-deoxy-2-*O*-methyl-D-*glycero*-4-pentulosuronic acid, **11**, which, by (*a*) enolization, (*b*) acid-catalyzed, enol ether hydrolysis, and (*c*) further periodate cleavage, can presumably give 3-formylpyruvic acid. The color yield in this application of the TBA assay is only ~5%.

(20) J. Preiss and G. Ashwell, *J. Biol. Chem.*, 237 (1962) 317–321.
(21) D. Charon and L. Szabó, *Carbohydr. Res.*, 34 (1974) 271–277.

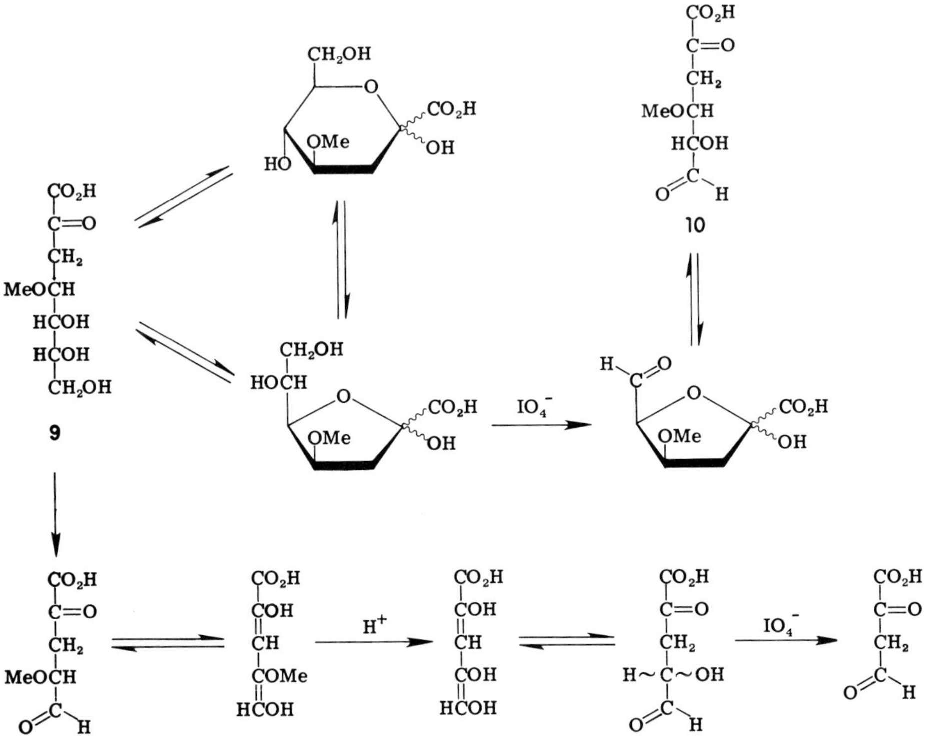

Scheme 5

Szabó's group[22] gave a detailed procedure for estimating KDO by the TBA method. They recommended conducting the periodate cleavage at 0° in order to avoid overoxidation of 3-formylpyruvate.[17] Practically quantitative color yields are obtained by this technique. It should be emphasized, however, that many different TBA procedures have been applied in the literature (compare Section I,2) for the qualitative and quantitative estimation of free and polysaccharide-bound KDO. As the TBA assay is unusually sensitive[17] to changes in such conditions as pH, temperature, or nature of the extracting solvent, the values obtained by different procedures are not comparable, unless standardized.[22] To appreciate more fully the results discussed in Section I,2, the following generalizations may be helpful.

(i) When the periodate–TBA reaction is applied to unhydrolyzed polysaccharide fractions under the strongly acidic conditions given by

(22) R. Chaby, D. Charon, R. S. Sarfati, L. Szabó, and F. Trigalo, *Methods Enzymol.*, 41B (1975) 32–34.

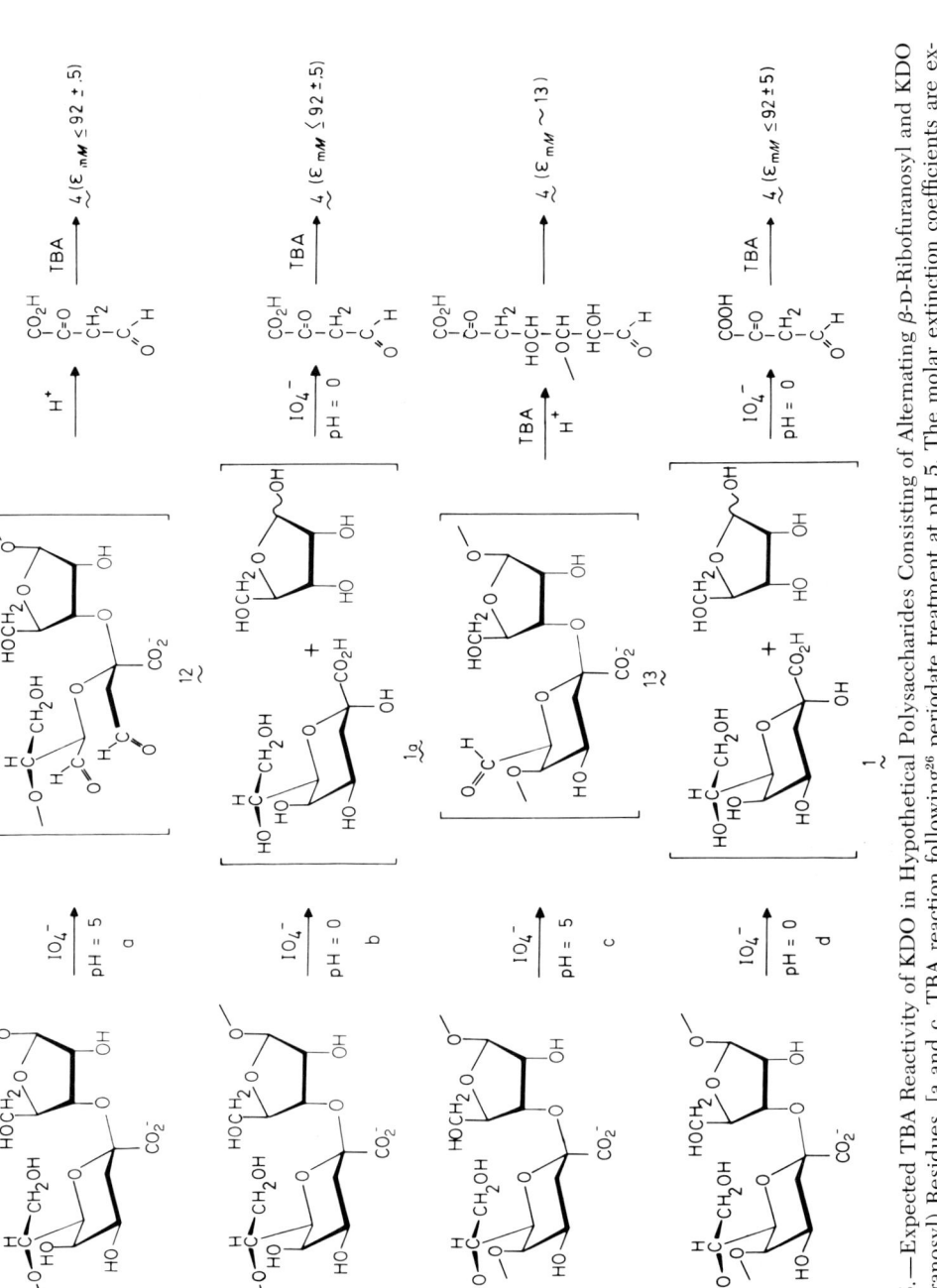

FIG. 5.—Expected TBA Reactivity of KDO in Hypothetical Polysaccharides Consisting of Alternating β-D-Ribofuranosyl and KDO (Ketopyranosyl) Residues. [a and c, TBA reaction following[26] periodate treatment at pH 5. The molar extinction coefficients are expected to depend on the linkage pattern at KDO. b and d, TBA reaction following periodate treatment under the strongly acidic conditions of Warren.[13] The molar extinction coefficients are expected to be independent of the linkage pattern at KDO.]

Warren,[13] the KDO moieties usually give a TBA response equivalent to that obtained after acid hydrolysis. Apparently, both the ketosidic and aglyconic linkages of KDO are cleaved during such treatment, so that KDO reacts essentially as the free 3-deoxyglyculosonic acid. Due to partial destruction of KDO by acid,[23-25] the color yield is less than optimal (see Fig. 5b, d).

(ii) An alternative procedure for the analysis of KDO moieties in whole polysaccharides employs periodate treatment at pH 5 (compare Method B of Dröge and coworkers[26]). When the polysaccharide contains KDO residues linked only through the side chain (with a free diol-grouping at C-4–C-5), an optimal TBA response is obtained. The intermediate dialdehyde **12** (see Fig. 5a) contains a 3-formylpyruvic acetal which is readily hydrolyzed[15] under the strongly acidic conditions that prevail during the reduction of the excess of periodic acid by arsenite, and during condensation with TBA. If OH-5 is substituted (see Fig. 5d) and the diol grouping of the side chain is free, the KDO moieties will give rise to a "weak" TBA response (ϵ_{mM} 13) by way of the intermediate **13**, which is analogous to **8** (see Scheme 4) and can undergo conversion into **4** by the "fragmentation" mechanism (see Scheme 3). When the periodate treatment is conducted at pH 5, KDO moieties in which OH-4 is substituted would not be expected to respond in the TBA assay,[21] as they cannot give rise to the 3-formylpyruvate-derived chromophore **4**, either by periodate cleavage at C-4–C-5, or by the "fragmentation" mechanism (see Scheme 6). The same holds true for KDO moieties that are substituted both in the side chain and at either C-4 or C-5.

2. Occurrence and Linkages of KDO in Lipopolysaccharides from Gram-Negative Bacteria

An early report that related to the presence of KDO in Gram-negative LPS was that of T. A. Sundarajan and coworkers[27] on the LPS from *Escherichia coli* K12. Among the products released from LPS by hy-

---

(23) C. S. Hershberger, M. Davis, and S. B. Binkley, *J. Biol. Chem.*, 243 (1968) 1585–1588.
(24) B. A. Dmitriev, L. V. Backinowsky, and N. K. Kochetkov, *Dokl. Akad. Nauk SSSR*, 193 (1970) 1304–1307; *Chem. Abstr.*, 74 (1971) 23,084x.
(25) D. Charon and L. Szabó, *J. Chem. Soc. Perkin Trans. 1*, (1973) 1175–1179.
(26) W. Dröge, V. Lehmann, O. Lüderitz, and O. Westphal, *Eur. J. Biochem.*, 14 (1970) 175–184.
(27) T. A. Sundarajan, A. M. C. Rapin, and H. M. Kalckar, *Proc. Natl. Acad. Sci. U. S. A.*, 48 (1962) 2187–2193.

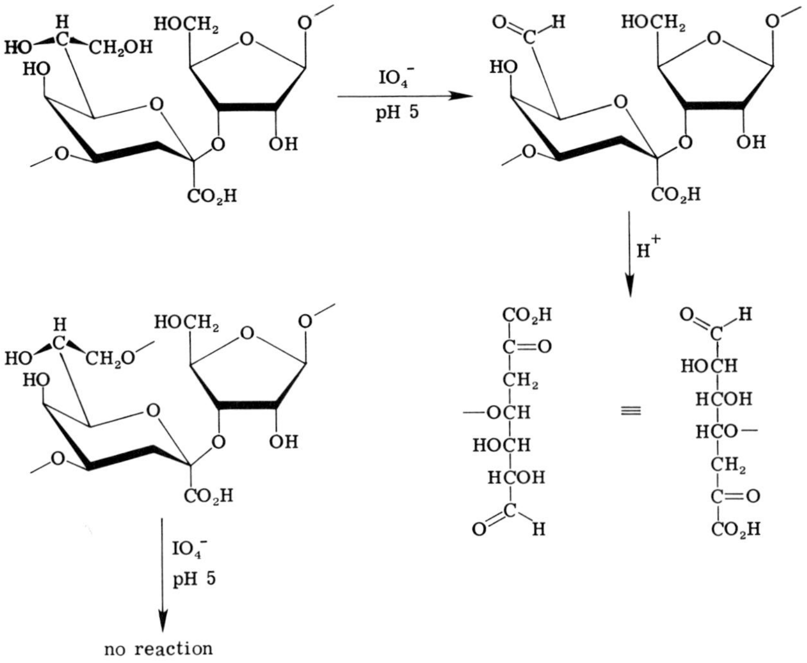

Scheme 6

drolysis with 0.05 M sulfuric acid for 45 min, the authors detected a compound that could be adsorbed on a column of Dowex-2 (acetate) ion-exchange resin and eluted with acetate buffer, pH 4.6. The substance gave a negative response in the orcinol, Ehrlich, and diphenylamine reactions,[28] but a positive response in the TBA reaction of Warren.[13] The compound occurred in the LPS from all of the several K-12 strains investigated. As KDO and NeuAc are the only components of Gram-negative, cell-surface structures to give a TBA chromophore having λ_{max} = 549 nm, and as NeuAc would have given positive orcinol and Ehrlich reactions, the unknown constituent described by Sundarajan and coworkers[27] is very likely[29] to be KDO.

The first, definitive identification of KDO as a constituent of the LPS from *Escherichia coli* 0111-B$_4$ (and from its UDP-galactose-4-epimerase-less mutant, J-5) was reported by Heath and Ghalambor.[29] When analyzing their LPS preparations for 3,6-dideoxy-L-*xylo*-hex-

(28) Z. Dische, *Methods Carbohydr. Chem.*, 1 (1962) 477–514.
(29) E. C. Heath and M. A. Ghalambor, *Biochem. Biophys. Res. Commun.*, 10 (1963) 340–345.

ose (colitose) by the TBA method of Cynkin and Ashwell,[30] Heath and his colleagues[29,31] detected an alkali-labile,[11] TBA chromophore of λ_{max} 549 nm, besides the alkali-stable, malonaldehyde-derived, TBA chromophore 3 (λ_{max} 532 nm). The authors isolated a 3-deoxyglyculosonic acid (later identified as KDO) from mild-acid hydrolyzates of the LPS, and demonstrated that this compound was identical to the one obtained by enzymic dephosphorylation of KDO 8-phosphate (**2**). Moreover, they showed that the formation of **2** from D-arabinose 5-phosphate and enolpyruvate phosphate, as catalyzed by crude extracts from *Pseudomonas aeruginosa*[9] (see Scheme 1), was also catalyzed by crude extracts of *E. coli* 0111-B_4. The contributions by Heath's group to the structural elucidation of KDO will be discussed in Section III,1.

Specific evidence concerning the location of KDO in LPS from a UDP-galactose-4-epimerase-less mutant* of *Salmonella typhimurium* was presented by Osborn.[33] She subjected the insoluble LPS to mild, acid-catalyzed hydrolysis (pH 3.4, at 100°), whereby polysaccharide and KDO were quantitatively released into the soluble fraction. Chromatography of this fraction on DEAE-cellulose, using gradient elution, yielded 75% of the total KDO as free KDO, and the remaining 25%, distributed over several peaks, as polysaccharide-bound KDO. These polysaccharide fractions reacted in the TBA assay without prior, acid-catalyzed hydrolysis, and no increase in their TBA reactivity was observed following acid-catalyzed hydrolysis. Treatment of the fractions with sodium borohydride abolished their reactivity in the TBA assay. These observations, together with the low reducing power of the polysaccharide fractions (which was only ~10% of that expected for glucose, by Nelson's procedure[34]) were taken to indicate that KDO residues constitute the reducing ends of the polysaccharide fractions. Because borohydride reduction prior to acid hydrolysis did not alter the TBA response of the whole LPS, Osborn[33] suggested that, in whole LPS, the polysaccharide portions might be linked to lipid A by ketosidic linkages of the reducing-terminal[35] KDO (see Fig. 6).

Sugimoto and Okazaki[36] found that mild, acid hydrolysis of LPS

* Such mutants synthesize a LPS that consists only of the lipid A and "core" structures, and lacks the O-specific side-chains[32] (compare Fig. 2).

(30) M. A. Cynkin and G. Ashwell, *Nature*, 186 (1960) 155–156.
(31) M. A. Ghalambor, E. M. Levine, and E. C. Heath, *J. Biol. Chem.*, 241 (1966) 3207–3215.
(32) H. Nikaido, *Biochim. Biophys. Acta*, 48 (1961) 460–469.
(33) M. J. Osborn, *Proc. Natl. Acad. Sci. U. S. A.*, 50 (1963) 499–506.
(34) N. Nelson, *J. Biol. Chem.*, 153 (1944) 375–380.
(35) B. L. Horecker, in Ref. 5, pp. 72–75.
(36) K. Sugimoto and R. Okazaki, *J. Biochem. (Tokyo)*, 62 (1967) 373–383.

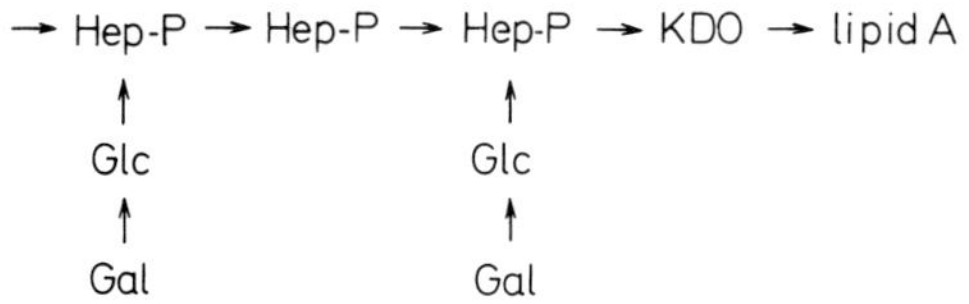

```
→ Hep-P → Hep-P → Hep-P → KDO → lipid A
      ↑              ↑
      Glc            Glc
      ↑              ↑
      Gal            Gal
```

FIG. 6.—Suggested Structure of the Core Region of LPS, Following the Investigation by Osborn.[33]

from *Escherichia coli* K-12 or JE 1538 (a smooth strain having the O-100 antigen) released an L-rhamnosyl-(1 → 5)-KDO (**14**; see Scheme 7). The disaccharide showed low reducing power (compare, the KDO-containing polysaccharide fractions observed by Osborn[33]) and gave a "weak" response in the TBA assay,[17] indicating that a (nonreducing) rhamnosyl group is linked to O-5 of a reducing, KDO residue. Hydrolysis of the compound in 0.1 M hydrochloric acid at 100° resulted in concomitant increases in reducing power and periodate–TBA reactivity. Following hydrolysis, both rhamnose and KDO were identified by paper chromatography in several solvent systems, using authentic standards. Treatment of the disaccharide with sodium borohydride abolished its reponse in the periodate–TBA reaction,[11] but did not alter its reactivity in the cysteine–sulfuric acid reaction for 6-deoxyhexoses.[37] To distinguish between substitution at C-4 or C-5 of KDO, periodate consumption by rhamnosyl-KDO was measured. One mole of the compound was found to consume four moles of periodate, with formation of one mole of formaldehyde. As the rhamnosyl group consumes two molecules of periodate, two molecules are used up by the KDO residue, corresponding to cleavage at C-8–C-7 and C-7–C-6. This was interpreted to indicate substitution at O-5 of KDO (see Scheme 7).

Following the identification of KDO in *Escherichia coli*,[29,31] *Pseudomonas aeruginosa*,[9] and *Salmonella typhimurium*,[33] several studies were made in order to determine the possible presence of KDO in

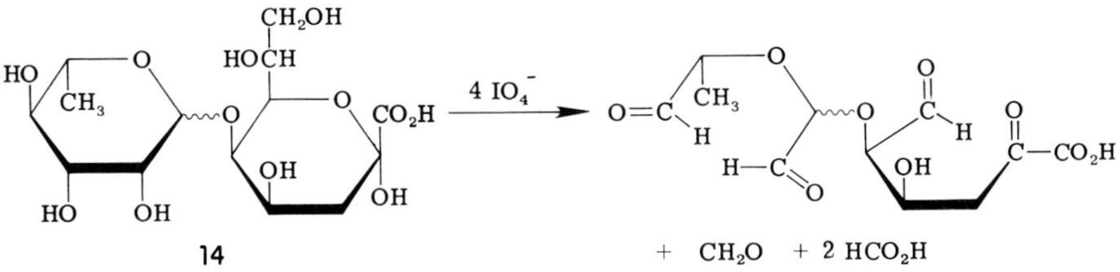

Scheme 7

(37) Z. Dische and L. B. Shettles, *J. Biol. Chem.*, 192 (1951) 579–582.

micro-organisms. Ellwood[38,39] isolated KDO from cell-wall preparations of Gram-negative bacteria, and characterized the compound in two paper-chromatographic systems that distinguish it from NeuAc. He was thus able to demonstrate the presence of KDO in the cell walls of almost all Gram-negative, bacterial species examined, including *Pasteurella*.[40] The compound was not detectable in Gram-positive organisms. NeuAc was present, together with KDO, in preparations from some Gram-negative strains, a finding that was in keeping with earlier reports showing the presence of NeuAc in *E. coli*[41,42] and[43] *Neisseria meningitidis* group C. Using a less specific methodology, Vincent and Cameron[44] found that Gram-negative, bacterial cells contain about eight times as much TBA-reactive (λ_{max} 549 nm) material as Gram-positive bacterial, or yeast, cells. The authors attributed this difference to the presence of KDO in the Gram-negative organisms.

Earlier reports[45,46] claiming the widespread occurrence of NeuAc in Gram-negative bacteria must be considered ill-founded, as the techniques employed did not discriminate[39] between NeuAc and KDO. Presumably, the TBA reactivity observed by the authors[45,46] is due to the presence of KDO (rather than of NeuAc) in most of the Gram-negative bacteria examined. Several techniques are now available to draw a distinction between NeuAc and KDO from bacterial, cell-wall preparations. Ellwood[39] carried out the TBA reaction under the conditions given by Weissbach and Hurwitz,[11] whereby NeuAc gives only 0.5% of the extinction obtained for equimolar solutions of KDO. The reliability of Ellwood's procedure was confirmed by Taylor.[47]

Hackenthal[48] studied the contents of KDO and NeuAc in 19 strains of *Neisseria meningitidis*. Following acid hydrolysis of acetone-dried cells, both KDO and NeuAc were separated from the mixtures by adsorption on Dowex-1 X-8 ion-exchange resin, and were eluted together by 0.6 M formic acid. The sum of periodate–TBA reactivity was determined by Warren's procedure.[13] The content of NeuAc was de-

(38) D. C. Ellwood, *Biochem. J.*, 99 (1966) 55P.
(39) D. C. Ellwood, *J. Gen. Microbiol.*, 60 (1970) 373–380.
(40) D. C. Ellwood, *Biochem. J.*, 106 (1968) 47P.
(41) G. T. Barry and W. F. Goebel, *Nature*, 179 (1957) 206.
(42) G. T. Barry, *J. Exp. Med.*, 107 (1958) 507–521.
(43) R. G. Watson, G. V. Marinetti, and H. W. Scherp, *J. Immunol.*, 81 (1958) 337–344.
(44) W. F. Vincent and J. A. Cameron, *J. Bacteriol.*, 93 (1967) 156–158.
(45) S. Aaronson and T. Lessie, *Nature*, 186 (1960) 719.
(46) R. J. Irani and K. Ganapathi, *Nature*, 195 (1962) 1227.
(47) P. W. Taylor, *Biochem. Biophys. Res. Commun.*, 61 (1974) 148–153, and personal communication.
(48) E. Hackenthal, *J. Immunol.*, 102 (1969) 1099–1102.

termined by the resorcinol reaction,[49] in which KDO does not react. The KDO content was then estimated by subtracting the TBA extinction for NeuAc (as calculated from the resorcinol value) from the total, TBA extinction. In this way, Hackenthal[48] found that the *N. meningitidis* strains of serogroup A contain KDO and not NeuAc, whereas the strains of serogroups B, C, Y, Z, and W-135 contain both KDO and NeuAc.

Gram-negative-bacterial polysaccharides containing NeuAc have since been the objects of structural and immunological studies.[50] A straightforward TBA method, suitable for the determination, in the presence of NeuAc, of the content of KDO in hydrolyzates, is used in the present author's laboratory.[51] Solutions containing both KDO and NeuAc are subjected to reduction with borohydride, whereby KDO is converted into a mixture of 3-deoxy-D-*glycero*-D-*galacto*- and 3-deoxy-D-*glycero*-D-*talo*-octonic acids (**15**; Scheme 8). This mixture, by an oxidation–decarboxylation mechanism similar to the Ruff degradation,[52] yields chromophore **3** (λ_{max} 532 nm, ϵ_{mM} ~15), under Warren's conditions[13] for the periodate–TBA reaction (see Scheme 8). The borohydride-reduction products formed from NeuAc (a mixture of 5-acetamido-3,5-dideoxy-D-*erythro*-L-*manno*- and 5-acetamido-3,5-dideoxy-D-*erythro*-L-*gluco*-nononic acids, **16**) do not contain a diol grouping at C-4–C-5, and hence do not yield the malonaldehyde-derived chromophore **3** in this reaction.

Following Osborn's suggestion[33] that KDO may be located at the junction of the polysaccharide and lipid A segments of LPS, several groups of authors have undertaken detailed studies to elucidate the structure of the "inner core" (see Fig. 2) in various organisms. Investigations of this kind are difficult for the following reasons. (*i*) The ketosidic linkages of KDO are extremely acid-labile. (*ii*) No entirely satisfactory procedure exists for the quantitative determination of KDO in polysaccharides of unknown structure and substitution pattern. (*iii*) KDO undergoes side reactions, leading to unknown or unstable products, under the usual conditions of hydrolysis of polysaccharides.

In a very elegant study, Dröge and coworkers[26] investigated the structures of the LPS from four *Salmonella minnesota* R-mutants designated mR595, mR3, mR7, and mR5. These mutants synthesize LPS that contains only lipid A and the "inner core" oligosaccharides,

(49) L. Svennerholm, *Biochim. Biophys. Acta*, 24 (1957) 604–611.
(50) J. B. Robbins and J. C. Hill (Eds.), Current Status and Prospects for Improved and New Bacterial Vaccines, *J. Infect. Dis., Suppl.*, 136 (1977) s1–s253.
(51) F. Hammerschmid, P. W. Taylor, and F. M. Unger, unpublished results.
(52) O. Ruff, *Ber.*, 31 (1898) 1573–1577.

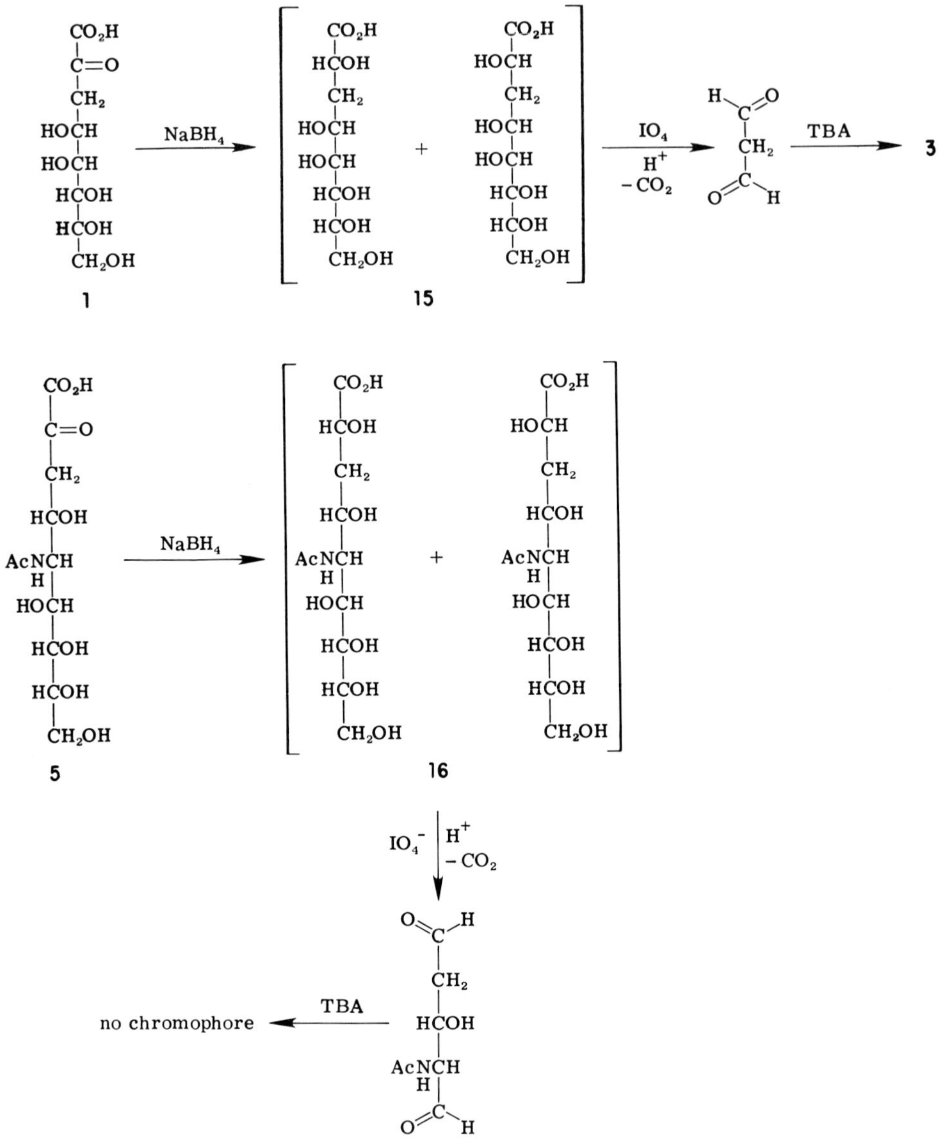

Scheme 8

whereas the "outer core" and O-antigenic chains are lacking (see Fig. 2). The sugar compositions of these LPS preparations are listed in Table II.

Samples of the four LPS were subjected to hydrolysis (0.1 M acetic

TABLE II

Sugar Composition of LPS from Four Strains of *Salmonella minnesota*
(Data as % of Sugar and, in Parentheses, as Molar Ratios)[26]

LPS from strain	KDO (semicarbazide)	Heptose	Glucose
mR595	18–20 (3)	0	0
mR3	18–20 (3)	5–7 (1)	0
mR7	16–20 (3)	10–13 (2)	0
mR5	16–18 (3)	10–11 (2)	5–6 (1)

acid, pH 3.4, for 1 h at 100°), and the hydrolyzates were analyzed by paper electrophoresis. All four hydrolyzates gave a spot that corresponded to authentic KDO, and an additional spot that could be stained with ninhydrin and also contained KDO. The compound corresponding to the latter spot was isolated in sufficient amounts by preparative, paper electrophoresis, and was found to contain phosphate, KDO, and amino groups in the molar ratios of 1:1:1. It gave in the TBA assay[10,11] a positive response that became negative following borohydride reduction.[11] No formaldehyde was released on periodate oxidation. These data suggested that the unknown compound was KDO, substituted at C-7 or C-8. The consumption of 2 moles of periodate per mole of KDO indicated that substitution was on C-7 (see Scheme 9).

No phosphate was released when the fraction was treated with phosphomonoesterase, indicating the presence of a phosphoric diester. Hydrolysis of the unknown KDO derivative (1 M HCl for 5 h at

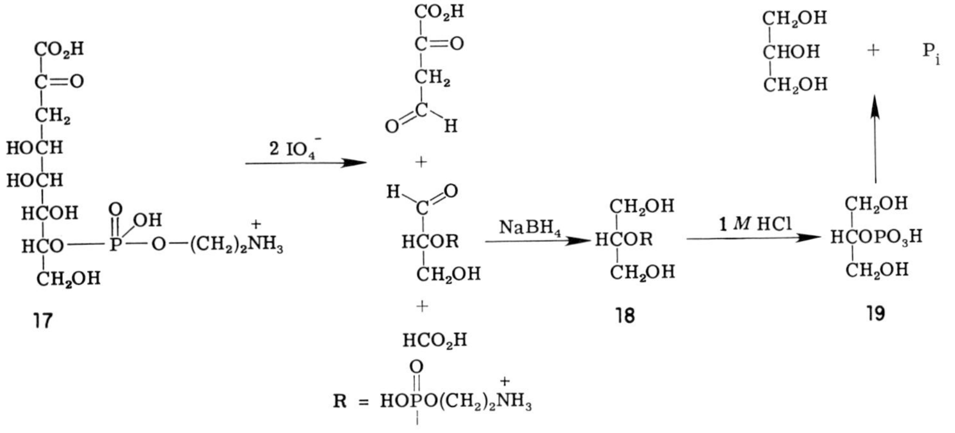

Scheme 9

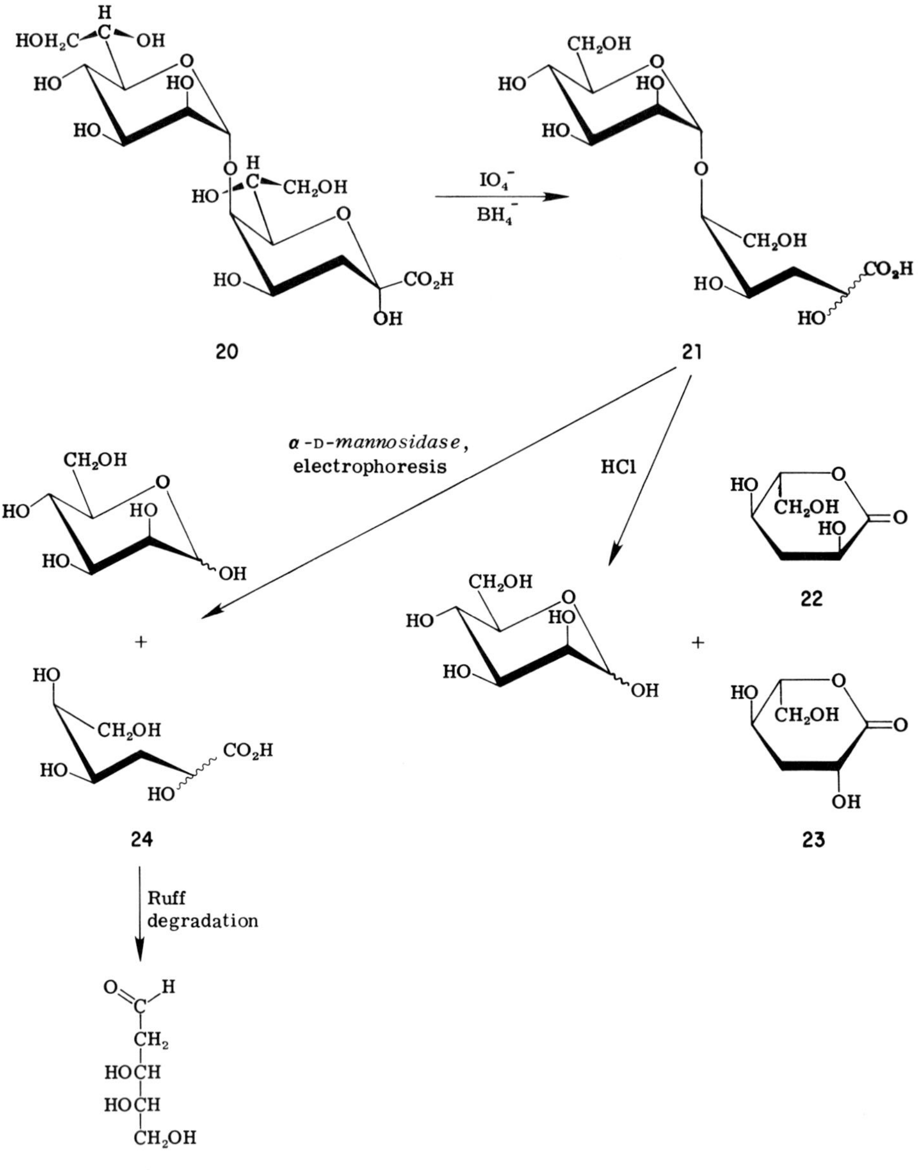

Scheme 10

100°) and paper electrophoresis of the hydrolyzate showed that mainly 2-aminoethanol had been released, with only traces of O-phosphono-2-aminoethanol present. Preparative-scale, periodate treatment, followed by borohydride reduction and paper electrophoresis, gave an almost neutral compound (**18**). 2-Aminoethanol was cleaved from **18** by hydrolysis with a strong acid, to give **19**, from which inorganic phosphate and glycerol were obtained by treatment with alkaline phosphatase (see Scheme 9). From these data, Dröge and coworkers[26] concluded that the unknown compound was KDO 7-(2-aminoethyl phosphate) (**17**).

A further fraction, obtained by paper electrophoresis of the hydrolyzate from mR3 LPS, contained residues of KDO and L-*glycero*-D-*manno*-heptose[53] in the ratio of 1:1 (compound **20**; Scheme 10). Periodate treatment of **20**, followed by borohydride reduction and electrophoresis, gave the diastereomeric pair of compounds, **21**, which contain glycosidically bound D-mannose. This sugar was identified following acid hydrolysis of **21**, as were 3-deoxy-L-*ribo*- and -*arabino*-hexono-1,5-lactone (the "α"- and "β"-L-gluco-metasaccharinic acid lactones),[54–56] **22** and **23**, which were also formed from **21** under these conditions. Mannose and "α"- and "β"-metasaccharinic acids[54–56] (**24**) were released from **21** by treatment with α-D-mannosidase. Ruff degradation[52] of **24** gave 2-deoxy L-*erythro*-pentose (**25**), which was identified, presumably by using the D sugar (see Scheme 10).

In the same study, Dröge and coworkers[26] investigated the KDO contents of the four LPS samples by three different methods, without prior hydrolysis (see Table III). (*i*) The semicarbazide assay revealed equal proportions of KDO (equivalent to three KDO residues each) in all four LPS samples. (*ii*) When analyzed by the TBA assay involving acidic periodate treatment (referred to by the authors as method A), the LPS from strain mR595 gave a TBA response equivalent to three KDO residues, whereas the heptose-containing LPS from the other three strains gave a TBA response indicative of only two KDO residues each. (*iii*) Following periodate treatment near neutrality (pH 5) (method B), the TBA response of the LPS from strain mR595 corresponded to two KDO residues. The values obtained for the heptose-containing LPS samples from strains mR3, mR7, and mR5 were indicative of only one KDO residue (see Table III).

(53) G. Bagdian, W. Dröge, K. Kotelko, O. Lüderitz, and O. Westphal, *Biochem. Z.*, 344 (1966) 197–211.
(54) H. Kiliani and H. Naegell, *Ber.*, 35 (1902) 3528–3533.
(55) *Methods Carbohydr. Chem.*, 2 (1963) 477–485.
(56) J. C. Sowden, *Adv. Carbohydr. Chem.*, 12 (1957) 35–79.

TABLE III

Determinations[a] of KDO Content in *Salmonella minnesota* LPS, Using the Semicarbazide Assay[88] and the TBA Assay[11]

LPS from strain	Semicarbazide	Method A	Method B
mR595	18–20 (3)	16–20 (3)	11 (2)
mR3	18–20 (3)	9–12 (2)	5–7 (1)
mR7	16–20 (3)	9–12 (2)	5–7 (1)
mR5	16–18 (3)	11 (2)	—

[a] For the TBA assay, periodate treatment was conducted in the presence of 125 mM sulfuric acid (method A), or at pH 5 (method B).[26] The data are given in percent, and as molar ratios (in parentheses).

The partial protection of individual KDO residues from periodate degradation, depending on the origin of the LPS and the method of TBA analysis, led the authors to suggest that a common, branched, trisaccharide structure, consisting of KDO residues, might actually constitute the linkage region between core oligosaccharides and lipid A in LPS from *Salmonella minnesota* (see Fig. 7). To test this hypothesis, LPS from strains mR595 and mR3 were subjected to "neutral" (pH 5) periodate oxidation, and the products reduced with sodium borohydride. Mild, acid-catalyzed hydrolysis of the products gave free KDO in each case, indicating that, in the intact LPS, one KDO residue must have been substituted both at O-4 (or O-5) *and* in the side chain (at O-7 or O-8). The release, following periodate and borohydride treatments, of this KDO residue by mild, acid hydrolysis indicated that both its anomeric and aglyconic linkages had been ketosidic. The

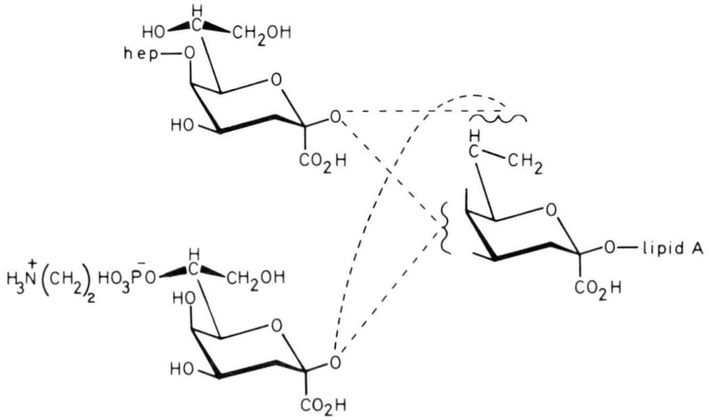

Fig. 7.—Representation of the Possible Linkages of the KDO Trisaccharide, Following the Investigation of Dröge and Coworkers.[26]

aglyconic linkages must, therefore, have involved the other two KDO residues present in the intact LPS.

Finally, Dröge and coworkers[26] showed that the 2-aminoethyl phosphate-substituted KDO group is the "lateral" KDO unit of the branched trisaccharide (see Fig. 7), as follows. LPS from *Salmonella minnesota* mR3 was subjected to periodate oxidation. This sample, together with a control that had not been oxidized, was then mildly hydrolyzed with acid (pH = 3.4) during 1 h at 100°. Following removal of lipid A, both samples were analyzed by gel-filtration on Sephadex G-10, and paper electrophoresis. As expected, the ninhydrin-positive material obtained from the control sample was identical with KDO 7-(2-aminoethyl phosphate) (**17**) as previously identified. This spot was absent from the periodate-treated sample. Instead, an almost neutral, ninhydrin-positive spot was observed. This material (compound **26**) was eluted, subjected to reduction with sodium [^{3}H]borohydride, and hydrolyzed under strongly acidic conditions (see Scheme 11). Fol-

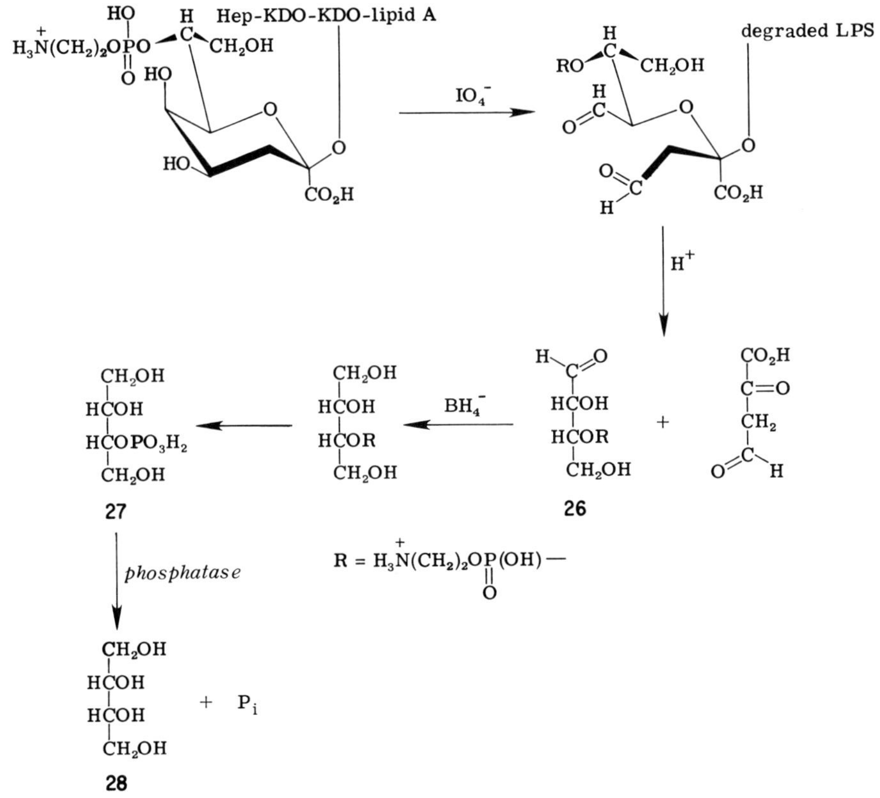

Scheme 11

Hep—(1⟶3)—Hep—(1⟶7 or 8)—KDO

FIG. 8.—Composition of the Core Oligosaccharide from a Galactose-epimeraseless mutant[32] of *Salmonella typhimurium*.[57]

lowing hydrolysis, free 2-aminoethanol was detected, together with an acidic compound (**27**). Treatment of **27** with phosphomonoesterase gave inorganic phosphate and **28**, which was chromatographically indistinguishable from erythritol. The authors concluded that the KDO 7-(2-aminoethyl phosphate) residue had been cleaved by periodate at C-4–C-5 (see Scheme 11). The lack of substitution at these atoms indicated that the KDO 7-(2-aminoethyl phosphate) is the "lateral" KDO group of the branched KDO trisaccharide in LPS from *Salmonella minnesota*.

Cherniak and Osborn[57] studied the composition of the core oligosaccharide from a galactose-epimerase-less mutant of *Salmonella typhimurium*. The structure of this species was reported as given in Fig. 8. The linkage of heptose to O-7 or O-8 of KDO was assumed on the basis of the direct reactivity of the reducing KDO residue in the TBA assay. On studying similar core-oligosaccharides from the galactose-epimerase-less strain *Escherichia coli* J5, Heath and coworkers[58,59] also found reducing-terminal KDO residues in three different oligosaccharides (see Fig. 9). They concluded, however, that the heptose is linked at O-5 of the reducing-terminal KDO residue, from the following results (compare Dröge and coworkers[26]). Smith degradation[60]

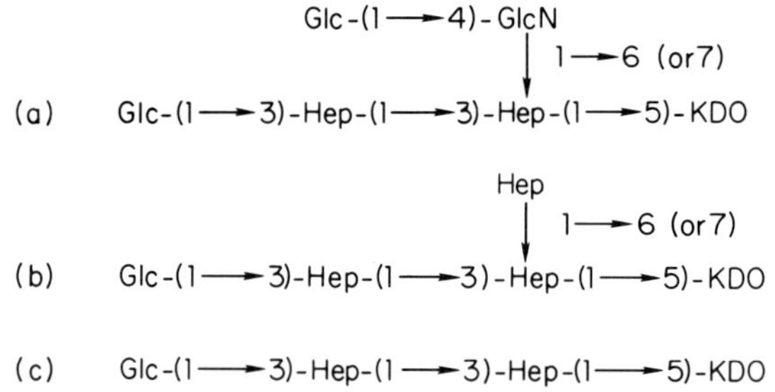

```
                    Glc-(1⟶4)-GlcN
                         │ 1⟶6 (or 7)
                         ↓
    (a)    Glc-(1⟶3)-Hep-(1⟶3)-Hep-(1⟶5)-KDO

                         Hep
                         │ 1⟶6 (or 7)
                         ↓
    (b)    Glc-(1⟶3)-Hep-(1⟶3)-Hep-(1⟶5)-KDO


    (c)    Glc-(1⟶3)-Hep-(1⟶3)-Hep-(1⟶5)-KDO
```

FIG. 9.—Composition of Three Oligosaccharides (a, b, and c) Identified[58,59] after Hydrolysis of LPS from *Escherichia coli* J5.

(57) R. Cherniak and M. J. Osborn, *Fed. Proc. Fed. Am. Soc. Exp. Biol.*, 25 (1966) 410 (Abstr. 1243).
(58) N. A. Fuller and E. C. Heath, *Fed. Proc. Fed. Am. Soc. Exp. Biol.*, 29 (1970) 337 Abstr.
(59) N. A. Fuller, M.-C. Wu, R. G. Wilkinson, and E. C. Heath, *J. Biol. Chem.*, 248 (1973) 7938–7950.
(60) G. W. Hay, B. A. Lewis, and F. Smith, *Methods Carbohydr. Chem.*, 5 (1965) 361–370.

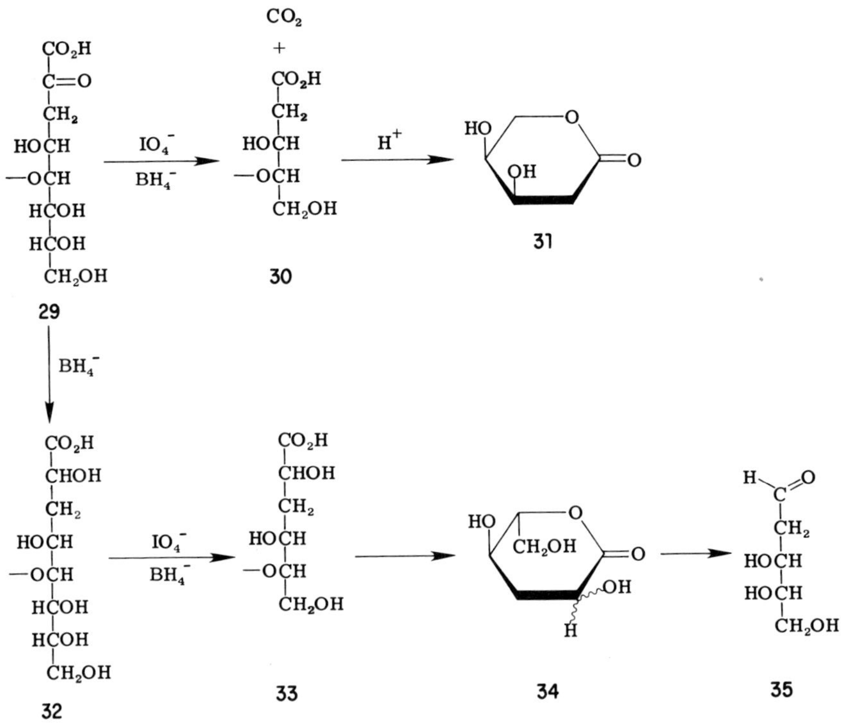

3-DEOXY-D-*manno*-2-OCTULOSONIC ACID (KDO)

Scheme 12

(that is, periodate treatment, followed by borohydride reduction) of oligosaccharides containing reducing-terminal KDO residues (for example, **29**; see Scheme 12) gave a residual oligosaccharide (**30**) containing a negatively charged group. Acid cleavage of **30** was accompanied by the formation of 2-deoxy-L-*erythro*-pentono-1,5-lactone (**31**), identified chromatographically by using a synthetic sample of the D enantiomer.[61]

When oligosaccharides containing reducing-terminal KDO residues were subjected to borohydride reduction prior to Smith degradation, to give **32**, the epimeric pair of oligosaccharides, **33**, was obtained, containing 3-deoxy-L-*ribo*- and -*arabino*-hexonic acid (the "α"- and "β"-L-glucometasaccharinic acids[54–56]) at the end that had formerly been reducing. Acidic cleavage of **33** resulted in formation of the epimeric pair of the corresponding 1,5-lactones, **34**. The identification of **34** was made by using standards belonging to the D series, as follows.

(61) S. Moore and K. P. Link, *J. Biol. Chem.*, 133 (1940) 294.

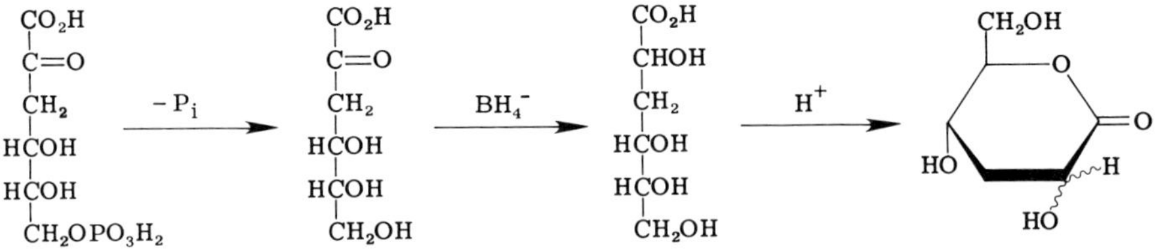

```
CO₂H              CO₂H              CO₂H              CH₂OH
|                 |                 |                     ___O
C=O               C=O               CHOH                /    \
|        -Pᵢ      |        BH₄⁻     |         H⁺       /      \
CH₂    ────→      CH₂     ────→     CH₂     ────→    |        |═O
|                 |                 |                  \      /
HCOH              HCOH              HCOH           HO  \~~~~/  H
|                 |                 |                    \  /
HCOH              HCOH              HCOH                 HO
|                 |                 |
CH₂OPO₃H₂         CH₂OH             CH₂OH

36                                  37                  38
```

Scheme 13

3-Deoxy-D-*erythro*-2-hexulosonate 6-phosphate[62] (**36**) was dephosphorylated by the action of alkaline phosphatase, and the product was reduced with sodium borohydride, to give **37**, an epimeric pair of 3-deoxyhexonic acids (metasaccharinic acids). Treatment of **37** with acid gave the corresponding epimeric pair of D-metasaccharinolactones, **38** (see Scheme 13) which are enantiomers of **34**, and hence chromatographically indistinguishable from **34**. The identity of **38** with **34**, in all physical properties except optical rotation, was further confirmed by gas–liquid chromatographic analysis of the per(trimethylsilyl) ether derivatives,[63] and mass spectrometry, and by degradation[31] of the metasaccharinolactones with ceric sulfate to give 2-deoxy-L-*erythro*-pentose, **35** (see Scheme 12). The nature of **35** was verified by its reactivity in the TBA assay for 2-deoxy sugars[10] (λ_{max} 532 nm) and by paper chromatography. The study by Heath and coworkers[58,59] on LPS from *Escherichia coli* complements the results obtained by Dröge and coworkers[26] on LPS from *Salmonella minnesota*.

On studying the KDO content of LPS from *Xanthomonas sinensis*, Volk and coworkers[64] found that only one KDO residue per LPS chain is present in this organism. Moreover, the authors observed the formation of 4,7- and 4,8-anhydro derivatives (**39** and **40**) of KDO (see Scheme 14) under the conditions used for mild hydrolysis of the LPS with acetic acid. Free KDO was also shown to yield the anhydro derivatives under analogous conditions. The structures of these unusual side-products were identified by mass spectrometry, as follows. Reduction of **39** and **40** with sodium borohydride gave the two epimeric mixtures, **41** and **42**. These were subjected to periodate treatment followed by borohydride reduction, the products separated by paper electrophoresis, and their trimethylsilyl ether derivatives analyzed by

(62) J. McGee and M. Doudoroff, *J. Biol. Chem.*, 210 (1954) 617–626.
(63) C. C. Sweeley, R. Bentley, M. Makita, and W. W. Wells, *J. Am. Chem. Soc.*, 85 (1963) 2497–2507.
(64) W. A. Volk, N. L. Salomonsky, and D. Hunt, *J. Biol. Chem.*, 247 (1972) 3881–3887.

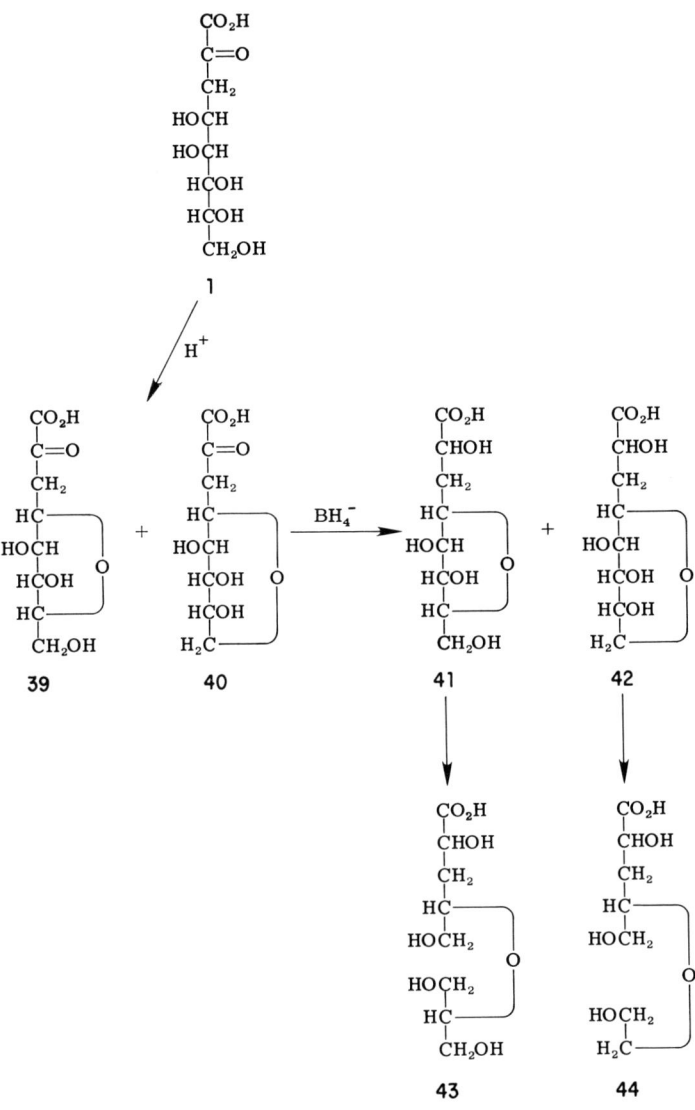

Scheme 14

high-resolution, mass spectrometry. Thereby, it was seen that periodate–borohydride treatment of **41** had resulted in a gain of 2 atomic mass units, to give compound **43**, whereas, following the same treatment, 100 atomic mass units had been lost from **42**, to give **44**. A mechanism for this side-reaction, involving 1,4-elimination of water followed by 1,4-addition of one of the hydroxyl groups (OH-7 or OH-8)

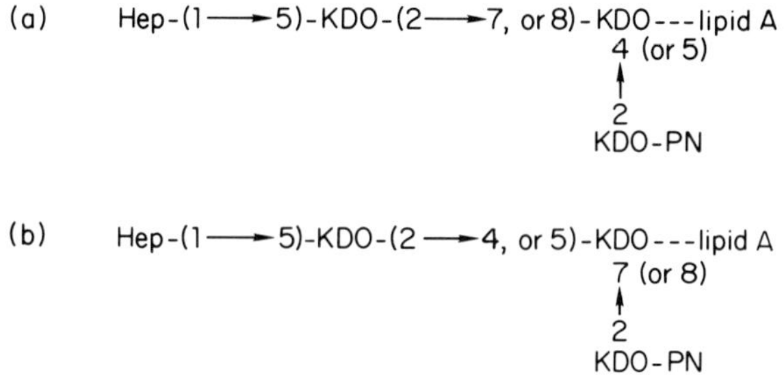

(a)   Hep-(1⟶5)-KDO-(2⟶7, or 8)-KDO---lipid A
                                4 (or 5)
                                  ↑
                                  2
                                KDO-PN

(b)   Hep-(1⟶5)-KDO-(2⟶4, or 5)-KDO---lipid A
                                7 (or 8)
                                  ↑
                                  2
                                KDO-PN

FIG. 10.—Experimentally Distinguishable, Linkage Arrangements (a and b) in the KDO Trisaccharide from the LPS of *Escherichia coli* B, According to Prehm and Coworkers[65] (Compare Fig. 7).

was proposed.[64] It was also suggested that the reaction may have occurred, but gone unnoticed, in similar investigations.

Prehm and coworkers[65] studied the content and linkages of KDO in several R mutants of *Escherichia coli* BB. Initially using the approach of Dröge and coworkers,[26] they obtained evidence for the existence, in these strains, of a branched KDO trisaccharide, similar to the one demonstrated in *Salmonella minnesota*.[26] However, in the work of Prehm and coworkers,[65] two additional refinements were introduced, yielding more-detailed information on the nature of the KDO region in LPS. First, the authors recognized that two, experimentally distinguishable, possibilities exist regarding the substitution pattern of the branch-point KDO (see Fig. 10). In one of these (see Fig. 10a), the KDO residue carrying the core oligosaccharide is linked at C-7 or C-8, and the "lateral" KDO 7-(2-aminoethyl phosphate) is attached at C-4 or C-5 of the branch-point KDO residue. In the other possible structure (see Fig. 10b), this order is reversed.

As may be seen in Schemes 15 and 16, the following sequence of reactions was conducted in order to decide this point. Smith degradation[60] (giving **44** or **48**), followed by very mild hydrolysis with 1% acetic acid during 5 h at 37°, would result in favored cleavage of the acetal bond by which the acyclic remnants of the "lateral" KDO 7-(2-aminoethyl phosphate) residue would be attached to the branch-point KDO residue (giving **45** or **49**); the ketosidic linkage of the KDO residue carrying the core oligosaccharide would be essentially unaffected. In this way, the diol grouping of the branch-point KDO, which had initially been resistant to periodate due to attachment of the "lateral" KDO residue, would now become susceptible to periodate cleavage (at C-4–C-5 in Scheme 15, or at C-7–C-8 in Scheme 16). If **45** is the

(65) P. Prehm, S. Stirm, B. Jann, and K. Jann, *Eur. J. Biochem.*, 56 (1975) 41–55.

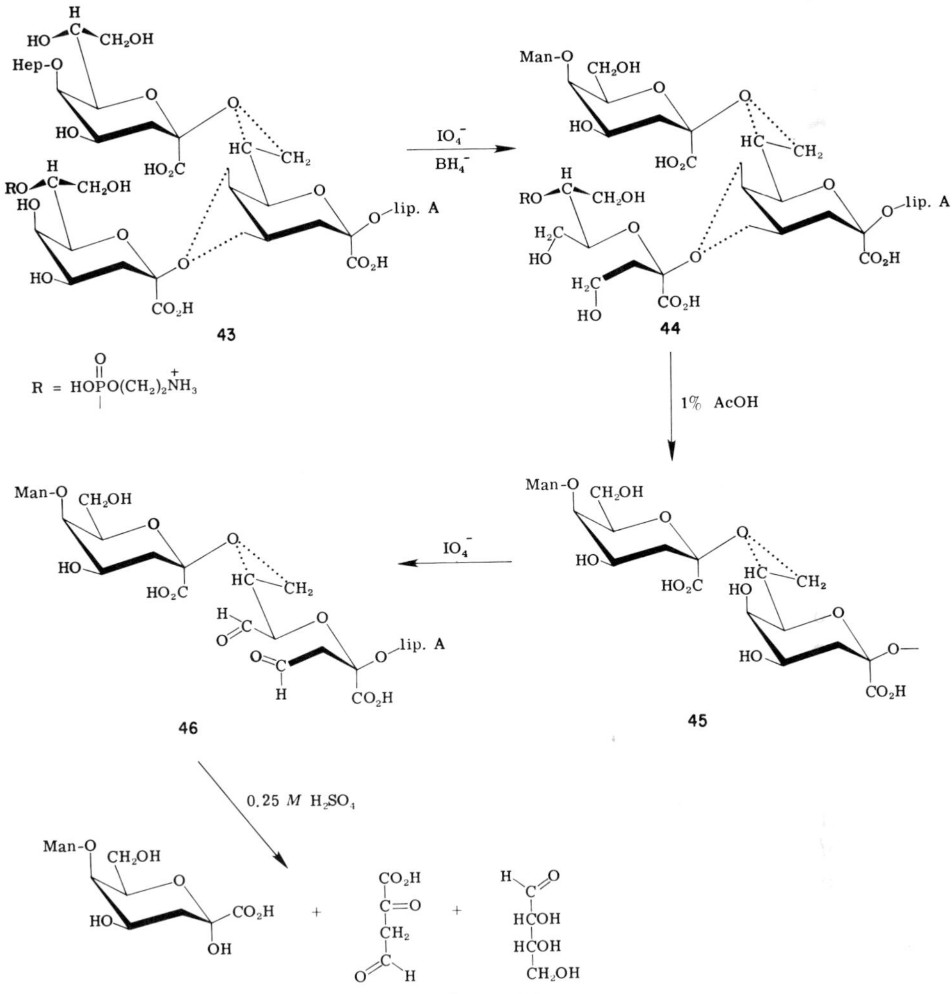

Scheme 15

intermediate, such cleavage would result in structure **46**, which gives 3-formylpyruvate directly on acid treatment. Under analogous conditions, formation of 3-formylpyruvate was not expected had **49** been the intermediate, leading to **50** on periodate treatment. In fact, the TBA reactivity measured was somewhat weaker (see Table IV) than was observed under conditions where the branch-point KDO residue would be fully TBA-reactive (hydrolysis with 125 mM sulfuric acid, instead of 1% acetic acid). Nonetheless, the result indicates that the "lateral" KDO residue is attached at C-4 or C-5 of the branch-point

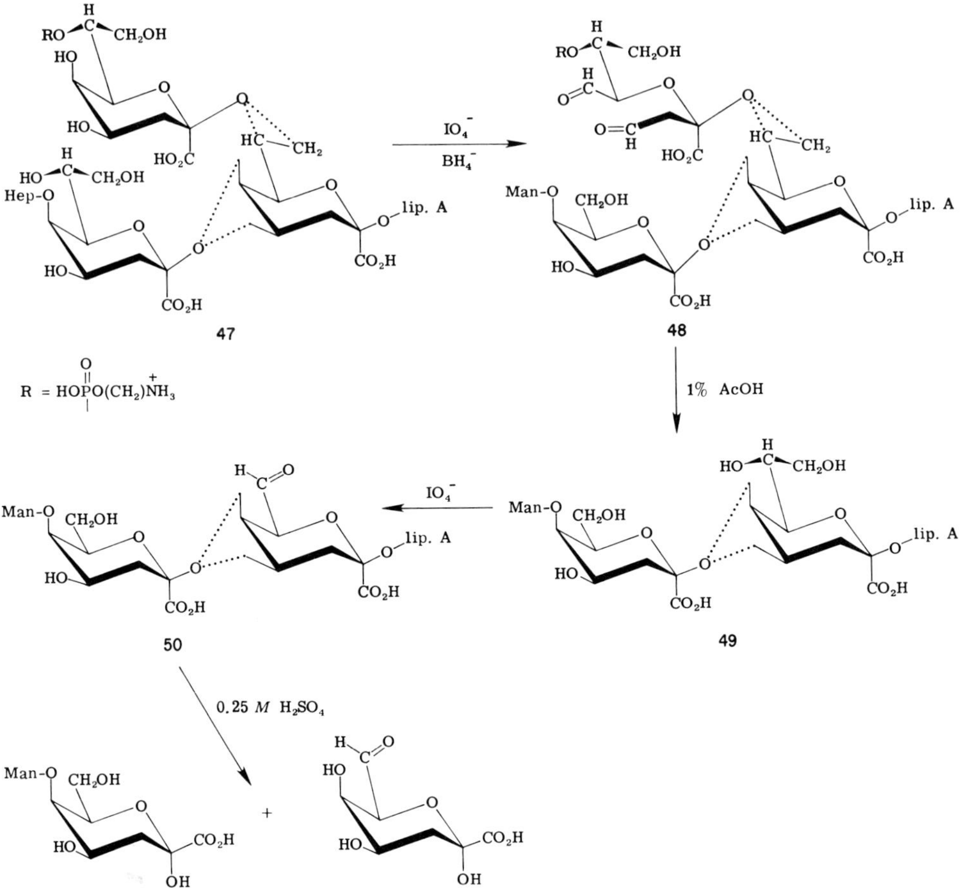

Scheme 16

KDO, that is, that the structure of the KDO trisaccharide in *Escherichia coli* BB corresponds to **43** (see Scheme 15) and not to **47** (see Scheme 16).

The second point of refinement introduced by Prehm and coworkers[65] concerns the methylation analysis[66] of LPS fractions containing KDO residues. As demonstrated for the LPS (**51**) from *E. coli* BB 12 (**51** has the shortest glycosyl extension reported so far), their procedure was as follows (see Scheme 17). Methylation by a modification[67]

(66) W. N. Haworth, *J. Chem. Soc.*, 107 (1915) 8.

TABLE IV

KDO Determination[a] in *Escherichia coli* BB LPS after Oxidation and Reduction[65]

Hydrolysis conditions	KDO released
No hydrolysis	0
125 mM Sulfuric acid for 8 min at 100°	12.4
1% Acetic acid for 5 h at 37°	8.5

[a] The LPS of *E. coli* BB was subjected to Smith degradation and then hydrolyzed as listed (compare, Schemes 15 and 16). This hydrolysis was followed by a KDO determination.[65]

of the Hakomori procedure[68] converted **51** into the methyl ester derivative (**52**). Reduction with calcium hydride in oxolane[69] gave **53** (wherein the carboxyl groups of the KDO residues had been converted into primary alcohol groups). Acid-catalyzed hydrolysis of **53**, followed by borohydride reduction, and acetylation, gave two epimeric mixtures (**54** and **55**) which, by mass-spectrometric analysis, were found to be distinct. The data indicated that one (**54**) is 1,2,6-tri-*O*-acetyl-3-deoxy-4,5,7,8-tetra-*O*-methyloctitol, and the other (**55**) is 1,2,4,6-tetra-*O*-acetyl-3-deoxy-5,7,8-tri-*O*-methyloctitol. From the methylation pattern, and from the absence of a di-*O*-methyloctitol, the authors concluded that only two KDO residues are present in *Escherichia coli* BB 12. One of these constitutes the nonreducing terminal, and is attached to the other by a (2 → 4)-linkage as shown in **51**. Interestingly, studies by Osborn's group[70] concerning *in vitro* transfer of KDO to lipid A acceptors led to the conclusion that two KDO residues had been enzymically transferred, and that they were (2 → 4)-linked, as in **51**. Moreover, the "lateral" KDO residue is apparently transferred before the one that eventually carries the core oligosaccharide. A possible explanation for this behavior has been given.[70]

The elegant and extensive studies of Szabó's group in Paris on LPS from *Bordetella pertussis* have been summarized in a plenary lecture that has appeared in print.[71] From their studies, the authors concluded

(67) K. Reske and K. Jann, *Eur. J. Biochem.*, 31 (1972) 320–328.
(68) S.-I. Hakomori, *J. Biochem. (Tokyo)*, 55 (1964) 205–208.
(69) M. Abdel-Akher and F. Smith, *Nature (London)*, 166 (1950) 1037–1038.
(70) R. S. Munson, Jr., N. S. Rasmussen, and M. J. Osborn, *J. Biol. Chem.*, 253 (1978) 1503–1511.
(71) L. Szabó, *Pure Appl. Chem.*, 49 (1977) 1187–1199.

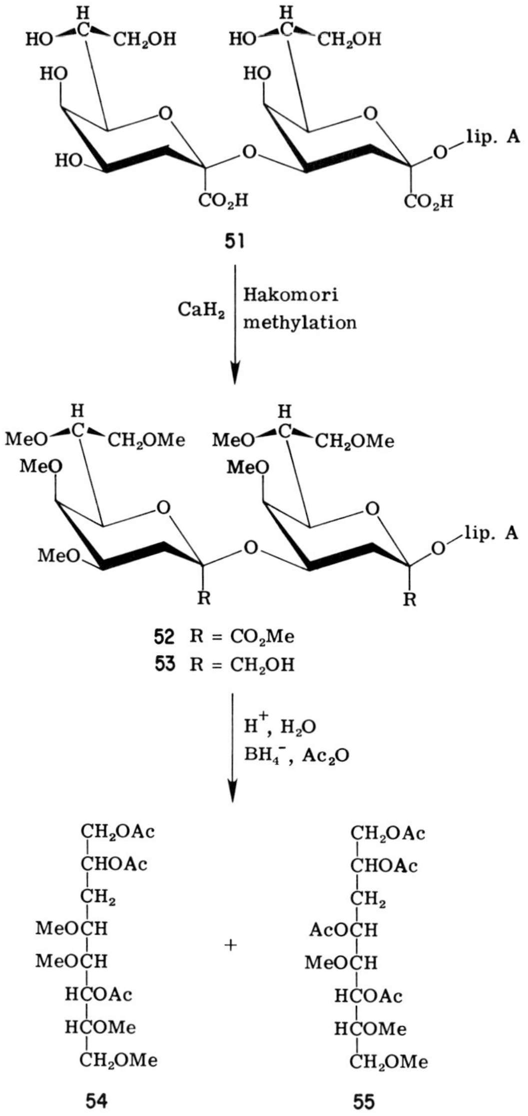

Scheme 17

51

$\downarrow$ CaH$_2$ | Hakomori methylation

52 R = CO$_2$Me
53 R = CH$_2$OH

$\downarrow$ H$^+$, H$_2$O; BH$_4^-$, Ac$_2$O

54

CH$_2$OAc
CHOAc
CH$_2$
MeOCH
MeOCH
HCOAc
HCOMe
CH$_2$OMe

+

55

CH$_2$OAc
CHOAc
CH$_2$
AcOCH
MeOCH
HCOAc
HCOMe
CH$_2$OMe

that the *Bordetella* LPS contains two polysaccharide chains attached to lipid A, as distinct from only one seen in the *Enterobacteriaceae*. One of the chains (polysaccharide I) in *Bordetella* LPS was released from the LPS by mild, acid-catalyzed hydrolysis, and contained a reducing-terminal KDO residue glycosylated at O-5. The other (polysaccharide II) required more drastic conditions for its acid-catalyzed, hydrolytic cleavage from the LPS, and contained a reducing-terminal KDO 5-phosphate residue. The attachment of polysaccharide I to lipid A, rather than to some site on polysaccharide II, was ascertained by treatment of the LPS with 2 M potassium hydroxide. This treatment released, from the LPS, fragments that contained polysaccharide I and constituents of the lipid A portion, but not the chain having the phosphorylated KDO residue. The substitution patterns of the KDO residues were established by using the periodate–borohydride techniques already discussed in this Section.[26,59] Many of the compounds needed for identification of degradation products, as well as other reference compounds, were synthesized by the Paris group (compare Section IV, 1 and 3).

A collection of LPS structures, including some containing KDO residues of known linkage, has been assembled by Nikaido.[72] Noticeable among these is the "R-2-type core" in *Escherichia coli* O-100 studied by Hämmerling and coworkers.[73] The LPS from this strain contains a KDO residue that is substituted by a galactosyl group on O-7 or O-8.

The presence of KDO in the LPS of some strains of anaerobic, Gram-negative, non-sporeforming bacilli has been suggested as a taxonomic criterion.[74]

Some studies of KDO in Gram-negative-bacterial LPS have utilized the color reaction of KDO with diphenylamine[22,71,75] (λ_{max} 425 nm). Other sugars interfere with this assay, and a somewhat complex formula[22,75] has been employed in calculation of the KDO content of a sample. Another rarely used test for the estimation of KDO is based on its reaction with *o*-phenylenediamine to afford 2-hydroxyquinoxalines.[76] Determinations of KDO content by gas–liquid chromatography (g.l.c.) seem to have been infrequently reported in the literature, mostly due to lack of generally applicable, quantitative, derivatization procedures. Two derivatives, **56** and **57**, have been recommended[77] for the g.l.c. of KDO. Kochetkov and coworkers[78] have studied the

(72) H. Nikaido, in G. D. Fasman (Ed.), *Handbook of Biochemistry and Molecular Biology*, 3rd edn., CRC Press, Cleveland, Ohio, pp. 409–415.
(73) G. Hämmerling, O. Lüderitz, O. Westphal, and P. H. Mäkelä, *Eur. J. Biochem.*, 22 (1971) 331–344.
(74) T. Hofstad, *J. Gen. Microbiol.*, 85 (1974) 314–320.
(75) R. Chaby, S. R. Sarfati, and L. Szabó, *Anal. Biochem.*, 58 (1974) 123–129.
(76) M. C. Lanning and S. S. Cohen, *J. Biol. Chem.*, 189 (1951) 109–114.
(77) G. A. Adams, C. Quadling, and M. B. Perry, *Can. J. Microbiol.*, 13 (1967) 1605–1613.
(78) N. K. Kochetkov, B. A. Dmitriev, and L. V. Backinowsky, *Carbohydr. Res.*, 11 (1969) 193–197.

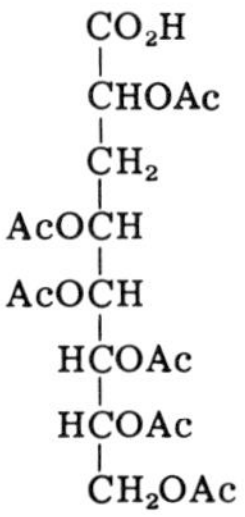

56 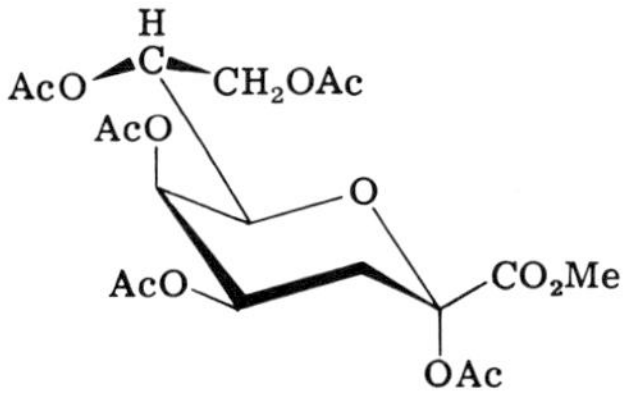 **57**

g.l.c. separation of KDO derivatives obtained from whole LPS by methanolysis and per(trimethylsilyl)ation. These authors reported the occurrence in their chromatograms of three peaks due to KDO. Charon and Szabó[79] reported on the formation of the methyl esters of two furanosides and one pyranoside by the action of methanol, in the presence of an acidic catalyst, on ammonium KDO. The two furanosides, following acetylation to give **58** and **59**, had shorter retention times than the pyranoside (presumably the thermodynamically stable, α-D-ketoside[80–82] **60**).

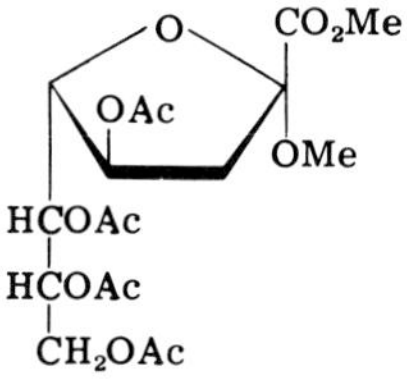

58 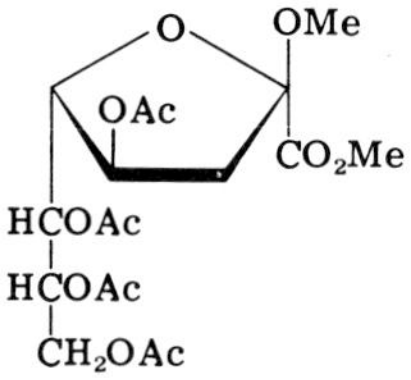 **59** 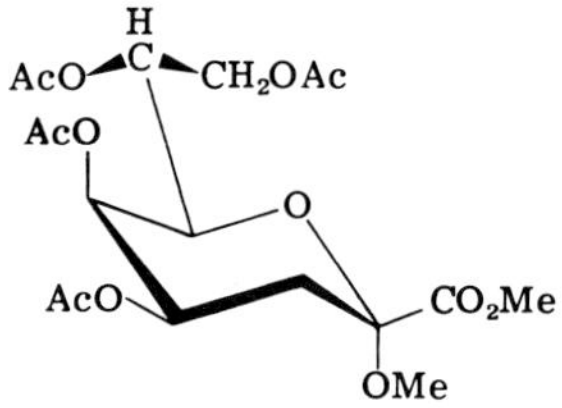 **60**

3. KDO in Acidic Exopolysaccharides (K-Antigens)

The first report of the occurrence of KDO in a structure other than LPS was that of Taylor[47] on the capsular polysaccharide (K-antigen) of a clinical isolate, *Escherichia coli* LP1092. Almost simultaneously, Bhattacharjee and coworkers[83] described an exopolysaccharide from *Neisseria meningitidis* serogroup 29e. This material contains KDO

(79) D. Charon and L. Szabó, *J. Chem. Soc., Perkin Trans. 1*, (1979) 2369–2374.
(80) F. M. Unger, D. Stix, and G. Schulz, *Carbohydr. Res.*, 80 (1980) 191–195.
(81) F. M. Unger, D. Stix, E. Schwarzinger, G. Schulz, and F. Hammerschmid, *Carbohydr. Res.*, in press.
(82) C. Kratky, D. Stix, and F. M. Unger, *Carbohydr. Res.*, in press.
(83) A. K. Bhattacharjee, H. J. Jennings, and C. P. Kenny, *Biochem. Biophys. Res. Commun.*, 61 (1974) 489–493.

and 2-acetamido-2-deoxygalactose in the ratio of 1:1. Vann and Jann[84] published their results on the linkage analysis of the exopolysaccharide from *Escherichia coli* 06:K13:H1. The K-13 antigen contains KDO and ribose in the ratio of 1:1. Methylation analysis[67,68] revealed the structure to be Rib*f*-(1 → 7)-KDO*p*-(2 → 3)-Rib*f*. The native polysaccharide contains one acetyl group linked to O-4 or O-5 of each KDO residue. Preliminary studies of the KDO-containing exopolysaccharide from *E. coli* LP1092 have been carried out in the present author's laboratory.[85,86] The data indicate that the material contains ribose and KDO in the ratio of 2:1 (by TBA assay and orcinol test[87]). The OH-4 and OH-5 groups of the KDO residue are unsubstituted, as the material responds fully in the TBA assay involving periodate cleavage near neutrality, as described by Dröge and coworkers[26] (method B). One of the ribose residues in the intact polysaccharide is protected from periodate degradation. These findings are compatible with the polysaccharide's being similar to the K-13 antigen, except for the presence, in the former, of a ribose side-branch.

III. Elucidation of the Structure of KDO

1. The KDO Molecule

The initial, structural information on KDO comes from the work of Levin and Racker,[9] who observed the enzyme-catalyzed formation of KDO 8-phosphate (**2**) from D-arabinose 5-phosphate and enolpyruvate phosphate (see Scheme 1). The evidence that led these authors to suggest structure **61** for **2** is as follows. (*i*) Compound **2** forms a chromophore of λ_{max} 549 nm in the TBA assay[11] for 3-deoxyglyc-2-ulosonic acids. (*ii*) Compound **2** reacts with semicarbazide, giving a semicarbazone (λ_{max} 250 nm, ϵ_{mM} 10.20) analogous in its u.v.-spectroscopic properties to the semicarbazone obtained from pyruvic acid.[88] (*iii*) Compound **2** and an equivalent amount of inorganic phosphate were formed from D-arabinose 5-phosphate and enolpyruvate phosphate, catalyzed by crude and by partially purified extracts from *P. aeruginosa*. Levin and Racker[9] assigned the configurations of C-5–C-7 of **61** as *arabino* (these correspond to C-2–C-4 of D-arabinose 5-phosphate). The configuration of C-4, and the size of the sugar ring, remained to be

(84) W. F. Vann and K. Jann, *Infect. Immun.*, 25 (1979) 85–92.
(85) P. W. Taylor, *Int. Symp. Carbohydr. Chem.*, 9th, (1978) F27.
(86) P. Messner and F. M. Unger, submitted for publication.
(87) Z. Dische, *Methods Carbohydr. Chem.*, 1 (1962) 484–488.
(88) J. DeLey and M. Doudoroff, *J. Biol. Chem.*, 227 (1957) 745–757.

$$\begin{array}{c}\text{CO}_2\text{H}\\|\\\text{C}=\text{O}\\|\\\text{CH}_2\\|\\\text{CHOH}\\|\\\text{HOCH}\\|\\\text{HCOH}\\|\\\text{HCOH}\\|\\\text{CH}_2\text{OPO}_3\text{H}_2\end{array}$$

61

established. Heath and Ghalambor[29,31] demonstrated that KDO, as released from LPS by mild, acid-catalyzed hydrolysis, is identical with the product of enzyme-catalyzed dephosphorylation of **2**. The authors tentatively assigned to KDO the D-*manno* configuration, because its rate of periodate consumption suggested that OH-4 and OH-5 are disposed *cis* to each other. This argument follows the one used by Preiss and Ashwell[20] to distinguish between 3-deoxy-D-*erythro*-2-hexulosonate and 3-deoxy-D-*threo*-2-hexulosonate. Ghalambor and Heath[89] verified the D-*arabino* configuration at C-5–C-7 of KDO by demonstrating its cleavage, catalyzed by KDO aldolase from *Aerobacter cloacae*, to give D-arabinose and pyruvate.

Degradation[90] of the borohydride-reduction products of KDO (**15**) with ceric sulfate gave a 2-deoxyheptose of unknown configuration.[29,31] Perry[91] synthesized an identical sugar (**62**, see Scheme 18) from D-mannose by the Sowden–Fischer[92,93] procedure. This finding established the configuration of KDO as D-*manno*.[94] 2-Deoxy-D-*manno*-heptose has been utilized for a synthesis of KDO by the Kiliani–Fischer cyanohydrin procedure.[95,96] The D-*manno* configuration of KDO was subsequently confirmed by a Wittig synthesis of KDO, starting from *aldehydo*-D-mannose pentaacetate,[97,98] by n.m.r. analysis of a

(89) M. A. Ghalambor and E. C. Heath, *Biochem. Biophys. Res. Commun.*, 11 (1963) 288–293.
(90) A. Meister, *J. Biol. Chem.*, 197 (1952) 309–317.
(91) M. B. Perry, *Can. J. Chem.*, 45 (1967) 1295–1297.
(92) J. C. Sowden and H. O. L. Fischer, *J. Am. Chem. Soc.*, 69 (1947) 1048–1050.
(93) J. C. Sowden and R. Schaffer, *J. Am. Chem. Soc.*, 73 (1951) 4662–4664.
(94) M. B. Perry and G. A. Adams, *Biochem. Biophys. Res. Commun.*, 26 (1967) 417–422.
(95) D. T. Williams and M. B. Perry, *Can. J. Biochem.*, 47 (1969) 691–695.
(96) D. Charon, R. S. Sarfati, D. R. Strobach, and L. Szabó, *Eur. J. Biochem.*, 11 (1969) 364–369.
(97) E. Hellwig, H. Berner, and F. M. Unger, *Int. Symp. Carbohydr. Chem., 8th*, (1976) Abstr. 4D-7.
(98) E. Hellwig, D. Stix, H. Berner, F. Franke, and F. M. Unger, submitted for publication.

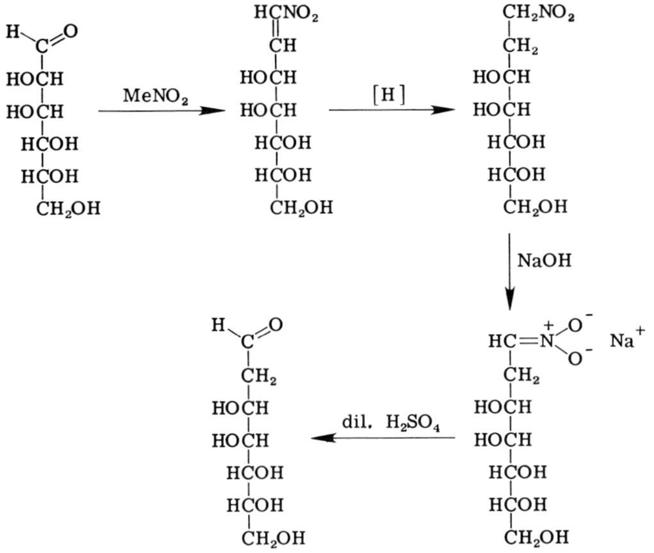

Scheme 18

[Scheme 18 showing synthesis with structures and reagents: MeNO₂, [H], NaOH, dil. H₂SO₄, compound **62**]

lactone intermediate in the Kuhn synthesis[99] of KDO,[100] and by X-ray crystallography[82] of the α-D-ketosidic KDO derivative **60**.

2. Spectroscopic Analysis of KDO and of Its Natural Derivatives

Studies of KDO-containing, natural polysaccharides using classical color reactions and degradation procedures have been discussed in Section II. Although these investigations yielded fundamental information regarding the occurrence and linkages of KDO moieties, they cannot provide an understanding of (i) the anomeric configurations of natural KDO derivatives, and (ii) the conformations of KDO residues in aqueous solutions. These features are conveniently studied by the various techniques of nuclear magnetic resonance (n.m.r.) spectroscopy. The few articles that have been published in this area will next be discussed.

(i) **The KDO Molecule.**—The α-D configuration was assigned to the preponderant form of KDO in aqueous solution,[101] based on the similarity of the corresponding ^{13}C-n.m.r. spectrum to that of a methyl ke-

(99) R. Kuhn and G. Baschang, *Justus Liebigs Ann. Chem.*, 659 (1962) 156–163.
(100) C. Hershberger and S. B. Binkley, *J. Biol. Chem.*, 243 (1968) 1578–1584.
(101) A. K. Bhattacharjee, H. J. Jennings, and C. P. Kenny, *Biochemistry*, 17 (1978) 645–651.

toside, **63**, of KDO which was inferred,[101] and later proved,[80] to be a methyl α-ketoside. Additional lines observed in the ^{13}C-n.m.r. spectra of KDO in aqueous solution were attributed to the possible presence of five- or six-membered-ring lactones in the solutions.[101] The latter as-

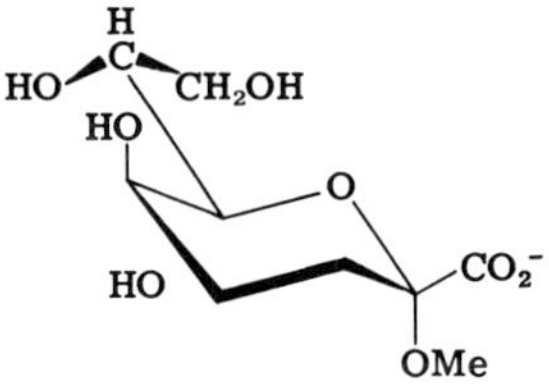

63

sumption was shown to be at least partially untenable by Cherniak and coworkers[102] who, on the basis of similar experiments, concluded that the distribution of the tautomeric forms of ammonium KDO in an aqueous solution is as follows: α-pyranose, ~60; β-pyranose, ~11; α-furanose, ~20; and β-furanose, ~9% (the assignment of the anomeric configurations to the furanoses being tentative). Using the 3,5-dideoxy-octulosonate derivative **64** (configuration at C-4 unknown), which they had prepared from 2-deoxy-D-*erythro*-pentose by the Cornforth reaction[18] (see Scheme 19), the authors demonstrated that the model compound, in aqueous solution, exhibits no ^{13}C-n.m.r. signals attributable to the presence of the 1,4-lactone **65**. It should also be pointed out that crystalline ammonium KDO can usually be recovered in excellent yields from its aqueous solutions.[103] This experience makes improbable the presence of significant concentrations of lactones, unless hydrolysis of the lactones should occur during crystallization.

A study of the tautomeric forms of ammonium KDO in aqueous so-

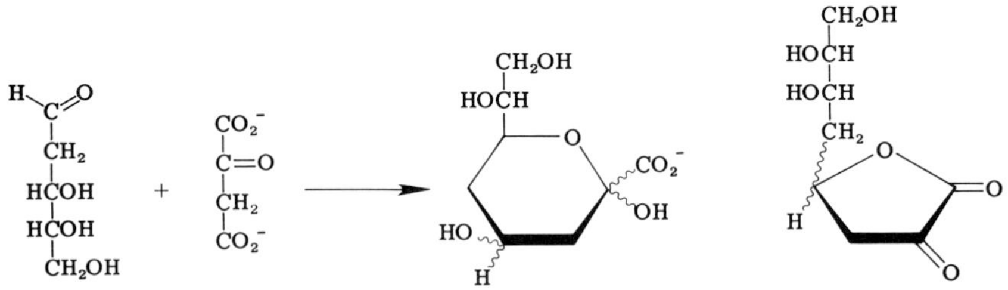

Scheme 19

(102) R. Cherniak, R. G. Jones, and D. S. Gupta, *Carbohydr. Res.*, 75 (1979) 39–49.
(103) F. M. Unger, unpublished observation.

lutions, employing 360-MHz, ^{1}H-n.m.r. spectroscopy, and using such model compounds as **63** and **66**, is currently being completed by Vliegenthart's group at Utrecht.[104]

(ii) **KDO-Containing Acidic Exopolysaccharides (K-Antigens).**—Bhattacharjee and coworkers[101] reported on an extensive, structural study, involving ^{13}C-n.m.r. spectroscopy, of the capsular polysaccharide antigen from *Neisseria meningitidis* serogroup 29-e (compare[83] Section II,3). These authors concluded that the structure of the polysaccharide is α-Gal*p*NAc-(1 → 7)-β-KDO*p*-(1 → 3)-Gal*p*NAc. The native, 29-e polysaccharide is acetylated at O-4 or O-5 of the KDO residues. Some KDO residues lacking O-acetyl groups are also present. The KDO residues in the polysaccharide are of the β-anomeric configuration, and adopt the $^5C_2$(D) conformation in aqueous solution. This is the only instance where the anomeric configuration of a naturally occurring, KDO ketoside has been assigned, and, as this conclusion is of great importance for considerations of KDO biosynthesis (compare Sections V,4 and VI), detailed consideration to its experimental basis is given here.

The assignment of the β-anomeric configuration to the KDO residues in the *Neisseria* polysaccharide[101] was made because of chemical-shift similarities between the ^{13}C-n.m.r. resonances observed for the β-ketosidic model compound, **66**, and those attributable to KDO in the spectrum of the polysaccharide (C-1 of compound **66**, 174.8 p.p.m.; of KDO in the polysaccharide, 174.5 p.p.m.; of the α anomer, **63**, 176.5 p.p.m.). It should be noted, however, that the C-1 resonance in the ^{13}C-n.m.r. spectrum of the exopolysaccharide from *Escherichia coli* LP 1092 occurs at 176.2 p.p.m. and would, by analogous reasoning, indicate the α-anomeric configuration of the KDO residues in this polysaccharide. As many of the biochemical features of KDO are similar, if not analogous, to those of N-acetylneuraminic acid[105] (NeuAc; see Section V), it would seem unlikely that natural KDO derivatives occur in more than one anomeric configuration. Thus, although the assignment of the β-anomeric configuration to the KDO residues in the *Neisseria* polysaccharide is likely to be correct, its experimental foundation is at present somewhat tenuous. The element of inference contained in the argument of Bhattacharjee and coworkers[101] calls for a definitive study aimed at assigning the anomeric configurations of the KDO residues in LPS and in K-antigens. The many influences of linkage patterns and configurational features on the chemical shifts of ^{13}C-

(104) J. F. G. Vliegenthart, personal communication.
(105) R. Schauer, *Angew. Chem.*, 85 (1973) 128–140.

n.m.r. resonances make it probable, in the view of the present author, that ultimate assignments will be based on the interpretation both of ^{13}C- and ^{1}H-n.m.r. data, especially in view of advances in the ^{1}H-n.m.r. spectroscopy of oligosaccharides at very high, magnetic-field strengths.[106]

In a preliminary investigation, Unger and coworkers[107] showed that

TABLE V

Proton Chemical Shifts and First-Order Coupling-Constants for Selected KDO Derivatives[a] (Ref. 81)

	Compound number						
	104	70	112	113	116	117	118
	Solvent						
	C_6D_6	$CDCl_3$	$CDCl_3$	$CDCl_3$	C_6D_6	$CDCl_3$	$CDCl_3$
Protons	Chemical shift (δ)						
3a	2.38	~2.30	2.39	2.10	2.25	δ H-3~ 5.90[b]	2.1–2.2
3e	2.46		2.56	2.36	2.29		
4	5.44	~5.3[c]	5.47	4.89	5.56	5.93	~5.10[c]
5	5.62	5.44	5.42	5.26	5.49	5.49	~5.35[c]
6	4.08	4.22	4.48	4.19	3.89	4.37	3.82
7	5.42	~5.3[c]	5.21	5.17	5.52	5.29	~5.20[c]
8	4.55	4.52	4.45	4.38[d]	4.67	4.64	4.56
8'	4.20	4.16	4.16		4.14	4.24	4.16
	Coupling constant (J, Hz)						
3a,3e	13.5	—[e]	13.5	12.5	12.8	—	—[e]
3a,4	13	—[e]	13.5	12.5	12.5	$J_{3,4\sim}$	—[e]
3e,4	4	—[e]	6.5	5.5	5	1.8[c]	—[e]
4,5	3	—[e]	3.5	2	3	4.5	—[e]
5,6	1.5	1	1	1	1.5	1	
6,7	9.7	10	10	10	9.7	9.5	9.5
7,8	2.5	2	2	—[d,e]	2.5	2.5	2.5
7,8'	4.5	4	4	—[d,e]	5	4.5	4.5
8,8'	12.5	12	12	—[d,e]	12.5	12	12

[a] Spectra were recorded at 90 MHz with tetramethylsilane as the internal standard. [b] 117 contains only one (alkenic) proton at C-3. [c] Signals overlap with others. [d] The signals for H-8,8' almost coalesce, but are still noticeably different. [e] Not determined.

(106) L. D. Hall, S. Sukumar, and G. R. Sullivan, *J. Chem. Soc. Chem. Commun.*, (1979) 292–294.
(107) F. M. Unger, D. Stix, P. W. Taylor, and G. Schulz, *Proc. Int. Symp. Glycoconjugates, 5th, Kiel, 1979*, Thieme, Stuttgart, 1979, pp. 124–125.

TABLE VI

Chemical Shifts and First-Order Coupling-Constants of the Protons
at C-3 of Selected KDO Derivatives (90 MHz)^a (Ref. 81)

Compound number	Chemical shift (δ, p.p.m.)		First-order coupling-constants (Hz)		
	H-3a	H-3e	$J_{3a,3e}$	$J_{3a,4}$	$J_{3e,4}$
1	1.97	1.87	15	15	6
114	1.99	2.43	13.5	14	5.3
66	1.74	2.38	13.5	13.5	5.3
115	1.90	2.10	15	15	6
63	1.79	2.06	15	14	6
119	1.68	2.05	14	13.5	5.5

^{a} Spectra were recorded for solutions in D$_2$O, with tetramethylsilane as the internal standard. ^{b} $J_{3a,2}$ ~14 Hz, $J_{3e,2}$ ~3.5 Hz. ^{c} $J_{3a,2}$ ~13.5 Hz, $J_{3e,2}$ ~3 Hz.

the 90-MHz, ^{1}H-n.m.r. spectrum of a solution of the KDO-containing exopolysaccharide from *Escherichia coli* LP 1092 in deuterium oxide contains four resonances that are distinct from the bulk of signals. These are due to the protons at C-3 of the KDO residues (H-3a: dd at δ 1.80, $J_{3a,3e}$ = $J_{3a,4}$ = ~14 Hz; H-3e: dd at δ 2.25, $J_{3e,3a}$ ~14, $J_{3e,4}$ ~5 Hz) and to the anomeric protons of the two ribofuranosyl residues (H-1': d at δ ~5.31, $J_{1',2'}$ ~1 Hz; H-1'': d at δ ~5.44, $J_{1'',2''}$ ~1 Hz). The difference in chemical shifts between the H-3a and H-3e resonances is ~0.45 p.p.m. In β-ketosidic, model compounds[81] of KDO, this difference is similar (0.26–0.64 p.p.m.; compare Tables V and VI). The corresponding, chemical-shift difference is usually much smaller in α-ketosidic derivatives of KDO (0–0.27 p.p.m., see Tables V and VI). In an aqueous solution of ammonium KDO, the signal due to (equatorial) H-3 occurs[104] at slightly higher magnetic field than that due to axial H-3. A similar effect, involving differences that are more pronounced, has been reported for NeuAc derivatives.[108]

Bhattacharjee and coworkers[101] reported the first synthesis of the methyl α- and β-ketoside carboxylates, **63** and **66**, to be used as model compounds for the n.m.r.-spectroscopic analysis of the *Neisseria meningitidis* exopolysaccharide. The method used was that employed by Kuhn and coworkers[109] for the synthesis of the analogous methyl β- and α-ketosides of NeuAc (compounds **67** and **68**; compare Section IV,5).

(108) J. Haverkamp, L. Dorland, J. F. G. Vliegenthart, and R. Schauer, *Int. Symp. Carbohydr. Chem., 9th*, (1978), Abstr. D-7.
(109) R. Kuhn, P. Lutz, and D. L. MacDonald, *Chem. Ber.*, 99 (1966) 611–617.

The anomeric configurations of compounds **63** and **66** were inferred by the authors[101] from a comparison of their ^{13}C-n.m.r. spectra with those of compounds **67** and **68**. This assignment was confirmed by Unger and coworkers,[80] who determined the heteronuclear coupling-constants $^3J_{\text{H3a,CCC-1'}}$ in the proton-spin-coupled, ^{13}C-n.m.r. spectra of these ketosides as <1 Hz and ~4 Hz, respectively. Ultimate proof for the α-anomeric configuration of **63** came from the X-ray crystallographic structure determination of compound **60**, carried out by Kratky and coworkers.[82]

The $^5C_2(\text{D})$ conformation of the model compound, **66**, was assigned by Bhattacharjee and coworkers,[101] based on the value of $J_{3a,4}$ (~14.5 Hz), which is indicative of a *trans*-antiperiplanar relationship of H-3*a* and H-4 in **66**. Unger and coworkers,[80,89] using ^{1}H-n.m.r. spectroscopy, confirmed that all pyranoside derivatives of KDO thus far reported are present in the $^5C_2(\text{D})$ conformation (see Tables V and VI). This conformation is also adopted by **60** in the crystalline state,[82] and is obviously determined by the β-equatorial orientation of the C-7–C-8 side-chain. Interestingly, methyl penta-*O*-acetyl-3-deoxy-D-*gluco*-2-octulosonate (**69**) also appears[98] to adopt the $^5C_2(\text{D})$ conformation, despite the presence of three, axially oriented, acetoxyl groups.

IV. CHEMICAL SYNTHESIS AND MONOSACCHARIDE CHEMISTRY OF KDO

1. Synthesis of KDO by the Cornforth Reaction

Following the synthesis by Cornforth and his associates[18] of NeuAc from oxalacetate and 2-acetamido-2-deoxy-D-mannose, Ghalambor and Heath[29,31] prepared KDO (isolated as the crystalline methyl 2,4,5,7,8-penta-*O*-acetyl-3-deoxy-D-*manno*-2-octulopyranosonate, **70**) from D-arabinose and oxalacetate by an analogous reaction (see Scheme 20).

The Cornforth reaction may be considered to proceed by the attack of an enolate species, such as **71**, on the carbonyl group of the sugar, to give the dicarboxylic acid salt, **72**. Decarboxylation gives the 3-deoxy-glyc-2-ulosonic acid salt, **73** (see Scheme 21). The reaction seems to be applicable to a wide range of aldoses, although yields are always low (~30%, maximum). As the carbonyl group of the sugar is diastereotopic, two isomers are formed in the reaction (if D-arabinose is the substrate, these are the D-*manno* and the D-*gluco* diastereoisomers). It is of interest that the *manno* isomer is preponderant where derivatives of D-arabinose have been subjected to the Cornforth reaction, and the products have been analyzed regarding their configuration at C-4. Detailed mechanistic studies of the Cornforth reaction have not come to the attention of the present writer.

In the original work of Ghalambor and Heath,[29,31] attempts at opti-

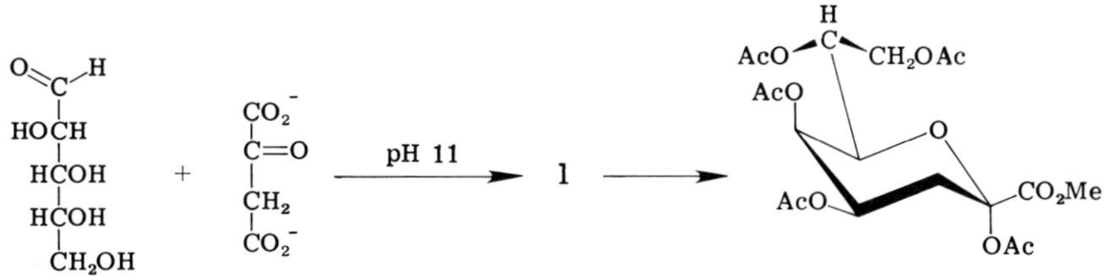

Scheme 20

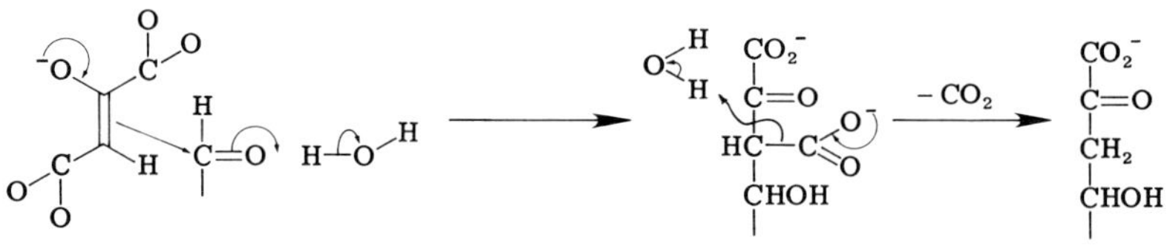

Scheme 21

mization of yields were apparently not made; in fact, the yields reported[29,31] are somewhat higher (~30%) if the calculation is based on the formation of TBA-reactive[11] materials. The low yields may thus be due to losses during processing and derivatization. The synthesis of KDO by the Cornforth reaction was reinvestigated by Hershberger and coworkers,[23] who modified the order of addition of the reagents, thereby obtaining greatly improved yields (~25%) of ammonium KDO, which they obtained, for the first time, in crystalline form. The procedure as given by Hershberger and coworkers[23] is used to prepare ammonium KDO in ~10-g quantities in the present author's laboratory, as follows.

Ice-water (200 mL) is placed in a beaker surrounded by an ice-water bath. By monitoring the pH value of the solution with the aid of a pH-meter, the pH is adjusted to 10, using 10 M sodium hydroxide solution. With constant stirring, small portions of oxalacetic acid (Aldrich; m.p. 155°; 24.4 g total, 184 mmoles) and 10 M sodium hydroxide solution are added alternately, so as to keep the pH value as close to 10 as possible. Following the addition of the oxalacetic acid, D-arabinose (Sigma; 75.0 g, 0.5 mole) is added all at once. The pH of the reaction mixture is adjusted to 11 with 10 M sodium hydroxide solution, and the mixture is kept for ~2 h at room temperature. The pH of the solution is then adjusted to 7 by using Dowex® AG-50 W-X4 (H⁺) cation-exchange resin (Bio-Rad, analytical grade; 200–400 mesh). The resin is removed by filtration, and the pH of the filtrate is adjusted to 5 with glacial acetic acid (Merck, analytical grade). The solution is then applied to a column (4.5 × 80 cm) of Dowex® AG-1 X-8 (HCO_3^-) anion-exchange resin (Bio-Rad). The column is washed with water (~2 L) to elute unreacted D-arabinose. KDO is eluted by using a linear, salt gradient [water (2 L) in the mixing vessel and 0.5 M ammonium hydrogencarbonate (2 L) in the reservoir], and fractions (25 mL) are collected. The KDO content of the fractions is monitored by thin-layer chromatography (t.l.c.) on plates of silica gel, with 10:10:3 (v/v) chloroform–methanol–water as the irrigant. Spots due to KDO (R_F 0.5–0.67, depending on the concentration) are identified by the bluish-grey coloration they give after spraying with the anisaldehyde–sulfuric acid reagent.[110] By contrast, D-arabinose gives a green color, and by-products eluting at higher salt-concentrations (possibly corresponding to the dicarboxylic acid intermediates, 73) give a brownish-orange coloration with this spray reagent. The fractions containing KDO are pooled, and lyophilized (twice if necessary) to remove most of the inorganic salt. The lyophilizate is taken up in a small volume of water, and the solution is

(110) E. Stahl and U. Kaltenbach, *J. Chromatogr.*, 5 (1961) 351–355.

passed through two columns (2.5 × 100 cm each, in tandem) of Sephadex-G10 (Pharmacia). The conductivity is measured on line, in order to assess the separation between ammonium KDO and inorganic salt. KDO-containing fractions are located by t.l.c. as before, pooled, and lyophilized. Ammonium KDO is crystallized from water–ethanol, and the crystals are air-dried. The yield is ~10 g (19.8%), m.p. 121–124°, $[\alpha]_D^{23}$ +40.3° (c 1.93, water); lit.[23] m.p. 121–123°, $[\alpha]_D^{27}$ +42.3° (c 1.7, water).

In the experience of the present author, minor deviations from this procedure may result in decreased yields. Oxalacetic acid of high quality is essential, and this should be verified by a melting-point determination prior to use. Decarboxylation of oxalacetate has been reported[111] to occur rapidly at pH ~7, and it should be kept to a minimum by maintaining the pH as close to 10 as possible when dissolving the oxalacetic acid. A modification of the Cornforth reaction is the cobalt(II)-ion-catalyzed condensation of D-erythrose 4-phosphate with oxalacetate to give 3-deoxyheptulosonic acid 7-phosphate[112] (as a mixture of the *arabino* and *ribo* isomers). Other procedures for the preparation of KDO will be discussed in subsections 3 and 4 of this Section.

The Cornforth reaction has frequently been applied for the synthesis of KDO derivatives, starting from appropriate derivatives of D-arabinose. Thus, Charon and Szabó[17] prepared 4-O-formyl-2-O-methyl-D-arabinose (**74**) by periodate degradation of 3-O-methyl-D-glucose. Application of the Cornforth reaction to **74** gave the crystalline compound, **75**, which presumably has the D-*manno* configuration, although this point was not established (see Scheme 22).

Compound **75** served as a model for a study of the TBA reactivity of 5-O-substituted KDO derivatives[17] (compare Section II,1). A synthesis of KDO 8-phosphate (**2**), which occurs in Nature as the product of the KDO 8-phosphate synthetase reaction[9] (Scheme 1; compare Section V,2 and 3), was performed by Charon and Szabó.[113] 2-O-Benzyl-D-

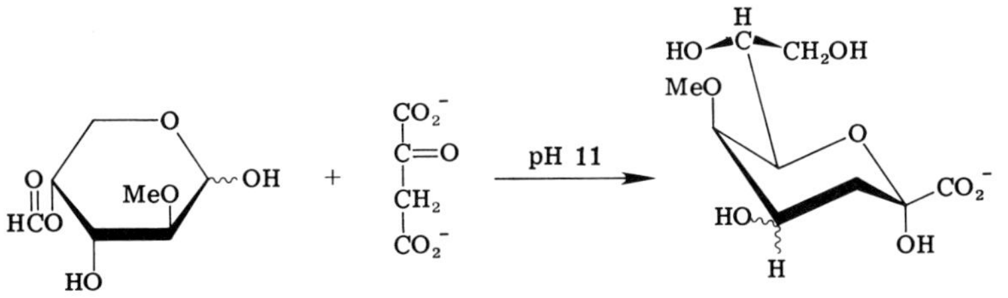

Scheme 22

(111) F. C. Kokesh, *J. Org. Chem.*, 41 (1976) 3593–3599.
(112) K. M. Herrmann and M. D. Poling, *J. Biol. Chem.*, 250 (1975) 6817–6821.
(113) D. Charon and L. Szabó, *J. Chem. Soc., Perkin Trans. 1*, (1976) 1628–1632.

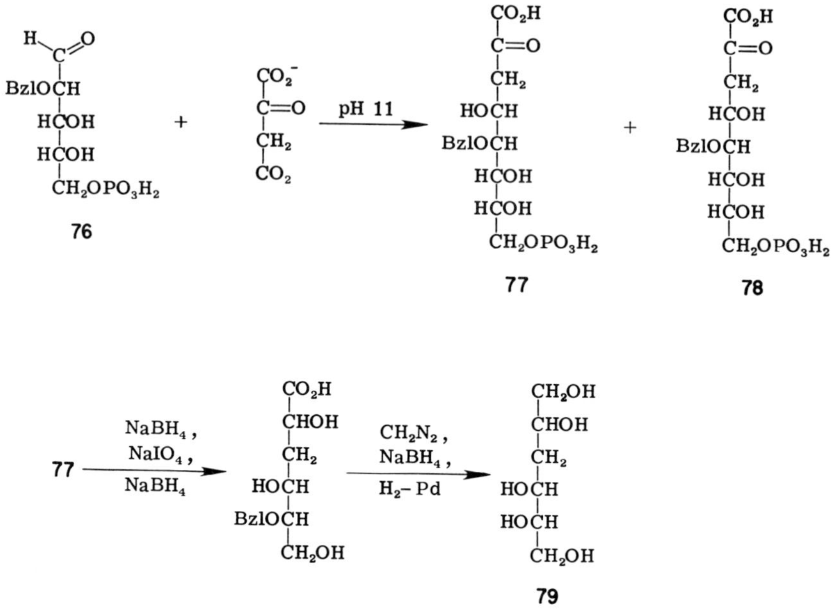

Scheme 23

arabinose 5-phosphate[114] (**76**) was condensed with oxalacetate, to give a mixture of epimers, **77** and **78**. The preponderant product, **77**, was shown to be of the D-*manno* configuration by relating it to the epimeric pair of alditols (**79**) derived from "α"- and "β"-L-glucometasaccharinic acids (see Scheme 23; compare Scheme 12).

L. Szabó and coworkers[115] prepared 2-*O*-phosphono-D-arabinose (**82**) by a sequence of reactions (see Scheme 24). Benzyl 3,4-*O*-isopropylidene-β-D-arabinopyranoside[116] (**80**) was esterified with 2-cyanoethyl phosphate in the presence of dicyclohexylcarbodiimide, to give **81**, from which the protecting groups were sequentially removed by use of sodium hydroxide, treatment with acid, and hydrogenation in the presence of palladium-on-charcoal, to give **82**. Cornforth reaction of **82** gave a mixture of the D-*manno* and D-*gluco* diastereoisomers, **83** and **84**, which were separated by ion-exchange chromatography. As in previous studies,[113] the D-*manno* isomer, **83**, preponderated by a factor of 3. Compound **83** was used as a model

(114) J. Stverteczky, P. Szabó, and L. Szabó, *J. Chem. Soc., Perkin Trans. 1*, (1973) 872–874.
(115) R. S. Sarfati, M. Mondange, and L. Szabó, *J. Chem. Soc., Perkin Trans. 1*, (1977) 2074–2077.
(116) C. E. Ballou, *J. Am. Chem. Soc.*, 79 (1957) 165–166.

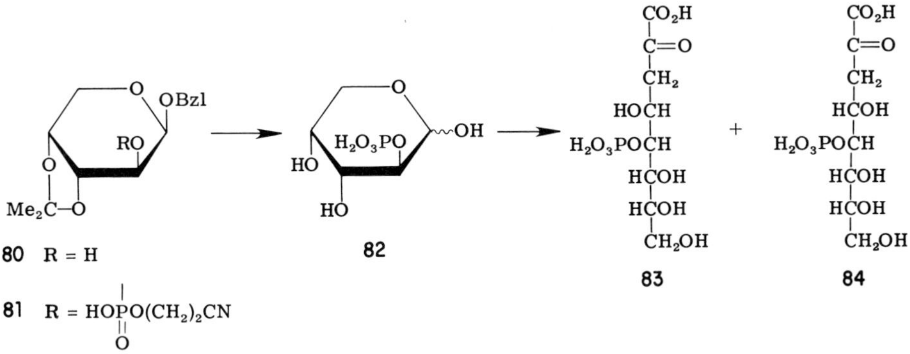

3-DEOXY-D-manno-2-OCTULOSONIC ACID (KDO)

80 R = H

81 R = HOPO(CH$_2$)$_2$CN, ‖O

82

83

CO_2H
$C=O$
CH_2
$HOCH$
H_2O_3POCH
$HCOH$
$HCOH$
CH_2OH

84

CO_2H
$C=O$
CH_2
$HCOH$
H_2O_3POCH
$HCOH$
$HCOH$
CH_2OH

Scheme 24

compound for the study of KDO derivatives contained[71] in *Bordetella pertussis* LPS (compare Section II,2).

Sarfati and Szabó[117] reported on synthesis of 5-O-α-L-rhamnopyranosyl-, 5-O-β-D-glucopyranosyl-, 5-O-(2-acetamido-2-deoxy-β-D-glucopyranosyl)-, and 5-O-(β-D-glucopyranosyluronic acid)-3-deoxyoctulosonic acids by Cornforth reactions of the appropriate 2-O-glycosyl-D-arabinose derivatives. The stabilities of these disaccharides towards acid hydrolysis, and their behavior in the TBA assay,[11] were studied. The Cornforth synthesis with 2-deoxy-D-*erythro*-pentose[102] has been discussed in Section III,2 (compare Scheme 19). Cornforth syntheses with azidodeoxy derivatives of D-arabinose,[118] to give azidodeoxy derivatives of 3-deoxyoctulosonic acids, will be discussed in Section VI.

2. Synthesis of KDO by the Kuhn Reaction

The procedure that Kuhn and Baschang[99] had reported for the synthesis of NeuAc was extended by Hershberger and Binkley[100] to a synthesis of KDO, as follows. Condensation of di-*tert*-butyl oxalacetate (**85**; see Scheme 25) with D-arabinose gave the epimeric mixture of lactone esters, **86** and **87**, which was separated by fractional recrystallization. When **86** was heated in aqueous solution, the enol lactone, **88**, was produced; from **87**, an enol lactone diastereomeric with **88** was obtained under these conditions. Compound **88** was converted into ammonium KDO by treatment with aqueous ammonia.

The enol lactone **88** is also an intermediate in a Wittig synthesis[98] of KDO (compare Section IV,4). Hershberger and Binkley[100] utilized cer-

(117) R. S. Sarfati and L. Szabó, *Carbohydr. Res.*, 65 (1978) 11–22.
(118) F. M. Unger, R. Christian, and P. Waldstätten, *Carbohydr. Res.*, 69 (1979) 71–77.

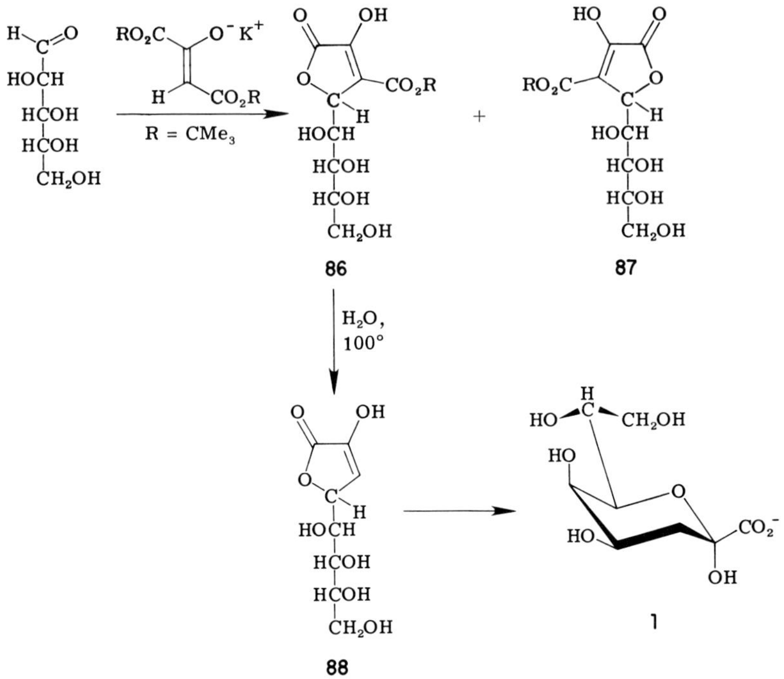

Scheme 25

tain of the intermediates in their synthesis of KDO in order to demonstrate the D-*manno* configuration of natural KDO. Ozonolysis of **86** gave only D-mannose, and analogous treatment of **87** yielded only D-glucose. On inspection of molecular models of the lactone esters, **86** and **87**, so constructed as to accommodate the *erythro* side-chains (R in Fig. 11) *trans* to the bulky *tert*-butoxycarbonyl groups, the authors found a dihedral angle of ~160° between H-4 and H-5 of **86**, and a dihedral angle of ~70° between H-4 and H-5 of **87**. These indications

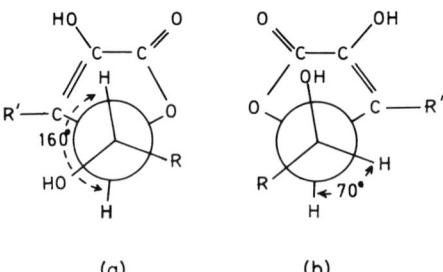

FIG. 11.—"Newman" Projection-Formulas, Viewed Along the C-4–C-5 Bond in a, Compound **86**, and b, Compound **87** (from Ref. 100, with permission).

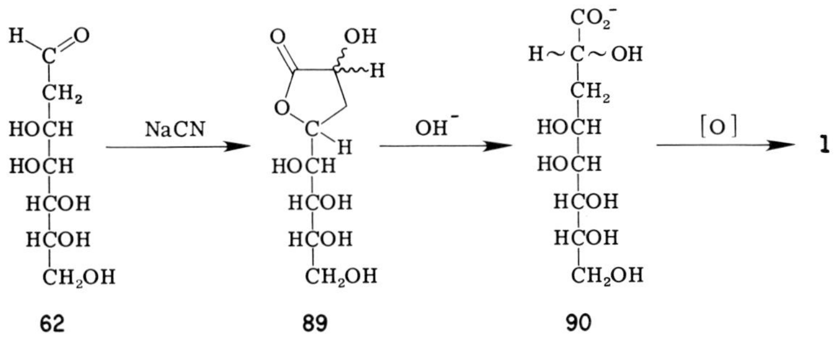

3-DEOXY-D-*manno*-2-OCTULOSONIC ACID (KDO)

Scheme 26

were borne out by the fact that $J_{4,5}$ of the lactone (which, following subsequent steps, gave a product identical with natural KDO) had a value of 4.45. Hence, this compound was assigned the D-*manno* configuration, and so it would have structure **86**. The value of $J_{4,5}$ of the epimeric lactone, **87**, was <0.5 Hz.

3. Cyanohydrin Synthesis of KDO

Heath and Ghalambor[29,31] had obtained a 2-deoxyheptose of unknown configuration by degradation[90] with ceric sulfate of **15** (see Scheme 8), the product obtained upon reduction of KDO with sodium borohydride. Perry and Adams[94] had shown this 2-deoxyheptose to be identical with 2-deoxy-D-*manno*-heptose (**62**; see Scheme 18), synthesized by Perry.[91]

When compound **62** was treated with sodium cyanide under the conditions of the Fischer–Kiliani cyanohydrin synthesis, the epimeric pair of the 3-deoxy-D-*glycero*-D-*galacto*- and 3-deoxy-D-*glycero*-D-*talo*-octonolactones, **89**, was obtained in good yield[95,96] (see Scheme 26). Compound **89** was converted into the epimeric mixture of the corresponding carboxylate salts, **90** (compare **15**, Scheme 8), and **90** was oxidized by the action of a vanadium chloride–potassium chlorate reagent[119,120] to give[95,96] KDO.

4. Wittig Syntheses of KDO

Kochetkov and coworkers[78,121] synthesized KDO by a Wittig route, starting from 2,3,4,5-tetra-*O*-acetyl-*aldehydo*-D-arabinose (**91**), as follows. The reaction of **91** with (*tert*-butoxyoxalyl)methylenetriphen-

---

(119) P. P. Regna and B. P. Caldwell, *J. Am. Chem. Soc.*, 66 (1974) 243–244.
(120) D. B. Sprinson, J. Rothschild, and M. Sprecher, *J. Biol. Chem.*, 238 (1963) 3170–3175.
(121) N. K. Kochetkov, B. A. Dmitriev, and L. V. Backinowsky, *Carbohydr. Res.*, 5 (1967) 399–405.

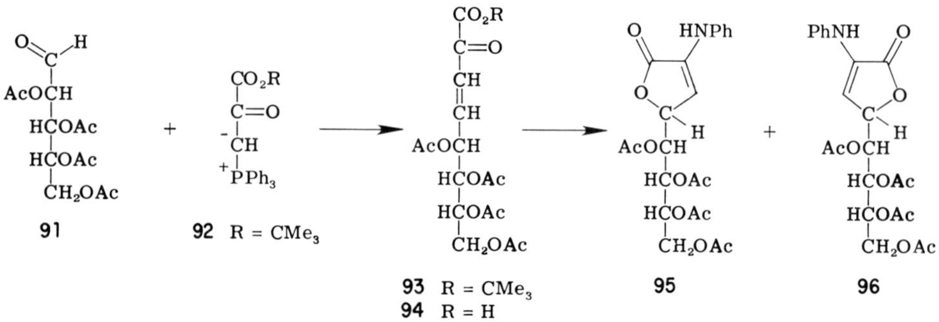

Scheme 27

ylphosphorane (**92**) gave the Wittig adduct **93** (see Scheme 27). Trifluoroacetic acid-catalyzed hydrolysis of the *tert*-butyl group of **93** gave the carboxylic acid, **94**, which was converted into the epimeric pair of enaminolactones, **95** and **96**, by the action of aniline hydrochloride; *N*-benzylideneaniline was added to the reaction mixture to remove water and hydrogen chloride. Compounds **95** and **96** were separated by fractional recrystallization. Methanolysis of **95**, followed by acid-catalyzed hydrolysis of the product, gave KDO. The L enantiomer of KDO[122] can be obtained by this reaction by using the L enantiomer of **91** as the starting material. Like the Cornforth and Kuhn syntheses of KDO (see this Section, 1 and 2), the synthesis reported by the Kochetkov group involves the formation of a chiral center at C-4, whereupon the separation of epimers (with concomitant loss in yields) becomes necessary.

Hellwig and coworkers[97,98] reported on a Wittig synthesis of KDO, starting from 2,3,4,5,6-penta-*O*-acetyl-*aldehydo*-D-mannose (**97**). The reaction of **97** with methoxy(methoxycarbonyl)methylenetriphenylphosphorane (**98**) gave the Wittig adduct, **99** (see Scheme 28). The enol ether grouping in **99** is unusually resistant to acid-catalyzed hydrolysis, and conversion of **99** into KDO was therefore achieved by an indirect route. Ethoxymercuration[123,124] of **99**, followed by treatment with sodium borohydride, gave the deoxyacetal ester, **100**, which was converted into the lactone, **101**, by the action of sodium methoxide in methanol. Trifluoroacetic acid-catalyzed hydrolysis of **101** yielded the enol lactone, **88** (compare Scheme 25), from which KDO was obtained

(122) N. K. Kochetkov, B. A. Dmitriev, and L. V. Backinowsky, *Izv. Akad. Nauk SSSR, Otd. Khim. Nauk,* (1968) 2341.
(123) G. R. Inglis, J. C. P. Schwartz, and L. McLaren, *J. Chem. Soc.,* (1962) 1014–1019.
(124) J. M. Beau, P. Sinaÿ, J. P. Kamerling, and J. F. G. Vliegenthart, *Carbohydr. Res.,* 67 (1978) 65–77.

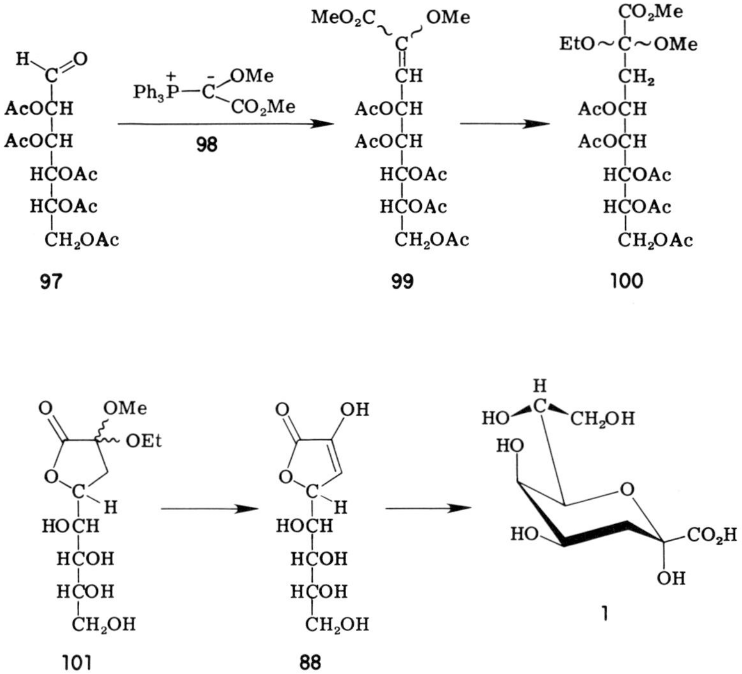

Scheme 28

by treatment with aqueous ammonia, as described by Hershberger and Binkley.[100] The yield of KDO is ~45%, based on 97. The D-*gluco* epimer of KDO was prepared by an analogous procedure.[97,98] NeuAc[125] and its 4-methyl ether derivative[124] had previously been synthesized by similar routes.

5. Monosaccharide Chemistry of KDO

Mild treatment of KDO with an acid results[25] in the formation of the enol lactone 88, which is readily hydrolyzed to give KDO. Through formation of 88, some KDO may escape analysis by the TBA assay[11] following hydrolysis with acetic acid (compare Refs. 25 and 26). The formation of 6-(D-*erythro*-glycerol-1-yl)-3-hydroxypyran-2-one (102) during exposure of KDO to acid has been claimed,[24] but not confirmed.[25] Another product arising from acid treatment of KDO is 5-(D-*erythro*-glycerol-1-yl)-2-furancarboxylic acid[24,25] (103); see Scheme 29.

(125) M. N. Mirzayanova, L. P. Davydova, and G. I. Samokhvalov, *J. Gen. Chem. USSR*, 40 (1970) 663–666; *Chem. Abstr.*, 73 (1970) 25,797k.

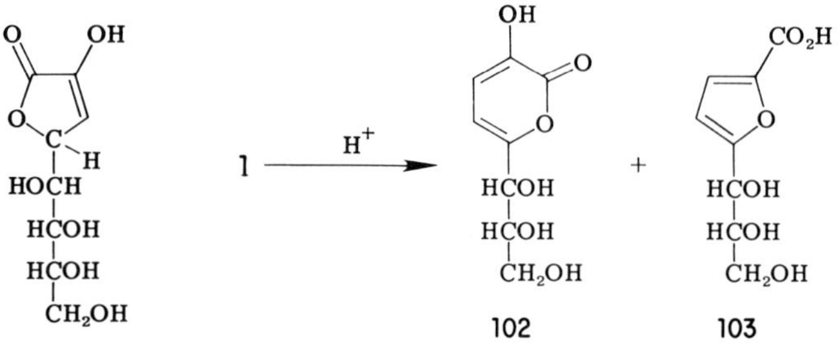

Scheme 29

Deacetylation of the pentaacetate (**104**) of KDO with 0.1 M sodium hydroxide under reflux was reported[23] to give rise to a crystalline product, of unknown structure, that did not give a response in the TBA assay[11] at 550 nm. Treatment of the mixture of the D-*manno* and D-*gluco* epimers of KDO (as obtained by the Cornforth reaction[18] of D-arabinose) with ethanethiol and conc. hydrochloric acid[126] gave compounds **105** and **106** in the ratio of 5:1.

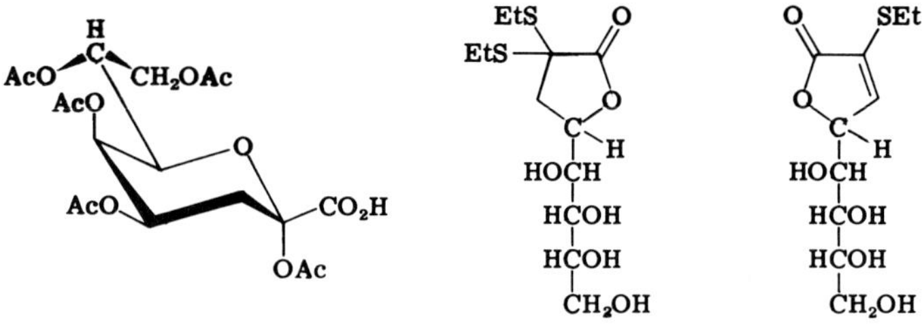

Acetylation of crystalline ammonium KDO with acetic anhydride in pyridine in the presence of 4-(dimethylamino)pyridine[127] gives[79,80] a quantitative yield of the crystalline α-pentaacetate **104**. Attempts at acetylation of KDO without the addition of 4-(dimethylamino)pyridine led to poor yields[29,31] or poorly defined products.[100] Benzoylation of the methyl ester (**107**) of KDO, with benzoyl chloride in pyridine appears to give mainly furanose derivatives (**108**) of unidentified anomeric configuration.[79] Treatment of **108** with hydrogen bromide in glacial acetic acid gave a crystalline ketofuranosyl bromide (presumably **109**) which reacted with methanol in the presence of silver carbonate to give a single methyl furanoside (**110**) of undetermined anomeric configuration (see Scheme 30).

(126) B. A. Dmitriev and L. V. Backinowsky, *Carbohydr. Res.*, 13 (1970) 293–296.
(127) W. Steglich and G. Höfle, *Angew. Chem.*, 81 (1969) 1001.

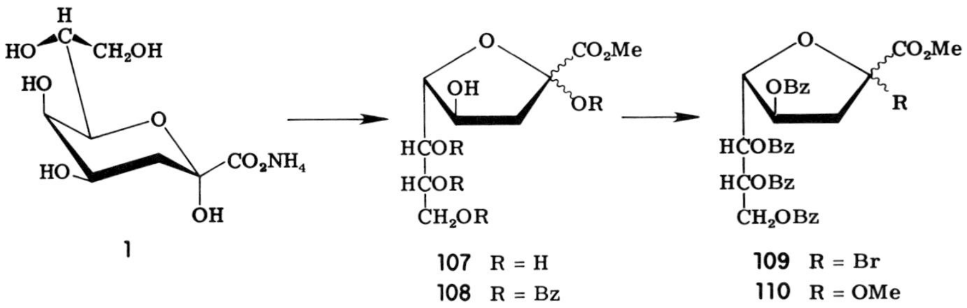

Scheme 30

107 R = H
108 R = Bz

109 R = Br
110 R = OMe

The methyl ester **107** was obtained by treatment of ammonium KDO with methanol in the presence of an acidic catalyst,[79] and it is likely that these esterification conditions (rather than the presence of the ester group of itself) favor the furanoid form. It would be of interest to ascertain whether the methyl ester (**111**) of KDO, when prepared by Zemplén deacetylation of **70** (see Scheme 31), would also yield furanoid derivatives upon benzoylation with benzoyl chloride in pyridine.

Esterification of **104** with diazomethane gives **70** in quantitative yield.[80] The low yield previously obtained[29,31] may have been due to impure starting-material. Compound **70** is crystalline,[29,31] and may be used for the preparation of the ketopyranosyl chloride **112** by the method that Kuhn and coworkers[109] reported for the synthesis of the analogous halide of NeuAc. In this procedure, compound **70** is dissolved in acetyl chloride, and the solution is cooled to $-70°$, saturated with hydrogen chloride gas, and allowed to warm to room temperature in a sealed container. The preparation of **112** in this way was first reported by Bhattacharjee and coworkers,[101] and, shortly thereafter, by Unger and coworkers.[128] The Canadian authors synthesized the

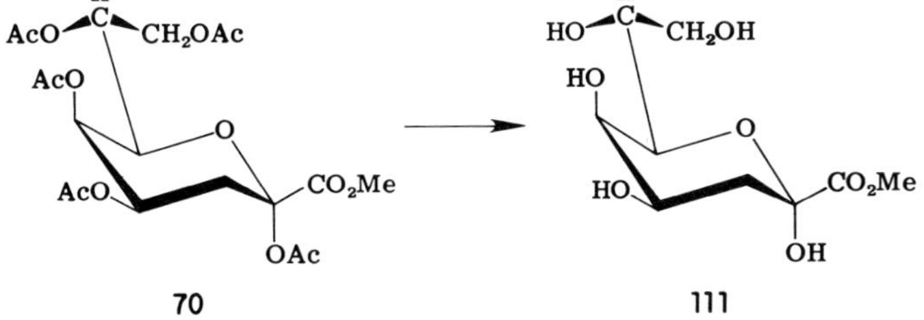

Scheme 31

(128) F. M. Unger, D. Stix, E. Schwarzinger, and G. Schulz, *Int. Symp. Carbohydr. Chem.*, 9th, London, 1978, Abstr. B49, p. 173.

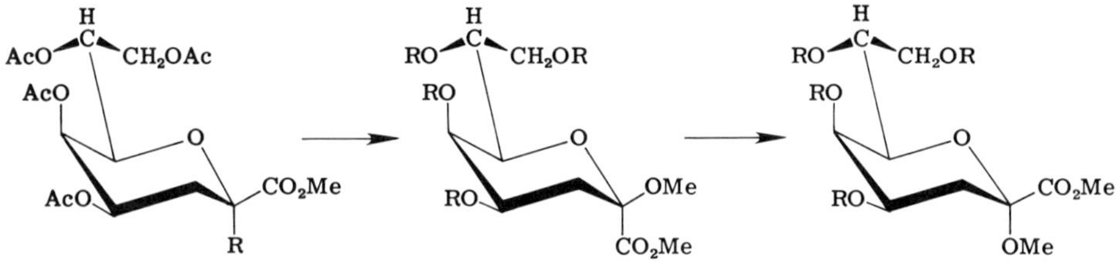

Scheme 32

- 70 R = OAc
- 112 R = Cl
- 113 R = Ac
- 114 R = H
- 115 R = H
- 116 R = Ac

methyl (methyl β-ketopyranosid)ate, **113**, by treatment of **112** with methanol in the presence of silver carbonate (compare Refs. 80 and 109). Zemplén deacetylation of **113** gave the methyl (methyl ketopyranosid)ate **114**, which was converted into its α anomer (**115**) by acid-catalyzed tautomerization in anhydrous methanol (see Scheme 32). The methyl α- and β-ketopyranosidonate salts, **63** and **66**, were prepared by deesterification of **115** and **114**, respectively, with 0.1 M sodium hydroxide.[101] The assignment of anomeric configurations to **63** and **66**, using ^{13}C-n.m.r. spectroscopy,[80,101] was discussed in Section III,2. Acetylation of **115** gave the crystalline derivative **116**, which was used for an X-ray structure determination.[82]

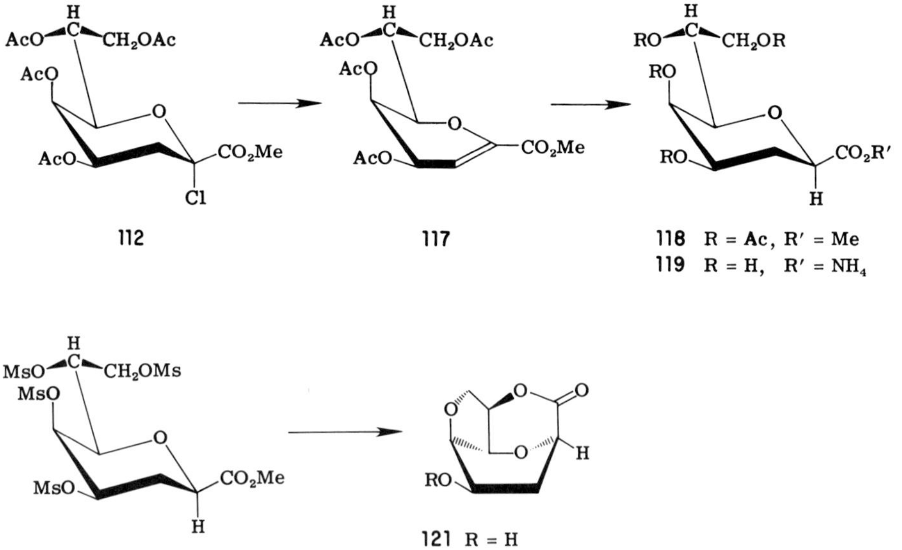

- 112
- 117
- 118 R = Ac, R′ = Me
- 119 R = H, R′ = NH$_4$
- 120
- 121 R = H
- 122 R = COC$_6$H$_4$NO$_2$-p

Scheme 33

Unger and coworkers[81] prepared the dehydro derivative **117** from the halide **112** by dehydrohalogenation with silver (diphenyl phosphate) (see Scheme 33). Catalytic hydrogenation of **117** gave the crystalline tetrahydropyrancarboxylate **118**. Alkaline deesterification of **118** gave the ammonium carboxylate salt (**119**), which is of interest as an enzyme inhibitor (see Section VI). The tetra-*O*-(methylsulfonyl) derivative[129] **120**, derived from **118** by Zemplén deacetylation, and treatment of the product with methanesulfonyl chloride in pyridine in the presence of 4-(dimethylamino)pyridine, was converted into a tricyclic lactone (**121**) by treatment with sodium methoxide in methanol. Compound **121** was identified[129] as its crystalline 4-*p*-nitrobenzoate (**122**). The formation of **121** confirms the (*R*) configuration at C-2 of compounds **118–122**, as deduced[81] from the ^{1}H-n.m.r. spectrum of **118**.

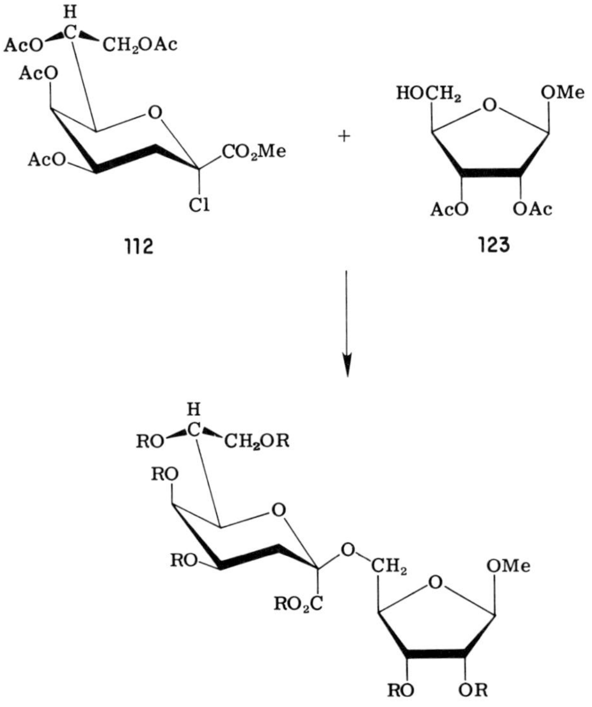

124 R = Ac, R' = Me
125 R = H, R' = Me
126 R = H, R' = NH$_4$

Scheme 34

(129) D. Stix, G. Schulz, and F. M. Unger, manuscript in preparation.

The Koenigs–Knorr reaction[109,130] of the halide **112** with methyl 2,3-di-*O*-acetyl-β-D-ribofuranoside (**123**) in 3:1 (v/v) benzene–1,4-dioxane in the presence of silver carbonate gave a low yield of the disaccharide **124** (see Scheme 34). Compound **124** and its deacetylation and deesterification products, **125** and **126**, were used as model compounds for ¹H- and ¹³C-n.m.r. studies[47,107] of the KDO-containing exopolysaccharides from *Escherichia coli* strains LP 1092 and 06:K13:H1 (refs. 84 and 86).

V. ENZYMES OF KDO METABOLISM

The enzyme-catalyzed isomerization of D-*erythro*-2-pentulose 5-phosphate to D-arabinose 5-phosphate diverts a substrate from the intermediary metabolism of the Gram-negative, bacterial cell for exclusive utilization in the biosynthetic pathway that leads to formation of KDO and to its incorporation into LPS. The second step in this enzyme pathway is the condensation of D-arabinose 5-phosphate with enolpyruvate phosphate to give KDO 8-phosphate (**2**; compare Scheme 1) and inorganic phosphate. The 8-phosphate of KDO is cleaved by a specific phosphatase, to give KDO and inorganic phosphate. In the fourth enzyme-catalyzed step of the pathway, KDO reacts with cytidine 5′-triphosphate (CTP) to give cytidine 5′-(3-deoxy-D-*manno*-2-octulosylonic phosphate) (CMP-KDO) and inorganic pyrophosphate. CMP-KDO is the "activated sugar nucleotide" that serves as a donor molecule during the incorporation of KDO groups into LPS and, probably, into KDO-containing exopolysaccharides. The conversions involved in the "KDO-pathway," summarized in Scheme 35, will now be briefly discussed.

1. D-Arabinose 5-Phosphate Isomerase (EC 5.3.1.13)

Studies of this enzyme (EC 5.3.1.13) were first reported by Volk,[131–133] and later by Lim and Cohen.[134] It is noteworthy that this enzyme reaction is reversible. The 41-fold purification of the enzyme from cells of *Propionibacterium pentosaceum*, using conventional procedures, has been reported.[133]

The enzyme is susceptible to heavy-metal inactivation, so that the biochemical manipulations were performed in the presence of (ethylenedinitrilo)tetraacetate (EDTA). The K_M for D-arabinose 5-phos-

(130) W. Koenigs and E. Knorr, *Ber.*, 34 (1901) 957–981.
(131) W. A. Volk, *J. Biol. Chem.*, 234 (1959) 1931–1936.
(132) W. A. Volk, *J. Biol. Chem.*, 235 (1960) 1550–1553.
(133) W. A. Volk, *Methods Enzymol.*, 9 (1966) 585–588.
(134) R. Lim and S. S. Cohen, *J. Biol. Chem.*, 241 (1966) 4304–4315.

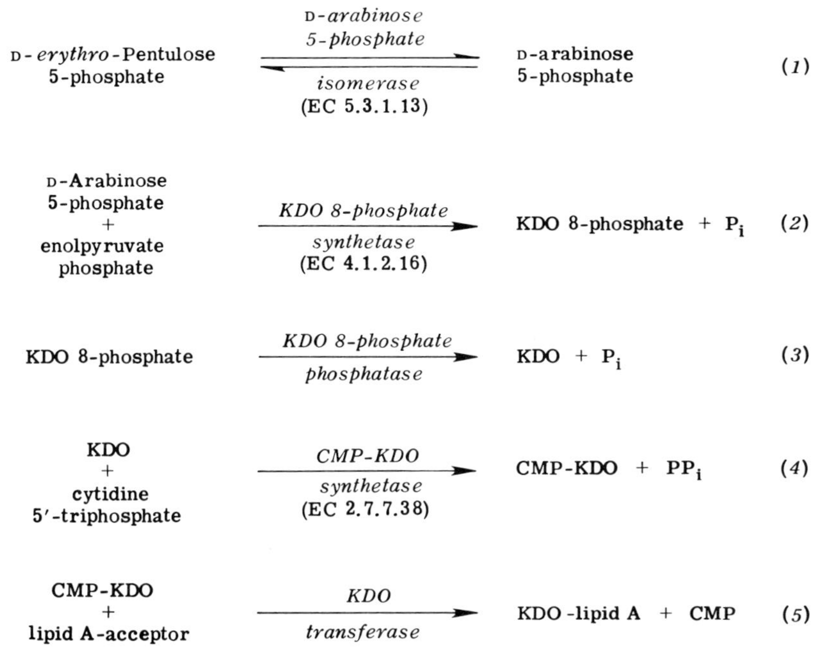

Scheme 35

$$\text{D-}erythro\text{-Pentulose 5-phosphate} \xrightleftharpoons[\text{isomerase (EC 5.3.1.13)}]{\text{D-}arabinose\text{ 5-phosphate}} \text{D-arabinose 5-phosphate} \quad (1)$$

$$\text{D-Arabinose 5-phosphate} + \text{enolpyruvate phosphate} \xrightarrow[\text{synthetase (EC 4.1.2.16)}]{\text{KDO 8-phosphate}} \text{KDO 8-phosphate} + P_i \quad (2)$$

$$\text{KDO 8-phosphate} \xrightarrow[\text{phosphatase}]{\text{KDO 8-phosphate}} \text{KDO} + P_i \quad (3)$$

$$\text{KDO} + \text{cytidine 5'-triphosphate} \xrightarrow[\text{synthetase (EC 2.7.7.38)}]{\text{CMP-KDO}} \text{CMP-KDO} + PP_i \quad (4)$$

$$\text{CMP-KDO} + \text{lipid A-acceptor} \xrightarrow[\text{transferase}]{\text{KDO}} \text{KDO-lipid A} + \text{CMP} \quad (5)$$

phate has been determined to be 1.98 mM. The pH optimum was found to be at pH 8.0. The enzyme was not inhibited by 2.5 mM iodoacetate, but 1.2 mM p-(chloromercuri)benzoate caused a 31% inhibition. At 37°, equilibrium is reached when 22.8% of D-arabinose 5-phosphate has been converted into D-*erythro*-2-pentulose 5-phosphate. The equilibrium constant is 0.295. The equilibrium lies more toward the formation of D-*erythro*-2-pentulose 5-phosphate as the reaction temperature is increased.

2. 3-Deoxyoctulosonate 8-Phosphate Synthetase (KDO 8-Phosphate Synthetase; EC 4.1.2.16)

The reaction catalyzed by KDO 8-phosphate synthetase (reaction 2, Scheme 35) was first observed by Levin and Racker[9] in extracts from *Pseudomonas aeruginosa* (see Scheme 1), and later by Ghalambor and Heath[29] in extracts from *Escherichia coli* 0111 B$_4$ and J-5. In the initial experiments of Levin and Racker,[135] the fate of D-ribose 5-phosphate in crude bacterial extracts was studied, and the KDO 8-phosphate discovered by the authors is really derived from D-ribose 5-phosphate by three, sequential, enzyme-catalyzed reactions (see Scheme 36).

(135) D. H. Levin and E. Racker, *Arch. Biochem. Biophys.*, 79 (1959) 396–399.

D-Ribose 5-phosphate ⟶ D-*erythro*-2-pentulose 5-phosphate   (*1*)

D-*erythro*-2-Pentulose 5-phosphate ⟶ D-arabinose 5-phosphate   (*2*)

D-Arabinose 5-phosphate + enolpyruvate phosphate ⟶ KDO 8-phosphate + P_i (*3*)

Scheme 36

These circumstances became apparent to the authors when they attempted to study the formation of KDO 8-phosphate as catalyzed by purified bacterial extracts. These extracts did not catalyze the formation of KDO 8-phosphate from D-ribose 5-phosphate, but required D-arabinose 5-phosphate as the substrate.[9] Heath and Ghalambor[29] showed that the KDO 8-phosphate synthetase reaction, observed in *Pseudomonas* extracts by Levin and Racker,[9] is also catalyzed by extracts from *Escherichia coli* strains 0 111 B_4 and J-5. Rick and Osborn[136] showed that the KDO 8-phosphate synthetase from a *Salmonella typhimurium* mutant conditionally defective in cell-wall synthesis had a K_M of ~6 mM as compared to a K_M of 170 μM for the enzyme from wild-type cells.

A comprehensive study of KDO 8-phosphate synthetase has been reported by Ray.[137] The author purified the enzyme 450-fold from crude extracts of *Escherichia coli* B cells. The synthetase has a molecular mass of 90,000 ±6,000 daltons and is composed of three identical subunits having an apparent molecular mass of 32,000 ±4,000 daltons. Two pH optima were observed, one being at pH 4.0–6.0 in succinate buffer, and the other, at pH 9.0 in glycine buffer. The isoelectric point of the enzyme is 5.1. The enzyme has an apparent K_M for D-arabinose 5-phosphate of 20 μM and an apparent K_M for enolpyruvate phosphate of 6 μM.

3. 3-Deoxyoctulosonate 8-Phosphate Phosphatase (KDO 8-Phosphate Phosphatase)

In a preliminary study, Berger and Hammerschmid[138] showed that extracts from *Escherichia coli* 0 111 B_4 contain an enzyme that cata-

---

(136) P. D. Rick and M. J. Osborn, *Proc. Natl. Acad. Sci. U. S. A.* 69 (1972) 3756–3760.
(137) P. H. Ray, *J. Bacteriol.*, 141 (1980) 635–644.
(138) H. Berger and F. Hammerschmid, *Biochem. Soc. Trans.*, 3 (1975) 1096–1097.

lyzes the hydrolysis of the phosphoric ester group in KDO 8-phosphate (see reaction 3, Scheme 35), but not that in D-arabinose 5-phosphate or p-nitrophenyl phosphate. Later, Ray and Benedict[139] reported on a definitive investigation of KDO 8-phosphate phosphatase. The authors purified the enzyme ~400-fold from crude extracts of *E. coli* B. The activity of the enzyme was stimulated 3–4-fold by the addition of Co^{2+} or Mg^{2+} to a final concentration of 1 mM. The pH optimum was found at pH 5.5–6.5, and the isoelectric point at 4.7–4.8. The enzyme has a molecular mass of 80,000 ±6,000 daltons and appears to be composed of two identical subunits of molecular mass 40–43.000. The K_M for KDO 8-phosphate was determined to be 58 ±9 μM in the presence of 1.0 mM Co^{2+}, and 180 ±20 μM in the absence of Co^{2+}.

4. CTP:CMP-3-deoxyoctulosonate Cytidylyltransferase (CMP-KDO-synthetase; EC 2.7.7.38)

Ghalambor and Heath[140] were the first to demonstrate the presence of this enzyme in extracts from *Escherichia coli* 0 111 B_4, and they subsequently reported[141] on its 170-fold purification from crude extracts of the same organism. To assay for the enzyme, a special modification of the TBA test is required, as follows. After incubation of KDO and CTP in the presence of the enzyme, the carbonyl groups of the remaining, unreacted KDO are reduced with sodium borohydride, the CMP-bound KDO being, of course, resistant to reduction. The products formed upon borohydride reduction of KDO (**15**, see Scheme 8) do not give rise to formation of a TBA chromophore having[11] λ_{max} 550 nm. Following the decomposition of the excess of borohydride with acetone, mild acid-catalyzed hydrolysis of CMP-KDO yields an amount of KDO that is stoichiometrically equivalent to the amount of CMP-KDO formed in the enzyme reaction, which is determined by application of the TBA test.[11] It should be pointed out that the borohydride-reduction products of KDO (**15** in Scheme 8) give rise to the 532-nm chromophore under the conditions of Warren[13] (compare Section II,1 and Ref. 51). This color formation can disturb the assay for CMP-KDO synthetase unless conditions are chosen wherein the 532-nm chromophore is not formed.

The principle of the assay procedure for CMP-KDO synthetase had been developed previously, to assay for CMP-NeuAc synthetase in ex-

(139) P. H. Ray and C. D. Benedict, *J. Bacteriol.*, 142 (1980) 60–68.
(140) M. A. Ghalambor and E. C. Heath, *Biochem. Biophys. Res. Commun.*, 10 (1963) 346–351.
(141) M. A. Ghalambor and E. C. Heath, *J. Biol. Chem.*, 241 (1966) 3216–3221.

tracts from hog submaxillary gland[142] and from *Neisseria meningitidis*.[143] For CMP-KDO synthetase from *E. coli* 0 111 B$_4$, Ghalambor and Heath[141] reported a K$_M$ (CTP) of 220 μM and a K$_M$ (KDO) of 800 μM. The enzyme exhibits an absolute requirement for Mg^{2+} ions. The pH optimum was found at pH 7.8 in Tris buffer. The CMP-KDO was isolated by chromatography on DEAE-cellulose, elution being performed by application of a linear gradient of lithium chloride. The "nucleotide sugar" was characterized as follows. (*i*) The ratios of cytidine to phosphorus to borohydride-resistant KDO were found to be 1.00:1.05:0.95. (*ii*) The ultraviolet absorption spectrum of CMP-KDO was indistinguishable from that of CMP. (*iii*) Following exposure of CMP-KDO to pH 2 for 5 min at room temperature, borohydride-resistant KDO was no longer detectable, and all of the material had been converted into CMP and KDO. (*iv*) The nucleotide was resistant to hydrolysis catalyzed by snake venom 5'-nucleotidase, indicating the presence of a phosphoric diester.

For considerations of KDO biosynthesis and transfer into macromolecules, it would be important to know the anomeric configuration of KDO in CMP-KDO. However, due to the extreme susceptibility of CMP-KDO to hydrolysis,[140,141] appropriate n.m.r. measurements have not been reported. As a hypothesis, the present author assumes that the KDO group in CMP-KDO has the α-D (axial) anomeric configuration (structure **127**, Scheme 37), for the following reasons. (*i*) The biochemical reactions leading to "activation" and macromolecular incorporation of KDO are similar to those of the corresponding pathway for NeuAc.[144] (*ii*) As may be seen in Scheme 37, free NeuAc (**5**) is mainly present as the β-D (axial), pyranose form in the $^2C_5$ conformation.[145] Its enzyme-catalyzed reaction with CTP may be viewed as being a nucleophilic displacement, by the anomeric hydroxyl group, of pyrophosphate, at the α-phosphorus atom of CTP. This reaction would occur with retention of the β-D-anomeric configuration of NeuAc. Indeed, NeuAc in CMP-NeuAc (**128**) has been shown[146] to have the β-D (axial) anomeric configuration. (*iii*) Transfer of NeuAc residues from CMP-NeuAc to glycoconjugates proceeds with inversion of configura-

(142) S. Roseman, *Proc. Natl. Acad. Sci. U. S. A.*, 48 (1962) 437–441.
(143) L. Warren and R. S. Blacklow, *Biochem. Biophys. Res. Commun.*, 7 (1962) 433–438.
(144) N. Sharon, *Complex Carbohydrates*, Addison–Wesley, Reading, Massachusetts, 1975, pp. 345–368.
(145) R. W. Ledeen and R. K. Yu, in A. Rosenberg and C. L. Schengrund (Eds.), *Biological Roles of Sialic Acid*, Plenum, New York, 1976, pp. 1–59.
(146) J. Haverkamp, T. Spoormaker, L. Dorland, J. F. G. Vliegenthart, and R. Schauer, *J. Am. Chem. Soc.*, 101 (1979) 4851–4853.

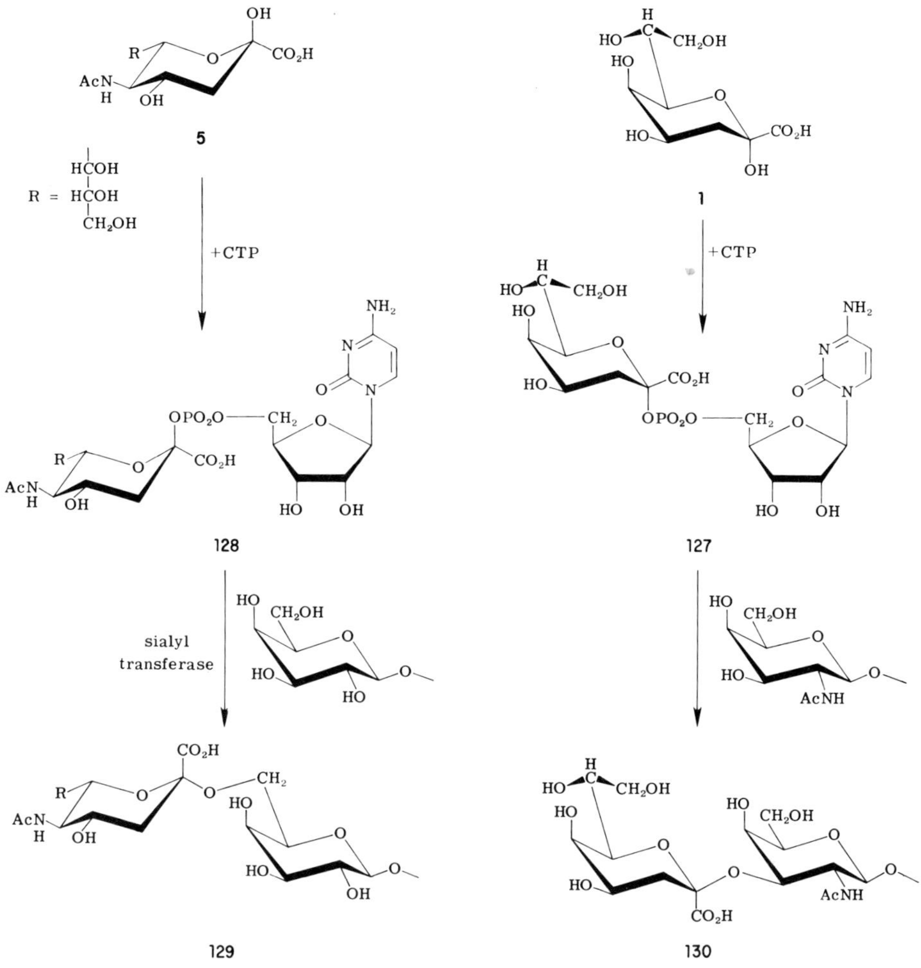

Scheme 37

tion at the anomeric center of NeuAc, so that the acid is present in the α-D (equatorial) anomeric configuration in its natural, ketosidic derivatives[109] (**129**). (*iv*) KDO has been shown to be present in aqueous solutions preponderantly as the α-D (axial), pyranose form in the $^5C_2$ conformation[80,102] (**1**, Scheme 37). Assuming that the formation of CMP-KDO occurs similarly, with retention of the anomeric configuration of the KDO group, this would have the α-D (axial) anomeric configuration in CMP-KDO (**127**). (*v*) Transfer of KDO groups from CMP-KDO with inversion would result in the β-D (equatorial) anomeric configuration of KDO residues, as deduced by Bhattacharjee and coworkers[101] for the KDO-containing exopolysaccharide from *Neisseria meningitidis* serogroup 29-e (**130**, Scheme 37).

Whereas *Escherichia coli* O 111 B_4 contains only one, cytoplasmically located, CMP-KDO synthetase activity, *E. coli* LP 1092 appears to contain two, chromatographically distinct, CMP-KDO synthetase activities. One of these is chromatographically similar to the enzyme from *E. coli* O 111 B_4, whereas the other can be eluted from DEAE-cellulose at significantly lower salt concentration.[85] The function, if any, of the additional CMP-KDO-synthetase activity is unknown. It has been suggested[85] that the enzyme plays a role in the biosynthesis of the KDO-containing exopolysaccharide from *E. coli* LP 1092.

5. CMP-3-deoxyoctulosonate:3-deoxyoctulosylono-lipid A 3-deoxyoctulosylonotransferase (KDO-transferase)

Enzyme-catalyzed transfer of KDO from CMP-KDO (**127**) into a lipid-A acceptor has been studied by Heath and coworkers,[1] using a cell-free system from *Escherichia coli* O 111 J-5. LPS from the organism did not function as an acceptor, and only weak acceptor-activity was displayed[147,148] by lipid A preparations obtained by mild, acid-catalyzed hydrolysis of LPS. Base-catalyzed, hydrolytic removal of the ester-linked, but not the amide-linked, fatty acid residues from lipid A resulted in an acceptor of maximal activity[1] (see Scheme 38; compare Scheme 39).

The study of KDO transfer was greatly facilitated through the discovery, by Osborn's group, of a lipid acceptor molecule, probably corresponding to the natural KDO acceptor, in a temperature-sensitive

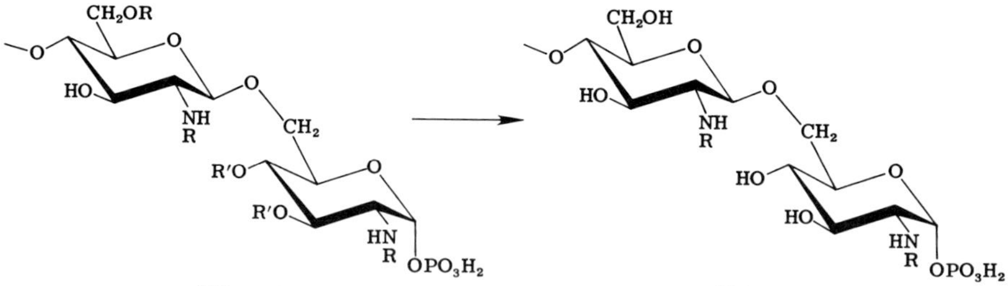

R = (R)-3-hydroxytetradecanoyl,
R' = long-chain acyl

Scheme 38

(147) O. Westphal and O. Lüderitz, *Angew. Chem.*, 66 (1954) 407–417.
(148) O. Lüderitz, C. Galanos, V. Lehmann, H. Mayer, E. T. Rietschel, and J. Weckesser, *Naturwissenschaften*, 65 (1978) 578–585.

Salmonella typhimurium mutant.[136,149] The defect of this mutant lies in the apparent K_M (D-arabinose 5-phosphate) of its KDO 8-phosphate synthetase (compare this Section, 2). This $\bar{K}_M$ increases more than 25-fold between 29 and 42°, so that the cells become increasingly dependent on exogenously supplied D-arabinose 5-phosphate as the growth temperature is raised. Cessation of LPS biosynthesis under nonpermissive conditions is accompanied by the accumulation of a KDO-deficient, precursor molecule.[149] Lehmann[150] and Rick and coworkers[151] described studies directed at the isolation and chemical characterization of such lipid A precursors (for example, **133**, Scheme 39). The

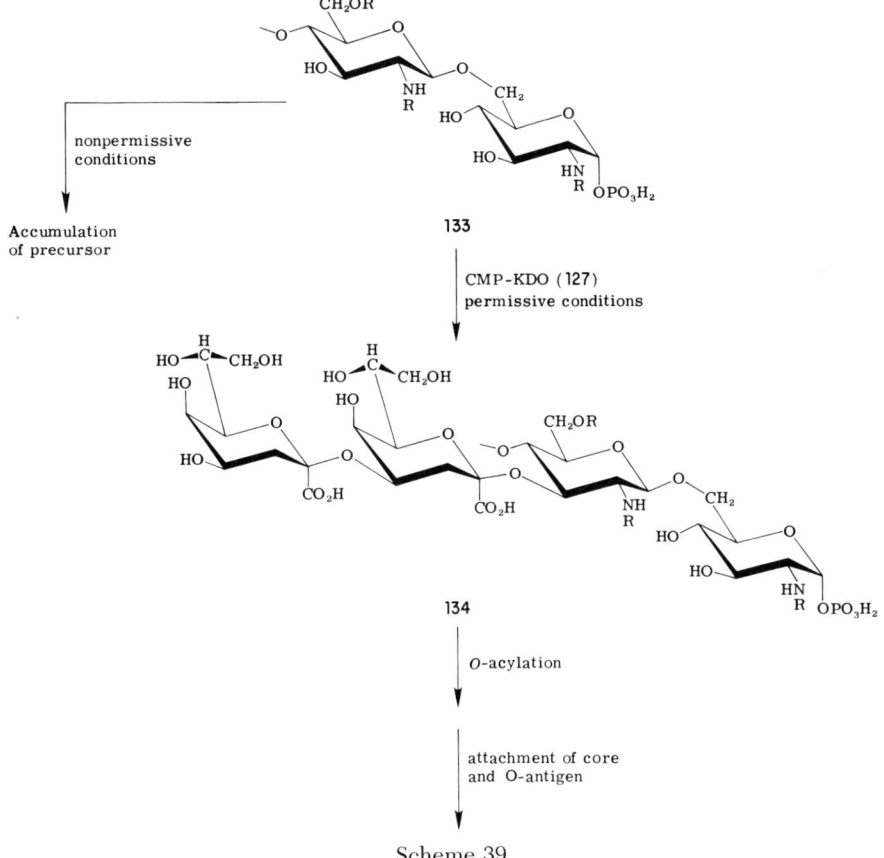

Scheme 39

(149) P. D. Rick and M. J. Osborn, *J. Biol. Chem.*, 252 (1977) 4895–4903.
(150) V. Lehmann, *Eur. J. Biochem.*, 75 (1977) 257–266.
(151) P. D. Rick, L. W.-M. Fung, C. Ho, and M. J. Osborn, *J. Biol. Chem.*, 252 (1977) 4904–4912.

molecule described by Osborn's group[151] is a 2-amino-2-deoxyglucosyl disaccharide containing one ester-linked and two amide-linked 3-hydroxytetradecanoyl residues. The 3-hydroxytetradecanoyl groups in lipid A from various species have been shown to be of the D configuration.[152] The disaccharide is further substituted by two phosphoric monoester groups, but lacks ester-linked 12:0 and 14:0 fatty acids, and KDO. A similar acceptor, containing two ester-linked 3-hydroxytetradecanoyl groups, was described by Lehmann,[150] who also showed, by pulse-chase experiments, that molecules of the type of **133** are, indeed, the precursors that accept KDO *in vivo*. A "neutral precursor," which differs from the "acidic precursors" (**133**) by the additional presence of glycosidically linked 4-amino-4-deoxy-L-arabinose, O-phosphono-2-aminoethanol, and polyamines, has also been described.[153]

Munson and coworkers[70] isolated from *Salmonella typhimurium* a membrane-bound enzyme that catalyzes the transfer of KDO from CMP-KDO to the acceptor molecule **133**. The product (**134**) was shown to contain two KDO residues, and periodate-oxidation studies indicated that the linkage of these is KDOp-(2 → 4)-KDOp. Interestingly, LPS from *Escherichia coli* BB 12 contains only two KDO residues, having the same linkage pattern,[65] and the question has been raised[70] as to why the third KDO residue is not added on in the mutant BB 12 or under *in vitro* conditions. Osborn and coworkers[70] suggested that the incorporation of the O-phosphono-2-aminoethanol group (which normally occurs on the "lateral," (2 → 4)-linked KDO residue[65]) may be required to precede the transfer of the third, "main chain" residue. The transfer of KDO onto lipid A acceptor (**133**), as elucidated by the work of Osborn and her associates, is presented in Scheme 39.

6. 3-Deoxyoctulosonate Aldolase (KDO Aldolase; EC 4.1.2.23)

When grown in a mineral medium containing KDO as the only source of carbon, cells of *Aerobacter cloacae* can be induced to produce an enzyme that catalyzes the cleavage of KDO to give D-arabinose and pyruvic acid.[89] This enzyme was purified ~60-fold by Ghalambor and Heath.[154] It has a pH optimum of 7, a K_M = 6 mM, and an equilibrium constant of 77 mM. The reversible nature of the enzyme reaction can be utilized to synthesize ^{14}C-labelled KDO from D-arabinose plus ^{14}C-pyruvic acid. Cleavage of KDO as catalyzed by KDO aldolase has

(152) E. T. Rietschel, *Eur. J. Biochem.*, 64 (1976) 423–428.
(153) V. Lehmann and E. Rupprecht, *Eur. J. Biochem.*, 81 (1977) 443–452.
(154) M. A. Ghalambor and E. C. Heath, *J. Biol. Chem.*, 241 (1966) 3222–3227.

been used to distinguish chemically synthesized KDO from its *gluco* epimer.[100]

VI. THE DESIGN OF INHIBITORS OF KDO METABOLISM

The cessation, at the nonpermissive temperature, of LPS biosynthesis in conditionally lethal mutants of *Salmonella typhimurium* appears to cause the accumulation of incomplete lipid A precursors in the inner membrane (see Fig. 1), and is followed by a halt in the synthesis of DNA, RNA, and protein. Although the mechanism by which the latter effect is exerted is not clear, it has been speculated, on this basis, that inhibitors of KDO biosynthesis, or of its incorporation into LPS, may eventually be useful as antibacterial agents having specific action against Gram-negative organisms.[102,155,156] Indeed, several groups of workers, mostly in the laboratories of the pharmaceutical industry, are engaged in a search for such inhibitors.

Several observations regarding this aspect have been published, and are briefly mentioned here. 5,6-Dideoxy-6-C-phosphono-D-*arabino*-hexofuranose (**135**), an isosteric phosphonate analog of D-arabinose 5-phosphate, is apparently converted, in the presence of enolpyruvate phosphate, into 3,8,9-trideoxy-9-C-phosphono-D-*manno*-2-nonulosonic acid (**136**) under catalysis by KDO 8-phosphate synthetase from *Escherichia coli* K 235. Compound **136**, an isosteric phosphonate analog of KDO 8-phosphate, is a product inhibitor of the synthetase, and, by the nature of the phosphonate group, is not subject to dephosphorylation as catalyzed by KDO 8-phosphate phosphatase[156] (see Scheme 40). Compound **119** (see Scheme 33) is a weak inhibitor of KDO 8-phosphate synthetase.[81] KDO inhibits KDO 8-phosphate phosphatase,[139] and D-ribose 5-phosphate has an inhibitory

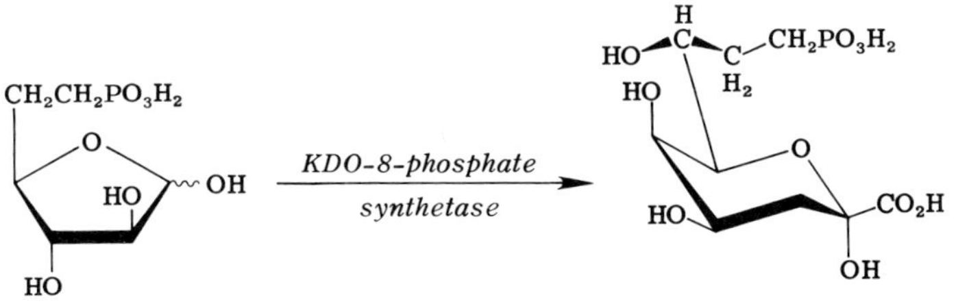

Scheme 40

(155) J. Drews, *Triangle Engl. Ed.*, 16 (1977) 141–149.
(156) F. M. Unger, D. Stix, E. Möderndorfer, and F. Hammerschmid, *Carbohydr. Res.*, 67 (1978) 349–356.

effect upon KDO 8-phosphate synthetase.[137] Several Cornforth reaction-products related to KDO (and of undesignated configuration at C-4) function as substrate analogs of KDO in the CMP-KDO synthetase reaction. These were obtained by condensation of oxalacetate with D-threose, D-altrose, and 5-azido-5-deoxy-D-arabinose, and are represented by structures[157] **137**, **138**, and **139**. Cornforth condensation-

products of azidodeoxy and diazidodideoxy derivatives of D-arabinose were also examined in the CMP-KDO assay.[118] Although most of these products (presumably **140–143**) are unstable and poorly characterized, it appears that the products obtained from 3-azido-3-deoxy- or 3,5-diazido-3,5-dideoxy-D-arabinose (**140** and **141**) are inert in the CMP-KDO synthetase reaction, whereas the products obtained from 2-azido-2-deoxy-D-arabinose (**142**) or 2,5-diazido-2,5-dideoxy-D-arabinose (**143**) exert an inhibitory effect upon the enzyme.[118,157]

(157) F. M. Unger, R. Christian, and P. Waldstätten, unpublished results.

METHYLATION TECHNIQUES IN THE STRUCTURAL ANALYSIS OF GLYCOPROTEINS AND GLYCOLIPIDS

By Heikki Rauvala, Jukka Finne, Tom Krusius, Jorma Kärkkäinen, and Johan Järnefelt

Department of Medical Chemistry, University of Helsinki, Siltavuorenpenger 10 A, SF-00170 Helsinki 17, Finland

I. Introduction 389
II. Permethylation of the Sample 390
   1. Detection of the Completeness of Methylation 391
   2. Use of Potassium *tert*-Butoxide to Generate the
      Methylation Reagent 392
III. Analysis of Permethylated Carbohydrates without
     Depolymerization of the Sample 392
   1. Oligosaccharides 392
   2. Glycolipids and Glycopeptides 394
IV. Degradation of the Permethylated Sample, and Analysis of the
     Partially Methylated Sugars 396
   1. Degradation of the Sample 396
   2. Identification of the Partially Methylated Monosaccharides 398
   3. Quantitation of the Methylated Monosaccharides 402
   4. Interpretation of the Quantitative Data 406
V. Methylation Analysis Combined with Specific Degradation 407
   1. Smith Degradation 407
   2. Degradation with Acid 408
   3. *N*-Deacetylation Followed by Acid Degradation 410
   4. β-Elimination 411
   5. Degradation with the Aid of Exoglycosidases 412
   6. Degradation with the Aid of Endoglycosidases 415

I. Introduction

The carbohydrate chains of glycoproteins[1] and glycolipids are commonly found as essential structures of such biologically important

(1) J. Montreuil, *Adv. Carbohydr. Chem. Biochem.*, 37 (1980) 157–223.

molecules as membrane receptors[1a,2] and antigens.[3,4] Structural information on the protein- and lipid-linked saccharide chains is, therefore, important for understanding of the function of these molecules.

Methylation analysis (and its different applications) is one of the most widely used methods in the structural characterization of carbohydrate chains (reviewed in refs. 5–8). In the present article, the progress achieved, and the problems encountered, in the analysis of animal glycolipids and glycoproteins are discussed, although many of the topics also concern the analysis of carbohydrate chains in general. Emphasis is laid on the efforts made to find ways to lessen the amounts of sample needed for the analysis, because only small amounts of glycan samples isolated from biological sources are often available.

II. Permethylation of the Sample

Methylation of a carbohydrate is most practically achieved by using the Hakomori procedure,[9] which has largely replaced the Purdie, Haworth, and Kuhn methylation methods.

The methylation reaction can be directly applied to the analysis of glycolipids, whereas glycoproteins are not generally methylated directly, due to their low solubility in dimethyl sulfoxide. Therefore, the carbohydrate moiety of the glycoprotein is first isolated as a glycopeptide after extensive proteolytic digestion, or as a reduced oligosaccharide after treatment with alkaline sodium borohydride.[9a] Another reason for the degradation is that glycoproteins often contain more than one carbohydrate chain. The approach employing glycopeptides can be used in the analysis of the alkali-stable (N–glycosyl) chains, whereas it is advisable to methylate the alkali-labile (O–glycosyl;

(1a) P. H. Fishman and R. O. Brady, *Science*, 194 (1976) 906–915.
(2) S. W. Craig and P. Cuatrecasas, *Proc. Natl. Acad. Sci. USA*, 72 (1976) 3844–3848.
(3) S.-I. Hakomori and A. Kobata, in M. Sela (Ed.), *The Antigens*, Vol. 2, Academic Press, New York, 1974, pp. 79–140.
(4) K. W. Talmadge and M. M. Burger, *MTP Int. Rev. Sci: Biochem. Ser. One*, 5 (1975) 43–93.
(5) G. O. Aspinall, *Int. Rev. Sci.: Org. Chem. Ser. Two, Carbohydrates*, 7 (1976) 201–222.
(6) K.-A. Karlsson, in S. Abrahamsson and I. Pascher (Eds.), *Structure of Biological Membranes*, Plenum Press, New York, 1977, pp. 245–274.
(7) B. Lindberg and J. Lönngren, *Methods Enzymol.*, 50 (1978) 3–33.
(8) S. Svensson, *Methods Enzymol.*, 50 (1978) 33–38.
(9) S.-I. Hakomori, *J. Biochem. (Tokyo)*, 55 (1964) 205–208.
(9a) R. N. Iyer and D. M. Carlson, *Arch. Biochem. Biophys.*, 142 (1971) 101–105.

that is, glycosidic) chains as reduced oligosaccharides, in order to avoid "peeling" reactions during the procedure.

1. Detection of the Completeness of Methylation

Completeness of the methylation reaction is a prerequisite for successful methylation analysis. It can be directly detected from the disappearance of the infrared absorption of the hydroxyl groups,[9] but this is often not possible because of the small amount of sample available.

The critical step in the permethylation procedure is the generation of the sugar alkoxides catalyzed by methylsulfinyl carbanion.[9] It is to be expected that extensive formation of the alkoxide species has occurred if an excess of the carbanion can still be detected in the mixture after the carbohydrates have reacted. For detection of the carbanion, the formation of a color with triphenylmethane[10] has been exploited.[11] No undermethylation, detectable by analysis of the methylated monosaccharide derivatives by gas–liquid chromatography–mass spectrometry (g.l.c.–m.s.), occurs when the mixture gives a positive triphenylmethane reaction after agitation of glycolipid or glycopeptide samples with the carbanion reagent. When the triphenylmethane reaction is negative, extensive undermethylation often occurs.[11] It should be emphasized that, although the amounts of carbohydrate used for the methylation are usually small, and only small amounts of the carbanion are needed for the sample itself, insufficient amounts of the reagent may be present in routine procedures. This is due to the fact that the anion is very reactive with proton donors other than carbohydrates, such as moisture present.[10] Use of the triphenylmethane method of detection thus makes it possible to lessen the reaction volumes needed for methylation.

A prerequisite for complete methylation is that the compounds shall be soluble under the reaction conditions, which is usually the case for glycolipids, glycopeptides, and oligosaccharides. Diminution of the reaction volumes from those generally used has the advantage that it results in lessening of the backround peaks in g.l.c., which is important when dealing with small samples. The methylation can be successfully conducted, even in the presence of 1–3% of water, provided that the amount of carbanion reagent is adjusted by means of the triphenylmethane test. It is also advantageous to carry out the procedure in tubes flushed with nitrogen, as the presence of water tends to in-

(10) E. J. Corey and M. Chaykovsky, *J. Am. Chem. Soc.*, 87 (1965) 1345–1353.
(11) H. Rauvala, *Carbohydr. Res.*, 72 (1979) 257–260.

crease the intensity of the non-sugar peaks arising from the reagent mixture.[11]

2. Use of Potassium *tert*-Butoxide to Generate the Methylation Reagent

The reaction of sodium hydride with dimethyl sulfoxide is generally used for the preparation of the carbanion reagent.[10] It has been shown that potassium *tert*-butoxide in dimethyl sulfoxide also generates methylsulfinyl carbanion, which is in equilibrium with the butoxide ion.[12] Use of this reagent for methylation has been studied, but its value for the methylation of carbohydrates has been unclear.[13] It has been observed that this reagent is equally as effective in catalyzing the methylation of carbohydrates as the sodium hydride–dimethyl sulfoxide reagent.[14] The advantage of using potassium *tert*-butoxide is that the preparation of the reagent is very simple and rapid. Commercial *tert*-butoxide salt can be dissolved in dimethyl sulfoxide without any further steps. When this method is used, the intensities of the nonsugar signals in g.l.c. tend to be lower than those seen when the reagent is prepared with the aid of sodium hydride.[14]

III. ANALYSIS OF PERMETHYLATED CARBOHYDRATES WITHOUT DEPOLYMERIZATION OF THE SAMPLE

1. Oligosaccharides

Analysis of permethylated oligosaccharides by g.l.c.–m.s. is generally possible for carbohydrates containing up to four or five monosaccharide residues. For higher polymers, direct-inlet m.s. can be used, as has also been shown in the analysis of permethylated glycolipids and glycopeptides (see later). When the oligosaccharide contains different monosaccharide species, such as hexoses, deoxyhexoses, and acetamidodeoxyhexoses, the sequence of sugar residues can be ascertained from primary fragments formed by fission of the glycosidic bonds and from the secondary fragments. The method has been applied to methyl glycosides of oligosaccharides[15] and to oligosac-

---

(12) J. I. Brauman, J. A. Bryson, D. C. Kahl, and N. J. Nelson, *J. Am. Chem. Soc.*, 92 (1970) 6679–6680.
(13) J. Eagles, W. M. Laird, R. Self, and R. L. M. Synge, *Biomed. Mass Spectrom.*, 1 (1974) 43–48.
(14) J. Finne, T. Krusius, and H. Rauvala, *Carbohydr. Res.*, 80 (1980) 336–339.
(15) J. Kärkkäinen, *Carbohydr. Res.*, 17 (1971) 1–10.

charide alditols.[16,17] Reduction to the alditol is to be recommended, because the linkage to the reduced sugar can then be determined from fragments formed by fissions of the alditol moiety, as shown for structure[18] **1**. Use of reduced oligosaccharides also has the advantage that possible "peeling" reactions which can occur under the alkaline conditions used for permethylation[14] can be avoided. Borodeuteride should, in general, be used in the reduction (instead of borohydride) in order to avoid possible symmetry in the alditol moiety.[16,17] In the analysis of aminohexitol-containing oligosaccharides,[18] the structure can also be concluded from borohydride-reduced samples, as the alditol moiety itself is asymmetrical, as in Gal-(1→3)-GalNAc-ol (**1**) obtained from the glycosidic chains of glycoproteins and from brain gangliosides.

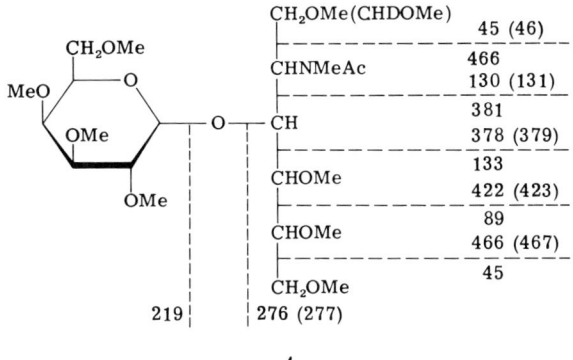

1

Investigation of the structure of neuraminic acid-containing oligosaccharides by g.l.c.–m.s. also seems promising.[19] For this purpose, reduction with lithium aluminum hydride according to Karlsson and coworkers[6] has been used, in order to increase the volatility of the sample used in the g.l.c. separation. The sample obtained from glycoprotein by degradation with alkaline sodium borohydride was permethylated, reduced with lithium aluminum hydride, and remethylated with CD_3I. By using this approach, the common glycosidic oligosaccharide[20,21] obtained from glycoproteins, namely, α-AcNeu-(2→3)-β-Gal-(1→3)-[α-AcNeu-(2→6)]-GalNAcol (**2**), was volatile

(16) J. Kärkkäinen, *Carbohydr. Res.*, 14 (1970) 27–33.
(17) J. Kärkkäinen, *Carbohydr. Res.*, 17 (1971) 11–18.
(18) I. Mononen, J. Finne, and J. Kärkkäinen, *Carbohydr. Res.*, 60 (1978) 371–375.
(19) T. Krusius, J. Finne, and H. Rauvala, unpublished results.
(20) J. Finne, *Biochim. Biophys. Acta*, 412 (1975) 317–325.
(21) J. Finne and T. Krusius, *FEBS Lett.*, 66 (1976) 94–97.

enough for g.l.c. separation. Some prominent ions in the mass spectrum of this disialosyl oligosaccharide as shown in **2** (*m/e* 115, 100%; *m/e* 351, 15%).

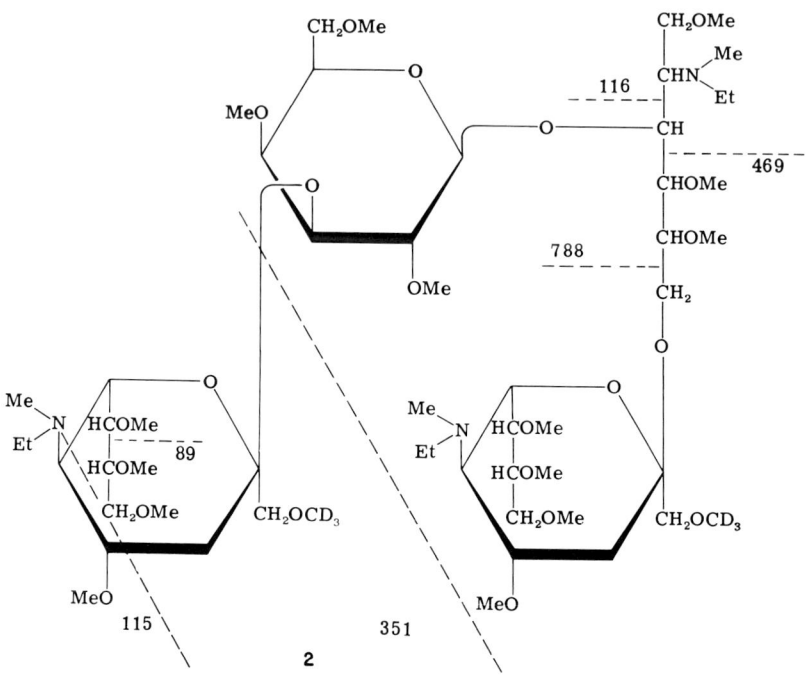

2. Glycolipids and Glycopeptides

Analysis of the oligosaccharide structures by mass spectrometry has been especially valuable in glycolipid studies.[22–24] This method has been extensively developed by Karlsson and coworkers.[6,25,26] These authors have studied several permethylated glycolipids, and shown that reduction of the permethylated sample with lithium aluminum hydride (effecting the reduction of the acetamido and the carboxyl groups of sugars) greatly stabilizes the fragments of high molecular

(22) C. C. Sweeley and G. Dawson, *Biochem. Biophys. Res. Commun.*, 37 (1969) 6–14.
(23) K. Samuelsson and B. E. Samuelsson, *Biochem. Biophys. Res. Commun.*, 37 (1969) 15–21.
(24) K.-A. Karlsson, B. E. Samuelsson, and G. O. Steen, *Biochem. Biophys. Res. Commun.*, 37 (1969) 22–27.
(25) K.-A. Karlsson, *FEBS Lett.*, 32 (1973) 317–320.
(26) K.-A. Karlsson, *Adv. Exp. Med. Biol.*, 71 (1976) 15–25.

weight. By use of this kind of derivatization, the composition and sequence of sugars in glycolipids containing up to nine monosaccharide residues can be analyzed by direct-inlet m.s. Terminal-sugar sequences, and sequences linked to the ceramide moiety, can be identified for even larger molecules. Use of mass-fragmentographic scanning in direct-inlet m.s. has provided useful structural data, even from mixtures of glycolipids.[27]

It has been shown that the method using direct-inlet m.s. is also valuable in the characterization of glycopeptides.[28] When permethylation, and reduction with lithium aluminum hydride, followed by (trimethylsilyl)ation, were used as in glycolipid studies, analysis of the N-glycosyl glycopeptide from transferrin by direct-inlet m.s. gave ions up to the complete nonasaccharide moiety of the desialosylated structure. Thus, the number of monosaccharide residues in the glycopeptide molecule was revealed by the mass spectrum. This finding should be of importance, as the fractionation of glycopeptides to give molecules having homogeneous carbohydrate structures is an extremely difficult task. The sugar sequence could also be deduced from the mass spectrum, as illustrated in 3.

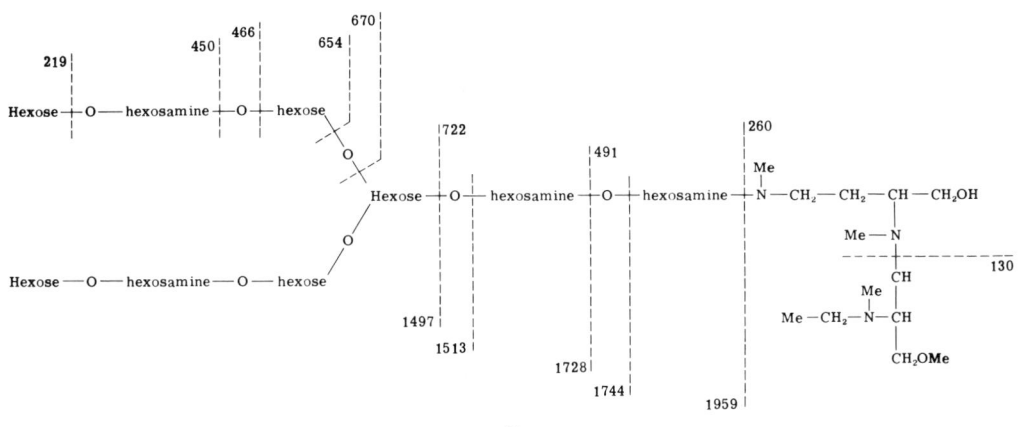

3

(27) M. E. Breimer, G. C. Hansson, K.-A. Karlsson, H. Leffler, W. Pimlott, and B. E. Samuelsson, *FEBS Lett.*, 89 (1978) 42–46.
(28) K.-A. Karlsson, I. Pascher, B. E. Samuelsson, J. Finne, T. Krusius, and H. Rauvala, *FEBS Lett.*, 94 (1978) 413–417.

IV. Degradation of the Permethylated Sample, and Analysis of the Partially Methylated Sugars

1. Degradation of the Sample

Depolymerization of the permethylated carbohydrate is achieved by hydrolysis with acid. Under these conditions, the amino sugar residues are N-deacetylated, and the aminohexosidic linkages become resistant to hydrolysis. Stellner and coworkers[29] showed that, when the acid degradation is conducted in 95% acetic acid, the amino sugar residues are also liberated, and can be analyzed by the methylation technique.[29] Therefore, acetolysis followed by acid hydrolysis is now commonly used, as it allows the analysis both of hexose and hexosamine residues.

Concerning the protein- and lipid-linked saccharides from animal cells, there are two major exceptions, the aminohexitol and the neuraminic acid residues, for which methanolytic cleavage is to be preferred over acetolysis–acid hydrolysis. Aminohexitol residues occur in the alkali-labile, glycosidic chains of glycoproteins, which are analyzed after treatment with alkaline sodium borohydride. They are also derived from N-glycosyl chains, and from glycolipids after treatment with endo-N-acetylhexosaminidase, a procedure that is becoming an important technique in structural analysis.[30] It was shown that the main product of the terminal aminohexitol residue in acetolysis–acid hydrolysis probably contains a free amino group,[31] as shown in **4** for the aminogalactitol derivative.

A solution of a sample of **4** in CH_3CO_2D, injected into the column for g.l.c.–m.s., showed a shift from m/e 74 to m/e 75 and 76, demonstrating that the compound contains freely exchangeable hydrogen due to a free amino group. In subsequent acetylation, the aminohexitol residue mainly gives the N-acetylacetamido derivative[31,32] **5**, instead of the N-methylacetamido derivative normally obtained[29] from amino sugars. 3-O-Substituted aminohexitols do not show this "anomalous" behavior, whereas aminohexitols substituted in other positions give various proportions of the N-acetylacetamido and N-methylacetamido derivatives, depending on the site of substitution. Glycosidically linked hexosamines do not produce N-acetylacetamido deriva-

---

(29) K. Stellner, H. Saito, and S.-I. Hakomori, *Arch. Biochem. Biophys.*, 155 (1973) 464–472.
(30) T. Tai, K. Yamashita, M. Ogata-Arakawa, N. Koide, T. Muramatsu, S. Iwashita, Y. Inoue, and A. Kobata, *J. Biol. Chem.*, 250 (1975) 8569–8575.
(31) J. Finne and H. Rauvala, *Carbohydr. Res.*, 58 (1977) 57–64.
(32) S. Hase and E. T. Rietschel, *Eur. J. Biochem.*, 63 (1976) 93–99.

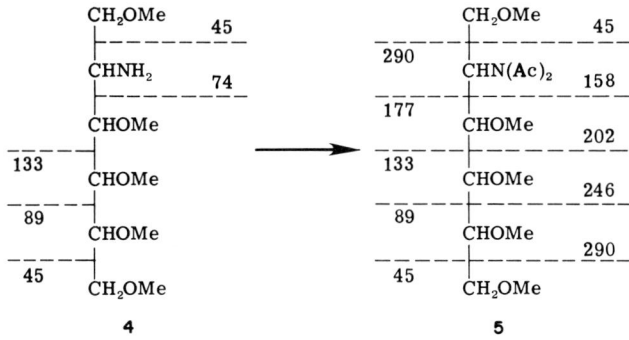

tives, probably because of the presence of a free hydroxyl group at C-1 after hydrolysis and reduction. Caroff and Szabó[32a] have reported that acetolysis–hydrolysis causes O-demethylation at O-1 of the aminohexitol, leading to a 1-O-acetyl derivative. As, however, the mass spectra reported[32a] were different from those given for the N-acetylacetamido derivative,[31,32] the two sets of investigations obviously concern different compounds. Why different products should have arisen under apparently similar conditions is not yet resolved. The partial 1-O-demethylation observed by Caroff and Szabó under very unusual methanolytic conditions (6 M hydrochloric acid in methanol for 2 h at 80°) does not occur when methanolysis is performed under "normal" conditions, namely, 0.5 M hydrochloric acid in methanol for 18 h at 80°. Considering the foregoing ambiguities arising after acetolysis–hydrolysis, it is recommended that they be avoided by the use of methanolysis.

Methanolysis is also to be preferred in the analysis of N-acetyl- and N-glycolylneuraminic acid residues, which are commonly found in animal glycolipids and glycoproteins. These compounds are normally decomposed under the conditions of acid hydrolysis used for the liberation of neutral and amino sugar derivatives. The terminal and the 8-O-substituted neuraminic acids are conveniently analyzed after methanolysis, giving the methyl glycoside methyl ester derivatives of these sugars.[33,34] It has been shown[33,34a] that very mild methanolytic conditions (0.05 M hydrochloric acid in methanol, for 16 h at 80°) permit the simultaneous identification of differently substituted N-glycolyl- and N-acetyl-neuraminic acids. Because of these con-

(32a) M. Caroff and L. Szabó, Biochem. Biophys. Res. Commun., 89 (1979) 410–413.
(33) H. Rauvala and J. Kärkkäinen, Carbohydr. Res., 56 (1977) 1–9.
(34) J. Haverkamp, J. P. Kamerling, J. F. G. Vliegenthart, R. W. Veh, and R. Schauer, FEBS Lett., 73 (1977) 215–219.
(34a) S. Inoue and G. Matsumura, Carbohydr. Res., 74 (1979) 361–368.

siderations, it is generally advisable to divide the permethylated sample into two portions: acetolysis–acid hydrolysis is used for the analysis of neutral sugars and of hexosamines, whereas aminohexitols and neuraminic acids are analyzed after methanolysis. It should, however, be noted that the neutral sugars[35–38] and the hexosamines[39,40] can also be analyzed after methanolysis, but the g.l.c. separation then becomes more complicated.

2. Identification of the Partially Methylated Monosaccharides

The partially methylated monosaccharides obtained on depolymerization of the permethylated sample are preferably analyzed as acetates by g.l.c.–m.s., as shown by Björndal and coworkers.[41,42] The neutral sugars and the amino sugars obtained in acetolysis–acid hydrolysis are reduced, and acetylated for the analysis, and the aminohexitol and the neuraminic acid residues are acetylated after methanolysis. Identification with the aid of g.l.c.–m.s. has been described for all of the common components of protein- and lipid-linked glycans and oligosaccharides from animal cells, namely, the neutral sugars,[41–43] hexitols,[44] hexosamines,[29,43,45,46] aminohexitols,[31,32] and neuraminic acids.[33,34,47]

Separation of the partially methylated sugars by g.l.c. gives evidence on the methyl-substitution pattern complementary to that obtained from the mass spectrum. The electron-impact, mass spectra do not differentiate between the sugar isomers (such as the galactose,

(35) J. Lönngren and S. Svensson, *Adv. Carbohydr. Chem. Biochem.*, 29 (1974) 41–106.
(36) N. K. Kochetkov and O. S. Chizhov, *Adv. Carbohydr. Chem.*, 21 (1966) 39–93.
(37) K. Heyns, H. F. Grützmacher, H. Scharmann, and D. Müller, *Fortschr. Chem. Forsch.*, 5 (1966) 448–490.
(38) B. Fournet, Y. Leroy, and J. Montreuil, *Méthodologie de la Structure et du Métabolisme des Glycoconjugués (Glycoprotéines et Glycolipides)*, Colloques Internationaux du Centre National de la Recherche Scientifique, No. 221, Villeneuve D'Ascq, 1973, pp. 111–130.
(39) S. K. Kundu and R. W. Ledeen, *Carbohydr. Res.*, 39 (1975) 179–191.
(40) S. K. Kundu and R. W. Ledeen, *Carbohydr. Res.*, 39 (1975) 329–334.
(41) H. Björndal, B. Lindberg, and S. Svensson, *Carbohydr. Res.*, 5 (1967) 433–440.
(42) H. Björndal, B. Lindberg, Å. Pilotti, and S. Svensson, *Carbohydr. Res.*, 15 (1970) 339–349.
(43) H. Björndal, C. G. Hellerqvist, B. Lindberg, and S. Svensson, *Angew. Chem.*, 82 (1970) 643–652.
(44) H. Yamaguchi and K. Okamoto, *J. Biochem. (Tokyo)*, 82 (1977) 511–518.
(45) G. O. H. Schwartzmann and R. W. Jeanloz, *Carbohydr. Res.*, 34 (1974) 169–173.
(46) T. Tai, K. Yamashita, and A. Kobata, *J. Biochem. (Tokyo)*, 78 (1975) 679–686.
(47) H. van Halbeek, J. Haverkamp, J. P. Kamerling, J. F. G. Vliegenthart, C. Versluis, and R. Schauer, *Carbohydr. Res.*, 60 (1978) 51–62.

glucose, and mannose derivatives), but this information is obtained from the retention times in g.l.c. Columns of OV-225, ECNSS-M, and OV-210 are useful in the separation of the neutral sugars. The retention times of the different, methylated, neutral sugars have been compiled in a manual of methylation analysis by Jansson and coworkers.[48] It may be noted that the separation of partially methylated alditol acetates derived from 2,6- and 3,6-di-O-substituted hexoses is successful on OV-210, but not on OV-225. Using the latter phase, the coexistence of these two species may easily be overlooked. The most efficient separations have been achieved in capillary columns, the available selection of which now comprises commercially available phases suitable for methylation analysis.[48] Less-polar phases, such as SE-30 and OV-101, are preferred for the N-acetylacetamido sugars, because more-intense peaks are observed (compared to the separations performed on the other phases mentioned). The g.l.c. separation of the methylated derivatives of hexosamines,[45,46] aminohexitols,[31,32] and neuraminic acids[33,34,47] on this type of phase has been described.

The methyl-substitution pattern of the neutral sugar residues can be reliably identified from their mass spectra.[41–43] Identification of the substitution pattern from the electron-impact, mass spectra rarely presents any major problems. It should, however, be noted that, in some cases, the occurrence of symmetrical molecules jeopardizes the identification. An example of this is the symmetrical, 3,4-di-O-methyl derivative (6) of mannose, which is commonly found in the analysis of

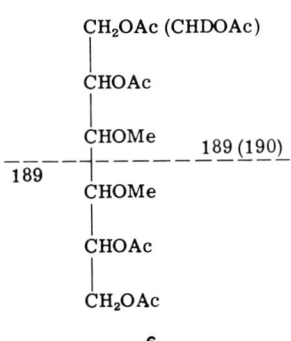

6

the animal glycoprotein-derived carbohydrates.[49,50] The mass spectrum of this sugar derivative resembles that of the 3,6-di-O-methyl de-

(48) P.-E. Jansson, L. Kenne, L. Liedgren, B. Lindberg, and J. Lönngren, *Chem. Commun., Univ. Stockholm*, 8 (1976).
(49) T. Krusius, J. Finne, and H. Rauvala, *FEBS Lett.*, 71 (1976) 117–120.
(50) T. Krusius and J. Finne, *Eur. J. Biochem.*, 78 (1977) 369–379.

```
        CH₂OAc (CHDOAc)
        |
        CHOAc
        |
        CHOMe          189 (190)
       |---------------------
        CHOAc
        |
        CHOAc
      ----|
       45  CH₂OMe

            7
```

rivative (**7**) of mannose. There are small differences in the ion intensities, such as the higher intensity at m/e 45 for **7**, due to the presence of the primary methoxyl group, and there is a prominent secondary fragment at m/e 113 in **7**. However, on reduction with sodium borodeuteride, the difference becomes clear-cut, as, in **6**, half of the intensity at m/e 189 is not shifted to m/e 190. Therefore, the use of sodium borodeuteride (instead of sodium borohydride) is to be recommended in the reduction step. Another method for introducing asymmetry into the hexose derivatives is by conversion into the aldononitriles.[50a] The fragmentation patterns of the partially methylated aldononitrile acetates are analogous to those of alditol acetates.

Identification, from their mass spectra, of the methyl-substitution pattern of hexosamines is more difficult than that of hexoses, because of a high tendency for cleavage between C-2 (containing the N-methylacetamido group) and C-3, for which reason the signals for other primary fragments tend to be of low intensity.[7] This is also the case for the aminohexitol derivatives. The type of mass spectrometer used seems to be of importance in the identification of these compounds. The intensities of the diagnostic fragments of larger molecular weight tend to be higher in the mass spectra of magnetically scanning instruments than in the mass spectra obtained with quadrupole instruments. Using a magnetically scanning instrument, the diagnostic ions in the mass spectra of the terminal, 3-*O*-, 4-*O*-, 6-*O*-, and 3,6-di-*O*-substituted aminohexitols are intense enough for reliable identification of the substitutions. Two common aminohexitol derivatives (after methanolysis and acetylation) from the glycosidic chains of glycoproteins, namely, the 3-*O*- and 3,6-di-*O*-substituted aminogalactitols,[31] and the relative intensities (% of base peak) of the primary fragments, are shown in **8** and **9**.

(50a) B. A. Dmitriev, L. V. Backinowsky, O. S. Chizhov, B. M. Zolotarev, and N. K. Kochetkov, *Carbohydr. Res.*, 19 (1971) 432–435.

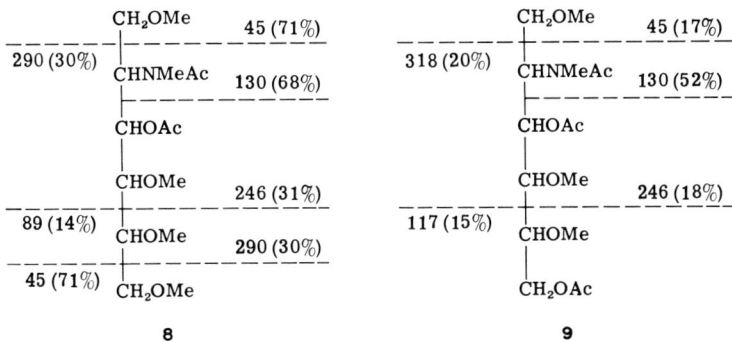

It has been observed that the neuraminic acid linkages can also be conveniently analyzed by the methylation technique.[33,34] Some important fragments, the origins of which were confirmed by deuterium-labelling experiments,[33] are shown for two common, N-acetylneuraminic acid derivatives, namely, the terminal sugar derivative 10 and the 8-O-substituted N-acetylneuraminic acid (11).

The chemical-ionization (c.i.), mass spectra of some methylated alditol acetates have been reported.[51,52] The main intensities in the c.i.

(51) M. McNeil and P. Albersheim, *Carbohydr. Res.*, 56 (1977) 239–248.
(52) R. A. Laine, L. C. Hodges, and M. C. Allen, *J. Supramol. Struct.*, 5, Suppl. 1, (1977) 31.

mass spectra occur near the molecular weight, and virtually no fragmentation of the alditol chains occurs. Although a proportion of the information as to the methyl-substitution pattern of the alditol chain is therefore lost, the c.i. mass spectra have the advantage of differentiating between different species of sugars. Thus, the methylated alditol acetates from the terminal glucose, galactose, and mannose residues can be differentiated by means of the difference in the intensities at (M + 1), (M + 1 − 32), and (M + 1 − 60) in the isobutane, c.i. mass spectra.[51] This finding is significant, as it is sometimes difficult to separate certain methylated alditol acetates, such as those arising from terminal mannose and glucose residues, from each other on any ordinary, g.l.c. phase.

Chemical-ionization mass spectra of partially methylated aldononitrile acetates have been published.[52a] Intense peaks appear at M + 1 and M + 1 − 60, permitting determination of the number of substitutions.

3. Quantitation of the Methylated Monosaccharides

The molar ratio of the different methylated derivatives of neutral sugars is generally calculated from the g.l.c. responses (using flame-ionization detection). In this kind of quantitation, the molar response of the different methylated sugars is assumed to be the same,[7] or effective carbon-response is used.[53]

Selective ion-monitoring offers more specificity and sensitivity for quantitation, but cannot yet be used generally due to lack of standardization for all types of sugars. The difficulty of the calibration task is illustrated in **12** and **13**. The ion at m/e 118 is useful for the detection and quantitation of the hexose species containing a methoxyl group at C-2, that is, for a wide variety of the hexose derivatives from protein- and lipid-linked glycans. Although, for the different stereoisomers (for example, the galactose, glucose, and mannose derivatives) having the same methyl-substitution pattern, the relative intensity of this ion is equal, this is not the case for hexoses having different methyl-substitution patterns. Thus, for the 2,3-di-O-methyl derivative **12**, the intensity of this ion would be much higher than for the 2,4-di-O-methyl derivative **13**. The reason for this is that the fragment m/e 118 is formed by fission between two methoxylated carbon atoms in **12**, and is therefore much more intense than for **13**, from which the ion is formed by

(52a) B. W. Li, T. W. Cochran, and J. R. Vercellotti, *Carbohydr. Res.*, 59 (1977) 567–570.
(53) D. P. Sweet, R. H. Shapiro, and P. Albersheim, *Carbohydr. Res.*, 40 (1975) 217–225.

```
        CHDOAc              CHDOAc
          |                   |
        CHOMe    118        CHOMe    118
     ---|-------         ---|-------
  261 CHOMe              CHOAc
        |                   |
        |               189 |
        CHOAc           ----|
        |                   CHOMe    234
        CHOAc              |-------
        |                   CHOAc
        CH₂OAc              |
                            CH₂OAc

          12                  13
```

fission between a methoxylated and an acetoxylated carbon atom.[41,42] Generally, the methyl-substitution pattern in the part of the alditol chain which is not present in the fragment monitored greatly influences the response factor. In addition, there tend to be differences in the ion intensities in the mass spectra obtained with different instruments.

The ion, m/e 117 or 118 (deuterated) just discussed is, however, useful for sugar detection to only a limited degree. The samples introduced into the g.l.c.–mass spectrometer generally contain contaminating material giving rise to fragments at these m/e values. For specific, sugar detection and measurement, it is advisable to use other ions that can be assigned to sugars explicitly. Moreover, by selecting m/e values corresponding to mass fragments of some specific derivatives, it is possible to distinguish between them, even if they are not separated in g.l.c. The ions at m/e 161 and 189 have been found very useful for this purpose, as shown in Fig. 1. The specificity is increased if the samples are deuterioreduced and are monitored at m/e 162 and 190 also.

Multiple-ion monitoring is, however, of considerable value in structural studies, but only if model compounds of known structure are available for comparison. Such an approach has been used in the study of the carbohydrate structures of glycoproteins from different tissues.[50] Separation of glycopeptides obtained from various tissues was performed on columns of concanavalin A–Sepharose. Structural analysis by multiple-ion monitoring of partially methylated, alditol acetates derived from the various fractions indicated that the glycopeptides were separated according to the linkage pattern of mannose (see Fig. 1).

Chemical-ionization, mass fragmentography was used for quantitative determination of variously linked monosaccharides in a study of the high-molecular-weight glycopeptide of erythrocytes.[75] It appears

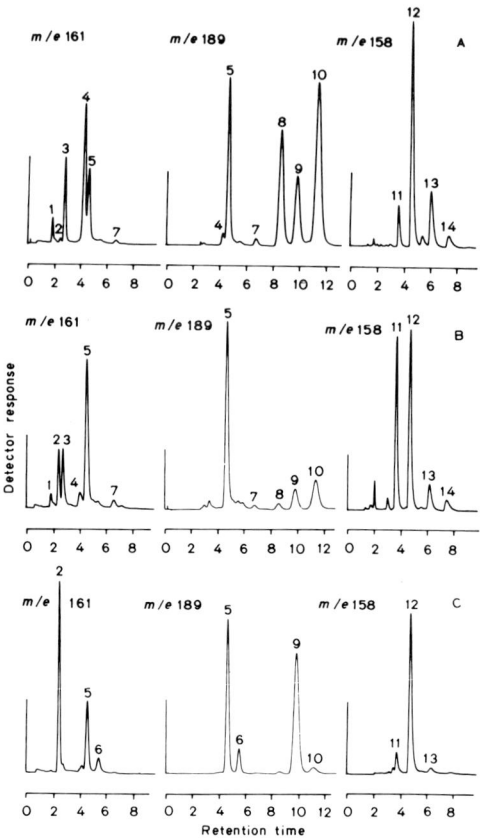

Fig. 1.—Mass Fragmentograms of Partially Methylated Alditol Acetates Obtained from Rat-brain Glycopeptides. [(A) Fraction A glycopeptides, (B) fraction B glycopeptides, and (C) fraction C glycopeptides. Peak 1, 2,3,4-tri-O-methylfucitol; peak 2, 2,3,4,6-tetra-O-methylmannitol and 2,3,4,6-tetra-O-methylglucitol; peak 3, 2,3,4,6-tetra-O-methylgalactitol; peak 4, 2,4,6-tri-O-methylgalactitol; peak 5, 3,4,6-tri-O-methylmannitol; peak 6, 2,3,4-tri-O-methylmannitol; peak 7, 2,3,4-tri-O-methylgalactitol; peak 8, 3,6-di-O-methylmannitol; peak 9, 2,4-di-O-methylmannitol; peak 10, 3,4-di-O-methylmannitol; peak 11, 2-deoxy-3,4,6-tri-O-methyl-2-(N-methylacetamido)glucitol; peak 12, 2-deoxy-3,6-di-O-methyl-2-(N-methylacetamido)glucitol; peak 13, 2-deoxy-6-O-methyl-2-(N-methylacetamido)glucitol; peak 14, 2-deoxy-3-O-methyl-2-(N-methylacetamido)-glucitol. Abscissa: retention time, in minutes. Conditions: 3% of QF-1, at 185°, for neutral sugars, and detection by mass fragmentography at m/e values 161 and 189; 2.2% of SE-30, at 205°, for amino sugars, and detection by mass fragmentography at m/e 158. Reproduced, by permission, from Ref. 50.]

that the $(M + 1 - 60)$ ion, because of its high intensity, provides a good basis for quantitative determination. The response factors for this ion were found to be approximately equal for the different methylated derivatives. It may turn out that the $(M + 1 - 60)$ ion of the partially methylated aldononitrile acetates[52a] also provides a good basis for quantitation.

Quantitation of the ratios of the different methylated monosaccharides in protein- and lipid-linked glycans is complicated by the occurrence of the acetamido sugars, because their response factors relative to those of the neutral sugars have not yet been determined. Therefore, the ratio of the total amount of the acetamido sugars to the neutral sugars should preferably be taken from the total sugar composition. The relative proportions of the different, partially methylated hexosamines can be determined by mass fragmentography, by using the intense, C-1–C-2 fragment at m/e 158 (159) (see **14**) or the second-

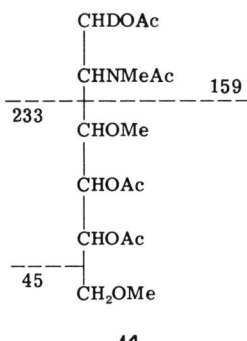

14

ary fragment at m/e 116 (117) formed from this by elimination of ketene. These fragments are common to all of the different, methylated hexosamines, and it has been reported[46] that, by using the m/e values 158 and 116, the relative proportion of the different methylated hexosamines may be directly calculated from the peak areas. However, as the fragmentation and the response factors may vary according to the instrument used, it is advisable to check the response for various methylated hexosamines before using the data for quantitative estimations. More-specific ions from the higher-mass region can be selected for identification, as, for example, the m/e 233 for the 3,6-di-O-methyl derivative **14**, which is common in the analysis of glycolipids and glycopeptides. Similarly, from the methylated aminohexitols, only the N-methylacetamido derivatives are formed in methanolysis, and these derivatives thus have in common[31] the intense C-1–C-2 fragment at m/e 130 (see **8** and **9**). This ion is therefore useful for mass fragmentography of the methylated aminohexitols, but detailed, quantitative evaluation with differentially substituted derivatives is still lacking.

Quantitation of the ratio of the terminal to the 8-O-substituted neuraminic acid may also be achieved by analysis using the methylated derivatives, whereas the proportion of total neuraminic acid relative to other sugars can be taken from the total sugar composition.[54,55] The fragment at m/e 274, formed by elimination of the side chain of the neuraminic acid (see **10** and **11**), is common for these derivatives, and is useful for quantitation. The ion at m/e 330 is also formed in about equal intensity from the terminal neuraminic acid (M − 45 − 32) and from the 8-O-substituted derivative (M − 45 − 60). The ion at m/e 376 is preferably used if small proportions of the 8-O-substituted derivative (relative to the terminal sugar) are found, as it is formed by a more-intense elimination from the 8-O-substituted derivative than from the terminal one.

4. Interpretation of the Quantitative Data

One special feature in the interpretation of the quantitative results seems not to have been exploited in practical analysis, and it certainly deserves attention. The result of the methylation analysis is sometimes complex, and can reveal the occurrence of 10 to 20 different methylated sugars. In this situation, it is not easy to decide, by simple inspection of the analytical data, whether the result could be caused by one complex, carbohydrate chain, or by a structure containing several saccharide chains bound to a common aglycon. In addition, it is not always easy to decide whether the result could fit any natural structure, or mixture of structures, or whether the complex result is attributable to undermethylation.

In such cases, it is practical to apply the following considerations. The analysis usually provides data that can be expressed as numbers of moles of each monosaccharide derivative per mole of sample. The expression [NS] − [DS] − 2 [TS] − 3 [TeS] (where NS = nonsubstituted (terminal), DS = disubstituted (branched), TS = trisubstituted, and TeS = tetrasubstituted sugars) should theoretically give a value of unity. If values far below unity are encountered, several sources of analytical error must be considered. Undermethylation must be considered, as well as erroneously low values for terminal sugars caused by the volatility of their methylated derivatives. Nonglycosidic substituents, such as sulfate,[56] phosphate, and acetal groups must also be considered.[48]

It should be emphasized that quantitative determination of differ-

(54) J. Finne, T. Krusius, and H. Rauvala, *Biochem. Biophys. Res. Commun.*, 74 (1977) 405–410.
(55) J. Finne, T. Krusius, H. Rauvala, and K. Hemminki, *Eur. J. Biochem.*, 77 (1977) 319–323.
(56) A. Stoffyn, P. J. Stoffyn, and E. Mårtensson, *Biochim. Biophys. Acta*, 152 (1968) 353–357.

ent methylated sugars is a prerequisite for the calculation shown. At present, the accuracy of such determinations leaves much to be desired, and it seems that applications using isotope-labelled, internal standards[57] are needed in order to improve the quantitative analysis. However, the data from methylation analysis should not be strikingly discrepant with the structures postulated.

V. METHYLATION ANALYSIS COMBINED WITH SPECIFIC DEGRADATION

Analysis of permethylated, oligosaccharidic alditols, glycolipids, and glycopeptides without degradation of the permethylated sample (see Section II) gives valuable information concerning the sugar sequence. Data on sugar sequence are also obtainable by applying the methylation technique to monitoring of the effects of selective or specific degradation on the saccharide structure. This kind of approach is especially important in the analysis of higher polymers, which cannot be directly analyzed by g.l.c.–m.s. or direct-inlet m.s. because of the low volatility of the sample.

Specific degradation of the carbohydrate chain has been extensively discussed in a few articles (for reviews, see Refs. 5, 7, 8, 35, and 58). In the following discussion, we shall briefly present some applications that have been valuable in the sequence study of glycoprotein and glycolipid saccharides.

1. Smith Degradation

Monosaccharide residues containing vicinal hydroxyl groups are oxidized by periodate, and are subsequently removed in the reduction–hydrolysis step. Therefore, the positions to which such monosaccharide residues are linked can be located by methylation analysis performed before, and after, Smith degradation. Alternatively,[59] the oxidized and reduced sample is methylated, the ether hydrolyzed, and the product realkylated with CD_3I or CH_3CH_2I. This kind of procedure can have advantages over that first described. For example, methylation before the hydrolysis step hinders the acetal protection of hydroxyl groups that can occur in acid hydrolysis.[7]

An example of the use of Smith degradation for carbohydrate sequencing is a study on the sites of AcNeu–Gal–GlcNAc linkages to

(57) I. Mononen and J. Kärkkäinen, *FEBS Lett.*, 59 (1975) 190–193.
(58) J. J. Marshall, *Adv. Carbohydr. Chem. Biochem.*, 30 (1974) 257–370.
(59) B. Lindberg, J. Lönngren, and S. Svensson, *Adv. Carbohydr. Chem. Biochem.*, 31 (1975) 185–240.

the mannotriaosyl residue of the glycopeptide from fetuin and other glycoproteins containing similar, branched-carbohydrate chains.[60] In the carbohydrate structure of fetuin, three peripheral branches are attached to the mannotriaosylchitobiose core,[61,62] as shown in **15**. Smith

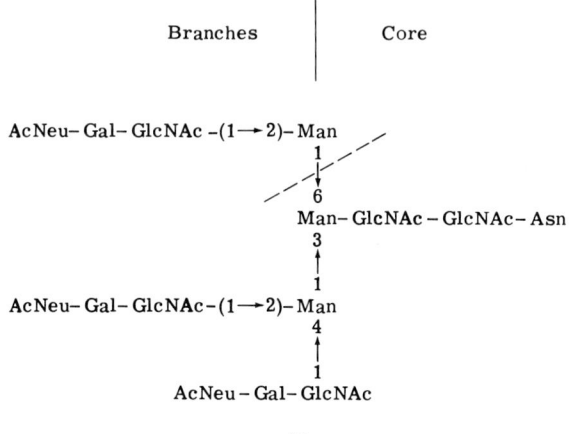

15

degradation and methylation analysis can be used to study whether the three branches are attached, without distinction, to the 3-O-substituted and 6-O-substituted sides of the innermost mannosyl residue, or whether there exists some selectivity. Of the two more-peripheral, mannose residues, that carrying two branches should be resistant to periodate, whereas the other should be oxidized. It can be shown by methylation analysis that the 6-O-substituent is almost quantitatively cleaved in the Smith degradation, whereas the 3-O-substituent is stable.[60] Therefore, the 3-O-substituted mannose carries two peripheral branches, as shown in **15**.

2. Degradation with Acid

Acid-labile linkages of carbohydrates are commonly those of furanosidic sugar residues or of deoxy monosaccharides.[59,63] Consequently, neuraminic acid and fucose residues, which occur as terminal monosaccharides of protein- and lipid-linked glycans, are removed by hydrolysis with a dilute acid. The linkages formed by these two sugar residues can conveniently be differentiated by subjecting a portion of

---

(60) T. Krusius and J. Finne, *Carbohydr. Res.*, in press.
(61) J. Baenziger and D. Fiete, *J. Biol. Chem.*, 254 (1979) 789–795.
(62) B. Nilsson, N. E. Norden, and S. Svensson, *J. Biol. Chem.*, 254 (1979) 4545–4553.
(63) J. N. BeMiller, *Adv. Carbohydr. Chem.*, 22 (1967) 25–108.

the sample to hydrolysis by a dilute acid, and another portion to digestion with neuraminidase. Methylation analysis before and after these degradations reveals the sites to which fucose and neuraminic acid are linked.

The saccharide chains of glycoproteins and glycolipids are commonly terminated by hexosyl-aminohexosyl sequences that are substituted by neuraminic acid or fucose residues.[64] The hexosyl-hexosamines can also form repeating units. As the acetamidohexosidic linkage is quite acid-labile before N-deacetylation, the hexosyl-hexosamine is usually liberated in a yield of ~10–20% of the theoretical maximum, when the glycopeptide or glycolipid is hydrolyzed with a dilute acid (for example, in 0.1 M hydrochloric acid for 1 h at 100°). Any neuraminic acid and fucose substituents that may be present are also cleaved off in the hydrolysis. Reduction of the sample with sodium borohydride or deuteride, followed by permethylation, and analysis of the oligosaccharide fragments by g.l.c.–m.s. provides information on the sugar sequence. For the reasons already given, hexosyl-hexosamine fragments are usually found in amounts sufficient for practical analysis. The analysis of these fragments is conveniently performed by g.l.c.–mass-fragmentographic detection at m/e 276 (the aminohexitol unit), or at other specific m/e values (see 1). An example of this kind of analysis of the N-glycosyl carbohydrates of rat brain[65] is shown in Fig. 2.

The glycosyl–N linkage of 2-acetamido-2-deoxy-D-glucose to L-asparagine, usually found in the alkali-stable carbohydrates of glycoproteins, is also rather stable to acid hydrolysis. This stability can be exploited in order to demonstrate the 2-acetamido-2-deoxy-D-glucose-L-asparagine fragment from the ovalbumin glycopeptides after hydrolysis and permethylation.[66]

Acetolysis often gives structural information complementary to that obtained by acid hydrolysis, as the rates of cleavage of the glycosidic linkages are different and, in some cases, even reversed.[59] The neuraminic acid residues are relatively stable, and it has been possible to isolate neuraminic acid-containing fragments after acetolysis of gangliosides,[67] oligosaccharides,[68] and glycopeptides.[69]

(64) H. Rauvala and J. Finne, FEBS Lett., 97 (1979) 1–7.
(65) T. Krusius and J. Finne, Eur. J. Biochem., 84 (1978) 395–403.
(66) P. Maury and J. Kärkkäinen, Clin. Chim. Acta, 91 (1979) 75–79.
(67) R. Kuhn and H. Wiegandt, Chem. Ber., 96 (1963) 866–880.
(68) L. Grimmonperez, S. Bouquelet, B. Bayard, G. Spik, M. Monsigny, and J. Montreuil, Eur. J. Biochem., 13 (1970) 484–492.
(69) B. Bayard and J. Montreuil, Carbohydr. Res., 24 (1972) 427–443.

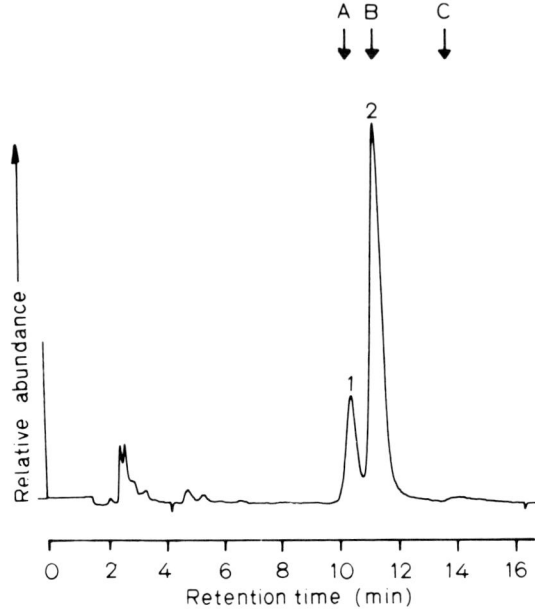

Fig. 2.—Mass Fragmentogram of Hexosyl-aminohexitols Obtained by Partial, Acid Hydrolysis of Rat-brain Glycopeptides. [Peak 1, β-Gal-(1→3)-GlcNAcol; peak 2, β-Gal-(1→4)-GlcNAcol. A, retention time of β-Gal-(1→3)-GlcNAcol; B, retention time of β-Gal-(1→4)-GlcNAcol; and C, retention time of β-Gal-(1→6)-GlcNAcol. Abscissa, retention time, in minutes. Conditions: 1% of OV-225, at 245°, and detection by mass fragmentography at m/e 276. Reproduced, by permission, from Ref. 65.]

3. N-Deacetylation Followed by Acid Degradation

A feasible way of introducing acid-stable linkages into carbohydrates is N-deacetylation. This can be achieved with hydrazine.[59,70,71] The use of sodium hydroxide–sodium benzenethioxide in aqueous dimethyl sulfoxide for this purpose has also been described.[72]

The difference in the acid hydrolysis of N-acetylhexosamine-containing carbohydrates before and after N-deacetylation was used in the study of complex glycoprotein saccharides from human erythrocyte membranes.[73–75] Methylation analysis of the glycopeptides prepared

(70) Z. Yosizawa, T. Sato, and K. Schmid, *Biochim. Biophys. Acta*, 121 (1966) 417–420.
(71) J. Montreuil, *Pure Appl. Chem.*, 42 (1975) 431–477.
(72) C. Erbing, K. Granath, L. Kenne, and B. Lindberg, *Carbohydr. Res.*, 47 (1976) c5–c7.
(73) J. Finne, T. Krusius, H. Rauvala, R. Kekomäki, and G. Myllylä, *FEBS Lett.*, 89 (1978) 111–115.
(74) T. Krusius, J. Finne, and H. Rauvala, *Eur. J. Biochem.*, 92 (1978) 289–300.
(75) J. Järnefelt, J. Rush, Y.-T. Li, and R. A. Laine, *J. Biol. Chem.*, 253 (1978) 8006–8009.

by extensive, proteolytic digestion revealed that these chains are highly branched, and are mainly composed of 3-O- and 6-O-substituted galactose and 4-O-substituted 2-amino-2-deoxyglucose residues.[74,75] Hydrolysis of the glycopeptides with 0.1 M hydrochloric acid for 1 h at 100° produced β-Gal-(1→4)-GlcNAc as the only disaccharide fragment, identified (as the reduced and then permethylated derivative) in g.l.c., using mass-fragmentographic detection.[74] After N-deacetylation, hydrolysis with 0.5 M hydrochloric acid for 16 h at 100°, reduction, N-reacetylation, and permethylation, one major peak in the oligosaccharide region was observed in g.l.c. The mass spectrum of this product suggested the structure 2-acetamido-2-deoxy-hexosyl-(1→3 or 4)-hexitol. Because of the result of the methylation analysis of the unhydrolyzed sample (see the preceding), it was concluded that the disaccharide was GlcNAc-(1→3)-Gal. Combination of the results with those from the determination of the monosaccharide composition, and a study of the anomeric nature of the glycosidic linkages, suggested[74] that the two disaccharides identified constitute overlapping fragments from the repeating sequence →3)-β-Gal-(1→4)-β-GlcNAc-(1→. On the basis of the results of methylation analysis, every second or third galactose residue carries another substituent at O-6. The chains are heterogeneous, varying in size from 20 to 70 monosaccharide residues.

Another way in which selective degradation, after N-deacetylation, may be effected is nitrous acid deamination. In the treatment with nitrous acid, the hexosamine residues are rearranged to give 2,5-anhydro sugars, and the aminohexosidic linkages are cleaved.[7,71] This reaction was exploited in the study, already mentioned, on the complex glycoprotein saccharides from the erythrocyte membrane. The carbohydrate structure was essentially depolymerized upon N-deacetylation–deamination, giving a disaccharide fraction composed of galactose and 2,5-anhydromannose.[75] Methylation analysis revealed that the 3-O- and 6-O-substituents of the galactose residues had been almost quantitatively cleaved. These experiments suggested the repeating sequence already given on the basis of a different set of experiments. The results also suggested that the substituents at O-6 of the galactose residues are amino sugar groups.

4. β-Elimination

The carbohydrate chains of glycoproteins linked through a glycosidic linkage to serine or threonine undergo β-elimination under alkaline conditions. The carbohydrate chain becomes degraded until a point is reached where a stable structure is formed. Degradation of the carbohydrate can be avoided by performing the reaction under reducing conditions. This kind of treatment is generally used in the isolation of the alkali-labile (glycosidic) glycoprotein carbohydrates.[58] As already mentioned in Section I, it is recommended that treatment with alkaline sodium borohydride be performed prior to methylation, to avoid degradation of the glycosidic chains during the derivatization.

Except for treatment with alkaline sodium borohydride to release the alkali-labile chains of glycoproteins, the wide range of β-elimination reactions[7,8,59] has found only limited application in studies on protein- and lipid-linked glycans. One possibility for exploiting these reactions is the elimination of terminal uronic acid groups by successive treatments with base and a dilute acid.[7] This degradation method was used for studying a galactosidic linkage of glycopeptides; uronic acid residues were introduced by oxidation with D-galactose oxidase and iodine. After permethylation, the compound was treated with base, producing the 4,5-unsaturated uronic acid, which is readily removed by hydrolysis with mild acid. The site of the uronic acid linkage was studied by remethylation.[65] The linkage was found to be to O-4 of 2-acetamido-2-deoxyglucose. In the case studied, treatment with D-galactosidase was ineffective, probably due to the fucose substitution at O-3 of the 2-amino-2-deoxyglucose residue.[65]

5. Degradation with the Aid of Exoglycosidases

Detailed information on the sugar sequence is obtained from specific treatment with an exoglycosidase in conjunction with methylation analysis. The terminal monosaccharide residues available to exoglycosidases are identified by methylation analysis before digestion with the glycosidase. The substituent removed is determined, after the degradation, by a second methylation analysis, which also identifies the new terminal monosaccharide available to a glycosidase in a succeeding degradation-step. As the glycosidases are normally specific for α, or β, configurations, the anomeric nature of the linkages may also be concluded from the susceptibility of the different sugars to the enzymes.

The carbohydrate structure can, in principle, be unambiguously determined by use of enzymic hydrolysis. However, the method is quite laborious, and usually requires large amounts of sample. It should also be noted that, unless the samples are purified after each step, the amount of the carbanion needed for the methylation tends to become increased, and must, therefore, be assessed by means of the triphenylmethane test in order to avoid undermethylation (see Section I). Blanks containing the enzyme used must also be analyzed in order to

STRUCTURAL ANALYSIS OF GLYCOPROTEINS AND GLYCOLIPIDS   413

avoid contamination by sugars present in the enzymes. With these precautions, a simple purification step, such as precipitation of the enzyme with methanol or ethanol, is sufficient before permethylation.

Exoglycosidases that are useful in carbohydrate degradations include neuraminidases, fucosidases, galactosidases, mannosidases, and aminohexosidases.[76,77] An example of the use of degradation by a glycosidase in conjunction with methylation analysis is a study on the oc-

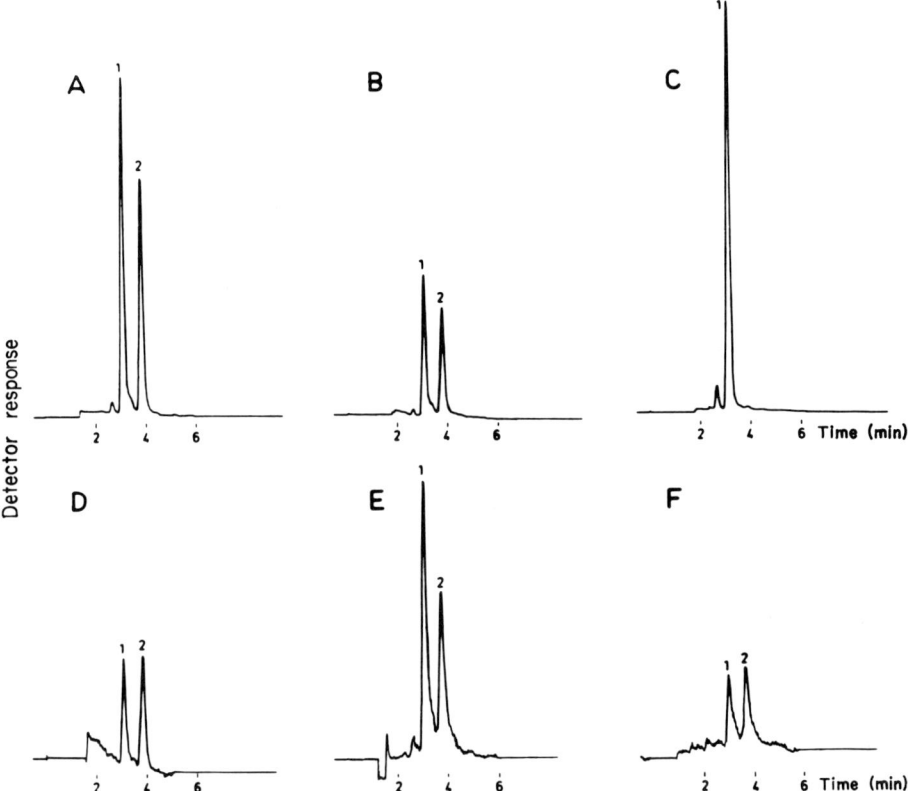

FIG. 3.—Mass Fragmentography of Methylated Neuraminic Acids. [Detection at m/e 274: A, disialosyl-lactosylceramide; B, the disialosyl ganglioside isolated; C, B after treatment with neuraminidase; D, the fraction containing the trisialosyl ganglioside. Detection at m/e 330: E, the disialosyl ganglioside; F, the trisialosyl fraction. The peaks identified are: (1) the permethylated (terminal) N-acetylneuraminic acid, and (2) the 8-O-acetyl (8-O-substituted) derivative of the methylated N-acetylneuraminic acid. Conditions: 1% of SE-30, at 240°. Reproduced, by permission, from Ref. 80.]

(76) S. Gatt, *Chem. Phys. Lipids*, 5 (1970) 235–249.
(77) Y.-T. Li and S.-C. Li, in M. I. Horowitz and W. Pigman (Eds.), *The Glycoconjugates*, Vol. 1, Academic Press, New York, 1977, pp. 52–68.

currence of AcNeu–AcNeu sequences in glycolipids and glycoproteins.[33,54,55] These sugar sequences had previously been found as terminal structures of glycolipids,[78,79] but their occurrence in glycoproteins was then not known. Methylation analysis before and after treatment with *Vibrio cholerae* neuraminidase revealed the presence of neuraminidase-labile, 8-*O*-substituted neuraminic acid residues. Therefore, it was concluded that the sequence α-AcNeu-(2→8)-AcNeu is present in glycoprotein carbohydrates.[54,55] Determination of the structure of a novel ganglioside from human kidney revealed that an α-AcNeu-(2→8)-AcNeu sequence [and, probably, an α-AcNeu-(2→8)-AcNeu-α-(2→8)-AcNeu sequence also (see Fig. 3)] occurs and is linked to the terminal β-Gal-(1→4)-GlcNAc structures in these glycolipids.[80] The site of linkage to the neutral-carbohydrate chain was shown to be O-3 of the terminal galactose residue (see Fig. 4). Thus, in all structures containing the AcNeu–AcNeu sequence that have thus far been found, this disaccharide is linked to O-3 of a galactose residue. The next sugar in the chain can be either 2-acetamido-2-deoxygalactose, glucose, or 2-acetamido-2-deoxyglucose.[78-80]

Sequential degradation with the aid of *Vibrio cholerae* neuraminidase, beef-kidney α-L-fucosidase, and *Aspergillus niger* β-D-galactosidase was used in a study of the complex, protein-linked glycans from human erythrocyte membrane.[74] By using these degradations, followed, after each step, by purification of the polymeric reaction-product by gel filtration for the methylation study, the structures of the terminal sugar sequences were identified as the H blood-group determinant (**16**) and the sialosylated branches (**17** and **18**). The pre-

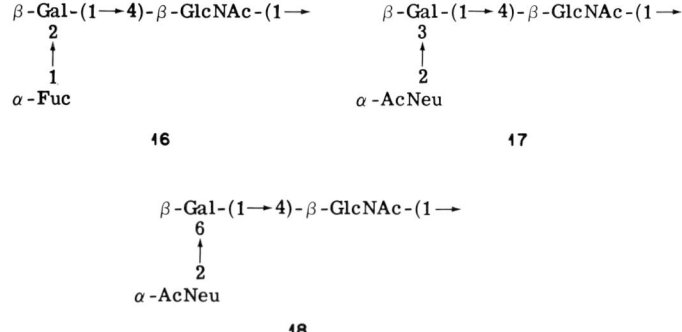

(78) R. Kuhn and H. Wiegandt, *Z. Naturforsch., Teil B*, 18 (1963) 541–543.
(79) I. Ishizuka and H. Wiegandt, *Biochim. Biophys. Acta*, 260 (1972) 279–289.
(80) H. Rauvala, T. Krusius, and J. Finne, *Biochim. Biophys. Acta*, 531 (1978) 266–274.

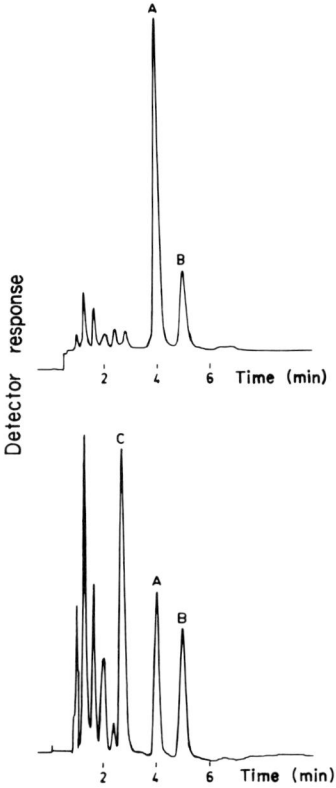

FIG. 4.—Mass Fragmentography (m/e 161) of Methylated Hexose Derivatives from the Disialosyl Ganglioside. [Top, before treatment with neuraminidase; bottom, after treatment with *Vibrio cholerae* neuraminidase. The methylated hexose derivatives identified are: (A) 2,4,6-tri-O-methylgalactose; (B) 2,3,6-tri-O-methylglucose; (C) 2,3,4,6-tetra-O-methylgalactose. The peaks eluting before C are unrelated signals that were also detected in a blank experiment employing the neuraminidase treatment. Conditions: 3% of QF-1, at 190°. Reproduced, by permission, from Ref. 80.]

cursor structures of these sequences were also identified.[74] The terminal-sugar sequences are linked to the repeating saccharide structure identified with the aid of acid degradation before and after N-deacetylation (see earlier).

6. Degradation with the Aid of Endoglycosidases

Endo-2-acetamido-2-deoxy-β-D-glucosidase has been used in structural studies of glycopeptides.[30] The enzyme cleaves the linkage between the two 2-amino-2-deoxyglucose residues that constitute the chitobiose residue linked to L-asparagine in glycoproteins containing

N-glycosylically-linked saccharide chains. The endo-2-amino-2-deoxyglucosidases from different sources have different structural specificities towards the substitution of the mannotriose structure linked to the chitobiose unit.[30,81]

The value of endogalactosidases in structural studies is becoming evident. The endo-β-D-galactosidase from *Escherichia freundii* hydrolyzes the galactosidic linkage of the →3)-β-Gal-(1→4)-β-GlcNAc-(1→ residue from protein-linked glycans and from oligosaccharides.[82] Thus, after hydrolysis, reduction, and methylation, the 3-O-substituted galactitol derivative could be identified.[82] The enzyme was shown to be effective upon glycolipids as well.[83] Oligosaccharide fragments terminating at the galactitol residue could be analyzed, after hydrolysis, reduction, and permethylation, by using direct-inlet m.s. for structural identification. Branching at O-6 of the galactose residue was shown to retard the reaction.[83] This was also shown to be the case in the analysis of the complex saccharide chains from the erythrocyte membrane, which contain the highly branched Gal–GlcNAc repeating unit. Because of the branching, only ~40% of the galactosidic linkages could be hydrolyzed, as was shown by methylation analysis.[75]

(81) T. Tai, K. Yamashita, S. Ito, and A. Kobata, *J. Biol. Chem.*, 252 (1977) 6687–6694.
(82) M. N. Fukuda and G. Matsumura, *J. Biol. Chem.*, 251 (1976) 6218–6225.
(83) M. N. Fukuda, K. Watanabe, and S.-I. Hakomori, *J. Biol. Chem.*, 253 (1978) 6814–6819.

BIBLIOGRAPHY OF CRYSTAL STRUCTURES OF CARBOHYDRATES, NUCLEOSIDES, AND NUCLEOTIDES* 1977 AND 1978

By George A. Jeffrey and Muttaiya Sundaralingam

Department of Crystallography, University of Pittsburgh, Pittsburgh, Pennsylvania 15260; Department of Biochemistry, University of Wisconsin, Madison, Wisconsin 53706

I. Introduction 417
II. Data for Carbohydrates 418
III. Data for Nucleosides and Nucleotides 485
IV. Preliminary Communications 526
 1. Carbohydrates 526
 2. Nucleosides and Nucleotides 527

I. Introduction

This bibliography is similar to those of previous years.[1] Perspective drawings for the structures are shown by using tapered bonds where necessary, with the hetero-atoms indicated by appropriate symbols. Where hydrogen-atom coordinates are not given, the atom is indicated by H. For the carbohydrate structures, an MMS-X graphics-NOVA 800 system was used. For the nucleosides and nucleotides, a Vector General graphics system was used to obtain the desired view for each molecule and the drawings were made by using a modified ORTEP program on a PDP 11/35 computer.

---

* Work supported by NIH Grants GM-17378 and GM-24526, and the College of Agricultural and Life Sciences, University of Wisconsin, Madison. The authors express their gratitude to Chizu Shiono, Mr. D. Blyler, and Mr. Tuli Haromy for assistance with the preparation of the Figures.

(1) G. A. Jeffrey and M. Sundaralingam, *Adv. Carbohydr. Chem. Biochem.*, 30 (1974) 445–466; 31 (1975) 347–371; 32 (1976) 353–384; 34 (1977) 345–378; 37 (1980) 373–436.

The alphabetic code, given in parentheses after the chemical name, is the reference code name in the Cambridge Crystallographic Data Bank, which provided most of the source data for this bibliography. Because these data are checked prior to insertion into the file, they are preferable to the original source and should be used when available. The calculated densities D_x or D_n are in g. cm^{-3}. The Cremer and Pople puckering parameters[2,3] are reported. For pyranose rings, θ, φ, and q are given. As θ approaches 0° [in the ideal, $^4C_1$(D) form] or 180° [in the ideal $^1C_4$(D) form], the value of the φ parameter becomes indeterminate. Therefore, when θ is $\pm 5°$ or $\pm 175°$, φ is not reported, because the value is not significant. For furanose rings, ψ and q are reported; $\psi = P - 90 + \epsilon$, where P is the pseudorotation angle of Altona and Sundaralingam,[4] and ϵ is a small discrepancy arising from an approximation in the Altona–Sundaralingam treatment. The value of ϵ is generally[5] less than 3°.

The crystal-structure analyses reported were carried out with computer-controlled diffractometers, unless otherwise noted. Reasonably accurate analysis requires anisotropic, thermal-parameter refinement of all atoms except the hydrogen atoms, which can be treated as having isotropic motion (except in the case of neutron diffraction). The number of variable parameters is, therefore, 9X + 4H, where X is the number of non-hydrogen atoms and H is the number of hydrogen atoms. Consequently, the accuracy of an analysis can be judged qualitatively by the ratio of observations to parameters, as well as by the residual, disagreement factor, R. An analysis with R < 0.04, and observation to parameter ratio > 10, is likely to be accurate; an analysis with R > 0.06 and observation to parameter ratio < 5 is not sufficiently accurate to justify discussion of the fine details, such as bond lengths and valence angles, although the configuration and general conformation will be correct. Analyses with R > 0.10, or observation to parameter ratio < 2, are suspect.

Most of the crystal-structure analyses report hydrogen-atom positions; these are less accurate than those of the non-hydrogen atoms by a factor of ten, except in the case of neutron diffraction, where the accuracies are comparable.

II. Data for Carbohydrates

$C_5H_{10}O_5$ β-L-Arabinopyranose (ABINOS01)[6]

(2) D. Cremer and J. A. Pople, *J. Am. Chem. Soc.*, 97 (1975) 1354–1358.
(3) G. A. Jeffrey and J. H. Yates, *Carbohydr. Res.*, 74 (1979) 319–322.
(4) C. Altona and M. Sundaralingam, *J. Am. Chem. Soc.*, 94 (1972) 8205–8212.
(5) G. A. Jeffrey and R. Taylor, *Carbohydr. Res.*, 81 (1980) 182–183.
(6) S. Takagi and G. A. Jeffrey, *Acta Crystallogr. Sect. B*, 33 (1977) 3033–3040.

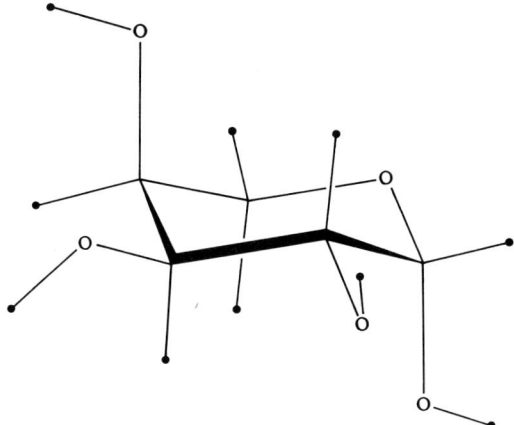

$P2_12_12_1$; $Z = 4$; $D_n = 1.627$; $R = 0.035$ for 1,004 neutron intensities. This is a neutron-diffraction refinement of a structure previously determined by X-ray diffraction.[7] The pyranose has the $^4C_1(\text{L})$ conformation ($\theta = 1°$, $q = 57$ pm). The C-5–O-5–C-1–O-1 bond-lengths are 143.2, 141.8, and 139.8 pm. The hydrogen bonding consists of infinite chains, with side chains and isolated links from the anomeric hydroxyl groups to ring-oxygen atoms. The hydrogen-bond lengths range over an unusually wide range, with H · · · O distances ranging from 173.5 to 220.1 pm.

$C_5H_{10}O_5$ β-L-Lyxopyranose (LYXOSE01)[8]

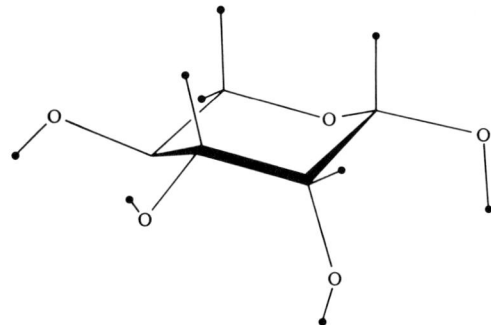

$P2_12_12_1$; $Z = 4$; $D_n = 1.538$; $R = 0.048$ for 1,463 neutron intensities. This is a neutron-diffraction refinement of a structure previously determined by X-ray diffraction.[9] The pyranose has the $^1C_4(\text{L})$ conforma-

(7) A. Hordvik, *Acta Chem. Scand.*, 15 (1961) 16–30.
(8) S. Nordenson, S. Takagi, and G. A. Jeffrey, *Acta Crystallogr. Sect. B*, 34 (1978) 3809–3811.
(9) A. Hordvik, *Acta Chem. Scand.*, 20 (1966) 1943–1954.

tion ($\theta = 177°$, $q = 57$ pm). The C-5–O-5–C-1–O-1 distances are 142.8, 143.1, and 138.6 pm. The hydrogen bonding consists of finite chains, with two bifurcated interactions having H $\cdots$ O distances of 195.7 and 250.0 pm (intramolecular), and 211.3 and 263.4 pm. The other hydrogen bonds are normal, with H $\cdots$ O distances of 171.8 and 179.3 pm.

$C_5H_{10}O_5 \cdot CaCl_2 \cdot 4\ H_2O$ α-L-Arabinopyranose·calcium chloride, tetrahydrate (ALARCA)[10]

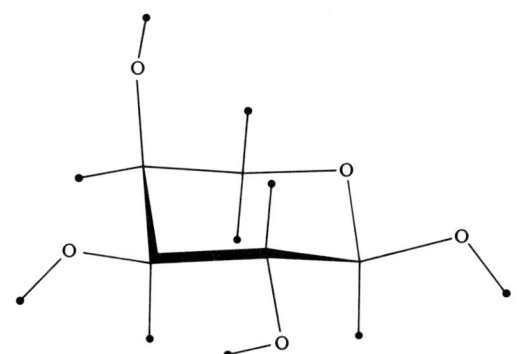

C2; $Z = 4$; $D_x = 1.62$; $R = 0.022$ for 1,840 intensities. The pyranose has the $^4C_1$ conformation ($\theta = 3°$, $q = 57$ pm). The C-5–O-5–C-1–O-1 bond-lengths are 143.8, 143.1, and 139.9 pm. The Ca^{2+} ion is eight-coordinated to four hydroxyl groups and four water molecules in a distorted antiprism, with $Ca^{2+} \cdots$ O distances ranging from 232.8 to 270.3 pm. The Cl^- ions are five-coordinated in distorted square pyramids, with H $\cdots$ Cl^- hydrogen-bond distances of 228 to 263 pm.

$C_6H_7Na^+O_8^- \cdot H_2O$ Sodium D-*erythro*-hex-2-enono-1,4-lactone, monohydrate (sodium D-isoascorbate monohydrate) (SIASCB)[11]

$P2_12_12_1$; $Z = 4$; $D_x = 1.723$; $R = 0.042$ for 809 intensities. This structure is of interest because it provides a comparison of the dimensions of the ion with that of the free acid.[12] There are small, but significant, differences in the conformation of the side-chain (the respective O-4–C-4–C-5–O-5 torsion angles are +51 and +86°) and in bond lengths,

(10) A. Terzis, *Cryst. Struct. Commun.*, 7 (1978) 95–99.
(11) J. A. Kanters, G. Roelofsen, and B. P. Alblas, *Acta Crystallogr. Sect. B*, 33 (1977) 1906–1912.
(12) N. Azarnia, H. M. Berman, and R. D. Rosenstein, *Acta Crystallogr. Sect. B*, 28 (1972) 2157–2161; Errata, 29 (1973) 1170.

of less than 3.5 pm. Similar differences were observed between sodium L-ascorbate[13] and L-ascorbic acid.[14] The hydrogen bonding is unusual, because it excludes one of the carbonyl oxygen atoms, O-1, in addition to the ring-oxygen atom. The other carbonyl oxygen atom, O-3, accepts four hydrogen bonds having H · · · O distances ranging from 190 to 239 pm.

$C_6H_8O_4$ 1,4:3,6-Dianhydro-α-D-glucopyranose[15]

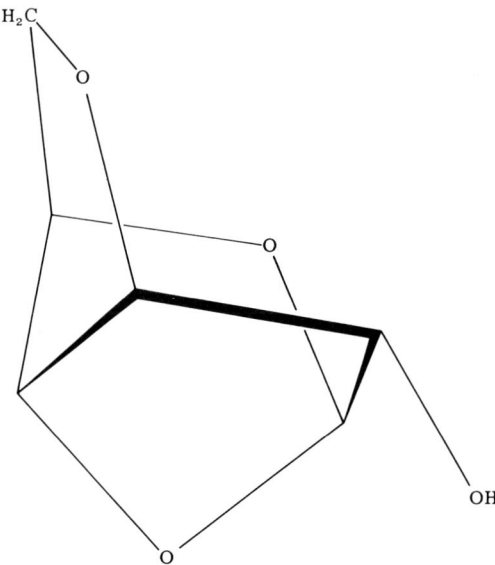

(13) J. Hvoslef, *Acta Crystallogr. Sect. B*, 25 (1969) 2214–2223.
(14) J. Hvoslef, *Acta Crystallogr. Sect. B*, 24 (1968) 1431–1440.
(15) V. J. Kopf and P. Köll, *Acta Crystallogr. Sect. B*, 34 (1978) 2502–2507.

$P2_12_12_1$; $Z = 4$; $D_x = 1.56$; $R = 0.050$ for 507 intensities. The pyranose has the $B_{1,4}$ conformation ($\theta = 95°$, $\varphi = 60°$, $q = 101$ pm). The furanoid part, containing O-3, has the ^4E conformation ($\psi = 71°$, $q = 40$ pm), and that containing O-4, the $^1T_0$ ($\psi = 188°$, $q = 55$ pm). The inter-ring torsion angles are C-2–C-3–O-3–C-6, $-84°$ and C-2–C-1–O-4–C-4, $-59°$. The rings are strained, with ring-oxygen valence-angles of 95, 105, and 109°. The ring-carbon valence-angles range from 99 to 114°. The C–O bond distances range from 141.8 to 144.1 pm. Hydrogen atom positions were not reported.

$C_6H_8O_5$ 3-Methoxyglutaconic acid[16]

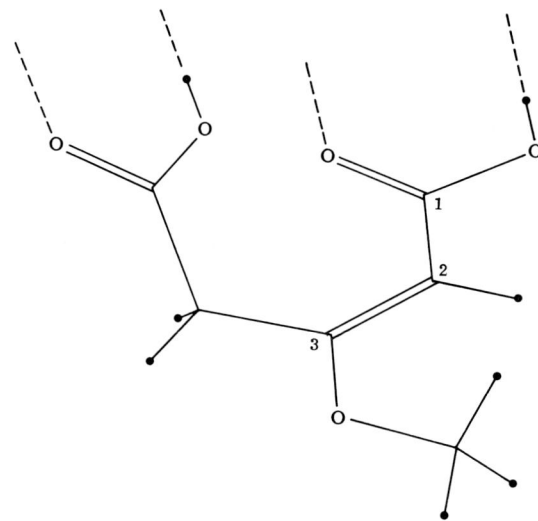

$P\bar{1}$; $Z = 2$; $D_x = 1.452$; $R = 0.066$ for 2,097 intensities. The carbon chain has a sickle conformation, being cis with respect to the double bond. The carbon and oxygen atoms lie in two planes, with C-2–C-3–C-4–C-5 = $+71°$. One of the carboxyl groups is eclipsed to the C–C bonds, with O-1=C-1–C-2—C-3 = $+1°$, and the other is so inclined that O-5=C-5–C-4–C-3 = $+22°$. The carboxyl groups are hydrogen bonded to afford dimers, with one of the hydrogen atoms disordered.

$C_6H_9K^+O_8^-$ Potassium D-glucarate[17]

$P2_1$; $Z = 2$; $D_x = 1.807$; $R = 0.123$ for 675 intensities (film data). The glucarate ion has a sickle conformation, with torsion angles at C-2–C-

---

(16) A. L. Spek and J. A. Kanters, *Cryst. Struct. Commun.*, 7 (1978) 251–254.
(17) T. Taga, Y. Kuroda, and K. Osaki, *Bull. Chem. Soc. Jpn.*, 50 (1977) 3079–3083.

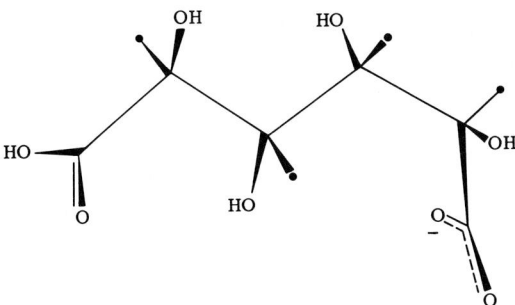

3–C-4–C-5 and C-3–C-4–C-5–C-6 of +67 and +69°, respectively. This is different from the bent-chain conformation of the Ca salt, tetrahydrate,[18] where the synclinal torsion angle, +56°, occurs at C-1–C-2–C-3–C-4. The K$^+$ ion is eight-coordinated, in a distorted, cubic-prism arrangement, to four carbonyl oxygen atoms and four hydroxyl oxygen atoms. Each cation binds to six anions. The hydroxyl groups are hydrogen bonded in chains. The hydrogen atom positions were not reported.

$C_6H_{10}O_4S$ 1,6-Anhydro-1(6)-thio-β-D-glucopyranose;   1(6)-thiolevoglucosan (TLEVGL)[19]

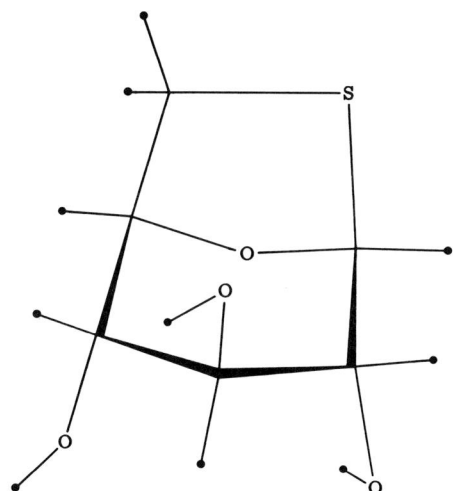

$P2_12_12_1$; Z = 4; D_x = 1.635; R = 0.026 for 909 intensities. The pyranoid conformation is a distorted $^1C_4$ (θ = 158°, φ = 172°, q = 59 pm). It is very similar to that observed in the crystal structure of levogluco-

(18) T. Taga and K. Osaki, *Bull. Chem. Soc. Jpn.*, 49 (1976) 1517–1520.
(19) S. Takagi and G. A. Jeffrey, *Acta Crystallogr. Sect. B*, 34 (1978) 816–820.

san.[20] The conformation of the anhydride is close to ^0E. The C–S bond lengths are 181.5 and 182.7 pm. The hydrogen bonding consists of short, finite chains. One of these includes a bifurcated interaction with a strong bond to the sulfur atom and a weak intramolecular bond, with hydrogen bond distances of H $\cdots$ S = 238 pm and H $\cdots$ O = 287 pm.

$C_6H_{11}NO_4$ 3-Amino-1,6-anhydro-3-deoxy-β-D-glucopyranose (AHXGLP)[21]

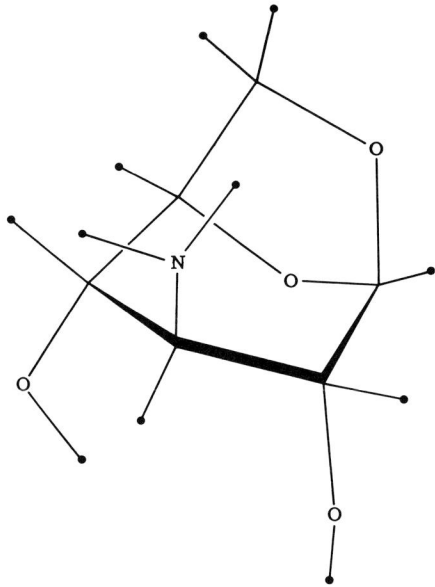

$P2_12_12_1$; Z = 4; D_n = 1.50; R = 0.056 for 1,170 neutron intensities. The pyranose has a distorted, $^1C_4$ conformation (θ = 153°, φ = 179°, q = 62 pm), very similar to that of 1(6)-thiolevoglucosan. The anhydride has a twist conformation. The hydrogen bonding consists of loops, and isolated links to the pyranose ring-oxygen atom. The only strong hydrogen bond is O-2–H $\cdots$ NH$_2$, with H $\cdots$ N = 176 pm. Other hydrogen-bond distances are O–H $\cdots$ O = 203 pm and N–H $\cdots$ O = 221 pm. There is a bifurcated bond from N–H to two oxygen atoms, with H $\cdots$ N distances of 249 and 226 pm; one of these bonds is intramolecular.

(20) Y. J. Park, H. S. Kim, and G. A. Jeffrey, *Acta Crystallogr. Sect. B*, 27 (1971) 220–227.
(21) J. H. Noordik and G. A. Jeffrey, *Acta Crystallogr. Sect. B*, 33 (1977) 403–408.

$C_6H_{12}Cl^-N^+O_4 \cdot H_2O$ 3-Amino-1,6-anhydro-3-deoxy-β-D-gluco-pyranose hydrochloride, monohydrate (AHGLCM)[22]

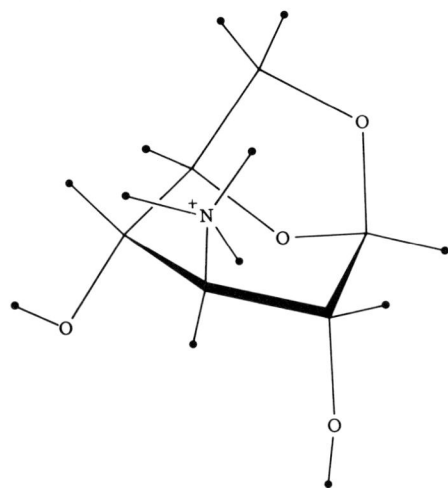

$P2_12_12_1$; Z = 4; D_x = 1.511; R = 0.029 for 1,163 intensities. The pyranose has a distorted, $^1C_4$ conformation (θ = 148°, φ = 186°, q = 62 pm). The anhydride has a conformation close to a twist. The bond lengths and valence angles agree well with those observed in the 3-amino compound.[21] The chloride ions are four-coordinated by hydrogen bonding to two water molecules and the N–H and O–H of the carbohydrate. The water molecules are hydrogen-bonded to two chloride ions and two N–H groups. The N^+–H · · · Cl hydrogen bonds are slightly shorter than the O–H · · · Cl bonds, 217 compared to 224 pm.

$C_6H_{12}O_4$ 2,6-Dideoxy-β-D-*ribo*-hexopyranose; β-D-digitoxose (BDDIGX)[23]

$P2_12_12_1$; Z = 4; D_x = 1.340; R = 0.043 for 794 intensities. The pyranose has the $^4C_1$ conformation (θ = 1.9°, q = 57 pm). The C-5–O-5–C-1–O-1 bond-lengths are 144.0, 144.5, and 140.4 pm. The C-1–C-2 bond is shorter than usual, 149.9 compared to 152.5 pm. The O-5–C-1–O-1 ring-valence angle of 104.8° is also smaller than the usual value

(22) H. Maluszynska, S. Takagi, and G. A. Jeffrey, *Acta Crystallogr. Sect. B*, 33 (1977) 1792–1796.
(23) J. A. Kanters, L. M. J. Batenburg, W. P. J. Gaykema, and G. Roelofsen, *Acta Crystallogr. Sect. B*, 34 (1978) 3049–3053.

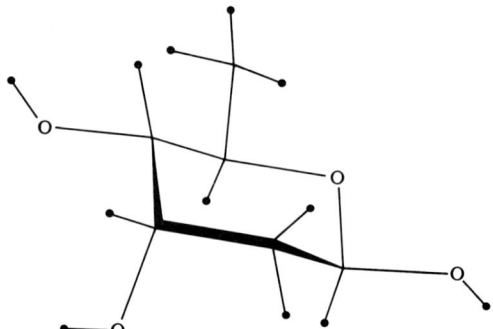

for a β-pyranose ring. The hydrogen bonding consists of infinite chains, and single links to the ring-oxygen atoms.

$C_6H_{12}O_5$ 6-Deoxy-α-DL-galactopyranose; α-DL-fucopyranose (ADLFUC)[24]

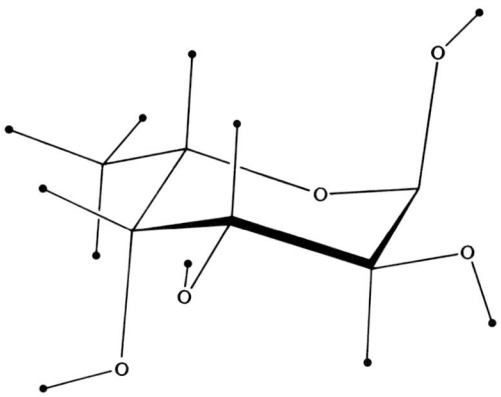

$P2_1/c$; Z = 4; D_x = 1.518; R = 0.045 for 1,583 intensities. Data were reported for the L enantiomer depicted, for which the pyranose conformation is $^1C_4$ (θ = 178°, q = 58 pm). The C-5–O-5–C-1–O-1 bond-lengths are 145.1, 143.1, and 139.4 pm. The hydrogen bonding consists of strong bonds from the anomeric hydroxyl groups to the ring-oxygen atoms, and weak, possibly bifurcated, interactions between the three other hydroxyl groups of the sugar molecules.

(24) F. Longchambon and H. Gillier-Pandraud, *Acta Crystallogr. Sect. B*, 33 (1977) 2094–2097.

$C_6H_{12}O_5$ Methyl α-L-arabinopyranoside (MALARA)[25]

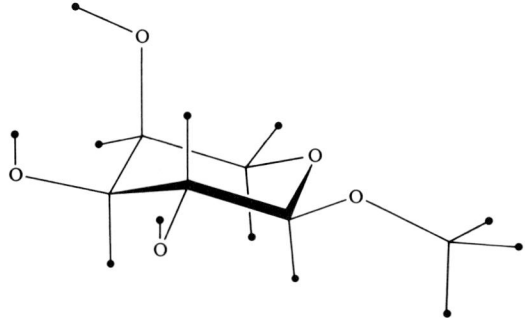

$P2_12_12_1$; $Z = 4$; $D_x = 1.450$; $R = 0.025$ for 953 intensities. The pyranoside has the $^4C_1$ conformation ($\theta = 1°$, $q = 58$ pm). The C-5–O-5–C-1–O-1–CH$_3$ bond-lengths are 143.2, 143.4, 137.5, and 143.4 pm, in good agreement with theory.[26] The glycosidic torsion-angle is $-79°$. The hydrogen bonding consists of infinite chains, and isolated links to the ring-oxygen atoms. The glycosidic oxygen atoms are excluded from the hydrogen bonding.

$C_6H_{12}O_5$ Methyl β-L-arabinopyranoside (MBLARA)[25]

(25) S. Takagi and G. A. Jeffrey, *Acta Crystallogr. Sect. B*, 34 (1978) 1591–1596.
(26) G. A. Jeffrey, J. A. Pople, J. S. Binkley, and S. Vishveshwara, *J. Am. Chem. Soc.*, 100 (1978) 373–378.

$P2_12_12_1$; $Z = 4$; $D_x = 1.435$; $R = 0.037$ for 964 intensities. The pyranoside has the $^4C_1$ conformation ($\theta = 2°$, $q = 58$ pm). The C-5–O-5–C-1–O-1–CH$_3$ bond-lengths are 143.4, 141.8, 139.3, and 142.3 pm, in good agreement with theory.[26] The glycosidic torsion-angle is $+69°$. The hydrogen bonding consists of infinite chains, which exclude both the ring and glycosidic oxygen atoms.

$C_6H_{12}O_5$ Methyl β-D-ribopyranoside (MDRIBP02)[27]

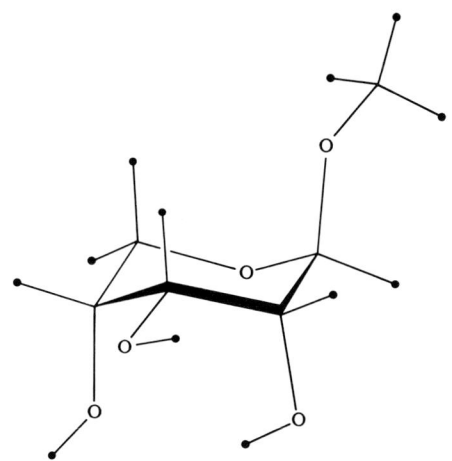

$P2_12_12_1$; $Z = 4$; $D_x = 1.475$; $R = 0.040$ for 865 X-ray intensities, and 0.043 for 1,048 neutron intensities. These are X-ray and neutron-diffraction refinements of a previously determined X-ray analysis.[28] The pyranoside has the $^1C_4$ conformation ($\theta = 176°$, $q = 57$ pm). This is in contrast to the 5-thio compound, for which the $^4C_1$ conformation was observed in the crystal structure.[29] The C-5–O-5–C-1–O-1–CH$_3$ distances are 143.3, 141.4, 139.4, and 141.8 pm. The glycosidic torsion-angle is $-67°$. There is an intramolecular hydrogen-bond between the syndiaxially oriented O-2–H and O-4, with a H · · · O distance of 195.9 pm and an O–H · · · O angle of 139°. This forms part of an infinite chain of hydrogen bonds which includes the other hydroxyl groups, but excludes the ring and glycosidic oxygen atoms.

(27) V. J. James, J. D. Stevens, and F. H. Moore, *Acta Crystallogr. Sect. B*, 34 (1978) 188–193.
(28) A. Hordvik, *Acta Chem. Scand. Ser. B*, 28 (1974) 261–263.
(29) R. L. Girling and G. A. Jeffrey, *Acta Crystallogr. Sect. B*, 29 (1973) 1102–1111.

$C_6H_{12}O_5$ Methyl α-D-xylopyranoside (MXLPYR)[30]

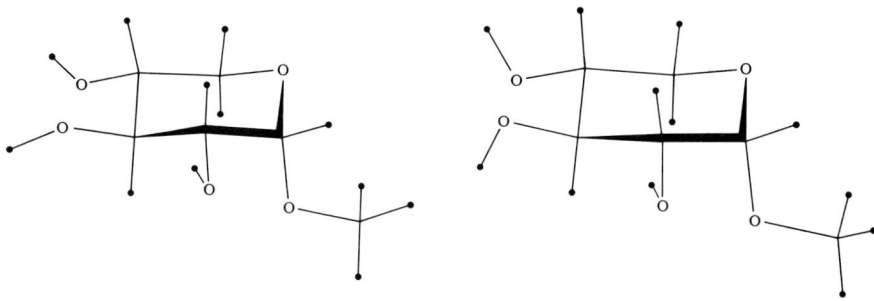

$P2_1$; $Z = 4$; $D_x = 1.402$; $R = 0.033$ for 1,777 intensities. The crystal structure contains two symmetry-independent molecules. The pyranoside conformations are $^4C_1$ and are almost identical in the two molecules ($\theta = 3°, 4°$; $q = 55$ pm, 57 pm). The mean C-5–O-5–C-1–O-1–CH_3 bond-lengths are 142.6, 141.4, 139.5, and 142.6 pm. The glycosidic torsion-angles are +71 and +63°. The largest difference in the two molecules is in the orientation of the O-3–H group. The hydrogen bonding consists of infinite chains, which exclude the ring and glycosidic oxygen atoms. The H · · · O distances lie in the narrow range of 170 to 182 pm.

$C_6H_{12}O_5$ Methyl β-D-xylopyranoside (XYLOBM01)[6]

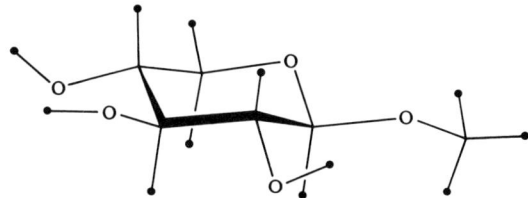

$P2_1$; $Z = 2$; $D_n = 1.403$; $R = 0.045$ for 1,059 neutron intensities. This is a neutron-diffraction refinement of a structure previously determined by X-ray diffraction.[31] The pyranoside has a slightly distorted, $^4C_1$ conformation ($\theta = 8°$, $\varphi = 36°$, $q = 58$ pm). The C-5–O-5–C-1–O-1–CH_3 bond-lengths are 142.1, 142.7, 138.1, and 142.6 pm. The glyco-

(30) S. Takagi and G. A. Jeffrey, *Acta Crystallogr. Sect. B*, 34 (1978) 3104–3107.
(31) C. J. Brown, E. G. Cox, and F. J. Llewellyn, *J. Chem. Soc.*, (1966) 922–927.

sidic torsion-angle O-5–C-1–O-1–CH$_3$ is $-72°$. The hydrogen bonding consists of finite chains which terminate at the ring-oxygen atoms and exclude the glycosidic oxygen atoms.

$C_6H_{12}O_5 \cdot H_2O$ 6-Deoxy-α-L-mannopyranose monohydrate; α-L-rhamnopyranose monohydrate (RHAMOH02)[32]

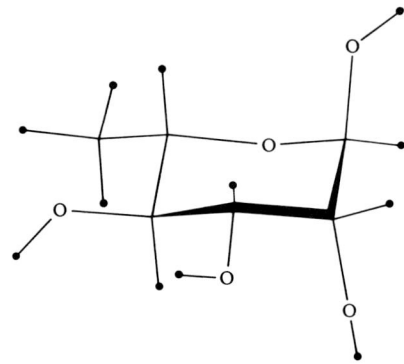

$P2_1$; Z = 2; D_n = 1.456; R = 0.042 for 1,727 neutron intensities. This is a neutron-diffraction refinement of a structure previously determined by X-ray diffraction.[33] The pyranose has a slightly distorted, $^1C_4$ conformation (θ = 174°, φ = 50°, q = 58 pm). The C-5–O-5–C-1–O-1 bond-lengths are 143.9, 141.9, and 140.4 pm. The hydrogen bonding consists of two infinite chains which intersect at the four-coordinated, water molecules. There is also a bifurcated interaction, one component of which is a weak, intramolecular bond between H–O-3 and O-4, with H · · · O distances of 194.9 and 271.5 pm.

$C_6H_{12}O_6$ β-D-Fructopyranose (FRUCTO02)[34,35]

$P2_12_12_1$; Z = 4; D_n = 1.601; R = 0.053 for 1,731 neutron intensities, and 0.040 for 954 X-ray intensities.[35] These are neutron and X-ray diffraction refinements of a crystal structure previously determined by X-ray diffraction.[36] The pyranose has the $^2C_5$ conformation (θ = 3°,

(32) S. Takagi and G. A. Jeffrey, *Acta Crystallogr. Sect. B*, 34 (1978) 2551–2555.
(33) C. G. Killean, J. L. Lawrence, and V. C. Sharma, *Acta Crystallogr. Sect. B*, 27 (1971) 1707–1710.
(34) S. Takagi and G. A. Jeffrey, *Acta Crystallogr. Sect. B*, 33 (1977) 3510–3515; Errata, 34 (1978) 1048.
(35) J. A. Kanters, G. Roelofsen, B. P. Alblas, and I. Meinders, *Acta Crystallogr. Sect. B*, 33 (1977) 665–672.
(36) R. D. Rosenstein, *Am. Crystallogr. Assoc. Meet., Buffalo, N. Y.*, (1968) Abstr. KK2.

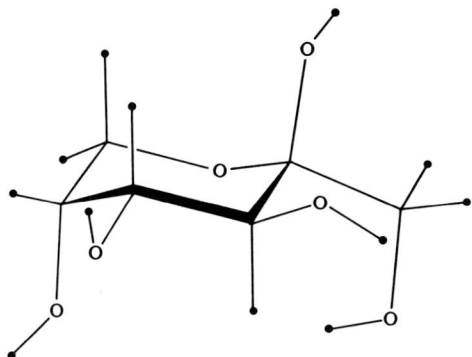

q = 56 pm). The orientation of the primary alcohol group is *gauche–gauche*. The C-6–O-6–C-2–O-2 bond-lengths are 143.0, 141.3, and 140.7 pm. The hydrogen bonding consists of infinite chains, with side chains and two bifurcated interactions involving the ring-oxygen atom, with H · · · O distances of 196.5, 234.9 pm and 197.7, 259.3 pm.

$C_6H_{12}O_6$ α-D-Talopyranose (ADTALO10)[37,38]

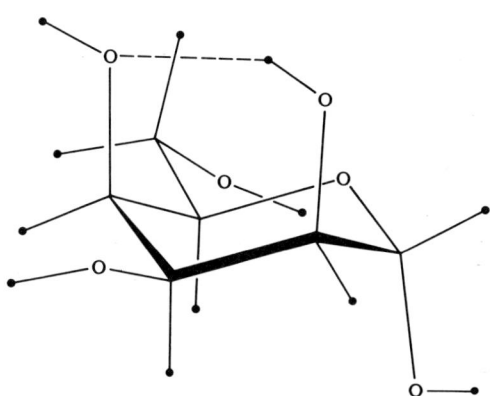

$P2_12_12_1$; Z = 4; D_x = 1.60; R = 0.051 for 2,605 intensities, and 0.047 for 1,556 intensities.[38] These two independent, X-ray analyses agree within experimental error. The pyranose has the $^4C_1$ conformation (θ = 3°, q = 59 pm), with three axial hydroxyl groups. The mean C-5–O-5–C-1–O-1 bond-lengths are 144.1, 143.6, and 140.1 pm. The hy-

---

(37) J. Ohanessian, D. Avenel, J. A. Kanters, and D. Smits, *Acta Crystallogr. Sect. B*, 33 (1977) 1063–1066.
(38) L. K. Hansen and A. Hordvik, *Acta Chem. Scand. Ser. A*, 31 (1977) 187–191.

drogen bonding consists of infinite chains which include a syndiaxial, O-2–H · · · O-4, intramolecular bond, and isolated links from the anomeric hydroxyl groups to the ring-oxygen atoms.

$C_6H_{12}O_6 \cdot SrCl_2 \cdot 5\ H_2O$ *epi*-Inositol · strontium chloride, pentahydrate (EPINSR)[39]

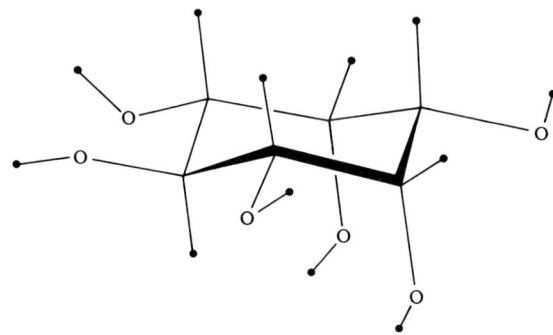

$P2_12_12_1$; Z = 4; D_x = 1.81; R = 0.020 for 1,730 intensities. The cyclohexane chair conformation, and the orientations of the hydroxyl groups are similar to those observed for the molecule in the crystal structure of *epi*-inositol.[40] The C–C bond-lengths range from 151.7 to 153.5 pm, and the C–O bond-lengths from 143.3 to 141.8 pm. The ring valence-angle at C-1 is strained to 114.6° in both structures; all other ring valence-angles are within ± 1° of tetrahedral. The strontium ion is nine-coordinated to four water molecules, three adjacent hydroxyl groups on one molecule, and two adjacent hydroxyl groups on an adjoining molecule. The Sr^+ · · · O distances range from 259.1 to 267.4 pm. The Cl^- ions are five-coordinated; one is hydrogen-bonded to four water molecules and one hydroxyl group, and the other, to three water molecules and two hydroxyl groups. The Cl · · · O distances range from 313.6 to 322.8 pm.

$C_6H_{14}O_6$ DL-Mannitol (DLMANT)[41]

$Pna2_1$; Z = 4; D_x = 1.504; R = 0.030 for 956 intensities at −150°C. The carbon chain of the molecule has the planar, zigzag conformation,

(39) R. A. Wood, V. J. James, and S. J. Angyal, *Acta Crystallogr. Sect. B*, 33 (1977) 2248–2251.
(40) G. A. Jeffrey and H. S. Kim, *Acta Crystallogr. Sect. B*, 27 (1971) 1812–1817.
(41) J. A. Kanters, G. Roelofsen, and D. Smits, *Acta Crystallogr. Sect. B*, 33 (1977) 3635–3640.

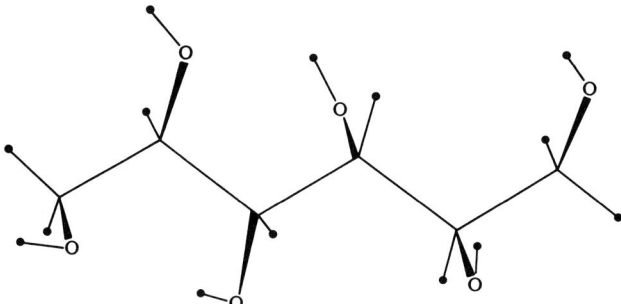

with the terminal hydroxyl groups in the $-sc$ orientation. The Figure depicts D-mannitol. The same conformation is observed for the polymorphs of D-mannitol.[42,43] Mean C–C and C–O distances are 152.5 and 149.2 pm, respectively; mean C–C–C and C–C–O angles are 112.6 and 110.1°. The hydrogen bonding involves all hydroxyl groups in infinite chains, and is very similar to that observed in DL-arabinitol,[44] which crystallizes in the same space-group. Atomic coordinates reported refer to the D enantiomer.

$C_7H_{14}O_6$ Methyl β-D-galactopyranoside (MBDGAL)[45,46]

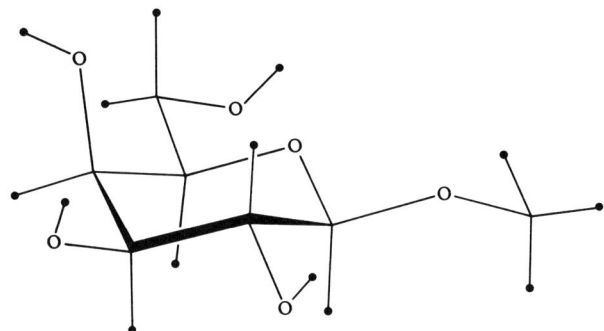

$P2_12_12_1$; $Z = 4$; $D_x = 1.482$; $R = 0.025$ for 1,086 intensities, and 0.040 for 917 intensities.[46] These two independent, X-ray analyses agree within their experimental errors. The pyranoside conformation is a

(42) H. M. Berman, G. A. Jeffrey, and R. D. Rosenstein, *Acta Crystallogr. Sect. B*, 24 (1968) 442–449.
(43) H. S. Kim, G. A. Jeffrey, and R. D. Rosenstein, *Acta Crystallogr. Sect. B*, 24 (1968) 1449–1455.
(44) F. D. Hunter and R. D. Rosenstein, *Acta Crystallogr. Sect. B*, 24 (1968) 1652–1660.
(45) S. Takagi and G. A. Jeffrey, *Acta Crystallogr. Sect. B*, 34 (1978) 2006–2010.
(46) B. Sheldrick, *Acta Crystallogr. Sect. B*, 33 (1977) 3003–3005.

slightly distorted $^4C_1$ ($\theta = 6°$, $\varphi = 348°$, q = 58 pm). The glycosidic bond-length (C-1–O-1) and the ring-oxygen valence-angle are smaller than in the α-pyranoside,[47] namely, 139.0 compared to 140.5 pm, and 111 compared to 114°. The glycosidic torsion-angle, O-5–C-1–O-1–CH_3, is $-78°$. The orientation of the primary alcohol group is *gauche – trans*. The hydrogen bonding consists of infinite chains and isolated, bifurcated interactions from O-3–H to the ring and glycosidic oxygen atoms, with H $\cdots$ O distances of 220 and 270 pm.

$C_7H_{14}O_6$ Methyl α-D-glucopyranoside (MGLUCP11)[48]

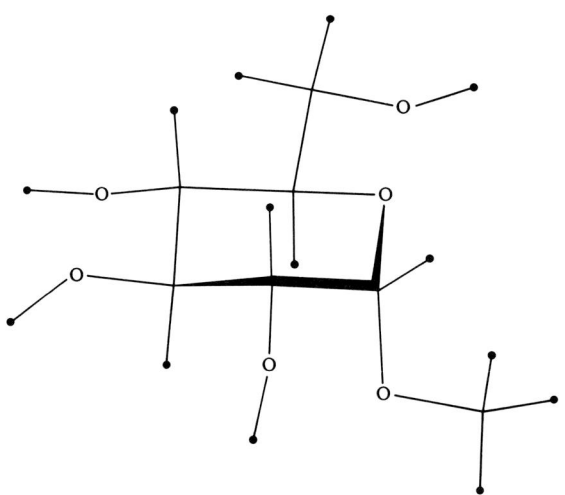

$P2_12_12_1$; Z = 4; D_n = 1.460; R = 0.046 for 1,123 neutron intensities. This is a neutron-diffraction refinement of a structure previously determined by X-ray diffraction.[49] The pyranoside has the $^4C_1$ conformation ($\theta = 2°$, q = 57 pm). The C-5–O-5–C-1–O-1–CH_3 bond-lengths are 142.8, 141.4, 140.1, and 142.2 pm, and the glycosidic torsion-angle is $+63°$. The primary alcohol group is *gauche –trans*. The hydrogen bonding consists of an infinite spiral of strong bonds and a very weak, bifurcated interaction, with H $\cdots$ O distances of 232 and 263 pm. The glycosidic oxygen atom is not hydrogen-bonded.

(47) B. M. Gatehouse and B. J. Poppleton, *Acta Crystallogr. Sect. B*, 27 (1971) 654–660.
(48) G. A. Jeffrey, R. K. McMullan, and S. Takagi, *Acta Crystallogr. Sect. B*, 33 (1977) 728–737.
(49) H. M. Berman and S. H. Kim, *Acta Crystallogr. Sect. B*, 24 (1968) 897–904.

$C_7H_{14}O_6$ Methyl α-D-mannopyranoside (MEMANP11)[48]

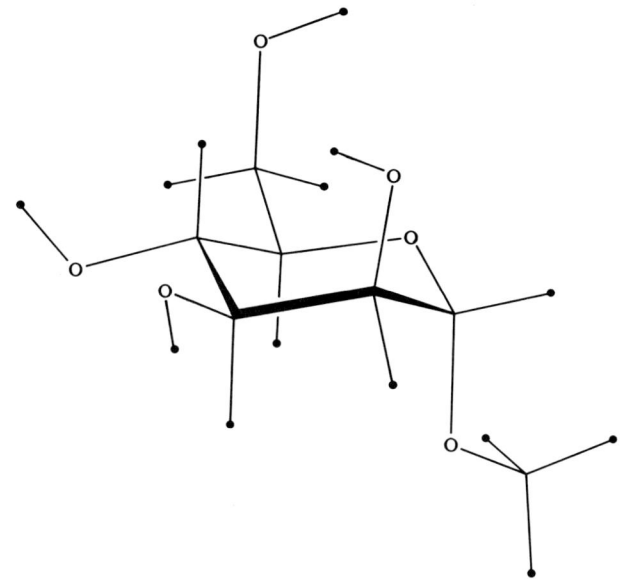

$P2_12_12_1$; Z = 4; D_n = 1.460; R = 0.030 for 1,304 neutron intensities. This is a neutron-diffraction refinement of a structure previously determined by X-ray diffraction.[50] This is a rare example of a pyranoside adopting the ideal, $^4C_1$ conformation (θ = 0°, q = 56 pm). The C-5-O-5-C-1-O-1-CH₃ bond-lengths are 143.9, 141.5, 140.1, and 142.2 pm, and the glycosidic torsion-angle is +61°. The primary alcohol group has the *gauche-gauche* orientation. The hydrogen bonding consists of finite chains and isolated bonds which involve all of the oxygen atoms in the molecule.

$C_7H_{14}O_6 \cdot 0.5\ H_2O$ Methyl β-D-glucopyranoside hemihydrate
(MBDGPH10)[51]

$P4_12_12$; Z = 8; D_x = 1.430; R = 0.035 for 1,269 intensities. The pyranoside has a slightly distorted, $^4C_1$ conformation (θ = 7°, φ = 38°, q = 60 pm). The C-5-O-5-C-1-O-1-CH₃ bond-lengths are 144.0, 143.3, 138.0, and 143.0 pm, and the glycosidic torsion-angle is −73°.

(50) B. M. Gatehouse and B. J. Poppleton, *Acta Crystallogr. Sect. B*, 26 (1970) 1761–1765.
(51) G. A. Jeffrey and S. Takagi, *Acta Crystallogr. Sect. B*, 33 (1977) 738–742.

The orientation of the primary alcohol group is *gauche–trans*. The hydrogen bonding consists of finite chains which intersect at four-coordinated water molecules on two-fold, crystallographic-symmetry axes.

$C_7H_{14}O_7$ D-*glycero*-β-D-*gulo*-Heptopyranose (β-D-"glucoheptose") (BDGHEP)[52]

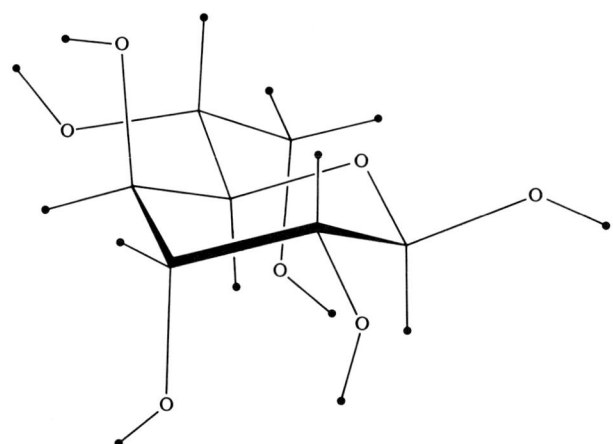

$P2_12_12_1$; Z = 4; D_x = 1.540; R = 0.029 for 855 intensities. The pyranose has a slightly distorted, $^4C_1$ conformation (θ = 6.2°, φ = 39.2°, q = 57 pm). The C-5–O-5–C-1–O-1 bond-lengths are 144.0, 142.9, and 140.1 pm. The orientation of the side chain is *gauche–trans*, and *gauche–gauche*. The hydrogen bonding consists of infinite chains which exclude the ring-oxygen atoms. The H · · · O distances range from 173 to 185 pm.

(52) J. A. Kanters, A. J. H. M. Kock, and G. Roelofsen, *Acta Crystallogr. Sect. B*, 34 (1978) 3285–3288.

$C_8H_{11}NO_5$ 2-Acetamido-2,3-dideoxy-D-*erythro*-hex-2-enono-1,4-lactone (ADEHXL10)[53]

$P2_12_12_1$; Z = 4; D_x = 1.472; R = 0.058 for 926 intensities. The C=C bond-length is 132.5 pm. The lactone ring is planar. The side-chain orientation is *gauche–gauche* with respect to C-3 and C-4. The hydrogen bonding involves three types of bonds, namely, N–H · · · O=C, O–H · · · O=C, and O–H · · · O–H, with H · · · O lengths of 221, 172, and 196 pm, respectively.

$C_8H_{14}O_7 \cdot H_2O$ 2,7-Anhydro-L-*glycero*-β-D-*manno*-octulopyranose, monohydrate[54]
Depiction on p. 438

$P2_1$; Z = 2; D_x = 1.55; R = 0.037 for 992 intensities. The pyranose has a distorted, $^1C_4$ conformation (θ = 155°, φ = 183°, q = 62 pm). The conformation of the anhydride is an envelope, with the pyranose ring-oxygen atom out of the plane. The seven-membered part has a boat conformation. Both CH_2OH groups are oriented *gauche* to the ring-oxygen atom of the anhydride. The ring C–O bond-lengths range from 140.3 to 145.2 pm. There is extensive hydrogen bonding involving all of the hydroxyl groups and ring-oxygen atoms.

(53) Z. Ružić-Toroš and F. Lazasini, *Acta Crystallogr. Sect. B*, 34 (1978) 854–858.
(54) H. J. Hecht, P. Luger, and R. Reinhardt, *J. Cryst. Mol. Struct.*, 7 (1977) 161–171.

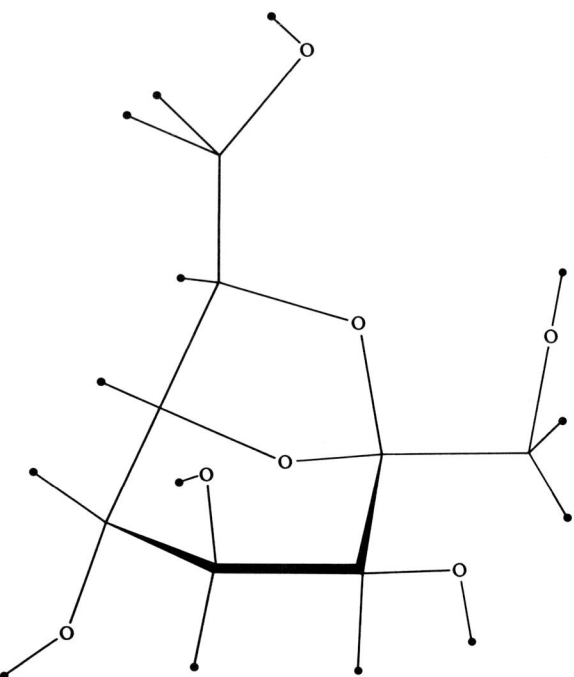

$C_8H_{16}ClNO_6$ N-(2-Chloroethyl)-D-gluconamide (CEGLCA01)[55]

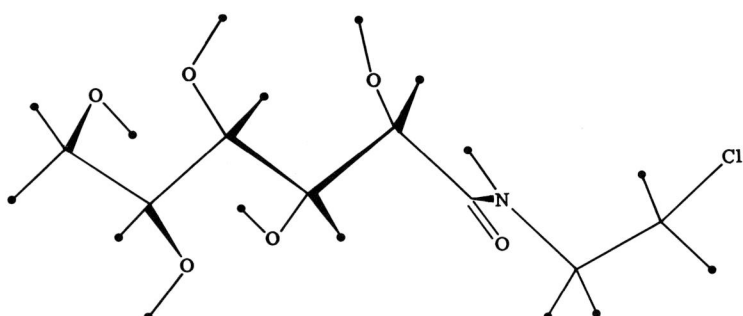

P1; Z = 1; D_n = 1.475; R = 0.05 for 961 neutron intensities. This is a neutron-diffraction refinement of a structure previously determined by X-ray diffraction.[56] All of the hydroxyl groups are involved in intermolecular hydrogen-bonds which form infinite chains, with H · · · O distances ranging from 175.5 to 200.2 pm, and O–H · · · O angles

(55) A. C. Sindt and M. F. Mackay, *Acta Crystallogr. Sect. B*, 33 (1977) 2659–2662.
(56) L. O. G. Satzke and M. F. Mackay, *Acta Crystallogr. Sect. B*, 31 (1975) 1128–1132; *Adv. Carbohydr. Chem. Biochem.*, 34 (1977) 356.

from 149 to 173°. The N–H forms a bifurcated bond, one component of which is intramolecular to a synclinal OH. The two H · · · O distances are 202.8 and 210.1 pm, making an O · · · H · · · O angle of 125°.

$C_{10}H_{11}NO_5$ 2,2'-Anhydro-(1-β-L-arabinofuranosyl-2-hydroxy-4-pyridone) (ARAFPY10)[57]

$P2_12_12_1$; Z = 8; D_x = 1.485; R = 0.05 for 1,476 intensities. There are two symmetry-independent molecules in the structure. The furanosyl groups have different conformations; ^4T_o (ψ = 156°, q = 36 pm), and

(57) W. L. B. Hutcheon and M. N. G. James, *Acta Crystallogr. Sect. B*, 33 (1977) 2228–2232; described the β-D form.

$^3T_4$ (ψ = 316°, q = 25 pm). There is also an orientational difference between the two primary alcohol groups; in one molecule it is *gauche–trans*, and in the other, *trans–gauche* (these being the two favored, staggered orientations). The 4-pyridone rings are planar, except that, in one of the molecules, the carbonyl oxygen atom is 1 pm out of the pyridine plane. No significant differences were noted in the bond lengths and valence angles between the two molecules. The molecules are linked by O–H · · · O hydrogen-bonds, one of which is unusually short, with H · · · O = 158 pm.

$C_{10}H_{16}NO_9S_2^-K^+\cdot H_2O$ C-Allyl-S-β-D-glucopyranosyl-O-sulfo(thiocarbohydroximidate), potassium salt, monohydrate ("sinigrin") (KMYRMH01)[58]

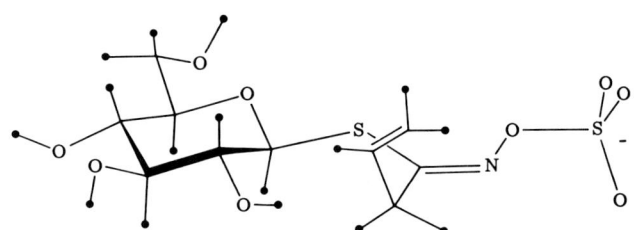

$P2_12_12_1$; Z = 4; D_x = 1.654; R = 0.056 for 1,825 intensities. The 1-thiopyranoside has a distorted, $^4C_1$ conformation (θ = 9°, φ = 43°, q = 58 pm). The O-5–C-1–S–C torsion-angle is −90°. The C–S bond-lengths are different, namely, 181 and 172 pm. The orientation of the primary alcohol group is *gauche–trans*. The K$^+$ ions are coordinated to three sulfate and two water oxygen atoms, with K · · · O distances ranging from 277 to 291 pm.

$C_{11}H_{15}NO_5$ 2-Acetamido-2,3-dideoxy-5,6-O-isopropylidene-D-*threo*-hex-2-enono-1,4-lactone (AIPTHL10)[59]

$P2_1$; Z = 4; D_x = 1.357; R = 0.048 for 1,989 intensities. The crystal structure contains two symmetry-independent molecules having almost the same conformation. The lactone rings are planar; the conformation of the dioxolane ring in one is close to an envelope, and that in the other has a twist conformation. The linkage between the rings, as defined by O-4–C-4–C-5–O-5, is −65 and −75°. The two independent molecules are hydrogen-bonded into dimers by the two N–

(58) W. H. Watson and Z. Taira, *Cryst. Struct. Commun.*, 6 (1977) 441–446.
(59) Z. Ružić-Toroš and I. Leban, *Acta Crystallogr. Sect. B*, 34 (1978) 1226–1230.

H · · · O=C hydrogen-bonds, with H · · · O distances of 212 and 220 pm. Some of the hydrogen atom positions were not reported.

$C_{11}H_{16}O_4$ 1-Deoxy-2-C-phenyl-D-arabinitol (DXPHAR)[60]

$P2_12_12_1$; Z = 4; D_x = 1.298; R = 0.056 for 1,045 intensities. The D-arabinitol residue has a sickle conformation, with a C-1–C-2–C-3–C-4 torsion-angle of −55°. The C-6 atom of the phenyl group forms a planar, zigzag chain with C-2 to C-5. The terminal hydroxyl group is +sc to the carbon chain. The plane of the phenyl ring is almost per-

(60) D. C. Rohrer, J. C. Fischer, D. Horton, and W. Weckerle, *Can. J. Chem.*, 56 (1978) 2915–2921.

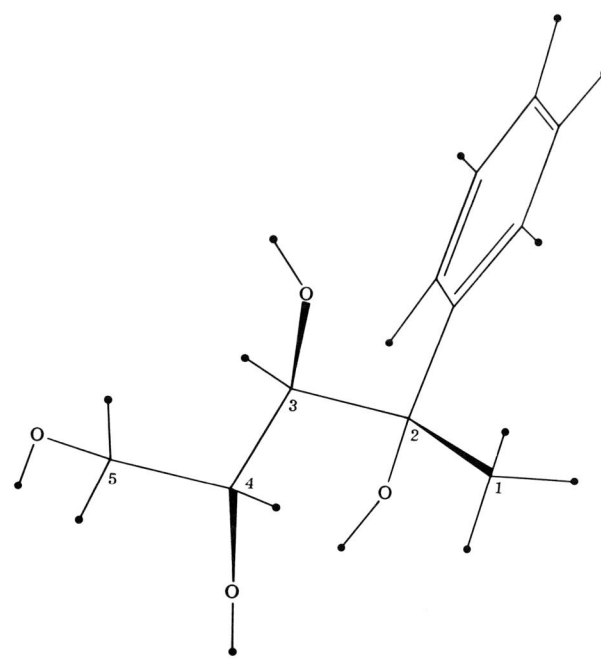

pendicular to that of the carbon chain, with C-3–C-2–C-6–C-7 = −94°. There is an intramolecular hydrogen-bond from O-2–H to O-4, with an H · · · O distance of 192 pm.

$C_{11}H_{16}O_7$ Methyl 3,4-di-O-acetyl-2,6-anhydro-β-D-talopyranoside (MATALP)[61]

Depiction on p. 443

$P2_1$; $Z = 2$; $D_x = 1.42$; $R = 0.037$ for 1,506 intensities. The pyranoside has a distorted conformation ($\theta = 90°$, $\varphi = 132°$, $q = 88$ pm) lying between $^{2,5}B$ ($\varphi = 120°$) and $^2S_0$ ($\varphi = 150°$). The conformation about the anhydro link is twisted so that C-5–C-6–O-2–C-2 = +20°. Both acetoxyl groups are equatorial. The methyl glycosidic torsion-angle is −66°. The C–O bond-lengths, excluding those in the acetoxyl groups, vary from 138.2 to 143.5 pm.

$C_{11}H_{18}O_4S_2$ Methyl 2,3,6-trideoxy-2-C-[1,1-(ethylenedithio)-2-hydroxyethyl]-α-L-*threo*-hexopyranosid-4-ulo-2^2,4-pyranose (MDETPP)[62]

Depiction at top of p. 444

(61) P. Köll and F. S. Tayman, *Chem. Ber.*, 110 (1977) 3297–3303.
(62) K.-H. Klaska, O. Jarchow, W. Koebernick, and H. Paulsen, *Carbohydr. Res.*, 56 (1977) 67–73.

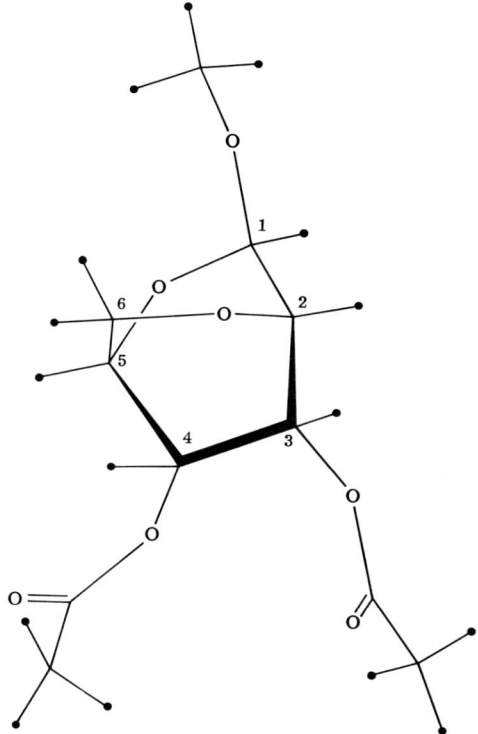

$P2_12_12_1$; $Z = 4$; $D_x = 1.442$; $R = 0.057$ for 2,009 intensities. The pyranoside has a distorted, $^1C_4$ conformation ($\theta = 167°$, $\varphi = 324°$, $q = 58$ pm). The conformation of the spirocyclic dithiolane is an envelope having one of the methylene carbon atoms out of the plane. The methyl glycosidic angle is $-59°$. The C–O bond-lengths range from 139.7 to 145.1 pm. The C–S bond-lengths are equal, in pairs, 183.4 and 181.3 pm.

$C_{11}H_{21}NO_4$ Methyl 3-acetamido-2,3,6-trideoxy-3-C-methyl-4-O-methyl-β-L-*xylo*-hexopyranoside (MADXHP)[63]
Depiction at bottom of p. 444

$P2_12_12_1$; $Z = 4$; $D_x = 1.220$; $R = 0.050$ for 699 intensities. The conformation of the pyranoside is $^1C_4$ ($\theta = 176°$, $q = 56$ pm), with the acetamido group equatorial. This result is inconsistent with an interpretation of some n.m.r. data which indicated axial attachment of a nitro group in a derivative of evernitrose.

(63) A. K. Ganguly, O. Z. Sarre, A. T. McPhail, and K. D. Onan, *J. Chem. Soc. Chem. Commun.*, (1977) 313–314.

$C_{12}H_{18}O_7 \cdot H_2O$ 2,3:4,6-Di-O-isopropylidene-α-L-xylo-2-hexulofuran-osonic acid, monohydrate (DIPKGA)[64]

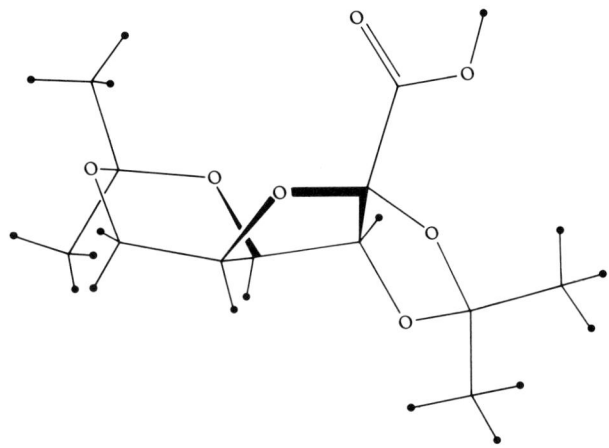

$P2_12_12_1$; $Z = 4$; $D_x = 1.319$; $R = 0.028$ for 1,789 intensities. The furanose has the $^4T_3$ conformation ($\psi = 128°$, $q = 36$ pm). The conformation of the dioxolane is an envelope, and that of the dioxane is a distorted chair. The molecule has a six-link sequence of C–O bonds, with the following lengths: 140.6, 142.7, 144.4, 140.8, 141.0, and 144.5 pm. These variations are similar to those observed in some oligosaccharides.[65] The C–C bond-lengths range from 150.0 to 154.1 pm. The hydrogen bonding consists of short, finite chains that link the carboxyl groups to the water molecules.

$C_{12}H_{22}O_{11}$ 3-O-α-D-Glucopyranosyl-β-D-fructopyranose; turanose (TURANS)[66,67]

$P2_12_12_1$; $Z = 2$; $D_x = 1.574$; $R = 0.031$ for 1,404 intensities, and 0.039 for 1,556 intensities.[67] These two independent, X-ray analyses agree within their experimental errors. The D-glucopyranosyl group has the $^4C_1$, and the D-fructopyranose residue, the $^5C_2$, conformation ($\theta = 4°$, 3°; $q = 56$ pm, 57 pm). The orientation of the linkage bonds is O-5–

(64) S. Takagi and G. A. Jeffrey, *Acta Crystallogr. Sect. B*, 34 (1978) 2932–2934.
(65) G. A. Jeffrey, J. A. Pople, and L. Radom, *Carbohydr. Res.*, 38 (1974) 81–95.
(66) J. A. Kanters, W. P. J. Gaykema, and G. Roelofsen, *Acta Crystallogr. Sect. B*, 34 (1978) 1873–1880.
(67) A. Neuman, D. Avenel, and H. G. Pandraud, *Acta Crystallogr. Sect. B*, 34 (1978) 242–248.

C-1–O-1–C-3′ = +98°, and C-1–O-1–C-3′–C-2′ = −128°. The orientations of the primary alcohol groups of the D-glucosyl group and D-fructose residue are *gauche–gauche* and *gauche–trans*, respectively. There is a weak, intramolecular hydrogen-bond linking O-4–H and O-2 of the two residues, with an H · · · O distance of 210 pm and an O–H · · · O angle of 166°. The hydrogen bonding consists of finite and infinite chains. It is unusual, in that it excludes one of the hydroxyl groups, O-2–H, which neither accepts nor donates a hydrogen bond. There is also a weak bond with H · · · O = 239 pm and O–H · · · O = 134°. These two weakly bonded, hydroxyl groups are identified with O–H stretching-vibration peaks at 3610 and 3550 cm^{-1}, respectively, in the infrared spectrum of the solid in a KBr pellet. Similar infrared identifications of weakly bonded hydroxyl groups in β-D-fructopyranose and other carbohydrates have been made.[35,68]

$C_{12}H_{22}O_{11}$ 4-*O*-α-D-Glucopyranosyl-α,β-D-glucopyranose; α,β-maltose (MALTOT)[69]

$P2_12_12_1$; Z = 4; D_x = 1.546; R = 0.043 for 1,379 intensities. The crystals contained 18% of the β anomer. Both the pyranosyl group and the pyranose residue have a slightly distorted, $^4C_1$ conformation (θ = 12°,

(68) J. A. Kanters, J. Kroon, A. F. Peerdeman, and J. A. Vliegenthart, *Nature*, 222 (1969) 370–371.
(69) F. Takusagawa and R. A. Jacobson, *Acta Crystallogr. Sect. B*, 34 (1978) 213–218.

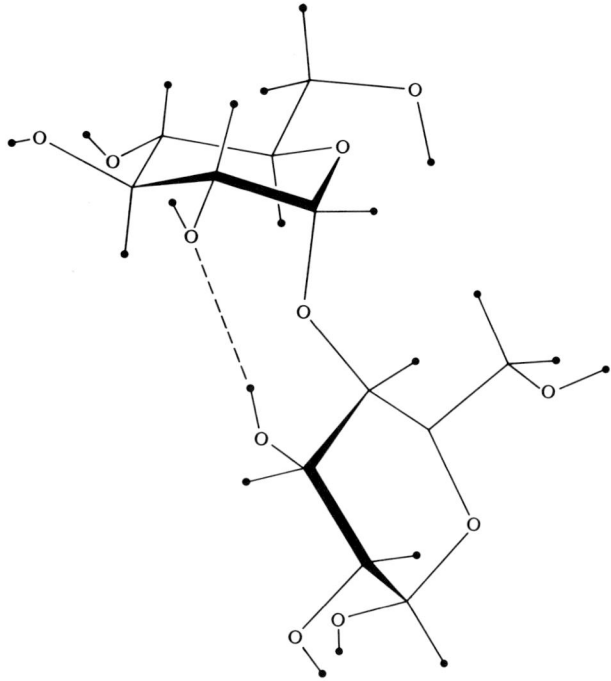

7°; φ = 57°, 53°; q = 52 pm, 55 pm). The linkage torsion-angles are O-5–C-1–O-1–C-4', +116°, and C-1–O-1–C-4'–C-3', +122°. The primary alcohol groups are both disposed *gauche–trans*. The glycosidic valence-angle, C-1–O-1–C-4', is 120°. The hydrogen bonding consists of finite chains which bifurcate and terminate at ring-oxygen atoms, and separate links from anomeric hydroxyl groups to ring-oxygen atoms. There is an intramolecular hydrogen-bond, namely, O-3–H · · · O-2'. The bifurcated H · · · O bond-lengths are 223 and 236 pm. Other H · · · O bond-lengths range from 172 to 217 pm.

$C_{12}H_{22}O_{11} \cdot 0.19\ H_2O$ 3-O-β-D-Glucopyranosyl-α,β-D-glucopyranose hydrate; laminarabiose hydrate (LAMBIO)[70]

P2; Z = 2; D_x = 1.555; R = 0.057 for 1,023 intensities. The crystal contains the α and β anomers in the ratio of 2:3. The β anomer is depicted. The pyranose conformations are $^4C_1$ (θ = 2°, 4°; q = 59 pm, 57 pm). The linkage angles are O-5–C-1–O-1–C-3' = −94°, and C-1–O-1–C-3'–C-4' = +79°. There is an intramolecular hydrogen-bond between O-4'–H and O-5. The fractional part of a water molecule is lo-

(70) H. Takeda, N. Yasuoka, and N. Kasai, *Carbohydr. Res.*, 53 (1977) 137–152.

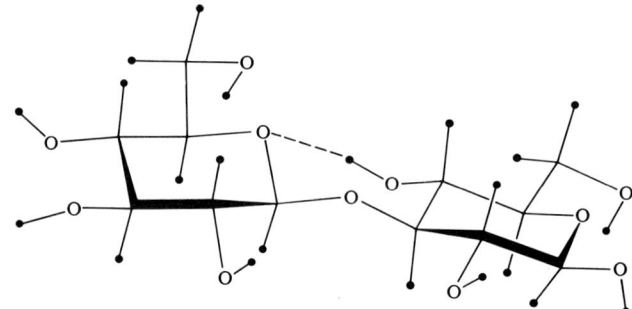

cated on the two-fold axis, and appears to be hydrogen-bonded primarily to the α anomers.

$C_{12}H_{22}O_{11} \cdot H_2O$ 6-O-α-D-Galactopyranosyl-α,β-D-glucopyranose monohydrate; α,β-melibiose monohydrate (MELIBM10)[71]

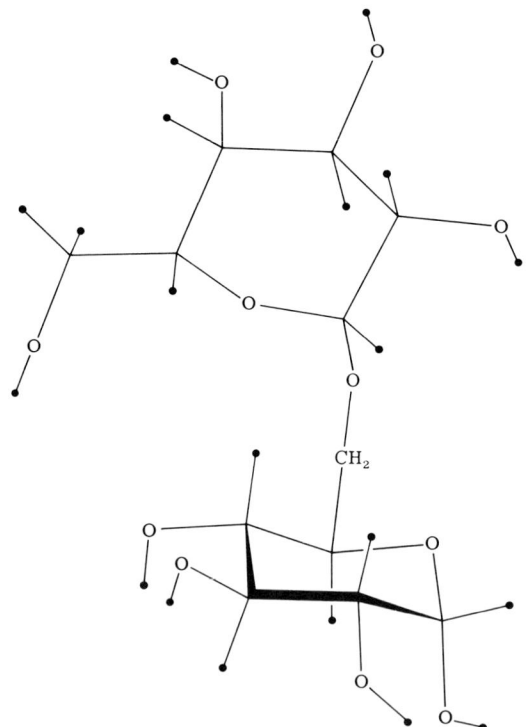

(71) M. E. Gress, G. A. Jeffrey, and D. C. Rohrer, *Acta Crystallogr. Sect. B*, 34 (1978) 508–512.

$P2_12_12_1$; $Z = 4$; $D_x = 1.554$; $R = 0.044$ for 1,805 intensities. The crystals contained 72% of the α and 28% of the β anomer. The α anomer is depicted. Other crystal-structure analyses have reported 85% of α and 15% of β (Ref. 72), and 80% of α and 20% of β (Ref. 73). Apart from the difference in anomeric composition, the agreement between the three crystal-structure results is excellent. The D-galactosyl group and D-glucose residue have the $^4C_1$ conformation ($\theta = 8°, 2°$; $q = 58$ pm, 56 pm, respectively). The linkage, bond torsion-angles are O-5'–C-1'–O-6–C-6 = +77°, C-1'–O-6–C-6–C-5 = −174°; they are very similar to those observed in the melibiose moiety of raffinose.[74] There is no direct, intramolecular hydrogen-bonding between residues, but the water molecule hydrogen-bonds intramolecularly to O-6–H and O-4' in the two moieties of the same molecule. The hydrogen bonding consists of loops and finite chains which intersect at the four-coordinated water molecules. There is a bifurcated interaction, with H · · · O distances of 247 and 256 pm. The hydroxyl groups of the α anomer form part of the infinite chain. The hydroxyl groups of the β anomer form a separate hydrogen-bond to one of the ring-oxygen atoms (which is not otherwise involved in the hydrogen bonding).

$C_{12}H_{22}O_{11} \cdot H_2O$ 4-O-α-D-Glucopyranosyl-β-D-glucopyranose monohydrate; β-maltose monohydrate (MALTOS11)[75]

$P2_1$; $Z = 2$; $D_n = 1.536$; $R = 0.044$ for 1,843 neutron intensities. This is a neutron-diffraction refinement of a structure previously determined by X-ray diffraction.[76] The pyranoid moieties have the $^4C_1$ conformation ($\theta = 1°, 5°$; $q = 57$ pm, 58 pm). The linkage torsion-angles are O-5–C-1–O-1–C-4', +122°, and C-1–O-1–C-4'–C-3', +133°. The orientations of the primary alcohol groups are *gauche–trans* and *gauche–gauche*. Except for the orientation of one of the primary alcohol groups, the molecular conformation is close to that observed in the structure of the anhydrous crystal.[69] There is an intramolecular hydrogen-bond from O-2–H · · · O-3', with an H · · · O distance of 183.5 pm. This forms part of a finite chain which intersects two infinite chains of hydrogen bonds, forming a three-dimensional net. In

(72) K. Hirotsu and T. Higuchi, *Bull. Chem. Soc. Jpn.*, 49 (1976) 1240–1245.
(73) J. A. Kanters, G. Roelofsen, H. M. Doesburg, and T. Koops, *Acta Crystallogr. Sect. B*, 32 (1976) 2830–2837.
(74) H. M. Berman, *Acta Crystallogr. Sect. B*, 26 (1970) 290–299.
(75) M. E. Gress and G. A. Jeffrey, *Acta Crystallogr. Sect. B*, 33 (1977) 2490–2495.
(76) G. J. Quigley, A. Sarko, and R. H. Marchessault, *J. Am. Chem. Soc.*, 92 (1970) 5834–5839.

the anhydrous crystal, the intramolecular hydrogen-bond is[69] from O-3–H · · · O-2'.

$C_{12}H_{22}O_{11} \cdot H_2O$ 2-O-β-D-Glucopyranosyl-α-D-glucopyranose monohydrate; sophorose monohydrate (SOPROS)[77]

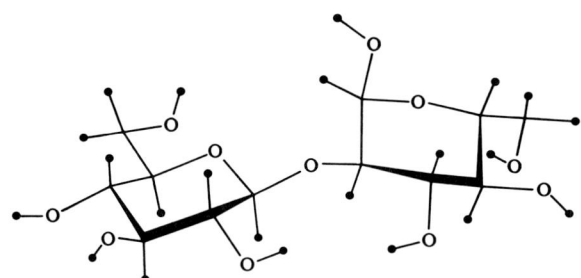

$P2_12_12_1$; Z = 4; D_x = 1.544; R = 0.072 for 1,315 intensities. The pyranose moieties have the $^4C_1$ conformation (θ = 4°, 9°; φ = 177°, 332°; q = 58 pm, 58 pm). The linkage, conformational angles, namely, O-5–C-1–O-1–C-2' and C-1–O-1–C-2'–C-3', are −79° and −140° (the conformational angles given in Table 5 of the paper[77] have

(77) J. Ohanessian, F. Longchambon, and F. Arene, *Acta Crystallogr. Sect. B*, 34 (1978) 3666–3671.

the incorrect sign, and refer to the L enantiomer). The C-5–O-5–C-1–O-1–C-2' bond-lengths are 143.6, 142.3, 140.3, and 144.0 pm. The primary alcohol orientations are *trans–gauche* and *gauche–gauche*. There is no intramolecular hydrogen-bond.

$C_{12}H_{22}O_{11} \cdot CaCl_2 \cdot 5\ H_2O$ α-D-Allopyranosyl α-D-allopyranoside · calcium chloride, pentahydrate (ALTRCA)[78]

$P2_1$; Z = 2; D_x = 1.83; R = 0.026 for 2,435 intensities. The pyranose moieties have a slightly distorted, $^4C_1$ conformation (θ = 7°, 9°; φ = 241°, 269°; q = 55 pm, 53 pm). The linkage-bond torsion-angles are O-5–C-1–O-1–C-1' = +45°, and C-1–O-1–C-1'–O-5' = +57°. [In α-D-glucopyranosyl α-D-glucopyranose (α,α-trehalose) dihydrate, these angles[79] are +62 and +75°.] The valence angles at the ring-oxygen atoms are unusually large, 117 and 116°; that at the linkage-oxygen atom is normal, 115°. The ring O-5–C-1 bond-lengths are also unusually short, 139.7 and 139.5 pm, whereas the anomeric bonds, C-1–O-1, are long, 143.3 and 142.5 pm. This is attributed to electron attraction by the cations, which results in a *reversal* of the usual trend for the sequence of acetal C–O bond-lengths. The orientations of the hydroxymethyl groups are *gauche–trans* and *gauche–gauche*. The cations are nine-coordinated to four hydroxyl oxygen atoms of different

(78) J. Ollis, V. J. James, S. J. Angyal, and P. M. Pljer, *Carbohydr. Res.*, 60 (1978) 219–228.
(79) G. M. Brown, D. C. Rohrer, B. Berking, C. A. Beevers, R. O. Gould, and R. Simpson, *Acta Crystallogr. Sect. B*, 28 (1972) 3145–3158.

molecules, a linkage oxygen atom, O-1, and four water oxygen atoms, with $Ca^{2+} \cdots O$ distances ranging from 241.0 to 261.4 pm.

$C_{12}H_{24}O_{10}$ 4-O-β-D-Galactopyranosyl-L-rhamnitol (GAPRHM)[80]

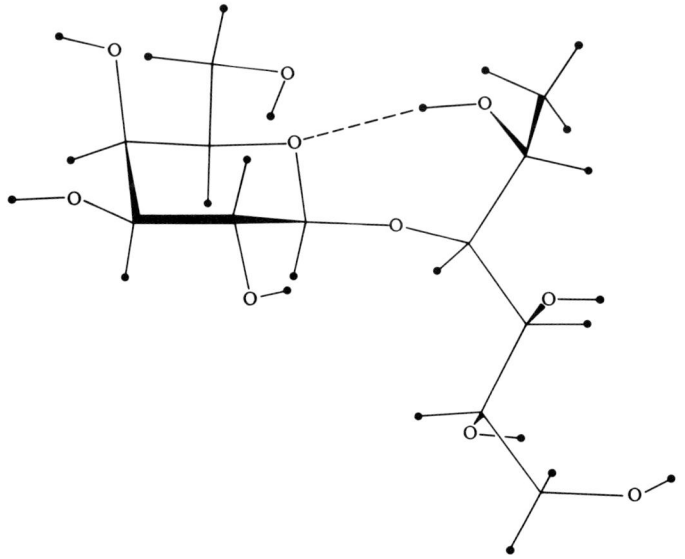

$P2_1$; $Z = 2$; $D_x = 1.464$; $R = 0.039$ for 1,620 intensities. The pyranosyl group has a distorted, $^4C_1$ conformation ($\theta = 12°$, $\varphi = 45°$, q = 58 pm). The C-5–O-5–C-1–O-1–C bond-lengths are 143.9, 142.6, 138.4, and 144.1 pm, and the glycosidic torsion-angle is −71°. The primary alcohol group has the *gauche-trans* orientation. The carbon chain of the L-rhamnitol residue has a planar, zigzag conformation from C-1′ to C-5′, but C-6′ is out of the plane because of the formation of an intramolecular hydrogen-bond from the terminal O-5′–H of the L-rhamnitol residue to the ring-oxygen atom, O-5, of the D-galactosyl group. The hydrogen bonding consists of infinite and finite chains which involve all of the hydroxyl groups and the ring-oxygen atoms.

$C_{13}H_{18}O_9$ 1,2,3,4-Tetra-O-acetyl-α-D-ribopyranose (AADRIB)[81]

$P2_1$; $Z = 4$; $D_x = 1.355$; $R = 0.053$; number of data not reported. The crystal structure contains two symmetry-independent molecules. Both

(80) S. Takagi and G. A. Jeffrey, *Acta Crystallogr. Sect. B*, 33 (1977) 2377–2380.
(81) V. J. James and J. D. Stevens, *Cryst. Struct. Commun.*, 6 (1977) 241–246.

pyranoses have the $^4C_1$ conformation, with two axial acetoxyl groups. One of the rings is significantly distorted (θ = 4°, 11°; φ = 309°, 261°; q = 54 pm, 54 pm). There are no other significant differences in the molecular dimensions, except for the orientation of the acetoxyl groups on C-2 and C-3.

$C_{13}H_{19}ClN_4O_5S$ 1-(6-Chloropurin-9-yl)-1-S-ethyl-1-thio-D-*glycero*-D-*ido*-hexitol (CLPTGL10)[82]

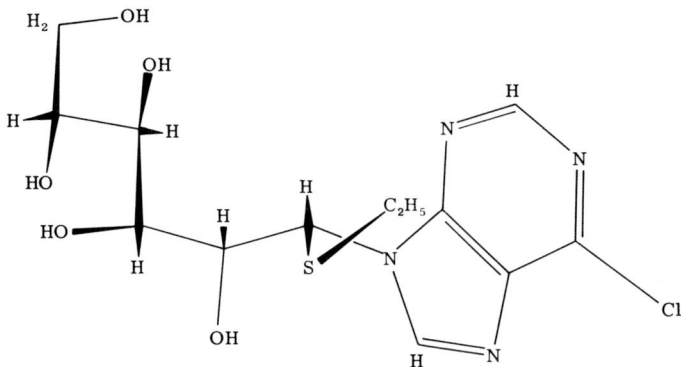

$P2_1$; $Z = 2$; $D_x = 1.49$; $R = 0.06$ for 854 intensities. The D-glucitol-1-yl group has a sickle conformation, with a C-1–C-2–C-3–C-4 torsion angle of $-46°$. The C–S bond-lengths are almost equal, namely, 181 and 182 pm. The purine ring is distorted from the planar shape, with displacements as great as 15 pm. The hydrogen atoms were not located in this low-accuracy analysis, and the hydrogen bonding is still uncertain.

$C_{13}H_{20}O_6$ (Z)-1-O-Acetyl-2,3:4,5-di-O-isopropylidene-D-*threo*-pent-1-enitol (APTPEN10)[83]
Depiction at top of p. 455

$P2_12_12_1$; $Z = 4$; $D_x = 1.21$; $R = 0.05$ for 1,605 intensities. The carbon chain has a sickle conformation, with C-1–C-2–C-3–C-4, C-2–C-3–C-4–C-5 torsion angles of $-60°$, $-50°$, respectively. The dioxolanes have a twist and an envelope conformation, and are related by an approximate, two-fold axis along C-3–C-4. The double-bond stereochemistry is Z.

$C_{13}H_{20}O_6$ (Z)-1-O-Acetyl-2,3:4,5-di-O-isopropylidene-D-*erythro*-pent-1-enitol (AIPEPN20)[83]
Depiction at bottom of p. 455

$P2_1$; $Z = 2$; $D_x = 1.25$; $R = 0.04$ for 1,387 intensities. Unlike that in the *threo* compound, the carbon chain has the zigzag conformation from C-2 to C-5. The C-1–C-2–C-3–C-4 torsion-angle is $-52°$. The

(82) A. Ducruix and C. Pascard, *Acta Crystallogr. Sect. B*, 33 (1977) 2501–2505; K. C. Blieszner, D. Horton, and R. A. Markovs, *Carbohydr. Res.*, 80 (1980) 241–262.
(83) A. Ducruix, C. Pascard-Billy, S. J. Eitelman, and D. Horton, *J. Org. Chem.*, 41 (1976) 2652–2653; A. Ducruix and C. Pascard-Billy, *Acta Crystallogr. Sect. B*, 33 (1977) 1384–1389.

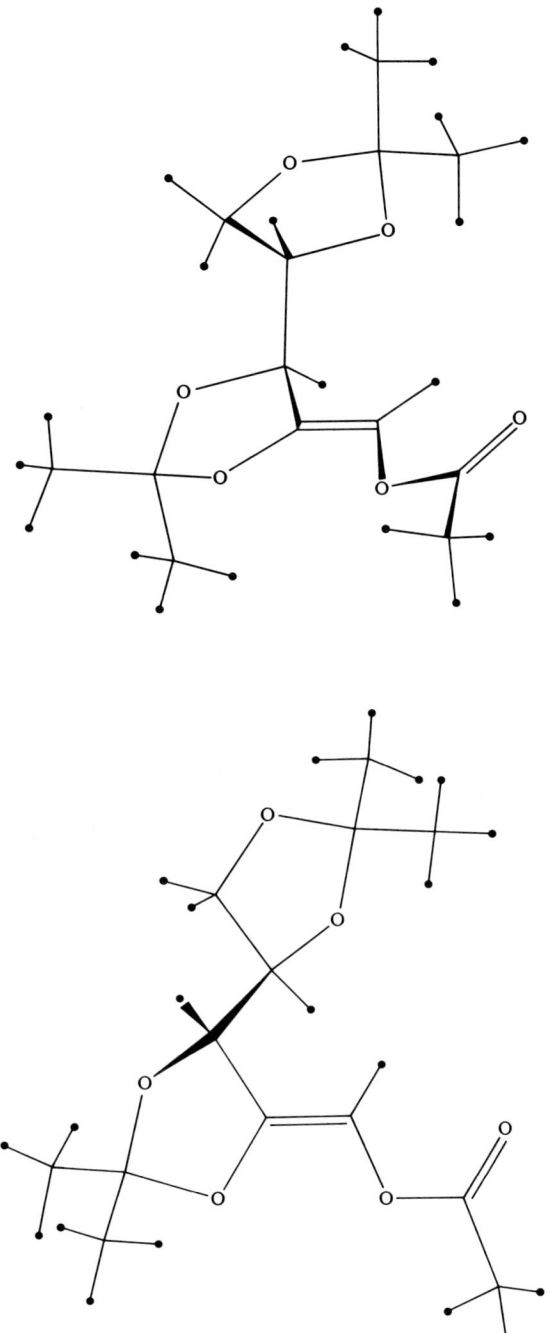

dioxolanes, which have a twist and an envelope conformation, are approximately centrosymmetrical about the midpoint of the C-3–C-4 bond linking them. The double-bond stereochemistry is Z.

$C_{13}H_{20}O_7$ 2,2^1:5,6-Di-O-isopropylidene-[2-C-(hydroxymethyl)-L-gulono-1,4-lactone] (IPHMGL)[84]

$P2_1$; $Z = 4$; $D_x = 1.32$; $R = 0.059$ for 1,985 intensities. The configuration of C-2 is (S). The two symmetry-independent molecules in the crystal structure have similar, but not identical, conformations. The gulonolactone and isopropylidene rings have envelope conformations. There appears to be a significant difference (of 5 pm) in two of the isopropylidene C–O bond-lengths in the two molecules.

$C_{14}H_{16}N_2O_3S$ 4-α-D-Erythrofuranosyl-1-p-tolylimidazoline-2-thione (TOEFIM10)[85]

(84) R. A. Anderson, F. W. B. Einstein, R. Hogi, and K. N. Slessor, *Acta Crystallogr. Sect. B*, 33 (1977) 2780–2783.
(85) I. Barragán, A. López-Castro, and R. Márquez, *Acta Crystallogr. Sect. B*, 33 (1977) 2244–2248.

$P2_12_12_1$; $Z = 4$; $D_x = 1.45$; $R = 0.074$ for 1,281 intensities. The furanosyl group has the $^0T_1$ conformation ($\psi = 29°$, q = 43 pm). The linkage to the planar imidazoline ring, as defined by O-1–C-1–C=C, is $-120°$. The angle between the phenyl and imidazoline rings is 53°.

$C_{14}H_{16}N_2O_3S$ 4-β-D-Erythrofuranosyl-1-p-tolylimidazoline-2-thione (BTEFIM10)[86]

(86) I. Barragán, A. López-Castro, and R. Márquez, *Acta Crystallogr. Sect. B*, 34 (1978) 295–298.

$P2_12_12_1$; Z = 4; D_x = 1.36; R = 0.040 for 1,541 intensities. The furanosyl group has the $^2T_3$ conformation (ψ = 83°, q = 42 pm). The linkage to the planar imidazoline ring, as defined by O-1–C-1–C=C, is +24°. The dihedral angle between the phenyl and imidazoline rings is 45°.

$C_{14}H_{18}N_2O_5S$ 1-Phenyl-(D-*glycero*-L-*gluco*-heptofurano)-[2,1-*d*]-imidazolidine-2-thione (PGHIMT10)[87]

$P2_1$; Z = 2; D_x = 1.47; R = 0.033 for 1,648 intensities. The furanoid part has the $^4T_0$ conformation (ψ = 151°, q = 38 pm). The carbon side-chain has a sickle conformation, with C-4–C-5–C-6–C-7 = +82°. The phenyl ring is inclined at 72° to the imidazolidine ring.

$C_{14}H_{19}Br_2ClO_8$ 3,4,5-Tri-*O*-acetyl-1,7-dibromo-6-(chloromethyl)-1,7-dideoxy-α-DL-*ido*-2-heptulopyranose (ABCMHP)[88]

$P2_1/c$; Z = 4; D_x = 1.704; R = 0.13 for 3,199 intensities. The analysis was carried out in order to confirm the configuration assigned to the molecule. The pyranose has the $^4C_1$ conformation (θ = 4°, q = 56 pm). ^{1}H-N.m.r. data suggested that, in CDCl$_3$ solution, the conformation is distorted toward a skew. The hydrogen atom positions were not reported.

(87) R. Jiménez-Garay, P. Villares, A. López-Castro, and R. Márquez, *Acta Crystallogr. Sect. B*, 34 (1978) 184–187.
(88) G. J. Abruscato, D. E. Kelly, W. J. Cook, and C. E. Bugg, *J. Org. Chem.*, 42 (1977) 3567–3571.

$C_{14}H_{21}NO_5$ 1-(Benzylmethylamino)-1-deoxy-β-D-*arabino*-2-hexulo-pyranose (BMAAHP10)[89]

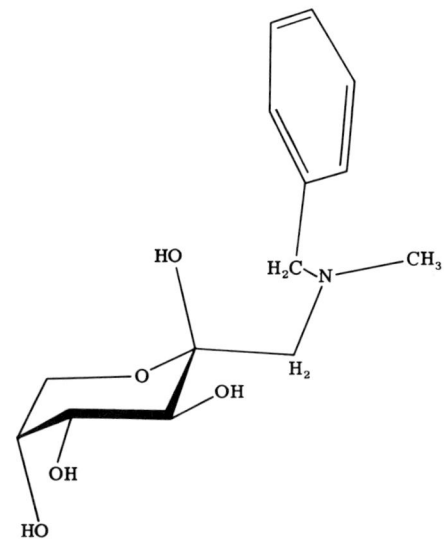

$P2_12_12_1$; Z = 4; D_x = 1.36; R = 0.079 for 1,236 intensities. The pyranose has the $^1C_4$ conformation (θ = 178°, q = 57 pm). The O-5–C-1–C-6–N and C-1–N–C–C torsion-angles are +87 and +166°. No hydrogen atom positions were reported.

(89) E. Moreno, S. Pérez-Garrido, P. Villares, and R. Jiménez-Garay, *Cryst. Struct. Commun.*, 7 (1978) 547–551.

$C_{14}H_{21}NO_5$ 1-(Benzylmethylamino)-1-deoxy-α-D-*lyxo*-2-hexulo-pyranose (BMALIX)[90]

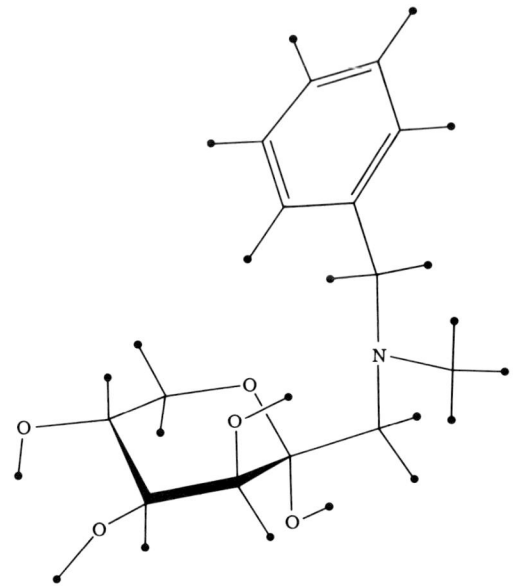

$P2_12_12_1$; Z = 4; D_x = 1.36; R = 0.047 for 4,243 intensities. The pyranose has a slightly distorted, $^4C_1$ conformation (θ = 8°, φ = 208°, q = 58 pm). The O-5–C-1–C-6–N torsion-angle is +62°. The benzyl ring is inclined at −49° to the plane of the methylamino group.

$C_{14}H_{24}N_4O_7S$ 5-Azido-5-deoxy-1,2-O-isopropylidene-3-O-(methyl-sulfonyl)-β-L-idofuranuronic N,N-diethylamide (AZDMIF10)[91,92]

$P2_1$; Z = 4; D_x = 1.32; R = 0.05 for 3,821 intensities. The configuration of C-5 is R. The two, symmetrically independent molecules in the crystal structure have different furanoid conformations, and orientations of the azido groups. One furanoid conformation is ^4E (ψ = 142°, q = 33 pm), and the other is $^4T_3$ (ψ = 124°, q = 35 pm). The orienta-

(90) S. Pérez, A. López-Castro, and R. Márquez, *Acta Crystallogr. Sect. B*, 34 (1978) 2341–2344.
(91) V. R. Klaska, O. Jarchow, C. Gunther, and H. Paulsen, *Acta Crystallogr. Sect. B*, 34 (1978) 226–232.
(92) H. Paulsen and C. Gunther, *Chem. Ber.*, 110 (1977) 2150–2157.

tions of the azido groups, as defined by the torsion angles C-4–C-5–N-1–N-2, are +78 and −56°. The diethylamide and methylsulfonyl groups have similar orientations, with C-4–C-5–C-6–N-6, −156° and −157°, and C-2–C-3–O-3–S, +87° and +93°. The isopropylidene groups have twist conformations, with O-1–C-1–C-2–O-2, +7 and +19°. The positions of the hydrogen atoms were not reported.

$C_{15}H_{22}O_{10}$ Methyl 2,3,4,6-tetra-O-acetyl-β-D-glucopyranoside (MTAGLV)[93]

$P2_12_12_1$; Z = 4; D_x = 1.293; R = 0.043 for 1,577 intensities. The pyranoside has a slightly distorted, $^4C_1$ conformation (θ = 8°, φ = 30°, q = 58 pm). The C-5–O-5–C-1–O-1–CH$_3$ bond-lengths are 142.7, 143.1, 138.5, and 144.3 pm. The glycosidic torsion-angle is −77°. The

(93) P. Zugenmaier and G. Rappenecker, *Acta Crystallogr. Sect. B*, 34 (1978) 164–167.

orientation of the acetoxyl groups is very similar to that observed on one of the rings of octa-*O*-acetyl-β-cellobiose.[94]

$C_{15}H_{22}O_{10}$ 1,3,4,6-Tetra-*O*-acetyl-2,5-*O*-methylene-D-mannitol (AMMANN)[95]

$P2_1$; $Z = 2$; $D_x = 1.281$; $R = 0.06$ for 1,338 intensities. The molecule contains a seven-membered, 1,3-dioxepane ring which has a conformation close to the twist-chair, with a two-fold axis through the mid-

(94) F. Leung, H. D. Chanzy, S. Pérez, and R. H. Marchessault, *Can. J. Chem.*, 54 (1976) 1365–1371.
(95) T. S. Cameron, R. E. Cordes, and T. B. Grindley, *Acta Crystallogr. Sect. B*, 33 (1977) 3718–3722.

point of C-2–C-3 and C-5. This is considered to be the preponderant conformation in solution.[96] There is a significant difference in the bond lengths of the two, ring C–O bonds, 139.3 and 143.4 pm, and there are some short, transannular, O · · · H contact distances of 225 and 236 pm.

$C_{15}H_{32}N_3O_9^+ Br^- \cdot 2\ H_2O$ 1,4-Diamino-1,4-dideoxy-3-O-(4-deoxy-4-propionamido-α-D-glucopyranosyl)-D-glucitol hydrobromide, dihydrate (DPGXHY)[97]

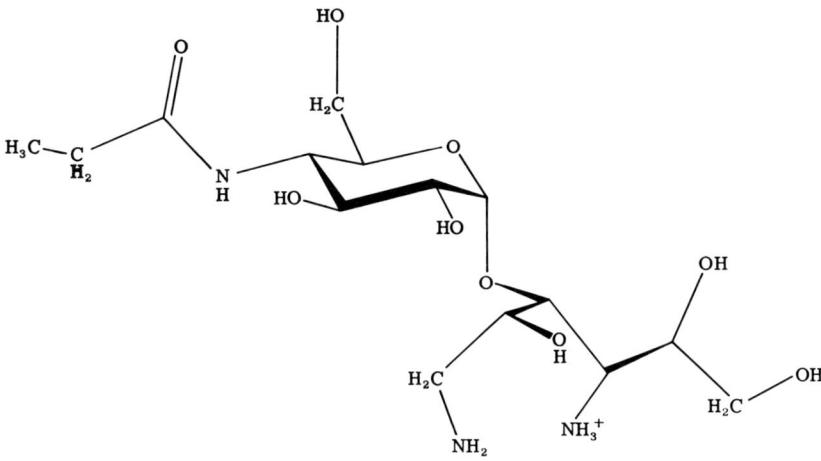

$P2_12_12_1$; $Z = 4$; $D_x = 1.21$; $R = 0.16$ for 2,306 intensities (partially photographic). This is the hydrobromide dihydrate of the aminoglycoside antibiotic P-2536(P). The conformation of the pyranosyl group is $^4C_1$ ($\theta = 3°$, $q = 60$ pm). The aglycon hexitol moiety has a bent carbon-chain conformation, with C-1–C-2–C-3–C-4 = $-75°$; C-2–C-3–C-4–C-5 = $-161°$. The results have low accuracy, and the hydrogen atoms were not located. The atomic coordinates reported were for the L enantiomer.

$C_{16}H_{21}NO_9$ 3,4,6-Tri-O-acetyl-2-(N-acetylacetamido)-1,5-anhydro-2-deoxy-D-*arabino*-hex-1-enitol (DXARAL)[98]

(96) T. B. Grindley and W. A. Swarez, *Can. J. Chem.*, 52 (1974) 4062–4071.
(97) K. Kamiya, Y. Wada, and N. Nara, *Chem. Pharm. Bull.*, 26 (1978) 2040–2045.
(98) B. Kojić-Prodić and V. Rogić, *Acta Crystallogr. Sect. B*, 34 (1978) 858–862.

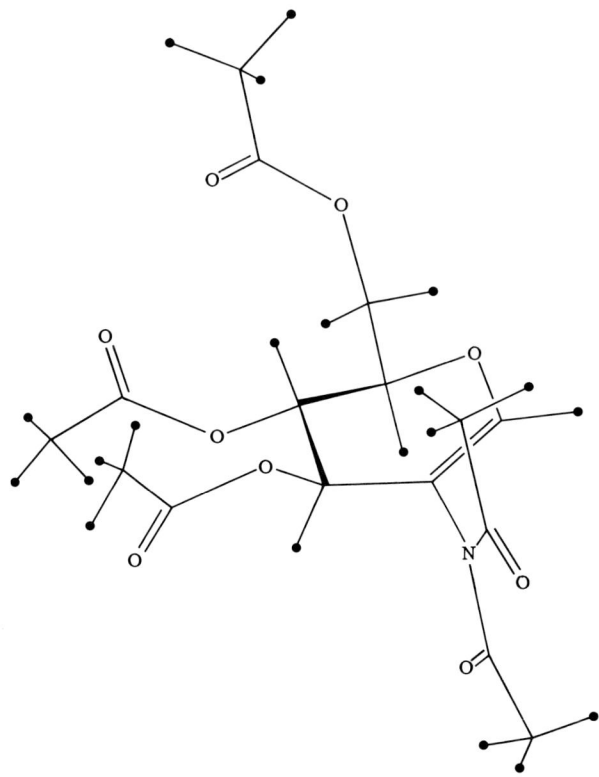

$P2_12_12_1$; $Z = 4$; $D_x = 1.361$; $R = 0.045$ for 2,151 intensities. The anhydro-enitol has the ^4E conformation ($\theta = 54°$, $\varphi = 292°$, $q = 48$ pm). The orientation of the acetylated, primary alcohol group is *gauche–gauche*, with an O-5–C-5–C-6–O-6 torsion-angle of $-65°$.

$C_{16}H_{21}NO_9$ 3,4,6-Tri-*O*-acetyl-2-(*N*-acetylacetamido)-1,5-anhydro-2-deoxy-D-*lyxo*-hex-1-enitol (ACALYX01)[99]

$P2_1$; $Z = 2$; $D_x = 1.346$; $R = 0.064$ for 2,191 intensities. The anhydro-enitol has the $^4H_5$ conformation ($\theta = 52°$, $\varphi = 269°$, $q = 54$ pm), with two of the acetoxyl groups axial. The O-5–C-1 and C-2–N bond-lengths adjacent to the C=C bond in the ring are 138.3 and 144.2 pm, indicating some π-bonding. The *N*-acetyl group is almost perpendicular to the plane of the C=C bond, with C-1–C-2–N–C-7 = $+104°$. The hydrogen atom positions of the methyl groups were not reported.

(99) V. Rogić, Ž. Ružić-Toroš, B. Kojić-Prodić, and N. Pravdić, *Acta Crystallogr. Sect. B*, 33 (1977) 3737–3743.

$C_{16}H_{21}NO_9$ 3,4,6-Tri-O-acetyl-2-(N-acetylacetamido)-1,5-anhydro-2-deoxy-D-*xylo*-hex-1-enitol (ACDXYL10)[99]

$P2_1$; $Z = 2$; $D_x = 1.357$; $R = 0.046$ for 1,753 intensities. The anhydro-enitol has the $^4H_5$ conformation ($\theta = 54°$, $\varphi = 252°$, $q = 48$ pm), with two axial acetoxyl groups. The lengths of the C–O and C–N bonds adjacent to the C=C bond are 136.1 and 144.1 pm, and the C-1–C-2–N–C-7 torsion-angle is +113°.

$C_{16}H_{22}O_{11}$ 1,2,3,4,6-Penta-O-acetyl-α-D-gulopyranose (PACDGP)[100]

$P2_12_12_1$; Z = 4; D_x = 1.325; R = 0.048; number of data not reported. The pyranose conformation is a distorted $^4C_1$ (θ = 7°, φ = 305°, q = 54 pm), with three axial acetoxyl groups. The bond lengths involving the ring-oxygen atom and the 1-O-acetyl group are C-5–O-5–C-1–O-1–C-7, 143.3, 138.9, 144.1, and 132.2 pm, respectively. The shortening of the ring O–C-1 bond and the lengthening of the anomeric C-1–O-1 bond relative to the mean, C–O bond-length agree with predictions from *ab initio*, molecular orbital calculations.[101]

$C_{16}H_{28}N_2O_{11}\cdot H_2O$ 2-Acetamido-4-O-[2-acetamido-2-deoxy-β-D-glucopyranosyl]-2-deoxy-D-glucopyranose monohydrate; N,N'-diacetylchitobiose monohydrate (ACHITM10)[102]

$P2_12_12_1$; Z = 4; D_x = 1.469; R = 0.054 for 2,202 intensities. The crystal structure contains about 10% of the α anomer. Both pyranoid moieties have the $^4C_1$ conformation, and one of them is ideal (θ = 3°, 0°;

(100) V. J. James and J. D. Stevens, *Cryst. Struct. Commun.*, 6 (1977) 119–122.
(101) G. A. Jeffrey and J. H. Yates, *Carbohydr. Res.*, 79 (1980) 155–163.
(102) F. Mo and L. H. Jensen, *Acta Crystallogr. Sect. B*, 34 (1978) 1562–1569.

q = 58 pm, 57 pm). The linkage-bond angles are O-5–C-1–O-1–C-4′, −80°, and C-1–O-1–C-4′–C-3′, +134°. (In cellobiose, these values are[103] −76 and +106°.) Both the primary alcohol groups and the N-acetyl groups have the *gauche–gauche* orientation. The hydrogen bonding consists of infinite chains with side links, and it excludes the two ring-oxygen atoms, the glycosidic oxygen atom, and one of the primary hydroxyl groups. Unlike the situation in the crystal structure of cellobiose, there is no intramolecular hydrogen-bonding from O-3′–H to the ring-oxygen atom, O-5 (unless it is a minor component of a bifurcated hydrogen-bond to O-6 of an adjacent molecule).

$C_{17}H_{18}BrNO_5$ [5-O-(Bromobenzoyl)-2,3-O-isopropylidene-α-D-ribofuranosyl]cyanomethane (IPBRCG)[104]

(103) S. S. C. Chu and G. A. Jeffrey, *Acta Crystallogr.*, 23 (1967) 1038–1049.
(104) H. Ohrui and S. Emoto, *J. Org. Chem.*, 42 (1977) 1951–1957.

P2$_1$2$_1$2$_1$; Z = 4; D$_x$ = 1.53; R not reported for 1,074 intensities. The D-ribofuranosyl group has the ^oE conformation (ψ = 3°, q = 31 pm), with the cyanomethyl group quasi-equatorial. The results were correlated with ^{1}H-n.m.r. data obtained for the compound in CDCl$_3$ solution (using the Karplus equations).

$C_{17}H_{24}O_{10}$ 2,3-(2R)-Epoxy-1-(2,3,4,6-tetra-O-acetyl-β-D-glucopyranosyl)propane (TAGPOX10)[105]

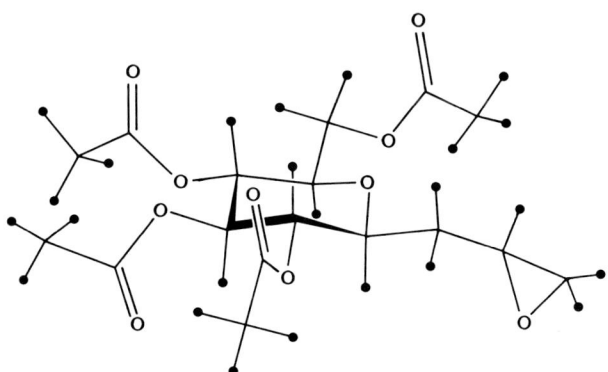

P2$_1$2$_1$2$_1$; Z = 4; D$_x$ = 1.320; R = 0.07 for 1,107 intensities. The pyranosyl group has the $^4C_1$ conformation (θ = 6°, φ = 93°, q = 57 pm). The configuration at the junction carbon atom of the epoxy ring is R. The glycosidic torsion-angle is −69°. The C–C and C–O bonds in the epoxy ring are 145 pm, and 139 and 144 pm.

$C_{17}H_{26}N_2O_7$ 3-(R)-Acetamido-spiro-3,4'-(R)-(3-deoxy-1,2:5,6-di-O-isopropylidene-α-D-ribo-hexofuranos-3-yl)-2-pyrrolidinone (SHFRAP)[106]

P2$_1$2$_1$2$_1$; Z = 4; D$_x$ = 1.241; R = 0.050 for 2,268 intensities. The furanose conformation is similar to that of the S isomer, reported next, being $^3T_4$, with ψ = 302°, q = 41 pm. The conformation of the pyrrolidinone is also the same in both compounds. The bond lengths in the furanose rings are the same within experimental error. The length of C-1–O-1 is significantly shorter than that of C-1–O-4 in both molecules, by 16 and 30 pm, respectively. There is strong, N–H · · · O, intermolecular hydrogen-bonding, and some evidence of weak, C–H · · · O bonding in both structures.

(105) A. D. Vasiliev, V. I. Andrianov, V. E. Ovchinnikov, and L. V. Gorbatii, *Kristallografiya*, 22 (1977) 526–533.
(106) S. J. Rettig and J. Trotter, *Can. J. Chem.*, 55 (1977) 1464–1462.

$C_{17}H_{26}N_2O_7$ 3-(S)-Acetamido-*spiro*-3,4'-(R)-(3-deoxy-1,2:5,6-di-O-isopropylidene-α-D-*ribo*-hexofuranos-3-yl)-2-pyrrolidinone (SHFSAP)[106]

$P2_12_12_1$; $Z = 4$; $D_x = 1.265$; $R = 0.070$ for 2,229 intensities. The furanose residue has the $^3T_4$ conformation ($\psi = 300°$, $q = 43$ pm). The pyrrolidinone has an envelope conformation, with the carbon atom that is linked to the furanose ring out of the plane.

$C_{18}H_{19}Br^-N_4O_4^+$ 5-α-D-Lyxofuranosyl-2,3-diphenyltetrazolium bromide (PLYXTZ)[107]

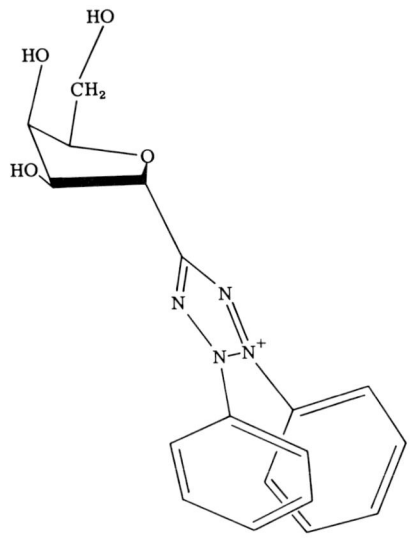

$P4_12_12$; $Z = 8$; $D_x = 1.54$; $R = 0.09$ for 2,090 intensities. The furanosyl group has the ^3E conformation ($\psi = 289°$, $q = 46$ pm). The feature of interest was the determination of the configuration of C-2, for correlation with chemical and spectroscopic data.[108] No hydrogen atom positions were reported.

$C_{18}H_{23}IO_9S$ Methyl   3,4-di-O-acetyl-6-deoxy-6-iodo-2-O-p-tolylsulfonyl-α-D-mannopyranoside (ADITSM)[109]

$P2_12_12_1$; $Z = 4$; $D_x = 1.606$; $R = 0.052$ for 1,820 intensities. The pyranoside has the $^4C_1$ conformation ($\theta = 5°$, $q = 58$ pm). The glycosidic, C–O bond is short, 139 pm, and the glycosidic torsion-angle is +69°. The C–I bond-length is 219 pm. The shortest intermolecular

---

(107) K. B. Lindberg, M. Czugler, V. Zsoldos-Mády, I. Pinter, and A. Messmer, *Acta Crystallogr. Sect. B*, 34 (1978) 3484–3486.
(108) V. Zsoldos-Mády, A. Messmer, I. Pinter, and A. Neszmélyi, *Carbohydr. Res.*, 62 (1978) 105–116.
(109) M. Ul-Haque and A. Freestone, *J. Chem. Soc. Perkin Trans. 2*, (1977) 1509–1513.

contact is iodine to sulfonyl oxygen atom, 327 pm. Some of the hydrogen atom positions were not reported.

$C_{19}H_{22}O_{10}S$ 3-O-Acetyl-4-C-carboxy-1,2-O-isopropylidene-5-O-p-tolylsulfonyl-α-L-gulofuranose-4¹,6-lactone (AITGFL)[110]

(110) S. E. V. Phillips and J. Trotter, *Acta Crystallogr. Sect. B*, 33 (1977) 1003–1007; described the α-D-form.

$P2_12_12_1$; $Z = 4$; $D_x = 1.438$; $R = 0.041$ for 2,016 intensities. The furanose has the $^3T_4$ conformation ($\psi = 114°$, $q = 31$ pm). The conformations of the lactone and dioxolane rings are close to envelopes. The dihedral angle between the furanose and dioxolane ring, as defined by O-4–C-1–C-2–O-2, is $+108°$. The angle between the furanose and lactone ring, as defined by O-4–C-4–C-5–C-6, is $-146°$. The C–S bond-length is 175.4 pm, and the C–O–S valence-angle is 119°.

$C_{19}H_{25}N_3O_{12}$ 3,5-Dimethyl-1-(2,3,4,6-tetra-O-acetyl-α-D-mannopyranosyl)isocyanuric acid (TAMCUR)[111]

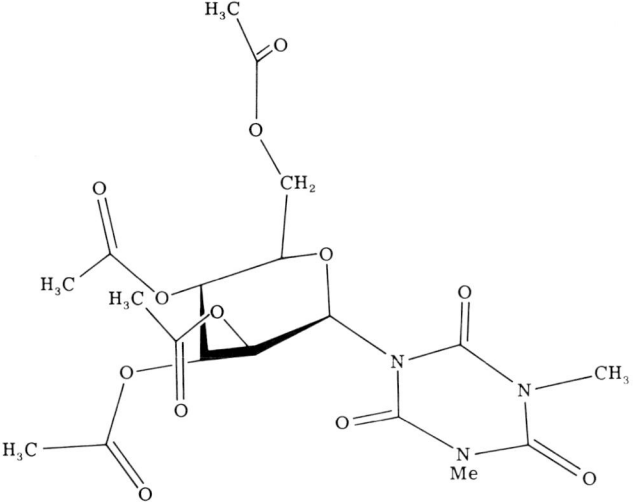

$P2_12_12_1$; $Z = 4$; $D_x = 1.40$; $R = 0.057$ for 1,386 intensities. The pyranosyl group has a distorted, $^4C_1$ conformation ($\theta = 18°$, $\varphi = 264°$, $q = 45$ pm). The axial, glycosidic torsion-angle, O-5–C-1–N–C, is $+97°$; the barrier to rotation as determined by ^{13}C-n.m.r. spectroscopy is 42 kJ.mol^{-1}. The hydrogen atom positions were not reported.

$C_{20}H_{22}O_4$ 2(S)-4-O-Acetyl-3,5-O-benzylidene-1,2-dideoxy-2-C-phenyl-D-*erythro*-pentitol (ABYPEP)[60]

C2; $Z = 4$; $D_x = 1.99$; $R = 0.059$ for 1,887 intensities. The S configuration at C-2 was confirmed. The 1,3-dioxane ring has a chair conformation. The carbon chain, C-1 to C-5, has the planar, zigzag conforma-

(111) W. Depmeier, H. V. Voithenberg, J. C. Jochims, and K.-H. Klaska, *Chem. Ber.*, 111 (1977) 2010–2020.

tion. The proton–proton couplings calculated from the dihedral angles for vicinal protons measured in the crystal, and measured by n.m.r. in chloroform solution, indicate no significant changes in conformation as between the solution and the crystal.

$C_{20}H_{24}O_{10} \cdot H_2O$ 8-[2-(β-D-Glucopyranosyloxy)isopropyl]-8,9-dihydro-9-hydroxyangelicin monohydrate; Apterin; Vaginidiol monoglucopyranoside[112]
    Depiction is at top of p. 474

$P2_12_12_1$; $Z = 4$; $D_x = 1.429$; $R = 0.054$ for 1,702 intensities. The pyranosyl group has the $^4C_1$ conformation ($\theta = 8°$, $\varphi = 177°$, $q = 58$ pm). The orientations of the β-linkage bonds to the aglycon are as follows: O-5–C-1–O-1–C-7, $-93°$, and C-1–O-1–C-7–C-10, $+61°$. The dihydrofuran ring has the $^2T_1$ conformation ($\psi = 47°$, $q = 20$ pm). The coumarin moiety is almost planar. The C–O bonds in the dihydrofuran ring are coplanar with the coumarin rings, and their lengths are 135.6 and 147.3 pm. There is an intramolecular hydrogen-bond from the hydroxyl group on the furan ring to the ring-oxygen atom of the D-glucosyl group; this is postulated to cause resistance to enzymic

(112) B. F. Pedersen and J. Karlsen, *Acta Crystallogr. Sect. B*, 34 (1978) 1893–1897.

474   GEORGE A. JEFFREY AND MUTTAIYA SUNDARALINGAM

action by β-D-glucosidase. There is also a close approach between the two ring-oxygen atoms, namely, 292.3 pm. The atomic coordinates reported referred to the L enantiomer.

$C_{22}H_{29}NO_9$ Ethyl 5-O-acetyl-3-O-benzyl-6-deoxy-6-(formamido)-1,2-O-isopropylidene-L-*glycero*-β-D-*talo*-heptofuranuronate (EDGLTH)[113]

Depiction is at bottom of p. 474

$P2_12_12_1$; Z = 4; D_x = 1.295; R = 0.05 for 2,975 intensities. The analysis was performed in order to establish the chiralities (S) at C-5 and C-6. The conformation of the furanose is close to $^4T_3$ (ψ = 139°, q = 36 pm), and that of the isopropylidene ring is close to an envelope. The only hydrogen bonding is intermolecular, N–H · · · O=C, with H · · · O = 220 pm. The article[113] names the compound as the D-*glycero*-L-*talo* isomer, but the structure given is that of the L-*glycero*-β-D-*talo* isomer.

$C_{22}H_{29}NO_9$ Ethyl 5-O-acetyl-3-O-benzyl-6-deoxy-6-(formamido)-1,2-O-isopropylidene-L-*glycero*-α-L-*allo*-heptofuranuronate (ELGDAH)[113]

$P2_1$; Z = 4; D_x = 1.315; R = 0.07 for 2,354 intensities. There are two symmetry-independent molecules in the crystal structure, and they

(113) J. C. A. Boeyens, A. J. Brink, R. H. Hall, A. Jordaan, and J. A. Pretorius, *Acta Crystallogr, Sect. B*, 33 (1977) 3059–3066.

have only minor conformational differences. The furanoses have E_4 and $^3T_4$ conformations (ψ = 316°, 300°, q = 39 pm, 39 pm). The isopropylidene rings have twist conformations. The largest conformational difference in the two molecules is in the orientation of the 3-O-benzyl group. The values of C-3–O-3–C–C(benzyl) are +174 and −148°. Some of the hydrogen atom positions were not reported.

$C_{23}H_{32}O_{10} \cdot H_2O$ Paucin monohydrate (PAUCIN)[114]

(114) P. J. Cox and G. A. Sim, *J. Chem. Soc. Perkin Trans. 2*, (1977) 259–262.

$P2_1$; $Z = 2$; $D_x = 1.31$; $R = 0.79$ for 1,929 intensities. The molecule is a pseudoguaianolide 6-O-acetyl-β-D-glucopyranoside. The β-D-glucopyranoside torsion-angle, O-5–C-1–O-1–C-2′, is −75°, and the C-1–O-1–C-2′–C-3′ angle is +83°. The conformation of the D-glucoside is $^4C_1$ ($\theta = 3°$, q = 58 pm). In the sesquiterpenoid moiety, the conformation of the cycloheptane ring is a twist-boat, with an approximate two-fold axis; that of the α-methylene γ-lactone ring is an envelope, and that of the cyclopentanone ring has a twist conformation having approximate C_{1h} symmetry. The water molecule participates in three hydrogen bonds, all of which are to carbonyl oxygen atoms.

$C_{24}H_{23}B_3O_6$ D-Mannitol 1,2:3,4:5,6-tris(phenylboronate) (MANBOS)[115]

$P2_12_12_1$; $Z = 4$; $D_x = 1.25$; $R = 0.102$ for 2,145 intensities. The molecule has three planar, 1,3,2-dioxaborolane rings, which are approximately parallel. Because of the ring formation, the D-mannitol residue has a sickle conformation, with C-2–C-3–C-4–C-5 = +136°. The C–O bond-lengths in the rings range from 144.0 to 147.1 pm, the B–O bond-lengths from 134.8 to 138.4 pm, and the B–C bond lengths from 154.6 to 157.9 pm. Two of the phenyl rings are coplanar with the dioxa-

(115) A. Gupta, A. Kirfel, and G. Will, *Acta Crystallogr. Sect. B*, 33 (1977) 637–641.

boralane ring to which they are attached, but the third is twisted at an angle of 28°.

$C_{24}H_{26}O_{10}$ 1-Naphthyl 2,3,4,6-tetra-O-acetyl-β-D-glucopyranoside (NAPAGQ)[116]

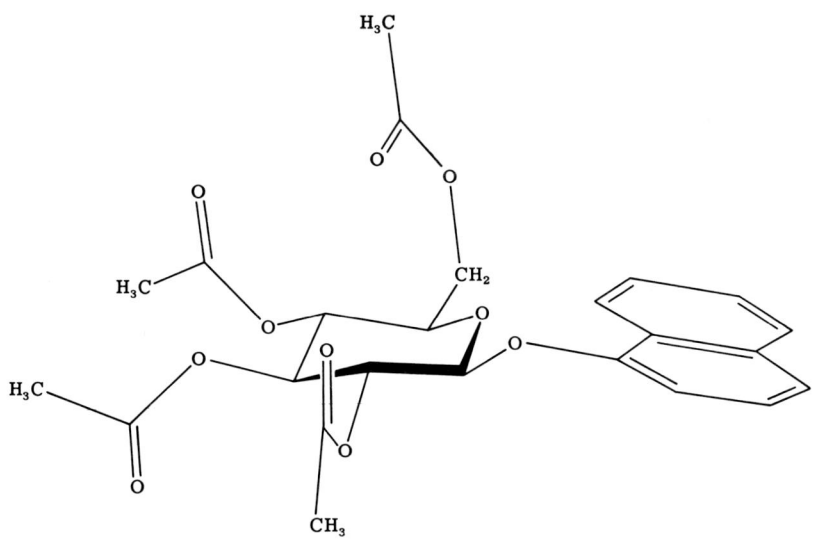

$P2_12_12_1$; $Z = 4$; $D_x = 1.28$; $R = 0.068$ for 1,997 intensities. The pyranoside has a distorted, $^4C_1$ conformation ($\theta = 9°$, $\varphi = 290°$, $q = 59$ pm). The glycosidic angles, O-5–C-1–O-1–C-1' and C-1–O-1–C-1'–C-2', are -79 and $+164°$, respectively. The C-5–O-5–C-1–O-1' bond-lengths are 144.0, 140.3, and 139.9 pm. The glycosidic valence-angle C-1–O-1–C is 119.7°. The naphthyl ring is slightly distorted from planarity. Hydrogen atom positions were not reported.

$C_{24}H_{50}Cl^-N^+O_8 \cdot H_2O$ β-D-Glucosylphytosphingosine hydrochloride, monohydrate (GLPHYC10)[117]

$P2_12_12_1$; $Z = 4$; $D_x = 1.19$; $R = 0.099$ for 522 intensities. This is, necessarily, a low-accuracy analysis and was aimed solely at determining the molecular packing. The D-glucosyl group has the $^4C_1$ conformation. The hydrocarbon chain is planar zigzag, from C-5' to C-18'. The section from C-4' to N-1' is also planar zigzag, and is inclined to the main section at an angle of 29°. The glycosidic torsion-angles, O-5–C-

(116) M. W. Makinen and N. W. Isaacs, *Acta Crystallogr. Sect. B*, 34 (1978) 1584–1590.
(117) S. Abrahamsson, B. Dahlen, and I. Pascher, *Acta Crystallogr. Sect. B*, 33 (1977) 2008–2013.

1-O-1-C-1' and C-1-O-1-C-1'-C-2', are −78 and +164°. The coordinates of the hydrogen atoms attached to the nitrogen and oxygen atoms were not determined.

$C_{26}H_{26}N_4O_4$ Methyl 4,6-O-benzylidene-2,3-dideoxy-2,3-di(phenylazo)-α-D-mannopyranoside (MBYZMN)[118]

I4; Z = 8; D_x = 1.21; R = 0.118 for 1,694 intensities. The pyranoside conformation is a slightly distorted $^4C_1$ (θ = 7°, φ = 270°, q = 57 pm), with the phenylazo groups respectively axial and equatorial (a diequatorial reaction-product was anticipated[119]). Hydrogen atom positions were not reported.

$C_{26}H_{36}O_{11} \cdot 2\ CH_4O$ Mascaroside di(methanolate) (MASCAR10)[120]

$P2_1$; Z = 4; D_x = 1.356; R = 0.06 for 2,006 intensities. This natural product is a β-D-glucopyranoside in the $^4C_1$ conformation. The glycosidic torsion-angle, O-5-C-1-O-1-C, is −81°. The aglycon is a diter-

---

(118) G. M. Sheldrick, B. E. Davison, and J. Trotter, *Acta Crystallogr. Sect. B*, 34 (1978) 1387–1389.
(119) A. J. Bellamy and R. D. Guthrie, *J. Chem. Soc.*, (1966) 1989–1993.
(120) A. Ducruix, C. Pascard, M. Hammonniere, and J. Poisson, *Acta Crystallogr. Sect. B*, 33 (1977) 2846–2850.

pene, very similar to cafestol,[121] the structure of which had been determined from that of a bromo derivative.[122] The primary hydroxyl group of the D-glucosyl group is disordered over the *gauche–gauche* and *gauche–trans* orientations. The coordinates of the hydroxyl hydrogen atoms were not reported.

$C_{27}H_{30}N_2O_{13}$ 1-Acetyl-3-benzamido-4-(2,3,4,6-tetra-*O*-acetyl-β-D-glucopyranosyloxy)-3-pyrrolin-2-one (ABGPON)[123]

(121) C. Djerassi, M. Cais, and L. A. Mitscher, *J. Am. Chem. Soc.*, 81 (1959) 2386–2394.
(122) A. I. Scott, G. A. Sim, G. Ferguson, D. W. Young, and F. McCapra, *J. Am. Chem. Soc.*, 84 (1962) 3197–3199.
(123) A. Mercer and J. Trotter, *Acta Crystallogr. Sect. B*, 34 (1978) 1596–1599.

P2$_1$; Z = 2; D$_x$ = 1.35; R = 0.079 for 1,838 intensities. The D-glucopyranosyl group has a slightly distorted, $^4C_1$ conformation (θ = 8°, φ = 317°, q = 62 pm). The pyrroline ring, the benzamido group, the acetyl groups, and the phenyl ring are all planar. The glycosidic torsion-angles are O-5–C-1–O-1–C, −78°, and C-1–O-1–C–C, −154°. As in the 1-O-acetyl derivatives of D-glucose, the O-5–C-1 bond is short, 139.2 pm, and the C-1–O-1 bond is long, 142.4 pm. The coordinates of the hydrogen atoms were not reported.

$C_{32}H_{31}NO_4P^+\cdot C_7H_7O_3S^-\cdot H_2O$ Methyl 4,6-O-benzylidene-2,3-dideoxy-2,3-[N-(triphenylphosphonio)-epimino]-α-D-allopyranoside p-toluene-sulfonate, monohydrate (MBYPAT)[124]

P2$_1$2$_1$2$_1$; Z = 4; D$_x$ = 1.29; R = 0.09 for 1,488 intensities. The analysis confirmed the structure having fused aziridine and pyranose rings that had been inferred from n.m.r. data. The pyranoside has a conformation lying between $^2H_5$ and a sofa conformation having one out-of-plane atom, with θ = 129°, φ = 134°, q = 55 pm. The aziridine ring makes an angle of 104° with C-1–C-2–C-3–C-4, which are coplanar. The p-toluenesulfonate anions are linked to the cations by hydrogen bonding through the water molecules to the pyranose ring-oxygen atoms and the sulfonyl oxygen atoms. Hydrogen atom positions were not reported.

$C_{34}H_{26}Cl_2O_9$ 2,3,4,6-Tetra-O-benzoyl-α-D-mannopyranosyl chloride (TBZMAC)[125]

(124) K. B. Lindberg, Cryst. Struct. Commun., 7 (1977) 607–612.
(125) F. W. Lichtenthaler, T. Sakakibara, and E. Oeser, Carbohydr. Res., 59 (1977) 57–61.

C2; Z = 4; D_x = 1.34; R = 0.044 for 3,400 intensities. The stereochemistry at C-2 was determined by this analysis. The conformation of the pyranosyl group is $^4C_1$ (θ = 5°, q = 54 pm), with the benzoyloxy group axial on C-2. The orientation about the primary alcohol group is *gauche–gauche*, and the 6-benzoyloxy group is axially oriented, with its plane almost perpendicular to that on C-2. The C–Cl bond-length is 181 pm, and the two ring C–O bonds are C-5–O-5, 144 pm, and C-1–O-5, 139 pm. These results are in good agreement with theoretical predictions.[126]

$C_{35}H_{42}O_6S_4$ 4,5,6-Tri-O-benzoyl-2,3-di-S-ethyl-2,3-dithio-D-allose diethyl dithioacetal (TBETHA)[127]

$P2_1$; Z = 2; D_x = 1.23; R = 0.09 for 1,343 intensities. The D-allose residue has a sickle conformation, with a C-2–C-3–C-4–C-5 torsion-angle of +93°. Atoms S-1a, C-1, C-2, C-3, and C-4 lie in one plane (within ±4 pm), and atoms C-3, C-4, C-5, and C-6 lie in a plane inclined at 88° thereto. The benzoyl groups are distorted from planar, with O=C–C=C torsion-angles ranging from 2 to 16°. The C–S bond-distances reported range from 169.2 to 207.3 pm, with some nonbonded S · · · S distances of 325 and 332 pm. However, this was

(126) G. A. Jeffrey and J. H. Yates, *J. Am. Chem. Soc.*, 101 (1979) 820–825.
(127) T. J. McLennan, W. T. Robinson, G. S. Bethell, and R. J. Ferrier, *Acta Crystallogr. Sect. B*, 33 (1977) 2888–2891.

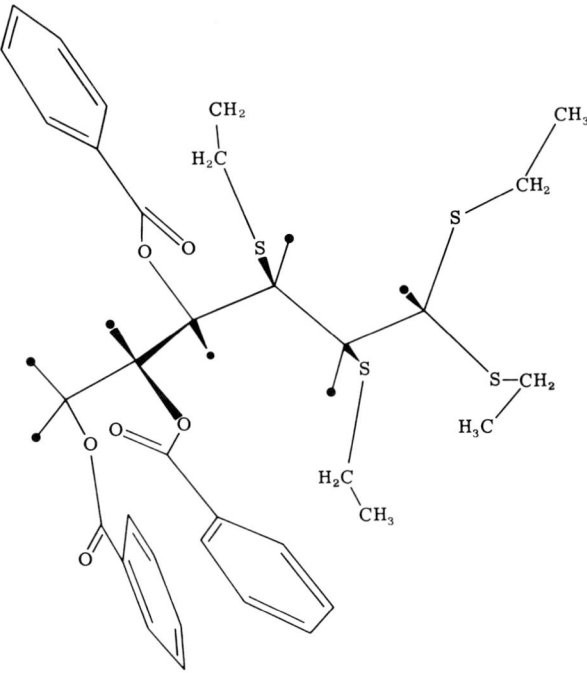

a low-accuracy structure-analysis, and only the coordinates of the hydrogen atoms on the main chain were reported.

$C_{39}H_{36}N_2O_{10}$ Methyl 3,6-di-*O*-benzoyl-2,4-di-(*N*-benzoylacetamido)-
2,4-dideoxy-α-D-idopyranoside (MABIDP)[128]
Depiction is at top of p. 484

$P2_12_12_1$; Z = 4; D_x = 1.25; R = 0.054 for 3,317 intensities. The pyranoside has the $^1C_4$ conformation (θ = 175°, q = 57 pm), with the benzoyl-substituted, primary alcohol group axial in the *gauche–gauche* orientation. The *N*-acetyl groups are so oriented that C-1–C-2–N–C(CH$_3$CO) is +70°, and C-5–C-4–N–C(CH$_3$CO) is +79°. The glycosidic torsion-angle is 74°, and the C-1–O-1 bond-length is 139.0 pm. Hydrogen atom positions were not reported.

$C_{40}H_{54}O_{27}$ 1,2,3,6-Tetra-*O*-acetyl-4-*O*-[2,3,6-tri-*O*-acetyl-4-*O*-(2,3,4,6-tetra-*O*-acetyl-β-D-glucopyranosyl)-β-D-glucopyranosyl]-β-D-glucopyranose; β-cellotriose undecaacetate (ACCELL10)[129]

(128) P. Luger and H. Paulsen, *Acta Crystallogr. Sect. B*, 34 (1978) 1254–1259.
(129) S. Pérez and F. Brisse, *Acta Crystallogr. Sect. B*, 33 (1977) 2578–2584.

$P2_1$; $Z = 2$; $D_x = 1.303$; $R = 0.091$ for 2,975 intensities. The pyranose residues have slightly distorted, $^4C_1$ conformations, with $\theta = 8°$, $11°$, $9°$; $\varphi = 294°$, $337°$, $311°$; q = 61 pm, 62 pm, 57 pm). The linkage torsion-angles are O-5–C-1–O-1–C-4′, $-98°$, C-1–O-1–C-4′–C-3′, $+102°$; and O-5′–C-1′–O-1′–C-4″, $-75°$, C-1′–O-1′–C-4″–C-3″, $+134°$. The orientation of the primary alcohol group on the nonreducing group and residue is *gauche–gauche*, and, on the reducing residue, *gauche–trans*. The valence angles of the two linkage oxygen atoms are 117 and 116°. The bond lengths and valence angles do not differ significantly from the values previously accepted.

III. Data for Nucleosides and Nucleotides
1977

$C_9H_{12}N_2O_5S$ 2-Thiouridine (TURIDN10)[130]

$P2_1$; $Z = 2$; $D_x = 1.619$; $R = 0.023$ for 1,334 intensities. The glycosyl disposition is *anti* (17.0°) and the D-ribosyl group has the $^3T_2$ conformation (13.4°, 41.5°). The orientation about the exocyclic, C-4'–C-5' bond is *trans* (−169°). The bases are linked by N-3-H · · · O-4 hydrogen bonds. The S-2 atom accepts a hydrogen bond from a neighboring O-5'-H group, O-5'-H · · · S = 335 pm.

$C_9H_{12}N_3O_7P$ β-D-Arabinofuranosylcytosine 2',5'-monophosphate (ARACYP)[131]

(130) S. W. Hawkinson, *Acta Crystallogr., Sect. B*, 33 (1977) 80–85.
(131) W.-J. Kung, R. E. Marsh, and M. Kainosho, *J. Am. Chem. Soc.*, 99 (1977) 5471–5477.

P2$_1$; Z = 2; D$_x$ = 1.698; R = 0.034 for 1,506 intensities. The molecule is a zwitterion with N-3 of the base protonated, and the phosphate group negatively charged. The seven-membered phosphate ring exhibits considerable strain, as seen by the greatly increased P–O–C angles, namely, C-2′–O-2′–P = 127.2° and C-5′–O-5′–P = 124.0°. The glycosyl disposition is *anti* (25.9°) and the conformation of the D-arabinosyl group is $^2T_3$ (170.7°, 41.6°). The exocyclic, C-4′–C-5′ torsion-angle is restricted to the *gauche*$^+$ (64.9°) range. The seven-membered phosphate group is in a chair conformation. One of the anionic phosphate oxygen atoms is simultaneously hydrogen-bonded to N-3 and N-4 of cytosine, while the other anionic oxygen atom is hydrogen-bonded to the second N-4 proton of a symmetry-related molecule. The carbonyl oxygen atom (O-2 of the base) is not hydrogen bonded.

$C_9H_{12}N_6O_4 \cdot H_2O$ 8-Azadenosine monohydrate (AZADEN2O)[132]

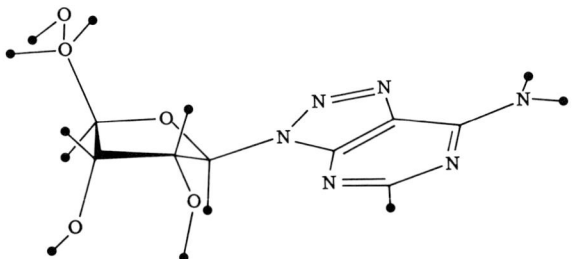

P2$_1$2$_1$2$_1$; Z = 4; D$_x$ = 1.521; R = 0.033 for 1,136 intensities. The crystal structure is isomorphous with that of formycin monohydrate.[133] The glycosyl disposition is *high anti* (103.7°) and the D-ribosyl group has the $^2T_1$ pucker (144.5°, 41.9°). The orientation about the exocyclic, C-4′–C-5′ bond is *trans* (179.1°). Errors: the O-1′ x-coordinate should be 0.3307, instead of 0.3007. The atom following O-4′ should be C-5′, not O-5′.

$C_9H_{13}N_3O_5$ α-Cytidine (ACYTID)[134]

C2; Z = 4; D$_x$ = 1.491; R = 0.033 for 1,002 intensities. The glycosyl disposition is *anti* (−28.4°) and bears an enantiomeric relationship to the usual *anti* range for β-nucleosides. The D-ribosyl group has the $^3T_2$

(132) P. Singh and D. J. Hodgson, *J. Am. Chem. Soc.*, 99 (1977) 4807–4815.
(133) P. Prusiner, T. Brennan, and M. Sundaralingam, *Biochemistry*, 12 (1973) 1196–1203; see *Adv. Carbohydr. Chem. Biochem.*, 31 (1975) 357.
(134) M. L. Post, G. I. Birnbaum, C. P. Huber, and D. Shugar, *Biochim. Biophys. Acta*, 479 (1977) 133–142.

pucker (357.2°, 40.1°), and the exocyclic, C-4'–C-5' torsion-angle is gauche⁺ (59.4°).

$C_{10}H_{11}BrN_4O_5$ 8-Bromoinosine (BRINOS10)[135]

P2₁2₁2₁; Z = 8; D_x = 1.931; R = 0.034 for 2,321 intensities. There are two molecules in the asymmetry unit, and both exhibit the *syn* disposition (−84.5°, +76.1°) for the base. The conformation of the D-ribosyl group is ³T₄ (28.5°, 39.4°) in molecule A and ³T₂ (359.9°, 36.2°) in molecule B. The exocyclic, C-4'–C-5' bond torsion-angle is *gauche⁺* for both molecules (55.2°, 59.7°). The purine bases of the crystallographically independent molecules are paired by N-1-H ··· O-6 hydrogen bonds across a pseudo-two-fold axis. The bases are stacked such that the Br atoms are tucked under the pyrimidine moiety of the adjacent

(135) H. Sternglanz, J. M. Thomas, and C. E. Bugg, *Acta Crystallogr. Sect. B*, 33 (1977) 2097–2102.

base, with contacts of 358 and 360 pm to N-3 and C-4, respectively, to the stacked base.

$C_{10}H_{11}NO_5$ 2,2'-Anhydro-(1-β-D-arabinofuranosyl-2-hydroxy-4-pyridone)[136]

$C_{10}H_{12}N_4O_5 \cdot HCl$ Formycin B hydrochloride[137]

$P2_12_12_1$; Z = 4; D_x = 1.60; R = 0.025 for 1,038 intensities. The glycosyl disposition is *syn* (−140.1°) and the D-ribosyl group has the $^2T_1$ conformation (154.9°, 36.6°). The orientation about the exocyclic, C-4'–C-5' bond is *gauche*⁺ (59.2°). The base is protonated at N-3, and this proton is involved in an intramolecular, donor hydrogen-bond to O-5' of the sugar, which acts as an acceptor. In contrast, in the usual intramolecular hydrogen-bond, the O-5'-H group acts as the donor to the unprotonated N-3 of the base. There is a C-H · · · O interaction, 300 pm, between the base C-2-H and the carbonyl O-6 atom of an adjacent base.

$C_{10}H_{13}NO_5$ 4-Deoxy-3-deazauridine (RFURPD)[138]

$P2_12_12_1$; Z = 4; D_x = 1.470; R = 0.049 for 749 intensities. The base disposition is *anti* (53.5°) and the conformation of the D-ribosyl group is $^2T_3$ (167.9°, 34.1°). The exocyclic C-4'–C-5' bond torsion-angle is

(136) See Ref. 57.
(137) P. Singh and D. J. Hodgson, *Acta Crystallogr.*, Sect. B, 33 (1977) 3189–3194.
(138) E. Egert and H. J. Lindner, *Acta Crystallogr.*, Sect. B, 33 (1977) 3704–3707.

gauche⁺ (50.6°). The ring-oxygen atom accepts a hydrogen bond from O-5'-H of a neighboring molecule; see S. Sprang and coworkers.[139]

$C_{10}H_{13}N_5O_3S \cdot H_2O$ 2'-Deoxy-6-thioguanosine monohydrate (DTGUOS)[140]

$C222_1$; $Z = 8$, $D_x = 1.443$; $R = 0.04$ for 1,330 intensities. The crystal structure is isomorphous with that of 2-thioguanosine monohydrate.[141] The glycosyl disposition is *anti* (58.6°), and the 2-deoxy-D-*erythro*-pentofuranosyl group has the $^2T_1$ pucker (149.4°, 38.8°). The orientation about the exocyclic, C-4'–C-5' bond is *trans* (− 174.4°). The bases are paired, through the N-1-H · · · N-7 and N-2-H · · · S hydrogen-

(139) S. Sprang, R. Scheller, D. Rohrer, and M. Sundaralingam, *J. Am. Chem. Soc.*, 100 (1978) 2867–2872.
(140) G. L. Gartland and C. E. Bugg, *Acta Crystallogr.*, Sect. B, 33 (1977) 3678–3681.
(141) U. Thewalt and C. E. Bugg, *J. Am. Chem. Soc.*, 94 (1972) 8892–8898.

bonds, to neighboring molecules, to form a planar ribbon of bases parallel to the *ac* plane.

$C_{10}H_{14}N_2O_5$ 3-Deazacytidine (DAZCYT10)[142]

$P2_12_12_1$; Z = 4; D_x = 1.49; R = 0.035 for 1,660 intensities. The glycosyl disposition is *anti* (67.8°) and the D-ribosyl group has the $^2T_1$ conformation (148.5°, 42.8°). The exocyclic, C-4'–C-5' bond torsion-angle is *trans* (179.5°). The 3-deazacytosine ring is bent into a shallow boat, with N-1 and C-4 showing the maximum displacements. There is no base stacking, or interbase hydrogen-bonding. The amino group N-4-H is hydrogen-bonded to the furanoid-ring O-1' of a neighboring molecule; see Refs. 138 and 139.

$C_{10}H_{15}N_3O_5$ 2'-*O*-Methylcytidine (OMCYTD20)[143]

$P2_12_12_1$; Z = 8; D_x = 1.428, R = 0.069 for 2,488 intensities. There are two independent molecules in the asymmetric unit, and they have similar conformations. The glycosyl dispositions are *anti* (47.4°, 43.6°), and the conformation of the D-ribosyl groups is $^2T_1$ (174.9°, 39.0°; 170.9°, 38.3°). The exocyclic, C-4'–C-5' bond torsion-angles are *gauche*$^+$ (55.9°, 51.6°). The O-2'-methyl groups adopt the staggered arrangement; in molecule A, the methoxyl group is disordered over two positions, with respective populations of 66 and 34%, where the torsion angle C-1'–C-2'–O-2'–C-O-2' is 95° and 156°, respectively, whereas, in molecule B, this angle is 168°. All intermolecular hydrogen-bonds

---

(142) W. L. B. Hutcheon and M. N. G. James, *Acta Crystallogr., Sect. B*, 33 (1977) 2224–2228.

(143) B. Hingerty, P. J. Bond, and R. Langridge, *Acta Crystallogr., Sect. B*, 33 (1977) 1349–1356.

are between bases and sugar residues. Error: bad coordinates for atom H-lm; fixed at 0.722, −0.429, and 0.462.

$C_{11}H_{14}N_2O_6$ 5-Acetyl-1-(2-deoxy-α-D-*erythro*-pentofuranosyl)uracil (ACDXUR)[144]

$P2_12_12_1$; Z = 4, D_x = 1.51; R = 0.039 for 1,535 intensities. The glycosyl disposition is *anti* (−11.7°) and the 2-deoxy-D-*erythro*-pentofuranosyl ring has the $^2T_3$ conformation (168.0°, 34.5°), a puckering rarely observed in α-nucleosides; see Ref. 145. The orientation about the ex-

(144) T. A. Hamer, M. K. O'Leary, and R. T. Walker, *Acta Crystallogr.*, Sect. B, 33 (1977) 1218–1223.
(145) M. Sundaralingam, *J. Am. Chem. Soc.*, 93 (1971) 6644–6647.

ocyclic, C-4′–C-5′ bond is in the unusual, *gauche*⁻ range (−79.0°). Thus, the rare orientation of the C-4′–C-5′ bond is correlated with the rare, C-2′-*endo*, sugar-ring pucker. The acetyl group is twisted at an angle of 13° to the pyrimidine ring. As is commonly observed for the uridine system, the carbonyl atom O-4 is involved in hydrogen bonding, whereas the carbonyl atom O-2 is not. The bases are stacked.

$C_{11}H_{14}N_2O_7$ 2′,3′-*O*-(Methoxymethylene)uridine (MXEURD)[146]

$P2_12_12_1$; $Z = 4$; $D_x = 1.523$; $R = 0.040$ for 1,416 intensities. The glycosyl disposition is *anti* (56.6°) and the furanoid ring has the ^2E confor-

(146) A. J. deKok, C. Romers, H. P. M. deLeeuw, C. Altona, and J. H. vanBoom, *J. Chem. Soc. Perkin Trans.* 2, (1977) 487–493.

mation (162.8°, 23.1°). The orientation about the exocyclic, C-4'–C-5' bond is *gauche*⁺ (52.9°). The dioxolane ring exhibits the C-2'-*endo*-C-2'-*exo* twist (138.6°, 32.3°). A prominent feature of the hydrogen-bonding scheme is the interbase, N-3-H · · · O-4, hydrogen-bonding spiral around a 2_1 axis. Error: bad coordinates for atoms H-81, H-82, and H-83.

$C_{11}H_{17}FN_2O_6S$ 1-S-Ethyl-1-(5-fluorouracil-1-yl)-1-thio-D-*gluco*-pentitol [(1-R)-1-S-ethyl-1-(5-fluorouracil-1-yl)-1-thio-D-arabinitol][147,148]

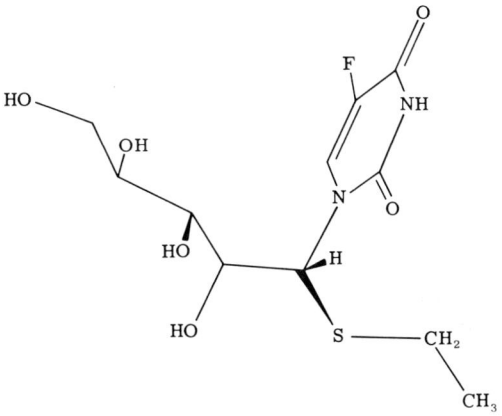

$P6_3$; $Z = 6$; $D_x = 1.37$; $R = 0.09$ for 1,141 intensities. The carbon chain of the sugar moiety is in the extended, zigzag conformation, so that the adjacent hydroxyl groups are in the *gauche*⁺, *trans*, *gauche*⁺ arrangement. The configuration of C-1 is (R). The ethylthio group also lies in the zigzag plane of the sugar moiety, giving an extended conformation to the molecule. There is an intramolecular hydrogen-bond in the structure. Error: the C-2' x-coordinate should be 0.7668, instead of 0.7868.

$C_{12}H_{19}ClN_4O_5S$ 1-(6-Chloropurin-9-yl)-1-S-ethyl-1-thio-D-*glycero*-D-*ido*-hexitol [(1-S)-1-(6-chloropurin-9-yl)-1-S-ethyl-1-thio-D-glucitol] (CLPTGL10)[149]

(147) A. Ducruix and C. Pascard-Billy, *Acta Crystallogr., Sect. B*, 33 (1977) 2505–2512.
(148) Compare D. Horton and R. A. Markovs, *Carbohydr. Res.*, 80 (1980) 263–275.
(149) A. Ducruix and C. Pascard-Billy, *Acta Crystallogr., Sect. B*, 33 (1977) 2501–2505; compare K. C. Blieszner, D. Horton, and R. A. Markovs, *Carbohydr. Res.*, 80 (1980) 241–262.

$P2_1$; $Z = 2$; $D_x = 1.49$; $R = 0.063$ for 857 intensities. The carbon chain of the sugar residue adopts a sickle conformation, with C-1 out of the zigzag plane defined by the other five atoms. The configuration of C-1 is (S). The ethylthio group and the glycosyl C-1 atom comprise a second plane, so that the dihedral angle between the two planes is $-52°$ (S–C-1–C-2–C-3). The purine ring exhibits the usual buckling about the C-4–C-5 bond, the dihedral angle between the two halves being 2.7°. The successive hydroxyl groups in the carbon chain exhibit the *gauche*$^+$, *gauche*$^+$, *trans*, *gauche*$^-$ sequence. These conformational features are comparable to those observed in D-glucitol[150] and pyridine–D-glucitol.[151]

$C_{13}H_{16}N_2O_8$ 3′,5′-Di-O-acetyluridine (at $-170°$)[152]

(150) Y. J. Park, G. A. Jeffrey, and W. C. Hamilton, *Acta Crystallogr., Sect. B*, 27 (1971) 2393–2401.

$P2_1$; $Z = 2$; $D_x = 1.434$; $R = 0.047$ for 1,867 intensities. The glycosyl disposition is *anti* (60.4°) and the D-ribosoyl group has the $^2T_3$ conformation (165.3°, 41.2°). The orientation about the exocyclic, C-4'–C-5' bond is *gauche*$^+$ (63.6°). The O atom of the 3'-acetyl group is disordered. There are only two hydrogen-bonds, and they involve the carbonyl oxygen atoms of the acetyl groups, one with N-3-H of the base and the other with O-2'-H of the sugar. Base stacking is absent.

$C_{13}H_{16}N_4O_6$ 6,7-Dimethyl-1-β-D-ribofuranosyl-lumazine (LUMZRF01)[153]

---

(151) H. S. Kim, G. A. Jeffrey, and R. D. Rosenstein, *Acta Crystallogr., Sect. B*, 27 (1971) 307–314.
(152) R. A. G. DeGraff, G. Admiraal, E. H. Loen, and C. Romers, *Acta Crystallogr., Sect. B*, 33 (1977) 2459–2464.
(153) W. Saenger, G. Ritzmann, and W. Pfleiderer, *Acta Crystallogr., Sect. B*, 33 (1977) 2989–2993.

$P2_1$; $Z = 4$; $D_x = 1.522$; $R = 0.059$ for 4,286 intensities. There are two independent molecules in the asymmetric unit. The glycosyl dispositions are *syn* (C-2–H-1–C-1'–O-4' = 72.1°, 60.8°), and they are stabilized by the intramolecular hydrogen-bonds O-5'-H · · · O-2. The D-ribosyl groups have the $^2T_3$ conformation (163.5°, 38.8°; 151.3°, 40.0°). The orientation about the exocyclic, C-4'–C-5' bond is *gauche*$^+$ (51.3°, 54.7°). The heterocycles are stacked 345 pm apart, and display base–D-ribose hydrogen-bonding.

$C_{15}H_{16}N_6O_7$ N^6-[(Carboxymethyl)aminocarbonyl]adenosine (PUCGLR10)[154]

$P2_1$; $Z = 2$; $D_x = 1.63$; $R = 0.046$ for 1,710 intensities. The glycosyl disposition is *anti* (4.0°), and the D-ribosyl group has the $^3T_2$ pucker (7.2°, 36.0°). The exocyclic, C-4'–C-5' bond torsion-angle is *gauche*$^+$ (50.9°). The carboxyl group and the ureido group lie almost in the plane of the adenine ring, and this orientation is stabilized by the bifurcated hydrogen-bond between the ureido N-H group, N-1 of the base, and the carbonyl oxygen atom of the carboxyl group. An analogous, bifurcated hydrogen-bond is present in the next molecule.[154]

$C_{15}H_{20}N_6O_8$ N^6-[(1-Carboxy-2-hydroxyethyl)aminocarbonyl]adenosine (PCTRIB10)[154]

$P2_1$; $Z = 2$; $D_x = 1.56$; $R = 0.041$ for 2,035 intensities. The glycosyl disposition is *anti* (33.0°), and the D-ribosyl group has the $^2T_1$ conformation (158.2°, 32.6°). The orientation about the exocyclic, C-4'–C-5' bond is *gauche*$^+$ (57.2°). Most of the amino acid side-chain lies in the

(154) R. Parthasarathy, J. M. Ohrt, and G. B. Chheda, *Biochemistry*, 16 (1977) 4999–5008.

plane of the adenine base, so that the N-H group of the ureido group is involved in a bifurcated hydrogen-bond, one to N-1 of the base and the other to the hydroxyl group of the carboxyl group. The N-6-H and the adjacent C-O groups engage in an eight-membered, hydrogen-bonded ring with the carboxyl group of a symmetry-related molecule.

$C_{15}H_{24}N_4O_7 \cdot 2.5\ H_2O$ 5-(L-Leucylamino)uridine (LEUARU10)[155]

I222; Z = 8; D_x = 1.236; R = 0.080 for 1,448 intensities. The glycosyl disposition is *anti* (50.5°), and the conformation of the D-ribosyl group is $^2T_3$ (170.8°, 40.5°). The O-5′ atom is statistically disordered, so that

(155) P. Narayanan and H. M. Berman, *Acta Crystallogr., Sect. B*, 33 (1977) 2047–2051.

the orientations about the exocyclic, C-4'–C-5' bond are gauche⁺ (41.3°) and gauche⁻ (−67.8°). The bases form layers of hydrogen-bonded ribbons: N-3 donates a hydrogen bond to O-2, and N-7 on the opposite end of O-6 donates a hydrogen bond to O-4.

$C_{18}H_{19}N_2O_8P$ 2'-O-Acetyluridine 3',5'-monophosphate, benzyl ester (AURCPB)[156]

$P2_1$; Z = 2; D_x = 1.50; R = 0.055 for 2,021 intensities. The glycosyl disposition is *syn* (−108.6°) and the conformation of the D-ribosyl group is $^3T_4$ (C-4'-*exo*) (48.7°, 47.3°). The 3',5' ring constrains the exocyclic, C-4'–C-5' bond orientation to the *gauche*⁻ range (−58.9°). The dioxophosphorinane ring has the usual chair conformation, and the benzyl group is in the equatorial orientation. There are short, intramolecular contacts from 278 pm to 204 pm, between the carbonyl oxygen atom (O-2) of the base and the four furanoid-ring atoms that lie in a plane, with the exception of C-4' as a result of the *syn* orientation. The only hydrogen bond connects the N-3-H group of the base to the oxygen atom of the acetyl group attached to O-2'. Errors: bad coordinates for H-5, H-6, and H-7; coordinates are not connected.

$C_{18}H_{20}Cl_2N_2O_7$ 1-(5-O-Acetyl-2,3-O-isopropylidene-β-D-ribofuranosyl)-5,6-(dichloromethylene)-5,6-dihydro-3-methylthymine (AIMCTY)[157]

$P2_12_12_1$; Z = 4; D_x = 1.449; R = 0.048 for 1,592 intensities. The glycosyl disposition is *anti* (51.8°). The D-ribosyl moiety has the $^3T_4$ con-

(156) W. Depmeier and J. Engels, *Acta Crystallogr., Sect. B*, 33 (1977) 2436–2440.
(157) J. Bode and H. Schenk, *Cryst. Struct. Commun.*, 6 (1977) 645–649.

formation (38.0°, 29.1°). The dioxolane ring is in a twist conformation, with C-2' and O-2' lying on opposite sides of the dioxolane ring. The orientation about the exocyclic, C-4'–C-5' bond is *gauche*⁺ (51.8°).

$C_{19}H_{25}FN_2O_{10}S \cdot C_2H_6O$ 2,3,4,5-Tetra-*O*-acetyl-1-*S*-ethyl-1-(5-fluoro-uracil-1-yl)-1-thio-D-*gluco*-pentitol [(1-*R*)-2,3,4,5-Tetra-*O*-acetyl-1-*S*-ethyl-1-(5-fluorouracil-1-yl)-1-thio-D-arabinitol], ethanolate (TASFUA)[147]

$P2_1$; $Z = 2$; $D_x = 1.32$; $R = 0.06$ for 2,108 intensities. The carbon chain of the sugar moiety maintains an almost planar, zigzag conformation, with the adjacent, ester oxygen atoms oriented in the *gauche*$^+$, *trans*, *gauche*$^-$ sequence. Atom O-5 of the terminal hydroxyl group and the terminal carbon atom of the ethylthio group are bent away from the zigzag chain. There is a molecule of ethanol of crystallization.

$(C_{13}H_{17}CuN_5O_8)_2 \cdot 2\ H_2O$ Copper(II) (glycylglycinato)cytidine, dihydrate (GLCYCV)[158]

$P2_1$; $Z = 4$; $D_x = 1.64$; $R = 0.127$ for 1,852 intensities. The two independent complexes have almost identical conformations. The glycosyl dispositions are *anti* (15.9°, 16.2°), and the sugar form is ^3E (17.9°, 46.9°) and (14.6°, 41.1°). The exocyclic, C-4'–C-5' bond torsion-angles are *gauche*$^+$ (59.8°, 55.0°). These conformational features are similar

(158) D. J. Szalda and T. J. Kistenmacher, *Acta Crystallogr. Sect. B*, 33 (1977) 865–869.

to those observed in uncomplexed cytidine.[159] The coordination of the copper is distorted square-pyramidal. The tridentate glycylglycinato occupies three of the equatorial coordination-sites, with N-3 of the base completing the fourth equatorial site. The carbonyl oxygen atom (O-2) occupies the axial position, with a weaker interaction. The dihedral angle between the cytosine ring and the equatorial, coordination plane is 101° (mean). The carbonyl oxygen atom (O-2) of cytosine is significantly displaced, by 19 pm, from the base plane, and the Cu atom lies 36 pm (mean) out of plane. The bases of both molecules show similar stacking, with interplanar separations of 351 and 342 pm.

$(C_{19}H_{23}N_7O_{12}P)_2(C_{21}H_{20}N_3)_2 \cdot 27\ H_2O$ Diethidium 5-iodouridylyl-(3′,5′)-adenosine, 27 hydrate (ETHIUA10)[160]

C2; Z = 4; D_x = 1.155; R = 0.20 for 2,017 intensities. The asymmetric unit of structure contains two ethidium molecules, two iodo-UpA molecules, and 27 water molecules. The two iodo-UpA molecules are held together by adenine–uracil, Watson–Crick-type, base pairing, and have a right-handed, helical conformation. The separation between the base pairs is ~670 pm, due to the presence of an intercalated ethidium molecule. Similarly, the separation between neighboring iodo-UpA dimers is ~670 pm, arising from the stacking (sandwiching) of the second ethidium molecule between the neighboring dimer helices. The pseudo-two-fold axis of symmetry of the

(159) S. Furburg, C. S. Petersen, and C. Romming, *Acta Crystallogr. Sect. B*, 18 (1965) 313–320; see *Adv. Carbohydr. Chem. Biochem.*, 31 (1975) 369.
(160) C.-C. Tsai, S. C. Jain, and H. M. Sobell, *J. Mol. Biol.*, 114 (1977) 301–315.

phenanthridinium ring system coincides with the approximate twofold symmetry relating the base-paired iodo-UpA dimer. The phenyl and ethyl groups of the intercalated ethidium molecule face the narrow groove of the dimer helix. Both the intercalated and sandwiched ethidium molecules are heavily stacked with the adjacent base-pairs. The adjacent base-pairs of the helical dimer are related by a twist of 8°; thus, the intercalation results in untwisting of the so-called miniature helix (base pairs of B-DNA are related by twist angles of 36°). The untwisting is accomplished by conformational changes in the sugar–phosphate chain, and, in this case, the iodouridine residues on the 5'-end have the $^3T_4$ and $^3T_2$ puckers (23.1°, 52.2°; 347.7°, 30.8°), while the adenosine residue on the 3'-end has the $^2T_1$ and $^2T_1$ puckers (146.8°, 40.0°; 156.9, 37.3°), unlike the favored C-3'-*endo* pucker found in RNA. The glycosyl torsion-angles adjust to the variation in the sugar pucker in such a way that the iodouridine residues are C-3'-*endo*, C-2'-*exo*, low *anti* (26°, 14°), and the adenosine residues are C-2'-*endo*, high *anti* (99°, 100°). The bond torsion-angles of the sugar phosphate chain also fall in the generally favored conformational domains for helices, namely, 5'→3': ψ(C-4'–C-5') 56°, 171°; ψ' (C-3'–C-4') 98°, 95°; ϕ' (C-3'–O-3') 207°, 218°; ω'(O-3'–P) 286°, 302°; ω(P–O-5') 291°, 276°; ϕ(O-5'–C-5') 236°, 230°; ψ(C-5'–C-4') 52°, 70°; ψ'(C-4'–C-3') 133°, 118°. The columns of ethidium iodo-UpA run parallel to the *b* crystallographic direction, and there are prominent water channels interleaving adjacent columns of the complex. The high temperature-factors reported for the water oxygen atoms indicate that they must be highly disordered (and this is supported by the high R factor) and not "relatively well-ordered," as stated in the paper.

$(C_{19}H_{24}N_8O_{12}P)_2(C_{21}H_{20}N_3)_2 \cdot (CH_3O)_4 \cdot 27\ H_2O$ Diethidium 5-iodocytidylyl-(3',5')-guanosine, 27 hydrate, 4 MeOH (ICYGET10)[161]

$P2_1$; $Z = 2$; $D_x = 1.206$; $R = 0.16$ for 3,180 intensities. The asymmetric unit of structure consists of two ethidium molecules, two iodoCpG molecules, 27 water molecules, and 4 methanol molecules. The two iodoCpG molecules are hydrogen-bonded by guanine–cytosine, Watson–Crick-type base-pairing. Adjacent, base-paired iodoCpG dimers of adjoining unit-cells are separated by 670 pm. Thus, in addition to an ethidium molecule intercalated between the base-paired, iodoCpG molecules, there is an ethidium sandwiched between the iodoCpG dimer helices. The phenyl and ethyl groups of the ethidium molecule

(161) S. C. Jain, C.-C. Tsai, and H. M. Sobell, *J. Mol. Biol.*, 114 (1977) 317–331.

lie in the narrow groove of the miniature, iodoCpG double-helix. The sandwiched ethidium lies in the opposite direction, so that the phenyl and ethyl groups lie on the side of the iodine substituents on the cytosine residues. The G–C base-pairs are related by a twist angle of ~8°. The sugar phosphate backbone has a conformation different from that found for the RNA double-helix; the iodocytidine residues on the 5'-end have the $^3T_2$ sugar pucker (12.6°, 33.8°; 10.9°, 34.2°), while the guanosine residues on the 3'-end have the opposite, $^2T_1$ pucker (152.3°, 41.1°; 159.4°, 33.5°). Thus, the intercalation here has wrought a mixed puckering of the sugars of the dinucleotide. The alternation in the pucker leads to the usual, coupled alterations in the base–glycosyl torsion-angle, that is, C-3'-*endo*, low *anti* (24°, 29°) and C-2'-*endo*, high *anti* (101°, 109°). The G–C base-pairs of the dimer helix have angular twists of 7° and 9°. The ethidium molecules show heavy stacking to the adjacent, G–C base-pairs. The ethidiumiodoCpG form columns parallel to the *a* axis, leaving hydrophilic channels between adjacent columns for the 27 water molecules and the 4 methanol molecules. The various, sugar phosphate, bond torsion-angles are ψ(C-5'–C-4') 51°, 90°; ψ'(C-4'–C-3') 87°, 84°; ϕ'(C-3'–O-3') 226°, 225°; ω'(O-3'–P) 281°, 291°; ω(P–O-5') 286°, 291°; ϕ(O-5'–C-5') 210°, 224°; ψ(C-5'–C-4') 72°, 55°; ψ'(C-4'–C-3') 131°, 134°.

1978

$C_9H_9N_3O_9 \cdot H_2O$ 5-Nitro-1-(β-D-ribosyluronic acid)uracil monohydrate (NRURAM01)[162]

(162) T. Srikrishnan and R. Parthasarathy, *Acta Crystallogr. Sect. B*, 34 (1978) 1363–1366.

$P2_1$; $Z = 2$; $D_x = 1.75$; $R = 0.056$ for 1,438 intensities. The pyrimidine ring is slightly bent into a boat shape, with N-1 and C-4 displaced by ~ 2 pm on the same side of the plane of the ring. The nitro group is twisted by 7–6° from the base ring. The glycosyl disposition is *anti* (53.9°), and the conformation is $^2T_3$ (172.9°, 38.6°). The orientation about the exocyclic, C-4′–C-5′ bond is *gauche*$^+$ (98.3°). An unusual feature of this structure is that a water molecule is sandwiched between neighboring pyrimidine bases that are 655 pm apart. There appears to be a significant C-H · · · O interaction, 310 pm, between the base C-6-H group and the 5′-hydroxyl group.

$C_9H_{12}N_2O_5 \cdot 2\ H_2O$ 1-β-L-Ribosyl-2(1*H*)-pyrimidinone dihydrate (RIBPYM)[163]

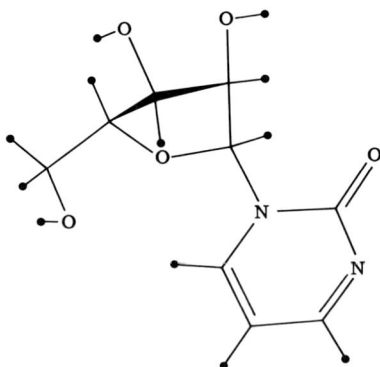

$P2_12_12_1$; $Z = 4$; $D_x = 1.46$; $R = 0.037$ for 1,167 intensities. The glycosyl disposition is *anti* ($-1.5°$) and the L-ribosyl group has the C-2′-*exo*,

(163) S. Furberg, *Acta Chem. Scand.*, 32 (1978) 478–482.

C-3′-endo, symmetrical-twist ($^3T_2$) pucker (179.7°, 39.0°). The exocyclic, C-4′–C-5′ bond torsion-angle is gauche⁻ (−52.3°). The low value of χ results in close contacts, between O-1′ of the furanoid ring and C-6 and H-6 of the base, of 268 and 218 pm, respectively, which, in part, accounts for the observed elongation in the glycosyl C-1′–N-1 bond-length of 151 pm. There is also a (short) C-6 · · · O-5′ contact of 305 pm, with an H-6 · · · O-5′ distance of 234 pm. Note: the author seems to have missed the point that this synthetic product is an L-nucleoside analog.

[($C_9H_{12}N_3O_8P$)Cd(H_2O)·H_2O]$_n$ Cytidine [cadmium(II) 5′-monophosphate] monohydrate[164]

$P2_12_12_1$; Z = 4; D_x = 2.11; R = 0.038 for 1,104 intensities. The glycosyl disposition is anti (16.1°), the sugar pucker is $^3T_2$ (14.2°, 36.4°), and the exocyclic, C-4′–C-5′ bond torsion-angle is gauche⁺ (58.7°). The Cd(II) ion forms a polymeric complex with the 5′-CMP. The Cd atom is in a distorted, square-pyramidal environment consisting of N-3 of the pyrimidine ring in the apical position and three phosphate oxygen atoms from different nucleotides in the equatorial position. The polymeric complex is characterized by cylindrical channels of cross-linked, -Cd–O–P–O–Cd- spirals.

[($C_9H_{12}N_3O_8P$)Co·H_2O]   Cytidine [cobalt(II) 5′-monophosphate], monohydrate[164]

$P2_1$; Z = 2; D_x = 1.89; R = 0.073 for 871 intensities. The glycosyl disposition is anti (69.7°), and the conformation of the D-ribosyl group is

(164) G. R. Clark and J. D. Orbell, Acta Crystallogr. Sect. B, 34 (1978) 1815–1822.

$^2T_1$ (137.2°, 34.2°). The exocyclic, C-4'–C-5' bond torsion-angle is gauche$^+$ (51.3°). The Co atom is tetrahedrally coordinated to N-3 of the cytosine ring, to a water molecule, and to two oxygen atoms from different phosphate groups, giving a two-dimensional, layer structure. The bases are stacked at an average separation of 373 pm. Errors: H-C-4' had bad coordinates (fixed at 0.563, 0.496, −0.173), and the y coordinate of H-C-5' probably equals 0.5834, instead of 0.3834.

$C_9H_{13}N_3O_5$ 6-Azathymidine (AZTYMO)[166]

$P2_12_12_1$; Z = 4; D_x = 1.57; R = 0.043 for 1,014 intensities. The glycosyl disposition is anti (88.0°) and the D-glycosyl group has the $^3T_2$ conformation (359.0°, 35.9°). The orientation about the exocyclic, C-4'–C-5' bond is gauche$^+$ (64.7°). Note: As published, the hydrogen atoms were not connected to heavier atoms.

(166) A. Banerjee and W. Saenger, Acta Crystallogr. Sect. B, 34 (1978) 1294–1298.

$C_9H_{13}N_3O_6$ 5-Aminouridine[167]

$P2_12_12_1$; $Z = 4$; $D_x = 1.597$; $R = 0.076$ for 849 intensities. The glycosyl disposition is *anti* (61.1°), and the conformation of the D-ribosyl group is $^2T_3$ (169.7°, 37.9°). The exocyclic, C-4'–C-5' bond torsion-angle is *gauche*$^+$ (49.9°). The bases are stacked, with overlaps between the carbonyl oxygen atom (O-2) and C-4 of the ring, 329 pm, and the amino nitrogen atom (N-5) and N-1 of the ring, 347 pm, of an adjacent base. There is an interbase hydrogen-bond, N-5-H · · · O-4.

$C_{10}H_{11}N_2O_7P$ Cytidine 2',3'-phosphate[168]

(167) E. Egert and H. J. Lindner, *Acta Crystallogr., Sect. B*, 34 (1978) 2204–2208.
(168) B. S. Reddy and W. Saenger, *Acta Crystallogr., Sect. B*, 34 (1978) 1520–1524.

$P2_12_12_1$; Z = 4; D_x = 1.62; R = 0.064 for 1,175 intensities. The orientation about the glycosyl bond is *anti* (59.1°), and the conformation of the D-ribosyl group is $^2T_3$ (puckering, 173.1°, 29.3°). The exocyclic, C-4′–C-5′ bond torsion-angle is *gauche*$^+$ (48.5°). The cytosine is protonated at N-3. Atoms N-4 and N-3 of the base act as hydrogen-bond donors to the pair of anionic oxygen atoms of an adjacent phosphate. As noted in other zwitterionic, nucleotide structures, the carbonyl oxygen atom (O-2) does not participate in hydrogen bonding. Errors: the N-3 z coordinate should be 0.6115 instead of 0.5115, and the atom labelled N-3 in Table 3 should be H-N-3.

$C_{10}H_{12}IN_5O_5 \cdot H_2O$ 8-Iodoguanosine monohydrate (IGUANM)[169]

$P2_1$; Z = 2; D_x = 1.886; R = 0.085 for 1,296 intensities. The glycosyl disposition is *syn* (−116.9°) and the conformation of the D-ribosyl group is $^2T_3$ (164.9°, 41.9°). The exocyclic, C-4′–C-5′ bond torsion-angle is *trans* (−177.9°). This conformation leads to an intermolecular hydrogen-bonding to N-7 of the base of a 2_1 axis-related molecule. An unusual feature is that neither of the hydrogen atoms of the amino group is involved in hydrogen bonding. Errors: (corrections given in Cambridge data file) the C-8 z-coordinate should be −0.0153, instead of 0.0153, and the N-9 x-coordinate should be −0.2957, instead of −0.1957.

$(C_{10}H_{12}N_5O_8P)_3Cu_3 \cdot 4\ H_2O$ Guanosine 5′-[copper(II) monophosphate], tetrahydrate[170]

(169) J. H. Al-Mukhtar and H. R. Wilson, *Acta Crystallogr., Sect. B*, 34 (1978) 337–339.
(170) G. R. Clark, J. D. Orbell, and K. Aoki, *Acta Crystallogr. Sect. B*, 34 (1978) 2119–2128.

$P2_12_12_1$; Z = 4; D_x = 1.81; R = 0.09, 0.11 for 2,316, 1,705 intensities. The preliminary report on the structure was jointly published[171] by two groups in 1976. The detailed, structural features[170] have now been presented. At about the same time as the preliminary work was reported, a third group[172] also reported the same structure. The structure is highly complicated, and the three independent, metal–nucleotide units form an infinite spiral consisting of -Cu–phosphate–sugar–base-linkages. The turns of the spirals are cross-linked by bridging through Cu and phosphate oxygen atoms, and by hydrogen bonding through coordinated water molecules and phosphate oxygen atoms. The guanine rings project radially from the outside of the spiral and intercalate between bases of neighboring chains. This base stacking probably leads to the differences in the sugar puckerings, wherein two are puckered C-3'-endo, $^3T_4$ (22.5°, 40.9°) and (14.2°, 40.3°) and one is C-2'-endo ($^2T_3$) (167.6°, 32.9°). All three nucleotides have their bases disposed in the *anti* orientation: 46.2, 60.1, and 41.9° for molecules A, B, and C, respectively. The exocyclic, C-4'–C-5' bond torsion-angles are *gauche*$^+$ (48.2°, 55.7°, 63.6°). The coordination about each Cu ion is distorted (4 + 1), square pyramidal. The apical sites are occupied by the N-7 atoms of the guanine bases. The four planar donors differ, in that, in Cu(A), it is a phosphate oxygen atom, whereas, in Cu(B) and Cu(C), they consist of two phosphate oxygen atoms and two water

(171) K. Aoki, G. R. Clark, and J. D. Orbell, *Biochim. Biophys. Acta*, 425 (1976) 369–371.
(172) E. Sletten and B. Lie, *Acta Crystallogr. Sect. B*, 32 (1976) 3301–3304.

molecules. Each nucleotide base forms a strong hydrogen-bond between the carbonyl oxygen atom (O-6) and a copper-coordinated, water molecule.

$C_{10}H_{13}Cl_2HgN_5O_5$ N^7-[Chloromercuri(II)]guanosine[174]

$P2_12_12_1$; $Z = 4$; $D_x = 2.45$; $R = 0.040$ for 1,410 intensities. The glycosyl disposition is *anti* (59.9°) and the D-ribosyl group is $^2T_1$ (151.8°, 48.3°). The orientation about the exocyclic, C-4′–C-5′ bond is *trans* (165.5°). The Hg is strongly coordinated to N-7 of the guanosine and to one Cl atom. The other Cl atoms are weakly coordinated to Hg, to complete an irregular, four-coordinate (tetrahedral) geometry. Infinite, –Cl–Hg–Cl–Hg–, zigzag (square wave) chains centered on the 2_1 axis run in the *c* direction. The structure exhibits good base-stacking.

$C_{10}H_{14}N_2NaO_9P \cdot CH_3OH$ 2,3-O-sodiouridine 5′-(methyl phosphate), methanolate (SUROMM)[175]

$P2_1$; $Z = 2$; $D_x = 1.566$; $R = 0.040$ for 1,197 intensities. The uracil residue has the *anti* (79.1°) orientation with respect to the sugar. The D-ribosyl group has the $^2T_3$ pucker (166.4°, 36.9°), and the orientation about the exocyclic, C-4′–C-5′ bond is *gauche*$^+$ (52.8°). The sodium atom is octahedrally coordinated by oxygen atoms at distances vary-

(174) M.-A. Martin, J. Hubert, R. Rivest, and A.L. Beauchamp, *Acta Crystallogr. Sect. B*, 34 (1978) 273–276.
(175) J. D. Hoogendorp and C. Romers, *Acta Crystallogr. Sect. B*, 34 (1978) 2724–2728.

ing between 231 and 241 pm. The ligands involve O-2' and O-3', the two anionic phosphate oxygen atoms, and the carbonyl group oxygen atoms (O-2 and O-4) of the surrounding molecules. Base stacking is absent, and there is base–phosphate hydrogen-bonding, N-3-H · · · O. The methanol links the 3'-hydroxyl group of one molecule and an anionic phosphate oxygen atom of an adjacent molecule.

$C_{10}H_{14}N_5O_5 \cdot H_2O$ 8-Azatubercidin monohydrate[176]

$P2_12_12_1$; Z = 4; D_x = 1.517; R = 0.040 for 1,112 intensities. The glycosyl disposition is in the high-*anti* range (102.5°). The D-ribosyl

(176) S. Sprang, R. Scheller, D. Rohrer, and M. Sundaralingam, *J. Am. Chem. Soc.*, 100 (1978) 2867–2872.

group exhibits the C-1'-*exo*-C-2'-*endo* ($^2T_1$), symmetrical-twist pucker (141.9°, 41.9°). The exocyclic, C-4'–C-5' bond torsion-angle is *trans* (179.5°). The bases are stacked on a 2_1 screw axis. There is an intermolecular hydrogen-bond between the amino group and the oxygen atom of the furanoid ring. Atom N-8 of the base is not involved in hydrogen bonding.

$C_{10}H_{16}N_2O_6$ (−)-(5-S)-5,6-Dihydro-5-hydroxythymidine[177]

$P2_1$; Z = 2; D_x = 1.420; R = 0.070 for 1,138 intensities. The orientation about the glycosyl bond is *anti* (−68.7°). The 2-deoxy-D-*erythro*-pentofuranosyl group has the O-1'-*endo* (^oE) pucker (257.7°, 37.6°), and the orientation about the exocyclic, C-4'–C-5' bond is *gauche*$^+$ (−49.3°). The pyrimidine ring has a twisted, half-chair conformation, with C-5 and C-6 displaced by 28.6 and 36.2 pm on opposite sides of the plane. The methyl group is equatorially attached to the base, and the hydroxyl group is axial. Error: all torsion angles published were the negative of the correct values.

$C_{11}H_{12}N_2O_5$ 2'-Deoxy-5-ethynyluridine (ETYXUR)[178]

$P2_1$; Z = 2; D_x = 1.474; R = 0.038 for 1,162 intensities. The glycosyl disposition is *anti* (18.6°), and the 2-deoxy-D-*erythro*-pentofuranosyl group is $^2T_1$ (156.4°, 42.7°). The orientation about the exocyclic, C-4'–C-5' bond is *trans* (−174.2°). The oxygen atom of the furanoid ring accepts an intermolecular hydrogen-bond from N-3-H of a base.

(177) A. Grand and J. Cadet, *Acta Crystallogr. Sect. B*, 34 (1978) 1524–1528.
(178) P. J. Barr, T. A. Hamor, and R. T. Walker, *Acta Crystallogr. Sect. B*, 34 (1978) 2799–2802.

$C_{11}H_{14}N_2O_5$ 2'-Deoxy-5-vinyluridine (VIDRID)[179]

$P2_12_12_1$; Z = 4; D_x = 1.485; R = 0.059 for 748 intensities. The base disposition is *anti* (38.8°). The 2-deoxy-D-*erythro*-pentofuranosyl group has the C-3'-*exo* pucker (190.0°, 38.0°). The orientation about the exocyclic, C-4'–C-5' bond is *trans* (170.3°). The carbonyl oxygen atom (O-2) makes a close contact to the vinyl methylene carbon atom, 319 pm.

(179) T. A. Hamor, M. K. O'Leary, and R. T. Walker, *Acta Crystallogr. Sect. B*, 34 (1978) 1627–1630.

$C_{11}H_{14}N_2O_8$ 5-(Carboxymethyl)uridine (CXMURD)[180]

$P2_12_12_1$; $Z = 4$; $D_x = 1.60$; $R = 0.047$ for 1,073 intensities. The glycosyl disposition is *anti* (49.4°), and the D-ribosyl group exhibits the $^2T_1$ pucker (159.9°, 40.9°). The exocyclic, C-4'–C-5' bond torsion-angle is *gauche*$^+$ (60.0°). The plane of the carboxymethyl group is roughly perpendicular to the pyrimidine base, thus placing the carbonyl oxygen atom below C-5 of the base, and on the *same side* of the ring-oxygen atom of the D-ribose. There are no interbase hydrogen-bonds.

$C_{11}H_{15}N_3O_7$ 5-(Carbamoylmethyl)uridine (CBMURO)[180]

$P2_12_12_1$; $Z = 4$; $D_x = 1.60$; $R = 0.036$ for 1,103 intensities. The glycosyl disposition is *anti* (5.3°), and the D-ribosyl group exhibits the $^3T_2$ pucker (4.2°, 39.7°). The orientation about the exocyclic, C-4'–C-5' bond is *trans* (179.9°). The carbamoylmethyl group at C-5 of the base is twisted away from the base plane in such a way that the carboxamide oxygen atom lies roughly over C-5 and points in a direction *opposite* to that of the D-ribose ring-oxygen atom. Interbase hydrogen-bonds, between N-3-H of one molecule and the carbonyl oxygen atom (O-2) of another, form an infinite, hydrogen-bonded spiral. The carbonyl oxygen (O-4) is hydrogen-bonded to the amide nitrogen atom.

(180) H. M. Berman, D. Marcu, and P. Narayanan, *Nucleic Acids Res.*, 5 (1978) 893–903.

The ring-oxygen atom of the D-ribosyl group points into the ring of the base of a neighboring molecule.

$C_{11}H_{15}N_5O_3S$ 5'-S-Methyl-5'-thioadenosine (TMSADS10)[181]

$P2_1$; Z = 4; D_x = 1.481; R = 0.075 for 2,612 intensities. There are two independent molecules in the asymmetric unit, and they show some significant, conformational differences. The glycosyl disposition is *anti* (molecule A, −24.7°; molecule B, +58.8°). The D-ribosyl group is C-2'-*exo*, C-3'-*exo* (216.5°, 35.2°) in molecule A, and $^2T_3$ (172.7°, 43.1°) in molecule B. The feature that is common to both molecules is the disposition of the methylthio side-group with respect to the rotation around the C-4'–C-5' bond, which is *gauche*⁻ (molecule A, −69.5°;

(181) N. Borkakoti and R. A. Palmer, *Acta Crystallogr. Sect. B*, 34 (1978) 867–874.

molecule B, $-73.7°$), an orientation rarely found in nucleosides and nucleotides. The bases are self-paired, involving N-6 and N-1. This, also, is an uncommon pairing-scheme for adenine systems. In addition, the bases are paired through N-6 and N-7 to the O-2'-H and O-3'-H groups of an adjacent molecule.

$C_{11}H_{15}N_5O_5 \cdot H_2O$ N^2-Methylguanosine monohydrate (MGUOSM)[182,183]

P1; Z = 1; D_x = 1.57. There were two independent investigations: R = 0.031 and 0.043 for 1,387 (MoK$_\alpha$) and 1,645 intensities, respectively. The glycosyl disposition is *syn* ($-114.9°$, $-114.6°$), and the D-ribosyl group has the $^2T_3$ pucker (168.0°, 36.3°; 168.1°, 36.2°). The orientation about the exocyclic, C-4'–C-5' bond is *gauche*$^-$ ($-77.2°$, $-76.9°$), and this precludes the formation of the usual, intramolecular hydrogen-bond to N-3 of the base.

$C_{11}H_{16}N_5O_7P$ Adenosine 5'-(methyl phosphate) (ADMOPM)[184]

I222; Z = 8; 1.39 < D_x < 1.51 (uncertain number of methanol molecules of crystallization); R = 0.038 for 1,372 intensities. The base disposition is *anti* (68.6°), and the D-ribosyl group is $^2T_3$ (165.9°, 31.9°). The orientation about the exocyclic, C-4'–C-5' bond is *gauche*$^+$ (68.3°).

(182) S. L. Ginell and R. Parthasarathy, *Biochem. Biophys. Res. Commun.*, 84 (1978) 886–894.
(183) R. A. G. deGraaff, F. B. Martens, and C. Romers, *Acta Crystallogr. Sect. B*, 34 (1978) 3012–3015.
(184) J. D. Hoogendorp, G. C. Verschoor, and C. Romers, *Acta Crystallogr. Sect. B*, 34 (1978) 3662–3666.

The torsion angle ϕ (C-4'–C-5'–O-5'-P) is 143.3° and is skewed from the ideal *trans* orientation. The P–O ester torsion-angles, ω (P-O-5') and ω' (P-O-3'), are 71.6° and 63.5°, respectively, and constitute a left-handed arrangement, in contrast to the right-handed arrangement for uridine 5'-(methyl phosphate). The molecule exists as a zwitterion, with the phosphate group negatively charged and N-1 of the base protonated. As is common, there is a strong hydrogen-bond, 263 pm, linking N-1-H and one of the anionic, phosphate oxygen atoms. The second anionic, phosphate oxygen atom of a neighboring molecule is hydrogen-bonded to the amino (N-6) atom. The bases are well stacked, and the dyad-related bases are paired through N-7 and N-6.

$C_{11}H_{19}Cl_2N_3O_6PtS$ *trans*-Dichloroplatinum(II)(dimethyl sulfoxide)cytidine[185]

(185) R. Melanson and F. D. Rochon, *Inorg. Chem.*, 17 (1978) 679–681.

$P2_12_12_1$; $Z = 4$; $D_x = 2.241$; $R = 0.043$ for 1,908 intensities. The glycosyl disposition is *anti* (13.5°), and the D-ribosyl group is $^3T_4$ (19.4°, 37.3°). The exocyclic, C-4'–C-5' bond torsion-angle is *gauche*$^+$ (60.8°). The Pt atom is coordinated to N-3 of the base, Pt–N-3 = 203 pm. The coordination around the Pt atom is planar. Besides N-3, two Cl atoms and the dimethyl sulfoxide group are liganded to the Pt. The plane of the pyrimidine ring is at 77.4° to the plane formed by the ligands around the Pt. Note: unconventional atom numbering was used in this paper.

$(C_{11}H_{21}ClN_5O_5Pt)^+Cl^-$ (Platinum ethylenediamine dichloride) 1-β-D-arabinofuranosylcytosine[186]

$P2_12_12_1$; $Z = 4$; $D_x = 2.209$; $R = 0.076$ for 1,449 intensities. The disposition of the base is *anti* (14.0°), and the D-arabinofuranosyl group is $^2T_3$ (171.3°, 44.7°). This feature, combined with the *gauche*$^+$ (69.9°) orientation around the exocyclic, C-4'–C-5' bond, allows the formation of an intra-sugar hydrogen-bond between O-2'–H and O-5'. One of the chlorine ligands in the (Pt ethylenediamine dichloride) complex is replaced by N-3 of the cytosine residue; thus, the Pt has the usual, square-planar coordination. The net positive charge in the molecule is neutralized by a Cl⁻ ion. The plane formed by the ligand atoms to the Pt is roughly perpendicular to the ring of the base. The intermolecular hydrogen-bond between the amino nitrogen atom (N-4) and the carbonyl oxygen atom (O-2) links equivalent molecules of adjacent unit cells, to form an infinite chain, as in the neutral, "ara-C" structure.[187]

(186) S. Neidle, G. L. Taylor, and A. B. Robins, *Acta Crystallogr. Sect. B*, 34 (1978) 1838–1841.
(187) A. K. Chwang and M. Sundaralingam, *Nature (London) New Biol.*, 243 (1978) 78–80.

$C_{12}H_{13}N_5O_4 \cdot H_2O$ Toyocamycin monohydrate (TOYOCM10)[188]

$P2_1$; $Z = 2$; $D_x = 1.503$; $R = 0.026$ for 1,120 intensities. The glycosyl disposition is *anti* (60.7°), and the conformation of the D-ribosyl group is $^2T_3$ (165.7°, 42.5°). The exocyclic, C-4′–C-5′ bond torsion-angle is *gauche*$^+$ (57.1°). The water molecule is hydrogen-bonded to the oxygen atom of the furanoid ring. The cyano group on C-7 is not involved in hydrogen bonding, but is tucked under a neighboring base, as is observed in halogenated-purine nucleosides. The amino hydrogen atom facing the cyano substituent is not engaged in hydrogen bonding, probably because it is shielded by the cyano group.

$C_{12}H_{18}N_7O_6P \cdot 4\ H_2O$ 8-(2-Aminoethylamino)adenosine 3′,5′-monophosphate, tetrahydrate[189]

(188) P. Prusiner and M. Sundaralingam, *Acta Crystallogr. Sect. B*, 34 (1978) 517–523.
(189) W. S. Sheldrick and E. Rieke, *Acta Crystallogr. Sect. B*, 34 (1978) 2324–2327.

$P2_12_12_1$; Z = 4; D_x = 1.55; R = 0.050 for 4,355 reflections. The molecule is a zwitterion. The disposition of the base is *syn* (− 107.0°), and the conformation of the D-ribosyl group is C-4'-*exo*, C-3'-*endo* ($^3T_4$) 51.0, 48.7°). The exocyclic, C-4'–C-5' bond torsion-angle is *gauche*⁻ (−62.9°). The 2_1 screw, symmetry-related molecules are stacked, with an interplanar separation between the bases of 317 pm. The (2-aminoethyl)amino group on C-8 exhibits the *gauche*⁺ orientation for the bond sequence N–C–C–N. There is a base–phosphate hydrogen-bond between the amino group (N-6) and an anionic, phosphate oxygen atom.

$C_{13}H_{17}N_5O_4$ 2',3'-O-Isopropylideneadenosine (IPADOS)[190]

$P2_12_12_1$; Z = 8; D_x = 1.430; R = 0.063 for 2,031 intensities. In the asymmetric unit, there are two independent molecules that differ substantially in the conformation of the sugar. In molecule A, the glycosyl disposition is *anti* (10.4°), the D-ribosyl group is planar (τ_m = 3.9), and the exocyclic, C-4'–C-5' bond torsion-angle is *gauche*⁺ (53.6°). In molecule B, the glycosyl disposition is *anti* (16.0°), the D-ribosyl group is C-3'-*exo*, C-4'-*endo* ($^4T_3$) (214.5°, 31.6°), and the orientation about the exocyclic, C-4'–C-5' bond is *trans* (174.5°). The dioxolane rings have different twist conformations. The bases are self-paired through the N-1, N-6 and N-6, N-7 sites, forming an infinite ribbon in the *ac* plane. There is a significant twist angle (27.6°) for the base pair.

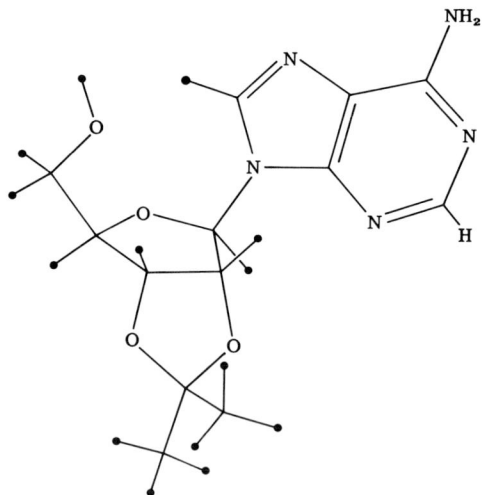

(190) S. Sprang, D. C. Rohrer, and M. Sundaralingam, *Acta Crystallogr. Sect. B*, **34** (1978) 2803–2810.

$C_{13}H_{19}N_5O_5 \cdot 2\ H_2O$ 8-(1-Hydroxyisopropyl)adenosine dihydrate (HIPADS)[191]

$P2_1$; $Z = 2$; $D_x = 1.436$; $R = 0.031$ for 1,467 intensities. The glycosyl disposition is *syn* (−118.6°), the D-ribosyl group has the $^2T_1$ pucker (160.4°, 40.3°), and the orientation about the exocyclic, C-4′–C-5′ bond is *gauche*$^+$ (51.7°). This conformational combination is stabilized by

(191) G. I. Birnbaum and D. Shugar, *Biochim. Biophys. Acta*, 517 (1978) 500–510.

the usual, intramolecular hydrogen-bond, O-5-H · · · N-3. One of the water molecules is hydrogen-bonded to N-7 and N-6 of the same base.

$C_{14}H_{14}BrN_7O_6$ Adenosine–5-bromouracil (ADOSBU)[192]

$P2_1$; $Z = 2$; $D_x = 1.78$; $R = 0.083$ for 2,003 intensities. The base disposition is *syn* ($-120.4°$) and N-3 of the base is involved in the usual, intramolecular hydrogen-bond to the oxygen atom of the hydroxymethyl group. The D-ribosyl group is close to the ^2E pucker (161.4°, 38.7°), and the exocyclic, C-4'–C-5' bond torsion-angle is *gauche*$^+$ (49.3°). The adenine and the uracil bases form a hydrogen-bonded pair by way of A(N-6) · · · U(O-2) and A(N-1) · · · U(N-1). Error: (corrected in the Cambridge Data File) the C-2'z-coordinate should be 0.5415 instead of 0.4515.

(192) K. Aiba, T. Hata, S. Sato, and C. Tamura, *Acta Crystallogr. Sect. B*, 34 (1978) 1259–1263.

$C_{14}H_{27}N_9O_{12}Pt$ [$(C_{10}H_{11}N_4O_5)\cdot C_4H_{13}N_3(NO_3)_2\cdot H_2O$] Inosine (diethylenetriamine)platinum(II) dinitrate, monohydrate[194]

$P2_1$; $Z = 2$; $D_x = 2.067$; $R = 0.040$ for 2,727 intensities. The disposition of the base is *anti* (8.1°), and the D-ribosyl group is $^3T_2$ (9.9°, 34.1°). The orientation about the exocyclic, C-4'–C-5' bond is *trans* ($-177.1°$). The N-7 atom of the base is coordinated to the Pt atom. The purine plane makes an angle of 74.5° to the Pt-coordination plane.

$(C_{10}H_{11}N_4NaO_8P)_2(C_4H_{13}N_3)Cu\cdot 10\ H_2O$ Bis(2,3-O-sodioinosine 5'-monophosphate) (diethylenetriamine) copper(II), decahydrate[195]

$P2_12_12_1$; $Z = 2$; $D_x = 1.61$; $R = 0.08$ for 2,051 intensities. The glycosyl disposition of both 5'-IMP ligands is *anti* (48.4°), and the D-ribosyl group is $^2T_3$ (168.0°, 34.9°). The exocyclic, C-4'–C-5' bond torsion-angle is in the *gauche*$^+$ range (60.3°). The Cu(II) ion is hexacoordinated, the equatorial ligands including the tridentate, triamine (dien) chelate and the N-7 atom of a 5'-IMP ligand. The axial ligands comprise

(194) R. Melanson and F. D. Rochon, *Acta Crystallogr. Sect. B*, 34 (1978) 3594–3598.
(195) C. C. Chiang, T. Sorrell, T. J. Kistenmacher, and L. G. Marzilli, *J. Am. Chem. Soc.*, 100 (1978) 5102–5110.

N-7 of a second 5'-IMP molecule and a water molecule. The equatorial, N-7–Cu binding (192 pm) is considerably stronger than the axial, N-7–Cu interaction (300 pm). The length of the Cu–N bond (central nitrogen atom of the triamine) is 184 pm, and the Cu–N (dien) terminal bond-lengths are both 200 pm. The ternary, Cu(II) complex is further stabilized by four interligand hydrogen-bonds: the dien chelate donates to the carbonyl oxygen atoms (O-6) of both 5'-IMP molecules, and is also involved in hydrogen bonding to one of the anionic, phosphate oxygen atoms of the equatorial 5'-IMP ligand. The axial water molecule is also involved in a hydrogen bond to a second phosphate oxygen atom of the equatorial 5'-IMP ligand. The coordination sphere about the Na$^+$ ion is a highly distorted octahedron. Two of the *cis* coordination sites are occupied by the D-ribosyl atoms, O-2' (243 pm) and O-3' (228 pm), and the four sites remaining are occupied by four water molecules (236, 248, 227, and 321 pm). The most weakly bound, water ligand donates a hydrogen bond to N-3 of a base.

[(C$_9$H$_{11}$N$_2$O$_9$P)(C$_{10}$H$_8$N$_3$)Cu(H$_2$O)]$_2$·5 H$_2$O   Uridine 5'-monophosphate copper(II) 2,2'-iminodipyridine, pentahydrate (CUMPAH10)[196]

P2$_1$; Z = 2; D$_x$ = 1.67; R = 0.086 for 2,692 intensities. This compound is a dimeric complex composed of two nucleotide, two iminodi-

(196) B. E. Fischer and R. Bau, *Inorg. Chem.*, 17 (1978) 27–34.

pyridine, and two water molecules, and two Cu^{2+} ions. The disposition of the base in nucleotide 1 is *syn* ($-77.6°$), and in the other, it is *anti* ($57.6°$). The conformations of the D-ribosyl groups are respectively $^3T_2$ ($346.0°$, $25.4°$) and $^oT_1$ ($100.7°$, $41.8°$). The exocyclic, C-4'–C-5' bond torsion-angles are *trans* ($-177.8°$, $-171.5°$) in both nucleotides. The nucleotide coordinates to the Cu^{2+} ion through the phosphate group only. The metal-ion coordination is square pyramidal, involving two oxygen ligands from two phosphate groups, two nitrogen ligands from two iminodipyridine groups, and one water molecule in an axial position. The apical water molecule is farther away from the metal ion than from the four equatorial ligands. The copper ions are bridged by two phosphate groups in such a way that an eight-membered ring is formed $-(Cu-O-P-O)_2$.

$C_{24}H_{29}N_5O_{10}S$ 2',3',5'-Tri-O-acetyl-6-O-(mesitylenesulfonyl)guanosine (AMSGOS)[197]

C2; Z = 4; D_x = 1.346; R = 0.110 for 1,102 intensities. The glycosyl disposition is *anti* ($23.7°$), and the D-ribosyl group is $^2T_1$ ($150.4°$, $37.6°$). The orientation about the exocyclic, C-4'–C-5' bond is *trans* ($173.9°$). The mesitylene group is disposed away from the guanine base, as the torsion angle C-6–O-6–S–C is in the *trans* range. There is no hydrogen bonding in this structure. Error: (corrected in the Cambridge Data File) the N-1 x-coordinate should be 0.2882 instead of 0.2282.

(197) S. Neidle and A. Achari, *Acta Crystallogr. Sect. B*, 34 (1978) 2794–2798.

IV. Preliminary Communications

1. Carbohydrates

$C_{19}H_{25}O_8P$ 1,2,4-Tri-*O*-acetyl-5,6-dideoxy-3-*O*-methyl-5-[(*S*)-phenyl-phosphinyl]-β-D-glucopyranose[198]

$C_{33}H_{26}O_9$ 1,2,3,4-Tetra-*O*-benzoyl-β-D-xylopyranose[199]

(198) M. Yamashita, Y. Nakatsukasa, S. Inokawa, K. Hirotsu, and J. Clardy, *Chem. Lett.*, (1978) 871–872.
(199) P. Luger, G. Kothe, and H. Paulsen, *Angew. Chem. Int. Ed. Engl.*, 16 (1977) 52–53.

2. Nucleosides and Nucleotides
(1977)

$C_9H_{11}CoN_2O_9P \cdot 7\ H_2O$ Uridine 5'-[cobalt(II) monophosphate], heptahydrate[200]

$C_9H_{11}N_3O_8 \cdot H_2O$ 5-Nitrouridine monohydrate[201]

$C_9H_{12}ClFN_4O_3$ 2,2'-Anhydro-(1-β-D-arabinofuranosyl-2,4-diamino-5-fluoropyrimidinium chloride)[202]

$Li(C_{22}H_{26}N_7O_{14}P_2) \cdot 2\ H_2O$ Lithium NAD⁺ dihydrate[203]

$(C_{19}H_{24}N_8O_{12}P)_2(C_{13}H_{12}N_3)_3 \cdot x\ H_2O$ 3:2 Cytidylyl-(3',5')-guanosine: proflavine, hydrate[204]

$(C_{19}H_{23}IN_8O_{12}P)_4(C_{13}H_{11}N_2)_4 \cdot 24\ H_2O$ 5-Iodocytidylyl-(3',5')-guanosine: 9-aminoacridine, 24 H_2O[205]

$[Pt(C_2H_8N_2)(C_9H_{12}N_3O_8P)]_2 \cdot 2\ H_2O$ (Platinum ethylenediamine cytidine 5'-phosphate)$_2$, dihydrate[206]

$[Cu(C_{10}H_{11}N_4O_8P)(C_{10}H_8N_2)(H_2O)_2] \cdot NO_3 \cdot H_2O$ Copper(II) inosine 5'-phosphate (2,2'-bipyridyl)(H$_2$O)$_2$, nitrate, monohydrate[207]

---

(200) B. A. Cartwright, D. M. L. Goodgame, I. Jeeves, and A. C. Skapski, *Biochim. Biophys. Acta*, 477 (1977) 195–198.
(201) E. Egert, H.-J. Lindner, W. Hillen, and H. G. Gassen, *Nucleic Acids Res.*, 4 (1977) 929–937.
(202) A. F. Cook, *J. Med. Chem.*, 20 (1977) 344–348.
(203) W. Saenger, B. S. Reddy, K. Mühlegger, and G. Weimann, *Nature*, 267 (1977) 225–229.
(204) H. M. Berman, H. L. Carrell, J. P. Glusker, and W. C. Stallings, *Nature*, 269 (1977) 304–307.
(205) T. D. Sakore, S. C. Jain, C.-C. Tsai, and H. M. Sobell. *Proc. Natl. Acad. Sci. U. S. A.*, 74 (1977) 188–192.
(206) S. Louie and R. Bau, *J. Am. Chem. Soc.*, 99 (1977) 3874–3876.
(207) K. Aoki, *J. Chem. Soc. Chem. Commun.*, (1977) 600–602.

[Cu($C_9H_{11}N_2O_9P$)($C_{10}H_9N_3$)(H_2O)]$_2$·5 H_2O [Copper(II) uridine 5'-phosphate (N-2-pyridyl-2-pyridinamine)H_2O]$_2$, pentahydrate[208]

[($C_5H_7O_2$)$_2$(NO_2)($C_{10}H_{13}N_5O_3$)Co]$_2$·7 H_2O Cobalt(III) bis(acetylacetonato)(nitro)(deoxyadenosine), heptahydrate[209]

(1978)

$C_9H_{12}N_2O_6$ 4-Hydroxy-1-β-D-ribofuranosyl-6-pyridazinone (3-deaza-6-azauridine)[210]

$C_{10}H_{13}N_2O_7$ 5-Methoxyuridine[211]

$C_{11}H_{14}N_2O_7S$·H_2O (Carboxymethyl)-2-thiouridine monohydrate[212] (TMAMUR)

$C_{11}H_{15}N_3O_6$ N^4-Acetylcytidine[213]

$C_{11}H_{21}N_3O_7S$ 5-(Methylaminomethyl)-2-thiouridine[214]

[($C_{10}H_8N_3$)($C_9H_{13}N_5O_7P$)Cu(H_2O)]$_2$(NO_3)$_2$·6 H_2O Adenosine 5'-monophosphate copper(II) iminodipyridine[215]

---

(208) B. E. Fischer and R. Bau, *J. Chem. Soc. Chem. Commun.*, (1977) 272–273.
(209) T. Sorrell, L. A. Epps, T. J. Kistenmacher, and L. G. Marzilli, *J. Am. Chem. Soc.*, 99 (1977) 2173–2179.
(210) B. J. Graves, D. J. Hodgson, D. J. Katz, D. S. Wise, and L. B. Townsend, *Biochim. Biophys. Acta*, 520 (1978) 229–232.
(211) W. Hillen, E. Egert, H. J. Lindner, H. G. Gassen, and H. Vorbrüggen, *J. Carbohydr. Nucleos. Nucleot.*, 5 (1978) 23–32.
(212) W. Hillen, E. Egert, H. J. Lindner, and H. G. Gassen, *Biochemistry*, 17 (1978) 5314–5320.
(213) R. Parthasarathy, S. Ginell, N. C. De, and G. Chheda, *Biochem. Biophys. Res. Commun.*, 83 (1978) 657–663.
(214) W. Hillen, E. Egert, H. J. Lindner, and H. G. Gassen, *FEBS Lett.*, 94 (1978) 361–364.
(215) K. Aoki, *J. Am. Chem. Soc.*, 100 (1978) 7106–7108.

[$(C_{12}H_8N_2)(C_{10}H_{12}N_5O_8P)Cu \cdot H_2O]_2$ Copper (II) guanosine 3'-monophosphate 1,10-phenanthroline[216]

$C_{15}H_{19}N_5O_6$ 2',3'-O-[(2-Carboxyethyl)ethylidene]adenosine[217]

$[(C_{40}H_{39}N_{14}O_{24}P_4)^{4-}(4NH_4^+)] \cdot xH_2O$ Tetra-ammonium 5'-p-adenylyl-(3',5')-thymidylyl-(3',5')-adenylyl-(3',5')-thymidine (d-pApTpApT), hydrate[218]

(216) C.-Y. Wei, B. E. Fischer, and R. Bau, *J. Chem. Soc. Chem. Commun.*, (1978) 1053–1055.
(217) D. Adamiak, M. Noltemeyer, W. Saenger, and F. Seela, *Z. Naturforsch.*, 33 (1978) 169–173.
(218) M. A. Viswamitra, O. Kennard, P. G. Jones, G. M. Sheldrick, S. Salisburg, L. Falvello, and Z. Shakked, *Nature*, 273 (1978) 687–688; *Curr. Sci.*, 47 (1978) 289–292.

AUTHOR INDEX

Numbers in parentheses are reference numbers and indicate that an author's work is referred to although his name is not cited in the text.

A

Aaronson, S., 338
Abdel-Akher, M., 353
Abrahamsson, S., 478
Abruscato, G. J., 458
Achart, A., 525
Achmatowicz, O., 177(176), 178(176), 183(176), 184(176), 185
Acton, E. M., 210(108), 211, 279(108)
Adamiak, D., 529
Adams, G. A., 48, 101(112), 355, 358, 371
Adamson, J., 196, 204(70), 231(2,68,69), 233(69), 254, 257(2,71), 261(69,71), 262(71), 263(68,69), 264(68,69,71), 265(71), 266(71), 268(2,185), 269(71), 276(71), 279(69)
Admiraal, G., 494(152), 495
Aerts, G., 279, 285(326)
Afanas'era, E. Ya., 242
Aiba, K., 522
Akada, H., 45
Akhrem, A. A., 242, 285
Albano, E. L., 204, 232(74), 233(74)
Albersheim, P., 401, 402(51)
Alblas, B. P., 420, 430, 446(35)
Allen, M. C., 401
Allerhand, A., 14, 25, 26(60), 27, 31, 40(60)
Al-Masudi, N. A., 110(17), 115
Al-Mukhtar, J. H., 508
Alpers, E., 307, 314
Al-Timari, U. S., 110(17), 115
Altona, C., 418, 492
Alviano, C. S., 67
Aminoff, D., 328, 331
Amit, B., 160(132), 165, 167(132), 169(141)
Anderson, R. A., 313, 456
Anderson, R. C., 113(30,32), 114(30), 115
Andrews, G. C., 297, 320

Andrianov, V. I., 468
Angyal, S. J., 19, 20, 432, 451
Anisuzzaman, A. K. M., 180(158), 185
Anjou, K., 319
Anno, K., 289
Ansari, A., 239
Anteunis, M. J., 279, 285(326)
Aoki, K., 508, 509(170), 527, 528
Araki, Y., 107(3,4,5,8,9), 108, 110(18,20), 111(23), 112(8,20,25,26), 115, 116(3, 4,9,20,36), 117, 121(4), 135(67), 140, 144(79), 146, 151
Araya, S., 45
Archibald, A. R., 82
Arene, F., 450
Ariaansz, R. F., 239
Arita, H., 205
Armitage, I. M., 31
Arnott, S., 80
Aseeva, N. N., 309
Ashwell, G., 296, 320(41), 324(4), 325, 331, 336, 358
Aspinall, G. O., 390, 407(5)
Atkins, E. D. T., 80
Atkins, G. R. S., 210
Atkins, H., 239
Audichya, T. D., 311
Austin, M. J., 41
Au-Young, Y., 310
Avenel, D., 431, 445
Axelson, D. E., 29
Ayer, D., 227, 228(172)
Azarnia, N., 420

B

Babcock, G. E., 53
Backinowsky, L. V., 334, 355, 371(78), 372, 373(24), 374, 400

AUTHOR INDEX

Bacon, B., 320
Baehler, B., 135(68), 140, 246, 265(251), 266(251), 272(251)
Baenziger, J., 408
Bagdian, G., 343
Baggett, N., 243
Baig, M. M., 320
Baker, B. R., 221
Baker, D. C., 181(162), 185
Balicki, R., 149
Ball, D. H., 210, 220(104)
Ballardie, F. W., 203, 206(67), 282(67), 283(67)
Ballou, C. E., 368
Balza, F., 20, 22(41), 23(41), 38(41b)
Bamford, C. H., 122
Bancher, E., 318
Banerjee, A., 506
Barbalat-Rey, F., 248, 249(254), 253(254), 257(254,264), 261(254), 268(254,264), 279(254), 285
Barber, P., 147
Barford, A. D., 162(120,121), 165, 217, 218(128), 254, 259(128), 262(128), 265(128), 266(128), 267(128), 269(128), 276(128), 279(128)
Bar-Guilloux, E., 212, 283(110)
Barker, R., 19
Barker, S. A., 295, 297
Barnett, J. E. G., 196, 199, 210, 228, 229(174), 233(3,4), 235, 237, 238, 244, 257(3), 279(4), 280(200), 281
Barnoud, F., 16, 70(18)
Barr, P. J., 512
Barragán, I., 456, 457
Barry, G. T., 338
Bartholomew, D. G., 160(133), 165, 167(133)
Barton, D. H. R., 230
Baschang, G., 359, 369
Bassieux, D., 22, 48(50), 49(50)
Batenburg, L. M. J., 425
Bau, R., 524, 527, 528, 529
Baugh, P. J., 165
Bayard, B., 409
Beau, J. M., 372, 373(124)
Beauchamp, A. L., 510
Beaudreau, C., 324(1), 325, 384(1)
Becerra, R., 234
Beck, M., 291
Becky, H. D., 255
Beecham, A. F., 298

Beevers, C. A., 451
Bekoe, A., 197
Bell, R. H., 157(128,129), 158(128), 164(128,129), 165, 187(128), 188(128), 189(184), 190
Bellamy, A. J., 479
Belleau, B., 310
BeMiller, J. N., 290, 294(12), 408
Bendich, A., 240
Benedict, C. D., 381, 387(139)
Bentley, R., 256, 319, 348
Berends, W., 288
Berger, H., 380
Berger, S., 227, 228(173)
Bergmann, E. D., 212, 218, 222(130), 243
Berking, B., 451
Berman, H. M., 296, 420, 433, 434, 449, 497, 514, 527
Bernacki, R. J., 285
Bernasconi, C., 146
Berner, H., 358, 369(98), 372(97,98), 373(97,98)
Berry, J. M., 26, 37(63), 96(63)
Bessell, E. M., 197, 204, 205(77), 207(77), 233(7), 254, 255, 270(77), 276(77), 279(77), 281(8), 282(7)
Bethell, G. S., 482
Beveridge, R. J., 142(81), 145(81), 146
Beving, H. F. G., 185
Bezsolitsen, V. P., 225
Bhat, H. B., 253
Bhattacharjee, A. K., 81, 89, 90(165), 91(166), 93(177), 356, 359, 360(101), 361(101), 363, 364(101), 375, 376(101), 383
Bhattacharjee, S. S., 59, 72, 77, 316
Bieber, M., 255
Biely, P., 240
Bilík, V., 295
Binkley, J. S., 427, 428(26)
Binkley, R. W., 129(69), 131(61,62), 132(61,62), 133(63,64), 134(63,64), 136(69), 139, 140, 175, 179(178,179), 185, 186, 187, 188(181,182,183), 189(183), 190
Binkley, S. B., 90, 238, 334, 359, 366(23), 367(23), 369, 373, 374(100), 387(100)
Binkley, W. W., 133(63,64), 134(63,64), 139, 179(178,179), 185, 186, 187, 188(181,182)
Birdsall, N. J. M., 220
Birnbaum, G. I., 486, 521

Bischofberger, K., 249, 250(255), 279(255), 321
Bishop, S. H., 25, 35(58), 42, 43(58), 72, 98(102)
Bister, W., 308, 309(109)
Björndal, H., 398, 399(41,42,43), 403(41,42)
Blackburne, I. D., 206, 207(81), 209(81)
Blacklow, R. S., 382
Blackwood, A. C., 48, 101(112)
Blaha, K., 300
Blieszner, K. C., 454, 493
Bluma, R., 200
Blunt, J. W., 20
Bobek, M., 226, 253(166), 257(166), 268(166)
Bochkov, A. F., 104
Bock, K., 17, 18(22), 31, 59, 200, 201(31,33,35,36), 203, 229(29), 230(27), 256, 258(27), 270(278), 271(278), 272(278), 273(278), 274(278)
Bode, J., 498
Boeyens, J. C. A., 251, 281(256), 475
Bogdanova, G. V., 307
Bognár, R., 308
Bond, P. J., 490
Bonenfant, A., 246, 247, 265(251,252), 266(251), 272(251), 279(252), 285
Bonner, T. G., 198
Borgegrain, R. A., 163(125), 165
Borkakoti, N., 515
Boswell, G. A., Jr., 225
Boullanger, P., 285
Bouquelet, S., 409
Bourne, E. J., 43, 295, 297
Bowers, A., 234
Boyd, J., 94
Brackenburry, J. M., 304
Bradbury, A. G. W., 136(70), 140
Brady, R. O., 390
Bragg, P. D., 315
Brauman, J. I., 392
Braun, W., 324(5), 326
Braunová, M., 225, 246(159)
Brauns, D. H., 199, 200(25), 203
Breimer, M. E., 395
Breitenbach, R., 301, 303(77), 320
Brennan, T., 486
Brewer, C. F., 78
Brezny, R., 311
Bridgman, P. W., 296
Brimacombe, J. S., 223, 224(145,146), 261(146), 263(146,150), 278(145, 146,150)
Brink, A. J., 246, 249, 250(249,255), 251, 279(255), 281(256), 475
Brisse, F., 483
Broom, A. D., 160(133), 165, 167(133)
Brown, C. J., 429
Brown, D. H., 298
Brown, G. M., 451
Bruckner, V., 121
Brunt, R. V., 212, 281
Bryce, T. A., 78
Bryson, J. A., 392
Buck, K. W., 219, 222(133), 243
Budesinsky, M., 216, 268(126)
Bugg, C. E., 458, 487, 489
Bundle, D. R., 81, 82, 85(169,172), 90(165)
Burger, M. M., 390
Burnett, L. J., 31
Burns, J. J., 320
Bursey, M. M., 297
Butchard, C. G., 204, 231, 232(73), 233(73), 283, 285
Bykhovskaya, E. G., 243
Bystrivky, S., 298

C

Cadet, J., 149, 512
Cais, M., 480
Caldwell, B. P., 371
Cameron, J. A., 338
Cameron, T. S., 462
Campbell, J. C., 203, 234(56,57), 235(56), 276(56,57)
Canet, D., 32
Capon, B., 203, 206(67), 282(67), 283(67)
Carlson, D. M., 390
Carlsson, B., 319
Caroff, M., 397
Carr, H. Y., 37
Carrell, H. L., 527
Carroll, P. M., 329
Cartwright, B. A., 527
Cary, L. W., 68
Casella, V. R., 239, 240(212)
Casu, B., 16, 17, 18(19), 77
Černý, M., 209, 216, 225, 242, 246 (159,225), 268(126), 279(122,123), 285
Cerutti, G., 318

Cesarotti, E., 285
Ceska, M., 45
Chabala, J. C., 317
Chaby, R., 332, 355(22)
Chambat, G., 16, 17, 21(28), 70(18), 96(28), 97(28)
Chan, A., 316
Chang, C. D., 163(127), 165
Chanzy, H. D., 462
Chapman, O. L., 174
Charon, D., 328(96), 329, 331(17), 332(17), 334(21), 337(17), 355(22), 356, 358, 367, 368(113), 371(96), 373(25), 374(79), 375(79)
Chatterjee, I. B., 320
Chauhau, M. S., 155
Chaykovsky, M., 391, 392(10)
Chemyshev, V. A., 298
Chen, E. C. M., 43
Chen, L.-Y., 181(163), 185
Cherniak, R., 346, 360, 369(102), 383(102), 387(102)
Chheda, G., 528
Chiang, C. C., 523
Chien, J. C. W., 35, 79
Chiklis, C. K., 102
Childers, R. F., 25, 26(60), 40(60)
Childs, R. F., 212
Chittenden, G. J. F., 294
Chiu, M. L., 74
Chizhov, O. S., 14, 19, 27(3), 68, 253, 254(265), 398, 400
Chmurny, G., 297
Choong, W., 197, 280, 285
Christensen, J. E., 210(108), 211, 279(108)
Christian, R., 369, 388(118)
Christl, M., 97
Christman, D. R., 239
Chu, S. S. C., 467
Chujo, R., 48
Chuvileva, A. I., 307
Chwang, A. K., 518
Clardy, J., 526
Clark, E. R., 298
Clark, G. R., 505, 508, 509(170)
Clark, J. C., 238
Cléophax, J., 285
Cleveland, P. G., 174
Clode, D. M., 177(164), 181(164), 182(169,170), 183(173,174)
Cluskey, J. E., 41

Cochran, T. W., 402, 405(52a)
Codington, J. F., 240, 279(222)
Cohen, S., 212
Cohen, S. G., 103
Cohen, S. S., 355, 378
Collins, P. M., 107(10,11), 108, 123, 125(46,47,51–53), 126(47,48), 127(48), 130(58,59), 131(58), 134(65), 137(75), 138(75), 139(75), 140, 141(75), 155, 161(116,118), 165, 168, 169(147–149), 170(147,148), 171(147–149), 172(147,149), 173(149)
Colonna, S., 285
Colson, P., 20, 22(43), 24, 25(55), 26, 35(55), 38(55), 40, 45(55), 46(55), 47(55), 48(55), 68, 98(55), 101(55)
Compton, J., 305
Conigliaro, P. J., 252
Conrad, H. E., 289, 297, 299(6), 315
Conway, T. F., 35, 71(86), 72(86)
Cook, A. F., 527
Cook, W. J., 458
Cooper, A. B., 173
Copeland, C., 321
Cordes, R. E., 462
Corey, E. J., 391, 392(10)
Cornforth, J. W., 329, 360(18), 365, 374(18)
Cottier, L., 146
Cox, D. D., 68
Cox, E. C., 429
Cox, P. J., 476
Coxon, B., 274, 294
Craciunescu, E., 168
Craig, D. C., 280
Craig, S. W., 390
Crawford, S. A., 288, 296(1)
Crawford, T. C., 288, 296(1), 297, 301, 303(77), 305(86), 320
Cremer, D., 418
Criddle, W. J., 147, 149
Cron, M. J., 311
Croon, I., 101
Cross, A. D., 227, 228(173)
Csaszar, J., 121
Cuadriello, D., 227, 228(173)
Cuatrecasas, P., 390
Cushley, R. J., 240, 279(222)
Cuthbert, M. W., 203, 206(67), 282(67), 283(67)
Cynkin, M. A., 336

Cyr, N., 16, 18, 20, 22(41), 23(41), 38(41b), 69(8), 77
Czarniecki, M. F., 32
Czugler, M., 470

D

Daane, W. M., 158(130), 165
D'Addieco, A. A., 295, 301(36)
Dässler, C.-G., 306, 311(99)
Dahlen, B., 478
Dale, J. K., 314
Daniel, P. F., 252
Daniels, P. J. L., 285
Danilov, S. N., 242
Danks, L. J., 230
Datta, R. K., 149
Davis, M., 334, 366(23), 367(23)
Davison, B. E., 479
Davydova, L. P., 373
Dawson, G., 255, 394
De, K. K., 309
De, N. C., 528
Dearie, W. M., 203, 206(67), 282(67), 283(67)
de Bie, M. J. A., 100
deBoer, W. R., 82, 88(168)
DeBruyn, A., 279, 285(326)
DeBruyn, C. K., 285(326)
Defaye, J., 189(184), 190, 212, 283(110)
DeGraff, R. A. G., 495(152), 495, 516
DeJongh, D. C., 255
de Kleijn, J. P., 238, 239
deKok, A. J., 492
deLeeuw, H. P. M., 492
DeLey, J., 357
DeMember, J. R., 102
Denot, E., 234
Depmeier, W., 472, 498
Descotes, G., 146, 206, 242(82), 254(82), 285
Deshayes, H., 142
DeVilliers, O. G., 246, 250(249)
Diana, G. D., 205
Dimitrijevich, S. D., 202, 204(49), 232(49)
Dische, Z., 335, 337, 357
Djerassi, C., 480
Dmitriev, B. A., 334, 355, 371(78), 372, 373(24), 374, 400
Dmytraczenko, A., 134(66), 140
Doddrell, D., 27, 31
Doerr, I. L., 240

Doesburg, H. M., 449
Doležalová, J., 246
Donaldson, B., 204, 265(76), 266(76), 269(76), 270(76)
Doner, L. W., 130(57), 139, 149, 181(165), 185
Dorland, L., 363, 382
Dorman, D. E., 19, 20, 24(47), 25(47), 37, 38, 49(35), 97
Doudoroff, M., 348, 357
Drašar, P., 216, 268(126)
Drew, J., 387
Driguez, H., 212, 283(110)
Dröge, W., 333(26), 334, 339, 343, 344(26), 345, 346, 348, 350, 355(26), 357
Drolshagen, G., 203
Ducruix, A., 121, 454, 479, 493, 499(147)
Dung, T., 285
Durette, P. L., 274
Dutton, G. G. S., 16, 17(17), 21(17), 26, 37(63), 96(17,63)
Dwek, R. A., 203, 204, 232(72), 233(72), 234(56,57), 235(56), 245, 257(72), 261(281), 262(281), 268(72), 276(56, 57), 283
Dworsky, P., 296, 320(40)
Dzhikiya, O. D., 68

E

Eagles, J., 392
Edstrom, R. D., 324(1), 325, 384(1)
Egert, E., 488, 490(138), 507, 527, 528
Einstein, F. W. B., 313, 456
Eitelman, S. J., 121, 454
El Ashmawy, A. E., 173
El Khadem, H. S., 311
Ellwood, D. C., 338
El Sawl, M. M., 205, 221(78)
Emoto, S., 467
Emsley, J. W., 256
Engels, J., 497
Epps, L. A., 528
Erbing, B., 203
Erbing, C., 410
Erlander, S. R., 37
Evans, M. E., 291
Eveleigh, D. E., 58
Evelyn, L., 207, 208(99), 227(89), 228(89), 229(89,170), 257(170), 276(89,170), 277(89)
Excoffier, G., 18

F

Faillard, H., 252
Falco, E. A., 117
Falvello, L., 529
Farkas, I., 308
Farmer, D. W. A., 220, 221(139)
Fauland, E., 308
Fay, I. W., 294
Feather, M. S., 209
Ferguson, G., 480
Ferrier, R. J., 307, 318, 482
Fields, R., 256
Fiete, D., 408
Filby, W. G., 147
Finlayson, A. J., 67
Finne, J., 392, 393, 395, 396, 397(31), 398(31), 399(31), 400(31), 403(50), 404(50), 405(31), 406, 408, 409, 410(65), 411(74), 412(65), 413(80), 414(54,55,74), 415(74,80)
Firth, M. E., 329, 360(18), 365(18), 374(18)
Fischer, B. E., 524, 528, 529
Fischer, E., 289, 292, 293, 294
Fischer, H., 308
Fischer, H. O. L., 290, 294(12), 358
Fischer, J. C., 441, 472(60)
Fisher, B. E., 41
Fisher, D., 227, 229(171), 234
Fishman, P. H., 390
Fletcher, H. G., Jr., 304, 315(89)
Fondy, T. P., 218, 222(131), 279(131)
Foster, A. B., 162(120,121), 165, 196, 197, 199, 204(70), 205(77), 207(77), 210(106), 211, 217, 218(128), 219(106), 220, 222(106,133), 223, 224(145,146), 231(2,68,69), 233(69), 243, 244(88), 253(137), 254(137,265), 255, 257(2,71,144), 259(128), 261(69, 71,146), 262(71,75,128), 263(68,69, 144,146,150), 264(68,69,71,137), 265(71,75,76,128,144), 266(71,75,76, 128,144,296), 267(128,137,296), 268(2), 269(71,75,76,128,296), 270(76,77), 276(71,75,77,128), 278(144,145,146,150), 279(69,77,128, 144), 281(8), 285
Foster, R. L., 203
Fournet, B., 398
Fourrey, J.-L., 155

Fowler, D. L., 219
Fowler, J. S., 233, 239, 240(212), 245, 285
Fox, J. J., 213, 224, 240, 245(151), 279(222)
Frank, N., 38, 41(95), 48(95), 98(95)
Franke, F., 358, 369(98), 372(98), 373(98)
Fraser, C. G., 47
Fraser, R. R., 243
Fraser-Reid, B., 113(27,28,29,30,31,32), 114(27,28,29), 115, 231, 235(186)
Fredericks, P. M., 206, 207(81), 209(8)
Freeman, M. R., 235
Freeman, R., 30, 31
Freestone, A., 470
Freibolin, H., 38, 41, 48(95), 98(95)
Fric, I., 300
Frush, H. L., 203, 315
Fu, Y. L., 226, 253(166), 257(166), 268(166)
Fujieda, K., 16, 17(9), 68(9), 71(9)
Fujimoto, M., 16, 68(14)
Fujimoto, Y., 131(60), 139
Fukuda, M. N., 416
Fuller, N. A., 346, 348(58,59), 355(59)
Fulmor, G. E., 196, 198(1)
Fung, W.-M., 385, 386(151)
Furburg, S., 501, 504
Furneaus, R. H., 307
Furuhata, K., 321

G

Gaertner, K., 207
Gagnaire, D. Y., 17, 18(26), 22(26), 23, 24(51), 25(51), 38(52b), 41, 48(50), 49(50), 98
Galanos, C., 384
Galkowski, T. T., 305
Gallagher, B. M., 239, 245
Ganapathi, K., 338
Ganguly, A. K., 230, 443
Gansow, O. A., 17
Gantt, R., 238
Garegg, P. J., 189(185), 190
Gartland, G. L., 489
Garzia, A., 318
Gassen, H. G., 527, 528
Gatehouse, B. M., 434, 435
Gatt, S., 413
Gatti, G., 77

Gaykema, W. P. J., 425, 445
Gelas, J., 212, 283(110)
Gelbard, G., 285
Gent, P. A., 223
Gentile, B., 253, 257(264), 268(264), 285
Gero, S. D., 285
Gerster, R., 192, 193(190)
Ghalambor, M. A., 335, 336(29), 337(29, 31), 348(31), 358, 364(89), 365, 366(29,31), 371, 374(29,31), 375(29, 31), 379, 380, 381, 382(140), 386(89)
Gibson, A. R., 180(160), 185
Gill, M., 37
Gillier-Pandraud, H., 426
Ginell, S. L., 516, 528
Ginocchio, S. D., 316
Girling, R. L., 428
Glaudemans, C. P. J., 71, 72(144)
Glushko, V., 27
Glusker, J. P., 527
Gnüchtel, A., 207
Godinho, L. S., 230
Goebel, W. F., 338
Goldstein, J. E., 259
Goodgame, D. M. L., 527
Goodman, L., 210(108), 211, 279(108)
Goodman, R. A., 25, 26(60), 40(60)
Gootz, R., 201
Gorbatii, L. V., 468
Gorbunov, B. N., 225
Gordon, H. T., 318
Gorin, P. A. J., 16, 17(15), 19(15), 20, 21(11,44,45), 22(15), 23, 24(44), 25(11,53), 33(57), 40, 48, 49(99), 52, 53, 54(12,39,44), 56(11,12,39,44,53, 61), 57(12), 58(57), 59(12), 60(12,16), 62(16), 64(12,53), 65(53), 66, 67, 68(15,125), 70(15), 77, 82(155), 86(155), 87(114), 95(11), 96(38), 97(125), 103(99)
Gorin, S. E., 68
Gotschlich, E. C., 84
Gottschalk, A., 329, 360(18), 365(18), 374(18)
Gould, R. O., 451
Goulding, R. W., 238
Graham, J. D., 24, 66(54)
Granath, K., 45, 410
Grand, A., 512
Grant, C. W. M., 283
Grant, D. M., 31

Graves, B. J., 528
Gray, B. E., 295
Green, B. S., 121
Green, M. J., 227
Greenwald, J., 192
Grenier, M. F., 146
Gress, M. E., 448, 449
Greve, W., 312, 313
Grieg, C. G., 252
Griffin, H. L., 37
Griffiths, W. C., 305
Grimmonperez, L., 409
Grindley, T. B., 462, 463
Grinsteins, V., 300
Grob, C. A., 329
Gross, B., 163(125), 165
Grützmacher, H. F., 398
Gschwendter, W., 256, 273(280)
Gürtler, P., 49
Guggenheim, B., 45
Guilloux, E. R., 189(184), 190
Gullyev, N., 68
Gulyas, E., 298
Gunther, C., 460
Gupta, A., 477
Gupta, D. S., 360, 369(102), 383(102), 387(102)
Gupta, P., 123, 125(46), 126(54), 127, 130(58,59), 131(58), 139
Guss, J. M., 80
Guthrie, R. D., 206, 207(81), 209(81), 479
Gutterson, N. I., 245

H

Haas, C., 103
Hachiya, K., 297, 304(51)
Hackenthal, E., 338, 339
Hadfield, A. F., 212, 219(112), 222(112), 253(111,112), 255(112), 283(111), 285
Hagaman, E. W., 21
Hagler, A. T., 121
Hahn, E. L., 37
Haines, A. H., 221
Hakomori, S.-I., 353, 357(68), 390, 391(9), 396, 398(29), 416
Halásková, J., 209
Hall, F. H., 263
Hall, L. D., 16, 20(7), 26, 31, 37(63), 96(63), 165, 202, 203(41,43), 204, 205(77), 207(77), 208(99), 217,

218(128), 221, 222(131), 223, 224(145,146), 227(89), 228(89), 229(89,170), 231(68,69), 233(69), 234(43,53), 235(43,53), 236(43), 237(43,53,196), 257(42,66,71, 144(170), 258(42,43,196,285), 259 (128), 260(42,43), 261(43,69,71,146), 262(42,66,71,75,128,283,286), 263(66, 68,69,141,144,146), 264(66,68,69, 71), 265(71,75,76,128,144,292,293), 266 (71,75,76,128,144,295,296), 267(128,296), 268(41), 269(70,71,75, 76,128,296), 270(76,77), 274(284,285), 275(42,284,285), 276(42,43,71,75,77, 89,128,170,196,283,285,286,292,295), 277(42,89), 278(141,144,145,146, 295), 279(69,77,128,131,144) 280(43), 283, 285, 362
Hall, R. H., 321, 475
Halmann, M., 192, 193(188,189)
Hamer, G. K., 17, 20, 22(41), 23(41), 38(41b), 76(25), 77(25)
Hamer, T. A., 491
Hamilton, W. C., 494
Hämmerling, G., 355
Hammerschmid, F., 339, 356, 363(81), 377(81), 380, 387(81)
Hammonniere, M., 479
Hamor, T. A., 512, 513
Hanessian, S., 199, 255
Hanninen, O., 318, 320
Hansen, L. K., 431
Hansson, G. C., 395
Hardegger, E., 162(122), 165
Hardingham, T. E., 80
Harrison, A. S., 204, 232(72), 233(72), 257(72), 268(72)
Harrison, R., 210(107), 211
Hascall, V. C., 35, 37(85), 80(85)
Hase, S., 396, 397(32), 398(32), 399(32)
Haskins, R. H., 23, 25(53), 52, 56(53), 64(53), 65(53)
Hasson, M. A., 35, 37(85), 80(85)
Hata, T., 522
Haverkamp, J., 100, 363, 382, 397, 398(34), 399(34,47), 401(34)
Hawkinson, S. W., 485
Haworth, W. N., 302, 352
Hay, G. W., 177(175), 183(175), 184(175), 185, 318, 346, 350(60)

Hayday, K., 146
Haynes, L. J., 199
Haynes, W. C., 41
Heath, E. C., 324(1), 325, 335, 336(29), 337(29,31), 346, 348(31), 355(59), 358, 364(89), 365, 366(29,31), 371, 374(29,31), 375(29,31), 379, 380, 381, 382(140), 384, 386(89)
Hecht, H. J., 437
Hecht, S. H., 291
Hehemann, D. G., 133(63,64), 134(63, 64), 139, 187, 188(183), 189(183)
Heidt, L. J., 147
Heins, A., 310
Heiske, D., 256, 273(280)
Helbig, R., 308
Helferich, B., 104, 107(7), 108, 201, 207, 220(84), 241, 307
Hellerqvist, C. G., 398, 399(43)
Hellwig, E., 358, 369(98), 372, 373(97, 98)
Hemmerich, P., 149
Hemminki, K., 406, 414(55)
Hems, R., 207, 210(106), 211, 219(106), 220, 222(106,133), 223, 224(145,146), 244(88), 253(137), 254(137), 257(144), 261(146), 263(144,146,150), 264(137), 265(144), 266(144,296), 267(137,296), 268, 269(296), 278(144, 145,146,150), 279(144)
Henbest, H. B., 219
Henrissat, B., 212, 283(110)
Hensley, D. E., 53
Hermann, K. M., 367
Hershberger, C. S., 334, 359, 366, 367(23), 369, 373, 374(100), 387(100)
Hess, K., 242
Hesse, R. H., 204, 230, 231(68,69), 233(69), 249(181), 261(69), 263(68, 69), 264(68,69), 279(69)
Heyns, K., 126(55), 127, 290, 291(13), 307, 314, 398
Hickman, J., 324(4), 325
Hicks, D. R., 113(27,29,30,32), 114(27,29, 30,32), 115
Higuchi, T., 449
Hill, H. D. W., 30, 31
Hill, J. C., 339
Hillen, W., 527, 528
Himmen, E., 241

Hingerty, B., 490
Hinoide, M., 45
Hirasaka, Y., 288, 298(3)
Hirose, K., 38, 97(93)
Hirotsu, K., 449, 526
Hirschmann, R., 253
Ho, C., 385, 386(151)
Hochmann, J., 29
Hockett, R. C., 294, 312
Hodges, L. C., 401
Hodgson, D. J., 486, 488, 528
Höfle, G., 374
Hoffman, J., 189(185), 190
Hoffmann-Ostenhof, O., 296, 319(40)
Hofstad, T., 355
Hoganson, E. D., 174
Hoge, R., 313, 456
Holder, N. L., 113(27,28), 114(27,28), 115
Holland, C. V., 274
Holton, S. L., 215
Honda, S., 75, 76, 79(151)
Honma, T., 202
Hoogendorp, J. D., 510, 516
Hoppen, V., 256, 273(280)
Hordvik, A., 419, 428, 431
Horecker, B. L., 336
Horitsu, K., 54
Horning, E. C., 319
Horsley, W. J., 30
Horton, D., 109(12,13), 115, 121, 151, 152(109), 153(109), 157(128,129), 158(128), 164(128,129), 165, 173, 177(164), 178(177), 180(157,159), 181(162,164), 182(169,170,171), 183(172–174), 185, 187(128), 188(128), 189(184), 190, 274, 441, 454, 472(60), 493
Hoski, Y., 299, 300
Hough, L., 211, 212, 222, 253(111), 283(111), 285, 315, 318
Howarth, G. B., 174
Hrinak, J., 242, 246(225)
Hribar, J. D., 255
Huber, C. P., 486
Huber, H., 31
Hubert, J., 510
Hudson, C. S., 299, 304, 315(89)
Hukins, D. W. L., 80
Hullar, T. L., 163(127), 165

Hulyalkar, R. K., 294, 302(26), 303(26), 314(26), 316
Hunt, B., 245
Hunt, D., 348, 350(64)
Hunter, F. D., 433
Hurwitz, J., 327, 328(11), 336(11), 337(11), 341(11), 357(11), 366(11), 369(11), 374(11), 381(11)
Hutcheon, W. L. B., 439, 488(57), 490
Hvoslef, J., 421

I

Ibanes, L. C., 234
Ido, T., 233, 239, 240(212)
Igarashi, K., 202
Igarashi, O., 49
Iitaka, Y., 321
Ikan, R., 243
Ikehara, M., 160(134–137), 165, 167(134), 168(135,136,137)
Iley, D. E., 113(29), 114(29), 115
Im, K. S., 131(60), 139
Imai, S., 45
Imai, Y., 288, 298(3)
Imakura, Y., 136(72,73), 140
Imanari, T., 319
Inai, T., 144(77,78), 146
Inch, T. D., 256
Inglis, G. R., 372
Inokawa, S., 109(15), 110(19), 111(19), 112(19), 115, 526
Inoue, Y., 48
Inouye, Y., 396, 397, 415(30), 416(30)
Irani, R. J., 338
Ireland, R. E., 306, 316(98)
Irisawa, J., 202
Isaac, D. H., 80
Isaacs, N. W., 478
Isbell, H. S., 290, 293, 294(12), 315, 316
Ishidate, M., 288, 298, 301, 302(76), 303(76), 304(76)
Ishido, Y., 107(3,4,5,8,9), 108, 110(18,20), 111(23), 112(8,20,25,26), 115, 116(3, 4,9,20,36), 117, 121(4), 135(67), 140, 144(79), 146, 151
Ishikura, K., 300
Ishizuka, I., 414
Ito, S., 416
Itoh, T., 302, 317(82)

Iwadare, K., 302
Iwashita, S., 396, 415(30), 416(30)
Iyer, R., 123, 125(46,47), 126(47), 130(58), 131(58), 139
Iyer, R. N., 390

J

Jackson, W. R., 219
Jacobsen, S., 200, 201(32), 244, 263(32), 266(32)
Jacobson, R. A., 446, 449(69), 450(69)
Jaenicke-Zauner, W., 124
Jain, S. C., 501, 502, 527
James, M. N. G., 439, 488(57), 490
James, V. J., 428, 432, 451, 452, 466
Jann, B., 350, 353(65), 386(65)
Jann, K., 350, 353(65), 357(67), 386(65)
Jansson, P.-E., 399, 406(48)
Jaques, L. B., 72
Jarchow, O., 442, 460
Jarman, M., 253, 254(265), 255, 285
Järnefelt, J., 403(75), 410, 411(75), 416(75)
Jarrell, H. C., 26, 35, 45, 71, 72(86), 182(166,167), 185
Jarý, J., 300, 316
Jeanes, A., 41, 43, 72
Jeanloz, R. W., 398, 399(45)
Jeeves, I., 527
Jefferis, R., 243
Jeffrey, G. A., 417, 418, 419, 423, 424, 425(21), 427, 428(26), 429(6), 430, 432, 433, 434, 435(48), 445, 448, 449, 452, 466, 467, 482, 494(151), 495
Jenkins, I. D., 180(161), 185, 198,236(9, 10), 237(9,10)
Jennings, H. J., 274, 14, 20, 22(43), 24, 25(4,55), 26(4), 35(55), 38(55), 40(55), 42, 45(55), 46(55), 47(55,94), 48(55), 71, 72(144), 81, 82, 85(169,172), 89, 90(165), 92(166), 93(177), 98(55), 101(55), 356, 359, 360(101), 361(101), 363(101), 364(101), 375(101), 376(101), 384(101)
Jensen, E. V., 242
Jensen, L. H., 466
Jensen, S. R., 200, 201(32), 263(32), 266(32)
Jewell, J. S., 109(14), 115, 119(14), 151, 152(109), 153(109), 274

Jiménez-Garay, R., 458, 459
Jochims, J. C., 308, 472
Johansson, I., 215, 305
Johnson, L. F., 16, 20(7), 72
Johnson, R. N., 162(121), 165, 204, 205(77), 207(77), 217, 218(128), 231(69), 233(69), 257(71), 259(128), 261(69,71), 262(71,75,128), 263(69), 264(69,71), 265(71,75,128), 266(71, 75,128), 267(128), 269(70,71,75,128), 270(77), 276(71,75,77,128), 279(69, 77,128)
Jones, B., 149
Jones, D. L., 235
Jones, J. K. N., 174, 294, 302(26), 303(26), 314(26)
Jones, L. B., 158(130), 165
Jones, P. G., 529
Jones, R. G., 360, 369(102), 383(102), 387(102)
Jonssen, E. K., 84
Jordaan, A., 246, 249, 250(249,255), 251, 279(255), 281(256), 475
Jorns, M. S., 149
Joseleau, J.-P., 16, 17(17), 21(17,28), 70(18), 96(17,28), 97(28)
Jouin, P., 155

K

Kadentsev, V. I., 254
Kärkkäinen, J., 392, 393, 397, 398(33), 399(33), 401(33), 407, 409, 414(33)
Kahl, D. C., 392
Kaindl, K., 318
Kainosho, M., 121, 485
Kalckar, H. M., 334, 335(27)
Kaltenbach, U., 366
Kalvoda, J., 227
Kamerling, J. P., 372, 373(124), 397, 398(34), 399(34,47), 401(34)
Kamiya, K., 463
Kan, L.-S., 149
Kanters, J. A., 420, 422, 425, 430, 431, 432, 436, 445, 446(35), 449
Kaplan, L., 240
Karabatsos, G. J., 24, 66(54)
Karabinos, J. V., 294
Karlsen, J., 473
Karlsson, K.-A., 390, 393, 394, 395

AUTHOR INDEX

Kasai, N., 447
Kass, E. H., 324(7), 326
Katz, D. J., 528
Katzin, L. I., 298
Kavai, I., 226, 253(166), 257(166), 268(166)
Kavanagh, L. W., 72
Kawabata, H., 311
Kawada, M., 310
Kefurt, K., 300
Kefurtova, Z., 300
Keilich, G., 38, 41(95), 48(95), 98(95)
Keiser, H., 78
Kekomäki, R., 410
Kelly, D. E., 458
Kelly, S., 320
Kennard, O., 529
Kenne, L., 399, 406(48), 410
Kenny, C. P., 81, 82, 89, 90(165), 91(166), 93(177), 356, 359, 360(101), 361(101), 363(101), 364(101), 375(101), 376(101), 384(101)
Kent, P. W., 196, 199, 202, 203, 204(49), 207, 208, 210(85,86), 219(94), 220, 221(139), 227, 229(171), 231, 232(49, 72,73), 233(3,4,72,73), 234(54,55,56, 57), 235(54,55,56), 237, 238, 243, 244, 245, 257(3,72), 261(281), 262(281), 268(72), 270(94), 276(56,57), 279(4, 94), 280(200), 281(54,55), 283, 285
Kershaw, K., 165
Keyanpour-Rad, M., 210(108), 211, 279(108)
Kharasch, N., 156
Khardin, A. P., 225
Khodair, A. I. A., 156
Khorlin, A. Ya., 208, 253(93), 276(93), 300(75), 301
Kiburis, J., 285
Kilbourne, H. W., 237
Kiliani, H., 343, 347(54)
Killean, C. G., 430
Kim, H. S., 424, 432, 433, 434, 494(151), 495
Kim, K. S., 142(81), 145(81), 146
King, C. G., 320
King, R. R., 68
Kirby, P. T., 204, 232(72), 233(72), 257(72), 268(72)
Kirfel, A., 477

Kirsanov, A. V., 225
Kirschenlohr, W., 308
Kisel, Ya. M., 243
Kissman, H. M., 207
Kistenmacher, T. J., 500, 523, 528
Kitagawa, I., 131(60), 136(71-73), 139
Klaska, K.-H., 442, 472
Klaska, V. R., 460
Klein, M. P., 30
Klemer, A., 199, 200(16), 202, 203
Kloss, J., 311
Knapp, R. D., 25, 35(58), 40, 42, 43(58), 72, 98(102), 104(98)
Knorr, E., 378
Knox, L. H., 227, 228(173)
Knunyants, I. L., 243, 244
Knutson, C. A., 318
Kobata, A., 390, 396, 398, 399(46), 405(46), 415(30), 416(30)
Kobayashi, M., 44
Koch, H. J., 16, 19, 20, 22, 23(41a), 38(41b), 69(8)
Koch, K. F., 21
Kochetkov, N. K., 68, 104, 334, 355, 371, 372, 373(24), 398, 400
Kock, A. J. H. M., 436
Kocon, R. W., 305
Koebernickj W., 114(33), 115, 442
Köll, P., 421, 442
Köllnerová, Z., 285
Koenigs, W., 378
Kohama, T., 16, 68(14)
Kohn, B. D., 301, 303(83), 316(83)
Kohn, P., 302, 303(83), 304, 316(83)
Koholic, D. J., 175, 190
Koide, N., 396, 415(30), 416(30)
Kojic-Prodić, B., 463, 464, 465(99)
Kokesh, F. C., 367
Komatsu, S., 299
Komoroski, R. A., 14, 27, 31
Konings, J., 288
Koops, T., 449
Kopf, V. J., 421
Koroteev, M. P., 244
Korsch, B., 16, 69(8)
Korytnyk, W., 226, 276(165), 285
Kosaza, K., 45
Kotelko, K., 343
Kothe, V. G., 202, 280(46), 285, 526
Kotō, S., 162(119), 165

Kotowycz, G., 256
Kouollik, G., 285
Kovać, P., 311
Kovollik, G., 207, 285
Krasyuk, E. I., 285
Kratky, C., 356, 359(82), 364, 365(82), 376(82)
Krátký, Z., 240
Kraus, C. A., 219
Kripach, N. B., 242
Kroon, J., 446
Kruk, C., 82, 88(168)
Krusius, T., 392, 393, 395, 399, 403(50), 404(50), 406, 408, 409, 410(65), 411(74), 412(65), 413(80), 414(54,55,74), 415(74,80)
Kruyssen, F. J., 82, 88(168)
Kucàr, S., 298
Kuge, T., 38, 97(93)
Kuhl, D. E., 239, 240(212)
Kuhlmann, K. F., 31
Kuhn, R., 308, 309(109), 327, 328(12), 331, 359, 363, 369, 375, 376(109), 378(109), 383(109), 409, 414
Kuhn, S., 243
Kumamoto, K., 110(19), 111(19), 112(19), 115
Kundu, S. K., 398
Kung, W.-J., 485
Kuniaka, A., 16, 68(14)
Kunieda, T., 312
Kuranari, M., 311
Kuroda, Y., 422
Kushida, K., 112(25), 115
Kvaratskheliya, L. D., 300(75), 301
Kwok, R. P., 210(108), 211, 279(108)

L

Lade, R. E., 285
LaForge, F. B., 244, 294
Laine, R. A., 401, 403(75), 410, 411(75), 416(75)
Laird, W. M., 392
Lambert, J. B., 259
Lamberts, B. L., 26, 45
Lancaster, J. E., 196, 198(1)
Lance, D. G., 174
Landy, M., 324(5), 326
Langen, P., 207, 285
Langridge, R., 490

Lanning, M. C., 355
Lapeyre, M., 16, 17(17), 21(17,28), 96(17,28), 97(28)
Lapokowski, M., 145(80), 146
Lapper, R. D., 85
Larsen, J. W., 198, 200(15), 233(15)
Larsson, K., 319
Lasker, S. E., 74
Lavrova, K. F., 309
Lawrence, J. L., 430
Lazasini, F., 437
Leaback, D. H., 252
Leban, I., 440
Ledeen, R. W., 382, 398
Lee, J. B., 205, 221(78)
Lefebvre, A., 17, 21(28), 96(28), 97(28)
Leffler, H., 395
Lehmann, V., 333(26), 334, 339(26), 343(26), 344(26), 345(26), 346(26), 348(26), 350(26), 355(26), 357(26), 384, 385, 386
Leigh, J. S., Jr., 30
Leive, L., 324(6), 326
Lemieux, R. U., 18, 231, 235(186), 256
Lemire, A. E., 239
Lenard, J., 198, 200(12), 201(12), 281(12)
Lerner, L. M., 301, 303(83), 304, 315(85), 316(83,90,91)
Leroy, Y., 398
Lessie, T., 338
Leung, F., 462
Levene, P. A., 209, 244, 294, 296, 299(30,45), 305, 311
Levin, D. H., 326, 336(9), 337(9), 357, 367(9), 379, 380(9)
Levine, E. M., 336, 337(31), 348(31), 358(31), 365(31), 366(31), 371(31), 374(31), 375(31)
Levy, D., 212
Levy, G. C., 14, 27(2), 29, 31, 32
Lewis, B. A., 318, 346, 350(60)
Li, S.-C., 413
Li, B. W., 402, 405(52a)
Li, Y.-F., 263
Li, Y.-T., 403(75), 410, 411(75), 413, 416(75)
Lie, B., 509
Liedgren, L., 399, 406(48)
Lim, R., 378
Lindberg, B., 189(185), 190, 203, 215, 305, 390, 398, 399(41,42,43), 400(7),

402(7), 403(41,42), 406(48), 407(7), 408(59), 409(59), 410(59), 411(7), 412(7,59)
Link, K. P., 347
Linn, C., 304
Linn, C. B., 310
Lichtenthaler, F. W., 481
Lindberg, K. B., 470, 481
Lindner, H. J., 488, 490(138), 507, 527, 528
Liska, F., 229
Littmann, O., 242
Liu, T. Y., 84
Live, D. H., 31
Llewellyn, F. J., 429
Lloyd, K. O., 16, 60, 61(16), 62(16)
Lloyd, W. J., 210(107), 211
Loebenstein, W. V., 219
Loen, E. H., 494(152), 495
Lönngren, J., 390, 398, 399, 400(7), 402(7), 406(48), 407(7,35), 408(59), 409(59), 410(59), 411(7), 412(7,59)
Loewus, F., 319, 320
London, R. E., 19
Longchambon, F., 426, 450
Lopes, D. P., 285
López-Castro, A., 456, 457, 458, 460
Louie, S., 527
Lüderitz, O., 324(2,3), 325, 333(26), 334, 339(26), 343(26), 344(26), 345(26), 346(26), 348(26), 350(26), 355(26), 357(26), 384
Luetzow, A. E., 180(159), 182(171), 183(172), 185
Luger, P., 202, 280(46), 285, 436, 483, 526
Lukacs, G., 33
Lundt, I., 17, 59(21), 200, 201(30), 202, 203(44,45), 229(30), 230(177), 275(45), 280(45)
Lutz, P., 327, 328(12), 331, 363, 375(109), 376(109), 378(109), 383(109)
Lyerla, J. R., 27

M

McBee, E. T., 237
MeCapra, F., 480
MacDonald, D. L., 363, 375(109), 376(109), 378(109), 383(109)
McDonald, G. G., 30
McGee, J., 348
MacGregor, R. R., 239, 245, 285
McGuire, E. J., 90
Mackay, M. F., 438
McLaren, L., 372
McKelvey, R. D., 146
Mackie, K. L., 26, 37(63), 96(63)
McKinnon, A. A., 78
McLennan, T. J., 482
McManus, S. P., 198, 200(15), 233(15)
McMullan, R. K., 434, 435(48)
McNeil, M., 401, 402(51)
McPhail, A. T., 227, 443
MacPherson, J., 298
Madison, R. K., 311
Maeda, S., 110(18), 115, 144(79), 146
Mäkelä, P. H., 355
Makinen, M. W., 478
Makita, M., 319, 348
Maluszynska, H., 425
Mancier, D., 23, 38(52b)
Mantsch, H. H., 85
Manville, J. F., 202, 203(43), 234(43,53), 235(43,53), 236(43), 237(43,53,196), 257(42,66), 258(42,43,196,285), 260(42,43), 261(43), 262(42,66,283, 286), 263(66), 264(66), 265(292,293), 266(296), 267(296), 269(296), 274(284,285), 275(42,284,285), 276(42,43,196,283,285,286,292), 277(42), 280(43)
Maradufu, A., 222
Marchessault, R. H., 449, 462
Marcu, D., 514
Marcus, D. M., 220, 231, 254(138), 267(138), 268(185), 279(138)
Marinetti, G. V., 338
Markley, J. L., 30
Markous, R. A., 454, 493
Markovskii, L. N., 225
Marniemi, J., 320
Marquez, R., 456, 457, 458, 460
Marsh, R. E., 485
Marshall, J. J., 407, 412(58)
Martens, F. B., 516
Mårtensson, E., 406
Martin, A., 81, 89, 90(165), 93(177)
Martin, J.-C., 206, 242(82), 254(82), 285
Martin, M.-A., 510
Martin, O. R., 247, 272(253), 285
Marzilli, L. G., 523, 528

Masuda, K., 40, 44(96), 49(96), 52(96)
Mathews, M. B., 80
Matsuda, A., 152(111), 154(111,112), 155
Matsuda, K., 16, 17(9), 21, 22(48), 23(48), 38(48), 39(48), 44(48), 46(48), 48(48), 49(48), 52(48), 68(9), 71(9)
Matsuda, Y., 149
Matsui, M., 301, 302(76), 303, 304(76), 319
Matsumura, G., 416
Matsunaga, I., 296, 319(46)
Matsushima, R., 144(77,78), 146
Matsushima, Y., 205
Matsuura, K., 106, 107(3,4,5,8,9), 108, 110(18,20), 111(23), 112(20,25,26), 115, 116(3,4,9,20,36), 117, 121(4), 144(79), 146, 151
Matsuura, T., 149
Matwiyoff, N. A., 19
Maury, P., 409
Mayer, H., 384
Mayer, R. M., 324(1), 325, 384(1)
Mazurek, M., 16, 17(15), 19(15), 22(15), 24, 26, 33(57), 40, 49(99), 58(57), 68(15), 70(15), 72, 77, 82(155), 86(155), 96(38), 103(99)
Meguro, H., 297, 304(51)
Meiboom, S., 37
Meinders, I., 430, 446(35)
Meinzer, J. L., 129(69), 136(69), 140
Meister, A., 358, 371(90)
Melanson, R., 517, 523
Melton, L. D., 180(160), 185
Melvin, E. H., 41
Mendonça, L., 16, 61(16), 62(16)
Mendonça-Previato, L., 16, 21(11), 23, 25(11,53,56(11,53,61), 62, 64(53), 65(53), 68(125), 70, 95(11), 97(125)
Mercer, A., 480
Mercier, D., 285
Meshreki, M. H., 182(169), 185
Messmer, A., 470
Messner, P., 357
Meth-Cohn, O., 174
Metzner, E. K., 68
Meyer, G. M., 294, 299(30)
Meyer, R. H., 49
Meyer, W. E., 176, 198(1)
Meyer zu Reckendorf, W., 293
Michaelis, E., 203

Micheel, F., 199, 200(16), 202, 203
Middleton, W. J., 225, 227(164), 230(164), 279(164)
Mihalov, V., 311
Mikhailopoulo, I. A., 242, 285
Miller, G. H., 173
Millington, J. E., 243
Millner, S., 238
Minami, K., 149
Minster, D. K., 291
Mirzayanova, M. N., 373
Mitscher, L. A., 480
Miyajima, G., 16, 17(9), 68(9), 71(9)
Miyazaki, S., 300
Mo, F., 466
Möderndorfer, E., 387
Moeller, H., 310
Moffatt, J. G., 180(161), 185, 198, 236(9, 10), 237(9,10)
Mofti, A. M., 224
Mondange, M., 368
Mononen, I., 393, 407
Monsigny, M., 409
Montreuil, J., 389, 398, 409, 410, 411(71)
Moody, G. J., 147
Moore, F. H., 428
Moore, S., 347
Moreno, E., 459
Moret, G., 253, 257(264), 268(264)
Morin, M. J., 285
Morio, S., 144(78), 146
Morris, A., 207, 210(86)
Morris, E. R., 78
Morton, G. O., 196, 198(1)
Moyna, P., 35, 71(86), 72(86)
Müller, D., 398
Müller, H., 209, 305
Muhlegger, K., 527
Muir, H., 80
Munasinghe, V. R. N., 134(65), 137(75), 138(75), 139(75), 140, 141(75), 168, 169(149), 171(149), 172(149), 173(149)
Munro, M. H. G., 20
Munson, R. S., Jr., 353, 386
Murakami, M., 300
Muramatsu, T., 396, 397, 415(30), 416(30)
Myers, T. C., 242
Myhre, D. V., 318
Myllylä, G., 410

AUTHOR INDEX 545

N

Naegell, H., 343, 347(54)
Nagabhushan, T. L., 18, 173
Nagasawa, J.-I., 135(67), 140
Nakajima, Y., 299, 300
Nakanishi, S., 242
Nakatsukasa, Y., 526
Nara, N., 463
Narayanan, P., 497, 514
Naumann, A., 296, 299(44)
Nef, J. U., 294, 299(31)
Neidle, S., 518, 525
Nelson, E. R., 296, 305(39)
Nelson, N., 336
Nelson, N. J., 392
Nelson, P. F., 296, 305(39)
Nelson, T. E., 40, 104(98)
Némec, J., 300, 316
Ness, R. K., 304, 315(89)
Neste, R., 126(55), 127
Neszmêlyi, A., 33, 470
Neukom, H., 319
Neuman, A., 445
Newth, F. H., 199
Ng, C. J., 224
Ng Ying Kins, N. M. K., 72
Nielsen, B., 244
Niemeyer, D. A., 311
Nifant'ev, E. E., 244
Nikaido, H., 324(3), 325, 336, 355
Nilsson, B., 408
Nishiyama, K., 110(20), 111(23), 112(20), 115, 116(20,36), 117
Nisizawa, T., 45
Nitta, Y., 299
Nolte, M., 202
Noltemeyer, M., 529
Noordik, J. H., 424, 425(21)
Norden, N. E., 408
Nordenson, S., 419
Nordick, H., 202
Nordin, R., 17
Norrish, R. G. W., 122
Norrman, B., 45

O

Oerte, G. W., 318
Oeser, E., 481
Ogata, T., 110(19), 111(19), 112(19), 115
Ogata-Arakawa, M., 396, 415(30), 416(30)
Ogura, H., 302, 317, 321
Ogura, M., 16, 68(14)
Ohanessian, J., 431, 450
Ohki, T., 24, 25(56), 35(59), 48(59), 49(56), 50(56)
Ohrt, J. M., 496
Ohrui, H., 467
Ohtsuka, E., 160(134–137), 165, 167(134), 168(135,136,137)
Ojha-Poncet, J., 253, 257(264), 268(264)
Okada, M., 301, 302(76), 303(76), 304(76), 319
Okamoto, K., 398
Okazaki, R., 336
Olah, G. A., 243
Oldfield, E., 25, 26(60), 40(60)
O'Leary, M. K., 491, 513
Ollis, J., 451
Olsson, L., 319
Onan, K. D., 227, 443
Ong, K.-S., 107(6), 108, 127
Oparaeche, N. N., 123, 126(48), 127(48), 134(65), 140, 168, 169(147–149), 170(147,148), 171(147–149), 172(147,149), 173(149)
Orbell, J. D., 505, 508, 509(170)
Orhanovic, Z., 239
Osaki, K., 422, 423
Osberghaus, R., 310
Osborn, M. J., 336, 337, 339(33), 346, 353, 380, 385(136), 386(70,151)
Othman, A. A., 110(17), 115
Otter, B. A., 117
Otto, A., 285
Ovchinnikov, V. E., 468

P

Paal, M., 126(55), 127
Pacák, J., 199, 209, 216, 225, 241, 246(159,225), 268(126), 279(122,123), 285
Paerels, G. B., 329, 330(15), 331(15), 334(15)
Pall, D. B., 219
Palmer, A. J., 238
Palmer, A. K., 211, 222
Palmer, R. A., 515
Palovcik, R., 311

Pandraud, H. G., 445
Parfondry, A., 18, 98, 101(184), 102
Park, Y. J., 424, 494
Parrish, F. W., 23, 210, 220(104), 291
Parthasarathy, R., 496, 503, 516, 528
Pascard-Billy, C., 121, 454, 479, 493, 499 (147)
Pascher, I., 395, 478
Pashinnik, V. G., 225
Patchornik, A., 160(131,132,138), 165, 167(131,132), 169(141)
Paterson, A. J., 20
Patin, D. L., 71
Pattison, F. L. M., 243
Paul, B., 18
Paulsen, H., 114(33), 115, 198, 200(13, 14), 202, 280(46), 285, 312, 313, 442, 460, 483, 526
Pearson, H., 31
Peat, I. R., 14, 27(2), 31, 32
Peat, S., 49
Pechet, M. M., 230
Peciar, C., 298
Pedersen, B. F., 473
Pedersen, C., 17, 18(22), 59(21,22), 200, 201(30,31,32,33,35,36), 202, 203 (44,45), 229(29,30), 230(27,28,177), 244, 256, 258(27), 263(32), 266(32, 295), 270(278), 271(278), 272(278), 273(278), 274(278), 275(45), 276(295), 278(295), 280(45)
Peerdeman, A. F., 446
Peng, J.-Y., 149
Penglis, A. A. E., 208, 220(95), 221(95), 240(95), 255(95)
Pérez, S., 460, 462, 483
Pérez-Garrido, S., 459
Perlin, A. S., 14, 16, 17, 18(19,20), 19, 20, 22(41), 23(41), 38, 58, 69(8), 72, 74, 76(25), 77(25), 98, 101(184), 102, 222, 316
Perry, M. B., 355, 358, 371(95)
Pervova, E. Ya., 244
Pete, J.-P., 137(74), 138(74), 140, 141(74), 142
Petersen, C. S., 501
Petersson, G., 319
Petrov, K. A., 244
Pfaffenberger, C. D., 319
Pfannemüller, B., 40, 104(98)
Pfeffer, P. E., 23

Pfleger, R., 242
Pfleiderer, W., 495
Phelps, D. E., 30
Philips, K. D., 173, 178(177), 185
Phillips, G. O., 106, 147, 165
Phillips, L., 256, 258, 259(287), 261(287), 262(287,291), 264(287), 267(287), 270(287), 276(287), 277(287), 277(287)
Phillips, S. E. V., 471
Pierce, O. R., 237
Pilotti, Å., 398, 399(42), 403(42)
Piloty, O., 289, 292
Pimlott, W., 395
Pinkard, R. M., 297
Pinter, I., 470
Pittman, C. U., Jr., 198, 200(15), 233(15)
Plattner, R. D., 163(126), 165
Plenkiewicz, J., 177(175,176), 178(176), 183(175,176), 184(175,176), 185
Pljer, P. M., 451
Podešva, J., 199, 216, 279(123)
Poisson, J., 479
Poling, M. D., 367
Polyakov, A. I., 309
Pople, J. A., 418, 427, 428(26), 445
Poppleton, B. J., 434, 435
Portella, C., 137(74), 138(74), 140, 141(74), 142
Post, M. L., 486
Powell, H. M., 197
Pravdić, N., 464, 465(99)
Prehm, P., 350, 353(65), 386(65)
Preiss, J., 331, 358
Pretorius, J. A., 475
Previato, J. O., 16, 21(11), 25(11), 52, 56(61), 68(125), 95(11), 97(125)
Přiklylová, V., 216
Privalova, I. M., 300(75), 301
Protopopov, P. A., 225
Prout, C. K., 203, 234(56,57), 235(56), 276(56,57)
Prusiner, P., 486, 519
Puhakainen, E., 318
Purcell, E. M., 37
Purvinas, R. M., 37

Q

Quadling, C., 355
Que, L., Jr., 218, 222(131), 279(131)

AUTHOR INDEX

Quichet, B., 285
Quigley, G. J., 449

R

Rabinsohn, Y., 121
Raboskaya, N. S., 244
Racker, E., 326, 336(9), 337(9), 357, 367(9), 379, 380(9)
Radatus, B. K., 177(176), 178(176), 183(176), 184(176), 185
Radchenko, S. S., 225
Radford, T., 255
Radom, L., 445
Raichle, M. E., 285
Rane, D. F., 285
Rankin, J. C., 41
Rao, G. V., 218, 222(131), 279(131)
Rao, K. N., 149
Rapin, A. M. C., 334, 335(27)
Rappenecker, G., 461
Rashka, M. A., 227
Rasmussen, N. S., 353, 386(70)
Rasmussen, P., 203
Ratcliffe, M., 109(16), 111(22,24), 112(16), 115
Ratenbergs, N., 300
Raunhardt, O., 319
Rauvala, H., 391, 392, 395, 396, 397(31), 398(31,33), 399(31,33), 400(31), 401(33), 405(31), 406, 409, 410, 411, 413(80), 414(33,54,55,74), 415(74,80)
Raunio, R., 320
Ray, P. H., 380, 381, 387(139), 388(137)
Re, A., 285
Reddy, B. S., 507, 527
Reed, M. F., 239
Rees, D. A., 78, 94
Refn, S., 200
Regna, P. P., 371
Rehorst, K., 296, 299(44)
Reich, H. J., 97
Reichman, U., 224, 245(151)
Reichstein, T., 209, 305
Reinhardt, R., 437
Reist, E. J., 68, 215
Reivich, M., 239, 240(212)
Rejtö, M., 121
Reske, K., 353, 357(67)
Rettig, S. J., 468, 469(106)

Rhoades, J. A., 21
Rice, W. F., 314
Richards, T., 147
Richardson, A. C., 210, 211, 212, 220(101), 222, 253(111), 283(111), 285
Richtmyer, N. K., 293
Rick, P. D., 380, 385(136), 386(151)
Ridder, I., 203
Riedl, H., 319
Rietschel, E. T., 384, 386, 396, 397(32), 398(32), 399(32)
Ripka, W. C., 225
Rist, C. E., 41
Ritchie, R. G. S., 16, 18, 162(123), 163(123), 165
Ritzmann, G., 495
Rivest, R., 510
Rivett, D. E., 237
Robbins, G. B., 299
Robbins, J. B., 339
Roberts, J. D., 19, 20, 24(47), 25(47), 31, 37, 38, 49(35), 97
Roberts, J. G., 49
Robins, A. B., 518
Robins, R. V., 204, 232(74), 233(74)
Robinson, W. T., 482
Robson, F. O., 203, 234(54,55), 235(54, 55), 281(54,55)
Rochon, F. D., 517, 523
Rodia, J., 309
Roeder, S. B. W., 31
Roelofsen, G., 420, 425, 430, 432, 436, 445, 446(35), 449
Rogić, V., 463, 464, 465(99)
Rohrer, D. C., 441, 448, 451, 472(60), 489, 490(139), 511, 520
Rolland, A., 285
Romers, C., 492, 494(152), 495, 510, 516
Romming, C., 501
Roseman, S., 382
Rosenstein, R. D., 296, 420, 430, 433, 494(151), 495
Rosenthal, A., 109(16), 110(21), 111(22, 24), 112(16), 115, 116(35), 117
Rothchild, J., 371
Rubin, L. E., 102
Ruff, O., 339, 343
Rupprecht, E., 386
Rush, J., 403(75), 410, 411(75), 416(75)
Ruskin, S. L., 294

Russell, C. R., 158(130), 165
Ruyle, W. V., 253
Ružić-Toroš, Z., 437, 440, 464, 465(99)

S

Saenger, W., 495, 506, 507, 527, 529
Safranski, M. J., 53
Saito, H., 24, 25(56), 35(59), 48(59), 49(56), 50, 396, 398(29)
Saito, M., 302
Sajdera, S. W., 80
Sakakibara, T., 481
Sakazawa, C., 149
Sakore, T. D., 527
Sakurai, Y., 49
Salisburg, S., 529
Salomonsky, N. L., 348, 350(64)
Salt, E., 295
Saman, E., 279
Samaritano, R. H., 304, 316(90,91)
Sammons, M. C., 297
Samokhualov, G. I., 373
Samuelson, O., 296, 305(39), 319
Samuelsson, B., 189(185), 190
Samuelsson, B. E., 394, 395
Samuelsson, K., 394
Sandstrom, W. M., 316
Sarel-Imber, M., 218, 222(130)
Sarfati, R. S., 328(96), 332, 355(22), 358, 368, 369, 371(96)
Sarko, A., 449
Sarre, O. Z., 443
Sasaki, T., 24, 25(56), 35(59), 48(59), 49(56), 50(56)
Saslaw, L. D., 326, 341(10), 348(10)
Sato, S., 522
Sato, T., 410
Satoh, S., 112(26), 115
Satzke, L. O. G., 438
Saunders, J. K., 115
Sawada, M., 147
Scanio, C. V., 320
Schaefer, F. C., 312
Schaffer, R., 358
Scharmann, H., 398
Schauer, R., 252, 361, 363, 382, 397, 398(34), 399(34,47), 401(34)
Sheldrich, G. M., 479, 529
Sheldrick, B., 433
Sheldrick, W. S., 519
Scheller, R., 489, 490(139), 511

Schenk, H., 498
Scherp, H. W., 338
Scherz, K., 148(100), 149, 318
Schiess, P. W., 329
Schittenhelm, W., 17
Schlessinger, R. H., 155
Schmid, K., 410
Schmidt, H. W. H., 319
Schmidt, M. F. G., 240
Schneider, H. J., 256, 273(280)
Schöllnhammer, G., 149
Schulte-Frohlinde, D., 124
Schulten, H. -R., 255
Schultz, A. G., 155
Schulz, G., 356, 360(80), 362, 363(81), 364(80), 374(80), 375(80), 376(80), 377(81), 378(107), 383(80), 387(81)
Schut, J., 329, 330, 331(15), 334(15)
Schwarcz, J. A., 17, 18(20), 22(48), 316
Schwartz, J. C. P., 372
Schwartz, R., 29
Schwarz, R. T., 240
Schwartzmann, G. O. H., 398, 399(45)
Schwarzenbach, D., 248, 249(254), 253(249), 257(254,264), 261(254), 268(254,264), 279(254)
Schwarzinger, E., 356, 363(81), 375, 377(81), 387(81)
Scott, A. I., 480
Scribner, R. M., 225
Seela, F., 529
Seifert, K.-G., 124
Self, R., 392
Sendo, Y., 300
Senna, K., 107(3,8,9), 108, 111(23), 112(8), 115, 116(3,9), 121
Sering, L., 300
Seto, S., 16, 17(9), 21, 22(48), 23(48), 38(48), 39(48), 40, 44(48,96), 45, 46(48), 48(48), 49(48,96), 52(48,96), 68(9), 71(9)
Seymour, F. R., 25, 35(58), 40, 42, 43(58), 72, 98(102), 104(98), 163(124, 126), 165
Shakked, Z., 529
Shapiro, R. H., 402
Sharma, M., 226, 276(165), 285
Sharma, R. A., 226, 253(166), 257(166), 268(166)
Sharma, V. C., 430
Sharon, N., 382
Shashkov, A. S., 14, 19, 27(3), 68

Shchegolev, A. A., 244
Shen, T. Y., 253
Sheppard, W. A., 225
Shettles, L. B., 337
Shigemasa, Y., 149
Shishniashvili, M. E., 298, 300(75), 301
Shiue, C., 285
Shoji, H., 182(169), 185
Shostakovskii, M. F., 309
Shreeve, W. W., 239
Shudo, K., 110(21), 115
Shue, H.-J., 227
Shugar, D., 486, 521
Shulman, M. L., 208, 253(93), 276(93)
Sidebotham, R. L., 43
Siefert, E., 38, 41(95), 48(95), 98(95)
Sim, G. A., 476, 480
Simms, H. S., 296, 299(45)
Simpson, R., 451
Sinaÿ, P., 372, 373(124)
Sindt, A. C., 438
Singh, P., 486, 488
Skapski, A. C., 527
Slessor, K. N., 20, 22(43), 180(160), 185, 293, 313, 456
Sletten, E., 509
Slodki, M. E., 53, 88, 163(126), 165
Smidst, O., 305
Smiley, J. D., 296, 320(41)
Smith, F., 318, 346, 350(60), 353
Smith, I. C. P., 14, 20, 22(43), 24, 25(4, 55), 26(4), 35(55), 36, 38, 40(55), 45(55), 46(55), 47(55,94), 48(55), 71(86), 72(86), 81, 82, 85(169,172), 89, 90(165), 93(177), 98(55), 101(55)
Smits, D., 431, 432
Sobell, H. M., 501, 502, 527
Sokolovskaya, T. A., 104
Som, P., 239
Sorochkin, I. N., 244
Sorrell, T., 523, 528
Southwick, J., 296
Sowa, W., 48, 101(112)
Sowden, J. C., 290, 294(12), 343, 347(56), 358
Spek, A. L., 422
Spencer, J. F. T., 20, 21(45), 25, 52, 53, 54, 56, 58, 59, 66, 67, 68
Sperber, N., 316
Spik, G., 409
Spoormaker, T., 382
Sprang, S., 489, 490(139), 511, 520

Sprecher, M., 371
Sprinson, D. B., 371
Srikrishuan, T., 503
Srivastava, H. S., 207, 208(90)
Srivastava, V. K., 207, 208(90)
Stacey, M., 295
Stafford, G. H., 82
Stahel, R., 293, 294
Stahl, E., 366
Stahl-Larivière, H., 259
Stallings, W. C., 527
Staub, A. M., 324(2,3), 325
Steen, G. O., 394
Steenken, S., 124
Steglich, W., 374
Stein, W., 290, 291(13)
Steiner, P. R., 202, 203(41), 221, 263(141), 266(295), 268(41), 276(295), 278(141,295)
Stellner, K., 396, 398(29)
Štěpán, V., 209
Stephenson, N. C., 197, 280
Sternglanz, H., 487
Stevens, J. D., 197, 224, 280, 428, 452, 466
Stewart, L. C., 293
Stick, R. V., 321
Sticzay, T., 298
Still, I. W. J., 155
Stirm, S., 350, 353(65), 386(65)
Stix, D., 356, 358, 359(82), 360(80), 362, 363(81), 364(80,82), 365(82), 369(98), 372(98), 374(80), 375(80), 376(80,82), 377(81), 378(107), 383(80), 387(81)
Stoeber, F., 307
Stoffyn, A., 406
Stoffyn, P. J., 406
Stout, E. I., 158(130), 165
Straatman, M. G., 239
Strachan, R. G., 253
Strobach, D. R., 328(96), 358, 371(96)
Štropová, D., 216, 225, 246(159), 268(126)
Stuart, R. S., 20, 22, 23(41a), 38(41b)
Stverteczky, J., 368
Sugimoto, K., 336
Sugiyama, H., 16, 17(9), 21, 22(48), 23(48), 38(48), 39(48), 40, 44(48,96), 45, 46(48), 48(48), 49(48,96), 52(48, 96), 68(9,14), 71(9)
Sukumar, S., 285, 362
Sullivan, G. R., 362
Sundarajan, T. A., 334, 335

Sundaralingam, M., 417, 418, 486, 489, 490(139), 491, 511, 518, 519, 520
Surfin, J. R., 285
Sutherland, I., 324(8), 326
Suzuki, M., 144(77), 146
Svennerholm, L., 339
Svensson, S., 390, 398, 399(41,42,43), 403(41,42), 407(8,35), 408(59), 409(59), 410(59), 412(8,59)
Svirdlov, A. F., 68
Swager, S., 263
Swarez, W. A., 463
Sweeley, C. C., 255, 319, 348, 394
Sweet, D. P., 402
Synge, R. L. M., 392
Szabó, I. F., 308
Szabó, L., 328(96), 329, 331(17), 332(17), 334(21), 337(17), 353, 355(22,71), 356, 358, 367, 368(113), 369, 371(96), 373(25), 374(79), 375(79), 397
Szafranek, J., 319
Szalda, D. J., 500
Szarek, W. A., 109(14), 115, 119(14), 134(66), 140, 142(81), 145(81), 146, 162(123), 163(126), 165, 174, 177(175,176), 178(176), 181(163), 182(166,167), 183(175,176), 184(175, 176), 185, 199
Szeja, W., 145(80), 146

T

Tabem, D. L., 300
Tachi-Dung, 206, 242(82), 254(82)
Taga, T., 422, 423
Tagiri, A., 297, 304(51)
Tai, T., 396, 398, 399(46), 405(46), 415(30), 416(30)
Taira, Z., 440
Takagi, S., 418, 419, 423, 425, 427, 429(6), 430, 433, 434, 435(48), 445, 452
Takahashi, H., 302, 317(82), 321
Takahata, H., 312
Takasuka, N., 24, 25(56), 49(56), 50(56)
Takeda, H., 447
Takeo, K., 38, 97
Taki, M., 147
Takiura, K., 75, 76, 79(151)
Takizawa, T., 312

Takusagawa, F., 446, 449(69), 450(69)
Talmadge, K. W., 390
Tam, S. Y.-K., 113(29), 114(29), 115
Tamura, G., 522
Tamura, Z., 296, 319(46)
Tanabe, M., 227
Tanaka, S., 160(134–137), 165, 167(134), 168(135,136,137)
Tanasesu, I., 168
Tanert, G., 147
Tanino, M., 311
Taravel, F. R., 17, 18
Tarelli, J. M., 190
Tarzia, G., 230
Tashima, S., 162(119), 165
Taylor, G. L., 518
Taylor, L. D., 102
Taylor, N. F., 207, 210(85,86,105), 211, 212, 213(14), 214, 220, 221(139), 237, 245, 257, 261(281), 262(281), 279(114), 281, 285
Taylor, P. W., 338, 339, 356, 357, 362, 378(47,107), 384(85)
Taylor, R. L., 289, 297, 299(6), 315
Taplor, R., 418
Tayman, F. S., 442
Terekhov, V., 244
Terzis, A., 420
Tewson, T. J., 224, 227(152), 239(152), 285
Tezuka, M., 152(111), 154(111,112), 155
Theander, O., 182(171), 183(172), 185, 305
Thede, L., 319
Thewalt, U., 489
Thierfelder, H., 289
Thom, D., 78
Thomas, F., 256, 273(280)
Thomas, J. M., 487
Thomas, M. T., 155
Thomas, P., 197, 233(7), 282(7)
Thompson, A., 316
Thornburg, W., 318
Thornton, E. R., 32
Tipson, R. S., 209, 210, 220(103)
Tjarks, L. W., 163(126), 165
Tobin, R., 37
Točik, F., 216, 279(122,123)
Tolman, R. L., 204, 232(74), 233(74)
Tomašić, J., 71, 72(144)

Tong, H. H., 210(108), 211, 279(108)
Torchia, D. A., 35, 37(85), 80
Tori, K., 33
Tork, L., 202, 203
Townsend, L. B., 528
Trabucchi, I., 318
Tracey, A. S., 293
Trachtman, M., 192, 193(189)
Travassos, L. R., 16, 23, 25(53), 52, 56(53), 60, 61(16), 62(16), 64(53), 65(53), 67, 68(125), 70, 97(125)
Travis, A. S., 123, 125(47), 126(47)
Triantaphylides, C., 192, 193(188,190)
Trigalo, F., 332
Trnka, T., 246
Tronchet, J., 253, 257(264), 268(264)
Tronchet, J. M. J., 135(68), 140, 246, 247, 248, 249(254), 253(254), 257(254), 261(254), 265(251,252), 266(251), 268(254), 272(251,253), 279(252,254), 285
Trotter, J., 468, 469(106), 471, 479, 480
Trummer, S., 102
Tsai, C.-C., 501, 502, 527
Tsuchiya, H. M., 41
Tsujimoto, S., 144(78), 146
Tsushima, S., 16, 17(9), 68(9), 71(9)
Türk, D., 307
Tullock, C. W., 225
Turner, W. N., 173
Turro, N. J., 147
Turvey, J. R., 94
Tuseev, A. P., 244
Tuzimura, K., 16, 17(9), 21, 22(48), 23(48), 38(48), 39(48), 40, 44(48,96), 46(48), 48(48), 49(48,96), 52(48,96), 68(9), 71(9), 297, 304(51)
Tyuleneva, V. V., 244

U

Ueda, T., 152(111), 154(111,112), 155
Ul-Haque, M., 470
Umemoto, K., 288, 298(3)
Unger, F. M., 339, 356, 357, 358, 359(82), 360(80), 362, 363(81), 364(82), 365(82), 369(98), 372(97,98), 373(97,98), 374(80), 375(80), 376(80, 82), 377, 378(107), 383(80), 387(81), 388(118)

Upson, F. W., 299, 304
Usov, A. I., 76
Usui, T., 16, 17(9), 21, 22(48), 23(48), 38, 39, 40, 44(48,96), 45, 46, 48, 49(48, 96), 52, 68(9), 71(9)

V

Valentekovic-Horvat, S., 285
Valentine, K. M., 23
VanBoom, J. H., 492
Vane, F. M., 24, 66(54)
van Halbeek, H., 398, 399(47)
Van Lear, J. E., 196, 198(1)
van Marle, T. W. J., 299
Vann, W. F., 357
Van Pragg, D., 240
van Zanten, B., 239
Vasayina, L. K., 76
Vasiliev, A. D., 468
Vass, G., 285
Vegh, L., 162(122), 165
Veh, R. W., 397, 398(34), 399(34), 400(34)
Velarde, E., 227, 228(173)
Vercellotti, J. R., 402, 405(52a)
Verheyden, J. P. H., 180(161), 185, 198, 236(9,10), 237(9,10)
Verschoor, G. C., 516
Veslius, C., 398, 399(47)
Vignon, M., 16, 17(17), 18(26), 21(17), 22(26), 41, 48(50), 49(50), 70(18), 96(17), 98
Villares, P., 458, 459
Vincendon, M., 22, 23, 24(51), 25(51), 38(52b), 48
Vincent, J. E., 317
Vincent, W. F., 338
Vishveshwara, S., 427, 428(26)
Viswamitra, M. A., 529
Vliegenthart, J. A., 446
Vliegenthart, J. F. G., 100, 361, 363(104), 372, 373(124), 382, 397, 398(34), 399(34,47), 401(34)
Vock, M., 207, 220(84)
Voelter, W., 295
Voithenberg, H. V., 472
Vold, R. L., 30
Volk, W. A., 348, 350(64), 378
von Biollaz, M., 227

von Gross, E., 107(8), 108
von Janta-Lipinski, M., 207
von Sonntag, C., 136(70), 140
von Sydow, E., 319
Vorbrüggen, H., 528
Vottero, P. J. A., 17, 98
Vyas, D. M., 162(123), 163(123), 165, 181(163), 185

W

Wada, Y., 463
Waddams, J. A., 298
Waldstein, P., 369, 388(118)
Walker, D. L., 113(27,29,30,31), 114(27, 29,30), 115
Walker, P. G., 252
Walker, R. T., 491, 512, 513
Walker, T. E., 19
Wan, C.-N., 233, 239, 240(212), 245, 285
Wander, J. D., 190
Wang, C.-C., 109(15), 115
Wang, S. Y., 149
Waravdekar, V. S., 326, 341(10), 348(10)
Ward, E., 149
Warren, L., 328, 329(13), 333, 334, 338, 339, 381, 382
Wasylishen, R. E., 256
Watanabe, K. A., 224, 245(151), 416
Watson, R. G., 338
Watson, W. H., 440
Waugh, J. S., 30
Wease, J. C., 180(159), 185
Webber, J. M., 210(106), 211, 219(106), 222(106,133), 243
Webber, M. G., 147
Weckerle, W., 180(157), 185, 441, 472(60)
Weckesser, J., 384
Weerman, R. A., 299
Wei, C.-Y., 529
Weibe, L., 200
Weidmann, H., 308
Weigel, H., 43
Weimann, G., 527
Weiser, D., 308
Weiss, M. J., 207
Weissbach, A., 327, 328(11), 336(11), 337(11), 341(11), 357(11), 366(11), 369(11), 374(11), 381(11)
Welch, J., 243

Welch, M. J., 224, 227(152), 239(152), 285
Welch, V. A., 203, 234(54,55), 235(54,55), 243, 244, 281(54,55)
Wells, W. W., 319, 348
Wenkert, E., 21
Werum, L. N., 318
Westermann, H., 202
Westphal, O., 324(2,3), 325, 333(26), 334, 339(26), 343(26), 344(26), 345(26), 346(26), 348(26), 350(26), 355(26), 357(26), 384
Westwood, J. H., 162(120,121), 165, 196, 197, 199, 204, 205(77), 207(77), 217, 218(128), 220, 223, 224(146), 231(2), 233(7), 253(137), 254(137,138,265), 255, 257(2), 259(128), 261(146), 262(75,128), 263(146,150), 264(137), 265(75,76,128), 266(75,76,128), 267(128,137,138), 268(2), 269(75,76, 128), 270(76,77), 276(75,77,128), 278(146,150), 279(77,128,138), 281(8), 282(7), 285
Weyer, J., 307, 314
Whaley, T. W., 19
Whelan, W. J., 49
Whiffen, D. H., 297
Whistler, R. L., 107(6), 108, 109(15), 115, 127, 130(57), 139, 149, 180(158), 181(165), 185, 209, 290, 294(12)
Whitton, B. R., 107(10,11), 108, 123, 126(48), 127(48), 155, 161(116,118), 165
Whyte, J. N. C., 319
Wiegandt, H., 409, 414
Wiggins, L. F., 302
Wilcox, C. S., 306, 316(98)
Wilham, C. A., 41
Wilkinson, R. G., 346, 348(59), 355(59)
Will, G., 477
Williams, D. M., 157(128,129), 158(128), 164(128,129), 165, 187(128), 188(128)
Williams, D. T., 358, 371(95)
Williams, R. J. P., 283
Williamson, K. L., 263
Wilson, E. R., 237
Wilson, H. R., 508
Winter, B., 180(157), 185
Winterbourne, D. J., 231
Winter-Mihaly, E., 157(128), 158(128), 164(128), 165, 187(128), 188(128)

Wirtz-Peitz, F., 252
Wise, E. B., 35
Wise, D. S., 528
Wise, W. B., 79
Wolf, A. P., 233, 239, 240(212), 245, 285
Wolff, S. M., 324(7), 326
Wolfrom, M. L., 71, 199, 252, 253, 289, 311, 316
Wood, K. R., 196, 227, 229(171), 233(3), 234, 244, 257(3)
Wood, R. A., 432
Woodward, B., 281
Woodward, R. B., 165
Wouter, J. T. M., 82, 88(168)
Wray, V., 256, 258, 259(287), 261(287), 262(287,291), 264(287), 265(279), 267(287,279), 270(287), 271(279), 272(279), 273(279,280), 274(279), 276(287), 277(287), 278(287)
Wright, J. A., 212, 213(14), 214, 279(114)
Wright, J. J., 173
Wright, J. R., 245
Wu, M.-C., 346, 348(59), 355(59)
Wulff, H., 203
Wysocki, J. R., 84

XY

Xavier, A. V., 283
Yamada, K., 110(20), 112(20), 115, 116(20)
Yamada, T., 147
Yamaguchi, H., 398
Yamaoka, N., 16, 17(9), 21, 22(48), 23(48), 38(48), 39(48), 40, 44(48,96), 46(48), 48(48), 49(48,96), 52(48,96), 68(9), 71(9)
Yamashita, K., 396, 398, 399(46), 405(46), 415(30), 416(15)
Yamashita, M., 526

Yaphe, W., 77
Yarotskii, S. V., 76
Yarovenko, N. N., 227
Yassuda, D. M., 227
Yasuoka, N., 447
Yates, J. H., 418, 466, 482
Yokoyama, M., 52
Yoshida, H., 110(19, 111(19), 112(19), 115
Yoshikawa, M., 136(71–73), 140
Yoshimoto, T., 149
Yoshino, H., 16, 68(14)
Yoshioka, M., 300
Yosioka, I., 136(71–73), 140
Yosizawa, Z., 410
Young, D. W., 480
Young, R. C., 208, 219(94), 270(94), 279(94)
Ysern, X., 25, 26(60), 40(60)
Yu, R. K., 382
Yuki, H., 75, 76, 79(151)
Yunker, M. B., 113(28,29), 114(28,29), 115

Z

Zanlungo, A., 116(35), 117
Zaugg, H. E., 316
Zehavi, U., 160(131,132,138), 165, 167(131,132), 169(141)
Zen, S., 162(119), 165
Zhdanov, Yu. A., 307
Zimmerman, H. K., 308
Zinner, H., 306, 311(99)
Zirner, J., 104
Zitrin, C. A., 316
Zolotarev, B. M., 254, 400
Zsoldos-Mády, V., 470
Zugenmaier, P., 461
Zweig, J. E., 72

SUBJECT INDEX

A

Acetals
  carbohydrate, photochemical reactions, 142–147, 170
  dithio, photolysis, 150–153
Acetolysis, structural analysis, 396, 409
Acetone
  photochemical cycloaddition with 3,4,6-tri-O-acetyl-D-glucal, 106, 117
  photochemical reaction with carbohydrate acetals, 143–145
Acyloxonium ion, in fluorination of carbohydrates, 201
Adenine, 9-[5-deoxy-2,3-O-isopropylidene-5-(phenylthio)-β-D-ribofuranosyl]-, irradiation, 152
Adenosine
  5'-methyl phosphate, crystal structure bibliography, 516, 517
  —, 8-(2-aminoethylamino)-, 3',5'-monophosphate tetrahydrate, crystal structure bibliography, 519, 520
  —, 2',3'-O-[(2-carboxyethyl)ethylidene]-, crystal structure bibliography, 529
  —, [N^6-(1-carboxy-2-hydroxyethyl)aminocarbonyl]-, crystal structure bibliography, 496, 497
  —, [N^6-(carboxymethyl)aminocarbonyl]-, crystal structure bibliography, 496
  —, diethidium 5-iodouridylyl-(3',5')-, 27 hydrate, crystal structure bibliography, 501, 502
  —, 8-(1-hydroxyisopropyl)-, dihydrate, crystal structure bibliography, 521, 522
  —, 2',3'-O-isopropylidene-, crystal structure bibliography, 520, 521
  —, 2',3'-O-isopropylidene-5'-S-phenyl-5'-thio-, photolysis mechanism, 153
  —, 5'-methyl-5'-thio-, crystal structure bibliography, 515, 516
Adenosine-5-bromouracil, crystal structure bibliography, 522

Adenosine 5'-monophosphate copper(II) iminodipyridine, crystal structure bibliography, 528
Agar, carbon-13 nuclear magnetic resonance spectra, 77, 78
Alcoholysis, gulono-1,4-lactones, 301
Aldehydes
  photochemical reactions with carbohydrates, 122–129
  reactions with gulono-1,4-lactones, 303
Alditol acetates
  methylated, chemical-ionization mass spectra, 401
  mass fragmentograms, 404
Alditols, irradiation of unprotected, 147–149
Aldohexoses, irradiation of unprotected, 147–149
Aldonic acids
  gas–liquid chromatography, 318
  high-performance liquid chromatography, 319
  ion-exchange chromatography, 319
  mass spectrometry, 318
  paper chromatography, 318
  thin-layer chromatography, 318
Aldononitrile acetates, chemical-ionization mass spectra of partially methylated, 402
Alginic acid, oligosaccharides, carbon-13 nuclear magnetic resonance spectra, 94
O-Alkylation, carbohydrate, chemical shifts, 20
Allofuranose, 3-deoxy-4-fluoro-3-iodo-1,2:5,6-di-O-isopropylidene-α-D-, preparation, 242
  —, 1,2:5,6-di-O-isopropylidene-α-D-, reaction with diethylaminosulfur trifluoride, 227
  —, 1,2:3,4-di-O-isopropylidene-3-O-nitro-α-D-, photolysis, 176

SUBJECT INDEX

Allono-1,4-lactone, 2,3:5,6-di-O-isopropylidene-D-, oxidation, 321
Allopyranose, 1,6-anhydro-2-deoxy-2-fluoro-β-D-, preparation, 246
—, 1,6-anhydro-2-deoxy-2-fluoro-3-C-methyl-β-D-, preparation, 246
—, 1,6-anhydro-2,4-dideoxy-2,4-difluoro-β-D-, preparation, 246
—, 2-deoxy-2-fluoro-D-, preparation, 216
—, 2,4-dideoxy-2,4-difluoro-D-, preparation, 246
Allopyranoside, α-D-allopyranosyl α-D-, calcium chloride pentahydrate, crystal structure bibliography, 451, 452
—, methyl 4,6-O-benylidene-2,3-dideoxy-2,3-[N-(triphenylphosphonio)epimino]-α-D-, p-toluenesulfonate monohydrate, crystal structure bibliography, 481
Allose, 4,5,6-tri-O-benzoyl-2,3-di-S-ethyl-2,3-dithio-D-, diethyl dithioacetal, crystal structure bibliography, 482, 483
Altropyranose, 1,6-anhydro-2-O-(methylsulfonyl)-β-D-, reaction with potassium fluoride, 220, 221
—, 2-deoxy-2-fluoro-D-, preparation, 216
Altropyranoside, methyl 3-benzamido-4,6-di-O-benzoyl-2,3-dideoxy-2-fluoro-α-D-, preparation, 240
—, methyl 2-deoxy-2-fluoro D-, synthesis, 241
Altropyranosyl fluoride, 2-deoxy-2-fluoro-D-, preparation, 215
Aminohexosidases, in carbohydrate degradations, 413
Amino sugars
 analysis, structural, 396
 ^{19}F labelling, 251
Amylopectin, carbon-13 nuclear magnetic resonance spectra of β-limit dextrins, 37, 39, 40
Amylose
 carbon-13 nuclear magnetic resonance spectra, 37–39
 carbon-13 signals, 23
—, O-D-glucopyranosyl-, carbon-13 nuclear magnetic resonance spectra, 104
Analgesics, N-(4-ethoxyphenyl)-L-gulonamide, and N-(4-methoxyphenyl)-L-gulonamide, 300

Angelicin, 8-[2-(β-D-glucopyranosyloxy)isopropyl]-8,9-dihydro-9-hydroxy-, monohydrate, crystal structure bibliography, 473–475
2,2′-Anhydro-(1-β-D-arabinofuranosyl-2,4-diamino-5-fluoropyrimidinium chloride), crystal structure bibliography, 527
2,2′-Anhydro-(1-β-L-arabinofuranosyl-2-hydroxy-4-pyridone), crystal structure bibliography, 439, 440
Antibiotics, aminocyclitol, 285
Antidepressants
 L-gulonic benzylhydrazide, 300
 L-gulonic (2-hydroxyethyl)hydrazide, 300
Antipyretics, N-(4-ethoxyphenyl)-L-gulonamide and N-(4-methoxyphenyl)-L-gulonamide, 300
Antitumor activity, lentinan, 51
Apterin, crystal structure bibliography, 473–475
Arabinal, di-O-acetyl-D-, bromofluorination, 237
Arabinans, per-O-methylated, carbon-13 nuclear magnetic resonance spectra, 71, 101
Arabinitol, 1-deoxy-2-C-phenyl-D-, crystal structure bibliography, 441, 442
—, (1-R)-1-S-ethyl-1-(5-fluorouracil-1-yl)-1-thio-D-, crystal structure bibliography, 493
—, (1R)-2,3,4,5-tetra-O-acetyl-1-S-ethyl-1-(5-fluorouracil-1-yl)-1-thio-D-, ethanolate, crystal structure bibliography, 499, 500
Arabinofuranose, 5-O-benzoyl-2-deoxy-2-fluoro-3-O-formyl-D-, preparation, 245
—, 1,2-di-O-acetyl-5-O-benzoyl-2-deoxy-2-fluoro-3-O-formyl-D-, preparation, 246
Arabinofuranosyl fluoride, per-O-benzoyl-2-bromo-2-deoxy-α-D-, synthesis, 203
Arabinogalactan, carbon-13 nuclear magnetic resonance spectra, 71, 97
Arabinopyranose
 α-L-, calcium chloride tetrahydrate, crystal structure bibliography, 420

β-L-, crystal structure bibliography, 418, 419
—, 2-deoxy-2,2-difluoro-D-, preparation, 196, 231
Arabinopyranoside, methyl α-L-, crystal structure bibliography, 427
—, methyl β-L-, crystal structure bibliography, 427, 428
—, methyl 3,4-O-ethylidene-β-L-, photochemical reaction mechanism with excited acetone, 143
—, trifluoromethyl 3,4-di-O-acetyl-2-deoxy-2-fluoro-β-D-, preparation, 233
Arabinopyranosyl fluoride, di-O-acetyl-2-bromo-2-deoxy-β-D-, preparation, 235
—, 3,4-di-O-acetyl-2-deoxy-2-fluoro-β-D-, preparation, 233
—, 3,4-di-O-benzoyl-2-O-methyl-D-, anomers, synthesis, 201
—, 2-deoxy-2-fluoro-D-, preparation, 214, 245
—, 3-deoxy-3-fluoro-D-, preparation, 213
—, 3-deoxy-3-fluoro-β-D-, preparation, 213
D-Arabinose 5-phosphate isomerase, metabolism of 3-deoxy-D-*manno*-2-octulosonic acid, 378, 379
Arbuzov reaction, fluorinated carbohydrate synthesis, 244
Ascorbic acid, L-, synthesis, 292, 314
8-Azaadenosine, monohydrate, crystal structure bibliography, 486
6-Azathymidine, crystal structure bibliography, 506, 507
8-Azatubercidin, monohydrate, crystal structure bibliography, 511, 512
Azides, irradiation of carbohydrate, 176-178, 180-184
Azines, irradiation of carbohydrate, 186

B

Barley glucan, carbon-13 nuclear magnetic resonance spectra, 49, 50
Bibliography
 crystal structure, of carbohydrates, nucleosides, and nucleotides, 417-529
 Emil Hardegger publications, 4-11
Biological activity
 of fluorinated carbohydrates, 281-284
 of gulonic acids and gulono-[1,4]-lactones, 320
Bis(2,3-O-sodioinosine 5'-monophosphate) (diethylenetriamine) copper(II), decahydrate, crystal structure bibliography, 523, 524
Borinates, *see* Sodium diphenylborinates.
Bovine nasal cartilage, carbon-13 nuclear magnetic resonance spectra and structure, 36, 78-81
Bromofluorination, of glycals, 234, 235
Brucine, L-gulonate, preparation, 299
1,3-Butanediol, 2,4-difluoro-, preparation, 237, 238

C

Cadmium(II) cytidine 5'-monophosphate, monohydrate, crystal structure bibliography, 505
Carbanion, detection, 391
Carbohydrates
 acetals, photochemistry, 142-147
 crystal structure bibliography, 417-485, 526
 dithioacetals, photolysis, 150-153
 excited derivatives, photochemical reactions, 125-131, 135
 fluorinated, 195-285
  biological applications, 281-284
  chemical shifts, 256-260
  ^{19}F-labelled, 285
  ^{19}F-nuclear magnetic resonance spectroscopy, 256-281, 285
  infrared spectra, 280
  mass spectrometry, 253-255, 285
  X-ray crystallography, 280, 285
 permethylated, degradation and structural analysis, 396-407
 photochemical reactions, 105-193
 structural analysis, methylation techniques, 389-416
 unsaturated, photochemical cycloaddition with carbonyl compounds, 106-108
 photochemical radical-addition reactions, 109-115
Carbohydrate esters, photochemical reactions, 136-142

Carbon-13–H coupling
  single-bond, 17
  three-bond, 17–19
Carbon-13 nuclear magnetic resonance signals
  monosaccharide, identification, 19–22
  oligosaccharide and polysaccharide, identification, 22–25
Carbon-13 nuclear magnetic resonance spectra
  3-deoxy-D-*manno*-2-octulosonic acid, 359–361
  polysaccharides, 13–104
  quantitation, 25–37
Carbon-13 parameters, configurationally dependent, 15–19
Carbonyl compounds, photochemical cycloaddition with unsaturated carbohydrates, 106–108
ι-Carrageenan, carbon-13 nuclear magnetic resonance spectra, 77–79
κ-Carrageenan, carbon-13 nuclear magnetic resonance spectra, 77, 78
β-Cellobioside, methyl, ^{13}C-signals, 23
β-Cellotriose, undecaacetate, crystal structure bibliography, 483–485
Cellulose
  carbon-13 nuclear magnetic resonance spectra, 48
  carbon-13 signals, 23
—, O-(carboxymethyl)-, carbon-13 nuclear magnetic resonance spectra, 98, 101, 102
—, O-(2-hydroxyethyl)-, carbon-13 nuclear magnetic resonance spectra, 98, 101, 102
—, O-methyl-, carbon-13 nuclear magnetic resonance spectra, 98, 100, 101
Cellulose acetate, carbon-13 nuclear magnetic resonance spectra, 97
Cephalosporin, hydrazone with L-gulonic phenylhydrazide, preparation, 300
Chemical-ionization mass spectra
  of methylated alditol acetates, 401
  of methylated monosaccharides, 402
Chemical shifts
  in fluorinated carbohydrates, 256–260
  polysaccharide, 15, 16
Chitobiose, N,N'-diacetyl-, monohydrate, crystal structure bibliography, 466, 467

β-Chitobiosyl fluoride, di-N-acetyl-
  biological activity, 282
  synthesis, 203
Chlorofluorination, of glycols, 237
N^7-[Chloromercuri(II)]guanosine, crystal structure bibliography, 510
(2-Chloro-1,1,2-trifluoroethyl)diethylamine, fluorinating agent, 227–229
Chondroitin sulfates, carbon-13 nuclear magnetic resonance spectra, 74–76, 80
Chondrosine, structure, 311
Chromatography
  gas–liquid, of aldonic acids, 318
  for methylation completeness, 391
  of partially methylated sugars, 398–402
  high-performance liquid, of aldonic acids, 319
  ion-exchange, of aldonic acids, 319
  paper, of aldonic acids, 318
  thin-layer, of aldonic acids, 318
Cinchonine, L-gulonate, preparation, 299
Circular dichroism, gulono-1,4-lactones, 297
Claisen condensation, fluorination of carbohydrates, 237
CMP-3-deoxyoctulosonate:3-deoxyoctulosylono-lipid A, metabolism of 3-deoxy-D-*manno*-2-octulosonic acid, 384–386
Cobalt(III) bis(acetylacetonate)(nitro)(deoxyadenosine), heptahydrate, crystal structure bibliography, 528
Cobalt(II) cytidine 5'-monophosphate, monohydrate, crystal structure bibliography, 505, 506
Colitose, periodate–thiobarbituric acid assay, 335, 336
Collagen, bovine nasal-cartilage, 79
Copper(II) (glycylglycinato)cytidine, dihydrate, crystal structure bibliography, 500, 501
Copper(II) guanosine 5'-monophosphate, tetrahydrate, crystal structure bibliography, 508–510
Copper(II) guanosine 3'-monophosphate 1,10-phenanthroline, crystal structure bibliography, 529
Copper(II) iminodipyridine adenosine 5'-monophosphate, crystal structure bibliography, 528

Copper(II) inosine 5′-phosphate (2,2′-bipyridyl)(H$_2$O)$_2$ nitrate, monohydrate, crystal structure bibliography, 527

Copper(II) uridine 5′-phosphate (2,2′-dipyridylamine)H$_2$O]$_2$, pentahydrate, crystal structure bibliography, 528

Cornforth reaction, 329
   3-deoxy-D-*manno*-octulosonic acid synthesis, 365–369

Crystal structure
   bibliography, of carbohydrates, nucleosides, and nucleotides, 417–529
   of fluorinated carbohydrates, 280, 285

CTP:CMP-3-deoxyoctulosonate cytidylyltransferase, metabolism of 3-deoxy-D-*manno*-2-octulosonic acid, 381–384

Cyclitol, amino-, antibiotics, synthesis, 285

Cycloaddition, photochemical, carbohydrate, 106–108

Cycloamyloses, carbon-13 nuclear magnetic resonance spectra, 37–39

Cyclobutanol, formation mechanism from penta-*O*-acetyl-*keto*-L-sorbose, 128

Cycloheptaamylose, carbon-13 signals, 23, 38

Cycloheptaamylose peracetate, carbon-13 nuclear magnetic resonance spectra, 97

Cyclohexaamylose, carbon-13 signals, 23, 38

Cyclohexaamylose peracetate, carbon-13 nuclear magnetic resonance spectra, 97

Cyclooctaamylose, carbon-13 signals, 38

Cyclooctaamylose peracetate, carbon-13 nuclear magnetic resonance spectra, 97

Cytidine, 2′,3′-phosphate, crystal structure bibliography, 507, 508

—, N^4-acetyl-, crystal structure bibliography, 528

—, (glycylglycinato)-, copper(II) complex, dihydrate crystal structure bibliography, 500, 501

—, 2′-*O*-methyl-, crystal structure bibliography, 490, 491

α-Cytidine, crystal structure bibliography, 486, 487

Cytidine [cadmium(II) 5′-monophosphate], monohydrate, crystal structure bibliography, 505

Cytidine [cobalt(II) 5′-monophosphate], monohydrate, crystal structure bibliography, 505, 506

Cytidine *trans*-dichloroplatinum(II)(dimethyl sulfoxide), crystal structure bibliography, 317, 318

Cytidine platinum ethylenediamine 5′-phosphate, crystal structure bibliography, 527

3:2 Cytidylyl-(3′,5′)-guanosine:proflavine, hydrate, crystal structure bibliography, 527

Cytosine, β-D-arabinofuranosyl-2′,5′-monophosphate, crystal structure bibliography, 485, 486
   platinum ethylenediamine dichloride, crystal structure bibliography, 518

D

N-Deacetylation, and acid degradation, 410, 411

3-Deaza-6-azauridine, crystal structure bibliography, 528

3-Deazacytidine, crystal structure bibliography, 490

3-Deazauridine, 4-deoxy-, crystal structure bibliography, 488, 489

Degradation, *see also* Depolymerization; β-Elimination; Smith degradation
   acid, methylation structural analysis, 408–410
   N-deacetylation and acid, 410, 411
   by endoglycosidases, 415, 416
   by exoglycosidases, 412–415

3-Deoxyoctulosonate aldolase, metabolism of 3-deoxy-D-*manno*-2-octulosonic acid, 386

3-Deoxyoctulosonate 8-phosphate phosphatase, metabolism of 3-deoxy-D-*manno*-2-octulosonic acid, 380, 381

3-Deoxyoctulosonate 8-phosphate synthetase, metabolism of 3-deoxy-D-*manno*-2-octulosonic acid, 379, 380

3-Deoxyoctulosylonotransferase, metabolism of 3-deoxy-D-*manno*-2-octulosonic acid, 384–386

SUBJECT INDEX

Depolymerization, of permethylated carbohydrates, 396–398
Dermatan sulfate, carbon-13 nuclear magnetic resonance spectra, 73, 74, 76
Deuteration, polysaccharide, 19, 20
Dextran
 carbon-13 nuclear magnetic resonance spectra, 98, 99
 of branched-chain, 41–46
 of linear, 40
—, per-*O*-benzyl-, carbon-13 nuclear magnetic resonance spectra, 98, 99
Dextran peracetate, carbon-13 nuclear magnetic resonance spectra, 98, 99
Diazo compounds, irradiation of carbohydrate, 178, 179
trans-Dichloroplatinum(II)(dimethyl sulfoxide)cytidine, crystal structure bibliography, 517, 518
Diethidium 5-iodocytidylyl-(3′,5′)-guanosine, 27 hydrate, crystal structure bibliography, 502, 503
Diethidium 5-iodouridylyl-(3′,5′)-adenosine, 27 hydrate, crystal structure bibliography, 501, 502
Diethylamine, *N*-(2-chloro-1,1,2-trifluoroethyl)-, fluorinating agent, 227–229
Diethylaminosulfur trifluoride, fluorinating agent, 225
Digitoxose, β-D-, crystal structure bibliography, 425, 426
1,3-Dioxolane, photochemical addition with unsaturated carbohydrate, 118, 119
Dithioacetals, *see* Acetals

E

β-Elimination, carbohydrate chains of glycoproteins, 411, 412
Endoglycosidases, degradation in structural analysis, 415, 416
Enzymes, 3-deoxy-D-*manno*-2-octulosonic acid metabolism, 378–387
Erythritol, 2-deoxy-2-fluoro-DL-
 structure, 197
 synthesis, 237
Esters, photochemical reactions with carbohydrates, 129–134

Exoglycosidases, degradation in structural analysis, 412–415
Exopolysaccharides, 3-deoxy-D-*manno*-octulosonic acid in acidic, 324, 356, 357, 361–365

F

Ferric L-gulonate, preparation, 299
Fetuin, Smith degradation and methylation analysis, 408
Fischer–Kiliani cyanohydrin synthesis, 3-deoxy-D-*manno*-2-octulosonic acid, 371
Fluorinated carbohydrates, *see* Carbohydrates
Fluorinating agents
 diethylaminosulfur [^{18}F]trifluoride, 239
 ^{18}F$_2$, 239
 silver [^{18}F]fluoride, 239
 tetraethylammonium [^{18}F]fluoride, 239
Fluorination
 by epoxide cleavage, 212–218
 by secondary sulfonyloxy group displacement, 218–225
 by sulfonyloxy group displacement, 204–212
Fluorine compounds, organic, 195–199
Fluorine-19 nuclear magnetic resonance spectroscopy
 carbohydrates, 197
 ^{19}F—^{13}C coupling constants, 270–274
 ^{19}F—^{1}H coupling constants, 260–268
 ^{19}F—^{19}F coupling constants, 268–270
 fluorinated carbohydrates, 256–281, 285
  applications, 274–280
 sugar binding to lysozymes, 283
(Fluoroacetyl)ation, amino sugars, 251
Fluorocarbons, 195
Fluoroformates, carbohydrate, decomposition, 242
Formycin B, hydrochloride, crystal structure bibliography, 488
Fourier transform, 14, 31, 198
Freudenberg rearrangement, thiocarbamates, 158
Fructofuranans, D-, carbon-13 nuclear magnetic resonance spectra, 71
Fructofuranosyl fluoride

α-D-, synthesis, 203
β-D-, synthesis, 203
Fructopyranose
   β-D-, crystal structure bibliography, 430, 431
—, 1-deoxy-1-fluoro-β-D-, synthesis, 209, 210
—, 2,3:4,5-di-O-isopropylidene-1-O-(methylsulfonyl)-β-D-, fluorination, 210
—, 1,2.4,5-di-O-isopropylidene-3-O-p-tolylsulfonyl-β-D-, fluorination, 222
—, 3-O-α-D-glucopyranosyl-β-D-, crystal structure bibliography, 445, 446
Fructopyranosyl fluoride, 3,4,5-tri-O-acetyl-1-O-methyl-α-D-, synthesis 203
—, 3,4,5-tri-O-acetyl-1-O-methyl-β-D-, synthesis, 203
Fructose
   D-, 6-(disodium phosphate), irradiation 193
—, 1,6-dideoxy-1,6-difluoro-D-, preparation, 209
Fucopyranose, α-DL-, crystal structure bibliography, 426
Fucose, 2-deoxy-2-fluoro-L-, preparation, 231
Fucosidases, in carbohydrate degradations, 413

G

Galactal, tri-O-acetyl-D-, bromination, 235
Galactitol, 1-deoxy-1-(ethylsulfinyl)-D-, photolysis, 156
—, 1-deoxy-1-fluoro-L-, preparation, 245
—, 1,6-dideoxy-1,6-difluoro-, preparation, 210
—, 1-S-ethyl-1-thio-D-, preparation, 151, 152
Galactofuranan, D-, carbon-13 nuclear magnetic resonance spectra, 68–70
Galactofuranose, 1,2:3,5-di-O-isopropylidene-6-(O-methylsulfonyl)-α-D-, fluorination, 210
—, 1,2:3,5-di-O-isopropylidene-6-O-p-tolylsulfonyl-α-D- fluorination, 210
Galactofuranosyl fluoride, per-O-benzoyl-β-D-, synthesis, 201
Galactokinase, specificity, fluorinated carbohydrate effect, 282
Galactomannan, carbon-13 nuclear magnetic resonance spectra, 62, 67, 68
Galactopyranose, 1,6-anhydro-3,4-O-isopropylidene-2-O-(methylsulfonyl)-β-D-, reaction with potassium fluoride, 220
—, 6-deoxy-α-DL-, crystal structure bibliography, 426
—, 6-deoxy-6,6-difluoro-1,2:3,4-di-O-isopropylidene-α-D-, preparation, 226
—, 6-deoxy-6-[^{18}F]fluoro-α-D-, synthesis, 239
—, 6-deoxy-6-fluoro-1,2:3,4-di-O-isopropylidene-α-D-, preparation, 244
—, 6-deoxy-6-iodo-1,2:3,4-di-O-isopropylidene-α-D-, irradiation, 187
—, 1,2:3,4-di-O-isopropylidene-α-D-, reaction with (2-chloro-1,1,2-trifluoroethyl)diethylamine, 227, 228
—, 1,2:3,4-di-O-isopropylidene-6-O-(methylsulfonyl)-α-D-, fluorination, 207
—, 1,2:3,4-di-O-isopropylidene-6-O-pyruvoyl-α-D-, photolysis, mechanism, 128
—, 1,2:3,4-di-O-isopropylidene-6-O-p-tolylsulfonyl-α-D-, fluorination, 219
—, 6-O-(fluoroformyl)-1,2:3,4-di-O-isopropylidene-α-D-, preparation and reactions, 242, 243
—, 6-O-o-iodobiphenylyl-1,2:3,4-di-O-isopropylidene-α-D-, irradiation, 187, 191
Galactopyranoside, methyl β-D-, crystal structure bibliography, 433, 434
—, methyl 2-acetamido-3,4-di-O-acetyl-2,-6-dideoxy-6-fluoro-α-D-, preparation, 208
—, methyl 2-(acetoxymercuri)-3,4,6-tri-O-acetyl-2-deoxy-β-D-, irradiation, 190, 191
—, methyl 2-benzamido-3-O-benzyl-2,4,-6-trideoxy-4,6-difluoro-D-, preparation, 222
—, methyl 2,3-di-O-methyl-4-O-(methylsulfonyl)-6-O-trityl-α-D-, fluorination, 220
Galactopyranosyl dipotassium phosphate, 6-deoxy-6-fluoro-α-D-, preparation, 245
Galactopyranosyl fluoride, 2-bromo-2-deoxy-α-D-, preparation, 235

SUBJECT INDEX

—, peracetyl-4-deoxy-4-fluoro-α-D-, synthesis, 204
—, peracetyl-4-deoxy-4-fluoro-β-D-, synthesis, 204
Galactose, D-, diethyl dithioacetal, photolysis, 151, 152
—, 2-deoxy-2-fluoro-D-, preparation, 231
—, 3-deoxy-3-fluoro-D-, preparation, 222
—, 4-deoxy-4-fluoro-D-, preparation, 220
—, 6-deoxy-6-fluoro-D-, preparation, 207
Galactosidases, in carbohydrate degradations, 413
Glucal, 4-O-benzyl-3,6-dideoxy-6-fluoro-D-, preparation, 209
—, 3,4-di-O-acetyl-6-O-p-tolylsulfonyl-D-, fluorination, 206
—, 3,4-di-O-benzoyl-6-O-p-tolylsulfonyl-D-, fluorination, 206
—, 3,4-di-O-benzyl-6-deoxy-6-fluoro-D-, preparation, 209
—, tri-O-acetyl-D-
  bromofluorination, 203, 235
  chlorofluorination, 237
  photochemical cycloaddition with acetone, 106, 117
  photochemical reaction with 2-propanol, 119, 120
  reaction with N-bromosuccinimide and hydrogen fluoride, 234
    with N-iodosuccinimide and hydrogen fluoride, 234
Glucans
  α-D-, carbon-13 nuclear magnetic resonance spectra, 37–48
  β-D-, carbon-13 nuclear magnetic resonance spectra, 48–52
Glucaric acid, D-, potassium salt, crystal structure bibliography, 422, 423
Glucitol
  D-, irradiation, 147, 148
—, (1-S)-1-(6-chloropurin-9-yl)-1-S-ethyl-1-thio-D-, crystal structure bibliography, 493, 494
—, 1-deoxy-1-(2,4-dinitroanilino)-D-, irradiation, 173
—, 1,4-diamino-1,4-dideoxy-3-O-(4-deoxy-4-propionamido-α-D-glucopyranosyl-D-, hydrobromide dihydrate, crystal structure bibliography, 463
—, 1,3:2,4-di-O-ethylidene-D-, oxidation, 301

Glucofuranose, 3,5-O-benzylidene-6-deoxy-6-fluoro-1,2-O-isopropylidene-α-D-, preparation, 207
—, 3,5-O-benzylidene-1,2-O-isopropylidene-6-O-(methylsulfonyl)-α-D-, fluorination, 207
—, 5,6-O-benzylidene-1,2-O-isopropylidene-3-O-(methylsulfonyl)-α-D-, reaction with potassium fluoride, 220
—, (E)-3-C-[2-(carboxamido)-1-(ethoxycarbonyl)ethylene]-3-deoxy-3-fluoro-1,2:5,6-di-O-isopropylidene-α-D-, preparation, 250, 252
—, 3-deoxy-3-C-ethoxalyl-3-fluoro-1,2:5,6-di-O-isopropylidene-α-D-, preparation, 249, 250
—, 3-deoxy-3-[(ethoxycarbonyl)(formylimino)methyl]-3-fluoro-1,2:5,6-di-O-isopropylidene-α-D-, preparation, 249, 250
—, 3-deoxy-3-fluoro-D-, preparation, 224
—, 3-deoxy-3-fluoro-1,2:5,6-di-O-isopropylidene-α-D-, preparation, 219, 227
—, 3-deoxy-3-[^{18}F]fluoro-1,2:5,6-di-O-isopropylidene-α-D-, preparation, 227, 239
—, 3-deoxy-3-fluoro-3-C-[ethoxy(ethoxycarbonyl)(formylimino)methyl]-1,2:5,6-di-O-isopropylidene-α-D-, preparation, 249, 250
—, 3-deoxy-3-fluoro-3-C-(hydroxymethyl)-1,2:5,6-di-O-isopropylidene-α-D-, preparation, 250, 251
—, 3-deoxy-3-iodo-1,2:5,6-di-O-isopropylidene-α-D-, irradiation, 187
—, 3-C-(difluoromethylene)-1,2:5,6-di-O-isopropylidene-α-D-, preparation, 249
—, 1,2:5,6-di-O-isopropylidene-α-D-, reaction with diethylaminosulfur trifluoride, 226, 227, 239
—, 1,2:3,4-di-O-isopropylidene-3-O-nitro-α-D-, photolysis, 176
—, 1,2:3,5-di-O-methylene-α-D-, reaction with (2-chloro-1,1,2-trifluoroethyl)diethylamine, 227
—, 1,2:3,5-di-O-methylene-6-O-p-tolylsulfonyl-α-D-, fluorination, 208
—, 3-O-o-iodobiphenylyl-1,2:5,6-di-O-isopropylidene-α-D-, irradiation, 187, 191

—, 1,2-O-isopropylidene-3,5,6-tri-O-(methylsulfonyl)-α-D-, reaction with potassium fluoride, 220
Glucofuranosyl fluoride, per-O-benzoyl-β-D-, synthesis, 201
Glucofuranurono-6,3-lactone
  D-, catalytic hydrogenation, 288
  reactions, 305, 308, 310
Glucoheptose, β-D-, crystal structure bibliography, 436
Glucomannan, carbon-13 nuclear magnetic resonance spectra, 66
Gluconamide, N-(2-chloroethyl)-D-, crystal structure bibliography, 438, 439
Gluconic acid, 2-deoxy-2-(2,4-dinitroanilino)-D-, sodium salt, irradiation, 173
—, 3-deoxy-3-fluoro-D-, preparation, 245
Glucopyranose, 2-acetamido-4-O-[2-acetamido-2-deoxy-β(α)-D-glucopyranosyl]-2-deoxy-D-, monohydrate, crystal structure bibliography, 466, 467
—, 2-acetamido-1,3,4-tri-O-acetyl-2-deoxy-6-O-(p-nitrophenylsulfonyl)-β-D-, fluorination, 206
—, 2-acetamido-1,3,4-tri-O-acetyl-2,6-dideoxy-6-fluoro-β-D-, preparation, 206
—, 3-amino-1,6-anhydro-3-deoxy-β-D-, crystal structure bibliography, 424
—, 2-amino-1,6-anhydro-2,4-dideoxy-4-fluoro-β-D-, preparation, 216
—, 1,6-anhydro-2,4-dideoxy-2,4-difluoro-β-D-, preparation, 216–218
—, 1,6-anhydro-1(6)-thio-β-D-, crystal structure bibliography, 423, 424
—, 3-deoxy-3-fluoro-D-, preparation, 216
—, 2-deoxy-2-(fluoroacetamido)-D-, preparation, 252
—, 2-deoxy-2-(trifluoroacetamido)-D-, preparation, 252
—, 1,4:3,6-dianhydro-α-D-, crystal structure bibliography, 421, 422
—, 6-O-α-D-galactopyranosyl-α,β-D-, monohydrate, crystal structure bibliography, 448, 449
—, 2-O-β-D-glucopyranosyl-α-D-, monohydrate, crystal structure bibliography, 450, 451
—, 3-O-β-D-glucopyranosyl-α,β-D-, 0.19-hydrate, crystal structure bibliography, 447, 448
—, 4-O-α-D-glucopyranosyl-α,β-D-, crystal structure bibliography, 446, 447
—, 4-O-α-D-glucopyranosyl-β-D-, monohydrate, crystal structure bibliography, 449, 450
—, 1,2,3,6-tetra-O-acetyl-4-O-[2,3,6-tri-O-acetyl-4-O-(2,3,4,6-tetra-O-acetyl-β-D-glucopyranosyl)-β-D-glucopyranosyl]-β-D-, crystal structure bibliography, 483–485
—, 1,2,4-tri-O-acetyl-5,6-dideoxy-3-O-methyl-5-[(S)-phenylphosphinyl]-β-D-, crystal structure bibliography, 526
Glucopyranose, 3-amino-1,6-anhydro-3-deoxy-β-D-, hydrochloride, monohydrate, crystal structure bibliography, 425
C-Allyl-S-β-D-glucopyranosyl-O-sulfo(thiocarbohydroximidate), potassium salt monohydrate, crystal structure bibliography, 440
—, 4-deoxy-4-fluoro-α-D-galactopyranosyl α-D-, preparation, 222
—, 6-deoxy-6-fluoro-α-D-glucopyranosyl α-D-
  biological activity, 283
  synthesis, 211
—, 6-deoxy-6-fluoro-4-O-(methylsulfonyl)-α-D-glucopyranosyl α-D-, biological activity, 283
—, 4,6-dideoxy-4,6-difluoro-α-D-galactopyranosyl α-D-, preparation, 222
—, methyl α-D-, crystal structure bibliography, 434
—, methyl β-D-, hemihydrate, crystal structure bibliography, 435, 436
—, methyl 2-acetamido-3,4-di-O-acetyl-2,4-dideoxy-4-fluoro-α-D-, preparation, 220
—, methyl 2-acetamido-3,4-di-O-acetyl-2,6-dideoxy-6-fluoro-α-D-, preparation, 208
—, methyl 2-benzamido-3-O-benzyl-2,4,6-trideoxy-4,6-difluoro-D-, preparation, 222
—, methyl 2-benzamido-2-deoxy-3-O-[D-1-(methoxycarbonyl)ethyl]-6-O-p-tolylsulfonyl-β-D-, fluorination, 205
—, methyl 4,6-O-benzylidene-2,3-di-O-(methylsulfonyl)-α-D-, reaction with potassium fluoride, 221
—, methyl 4,6-O-benzylidene-2,3-di-O-p-tolylsulfonyl-α-D-, reaction with potassium iodide, 221

SUBJECT INDEX

—, methyl 4,6-O-benzylidene-3-O-(methylsulfonyl)-α-D-, reaction with potassium fluoride, 221
—, methyl 4-deoxy-4-fluoro-α-D-, structure, 197
—, methyl 2-deoxy-2-(trifluoroacetamido)-α-D-, biological activity, 283
—, methyl 2,3-di-O-benzyl-4-O-(p-bromophenylsulfonyl)-6-O-trityl-β-D-, fluorination, 222
—, methyl 6-O-p-tolylsulfonyl-α-D-, fluorination, 205, 207
—, methyl 2,3,4,6-tetra-O-acetyl-β-D-, crystal structure bibliography, 461, 462
—, methyl 2,3,6-tri-O-benzoyl-4-O-(p-bromophenylsulfonyl)-β-D-, fluorination, 222
—, 1-naphthyl 2,3,4,6-tetra-O-acetyl-β-D-, crystal structure bibliography, 478
—, per-O-acetyl-4-acetamido-4,6-dideoxy-6-fluoro-α-D-galactopyranosyl α-D-, preparation, 212
—, phenyl 2,3-di-O-acetyl-6-O-(methylsulfonyl)-4-O-(2,3,4,6-tetra-O-acetyl-α-D-glucopyranosyl)-α-D-, fluorination, 205
—, trifluoromethyl 3,6-di-O-acetyl-2-deoxy-2-fluoro-4-O-(2,3,4,6-tetra-O-acetyl-β-D-galactopyranosyl)-α-D-, preparation, 232
—, trifluoromethyl 3,4,6-tri-O-acetyl-2-deoxy-2-fluoro-α-D-, preparation, 231
Glucopyranosiduronic acid, methyl, methyl ester, photochemistry, 140, 141
Glucopyranosyl bromide, 3,4,6-tri-O-acetyl-2-deoxy-2-(trifluoroacetamido)-α-D-, preparation, 253
Glucopyranosyl fluoride, 2-acetamido-2-deoxy-α-D-, synthesis, 203
—, 2-acetamido-2-deoxy-β-D-, synthesis, 203
—, 2-chloro-2-deoxy-β-D-, preparation, 237
—, 3-deoxy-3-fluoro-D-, preparation, 215
—, 6-deoxy-6-fluoro-α-D-, synthesis, 204
—, 6-deoxy-6-fluoro-β-D-, synthesis, 204
—, 2-deoxy-2-(p-tolylsulfonamido)-α-D-, synthesis, 203
—, 2-deoxy-2-(p-tolylsulfonamido)-β-D-, synthesis, 203
—, 3,6-di-O-acetyl-2-deoxy-2-fluoro-4-O-(2,3,4,6-tetra-O-acetyl-β-D-galactopyranosyl)-α-D-, preparation, 232
—, per-O-acetyl-α-D-, synthesis, 202
—, per-O-acetyl-β-D-, synthesis, 201, 202
—, per-O-acetyl-2-bromo-2-deoxy-β-D-, synthesis, 201
—, per-O-acetyl-6-bromo-6-deoxy-α-D-, synthesis, 203
—, per-O-acetyl-6-bromo-6-deoxy-β-D-, synthesis, 203
—, per-O-acetyl-2-chloro-2-deoxy-D-, synthesis, 202
—, per-O-acetyl-2-deoxy-2-fluoro-α-D-, synthesis, 204
—, per-O-acetyl-2-deoxy-2-fluoro-β-D-, synthesis, 204
—, per-O-acetyl-3-deoxy-3-fluoro-α-D-, synthesis, 204
—, per-O-acetyl-3-deoxy-3-fluoro-β-D-, synthesis, 204
—, per-O-acetyl-2-deoxy-2-fluoro-4-O-β-D-galactopyranosyl-α-D-, synthesis, 204
—, tetra-O-acetyl-α-D-, synthesis, 199
—, tri-O-acetyl-2-bromo-2-deoxy-α-D-, preparation, 234, 235
—, tri-O-acetyl-2-bromo-2-deoxy-β-D-, preparation, 234, 235
—, tri-O-acetyl-2-chloro-2-deoxy-α-D-, preparation, 237
—, 3,4,6-tri-O-acetyl-2-deoxy-2-fluoro-α-D-, preparation, 231
—, tri-O-acetyl-2-deoxy-2-iodo-α-D-, preparation, 234
—, tri-O-acetyl-2-deoxy-2-iodo-β-D-, preparation, 234
Glucopyranosyl phosphate, 2-acetamido-2-deoxy-α-D-, carbon-13 nuclear magnetic resonance spectra, 82, 85
Glucopyranosyl sulfone, 2,3,4,6-tetra-O-acetyl-β-D-, photolysis, 155–157
Glucoseptanoside, methyl 3,4-O-isopropylidene-2-O-(p-nitrophenylsulfonyl)-α-D-, fluorination, 224
(Glucosyl fluoride)urono-(6,3)-lactone, per-O-acetyl-β-D-, synthesis, 203
—, per-O-benzoyl-β-D-, synthesis, 203
Glucose
  D-, 6-(disodium phosphate), irradiation, 192, 193
  irradiation, 147–149

—, 2-acetamido-2,6-dideoxy-6-fluoro-D-, preparation, 208
—, 4-acetamido-2,4-dideoxy-2-fluoro-β-D-, preparation, 216
—, 6-amino-2,6-dideoxy-2-fluoro-D-, preparation, 246
—, 2-deoxy-2-fluoro-D-, preparation, 216, 231
—, 2-deoxy-2-fluoro-D-[^{14}C]-, preparation, 240
—, 2-deoxy-2-fluoro-D-[^{3}H]-, preparation, 239
—, 3-deoxy-3-fluoro-D-
  biological activity, 281
  preparation, 222
—, 3-deoxy-3-[^{18}F]fluoro-D-, synthesis, 239
—, 4-deoxy-4-fluoro-D-, preparation, 216, 218, 220
—, 6-deoxy-6-fluoro-D-, preparation, 207
—, 2-deoxy-2-(trifluoroacetamido)-α-D-, biological activity, 283
—, 2-deoxy-2-(trifluoroacetamido)-β-D-, biological activity, 283
—, 2,4-dideoxy-2,4-difluoro-D-, preparation, 216
—, 2,6-dideoxy-2,6-difluoro-D-, preparation, 216
Glucose 6-phosphate, 2-deoxy-2-[^{18}F]fluoro-D-, synthesis, 245
Glucosyl dipotassium phosphate, α-D-, irradiation, 190, 191
Glucosyluronorhamnomannan, carbon-13 nuclear magnetic resonance spectra, 64
Glucuronic acid, D-, catalytic hydrogenation, 289
Glucuronoxylan, carbon-13 nuclear magnetic resonance spectra, 95
Glutaconic acid, 3-methoxy-, crystal structure bibliography, 422
Glycals
  electrophilic additions, 234–237
  reaction with hydrogen fluoride, 229–237
Glycerol, 1-deoxy-1-fluoro-D-, preparation, 210
—, 2-deoxy-2-fluoro-, synthesis, 237
Glycogen, carbon-13 nuclear magnetic resonance spectra of β-limit dextrins, 39, 40

Glycolipids
  permethylated, mass spectrometry, structure analysis, 394, 395
  structural analysis, methylation techniques, 389–416
Glycopeptides, mass spectrometry, structure analysis, 394, 395
Glycoproteins, structural analysis, methylation techniques, 389–416
Glycosides, aryl, photolysis, 142, 147
Glycosyl fluorides
  preparation, from glycals, 229–237
  synthesis, 199–204
Glyc-2-ulosonic acids, 3-deoxy-, periodate–thiobarbituric acid assay, 327, 238
Gram-negative bacteria, 3-deoxy-D-manno-2-octulosonic acid constituent, 324, 325
Guanosine, N^7-[chloromercuri(II)]-, crystal structure bibliography, 510
—, 2'-deoxy-6-thio-, monohydrate, crystal structure bibliography, 489, 490
—, diethidium 5-iodocytidylyl-(3',5')-, 27 hydrate 4 MeOH, crystal structure bibliography, 502, 503
—, 8-iodo-, monohydrate, crystal structure bibliography, 508
—, N^2-methyl-, monohydrate, crystal structure bibliography, 516
—, 2',3',5'-tri-O-acetyl-6-O-(mesitylenesulfonyl)-, crystal structure bibliography, 525, 526
Guanosine 5'-[copper(II) monophosphate], tetrahydrate, crystal structure bibliography, 508–510
Gulitol, D-, preparation, 315
—, 3,5-O-benzylidene-L-, preparation, 315
—, 2,3-O-isopropylidene-L-, preparation, 315
Gulofuranose, 1-(benzothiazol-2-yl)-2,-3:5,6-di-O-isopropylidene-L-, preparation, 318
—, 1-(1-benzylbenzimidazol-2-yl)-2,3:5,6-di-O-isopropylidene-L-, preparation, 318
—, 1-(chloroethynyl)-2,3:5,6-di-O-isopropylidene-L-, preparation, 317
—, 1-(diethoxy-1-propynyl)-2,3:5,6-di-O-isopropylidene-L-, preparation, 317

—, 2,3:5,6-di-O-isopropylidene-1-(phenylethynyl)-L-, preparation, 317
—, 2,3:5,6-di-O-isopropylidene-1-(pyridin-2-yl)-L-, preparation, 318
—, 2,3:5,6-di-O-isopropylidene-1-[(tetrahydropyran-2-yloxy)-1-propynyl]-L-, preparation, 317
—, 1-(1,3-dithian-2-yl)-2,3:5,6-di-O-isopropylidene-β-L-, preparation, 321
—, 1-ethynyl-2,3:5,6-di-O-isopropylidene-L-, preparation, 317
—, 2,3,5,6-tetra-O-benzoyl-D-, preparation, 316
Gulofuranose-4[1],6-lactone, 3-O-acetyl-4-C-carboxy-1,2-O-isopropylidene-5-O-p-tolylsulfonyl-α-L-, crystal structure bibliography, 471, 472
Gulonamide
  D-, preparation, 299
  L-, preparation, 299
—, 2-acetamido-N-cyclohexyl-2-deoxy-3,4:5,6-di-O-isopropylidene-D-, preparation, 309
—, 4-O-(2-acetamido-2-deoxy-D-galactopyranosyl)-L-, preparation, 311
—, N-allyl-D-, preparation, 300
—, 3,5-O-benzylidene-L-, preparation, 301
—, 2,3:4,5-di-O-carbonyl-6,6-dichloro-6-deoxy-, preparation, 312
—, 2,3:5,6-di-O-isopropylidene-L-, preparation, 301
—, N,N-dimethyl-D-, preparation, 300
—, N-(4-ethoxyphenyl)-L-, preparation, 300
—, N-(methoxycarbonylmethyl)-D-, preparation, 300
—, N-(4-methoxyphenyl)-L-, preparation, 300
Gulonic acid
  biological role, 320
  D-, oxidation, 314
  L-, oxidation, 314
  preparation, 295
  preparation, and salts, 298, 299
—, N-acetylchondrosaminido-L-, methyl ester, preparation, 311
—, 6-amino-6-deoxy-L-, preparation, 308
—, 5-amino-5,6-dideoxy-DL-, preparation, 310

—, 2,6-anhydro-6-deoxy-6-phenyl-L-, preparation, 307
—, 2-anilino-2-deoxy-D-, ammonium salt, preparation, 308
—, 3-O-benzoyl-2,4,5,6-tetra-O-benzyl-L-, preparation, 302
—, 6-deoxy-6-[1,3-di(carboxypropyl)amino]-L-, preparation, 310
—, 6-deoxy-6-{p-[(2,6-dimethoxypyrimidin-4-yl)sulfamoyl]anilino}-6-sulfo-L-, preparation, 311
—, 6-deoxy-6,6-diphenyl-L-, preparation, 310
—, 3,5:4,6-di-O-benzylidene-L-, ethyl ester, preparation, 301
—, 2,3:4,5-di-O-carbonyl-6,6-dichloro-6-deoxy-, methyl ester, preparation, 312
—, 2,3:4,5-di-O-methylene-L-, methyl ester, preparation, 302
—, 3-O-β-D-glucopyranosyl-L-, preparation, 305
—, 2-O-methyl-L-, methyl ester, 311
—, 2,4,5,6-tetra-O-benzyl-L-, preparation of, and esters, 302
—, 3,4,5-tri-O-acetyl-2,6-anhydro-L-, methyl ester, preparation, 307
Gulonic phenylhydrazide, L-, preparation, 299
Gulono-1,6-lactam, 6-amino-6-deoxy-L-, preparation, 308
Gulono-1,4-lactone
  D-, crystal structure, 296
  mass spectra, 297
  synthesis, 287, 289, 294
  L-, hydrolysis, 296, 298
  spectroscopic properties, 297
  synthesis, 287, 288, 295
—, 5-acetamido-5-deoxy-L-, preparation, 308
—, 3,6-anhydro-L-, preparation, 307
—, 3,5-O-benzylidene-L-, preparation, 303
—, 3,5-O-benzylidene-6-O-trityl-L-, preparation, 305
—, 2-O-(tert-butyldimethylsilyl)-3,5-O-isopropylidene-L-, preparation, 306
—, 3,5-O-(2-chlorobenzylidene)-L-, preparation, 303
—, 6-deoxy-D-, preparation, 305
—, 6-deoxy-L-, preparation, 305, 306, 307

—, 6-deoxy-6,6-bis(4,4-dimethyl-2,6-dioxocyclohexyl)-3,5-O-ethylidene-L-, preparation, 311
—, 6-deoxy-6-(2-isonicotinoylhydrazino)-6-sulfo-D-, preparation, 310
—, 2,3-di-O-acetyl-5,6-O-isopropylidene-D-, preparation, 321
—, 2,6:3,5-di-O-benzylidene-L-, preparation, 304
—, 2,3:5,6-di-O-isopropylidene-D-
  oxidation, 321
  preparation, 302
—, 2,3:5,6-di-O-isopropylidene-L-
  preparation, 302
  reaction with lithiated 1,3-dithiane, 321
—, 2,2^1:5,6-di-O-isopropylidene[2-C-(hydroxymethyl)-L-
  crystal structure bibliography, 456
  preparation, 313
—, 2,3:5,6-di-O-isopropylidene-2-C-(hydroxymethyl)-L-, preparation, 313
—, 3,5-O-ethylidene-L-, preparation, 303, 304
—, 5,6-O-ethylidene-L-, preparation, 303
—, 3,5-O-isobutylidene-L-, preparation, 303
—, 2,3-O-isopropylidene-D-, preparation, 303
—, 3,5-O-isopropylidene-L-, preparation, 303
—, 5,6-O-isopropylidene-D-, preparation, 302, 321
—, 3,5-O-(2-methoxybenzylidene)-L-, preparation, 303
—, 3-O-methyl-L-, preparation, 305
—, 3,5-O-(2-methylbenzylidene)-L-, preparation, 303
—, 3,5-O-(3-methylbenzylidene)-L-, preparation, 303
—, 2,3,5,6-tetra-O-acetyl-D-, preparation, 304
—, 2,3,5,6-tetra-O-benzoyl-D-, preparation, 304
—, 2,3,5,6-tetra-O-(trimethylsilyl)-D-, preparation, 304
—, 2,5,6-tri-O-(*tert*-butyldimethylsilyl)-L-, preparation, 305
—, 3,5,6-tri-O-(*tert*-butyldimethylsilyl)-L-, preparation, 305

Gulono-1,4-lactones, 287–321
  alcoholysis, 301
  biological role, 320
  nucleophilic additions to carbonyl group, 317
  oxidation, 314
  reactions of hydroxyl groups, 302–305
  reduction, 315–317
Gulono-1,5-lactone, L-, formation, 296, 297
Gulononitrile, 2-amino-2-deoxy-D-, N-derivatives, preparation, 308
—, 2,5-anhydro-3,4,6-tri-O-(*p*-nitrobenzoyl)-D-, preparation, 311
—, 2-deoxy-3,5:4,6-di-O-ethylidene-2-C-methyl-L-, preparation, 312
—, 2,3:4,5-di-O-carbonyl-6,6-dichloro-6-deoxy-DL-, preparation, 312
Gulopyranose, 1,2,3,4,6-penta-O-acetyl-α-D-, crystal structure bibliography, 466
Gulose,
  D-, by D-gulono-1,4-lactone reduction, 315, 316
  preparation, 293, 294
  L-, preparation, 290, 291
—, 6-deoxy-L-, preparation, 316
—, 2,3:5,6-di-O-isopropylidene-D-, preparation, 316
—, 2,3-O-isopropylidene-D-, preparation, 316

H

Hardegger, Emil, obituary, 1–11
Heparin, carbon-13 nuclear magnetic resonance spectra, 72–74
Hept-5-enofuranurononitrile, E- and Z-6-fluoro-1,2-O-isopropylidene-3-O-methyl-α-D-*xylo*-, preparation, 247
D-*glycero*-L-*gluco*-Heptofurano[2,1-*d*]imidazolidine-2-thione, 1-phenyl-, crystal structure bibliography, 458
Heptofuranuronic acid, 5-O-acetyl-3-O-benzyl-6-deoxy-6-(formamido)-1,2-O-isopropylidene-L-*glycero*-α-L-*allo*-, ethyl ester, crystal structure bibliography, 475, 476
—, 5-O-acetyl-3-O-benzyl-6-deoxy-6-(formamido)-1,2-O-isopropylidene-L-gly-

cero-β-D-*talo*-, ethyl ester, crystal structure bibliography, 475
Heptopyranose, D-*glycero*-β-D-*gulo*-, crystal structure bibliography, 436
2-Heptulopyranose, 3,4,5-tri-O-acetyl-1,7-dibromo-6-(chloromethyl)-1,7-dideoxy-α-DL-*ido*-, crystal structure bibliography, 458, 459
2-Heptulosonic acid, 3-deoxy-4-O-methyl-D-*arabino*-, periodate–thiobarbituric acid assay, 331, 332
Hept-5-ynofuranose, 1,2-O-isopropylidene-3-O-methyl-6-phenyl-, preparation, 246, 247
Hex-1-enitol, 2-cyano-1,2-dideoxy-3,5:4,6-di-O-ethylidene-L-*xylo*-, preparation, 313
—, 1,2-dideoxy-3,5:4,6-di-O-ethylidene-1-C-nitro-L-*xylo*-, preparation, 312
—, 3,4,6-tri-O-acetyl-2-(N-acetylacetamido)-1,5-anhydro-2-deoxy-D-*arabino*-, crystal structure bibliography, 463, 464
—, 3,4,6-tri-O-acetyl-2-(N-acetylacetamido)-1,5-anhydro-2-deoxy-D-*lyxo*-, crystal structure bibliography, 464, 465
—, 3,4,6-tri-O-acetyl-2-(N-acetylacetamido-1,5-anhydro-2-deoxy-D-*xylo*-, crystal structure bibliography, 465, 466
Hex-5-enofuranose, 6-deoxy-6-fluoro-1,2-O-isopropylidene-3-O-methyl-6-(methylphenylamino)-α-D-*xylo*-, preparation, 246
—, 5,6-dideoxy-6,6-difluoro-1,2-O-isopropylidene-3-O-methyl-α-D-*xylo*-, preparation, 246
—, 5,6-dideoxy-1,2-O-isopropylidene-α-D-*xylo*-, photochemical addition with 1,3-dioxolane, 118, 119
Hex-2-enono-1,4-lactone, D-*erythro*-, sodium salt monohydrate, crystal structure bibliography, 420, 421
—, 2-acetamido-2,3-dideoxy-D-*erythro*-, crystal structure bibliography, 437
—, 2-acetamido-2,3-dideoxy-5,6-O-isopropylidene-D-*threo*-, crystal structure bibliography, 440, 441
Hex-2-enopyranoside, ethyl 4-O-acetyl-2,-3-dideoxy-6-O-*p*-tolylsulfonyl-α-D-*erythro*-, fluorination, 206
—, ethyl 4-O-acetyl-2,3,6-trideoxy-6-fluoro-α-D-*erythro*-, preparation, 241, 242
—, ethyl 4-O-benzyl-2,3,6-trideoxy-6-fluoro-α-D-*erythro*-, preparation, 209
—, ethyl 2,3,6-trideoxy-6-fluoro-α-D-*erythro*-, preparation, 242
Hex-4-enopyranoside, methyl 2-benzamido-3-O-benzyl-2,4,5,6-tetradeoxy-6-fluoro-β-L-*threo*-, preparation, 222
Hex-2-enopyranosyl fluoride, 2,4,6-tri-O-acetyl-3-deoxy-D-*ribo*-, preparation, 230
Hex-3-enopyranosylulose fluoride, 6-O-acetyl-3,4-dideoxy-D-*glycero*-, preparation, 230
Hexitol, 1-(6-chloropurin-9-yl)-1-S-ethyl-1-thio-D-*glycero*-D-*ido*-, crystal structure bibliography, 454, 493, 494
Hexofuranose, 3-C-(chlorofluoromethylene)-3-deoxy-1,2:5,6-di-O-isopropylidene-α-D-*ribo*-, *cis*- and *trans*-, preparation, 247, 248
—, 3-C-(chlorofluoromethylene)-3-deoxy-1,2:5,6-di-O-isopropylidene-α-D-*xylo*-, *cis*- and *trans*-, preparation, 247, 248
—, 3-deoxy-3-C-(difluoromethylene)-1,-2:5,6-di-O-isopropylidene-α-D-*ribo*-, preparation, 249
—, 3-deoxy-3-C-(difluoromethylene)-1,-2:5,6-di-O-isopropylidene-α-D-*xylo*-, preparation, 249
—, 3-deoxy-3-C-(fluoromethyl)-1,2:5,6-di-O-isopropylidene-α-D-*ribo*-, *cis*- and *trans*-, preparation, 247, 248
—, 3-deoxy-3-C-(fluoromethyl)-1,2:5,6-di-O-isopropylidene-α-D-*xylo*-, *cis*- and *trans*-, preparation, 247, 248
Hexokinase, specificity, fluorinated carbohydrate effect, 281
Hexonic acid, 3,4,5,6-tetra-O-acetyl-2-deoxy-2-diazo-D-*arabino*-, methyl ester, irradiation, 179
Hexopyranose, 2,6-dideoxy-β-D-*ribo*-, crystal structure bibliography, 425, 426
Hexopyranoside, methyl 3-acetamido-2,-

3,6-trideoxy-3-C-methyl-4-O-methyl-β-L-*xylo*-, crystal structure bibliography, 443, 444

Hexopyranosid-4-ulo-2²,4-pyranose, methyl 2,3,6-trideoxy-2-C-[1,1-(ethylenedithio)-2-hydroxyethyl]-α-L-*threo*-, crystal structure bibliography, 442, 443

Hexopyranos-3-ulose, 1,6-anhydro-2,4-dideoxy-2,4-difluoro-β-D-*ribo*-, reaction with phenylsulfur trifluoride, 225

Hexopyranosyl fluoride, 3,6-di-O-benzoyl-2-deoxy-α-D-*ribo*-, preparation, 230

—, 4,6-di-O-benzoyl-2,3-dideoxy-D-*erythro*-, preparation, 230

—, per-O-benzoyl-2-deoxy-α-D-*arabino*-, synthesis, 203

—, per-O-benzoyl-2-deoxy-α-D-*lyxo*-, synthesis, 201

—, tri-O-acetyl-2-deoxy-α-D-, preparation, 234

Hexose, 3,6-dideoxy-L-*xylo*-, periodate-thiobarbituric acid assay, 335, 336

Hexulofuranosonic acid, 2,3:4,6-di-O-isopropylidene-α-L-, monohydrate, crystal structure bibliography, 445

Hexulopyranose, 4-deoxy-4-fluoro-1,2-O-isopropylidene-β-D-*xylo*-, preparation, 218

—, 1,2-O-isopropylidene-β-D-*xylo*-, preparation, 218

2-Hexulopyranose, 1-(benzylmethylamino)-1-deoxy-β-D-*arabino*-, crystal structure bibliography, 459

—, 1-(benzylmethylamino)-1-deoxy-α-D-*lyxo*-, crystal structure bibliography, 460

3-Hexulopyranose, 1,6-anhydro-4-O-benzyl-2-deoxy-2-fluoro-β-D-*ribo*-, preparation, 246

—, 1,6-anhydro-2,4-dideoxy-2,4-difluoro-β-D-, hydrate, preparation, 246

4-Hexulose, 3-deoxy-1-fluoro-3-iodo-1,2:5,6-di-O-isopropylidene-D-*xylo*-, preparation, 242

2-Hexulosonic acid, L-*xylo*-, preparation, 288, 314

3-Hexulosonic acid, 2,4,5,6-tetra-O-benzyl-L-*xylo*-, benzyl ester, preparation, 314

2-Hexulosono-1,4-lactone, 3,6-anhydro-L-*xylo*-, preparation, 314

Hyaluronic acid, bovine nasal cartilage, 36, 79

[2-C-(Hydroxymethyl)-L-gulonolactone], 2,2¹:5,6-di-O-isopropylidene-, crystal structure bibliography, 456

I

Idofuranose, 3,6-anhydro-5-deoxy-5-fluoro-1,2-O-isopropylidene-β-L-, preparation, 221

—, 6-O-benzoyl-3-deoxy-3-fluoro-1,2-O-isopropylidene-β-L-, preparation, 224

—, 3-deoxy-3-fluoro-1,2:5,6-di-O-isopropylidene-β-L-, preparation, 223

—, 3-deoxy-3-fluoro-1,2-O-isopropylidene-β-L-, preparation, 224

—, 5,6-di-O-benzoyl-3-deoxy-3-fluoro-1,2-O-isopropylidene-β-L-, preparation, 223, 224

Idofuranoside, methyl 3,6-anhydro-5-deoxy-5-fluoro-1,2-O-isopropylidene-β-L-, preparation, 221

Idofuranosyl fluoride, 2-O-acetyl-3,6-anhydro-5-O-benzoyl-β-L-, synthesis, 203

Idofuranuronic N,N-diethylamide, 5-azido-5-deoxy-1,2-O-isopropylidene-3-O-(methylsulfonyl)-β-L-, crystal structure bibliography, 460, 461

Idopyranoside, methyl 3,6-di-O-benzoyl-2,4-di(N-benzoylacetamido)-2,4-dideoxy-α-D-, crystal structure bibliography, 483

Idose, 3-deoxy-3-fluoro-L-, preparation, 223, 224

Imidazoline-2-thione, 4-α-D-erythrofuranosyl-1-*p*-tolyl-, crystal structure bibliography, 456, 457

—, 4-β-D-erythrofuranosyl-1-*p*-tolyl-, crystal structure bibliography, 457, 458

Infrared spectra, of fluorinated carbohydrates, 280

Inhibitors, of 3-deoxy-D-*manno*-2-octulosonic acid metabolism, 387, 388

Inosine, 8-bromo-, crystal structure bibliography, 487, 488

SUBJECT INDEX

Inosine (diethylenetriamine) platinum(II) dinitrate, monohydrate, crystal structure bibliography, 523
epi-Inositol, strontium chloride pentahydrate, crystal structure bibliography, 432
Insulin, carbon-13 signals, 23
Inulin, carbon-13 nuclear magnetic resonance spectra, 72
o-Iodobiphenylyl ethers, irradiation of carbohydrate, 187, 190, 191
Iodo compounds, irradiation of carbohydrate, 186–190
5-Iodocytidylyl-(3′,5′)-guanosine:9-aminoacridine, 24 hydrate, crystal structure bibliography, 527
Iodofluorination, of glycals, 234, 235, 242
Isocyanuric acid, 3,5-dimethyl-1-(2,3,4,6-tetra-O-acetyl-α-D-mannopyranosyl)-, crystal structure bibliography, 472
Isomerization, photochemical, 121

K

K-antigens, see Exopolysaccharides
KDO, see 2-Octulosonic acid, 3-deoxy-D-manno-
Keratan sulfate, bovine nasal-cartilage, 36, 79
Ketones
  diazo, reaction with hydrogen fluoride, 242–244
  photochemical reactions with carbohydrates, 122–129
  reactions with gulono-1,4-lactones, 302
Kuhn reaction, 3-deoxy-D manno-2-octulosonic acid synthesis, 369–371

L

Lactal, hexa-O-acetyl-, fluorination, 232
Lactose, N-acetylneuraminyl-, structure, 32
Lactosyl fluoride, per-O-acetyl-α-, synthesis, 202
Laminarabiose, 0.19-hydrate, crystal structure bibliography, 447, 448
Laminaran, carbon-13 nuclear magnetic resonance spectra, 48
  carbon-13 signals, 23
Lentinan
  antitumor activity, 51
  carbon-13 nuclear magnetic resonance spectra, 51, 52
Levoglucosan, 1(6)-thio-, crystal structure bibliography, 423, 424
Lichenan, carbon-13 nuclear magnetic resonance spectrum, 24, 49, 50
Lipopolysaccharides
  3-deoxy-D-manno-2-octulosonic acid component, 324
  methylation analysis, 352
  occurrence and linkages of KDO, 334–356
  structure, 337, 339, 343–356
  sugar composition, 341
Lithium NAD$^+$ dihydrate, crystal structure bibliography, 527
Lumazine, 6,7-dimethyl-1-β-D-ribofuranosyl-, crystal structure bibliography, 495, 496
Lyxopyranose, β-L-, crystal structure bibliography, 419, 420
Lyxopyranosyl fluoride, per-O-acetyl-2-deoxy-2-fluoro-β-D-, preparation, 204
—, per-O-benzoyl-α-D-, preparation, 201
Lyxurono-1,4-lactone, 3,4-O-isopropylidene-L-, preparation, 314

M

α,β-Maltose, crystal structure bibliography, 446, 447
β-Maltose, monohydrate, crystal structure bibliography, 449, 450
Mannans
  branched-chain D-, carbon-13 nuclear magnetic resonance spectra, 56–60, 64
  linear D-, carbon-13 nuclear magnetic resonance spectra, 52–56
Mannitol
  D-, 1,2:3,4:5,6-tris(phenylboronate), crystal structure bibliography, 477, 478
  DL-, crystal structure bibliography, 432, 433
—, 2,5-anhydro-1-deoxy-1,1-difluoro-D-, preparation, 233
—, 3-O-benzoyl-1,2:5,6-di-O-isopropylidene-4-O-(methylsulfonyl)-D-, reaction with potassium fluoride, 221
—, 3,4-O-benzylidene-2,5-O-methylene-

1,6-di-O-p-tolylsulfonyl-D-, fluorination, 210
—, 1,6-dideoxy-1,6-difluoro-D-, preparation, 211
—, 1,3,4,6-tetra-O-acetyl-2,5-O-methylene-D-, crystal structure bibliography, 462, 463
Mannofuranosyl fluoride, per-O-benzoyl-α-D-, synthesis, 201
Mannono-1,4-lactone, 2,3:5,6-di-O-isopropylidene-D-, oxidation, 321
Mannopyranan, α-D-, carbon-13 nuclear magnetic resonance spectra, 54–60
Mannopyranose, 6-deoxy-α-L-, monohydrate, crystal structure bibliography, 430
—, 2-deoxy-2-fluoro-β-D-
crystal structure, 280
preparation, 224
—, 1,6:2,3-dianhydro-4-deoxy-4-fluoro-β-D-, preparation, 217
Mannopyranoside, methyl α-D-, crystal structure bibliography, 435
—, methyl 4,6-O-benzylidene-2,3-dideoxy-2,3-di(phenylazo)-α-D-, crystal structure bibliography, 479
—, methyl 3,4-di-O-acetyl-6-deoxy-6-iodo-2-O-p-tolylsulfonyl-α-D-, crystal structure bibliography, 470, 471
—, trifluoromethyl 3,6-di-O-acetyl-2-deoxy-2-fluoro-4-O-(2,3,4,6-tetra-O-acetyl-β-D-galactopyranosyl)-β-D-, preparation, 232
—, trifluoromethyl 3,4,6-tri-O-acetyl-2-deoxy-2-fluoro-β-D-, preparation, 231
Mannopyranosyl chloride, 2,3,4,6-tetra-O-benzoyl-α-D-, crystal structure bibliography, 481, 482
Mannopyranosyl fluoride, 3,6-di-O-acetyl-2-deoxy-2-fluoro-4-O-(2,3,4,6-tetra-O-acetyl-β-D-galactopyranosyl)-β-D-, preparation, 232
—, per-O-acetyl-α-D-, synthesis, 202
—, per-O-acetyl-2-bromo-2-deoxy-α-D-, synthesis, 202, 203
—, per-O-acetyl-2-chloro-2-deoxy-D-, synthesis, 202
—, per-O-acetyl-2-deoxy-2-fluoro-α-D-, synthesis, 204
—, per-O-acetyl-2-deoxy-2-fluoro-β-D-, synthesis, 204

—, per-O-acetyl-2-deoxy-2-fluoro-4-O-β-D-galactopyranosyl-β-D-, synthesis, 204
—, per-O-benzoyl-α-D-, preparation, 201
—, tri-O-acetyl-2-bromo-2-deoxy-α-D-, preparation 234, 235
—, tri-O-acetyl-2-chloro-2-deoxy-α-D-, preparation, 237
—, tri-O-acetyl-2-chloro-2-deoxy-β-D-, preparation, 237
—, 3,4,6-tri-O-acetyl-2-deoxy-2-fluoro-β-D-
F-2 chemical shift, 257
preparation, 231
—, tri-O-acetyl-2-deoxy-2-iodo-α-D-, preparation, 234
Mannose, 2-deoxy-2-fluoro-D-, preparation, 231, 233
—, 2-deoxy-2-fluoro-D-[^{3}H]-, preparation, 239
—, 3,4-di-O-methyl-, mass spectrum, 399
—, 3,6-di-O-methyl-, mass spectrum, 399, 400
Mannosidases, in carbohydrate degradations, 413
Mascaroside di(methanolate), crystal structure bibliography, 479, 480
Mass fragmentography, 405, 413, 415
Mass spectrometry
of aldonic acids, 318
chemical-ionization, of methylated alditol acetates, 401, 402
of fluorinated carbohydrates, 253–255, 285
for methylation completeness, 391
of partially methylated sugars, 398–402
of permethylated glycolipids, 394, 395
α,β-Melibiose, monohydrate, crystal structure bibliography, 448, 449
Metabolism
3-deoxy-D-$manno$-2-octulosonic acid, enzymes, 378–387
inhibitors, for 3-deoxy-D-$manno$-2-octulosonic acid, 387, 388
Methane, [5-O-(bromobenzoyl)-2,3-O-isopropylidene-α-D-ribofuranosyl]cyano-, crystal structure bibliography, 467, 468
—, trifluoro(fluoroxy)-, fluorinating agent, 230–233

SUBJECT INDEX

Methanolysis, structural analysis, 397
Methylation, see also Permethylation
  structural analysis, of glycoproteins and glycolipids, 389–416
  Smith degradation, 407, 408
Monosaccharides
  carbon-13 signals, identification, 19–22
  identification of partially methylated, 398–402
  methylated, quantitation, 402–407

N

Neuraminic acid
  mass fragmentography, 413–415
  oligosaccharides containing, structure analysis of permethylated, 393
—, 5-acetamido-3,5-dideoxy-3-fluoro-, preparation, 238
—, N-acetyl-
  detection, 338–340
  structure, 328, 329
—, N-(fluoroacetyl)-, preparation, 252
Neuraminic acid linkages
  structural analysis, by acid degradation, 408, 409
  by methylation, 401
Neuraminidases, in carbohydrate degradation, 413
Nitrites, carbohydrate, photolysis, 177
Nitro compounds, photolysis, of carbohydrates, 160, 165–176
Nonulosonic acid, 5-acetamido-3,5-dideoxy-3-fluoro-D-$glycero$-D-$galacto$-, preparation, 238
2-Nonulosonic acid, 3,8,9-trideoxy-9-C-phosphono-D-$manno$-, preparation, 387
Nuclear magnetic resonance spectroscopy, see Carbon-13 nuclear magnetic resonance spectroscopy; Fluorine-19 nuclear magnetic resonance spectroscopy
Nuclear Overhauser enhancement, 26–35
Nucleocidin
  structure, 196, 198
  synthesis, 237
Nucleosides
  crystal structure bibliography, 485–529

  photolysis of o-nitrophenyl ethers, 167, 168
  sulfur-containing, photolysis, 154, 155
Nucleotides, crystal structure bibliography, 485–529

O

Obituary, Emil Hardegger, 1–11
Oct-6-enose, (E)-6,7,8-trideoxy-1,2:3,4-di-O-isopropylidene-7-C-nitro-α-D-$galacto$-, irradiation, 174, 175
Octulopyranose, 2,7-anhydro-L-$glycero$-β-D-$manno$-, monohydrate, crystal structure bibliography, 437, 438
2-Octulopyranosonic acid, 2,4,5,7,8-penta-O-acetyl-3-deoxy-D-$manno$-
  deacetylation products, 374, 375
  methyl ester, synthesis, 365–367, 375
—, 4,5,7,8-tetra-O-acetyl-2-chloro-2,3-dideoxy-D-$manno$-, methyl ester, synthesis, 375, 376
2-Octulosonic acid, 5-O-(2-acetamido-2-deoxy-β-D-glucopyranosyl)-3-deoxy-, synthesis, 369
—, 7-(2-aminoethyl)-3-deoxy-D-$manno$-, phosphate, 343, 345
—, 3-deoxy-D-$manno$- (KDO)
  in acidic exopolysaccharides, 356, 357, 361–365
  biosynthesis, 325
  chemistry and biological significance, 323–388
  in lipopolysaccharides, 324
  metabolism, enzymes, 378–387
    inhibitors, 387, 388
  monosaccharide chemistry, 373–378
  occurrence and linkages in lipopolysaccharides, 334–356
  periodate–thiobarbituric acid assay, 326–334
  phosphate, synthesis, 326, 367
  polysaccharides containing, carbon-13 nuclear magnetic resonance spectra, 91–94
  semicarbazide assay, 341, 343, 344
  spectroscopic analysis, 359–365
  structure, 357–365
  synthesis, by Cornforth reaction, 365–369
    cyanohydrin, 371

by Kuhn reaction, 369–371
Wittig synthesis, 358, 369, 371–373
—, 3-deoxy-5-O-β-D-glucopyranosyl-, synthesis, 369
—, 3-deoxy-5-O-(β-D-glucopyranosyluronic acid)-, synthesis, 369
—, 3-deoxy-5-O-methyl-, synthesis, 329, 331
—, 3-deoxy-5-O-α-L-rhamnopyranosyl-, synthesis, 369
Oligosaccharides
  carbon-13 nuclear magnetic resonance signals, identification, 22–25
  neuraminic acid-containing, permethylated, structure analysis, 393
  permethylated, structure analysis, 392–394
Organometallic compounds, irradiation, 190, 191
Oximes, irradiation of carbohydrate, 179, 185

P

Paucin, monohydrate, crystal structure bibliography, 476, 477
Pent-1-enitol, (Z)-1-O-acetyl-2,3:4,5-di-O-isopropylidene-D-*erythro*-, crystal structure bibliography, 454–456
—, (Z)-1-O-acetyl-2,3:4,5-di-O-isopropylidene-D-*threo*-, crystal structure bibliography, 454
Pentitol, 2(S)-4-O-acetyl-3,5-O-benzylidene-1,2-dideoxy-2-C-phenyl-D-*erythro*-, crystal structure bibliography, 472, 473
—, 1-S-ethyl-1-(5-fluorouracyl-1-yl)-1-thio-D-*gluco*-, crystal structure bibliography, 493
—, 2,3,4,5-tetra-O-acetyl-1-S-ethyl-1-(5-fluorouracil-1-yl)-1-thio-D-*gluco*-, ethanolate, crystal structure bibliography, 499, 500
Pentofuranoside, methyl 2-deoxy-5-O-(methylsulfonyl)-D-*erythro*-, fluorination, 207
—, methyl 2-deoxy-5-O-*p*-tolylsulfonyl-D-*erythro*-, fluorination, 207
—, methyl 2-deoxy-5-O-*p*-tolylsulfonyl-3-O-trityl-D-*erythro*-, fluorination, 207

Pentofuranos-3-ulose, 5-deoxy-1,2-O-isopropylidene-β-L-*threo*-, irradiation, 123, 124
Pentopyranoside, methyl 2-deoxy-2,2-difluoro-3,4-O-isopropylidene-β-L-*erythro*-, preparation, 226
Pentopyranosid-2-ulose, *tert*-butyl 3,4-O-isopropylidene-α-L-*erythro*-, photolysis, 122, 123
2-Pentulose, L-*threo*-, preparation, 314
Periodate–thiobarbituric acid assay, 3-deoxy-D-*manno*-2-octulosonic acid, 326–334
Periodate–thiobarbituric acid chromaphores, 326, 328, 330, 336
Permethylation, completeness detection, 390–392
Phenylsulfur trifluoride, fluorinating agent, 225
Phosphates, irradiation of carbohydrate, 192, 193
O-Phosphonoglycans, carbon-13 nuclear magnetic resonance spectra, 82–88
O-Phosphonohexopyranans, carbon-13 nuclear magnetic resonance spectra, 81–88
O-Phosphonomannan, carbon-13 nuclear magnetic resonance spectra, 82, 86–88
Phosphorane, [5-(S)-(3-deoxy-3-fluoro-1,2:5,6-di-O-isopropylidene-α-D-glucofuranos-3-yl)-5-hydroxy-2,4-dioxopyrrolidin-3-ylidene]triphenyl-preparation, 250, 251
  X-ray crystallography, 281
—, [5-(R)-(3-deoxy-3-fluoro-1,2-O-isopropylidene-5-O-trityl-α-D-xylofuranos-3-yl)-5-hydroxy-2,4-dioxopyrrolidin-3-ylidene]triphenyl-, preparation, 251, 252
Phosphoric triamide, hexamethyl-, photochemical reaction with carbohydrate esters, 137–142
Photochemistry, carbohydrate, 105–193
Phytosphingosine, β-D-glucosyl-, hydrochloride monohydrate, crystal structure bibliography, 478, 479
Platinum ethylenediamine cytidine 5′-phosphate, dihydrate, crystal structure bibliography, 527
(Platinum ethylenediamine dichloride)

1-β-D-arabinofuranosylcytosine, crystal structure bibliography, 518
Poly(butyl methacrylate), spin–lattice relaxation-times and nuclear Overhauser enhancement values, 30
Polysaccharides
  carbon-13 nuclear magnetic resonance signals, identification, 22–25
  carbon-13 nuclear magnetic resonance spectroscopy, 13–104
  3-deoxy-D-*manno*-octulosylonic acid-containing, carbon-13 nuclear magnetic resonance spectra, 91–94
  deuteration, 19, 20, 22
  diphenylborinates, carbon-13 nuclear magnetic resonance spectra, 103
  furanose-containing, carbon-13 nuclear magnetic resonance spectra, 68–72
  mannose-containing, carbon-13 nuclear magnetic resonance spectra, 60–68
  quantitation of ^{13}C-signal intensities, 25–37
  serogroup, carbon-13 nuclear magnetic resonance spectra, 81–88, 96
  sialic acid-containing, carbon-13 nuclear magnetic resonance spectra, 89–91, 95
  sulfated, carbon-13 nuclear magnetic resonance spectra, 72–81
  D-xylose-containing, carbon-13 nuclear magnetic resonance spectra, 95
Potassium *tert*-butoxide, in methylations, 392
Potassium fluoride, fluorination agent, 220
Potassium D-glucarate, crystal structure bibliography, 422, 423
Propane, 2,3-(2R)-epoxy-1-(2,3,4,6-tetra-O-acetyl-β-D-glucopyranosyl)-, crystal structure bibliography, 468
2-Propanol, photochemical reaction with 3,4,6-tri-O-acetyl-D-glucal, 119, 120
Proteoglycans complex, carbon-13 nuclear magnetic resonance spectra, 36, 78–80
Pseudonigeran, carbon-13 signals, 23
Psicopyranose, 1,2:4,5-di-O-isopropylidene-3-O-*p*-tolylsulfonyl-, fluorination, 222
Pullulan, carbon-13 nuclear magnetic resonance spectra, 46–48, 101
Pustulan, carbon-13 nuclear magnetic resonance spectra, 49
6-Pyridazinone, 4-hydroxy-1-β-D-ribofuranosyl-, crystal structure bibliography, 529
4-Pyridone), 2,2'-anhydro-(1-β-D-arabinofuranosyl-2-hydroxy-, crystal structure bibliography, 439, 440, 488
Pyrimidinium chloride), 2,2'-anhydro-(1-β-D-arabinofuranosyl-2,4-diamino-5-fluoro-, crystal structure bibliography, 527
2(1H)-Pyrimidinone, 1-β-L-ribosyl-, dihydrate, crystal structure bibliography, 504, 505
2-Pyrrolidinone, 3-(R)-acetamido-*spiro*-3,4'-R-(3-deoxy-1,2:5,6-di-O-isopropylidene-α-D-*ribo*-hexofuranos-3-yl)-, crystal structure bibliography, 468, 469
—, 3-(S)-acetamido-*spiro*-3,4'-R-(3-deoxy-1,2:5,6-di-O-isopropylidene-α-D-*ribo*-hexofuranos-3-yl)-, crystal structure bibliography, 469, 470
3-Pyrrolin-2-one, 1-acetyl-3-benzamido-4-(2,3,4,6-tetra-O-acetyl-β-D-glucopyranosyloxy)-, crystal structure bibliography, 480, 481

Q

Quinine, L-gulonate, preparation, 299

R

Rhamnitol, 4-O-β-D-galactopyranosyl-L-, crystal structure bibliography, 452
Rhamnomannan,
  carbon-13 nuclear magnetic resonance spectra, 61–68
  carbon-13 signals, 23
Rhamnopyranose, α-L-, monohydrate, crystal structure bibliography, 430
Ribitol, 2,5-anhydro-1-deoxy-1,1-difluoro-D-, synthesis, 196
—, 2,5-anhydro-1,1-difluoro-, synthesis, 233
—, 2-deoxy-2-fluoro-DL-, synthesis, 238

Riboflavine, photochemistry, 149
Ribofuranoside, methyl 2,3-O-isopropylidene-5-O-(methylsulfonyl)-D-, fluorination, 207
Ribono-1,4-lactones, nucleophilic additions, 317
Ribopyranose, 1,2,3,4-tetra-O-acetyl-α-D-, crystal structure bibliography, 452, 453
Ribopyranoside, methyl β-D-, crystal structure bibliography, 428
—, trifluoromethyl 2,3,4-tri-O-acetyl-2-deoxy-2-fluoro-β-D-, preparation, 233
Ribopyranosyl fluoride, 3,4-di-O-acetyl-2-fluoro-α-D-, preparation, 233
—, per-O-acetyl-2-deoxy-2-fluoro-β-D-, synthesis, 204
—, 2,3,4-tri-O-acetyl-2-fluoro-β-D-, preparation, 233
Ribose, 2-deoxy-2-fluoro-D-, preparation, 240
—, 5-deoxy-5-fluoro-D-, preparation, 207
Ring contractions, in fluorinations, 200
Ring expansion, in glycosyl fluoride synthesis, 201

S

Sialic acid, polysaccharides containing, carbon-13 nuclear magnetic resonance spectra, 89–91, 95
Sinigrin, crystal structure bibliography, 440
Smith degradation, methylation structural analysis, 407, 408
2,3-O-Sodiouridine 5'-(methyl phosphate) methanolate, crystal structure bibliography, 510, 511
Sodium diphenylborinates, polysaccharide, carbon-13 nuclear magnetic resonance spectra, 103
Sodium L-gulonate, preparation, 299
Sodium D-isoascorbate, monohydrate, crystal structure bibliography, 420, 421
Sophorose, monohydrate, crystal structure bibliography, 450, 451
Sorbopyranose, 5-deoxy-5-fluoro-α-L-, preparation, 218
Sorbose, 4-deoxy-4-fluoro-α-D-, preparation, 218

—, 1,3,4,5,6-penta-O-acetyl-keto-L-, photolysis, cyclobutanol formation, 128
Spin-lattice relaxation time, 15, 26, 28–35
Stachyose, carbon-13 nuclear magnetic resonance spectrum, 31
Stereospecificity, fluorination of carbohydrates, 259, 266
Steroids, fluorinations, 219, 230
κ-Strophanthoside, structure, 33
Structural sequences, in polymers, 14
Strychnine, L-gulonate, preparation, 299
Sugars
  deoxyiodo, irradiation, 186–189
  2-deoxy-, periodate–thiobarbituric acid assay, 326
  deoxy-, preparation, 157
Sulfides, carbohydrate, photolysis, 152
Sulfonates, carbohydrate, photolysis, 153–157, 162, 163
Sulfones, carbohydrate, photolysis, 153–157, 161
Sulfonyloxy group
  displacement, of primary by fluorination, 204–212
  of secondary by fluorination, 218–225
Sulfoxides, carbohydrate, photolysis, 153–157

T

Tagatose, 4-deoxy-4-fluoro-D-, preparation, 218
Talofuranosyl fluoride, per-O-benzoyl-α-D-, synthesis, 201
Talopyranose, α-D-, crystal structure bibliography, 431, 432
Talopyranoside, methyl 3,4-di-O-acetyl-2,6-anhydro-β-D-, crystal structure bibliography, 442
Teichoic acid, carbon-13 nuclear magnetic resonance spectra, 82, 88
Tetraammonium 5'-p-adenylyl-(3',5')-thymidylyl-(3',5')-adenylyl-(3',5')-thymidine, hydrate, crystal structure bibliography, 529
Tetrabutylammonium fluoride, fluorinating agent, 219
Tetrazolium bromide, 5-α-D-lyxofurano-

SUBJECT INDEX

syl-2,3-diphenyl-, crystal structure bibliography, 470
Tetrulose, 3,4-di-*O*-benzoyl-1-bromo-1-deoxy-D-*glycero*-, preparation, 243
—, 3,4-di-*O*-benzoyl-1-chloro-1-deoxy-D-*glycero*-, preparation, 243
—, 3,4-di-*O*-benzoyl-1-deoxy-1-diazo-D-*glycero*-, reaction with hydrogen fluoride, 243
Thiocarbamates, carbohydrate, photolysis, 157, 158, 164
Thiocarbohydroximidic acid, *C*-allyl-*S*-β-D-glucopyranosyl-*O*-sulfo-, potassium salt, monohydrate, crystal structure bibliography, 440
Threitol, 2-deoxy-2-fluoro-DL-, synthesis, 237
Thymidine, (−)-(5*S*)-5,6-dihydro-5-hydroxy-, crystal structure bibliography, 512
Thymine, 1-(5-*O*-acetyl-2,3-*O*-isopropylidene-β-D-ribofuranosyl)-5,6-(dichloromethylene)-5,6-dihydro-3-methyl-, crystal structure bibliography, 498, 499
p-Toluenesulfonates, carbohydrate, photolysis, 162, 163
Toyocamycin, monohydrate, crystal structure bibliography, 519
Transferrin, *N*-glycosyl glycopeptide, mass spectrometric analysis, 395
α,α-Trehalose, 4,4′-dideoxy-4,4′-difluoro-, preparation, 222
—, 6,6′-dideoxy-6,6′-difluoro-, preparation, 211
—, 4,6,4′,6′-tetradeoxy-4,6,4′,6′-tetrafluoro-, preparation, 222
(Trifluoroacetyl)ation, amino sugars, 251
Trifluoro(fluoroxy)methane, fluorinating agent, 230–233
Turanose, crystal structure bibliography, 445, 446

U

Uracil, 5-acetyl-1-(2-deoxy-α-D-*erythro*-pentofuranosyl)-, crystal structure bibliography, 491, 492
—, 5-nitro-1-(β-D-ribosyluronic acid)-, monohydrate, crystal structure bibliography, 503, 504

Uridine, 2′-*O*-acetyl-, 3′,5′-monophosphate benzyl ester, crystal structure bibliography, 498
—, 5-amino-, crystal structure bibliography, 507
—, 5-(carbamoylmethyl)-, crystal structure bibliography, 514, 515
—, 5-(carboxymethyl)-, crystal structure bibliography, 514
—, (carboxymethyl)-2-thio-, monohydrate, crystal structure bibliography, 528
—, 2′-deoxy-, photochemistry, 149, 150
—, 2′-deoxy-5-ethynyl-, crystal structure bibliography, 512, 513
—, 2′-deoxy-5-vinyl-, crystal structure bibliography, 513
—, 3′,5′-di-*O*-acetyl-, crystal structure bibliography, 494, 495
—, 5-(L-leucylamino)-, crystal structure bibliography, 497, 498
—, 5-methoxy-, crystal structure bibliography, 528
—, 2′,3′-*O*-(methoxymethylene)-, crystal structure bibliography, 492, 493
—, 5-(methylaminomethyl)-2-thio-, crystal structure bibliography, 528
—, 5-nitro-, monohydrate, crystal structure bibliography, 527
—, 2-thio-, crystal structure bibliography, 485
Uridine 5′-[cobalt(II) monophosphate], heptahydrate, crystal structure bibliography, 527
Uridine 5′-monophosphate copper(II) 2,2′-iminodipyridine, pentahydrate, crystal structure bibliography, 524, 525
Uridine 2,3-*O*-sodio 5′-(methyl phosphate) methanolate, crystal structure bibliography, 510, 511

V

Vaginidiol monoglucopyranoside, crystal structure bibliography, 473–475

W

Walden inversion, in fluorinations, 218
Wittig reaction, fluorination of carbohydrates, 246

Wittig synthesis, of 3-deoxy-D-*manno*-2-octulosonic acid, 358, 369, 371–373

X

Xanthates, Xanthides, carbohydrate, photolysis, 157–160
Xylitol, 1,4-anhydro-5-deoxy-5-fluoro-, preparation, 242
Xylofuranose, 3-*C*-(acetoxymethyl)-5-*O*-acetyl-3-deoxy-3-fluoro-1,2-*O*-isopropylidene-α-D-, preparation, 250, 251
—, 3-*O*-benzyl-1,2-*O*-isopropylidene-5-*O*-*p*-tolylsulfonyl-α-D-, fluorination, 208
—, (*E*)-3-*C*-[2-(carboxamido)-(ethoxycarbonyl)ethylene]-3-deoxy-3-fluoro-1,2-*O*-isopropylidene-5-*O*-trityl-α-D-, preparation, 251, 252
—, 1,2-*O*-isopropylidene-α-D-, reaction with (2-chloro-1,1,2-trifluoroethyl)diethylamine, 229
Xylofuranosyl fluoride, per-*O*-acetyl-α-D-, synthesis, 201
—, per-*O*-acetyl-β-D-, synthesis, 201
Xylopyranan, carbon-13 nuclear magnetic resonance spectra, 95
Xylopyranose, per-*O*-acetyl-2-deoxy-2-fluoro-α-D-, synthesis, 202
—, 1,2,3,4-tetra-*O*-benzoyl-β-D-, crystal structure bibliography, 526
—, tri-*O*-acetyl-2-deoxy-2-fluoro-α-D-, crystallographic analysis, 280
Xylopyranoside, methyl α-D-, crystal structure bibliography, 429
—, methyl β-D-, crystal structure bibliography, 429, 430
—, methyl 3-deoxy-3-fluoro-β-L-, preparation, 212
Xylopyranosyl fluoride, per-*O*-acetyl-β-D-, synthesis, 202
—, per-*O*-acetyl-2-deoxy-2-fluoro-α-D-, synthesis, 204
—, per-*O*-acetyl-3-deoxy-3-fluoro-β-D-, synthesis, 204
—, per-*O*-benzoyl-α-D-, preparation, 201
—, per-*O*-benzoyl-β-D-, preparation, 201
Xylose, D-, D-gulono-1,4-lactone synthesis, 294, 295
—, 2-deoxy-2-fluoro-α-D-, preparation, 213
—, 3-deoxy-3-fluoro-D-, preparation, 212, 214, 244
—, 3,5-dideoxy-3,5-difluoro-D-, preparation, 244

Y

Yarovenko–Rashka reagent, fluorinating agent, 227–229

ERRATA

Volume 37

Page 118, formulas **64, 65,** *and* **67**. For "CH_2OH" read "CO_2H."
Page 318, paragraph 2, line 4. For superscript 279 read 179.
Page 355, line 7 up. For superscript 286 read 289.
Page 365, line 2. For superscripts 358 and 360 read 658 and 660, respectively.